Plant Abiotic Stress Tolerance

Mirza Hasanuzzaman • Khalid Rehman Hakeem
Kamrun Nahar • Hesham F. Alharby
Editors

Plant Abiotic Stress Tolerance

Agronomic, Molecular and Biotechnological Approaches

Springer

Editors
Mirza Hasanuzzaman
Department of Agronomy
Faculty of Agriculture
Sher-e-Bangla Agricultural University
Dhaka, Bangladesh

Kamrun Nahar
Department of Agricultural Botany
Faculty of Agriculture
Sher-e-Bangla Agricultural University
Dhaka, Bangladesh

Khalid Rehman Hakeem
Department of Biological Sciences
King Abdulaziz University
Jeddah, Saudi Arabia

Hesham F. Alharby
Department of Biological Sciences
King Abdulaziz University
Jeddah, Saudi Arabia

ISBN 978-3-030-06117-3 ISBN 978-3-030-06118-0 (eBook)
https://doi.org/10.1007/978-3-030-06118-0

Library of Congress Control Number: 2019932811

This Springer imprint is published by the registered company Springer Nature Switzerland AG
The registered company address is: Gewerbestrasse 11, 6330 Cham, Switzerland

This book is dedicated to:
Sher-e-Bangla Agricultural University
—The Center of Excellence for Agricultural Education and Research
and
King Abdulaziz University, Jeddah, Saudi Arabia, which is ranked No. 1 in the Arab World by the Times Higher Education Ranking in 2018

Foreword

The plant stresses are defined as responses describing a suite of molecular and cellular processes triggered by the detection by a plant of some form of stress. These can be abiotic such as water deficit, water-logging or flooding, extreme cold, frost, heat, salinity, sodicity, and metal and metalloid toxicity or biotic which are responsible for the damage done to an organism by other living organisms like herbivores or pathogens, bacteria, viruses, fungi, parasites, beneficial and harmful insects, weeds, and cultivated or native plants. It has been estimated that salinity and drought are expected to cause serious salinization of more than 50% of all available productive, arable lands by the year 2050. Extreme environmental events in the era of global climatic change further aggravate the problem and remarkably restrict the plant growth and development. We now have very high yielding crops, but these too are susceptible to abiotic stresses. Potential yield of economically important crops is drastically coming down every year just because of abiotic stresses. In view of this, improvement in crop stress responses is a big challenge. Understanding the mechanisms by which plants perceive and transduce the stress signals to initiate adaptive responses is essential for engineering stress-tolerant crop plants. Systems biology approaches facilitate a multi-targeted approach, which involves the molecular parts of an organism and attempts to fit them into functional networks or models designed to describe and predict the dynamic activities of that organism in different environments. Recent advances in biotechnology have changed our capabilities for gene discovery and functional genomics. While many of the functions of individual parts are unknown, their function can sometimes be inferred through association with other known parts, providing a better understanding of the biological system as a whole. High-throughput omics technologies facilitate the identification of new genes and gene function. The mechanisms underlying stress factors have long been the focus of research. Plants overcome environmental stresses by the development of tolerance, resistance, or avoidance mechanisms.

This book titled *Plant Abiotic Stress Tolerance: Agronomic, Molecular and Biotechnological Approaches*, edited by Dr. Mirza Hasanuzzaman, Professor of Agronomy in Bangladesh with a Ph.D. in Plant Stress Physiology and Antioxidant Metabolism from Japan; Dr. Khalid Rehman Hakeem, Associate Professor at King Abdulaziz University, Saudi Arabia, with specialization in Plant Ecophysiology, Biotechnology, and Molecular Biology; Dr. Kamrun Nahar, Associate Professor at Sher-e-Bangla Agricultural University, Bangladesh, with Ph.D. in Plant Abiotic Stress Physiology from Japan; and Dr. Hesham F. Alharby, Head of Plant Section, King Abdulaziz University, Saudi Arabia, presents a collection of 18 chapters. It presents the trends in plant abiotic stress tolerance: agronomic, molecular, and biotechnological approaches. The chapters included here provide detailed latest information. It will be a good guide for researchers working in the field of crop improvement, genetic engineering, and abiotic stress tolerance.

Chapter 1 deals with the maize production under salinity and drought condition: oxidative stress regulation by the antioxidant defense and glyoxalase systems. Authors have reviewed and discussed the present circumstances of maize production and recent progress of varietal improvement for drought and salt tolerance emphasizing how ROS and MG are being regulated in a plant cell by the antioxidant defense and glyoxalase pathways. This chapter also focused on the recent approaches in attenuating oxidative stress in maize plants grown under salinity and drought. Chapter 2 focuses on plants' behavior under soil acidity stress: insight into morphophysiological, biochemical, and molecular responses. This chapter reviews the mechanism of damage under acidity (H^+ rhizotoxicity) stress on plants and also the recent approaches to improve growth and productivity under acidic condition, from the available literature. Chapter 3, titled as salinity: a major agricultural problem—causes, impacts on crop productivity, and management strategies, uncovers decreased crop productivity due to salinity which is expected to elevate in the coming decades. It is expected to pose severe threats to global food security in the future if the challenge is not properly directed. Authors present sustainable agronomic practices, deployment of molecular and functional genomic approaches here which can boost our understanding of salinity stress and create salt-tolerant traits in major field crops. These will potentially contribute to production and yield enhancement under elevated saline conditions. In Chapter 4, plant salinity stress tolerance in plants—physiological, molecular, and biotechnological approaches—has been dealt with considering the advances made in recent decades. The breeding for increased tolerance through gene transfer and the production of transgenic plants is considered an excellent and low-cost method. Perhaps the most valuable outcome of the biotechnology program is to use molecular tools for the breeding programs. Identifying tightly linked molecular markers with the target gene and mapping on the chromosome is an important goal for cloning the genes and marker-assisted selection. Chapter 5 talks about water-deficit stress effects and responses in maize. This chapter describes the mechanism of drought resistance in plants on a morphological, physiological, and molecular basis. The development of crop varieties with increased tolerance to drought, both by conventional breeding methods and by genetic engineering, is given as an important approach to meet up global food demands with less water. Chapter 6 sheds light on the temperature extremes: impact on rice growth and development. In this chapter, authors have summarized the studies regarding the effect of temperature extremes on different growth stages of rice and discussed the possible strategies and opportunities for improving the rice tolerance to heat and cold stresses. Chapter 7 discusses submergence stress in rice: physiological disorders, tolerance mechanisms, and management. Authors mention that several transcription factors are involved in the negative regulation of genes to reduce the elongation. In escape strategy, ethylene-mediated factors are involved in elongation of internodal distance; they have also proposed the physiological and molecular approaches for enhancing the rice tolerance to flood-prone and rainfed lowland conditions. Chapter 8 deals with the oxidative stress and antioxidant defense mechanism in plants under salt stress. It presents studies emphasizing on the plant response to salinity stress through physical, biological, and DNA changes and its alterations to saline places by osmoregulation, ion homeostasis, apoplastic acidification, production of various antioxidants, several genes, hormonal conventions, and production of stress-responsive proteins. According to the authors, intensive exploration work on a combination of several control practices may lead to excellent crop yield in saline soils that might contribute significantly and efficiently to global food security. Chapter 9 titled as oxidative stress and antioxidant defense in plants under drought discusses the oxidative damage caused by the water deficit condition in plant and focuses on the production and scavenging system of ROS in plants. It also provides the details of production site of reactive oxygen species and their reaction with different cellular organelles. A comprehensive scavenging enzymatic and nonenzymatic types and their mode of action to neutralize the harmful effects imposed by drought stress are presented. Chapter 10 discusses the potential of reactive oxygen species metabolism and antioxidant defense in plants under metal/metalloid stress. It is gaining enormous research interest as it limits crop production by harshly altering the physiology and biochemistry of plants. Authors have reviewed the recent reports on different molecular approaches of metal-/metalloid-induced stress tolerance strategies. Chapter 11 covers reactive oxygen species signaling in plants. Various aspects of reactive oxygen species and enzymes in plant response to stress regulation and metabolism are discussed here. Chapter 12 deals with the role of selective exogenous elicitors in plant responses to abiotic stress tolerance. This chapter summarizes the role of elicitors during stressful environments. Some of the signaling aspects through which the cell metabolism is modulated by these elicitors have also been discussed. A brief crosstalk mechanism of some of these exogenous elicitors during these environmental perturbations has also been covered.

Chapter 13 uncovers calcium-mediated growth regulation and abiotic stress tolerance in plants. Authors have focused on the role of calcium against devastating effect of abiotic stresses in plant growth, development, physiology, and yield. Recent information focused on the calcium-induced stimulation of plant growth and physiology as well as abiotic stress tolerance in plants has been presented at length. Chapter 14 deals with silicon—a sustainable tool in abiotic stress tolerance in plants. Silicon fertilizer provides economic as well as ecological benefits to plant growers. Authors enlighten the fact that concerted efforts in the area of silicon research can lead to its accelerated and improved application in the form of fertilizer for sustainable agriculture. Chapter 15 deals with the response of gerbera plants to different salinity levels and leaching ratios on soilless culture. This study has been carried out in order to determine the effects of different salinity levels and leaching ratios on plant growth, yield and quality, and water consumption of gerbera grown by soilless culture. In Chapter 16, crosstalk of nitric oxide and reactive oxygen species in various processes of plant development: past and present, nitric oxide is discussed as a regulator of many physiological processes including cell wall biosynthesis, reactive oxygen species metabolism, stress-induced or constitutive gene expression, programmed cell death, ripening, and senescence. Chapter 17 evaluates the ameliorative capability of plant growth promoting rhizobacteria and arbuscular mycorrhizal fungi against salt stress in plants. Authors describe the causes of soil salinization and discuss potential impacts of salinity stress on plants as well as the action mechanisms of plant growth promotion and/or regulation. They are also highlighting their intrinsic traits that can be upscaled to increase their usefulness as a value-added product for stress agriculture. In Chapter 18, plant miRnome, miRNA biogenesis and abiotic stress response, has been discussed with current knowledge on miRNA biogenesis, mode of action, and the role of miRNA in abiotic stress response in plants.

This book includes a practical update on our knowledge on plant abiotic stress tolerance with special reference to agronomic, molecular, and biotechnological approaches. It will lead to new discussions and efforts to the use of various tools for the improvement of crops for abiotic stress tolerance.

Izmir, Turkey — Münir Öztürk

Preface

Plants have to experience a series of environmental stresses throughout the entire life-span in terms of biotic and abiotic stress. Among these, abiotic stress is the most detrimental one that is responsible for nearly 50% of crop yield reduction, and it appears to be a potential threat to global food security in coming decades. Plant growth and development reduces drastically due to adverse effects of abiotic stresses. It has been estimated that crop can exhibit only 30% of their genetic potentiality under abiotic stress condition. Therefore, this is a fundamental need to understand the stress responses, thus facilitating breeders to develop stress-resistant and stress-tolerant cultivars along with good management practices to withstand abiotic stresses. Also a holistic approach to understand molecular and biochemical interactions of plants is important to implement the knowledge of plant resistance mechanisms under abiotic stresses. Agronomic practices like selecting cultivars that are tolerant to a wide range of climatic condition, planting date, irrigation scheduling, and fertilizer management could be some of the effective short-term adaptive tools to fight against the abiotic stresses. In addition, for long-term adaptation changes and alternations in plant molecular level, "system biology" and "omics approaches" in recent studies could bring some tremendous revolutionary modification in realizing abiotic stresses. The genetic approach, for example, selection and identification of major conditioning genes by linkage mapping and quantitative trait loci (QTL), production of mutant genes, and transgenic introduction of novel genes, has imparted some tolerant characteristics in crop varieties from their wild ancestors. Recently, research has revealed the interactions between micro-RNAs (miRNAs) and plant stress responses exposed to salinity, freezing stress, and dehydration. Transgenic approaches to generate stress-tolerant plant are one of the most interesting researches until now.

The current book is presenting the recent development of agronomic and molecular approaches in conferring plant abiotic stress tolerance in an organized way. The abiotic stresses covered in this book include salinity, water deficiency, water submergence, and extreme temperatures. We have mentioned the strategies in use to mitigate these stresses by incorporating various approaches. These strategies include the application of silicon, AMF, and various exogenous elicitors. The book is also highlighting the mechanism of action of these stress busters in order to increase their usefulness as a value-added product for stressed agriculture. The role of antioxidant enzyme machinery as a defensive feature has been broadly explained in this book. Besides, the current knowledge on miRNA biogenesis, mode of action and the role of miRNA in abiotic stress response in plants.

This is our opportunity to thank the authors who have given their time unselfishly to meet the deadlines for each chapter. We greatly appreciate their commitment. Our profound thanks also to Mr. Abdul Awal Chowdhury Masud, Ms. Khursheda Parvin, Mr. Sayed Mohammad Mohsin, and Mr. MHM Borhannuddin Bhuyan for their critical review and valuable support in formatting and incorporating all editorial changes in the manuscripts. We are also thankful to Prof. Münir Öztürk for his suggestions and writing the foreword for this volume.

We also thank Springer International team for their generous cooperation at every stage of the book production.

Dhaka, Bangladesh — Mirza Hasanuzzaman
Jeddah, Saudi Arabia — Khalid Rehman Hakeem
Dhaka, Bangladesh — Kamrun Nahar
Jeddah, Saudi Arabia — Hesham F. Alharby

About the Book

Abiotic stress is one of the major constraints for crop production in the era of climate change. Therefore, this is a fundamental need to understand the stress responses, thus facilitating breeders to develop stress-resistant and stress-tolerant cultivars along with good management practices to withstand abiotic stresses. Also, a holistic approach to understand molecular and biochemical interactions of plants is important to implement the knowledge of plant resistance mechanisms under abiotic stresses. Agronomic practices like nutrient management could be some of the effective short-term adaptive tools to fight against the abiotic stresses. In addition, for long-term adaptation changes and alternations in plant molecular level, "system biology" and "omics approaches" in recent studies could bring some tremendous revolutionary modification in realizing abiotic stresses.

In the recent years, considerable progress has been made in improving crops for changing environments, and many reports have been published. This book contains 18 informative chapters about the up-to-date knowledge on wheat responses and tolerance to various abiotic stresses written by 74 experts aiming to become a useful information tool for agronomists, plant breeders, and plant physiologists as well as a guide for students in the field of plant science and agriculture. Importantly, this book will lead to new discussion and efforts toward plant abiotic stress tolerance using agronomic and molecular approaches.

Contents

Editors

Mirza Hasanuzzaman Department of Agronomy, Faculty of Agriculture, Sher-e-Bangla Agricultural University, Dhaka, Bangladesh

Khalid Rehman Hakeem Department of Biological Sciences, King Abdulaziz University, Jeddah, Saudi Arabia

Kamrun Nahar Department of Agricultural Botany, Faculty of Agriculture, Sher-e-Bangla Agricultural University, Dhaka, Bangladesh

Hesham F. Alharby Department of Biological Sciences, King Abdulaziz University, Jeddah, Saudi Arabia

Contributors

Sandeep B. Adavi Division of Plant Physiology, ICAR-Indian Agricultural Research Institute, New Delhi, Delhi, India

Bilal Ahmad Plant Physiology and Biochemistry Division, Department of Botany, Aligarh Muslim University, Aligarh, India

Niaz Ahmad Department of Soil Science, Bahauddin Zakariya University, Multan, Pakistan

Riaz Ahmad Department of Horticulture, Bahauddin Zakariya University, Multan, Pakistan

Shakeel Ahmad Department of Agronomy, Faculty of Agricultural Sciences and Technology, Bahauddin Zakariya University, Multan, Punjab, Pakistan

Ishtiaq Ahmed Department of Horticultural Sciences, University College of Agriculture and Environmental Sciences, The Islamia University of Bahawalpur, Bahawalpur, Punjab, Pakistan

Sadia Sabrina Alam Plant Breeding Division, Bangladesh Agricultural Research Institute, Joydebpur, Gazipur, Bangladesh

Basharat Ali Department of Agronomy, University of Agriculture, Faisalabad, Punjab, Pakistan

Muhammad Arif Ali Department of Soil Science, Bahauddin Zakariya University, Multan, Pakistan

Shafaqat Ali Department of Environmental Sciences and Engineering, Government College University, Faisalabad, Pakistan

Jubayer Al Mahmud Department of Agroforestry and Environmental Science, Faculty of Agriculture, Sher-e-Bangla Agricultural University, Dhaka, Bangladesh

Mohammad Amiruzzaman Plant Breeding Division, Bangladesh Agricultural Research Institute, Joydebpur, Gazipur, Bangladesh

Taufika Islam Anee Department of Agronomy, Faculty of Agriculture, Sher-e-Bangla Agricultural University, Dhaka, Bangladesh

Muhammad Akbar Anjum Department of Horticulture, Bahauddin Zakariya University, Multan, Pakistan

Muhammad Saleem Arif Department of Environmental Sciences and Engineering, Government College University, Faisalabad, Pakistan

Umair Ashraf Department of Botany, University of Education (Lahore), Faisalabad, Punjab, Pakistan

Department of Agronomy, University of Agriculture, Faisalabad, Punjab, Pakistan

Shafia Bashir Department of Botany, Punjab Agricultural University, Ludhiana, Punjab, India

Shahnewaz Begum Plant Breeding Division, Bangladesh Agricultural Research Institute, Joydebpur, Gazipur, Bangladesh

Tasnim Farha Bhuiyan Department of Agricultural Botany, Faculty of Agriculture, Sher-e-Bangla Agricultural University, Dhaka, Bangladesh

M. H. M. Borhannuddin Bhuyan Laboratory of Plant Stress Responses, Department of Applied Biological, Sciences, Faculty of Agriculture, Kagawa University, Takamatsu, Kagawa, Japan

Bangladesh Agricultural Research Institute, Joydebpur, Gazipur, Bangladesh

Citrus Research Station, Bangladesh Agricultural Research Institute (BARI), Jaintiapur, Sylhet, Bangladesh

Muhammad Adnan Bukhari Department of Agronomy, University College of Agriculture and Environmental Sciences, The Islamia University of Bahawalpur, Bahawalpur, Punjab, Pakistan

Shaghef Ejaz Department of Horticulture, Bahauddin Zakariya University, Multan, Pakistan

Shah Fahad Department of Agriculture, The University of Swabi, Swabi, Khyber Pakhtunkhwa, Pakistan

Masayuki Fujita Laboratory of Plant Stress Responses, Department of Applied Biological, Sciences, Faculty of Agriculture, Kagawa University, Takamatsu, Kagawa, Japan

Mirza Hasanuzzaman Department of Agronomy, Faculty of Agriculture, Sher-e-Bangla Agricultural University, Dhaka, Bangladesh

Ahmad Hassan Department of Horticulture, Bahauddin Zakariya University, Multan, Pakistan

Mamta Hirve School of Biochemistry, Devi Ahilya University, Indore, Madhya Pradesh, India

Hafiz Athar Hussain College of Agronomy and Biotechnology, Southwest University, Chongqing, China

Institute of Environment and Sustainable Development in Agriculture, Chinese Academy of Agricultural Sciences, Beijing, China

Sadam Hussain Department of Agronomy, University of Agriculture, Faisalabad, Punjab, Pakistan

Sajjad Hussain Department of Horticulture, Faculty of Agricultural Sciences and Technology, Bahauddin Zakariya University, Multan, Punjab, Pakistan

Shahid Iqbal College of Plant Protection, Yunnan Agricultural University, Kunming, China

Md. Robyul Islam Plant Breeding Division, Bangladesh Agricultural Research Institute, Joydebpur, Gazipur, Bangladesh

Meeta Jain School of Biochemistry, Devi Ahilya University, Indore, Madhya Pradesh, India

Shailendra K. Jha Division of Genetics, ICAR-Indian Agricultural Research Institute, New Delhi, Delhi, India

Riti Thapar Kapoor Plant Physiology Laboratory, Amity Institute of Biotechnology, Amity University, Noida, Uttar Pradesh, India

Sunita Kataria School of Biochemistry, Devi Ahilya University, Indore, Madhya Pradesh, India

Muhammad Fasih Khalid Department of Horticulture, Bahauddin Zakariya University, Multan, Pakistan

Abdul Khaliq Department of Agronomy, University of Agriculture, Faisalabad, Punjab, Pakistan

Mojtaba Kordrostami Department of Plant Biotechnology, Faculty of Agricultural Sciences, University of Guilan, Rasht, Iran

Yasemin S. Kukul Kurttaş Department of Agricultural Structures and Irrigation, E.U. Faculty of Agriculture, Ege University, Bornova-Izmir, Turkey

Tang Li College of Plant Protection, Yunnan Agricultural University, Kunming, China

Umer Mahmood Department of Agronomy, University of Agriculture, Faisalabad, Punjab, Pakistan

College of Agronomy and Biotechnology, Southwest University, Chongqing, China

Abdul Majeed Department of Botany, Government Degree College Naguman, Peshawar, Khyber Pakhtunkhwa, Pakistan

Chanchal Malhotra Plant Physiology Laboratory, Amity Institute of Biotechnology, Amity University, Noida, Uttar Pradesh, India

Maryam Department of Botany, Government Sadiq College Women University, Bahawalpur, Punjab, Pakistan

Hari Singh Meena Division of Plant Physiology, ICAR-Indian Agricultural Research Institute, New Delhi, Delhi, India

M. Kamil Meriç E.U. Bergama Vocational School, Ege University, Bergama-Izmir, Turkey

Zahir Muhammad Department of Botany, University of Peshawar, Peshawar, Khyber Pakhtunkhwa, Pakistan

Muhammad Nafees Department of Horticultural Sciences, University College of Agriculture and Environmental Sciences, The Islamia University of Bahawalpur, Bahawalpur, Punjab, Pakistan

Kamrun Nahar Department of Agricultural Botany, Faculty of Agriculture, Sher-e-Bangla Agricultural University, Dhaka, Bangladesh

Tahia Naznin Plant Breeding Division, Bangladesh Agricultural Research Institute, Joydebpur, Gazipur, Bangladesh

Muhammad Noman Department of Bioinformatics and Biotechnology, Government College University, Faisalabad, Pakistan

Sumaiya Haque Omy Plant Breeding Division, Bangladesh Agricultural Research Institute, Joydebpur, Gazipur, Bangladesh

Deepu Pandita Government Department of School Education, Jammu, Jammu and Kashmir, India

Khursheda Parvin Laboratory of Plant Stress Responses, Department of Applied Biological, Sciences, Faculty of Agriculture, Kagawa University, Takamatsu, Kagawa, Japan

Department of Horticulture, Faculty of Agriculture, Sher-e-Bangla Agricultural University, Dhaka, Bangladesh

Rajkumar Prajapati School of Biochemistry, Devi Ahilya University, Indore, Madhya Pradesh, India

Tauqeer Qadir Department of Agronomy, University of Agriculture, Faisalabad, Punjab, Pakistan

Babak Rabiei Department of Agronomy and Plant Breeding, Faculty of Agricultural Sciences, University of Guilan, Rasht, Iran

Muhammad Junaid Rao Key Laboratory of Horticultural Plant Biology (Ministry of Education), Huazhong Agricultural University, Wuhan, People's Republic of China

College of Horticulture and Forestry Sciences, Huazhong Agricultural University, Wuhan, People's Republic of China

Muhammad Riaz Department of Environmental Sciences and Engineering, Government College University, Faisalabad, Pakistan

Md. Motiar Rohman Plant Breeding Division, Bangladesh Agricultural Research Institute, Joydebpur, Gazipur, Bangladesh

Yawar Sadiq Plant Physiology and Biochemistry Division, Department of Botany, Aligarh Muslim University, Aligarh, India

Muhammad Saqib Department of Horticulture, Bahauddin Zakariya University, Multan, Pakistan

Özlem Akat Saraçoğlu E.U. Bayindir Vocational School, Ege University, Bayindir-Izmir, Turkey

Lekshmy Sathee Division of Plant Physiology, ICAR-Indian Agricultural Research Institute, New Delhi, Delhi, India

Adnan Noor Shah Department of Agronomy, Faculty of Agriculture, Gomal University, Dera Ismail Khan, Khyber Pakhtunkhwa, Pakistan

Sher Muhammad Shahzad Department of Soil and Environmental Sciences, University College of Agriculture, University of Sargodha, Sargodha, Pakistan

Mohsin Tariq Department of Bioinformatics and Biotechnology, Government College University, Faisalabad, Pakistan

Department of Integrative Biology, University of California, Berkeley, CA, USA

İ. Hakkı Tüzel Department of Agricultural Structures and Irrigation, E.U. Faculty of Agriculture, Ege University, Bornova-Izmir, Turkey

Shabir H. Wani Mountain Research Centre for Field Crops, Sher-e-Kashmir University of Agricultural Sciences and Technology of Kashmir, Anantnag, Jammu and Kashmir, India

Tahira Yasmeen Department of Environmental Sciences and Engineering, Government College University, Faisalabad, Pakistan

Key Laboratory of Economic Plants and Biotechnology, Kunming Institute of Botany, Chinese Academy of Sciences, Kunming, Yunnan, China

Abbu Zaid Plant Physiology and Biochemistry Division, Department of Botany, Aligarh Muslim University, Aligarh, India

Iqra Zakir Department of Agronomy, Bahauddin Zakariya University, Multan, Pakistan

Department of Agriculture, The University of Swabi, Swabi, Khyber Pakhtunkhwa, Pakistan

About the Editors

Mirza Hasanuzzaman is Professor of Agronomy at Sher-e-Bangla Agricultural University in Dhaka. He received his Ph.D. on "Plant Stress Physiology and Antioxidant Metabolism" from Ehime University, Japan, with a scholarship from the Japanese government (MEXT). Later, he completed his postdoctoral research at the Center of Molecular Biosciences, University of the Ryukyus, Japan, as a Recipient of the Japan Society for the Promotion of Science (JSPS) postdoctoral fellowship. He was also the Recipient of the Australian Government's Endeavour Research Fellowship for postdoctoral research as an Adjunct Senior Researcher at the University of Tasmania, Australia. Dr. Hasanuzzaman's current work is focused on the physiological and molecular mechanisms of environmental stress tolerance. Dr. Hasanuzzaman has published over 80 articles in peer-reviewed journals. He has edited 6 books and written 30 book chapters on important aspects of plant physiology, plant stress tolerance, and crop production. According to Scopus®, Dr. Hasanuzzaman's publications have received about 3300 citations with an *h*-index of 30. He is an Editor and Reviewer for more than 50 peer-reviewed international journals and was a Recipient of the "Publons Peer Review Award 2017 and 2018." He has been honored by different authorities for his outstanding performance in different fields like research and education, and he has received the World Academy of Science Young Scientist Award (2014).

Khalid Rehman Hakeem (Ph.D.) is Associate Professor at King Abdulaziz University, Jeddah, Saudi Arabia. After completing his doctorate (Botany; specialization in Plant Eco-physiology and Molecular Biology) from Jamia Hamdard, New Delhi, India, in 2011, he worked as a lecturer at the University of Kashmir, Srinagar, for a short period. Later, he joined Universiti Putra Malaysia, Selangor, Malaysia, and worked there as Postdoctorate Fellow in 2012 and Fellow Researcher (Associate Professor) from 2013 to 2016. Dr. Hakeem has more than 10 years of teaching and research experience in plant eco-physiology, biotechnology and molecular biology, medicinal plant research, plant-microbe-soil interactions as well as in environmental studies. He is the recipient of several fellowships at both national and international levels; also, he has served as the visiting scientist at Jinan University, Guangzhou, China. Currently, he is involved with a number of international research projects with different government organizations.

So far, Dr. Hakeem has authored and edited more than 35 books with international publishers, including Springer Nature, Academic Press (Elsevier), and CRC Press. He also has to his credit more than 80 research publications in peer-reviewed international journals and 55 book chapters in edited volumes with international publishers.

At present, Dr. Hakeem serves as an editorial board member and reviewer of several high-impact international scientific journals from Elsevier, Springer Nature, Taylor and Francis, Cambridge, and John Wiley Publishers. He is included in the advisory board of Cambridge Scholars Publishing, UK. He is also a fellow of Plantae group of the American Society of Plant Biologists, member of the World Academy of Sciences, member of the International Society for Development and Sustainability, Japan, and member of Asian Federation of Biotechnology, Korea. Dr. Hakeem has been listed in Marquis Who's Who in the World, since 2014–2019. Currently, Dr. Hakeem is engaged in studying the plant processes at eco-physiological as well as molecular levels.

Kamrun Nahar is Associate Professor in the Department of Agricultural Botany at Sher-e-Bangla Agricultural University, Dhaka, Bangladesh. She received her Ph.D. Degree on "Environmental Stress Physiology of Plants" in 2016 from the United Graduate School of Agricultural Sciences, Ehime University, Japan, with Japanese Government (MEXT) Scholarship. Dr. Nahar has been involved in research with field crops emphasizing stress physiology since 2006. She has completed several research work and also continuing research project funded by Sher-e-Bangla Agricultural University Research System and Ministry of Science and Technology (Bangladesh). She is supervising M.S. students. Dr. Nahar published 50 articles and chapters related to plant physiology and environmental stresses with Springer, Elsevier, CRC Press, Wiley, etc. Her publications reached about 2000 citations with h-index of 22 (according to Scopus®). She is involved in editorial activities and Reviewer of international journals. She is Active Member of about 20 professional societies. Dr. Nahar attended different international conferences and presented 10 papers and posters in national and international conferences in different countries (the USA, Australia, Japan, Austria, Russia, China, etc.).

Hesham F. Alharby (Ph.D.) is Associate Professor in the Department of Biological Sciences at King Abdulaziz University (KAU), Jeddah, Saudi Arabia. He received his Ph.D. from the School of Plant Biology at the University of Western Australia, Perth, Australia, in 2014. Dr. Alharby's current research works are related to various aspects of plant biology, particularly in eco-physiology and molecular biology. So far, Dr. Alharby has published more than 40 papers in peer-reviewed international journals and attended several international conferences. He was a Head of the Laboratories at Teachers College, Jeddah, Saudi Arabia, in 2005. At the moment, he is the Head of Plant Section in the Department of Biological Sciences at KAU.

Maize Production Under Salinity and Drought Conditions: Oxidative Stress Regulation by Antioxidant Defense and Glyoxalase Systems

Md. Motiar Rohman, Md. Robyul Islam, Tahia Naznin, Sumaiya Haque Omy, Shahnewaz Begum, Sadia Sabrina Alam, Mohammad Amiruzzaman, and Mirza Hasanuzzaman

Introduction

Abiotic stress factors, particularly drought and soil salinity, are the major stresses limiting crop yields worldwide. The forthcoming global climate changes have been increasing the possibilities of higher mean temperatures, extreme seasonal weather patterns, and the frequency, intensity and duration of drought, as well as heat waves. At the same time, increasing soil salinity in coastal regions has focused attention on the possibility of crop damage in fields located in sea regions worldwide. These problems due to salinity and drought will affect the production of agricultural crops in the upcoming years, particularly in arid and semiarid regions (IPCC 2014). This situation is a great threat to ensuring food security for densely populated countries, because crops that have a higher yield but have lower adaptability to salinity and drought will need to be replaced by crops that have higher adaptive potential but are likely to have a lower yield. As a result, there is an urgent need to develop highly adaptive crops that also have a higher yield, in order to address both salinity and drought.

The primary effects of drought stress are changes in key biochemical and physiological processes as a consequence of drought-induced osmotic stress, which may cause oxidative damage in most plants (Ashraf 2010). Salinity affects plant growth and development in two ways. First, it imposes osmotic stress by reducing the soil water potential, leading to limited water uptake. Second, it causes excessive uptake of ions, particularly Na^+ and Cl^-, that ultimately interfere with various metabolic

M. M. Rohman (✉) · M. R. Islam · T. Naznin · S. H. Omy · S. Begum
S. S. Alam · M. Amiruzzaman
Plant Breeding Division, Bangladesh Agricultural Research Institute,
Joydebpur, Gazipur, Bangladesh
e-mail: mrahman@bari.gov.bd

M. Hasanuzzaman
Department of Agronomy, Faculty of Agriculture, Sher-e-Bangla Agricultural University,
Dhaka, Bangladesh

M. Hasanuzzaman et al. (eds.), *Plant Abiotic Stress Tolerance*,
https://doi.org/10.1007/978-3-030-06118-0_1

processes. Plant responses to the osmotic and ionic components of salt stress are complicated and involve many gene networks and metabolic processes (Hasegawa et al. 2000; Munns and Tester 2008). Such responses depend mainly on the inherent salt tolerance of the plant, the severity of salt stress (the concentration of salt in the soil solution), and the duration of the plant roots' exposure to the salt. Both salinity and drought tolerance are complex traits, and plant breeders' efforts to produce crops with higher yields have largely been unsuccessful because of mutagenic adaptive responses to these traits.

Both salinity and drought stresses have negative consequences for gas exchange, resulting in low CO_2 assimilation for photosynthesis and consequently a significant reduction in electron transportation. As a result, reactive oxygen species (ROS) are generated, such as singlet oxygen ($^1O_2$), superoxide anions ($O_2^{\bullet-}$), hydrogen peroxide (H_2O_2), perhydroxy radicals ($HO_2^{\bullet}$), and alkoxy radicals ($RO^{\bullet}$) (Gill and Tuteja 2010; Moller et al. 2007). Normally, generation of ROS is balanced with scavenging by various antioxidants (Foyer and Noctor 2005). This balance between the generation and scavenging of ROS is broken down by various biotic and abiotic stresses, including salinity and drought. At higher concentrations, ROS are highly reactive and cause damage to proteins, DNA, lipids, and carbohydrates, resulting in cell death (Fig. 1). As a result, accumulation of ROS under environmental stresses is the foremost reason for reduced productivity of crops (Mittler 2002; Apel and Hirt 2004; Mahajan and Tuteja 2005). Methylglyoxal (MG) is a potentially cytotoxic compound, which can react with and modify other molecules, including DNA and proteins (Yadav et al. 2005a). It is an α-oxoaldehyde compound and is produced copiously under different abiotic stress via different enzymatic and nonenzymatic reactions (Singla-Pareek et al. 2008; Yadav et al. 2005a, b). Therefore, both ROS and MG must be detoxified, or dangerous increases of them must be prevented, to keep them below toxic levels for cellular survival under stressful conditions.

Plants possess efficient means for scavenging of ROS produced during various environmental stresses, including salinity and drought. This requires the defense mechanisms of both enzymatic and nonenzymatic antioxidants for cellular protec-

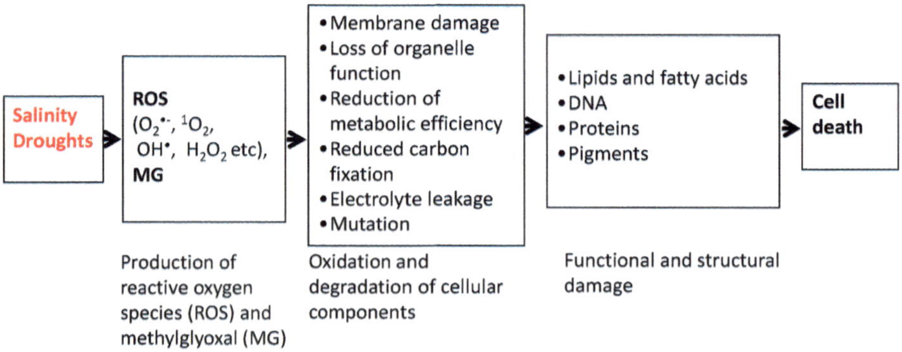

Fig. 1 General effects of reactive oxygen species (*ROS*) and methylglyoxal (*MG*) in plant cells under salinity and drought. $^1O_2$, singlet oxygen; $O_2^{\bullet-}$, superoxide anions and H_2O_2, hydrogen peroxide

tion (Choudhury et al. 2013). The enzymatic antioxidants include superoxide dismutase (SOD), catalase (CAT), ascorbate peroxidase (APX), monodehydroascorbate reductase (MDHAR), dehydroascorbate reductase (DHAR), glutathione reductase (GR), glutathione peroxidase (GPX), and glutathione *S*-transferase (GST). The nonenzymatic antioxidants include ascorbic acid (AsA), glutathione (GSH), phenolic compounds, alkaloids, nonprotein amino acids, and α-tocopherols (Apel and Hirt 2004; Gill and Tuteja 2010). Cytotoxic MG is detoxified and GSH homeostasis is maintained via the glyoxalase (Gly) system (Yadav et al. 2005a, b), which consists of two enzymes: Gly I and Gly II. It has been reported that coordinated induction or regulation of both antioxidant and glyoxalase pathway enzymes are necessary for the plant to achieve substantial tolerance of oxidative stress (Singla-Pareek et al. 2008; Saxena et al. 2011).

Maize (*Zea mays* L.) is a C_4 plant belonging to the Poaceae family. Naturally, it is a cross-pollinated crop, and it is assumed to have adaptability due to improved photosynthesis, in comparison with C_3 plants, but at the expense of reduced photorespiration (Kanai et al. 1999). As a result, maize has been thought to suffer the least oxidative stress. However, several studies have recently proved that maize suffers substantially from oxidative damage under abiotic stress, particularly under salinity and drought. At the same, efficient antioxidant defense, with an important role in ROS scavenging, has been reported in maize (Stepien and Klobus 2005). In 2015, Farooq et al. (2015) published a review highlighting osmoregulation and osmoprotection, ion homeostasis, apoplastic acidification, an antioxidant defense system, hormonal regulation, and molecular mechanisms in maize. They also recommended some management practices to reduce salinity-mediated damage. Since then, a good number of studies have demonstrated regulation of oxidative stress under salinity in maize, but they have been very scattered. At the same time, studies on antioxidant-mediated mitigation of oxidative damage under drought stress have yielded information that has improved our understanding of oxidative damage. Therefore, this chapter focuses on recent approaches to attenuation of oxidative stress in maize plants grown under salinity and drought.

General Situation of Maize Production Under Salinity and Drought

Globally, maize is the third most important crop. It is a versatile crop grown in a wide range of agroclimatic zones. In fact, the suitability of maize for diverse environments is unmatched by that of any other crop. It is grown from below sea level to altitudes higher than 3000 m, in areas with 250 mm to more than 5000 mm of rainfall per year, and with a growing cycle ranging from 3 to 10 months (Sheikh et al. 2017). According to the Food and Agriculture Organization of the United Nations (FAO 2016), in 2016 the total area of maize cultivation was 195.4 million hectares, with production of 1100.2 million tonnes and an average yield of 5.63 tonnes per hectare (Fig. 2).

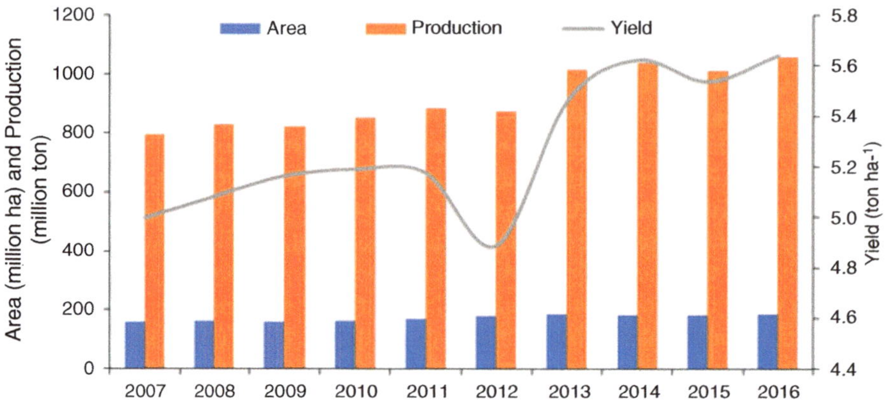

Fig. 2 Total area of production and yield of maize in the world

Abiotic stresses such as salinity, drought, and extreme temperatures are responsible for up to a 50–70% decline in major crop production (Mittler 2006). In the world, more than 800 million hectares of land is affected by either salinity (397 million hectares) or sodicity (434 million hectares) (Munns 2005). Drought affects agricultural crops more than any other stress and is becoming even more severe in the world as a result of global climate change. Statistically, drought stress doubled globally from 1970 to 2000 (Isendahl and Schmidt 2006). Therefore, it has become essential to develop crop varieties for those problematic soils. With this view, understanding of the mechanisms of salinity and drought tolerance has allowed significant achievements in the development of saline-tolerant and drought-tolerant maize globally. Drought-tolerant maize hybrids and open-pollinated varieties (OPVs), developed by the International Maize and Wheat Improvement Center (CIMMYT) in collaboration with other partner organizations, have been reported by Aslam et al. (2015). Recently, the Bangladesh Agricultural Research Institute (BARI) released two maize hybrids for cultivation in drought-prone areas of Bangladesh.[1] However, development of maize varieties for saline soil has not yet been successful, although numerous attempts have been undertaken.

Oxidative Stress in Maize

Both salinity and drought-mediated osmotic stress impel stomatal closure, resulting in low CO_2 assimilation. The osmotic effect of salt outside the roots induces stomatal responses (Munns and Tester 2008). As stress becomes prolonged, photosynthetic inhibition occurs because of severe water shortage, complete stomatal closure, ion toxicity, nutritional imbalance, and membrane disruption, affecting a range of physiological processes involved in cell metabolism (Munns 2002). The O_2 molecule has two impaired electrons. This spin restriction makes O_2 preferentially accept its

[1] Information collected from the Plant Breeding Division, BARI, Gazipur, Bangladesh.

electrons one at a time, leading to generation of ROS, which can damage cells (Gill and Tuteja 2010). ROS are also produced continuously as by-products of various metabolic pathways that are localized in different cellular compartments such as the chloroplast, mitochondria, and peroxisomes (del Rio et al. 2006; Navrot et al. 2007). Molecular oxygen is essentially a relatively stable molecule and nonreactive to living cells. However, when triplet oxygen receives extra energy or electrons under environmentally stressful conditions, it generates a variety of ROS, causing oxidative damage to cellular organs, including lipids, proteins, and nucleic acids. The most common ROS are $^1O_2$, H_2O_2, $O_2^{\bullet-}$, and $OH^{\bullet}$. Triplet oxygen has two unpaired electrons with parallel spins located in different orbitals. Upon receiving extra energy from a photosensitizer such as chlorophyll (chl), these two electrons show antiparallel spin, increasing the oxidizing power of oxygen (singlet oxygen) [reviewed by Krieger-Liszkay (2004)]. When triplet oxygen receives an electron, it produces $O_2^{\bullet-}$, which generates H_2O_2 and $OH^{\bullet}$ through a series of chemical conversions [reviewed by Apel and Hirt (2004)]. In photosynthesis, light energy is captured by photosystems I and II (PSI and PSII) and used to excite electrons, which go through a series of electron transport reactions. It is estimated that about 10% of the photosynthetic electrons leak from the photosynthetic electron transport chain (ETC) to oxygen as a final electron acceptor (the Mehler reaction), resulting in formation of $O_2^{\bullet-}$ (Foyer and Noctor 2000). When the terminal oxidases—cytochrome c oxidase and the alternative oxidase—react with O_2, four electrons are transferred and H_2O is released (Gill and Tuteja 2010). However, occasionally O_2 reacts with other ETC components and only one electron is transferred, resulting in formation of $O_2^{\bullet-}$. It has been noted that $O_2^{\bullet-}$ is the first ROS to be generated in plant tissues, accounting for 1–2% of O_2 consumption (Puntarulo et al. 1988). Usually, $O_2^{\bullet-}$ is produced during electron transport upon reduction of O_2 and also via the noncyclic pathway in the ETC of chloroplasts and other compartments of the plant cell. Reduction of O_2 to $O_2^{\bullet-}$ can occur in the ETC at the level of PSI. The $O_2^{\bullet-}$ may produce more reactive ROS such as $OH^{\bullet}$ and $^1O_2$ (Elstner 1987). These ROS are responsible for peroxidation of membrane lipids and cellular leakage. The protonation of the generated $O_2^{\bullet-}$ can produce a powerful oxidizing agent, perhydroxy radical ($HO_2^{\bullet}$), on negatively charged membrane surfaces, and the $HO_2^{\bullet}$ then directly attacks polyunsaturated fatty acid (PUFA) (Bielski et al. 1983). Furthermore, $O_2^{\bullet-}$ can produce H_2O_2 and $OH^{\bullet}$ through the Haber–Weiss reaction and the Fenton reaction (Apel and Hirt 2004).

Under normal conditions, the most common ROS ($O_2^{\bullet-}$ and H_2O_2) result from electron leakage from the photosynthetic and respiratory ETCs to oxygen. H_2O_2 is also produced through photorespiration resulting from the oxygenase activity of ribulose-1,5-bisphosphate carboxylase/oxygenase (Rubisco). Rates of photorespiration are controlled by the ratio of [CO_2] to [O_2] and temperature. The key feature of C_4 photosynthesis is the operation of a CO_2-concentrating mechanism in the leaves (Hatch 1987). C_4 plants such as maize use nicotinamide adenine dinucleotide phosphate (NADP)–malic enzyme–type photosynthesis (Omoto et al. 2012) and fix atmospheric CO_2 principally into oxaloacetate through phosphoenolpyruvate carboxylase in mesophyll cells. Oxaloacetate is then transported to mesophyll cell chloroplasts and reduced to malate by the NADP-dependent malate dehydrogenase enzyme. Malate is then shifted to bundle sheath cells of chloroplasts and decarboxylated by the NADP–malic enzyme to provide CO_2 and reducing power. This CO_2 is fixed via the Calvin cycle, as also happens in C_3 plants [reviewed by Farooq et al. (2015)]. In that way, higher CO_2 contents near Rubisco in the bundle sheath cells hold down oxygenase activity and help to improve photosynthesis in comparison with C_3 plants (Stepien and Klobus 2005).

In C_3 plants, photorespiration constitutes about 20–30% of photosynthesis under atmospheric conditions at 25 °C (Sage 2004). Contrarily, C_4 plants show lower rates of photorespiration (3.5–6% of photosynthesis) under various environmental conditions because of their CO_2-concentrating mechanism (Lacuesta et al. 1997; Carmo-Silva et al. 2008). However, under normal conditions, ROS do not vary largely between C_3 and C_4 plants. Stepien and Klobus (2005) have described similar lipid peroxidation through ROS activity in maize and wheat. Possible reasons for this might be either that the actual contribution of photorespiration to generation of ROS under normal conditions in C_3 plants is less than expected, or that C_3 plants have greater enzymatic activity [e.g., higher CAT activity] than C_4 plants, which immediately detoxifies H_2O_2. However, Stepien and Klobus (2005) found that maize has a better capacity to preserve the photosynthetic apparatus against overproduction of ROS than wheat. Vitkauskaitė and Venskaitytė (2011) reported higher water use efficiency and lower stomatal conductance, transpiration rates, intercellular CO_2 concentrations, and lipid peroxidation in the C_4 plant *Panicum miliaceum* L. Similarly, Uzilday et al. (2011) reported higher malondialdehyde (MDA) and ROS levels in *Cleome spinosa* (C_3) than in *C. gynandra* (C_4).

Production of lipid peroxidation has been reported in details by Gill and Tuteja (2010). Overproduction of ROS and PUFA in the cell wall causes membrane leakage with secondary damage to membrane proteins (Moller et al. 2007). Several aldehydes—such as 4-hydroxy-2-nonenal (HNE) and MDA, as well as hydroxyl and keto fatty acids—are formed as a result of PUFA peroxidation (Box 1). The breakdown products of aldehyde can form conjugates with proteins and DNA. In maize, cytoplasmic male sterility might be caused by aldehydes formed in the mitochondria because a restorer gene in this species encodes a mitochondrial aldehyde dehydrogenase (Moller et al. 2007). Higher MDA has been reported in seedlings of susceptible maize genotypes under salinity and drought conditions (Rohman et al. 2016a, b; AbdElgawad et al. 2016). Production of MDA in maize also depends on the types and intensity of stress. For example, higher MDA levels have been observed in sudden drought stress than in gradual stress (Zhang et al. 2014).

The lipoxygenase enzyme (LOX; International Union of Biochemistry and Molecular Biology (IUBMB) Enzyme Commission number (EC) 1.13.11.12) catalyzes the peroxidation of PUFAs to their corresponding hydroperoxides (Doderer et al. 1992). Increased LOX activity has been assumed to be a reason for increased

Box 1 Oxidation of Polyunsaturated Fatty Acid by Reactive Oxygen Species (ROS) in Plant Cells

$$\text{Linolenic acid} \xrightarrow{\text{ROS}} \text{Malondialdehyde (MDA)}$$

$$\text{Linoleic acid} \xrightarrow{\text{ROS}} \text{4 hydroxy 2 nonenal (HNE)}$$

lipid peroxidation of PUFAs, as reported in many plants (Demiral and Türkan 2004; Azooz et al. 2009; Sánchez-Rodríguez et al. 2012). Therefore, LOX is important for cell damage. An association between higher LOX and higher MDA production in maize has been reported in both inbreds and hybrids (Azooz et al. 2009; Rohman et al. 2016a, b). Lipid peroxidation has been reported to be higher in susceptible genotypes of different crops, including maize; therefore, it has become useful as a selection criterion for stress tolerance (Azooz et al. 2009; Rao et al. 2013). Recently, we found lower MDA with a higher yield in some tolerant maize genotypes than in susceptible genotypes (Rohman et al. 2018).

Antioxidative Defense Mechanisms

From different studies, it has been established that plants exposed to abiotic stresses—such as salinity, drought, temperature extremes, heavy metals, and nutrient deficiency—produce ROS (e.g., $^1O_2$, $O_2^{\bullet -}$, H_2O_2, and $OH^{\bullet}$). Plants possess an internal mechanism to protect themselves from these ROS in their cells and organelles—such as chloroplasts, mitochondria, and peroxisomes—and the cellular antioxidant machinery is important for protection against various stresses (Gill and Tuteja 2010). Salinity- and drought-mediated overproduction of ROS changes general metabolic processes, causing oxidative stress in maize (Farooq et al. 2015; Rohman et al. 2016b; Anjum et al. 2011). The defense system consists of both enzymatic and nonenzymatic antioxidants. Among the enzymatic antioxidants, SOD, CAT, POD (peroxidase), APX, GPX, MDHAR, DHAR, and GR are important. GSH, AsA (both water soluble), carotenoids, and tocopherols (lipid soluble) are nonenzymatic antioxidants (Stepien and Klobus 2005; Ahmad et al. 2016).

Superoxide Dismutase

Superoxide dismutase (SOD; EC 1.15.1.1), a metalloenzyme, plays a very important role against oxidative stress in all aerobic organisms and in all subcellular organelles sensitive to ROS. SOD deploys first-line protection and catalyzes the dismutation of $O_2^{\bullet -}$ to O_2 and H_2O_2. SOD enzymes are localized in different cellular compartments and, depending on their metal cofactor, they have three different isozymes: copper/zinc (Cu/Zn-SOD), manganese (Mn-SOD), and iron (Fe-SOD) (Mittler 2002). Mn-SOD is localized in the mitochondria of eukaryotic cells and in peroxisomes (del Río et al. 2003); some Cu/Zn-SOD is present in cytosolic fractions and also in chloroplasts of higher plants (del Rio et al. 2002). Fe-SOD is usually associated with the chloroplast compartment (Alscher et al. 2002). Mn-SOD and Fe-SOD are prokaryotic. Cu/Zn-SOD is eukaryotic and a dimer, whereas Mn-SOD in mitochondria is a tetramer. Increased activity of SOD is correlated with increased tolerance of plants against different environmental stresses. The activation of SOD in maize under salinity and drought conditions is shown in Table 1.

Table 1 Regulation of enzymatic antioxidants in maize under salinity and drought

Antioxidant enzyme	Stress	Regulation	Reference
SOD	Drought	Up	Yang et al. (2015)
	Recovery (from drought)	Down	
	Salinity	Up	Rios-Gonzalez et al. (2002)
	Salinity	Up	de Azevedo Neto et al. (2006)
		Down (sensitive cultivar)	
	Salinity	Up	Rohman et al. (2015)
	Salinity	Up	Rohman et al. (2016a)
	Drought	Up	Rohman et al. (2016b)
	Salinity	Up	Fu et al. (2017)
	Drought	Up	Moussa and Abdel-Aziz (2008)
	Drought	Up	Goodarzian-Ghahfarokhi et al. (2016)
	Salinity and drought	Up	Saed-Moocheshi et al. (2014)
	Salinity	Up	Jiang et al. (2017)
	Salinity	Up (moderate stress)	Wang et al. (2017)
		Down (severe stress)	
	Drought	Up (tolerant)	Chugh et al. (2011)
		Down (sensitive)	
	Salinity	Up	Kaya et al. (2013)
	Salinity	Up	AbdElgawad et al. (2016)
	Salinity	Up	Ahmad et al. (2016)
	Drought	Up	Naeem et al. (2018)
	Salinity	Up	Bustos et al. (2008)
	Salinity	Up	Ashraf et al. (2018)
	Drought	Up	Avramova et al. (2015)
	Drought	Up	Ye et al. (2016)
	Drought	Up	Avramova et al. (2017)
	Drought	Up	Noman et al. (2015)
	Drought	Up	Rajasekar et al. (2015)
	Drought	Up (Mn-SOD, Fe-SOD)	Shiriga et al. (2014)
		Unchanged (Cu/Zn-SOD)	
	Salinity	Up	Jiang et al. (2017)
	Salinity	Up	Stepien and Klobus (2005)
	Drought	Up	Abdelgawad et al. (2014)
	Drought	Up	Ashraf et al. (2015)
	Salinity	Up	Kaya et al. (2015)
	Drought	Up	Huo et al. (2016)
	Drought	Up	Ali and Ashraf (2011)
	Salinity	Down	Ashraf (2010)
	Salinity	Up	Estrada et al. (2013)

(continued)

Table 1 (continued)

Antioxidant enzyme	Stress	Regulation	Reference
	Salinity	Up	Gondim et al. (2012)
	Salinity	Up	Tuna et al. (2013)
	Salinity	Up	Kaya et al. (2010)
	Drought	Up	Anjum et al. (2011)
CAT	Drought	Up	Yang et al. (2015)
	Recovery (from drought)	Down	
POD	Salinity	Up	Rios-Gonzalez et al. (2002)
	Salinity	Down	de Azevedo Neto et al. (2006)
	Salinity	Varied	Fu ct al. (2017)
	Salinity	Down	Rohman et al. (2015)
	Salinity	Down	Rohman et al. (2016a)
	Drought	Up	Rohman et al. (2016b)
	Drought	Up	Moussa and Abdel-Aziz (2008)
	Salinity and drought	Up	Saed-Moocheshi et al. (2014)
	Salinity	Up (moderate stress)	Wang et al. (2017)
		Down (severe stress)	
	Drought	Up (tolerant)	Chugh et al. (2011)
		Down (sensitive)	
	Salinity	Up	Kaya et al. (2013)
	Salinity	Up	AbdElgawad et al. (2016)
	Salinity	Up	Latef and Tran (2016)
	Drought	Up	Naeem et al. (2018)
	Salinity	Up	Ashraf et al. (2018)
	Drought	Up	Avramova et al. (2015)
	Drought	Up	Ye et al. (2016)
	Drought	Up	Avramova et al. (2017)
	Drought	Up	Rajasekar et al. (2015)
	Drought	Up	Abdelgawad et al. (2014)
	Drought	Down	Ashraf et al. (2015)
	Salinity	Up	Kaya et al. (2015)
	Drought	Up	Huo et al. (2016)
	Salinity	Up	Akram et al. (2017)
	Salinity	Unchanged	Ashraf (2010)
	Salinity	Up	Estrada et al. (2013)
	Drought	Up	Sandhya et al. (2010)
	Salinity	Down	Gondim et al. (2012)
	Salinity	Down	Agami (2013)
	Drought	Up	Vardharajula et al. (2011)
	Salinity	Up	Tuna et al. (2013)
	Salinity	Up	Kaya et al. (2010)

(continued)

Table 1 (continued)

Antioxidant enzyme	Stress	Regulation	Reference
	Salinity	Up	Gholizadeh and Kohnehrouz (2010)
	Drought	Up	Anjum et al. (2011)
	Salinity	Up	Rios-Gonzalez et al. (2002)
	Salinity	Up	Fu et al. (2017)
	Salinity	Up	Rohman et al. (2015)
	Salinity	Up	Rohman et al. (2016a)
	Drought	Up	Rohman et al. (2016b)
	Drought	Up	Moussa and Abdel-Aziz (2008)
	Salinity and drought	Up	Saed-Moocheshi et al. (2014)
	Salinity	Up (moderate stress)	Wang et al. (2017)
		Down (severe stress)	
	Drought	Up (tolerant)	Chugh et al. (2011)
		Down (sensitive)	
	Salinity	Unchanged	Kaya et al. (2013)
	Salinity	Unchanged	AbdElgawad et al. (2016)
	Salinity	Up	Latef and Tran (2016)
	Drought	Up	Naeem et al. (2018)
	Salinity	Up	Ashraf et al. (2018)
	Drought	Up	Avramova et al. (2015)
	Drought	Up	Ye et al. (2016)
	Drought	Up	Avramova et al. (2017)
	Drought	Up	Noman et al. (2015)
	Salinity	Up	Jiang et al. (2017)
	Drought	Up	Abdelgawad et al. (2014)
	Drought	Up	Ashraf et al. (2015)
	Salinity	Up	Kaya et al. (2015)
	Drought	Up	Huo et al. (2016)
	Salinity	Up	Akram et al. (2017)
	Drought	Up	Ali and Ashraf (2011)
	Salinity	Unchanged	Ashraf (2010)
	Salinity	Down	Agami (2013)
	Salinity	Up	Tuna et al. (2013)
	Salinity	Up	Kaya et al. (2010)
	Salinity	Down	Gholizadeh and Kohnehrouz (2010)
	Drought	Up	Anjum et al. (2011)
APX	Salinity	Unchanged	de Azevedo Neto et al. (2006)
	Salinity	Up	Fu et al. (2017)
	Salinity	Down	Rohman et al. (2015)
	Salinity	Down	Rohman et al. (2016a)

(continued)

Table 1 (continued)

Antioxidant enzyme	Stress	Regulation	Reference
APX	Drought	Up	Rohman et al. (2016b)
	Drought	Down	Goodarzian-Ghahfarokhi et al. (2016)
	Salinity	Up	Jiang et al. (2017)
	Salinity	Up (moderate stress)	Wang et al. (2017)
		Down (Severe stress)	
	Drought	Up (tolerant)	Chugh et al. (2011)
		Down (sensitive)	
	Salinity	Up	AbdElgawad et al. (2016)
	Drought	Up	Naeem et al. (2018)
	Drought	Up	Ye et al. (2016)
	Drought	Up	Avramova et al. (2017)
	Drought	Up	Rajasekar et al. (2015)
	Salinity	Up	Jiang et al. (2017)
	Salinity	Up	Stepien and Klobus (2005)
	Salinity	Up	Jiang et al. (2017)
	Salinity	Up	Shan et al. (2014)
	Drought	Up	Sandhya et al. (2010)
	Salinity	Up	Gondim et al. (2012)
	Drought	Up	Vardharajula et al. (2011)
GPX	Salinity	Up	de Azevedo Neto et al. (2006)
		Down (sensitive)	
	Salinity	Up	Fu et al. (2017)
	Salinity	Up	Rohman et al. (2015)
	Salinity	Up	Rohman et al. (2016a)
	Drought	Up	Avramova et al. (2017)
	Drought	Up	Sandhya et al. (2010)
	Salinity	Up	Gondim et al. (2012)
	Drought	Up	Vardharajula et al. (2011)
GR	Salinity	Up (tolerant)	de Azevedo Neto et al. (2006)
		Down (sensitive)	
	Salinity	Up	Fu et al. (2017)
	Salinity	Up	Rohman et al. (2015)
	Salinity	Up	Rohman et al. (2016a)
	Drought	Up	Ahmad et al. (2016)
	Drought	Unchanged	Chugh et al. (2011)
	Salinity	Up	AbdElgawad et al. (2016)
	Drought	Up	Avramova et al. (2017)
	Salinity	Up	Stepien and Klobus (2005)
	Salinity	Up	Shan et al. (2014)

(continued)

Table 1 (continued)

Antioxidant enzyme	Stress	Regulation	Reference
DHAR	Salinity	Up	AbdElgawad et al. (2016)
	Drought	Up	Avramova et al. (2017)
	Salinity	Up	Rohman et al. (2015)
	Salinity	Up	Rohman et al. (2016a)
	Salinity	Up	Rohman et al. (2016b)
	Drought	Up	Shan et al. (2014)
MDHAR	Salinity	Up (severe)	AbdElgawad et al. (2016)
		Unchanged (mild)	
	Drought	Up	Avramova et al. (2017)
	Salinity	Up	Rohman et al. (2015)
	Salinity	Up	Rohman et al. (2016a)
	Salinity	Up	Rohman et al. (2016b)
	Salinity	Up	Shan et al. (2014)
GST	Salinity	Up	Rios-Gonzalez et al. (2002)
	Salinity	Up	AbdElgawad et al. (2016)
	Salinity	Up	Rohman et al. (2016a)

APX **ascorbate peroxidase,** ***CAT*** **catalase,** ***DHAR*** **dehydroascorbate reductase,** ***GPX*** **glutathione peroxidase,** ***GR*** **glutathione reductase,** ***GST*** **glutathione** ***S*****-transferase,** ***MDHAR*** **monodehydroascorbate reductase,** ***POD*** **peroxide,** ***SOD*** **superoxide dismutase**

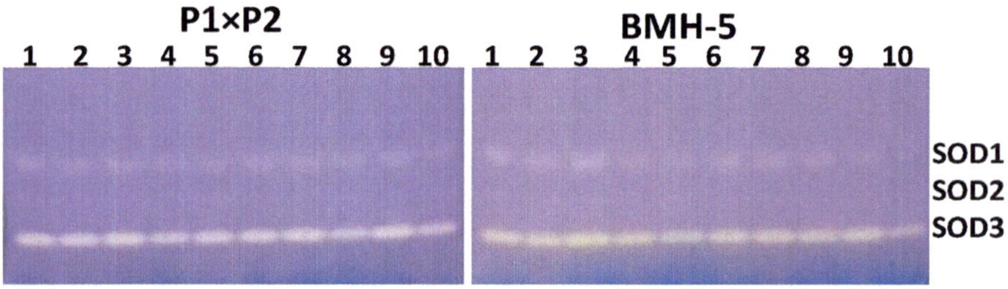

Fig. 3 Different isozymes of superoxide dismutase (*SOD*) under salinity stress

Different types of SOD under salinity conditions have not yet been described in maize. Recently, we analyzed the SOD band by negative staining in a tolerant hybrid (P1 × P2) and a sensitive hybrid (BMH-5) (Fig. 3), but the types have not yet been analyzed. Maize SOD was also found to increase under chilling stress (Fryer et al. 1998). Activation of SOD isozymes has been reported under salinity conditions in olive (Valderrama et al. 2006) and under drought conditions in rice (Sharma and Dubey 2005; Mishra et al. 2013), sugar beet (Sayfzadeh et al. 2011), and wheat (Sgherri et al. 2000).

Catalase

Catalase (CAT; EC 1.11.1.6) is a tetrameric heme-containing enzyme, responsible for metabolizing stress-provoked ROS (such as H_2O_2) and controlling their potential impacts on cellular metabolism and functions (Anjum et al. 2016). It is important for scavenging of H_2O_2 generated in peroxisomes during photorespiratory oxidation, β-oxidation of fatty acids, and purine catabolism (del Rio et al. 2006). It is considered to have the highest specificity for H_2O_2, with a very fast turnover rate. However, it is reluctant to show affinity for organic peroxides because in an enzymatic reaction, CAT is independent of other cellular reductants for instituting its activity (Scandalios 2005). Therefore, CAT is unique, as it does not require a cellular reducing equivalent. As a result, it shows a much lower affinity for H_2O_2 than APX. Three CAT classes have been proposed (Anjum et al. 2016): class I CATs are expressed in photosynthetic tissues and are regulated by light, class II CATs are expressed at high levels in vascular tissues, and class III CATs are highly abundant in seeds and young seedlings. CATs are reported to be changed by abiotic stresses, and their activity is either enhanced or decreased depending on the type and intensity of stress (Hasanuzzaman et al. 2014; Rohman et al. 2016a; Sharma and Dubey 2005). Three isoforms have been reported in maize (CAT1, CAT2, and CAT3) on separate chromosomes and are differentially expressed and independently regulated (Scandalias 1990). CAT1 and CAT2 are found in peroxisomes and the cytosol, whereas CAT3 is found in the mitochondria. As a C_4 crop, maize was thought not to show increased CAT activity. Recently, however, CAT has been reported to increase in tolerant maize genotypes (Rohman et al. 2016b; Chugh et al. 2011). Activation of CAT activity under salinity and drought is shown in Table 1.

Peroxidase

Peroxidase (POD; EC 1.11.1.7) is a heme-containing enzyme. It uses guaiacol and pyrogallol at the expense of H_2O_2 (Asada 1999). It is found in higher organisms and microbes. Many POD isoenzymes are localized in the vacuoles, cell walls, and cytosol of plant tissue (Asada 1992). POD is important for biosynthesis of lignin and ethylene, degradation of indole acetic acid, and healing of wounds, as well as for conferring plant tolerance under environmental stresses (Kobayashi et al. 1996). PODs are important for quenching reactive intermediary forms of O_2 and peroxy radicals under stress conditions (Vangronsveld and Clijsters 1994). Induced activity of POD under salinity has been reported in soybean (Weisany et al. 2012), licorice (Pan et al. 2006), and *Lepidium sativum* (Manaa et al. 2014), while drought stress has been shown to induce POD activity in rapeseed (Abedi and Pakniyat 2010) and licorice (Pan et al. 2006). Recently, Tayefi-Nasrabadi et al. (2011) observed increased tolerance achieved through induction of POD activity in sunflower. POD has been reported to play a significant role in ROS detoxification in maize

(Rohman et al. 2016a, b). The activity of maize POD is organ specific (Rios-Gonzalez et al. 2002). Activation of POD in maize under salinity and drought is shown in Table 1.

Ascorbate Peroxidase

Ascorbate peroxidase (APX; EC 1.1.11.1) plays the most important role in ROS scavenging and thus protects cells in all living organisms. APX is a multigenic family consisting of at least five different isoforms, including thylakoid (tAPX) and glyoxysome membrane forms (gmAPX), as well as a chloroplast stromal soluble form (sAPX) and a cytosolic form (cAPX) (Peng et al. 2005). APX scavenges H_2O_2 in the water–water and AsA–GSH cycles, where AsA acts as an electron donor (Apel and Hirt 2004). Enhanced activity of APX in response to abiotic stresses, including drought and salinity, has been established (Sharma and Dubey 2005, 2007; Han et al. 2009; Maheshwari and Dubey 2009; Hefny and Abdel-Kader 2009; Hasanuzzaman et al. 2014; Nahar et al. 2015; Ghaderi et al. 2015). Regulation of APX in different crop species under various stresses has also been reviewed (Gill and Tuteja 2010). Recently, Anjum et al. (2017) described enhanced APX activity, which varied with drought duration and severity, where high-yielder genotypes exhibited comparatively higher APX activity. Previously, we had also observed higher APX activity in tolerant inbreds and hybrids under both drought and salinity stress (Rohman et al. 2016a, b). AbdElgawad et al. (2016) reported higher APX activity in the root of maize under salinity. Activation and regulation of APX activity in maize under drought and salinity conditions are shown in Table 1. Six APX homologues with primary gene/protein features have been reported in maize (Ozyigit et al. 2016). Recently, we analyzed APX isozymes in two maize hybrids: P1 × P7 (a saline-tolerant hybrid) and BMH-5 (a saline-susceptible hybrid) (Fig. 4). The experiment was conducted under both aerobic and anaerobic saline conditions. In both cases, five isozymes were detected by sodium dodecyl sulfate polyacrylamide gel electrophoresis (SDS-PAGE). It is clear that in both hybrids, new isozymes—particularly APX1 and APX2—can play an important role in H_2O_2 scavenging.

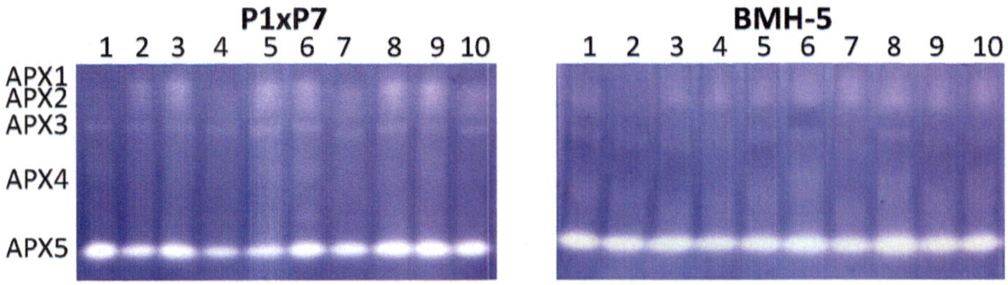

Fig. 4 Different isozymes of ascorbate peroxidase (*APX*)

Glutathione Peroxidase

Glutathione peroxidases (GPXs; EC 1.11.1.9) are a large and diverse family of isoenzymes, which use GSH to reduce H_2O_2 and organic and lipid hydroperoxides, thereby decreasing oxidative stress (Noctor et al. 2002). Ozyigit et al. (2016) identified different GPX homologues in 18 plant species; most of them were localized in the chloroplast, mitochondria, cytosol, and endoplasmic reticulum, where three GPXs were reported in maize. Plant GPXs have a cysteine residue at their active site (Koua et al. 2009), which is functional in both GSH and thiol peroxidase classes of the nonheme family. Many studies have described the significant stress-mitigating role of GPXs under various abiotic and biotic stress conditions such as oxidative stress, pathogen attack, metal, cold, drought, and salinity (Navrot et al. 2006; Diao et al. 2014; Fu 2014; Gao et al. 2014). Overexpression of GPX in soybean and tomato has been associated with higher tolerance of abiotic stress (Herbette et al. 2011; Ferreira Neto et al. 2013). In addition to stress responses, GPXs are also thought to regulate cellular redox homeostasis by modulating the thiol–disulfide balance (Bela et al. 2015). Upregulated GPX expression has been reported to maintain redox homeostasis in *Brassica rapa* under oxidative stress (Sugimoto et al. 2014). Introduction of a radish phospholipid hydroperoxide GPX gene (RsPHGPx) into yeast has been shown to be protective against membrane damage (Yang et al. 2005). The available information on regulation of GPX activity in maize under drought conditions is limited. Rohman et al. (2016a, b) reported the importance of GPX in both inbred and hybrid maize under drought and salinity. Recent works on GPX activation under salinity and drought are summarized in Table 1.

Glutathione Reductase

The changes in the ratio of GSH to glutathione disulfide (GSSG) and the ratio of AsA to dehydroascorbate (DHA) are essential for maintaining the most important nonenzymatic antioxidants GSH and AsA in response to oxidative stress. Glutathione reductase (GR; EC 1.8.1.7) is crucial for catalyzing the reduction of GSH and is involved in many metabolic regulatory and antioxidative processes in plants in the presence of reduced NADP (NADPH) (Apel and Hirt 2004). GR is located in the chloroplasts, cytosol, mitochondria, and peroxisomes, but about 80% of GR activity in photosynthetic tissues occurs in chloroplastic isoforms (Edwards et al. 1990). Both GSH and GR in chloroplasts are involved in detoxification of H_2O_2 generated by the Mehler reaction. GSH participates in maintaining sulfhydryl (–SH) groups, a substrate for GSTs (Noctor et al., 2012). Both GR and GSH play vital roles in maintaining plant tolerance under various stresses. The importance of GR in maintaining GSH and as an antioxidant has been confirmed by many studies (Noctor et al. 2012; Apel and Hirt 2004; Hasanuzzaman et al. 2014). Caverzan et al. (2016) reported upregulation of GR in different wheat genotypes. GR also provides tolerance in transgenic plants (Yousuf et al. 2012). The important of maize GR has been confirmed by many studies under salinity and drought (Table 1). Sometimes, the activity was tissue specific (Rohman et al. 2016a, b; AbdElgawad et al. 2016). Al Hassan et al. (2017) studied three species of *Juncus*—*J. maritimus*, *J. acutus* (both halophytes), and *J. articulatus* (salt-sensitive)—where GR activity was significantly higher in halophyte species than in salt-sensitive species.

Monodehydroascorbate Reductase

For ROS scavenging, high ratios of AsA to DHA and of GSH to GSSG are essential in cells. In the AsA–GSH cycle, MDHAR and DHAR are important for maintaining AsA (Apel and Hirt 2004). MDHAR is a flavin adenine dinucleotide (FAD) enzyme and is present as chloroplastic and cytosolic isozymes. It shows high specificity for monodehydroascorbate (MDHA) as an electron acceptor, using NAD(P)H as an electron donor (Hossain and Asada 1985). Hydrogen peroxide is also converted into water by the AsA–GSH cycle. The reducing agent in the first reaction, catalyzed by APX, is ascorbate, which oxidizes into MDHA. MHDAR reduces MDA into ascorbate with the help of NAD(P)H. Dehydroascorbate is produced spontaneously by MHDA and can be reduced to ascorbate by DHAR, using GSH to oxidize into GSSG. Sudan et al. (2015) reported a stress tolerance role of MDHA in finger millet under different abiotic stresses (drought, salt, and ultraviolet radiation). The role of the MDHAR gene is crucial for a mutualistic interaction between *Piriformospora indica* and *Arabidopsis* (Vadassery et al. 2009). Coexpression of the MDHAR and DHAR genes confers tolerance in *B. rapa* (Shin et al. 2013). Leterrier et al. (2005) reported that peroxisomal *MDAR1* in pea has a differential regulation, which could be indicative of its specific function in peroxisomes. High MDHAR activity has been reported in rice (Hasanuzzaman et al. 2014). MDHAR activity in maize was found to be regulated under drought and salinity (AbdElgawad et al. 2016; Rohman et al. 2016a, b). The activation of MDHAR in maize under salinity and drought is summarized in Table 1.

Dehydroascorbate Reductase

Dehydroascorbate reductase (DHAR; EC 1.8.5.1) maintains AsA by catalyzing the reduction of DHA, using GSH as the reducing substrate (Apel and Hirt 2004). It is very important to maintain AsA levels in plants under environmental stress. Overexpression of DHAR increases the AsA content in tobacco, maize, and potato (Chen et al. 2003; Qin et al. 2011). The role, purification, and overexpression of DHAR under different stresses in different crops have been described by various researchers (Hossain and Asada 1984; Dipierro and Borraccino 1991; Yoshida et al. 2006; Rubio et al. 2009; Wang et al. 1999; Hernandez et al. 2001; Sharma and Dubey 2005, 2007; Maheshwari and Dubey 2009; Eltayeb et al. 2011; Lu et al. 2007; Chang et al. 2017). Coexpression of the DHAR and MDHAR genes has been associated with better tolerance in *B. rapa* (Shin et al. 2013). Coordinated increases in the activities of MDHAR, DHAR, and GR have been associated with overexpression of Mn-SOD in maize leaves under cold stress (Kingston-Smith and Foyer 2000). Though drought-tolerant hybrid seedlings show higher DHAR activity under drought, saline-sensitive inbreds show higher DHAR activity under salinity (Rohman et al. 2016a, b). The activation of maize DHAR under salinity and drought is reported in Table 1.

Glutathione S-*Transferases*

Glutathione *S*-transferases (GSTs; EC 2.5.1.18) are multigenic family enzymes, which catalyze the conjugation of GSH to electrophilic substrates. GSTs can metabolize various toxic exogenous compounds (xenobiotics) by GSH conjugation (Cummins et al. 2011). Plant GSTs are commonly known for their role in herbicide detoxification and are also considered as glutathione peroxidases. Furthermore, GSTs function as nonenzymatic carriers (ligandins) in intracellular transport and catalyze anthocyanin–GSH conjugates, thereby allowing transport into vacuoles via a glutathione pump (Marrs 1996; Rohman et al. 2009). Plants GSTs are classified into eight classes: phi, tau, theta, zeta, lambda, glutathione-dependent DHARs, tetrachlorohydroquinone dehalogenase (TCHQD), and membrane-associated proteins in eicosanoid and glutathione metabolism (MAPEG) (Xu et al. 2015). Among these eight classes, phi and tau are the largest plant-specific classes and are often highly stress-inducible GSTs (Dixon and Edwards 2010). The role of GSTs in stress has been demonstrated in several transgenic studies (Roxas et al. 2000; Takesawa et al. 2002; Kumar et al. 2013; Yang et al. 2014; Yu et al. 2003; Chen et al. 2012). A total of 42 highly expressed GSTs have been found in maize (McGonigle et al. 2000). The Bronze2 (Bz2) gene in maize, encoding a GST, is involved in anthocyanin biosynthesis and transportation into the vacuole, and is highly expressed under stress (Marrs and Walbot 1997). Higher GST activity is associated with detoxification of hydroperoxides, as well as direct scavenging of H_2O_2, as shown by glutathione peroxidase activity under stress (Noctor et al. 2012). Remme et al. (2013) studied three homologues of maize GSTs, among which a GST showing higher activity was upregulated under salinity stress. Increased GST activity in maize under salinity and drought may be involved in leaf senescence (Rohman et al. 2016a, b). Higher GST activity under salinity and drought have also been reported by other research groups (Table 1).

Methylglyoxal Detoxification Enzymes

Methylglyoxal is a highly reactive alpha-ketoaldehyde, which interact with proteins and nucleic acids. It is produced primarily as a by-product of several metabolic pathways, such as glycolysis, lipid peroxidation, and oxidative degradation of glucose. Therefore, it must be detoxified or eliminated from the cell through biological systems. In plants, MG is detoxified mainly via the glyoxalase system, which comprises two enzymes: glyoxalase I (Gly I; EC 4.4.1.5) and glyoxalase II (Gly II; EC 3.1.2.6). Gly I converts MG to *S*-D-lactoylglutathione (SLG), utilizing GSH, while Gly II converts SLG to D-lactic acid (Yadav et al. 2005a, b). During the latter reaction, GSH is regenerated. Overexpression of glyoxalase enzymes in plants has been found to limit increases in ROS and MG levels under stress conditions by maintaining GSH homeostasis and antioxidant enzyme levels (Singla-Pareek et al. 2006, 2008). Increased activities of glyoxalases have also been reported under drought in other plant species (Hasanuzzaman et al. 2017; Alam et al. 2014; Hasanuzzaman and Fujita 2011). Decreased oxidative stress and enhanced salinity tolerance were observed in transgenic tomato with a highly expressed glyoxalase pathway (Viveros et al. 2013). Upregulation or overexpression of these enzymes increases tolerance of abiotic stresses in tobacco (Singla-Pareek et al. 2008; Saxena et al. 2011). A coordinated response of antioxidants and glyoxalase in increasing salinity tolerance was reported in rice (Hasanuzzaman et al. 2014) and in mung bean (Nahar et al. 2016). Though increased Gly I activity was observed in maize under salinity, Gly II activity was decreased (Rohman et al. 2016a), and increased activity of Gly I and Gly II under drought was observed only in tolerant maize genotypes (Rohman et al. 2016b). However, Mir et al. (2018) found that the activity of both enzymes was decreased under salinity. Therefore, the response of glyoxalase in maize is genotype dependent. Some external stimuli such as proline, betaine, and methyl jasmonate have been found to be useful to increase glyoxalases, conferring tolerance in maize under salinity (Mir et al. 2018; Rohman et al. 2016a).

Nonenzymatic Antioxidants

Nonenzymatic components of the antioxidative defense system include AsA and GSH, as well as tocopherol, carotenoids, and phenolic compounds. They play crucial roles in defense, interacting with numerous cellular components. In this chapter, only AsA and GSH are discussed with respect to salinity and drought stress in maize.

Ascorbic Acid

Ascorbic acid (AsA) is the most abundant and powerful ROS scavenger because it has the ability to donate electrons in several enzymatic and nonenzymatic reactions. It is found in all plant tissues, and its highest levels are found in mature leaves with fully developed chloroplasts and the highest chl levels. Under normal physiological conditions, AsA mostly remains in a reduced form in leaves and chloroplasts (Szarka et al. 2007). It protects cells and organelles from ROS, which overaccumulate under environmental stress, including salinity and drought (Latif et al. 2016; Mukhtar et al. 2016; Naz et al. 2016). It plays an important role in cell division and expansion, photosynthesis, hormone biosynthesis, and regeneration of antioxidants (Gallie 2012; Lisko et al. 2014). It protects membranes by directly scavenging $O_2^{\bullet-}$ and $OH^{\bullet}$ and by regenerating α-tocopherol from tocopheroxyl radicals (Gill and Tuteja 2010). AsA acts as a cofactor of violaxanthin de-epoxidase, thus sustaining dissipation of excess excitation energy in chloroplasts (Smirnoff 2000). In the AsA–GSH cycle, the AsA redox system consists of reduced AsA and oxidized MDHA and DHA, both oxidized forms of AsA being relatively unstable. MDHA is very transitory and disproportionates to either AsA or DHA (Apel and Hirt 2004). Therefore, it is very important to convert DHA into AsA for maintenance of cellular redox. MDHA is reduced to AsA in the presence of lactaldehyde reductase (NADPH), while DHA can be chemically reduced by GSH to ASH (Foyer and Halliwell 1976; Apel and Hirt 2004; Akram et al. 2017). In the AsH–GSH cycle, AsA not only preserves the activity of APX but also preserves the activities of enzymes containing prosthetic transition metal ions (Noctor and Foyer 1998). Under stressful conditions, redox homeostasis of AsA is lost as a result of increased GSSG formation. In these circumstances, exogenous application of AsA as a pretreatment with foliar application is considered advisable (Mukhtar et al. 2016; Naz et al. 2016). The results of exogenous AsA application, with improvements in tolerance under various abiotic stresses, have been reported by Akram et al. (2017), who found that both presowing and foliar spray were useful in maize, producing higher pigment levels, greater membrane stability, and higher antioxidant levels under drought and cold stresses. Tuna et al. (2013) applied 125 mM in irrigation water 25 days after sowing and measured biochemical parameters at the cob formation stage. They reported that AsA maintained lower Na^+/K^+ levels and higher antioxidant enzymatic antioxidant levels, as well as increasing the yield. Recently, we found increased antioxidant levels in maize seedlings treated with AsA in hydroponic culture (Billah et al. 2017). Saline-tolerant genotypes show higher AsA and AsA redox under salinity and drought at the seedling stage (Rohman et al. 2016a, b). Chugh et al. (2013) observed higher AsA content in drought-tolerant maize than in drought-sensitive maize at the reproductive stage.

Glutathione

Glutathione (GSH), a tripeptide of γ-glutamyl-cysteinyl-glycine, is the most important nonenzymatic antioxidant in the AsA–GSH cycle, and plays a central role in antioxidant defense through ROS scavenging and maintenance of redox homeostasis in plant tissue in a stressful environment (Noctor et al. 2012). It is localized in the cytosol, endoplasmic reticulum, vacuole, mitochondria, chloroplasts, and apoplast. In physiological processes, it plays an important role in detoxification of xenobiotics, transport, conjugation of metabolites, signal transduction, and stress-responsive gene expression (Mullineaux and Rausch 2005; Noctor et al. 2012). It is

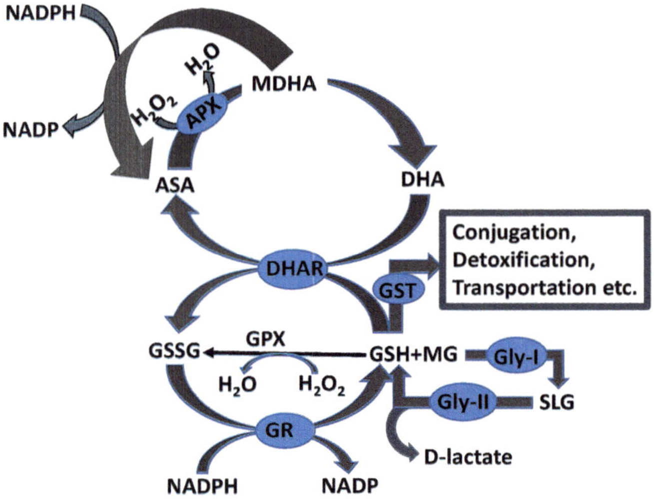

Fig. 5 Uses and maintenance of ascorbic acid (*AsA*) and glutathione (*GSH*) in the AsA–GSH cycle in plants

also important for growth and development of tissue, cell death and senescence, and enzymatic regulation for pathogen resistance (Sharma et al. 2012). Plants maintain a high cellular ratio (of about 20:1) between GSH and GSSG (its oxidized form) in unstressed conditions, and the balance between GSH and GSSG is a central component in maintaining cellular redox state (Foyer and Noctor 2005; Labudda and Azam 2014). GSH is necessary for activation of the functions of GPXs, GSTs, and glyoxalases, where GR plays a central role in recycling of GSH from GSSG in the presence of NADPH (Fig. 5). In the AsA–GSH cycle, regeneration of AsA is extremely important because fully oxidized DHA has a short half-life and will be lost unless it is reduced back. GSH is crucially important in maintaining AsA (Apel and Hirt 2004). Therefore, GSH plays a key role in ROS tolerance under oxidative stress, in association with other ROS-metabolizing enzymes. Drought tolerance has been correlated with higher GSH in tolerant maize genotypes (Chugh et al. 2013). Rohman et al. (2016a, b) found higher levels and redox of GSH in tolerant maize genotypes at the seedling stage. Herbicide safener has been reported to increase tolerance in a chilling-sensitive maize genotype by maintaining higher levels of GSH and its related enzymes (Kocsy et al. 2001).

Osmoregulation and Osmoprotection

Naturally, some common mechanisms and reactions are activated in the body of plants when they are affected by soil salinity, water deficit, and other abiotic stresses. These responses generally include growth inhibition, photosynthetic pigment degradation, regulation of ion transport, accumulation and biosynthesis of specific osmolytes for osmotic adjustment, or activation of antioxidant systems, since the aforesaid stressful conditions cause secondary oxidative stress in plants (Bartels and Sunkar 2005; Parida and Das 2005; Munns and Tester 2008; Szabados and Savoure 2010; Volkov 2015; Kumar et al. 2017).

Osmoregulation, or osmotic modification, is one of the key adaptations of plants to minimize the detrimental effects of salt and drought stress at the cellular level, exclusively during the first phase of salt stress (Farooq et al. 2015). Osmoprotection is primarily met with deposition of solutes (organic and inorganic) under salinity and/or drought to lower water potential without lessening the actual water content (Serraj and Sinclair 2002). Among the major osmolytes are soluble sugars, sugar alcohols, proline, glycine betaine, organic acids, and trehalose. Proline accumulation increases in maize plants facing salt stress (Kaya et al. 2010). In one study it was found that at 400 mM NaCl, the leaves of sweet corn accrued more than 600 μmol g^{-1} of proline (de Azevedo Neto et al. 2004). Likewise, salinity stress has been shown to result in a significant increase in proline content in maize (Kaya et al. 2013).

Saed-Moocheshi et al. (2014) reported that salinity and drought stress increased the amount of free proline, but the effect of drought was greater than that of salinity. In a salt-tolerant maize genotype (BR5033), soluble amino acid buildup significantly increased in leaves under salt stress, with the largest increase being 113%; BR5033 was the only genotype to also markedly increase amino acid content in the roots (de Azevedo Neto et al. 2004). With regard to carbohydrate levels in the leaves and roots in saline environments, all of the tested genotypes had the same or lower levels, except for the salt-tolerant maize genotype BR5033, in which the level was increased by 14% in the leaves (de Azevedo Neto et al. 2004). Moreover, in a comparison of root amino acid and carbohydrate content in salt-stressed plants of both salt-sensitive (BR5011) and salt-tolerant (BR5033) maize genotypes, the roots of the salt-tolerant (BR5033) maize genotype accumulated 132% and 122% more amino acids and carbohydrates, respectively, than those of the salt-sensitive (BR5011) maize genotype (de Azevedo Neto et al. 2004).

Hussein et al. (2007) observed decreased content of the amino acids arginine, lysine, serine, and glutamic acid; no change in glycine content; and increased proline content in response to salt stress in maize. Salt stress was also shown to induce accumulation of polyamine but not spermidine, possibly because of the fast turnover of the latter (Erdei et al. 1996). In one study, accumulation of glycine betaine and free proline increased under drought stress in Giza 2 (salt-tolerant) and Trihybrid 321 (salt-susceptible but adapted) maize genotypes; the magnitude of the increase in both osmolytes was higher in Giza 2 than in Trihybrid 321 (Moussa and Abdel-Aziz 2008). In another study the soluble sugar and proline levels of tested maize leaves and roots were higher in Con1 plants (control plants treated with 200 mL of the vehicle used in the study) than in Con2 plants (control plants grown in untreated meadow soil with a pH of 8.23). Moreover, the results showed a larger increase in the roots than in the leaves and also indicated that proline was more effective than soluble sugars in alleviating osmotic stress (Fu et al. 2017).

A recent investigation showed that accumulation of compatible osmolytes such as proline and soluble sugar in leaves of the drought-tolerant Dekalb-6525 and the drought-sensitive Yousafwala maize hybrids was substantially improved under drought stress (Naeem et al. 2018). However, the greater accumulation of soluble sugar in genotype Dekalb-6525 signified that it has greater genotypic tolerance of drought stress than the Yousafwala genotype, as these sugars help to maintain plant water relations, stabilizing membrane and protein structures through scavenging of hydroxyl radicals and adjustment of cytoplasmic pH under stress conditions (Guo et al. 2010). Correspondingly, accumulation of osmotically active substances (e.g., proline and soluble sugar) is noticeably greater in plants exposed to drought stress, possibly as a response to tissue damage or an approach to lessen oxidative damage in the plants (Aghaz et al. 2013).

The accumulation of soluble proteins under high alkaline pressure in maize plants may supply a storage form of nitrogen that is reutilized when the stress is over (Abdel Latef 2010; Abdel Latef and Chaoxing 2014; Zhang et al. 2014). It has been observed that the increase in the soluble protein content of maize under the highest alkaline level applied is accompanied by a marked reduction in growth. This finding suggests that under alkaline stress, maize plants divert most of the synthesized proteins from being used for growth to being used for osmoregulation to survive the stress. Thus, the strategy that maize plants use to cope with alkaline stress could be enhancement of nitrogen metabolism. However, the effects of alkaline stress on the content of osmoprotectants and antioxidants have been found to vary (Latef and Tran 2016).

In essence, the effects of salinity and drought-induced osmotic stress in maize plants can be minimized with the help of osmoregulation. Among the osmolytes responsible for osmoregulation in maize, proline and glycine betaine are the first-line ones in salt and/or drought stress conditions (Moussa and Abdel-Aziz 2008; Kumar et al. 2017). However, when administered as a foliar spray to maize seedlings under saline stress, exogenous proline was observed to provide better tolerance than betaine by upregulating both enzymatic and nonenzymatic antioxidants (Rohman et al. 2016a). Wang et al. (2017) observed partial tolerance under salinity stress with application of γ-aminobutyric acid (GABA) as a root drench, with improved photosynthesis and upregulated osmoprotectants and antioxidants. Exogenously applied nitric oxide conferred tolerance of salinity-induced oxidative stress in seedlings of maize cultivars that differed in salinity tolerance, by reducing ROS and MDA and by maintaining higher chl and activity of SOD, POD, and CAT enzymes (Kaya et al. 2015).

Salicylic acid is another important secondary metabolite that has been shown to induce salinity resistance in plants. Maize plants treated with exogenously applied salicylic acid have been reported to increase their dry biomass in both saline and nonsaline environments, but the effect was more pronounced in the saline environment (Gunes et al. 2007). In another study, foliar application of salicylic acid in a concentration of 200 parts per million (ppm) remedied the deleterious effects of salinity in a maize crop irrigated with 4000 ppm saline water, by improving the plant height; the number and area of green leaves; the stem diameter; and the stem, leaf, and whole-plant dry weights (Hussein et al. 2007). Moreover, salicylic acid application was shown to enhance proline accumulation and the content of amino acids such as arginine, lysine, serine, and glutamic acid in maize under stress conditions, and helped to overcome a salinity-induced nutrient imbalance (Hussein et al. 2007). Foliar spraying of hydrogen peroxide has also been reported to increase the activities of catalase, guaiacol peroxidase, ascorbate peroxidase, and superoxide dismutase in maize seedlings (Gondim et al. 2012). Presowing treatments with 28-homobrassinolide further enhanced the activities of antioxidative enzymes in addition to lowering lipid peroxidation and increasing the protein concentration, thus alleviating oxidative stress in salt-treated maize plants (Arora et al. 2008). Different rhizobacteria promoting growth in maize under salinity were reviewed by Farooq et al. (2015). However, exogenous protectants have seldom been shown to regulate glyoxalase activities in maize. Maize seed priming with 10 μM of jasmonic acid for 12 h was reported to increase the activities of glyoxalases, along with the AsA and GSH content, in saline-treated 25-day-old seedlings (Mir et al. 2018). Previously, a foliar spray of proline was found to be more effective than betaine in upregulating glyoxalases in saline-treated maize seedlings (Rohman et al. 2016a). In one study, exogenous trehalose (10 mM) in irrigation water improved photosynthesis, water relations, and oxidative stress tolerance, with upregulation of some key enzymes such as SOD, POD, and CAT (Ali and Ashraf 2011). Application of urea, along with the urease inhibitor *N*-(*n*-butyl) thiophosphoric triamide, in polyethylene glycol (PEG)–treated drought-stressed maize seedlings significantly enhanced their drought tolerance by protecting the photosynthetic apparatus, activating the antioxidant defense system, and improving osmoregulation (Gou et al. 2017). In this study, increased SOD, POD, and CAT reduced ROS and MDA. Hormones such as GA3 and cytokinin, applied as a foliar spray at the vegetative and reproductive stages, were found to improve growth and yield (Akter et al. 2014). Thus, a substantial number of studies have found that external osmoprotectants modulate oxidative stress in maize. However, most of the studies were limited to the seedling stage.

Conclusion

Cultivation of maize, as a high-potential crop, has increased substantially. Efforts to develop a drought-tolerant maize variety have been comparatively more successful than efforts to develop a saline-tolerant variety. The studies discussed in this chapter have revealed that maize suffers significant oxidative stress under salinity and drought. Among enzymatic antioxidants, superoxide dismutase (SOD), catalase (CAT), ascorbate peroxidase (APX), and glutathione peroxidase (GPX)—as well as POD—play important roles in scavenging of reactive oxygen species (ROS) in both types of stress. The AsA-maintaining enzymes, MDHAR and DHAR, sometimes do not show harmony in inducing activity under stress, indicating the probability of oxidation of ascorbic acid (AsA). In this circumstance, the maintenance of GSH by GR has revealed its important role in maintaining redox homeostasis of GSH under salinity and drought conditions. Reports of methylglyoxal (MG) detoxification in maize by glyoxalases are very limited. A significant number of studies have focused on experimental use of osmoprotectants to reduce oxidative stress. However, they were conducted mostly at the seedling stage and under very short-term salinity or drought stress. As a result, it is speculative to say that osmoprotectants will be able to lessen oxidative stress when plants are grown on a large scale in salinity-affected fields. Therefore, it is essential for a complete and economical solution package to be developed in order to reduce oxidative stress for sustainable maize production in drought-stressed and/or salinity-stressed environments. On the other hand, enzymatic antioxidant levels increase significantly under salinity and drought conditions, which provides a strong rationale for employment of molecular and biotechnological approaches for their further development and use.

Acknowledgement: The authors acknowledged the research grant of Sub-project CRG-389, NATP-2 for SOD and APX isozymes analyses.

References

Abdel Latef AA (2010) Changes of antioxidative enzymes in salinity tolerance among different wheat cultivars. Cereal Res Commun 38:43–55. https://doi.org/10.1556/CRC.38.2010.1.5

Abdel Latef AA, Chaoxing H (2014) Does the inoculation with *Glomus mosseae* improve salt tolerance in pepper plants? J Plant Growth Regul 33:644–653. https://doi.org/10.1007/s00344-014-9414-4

AbdElgawad ZA, Khalafaallah AA, Abdallah MM (2014) Impact of methyl jasmonate on antioxidant activity and some biochemical aspects of maize plant grown under water stress condition. Agric Sci 5:1077–1088

AbdElgawad H, Zinta G, Hegab MM, Pandey R, Asard H, Abuelsoud W (2016) High salinity induces different oxidative stress and antioxidant responses in maize seedlings organs. Front Plant Sci 7:276. https://doi.org/10.3389/fpls.2016.00276

Abedi T, Pakniyat H (2010) Antioxidant enzyme changes in response to drought stress in ten cultivars of oilseed rape (*Brassica napus* L.). Czech J Genet Plant Breed 46(1):27–34

Agami RA (2013) Alleviating the adverse effects of NaCl stress in maize seedlings by pre-treating seeds with salicylic acid and 24-epibrassinolide. S Afr J Bot 88:171–177

Aghaz M, Bandehagh A, Aghazade E, Toorchi M, Ghassemi-Gholezani K (2013) Effects of cadmium stress on some growth and physiological characteristics in dill (*Anethum graveolens*) ecotypes. Int J Agric Res Rev 3:409–413

Ahmad P, Abdel Latef AA, Hashem A, Abd_Allah EF, Gucel S, Tran LSP (2016) Nitric oxide mitigates salt stress by regulating levels of osmolytes and antioxidant enzymes in chickpea. Front Plant Sci 7:347. https://doi.org/10.3389/fpls.2016.00347

Akram NA, Shafiq F, Ashraf M (2017) Ascorbic acid—a potential oxidant scavenger and its role in plant development and abiotic stress tolerance. Front Plant Sci 8:613. https://doi.org/10.3389/fpls.2017.00613

Akter N, Islam MR, Karim MA, Islam T (2014) Alleviation of drought stress in maize by exogenous application of gibberellic acid and cytokinin. J Crop Sci Biotechnol 17(1):41–48

Al Hassan M, Chaura J, Donat-Torres MP, Boscaiu M, Vicente O (2017) Antioxidant responses under salinity and drought in three closely related wild monocots with different ecological optima. AoB Plants 9:plx009. https://doi.org/10.1093/aobpla/plx009

Alam MM, Nahar K, Hasanuzzaman M, Fujita M (2014) Trehalose-induced drought stress tolerance: a comparative study among different *Brassica* species. Plant Omics 7(4):271–283
Ali Q, Ashraf M (2011) Induction of drought tolerance in maize (*Zea mays* L.) due to exogenous application of trehalose: growth, photosynthesis, water relations and oxidative defense mechanism. J Agron Crop Sci 197(4):258–271. https://doi.org/10.1111/j.1439-037X.2010.00463.x
Alscher RG, Erturk N, Heath LS (2002) Role of superoxide dismutases (SODs) in controlling oxidative stress in plants. J Exp Bot 53(372):1331–1341
Anjum SA, Wang LC, Farooq M, Hussain M, Xue LL, Zou CM (2011) Brassinolide application improves the drought tolerance in maize through modulation of enzymatic antioxidants and leaf gas exchange. J Agron Crop Sci 197(3):177–185. https://doi.org/10.1111/j.1439-037X.2010.00459.x
Anjum NA, Sharma P, Gill SS, Hasanuzzaman M, Khan EA, Kachhap K, Mohamed AA, Thangavel P, Devi GD, Vasudhevan P, Sofo A (2016) Catalase and ascorbate peroxidase—representative H_2O_2-detoxifying heme enzymes in plants. Environmental science and pollution research 23(19):19002–29
Anjum SA, Ashraf U, Tanveer M, Khan I, Hussain S, Shahzad B, Zohaib A, Abbas F, Saleem MF, Ali I, Wang LC (2017) Drought induced changes in growth, osmolyte accumulation and antioxidant metabolism of three maize hybrids. Front Plant Sci 8:69. https://doi.org/10.3389/fpls.2017.00069
Apel K, Hirt H (2004) Reactive oxygen species: metabolism, oxidative stress, and signal transduction. Annu Rev Plant Biol 55:373–399
Arora N, Bhardwaj R, Sharma P, Arora HK (2008) 28-Homobrassinolide alleviates oxidative stress in salt treated maize (*Zea mays* L.) plants. Braz J Plant Physiol 20:153–157
Asada K (1992) Ascorbate peroxidase: a hydrogen peroxide scavenging enzyme in plants. Physiol Plant 85(2):235–241
Asada K (1999) The water–water cycle in chloroplasts: scavenging of active oxygens and dissipation of excess photons. Annu Rev Plant Physiol Plant Mol Biol 50:601–639
Ashraf M (2010) Inducing drought tolerance in plants: some recent advances. Biotechnol Adv 28:169–183
Ashraf MA, Rasheed R, Hussain I, Iqbal M, Haider MZ, Parveen S, Sajid MA (2015) Hydrogen peroxide modulates antioxidant system and nutrient relation in maize (*Zea mays* L.) under water-deficit conditions. Arch Agron Soil Sci 61(4):507–523. https://doi.org/10.1080/03650340.2014.938644
Ashraf MA, Akbar A, Parveen A, Rasheed R, Hussain I, Iqbal M (2018) Phenological application of selenium differentially improves growth, oxidative defense and ion homeostasis in maize under salinity stress. Plant Physiol Biochem 123:268–280
Aslam M, Maqbool MA, Cengiz R (2015) Drought stress in maize (*Zea mays* L.): effects, resistance mechanisms, global achievements and biological strategies for improvement. Springer, Cham, pp 5–17. https://doi.org/10.1007/978-3-319-25442-5_2
Avramova V, AbdElgawad H, Zhang Z, Fotschki B, Casadevall R, Vergauwen L, Knapen D, Taleisnik E, Guisez Y, Asard H, Beemster GTS (2015) Drought induces distinct growth response, protection, and recovery mechanisms in the maize leaf growth zone. Plant Physiol 169:1382–1396
Avramova V, AbdElgawad H, Vasileva I, Petrova AS, Holek A, Mariën J, Asard H, Beemster GTS (2017) High antioxidant activity facilitates maintenance of cell division in leaves of drought tolerant maize hybrids. Front Plant Sci 8:84. https://doi.org/10.3389/fpls.2017.00084
Azooz MM, Ismail AM, Elhamd MA (2009) Growth, lipid peroxidation and antioxidant enzyme activities as a selection criterion for the salt tolerance of maize cultivars grown under salinity stress. Int J Agric Biol 11:21–26
Bartels D, Sunkar R (2005) Drought and salt tolerance in plants. Crit Rev Plant Sci 24:23–58
Bela K, Horváth E, Gallé Á, Szabados L, Tari I, Csiszár J (2015) Plant glutathione peroxidases: emerging role of the antioxidant enzymes in plant development and stress responses. J Plant Physiol 176:192–201. https://doi.org/10.1016/j.jplph.2014.12.014

Bielski BH, Arudi RL, Sutherland MW (1983) A study of the reactivity of HO_2/O_2^- with unsaturated fatty acids. J Biol Chem 258:4759–4761

Billah M, Rohman MM, Hossain N, Uddin MS (2017) Exogenous ascorbic acid improved tolerance in maize (*Zea mays* L.) by increasing antioxidant activity under salinity stress. Afr J Agric Res 12(17):1437–1446. https://doi.org/10.5897/AJAR2017.12295

Bustos D, Lascano R, Villasuso AL, Machado E, Senn ME, Córdoba A, Taleisnik E (2008) Reductions in maize root-tip elongation by salt and osmotic stress do not correlate with apoplastic $O_2^{\bullet-}$ levels. Ann Bot 102:551–559. https://doi.org/10.1093/aob/mcn141

Carmo-Silva AE, Powers SJ, Keys AJ, Arrabac MC, Parry MAJ (2008) Photorespiration in C_4 grasses remains slow under drought conditions. Plant Cell Environ 31:925–940

Caverzan A, Casassola A, Brammer SP (2016) Antioxidant responses of wheat plants under stress. Genet Mol Biol 39(1):1–6

Chang L, Sun H, Yang H, Wang X, Su Z, Chen F, Wei W (2017) Over-expression of dehydroascorbate reductase enhances oxidative stress tolerance in tobacco. Electron J Biotechnol 25:1–8

Chen Z, Young TE, Ling J, Chang SC, Gallie DR (2003) Increasing vitamin C content of plants through enhanced ascorbate recycling. Proc Natl Acad Sci U S A 100(6):3525–3530

Chen JH, Jiang HW, Hsieh EJ, Chen HY, Chien CT, Hsieh HL (2012) Drought and salt stress tolerance of an *Arabidopsis* glutathione *S*-transferase U17 knockout mutant are attributed to the combined effect of glutathione and abscisic acid. Plant Physiol 158(1):340–351. https://doi.org/10.1104/pp.111.181875

Choudhury S, Panda P, Sahoo L, Panda SK (2013) Reactive oxygen species signaling in plants under abiotic stress. Plant Signal Behav 8(4):e23681. https://doi.org/10.4161/psb.23681

Chugh V, Kaur N, Gupta AK (2011) Evaluation of oxidative stress tolerance in maize (*Zea mays* L.) seedlings in response to drought. Indian J Biochem Biophys 48:47–53

Chugh V, Kaur N, Grewal MS, Gupta AK (2013) Differential antioxidative response of tolerant and sensitive maize (*Zea mays* L.) genotypes to drought stress at reproductive stage. Indian J Biochem Biophys 50:150–158

Cummins I, Dixon DP, Freitag-Pohl S, Skipsey M, Edwards R (2011) Multiple roles for plant glutathione transferases in xenobiotic detoxification. Drug Metab Rev 43:266–280. https://doi.org/10.3109/03602532.2011.552910

de Azevedo Neto AD, Prisco JT, Enéas-Filho J, de Lacerda CF, Silva JV, da Costa PHA, Gomes-Filho E (2004) Effects of salt stress on plant growth, stomatal response and solute accumulation of different maize genotypes. Braz J Plant Physiol 16:31–38

de Azevedo Neto AD, Prisco JT, Eneas J, de Abreu CEB, Gomes-Filho E (2006) Effect of salt stress on antioxidative enzymes and lipid peroxidation in leaves and roots of salt-tolerant and salt-sensitive maize varieties. Environ Exp Bot 56:87–94. https://doi.org/10.1016/j.envexpbot.2005.01.008

del Rio LA, Corpas FJ, Sandalio LM, Palma JM, Gomez M, Barroso JB (2002) Reactive oxygen species, antioxidant systems and nitric oxide in peroxisomes. J Exp Bot 53:1255–1272

del Río LA, Corpas FJ, Sandalio LM, Palma JM, Barroso JB (2003) Plant peroxisomes, reactive oxygen metabolism and nitric oxide. IUBMB Life 55(2):71–81

del Rio LA, Sandalio LM, Corpas FJ, Palma JM, Barroso JB (2006) Reactive oxygen species and reactive nitrogen species in peroxisomes. Production, scavenging, and role in cell signaling. Plant Physiol 141(2):330–335. https://doi.org/10.1007/s10535-015-0542-x

Demiral T, Türkan I (2004) Does exogenous glycine betaine affect antioxidative system of rice seedlings under NaCl treatment? J Plant Physiol 161(10):1089–1100

Diao Y, Xu H, Li G, Yu A, Yu X, Hu W (2014) Cloning a glutathione peroxidase gene from *Nelum bonucifera* and enhanced salt tolerance by overexpressing in rice. Mol Biol Rep 41:4919–4927. https://doi.org/10.1007/s11033-014-3358-4

Dipierro S, Borraccino G (1991) Dehydroascorbate reductase from potato tubers. Phytochemistry 30(2):427–429

Dixon DP, Edwards R (2010) Glutathione transferases. Arabidopsis Book 8:e0131. https://doi.org/10.1199/tab.0131. PMID: 22303257

Doderer A, Kokkelink I, van der Veen S, Valk BE, Schram A, Douma AC (1992) Purification and characterization of two lipoxygenase isoenzymes from germinating barley. Biochim Biophys Acta 1120(1):97–104

Edwards EA, Rawsthorne S, Mullineaux PM (1990) Subcellular distribution of multiple forms of glutathione reductase in leaves of pea (*Pisum sativum* L.). Planta 180(2):278–284

Elstner EF (1987) Metabolism of activated oxygen species. In: Davies DD (ed) Biochem plants. Academic, London, pp 253–315

Eltayeb AE, Yamamoto S, Habora MEE, Yin L, Tsujimoto H, Tanaka K (2011) Transgenic potato overexpressing *Arabidopsis* cytosolic *AtDHAR1* showed higher tolerance to herbicide, drought and salt stresses. Breed Sci 61(1):3–10

Erdei L, Szegletes Z, Barabás K, Pestenácz A (1996) Responses in polyamine titer under osmotic and salt stress in sorghum and maize seedlings. J Plant Physiol 147(5):599–603

Estrada B, Aroca R, Barea JM, Ruiz-Lozano JM (2013) Native arbuscular mycorrhizal fungi isolated from a saline habitat improved maize antioxidant systems and plant tolerance to salinity. Plant Sci 201:42–51

FAO [Food and Agriculture Organization of the United Nations] (2016) FAOStat: food and agriculture data. http://www.fao.org/faostat/en/#data/QC

Farooq M, Hussain M, Wakeel A, Siddique KHM (2015) Salt stress in maize: effects, resistance mechanisms, and management. A review. Agron Sustain Dev 35:461–481

Ferreira Neto JRC, Pandolfi V, Guimaraes FCM, Benko-Iseppon AM, Romero C, De Oliveira Silva RL (2013) Early transcriptional response of soybean contrasting accessions to root dehydration. PLoS One 8:83–66. https://doi.org/10.1371/journal.pone.0083466

Foyer CH, Halliwell B (1976) The presence of glutathione and glutathione reductase in chloroplasts: a proposed role in ascorbic acid metabolism. Planta 133(1):21–25

Foyer CH, Noctor G (2000) Oxygen processing in photosynthesis: regulation and signaling. New Phytol 146:359–388

Foyer CH, Noctor G (2005) Redox homeostasis and antioxidant signaling: a metabolic interface between stress perception and physiological responses. Plant Cell 17:1866–1875

Fryer MJ, Andrews JR, Oxborough K, Blowers DA, Baker NR (1998) Relationship between CO_2 assimilation, photosynthetic electron transport, and active O_2 metabolism in leaves of maize in the field during periods of low temperature. Plant Physiol 116(2):571–580

Fu JY (2014) Cloning of a new glutathione peroxidase gene from tea plant (*Camellia sinensis*) and expression analysis under biotic and abiotic stresses. Bot Stud 55:7. https://doi.org/10.1186/1999-3110-55-7

Fu J, Liu Z, Li Z, Wang Y, Yang K (2017) Alleviation of the effects of saline–alkaline stress on maize seedlings by regulation of active oxygen metabolism by *Trichoderma asperellum*. PLoS One 12(6):e0179617. https://doi.org/10.1371/journal.pone.0179617

Gallie DR (2012) The role of L-ascorbic acid recycling in responding to environmental stress and in promoting plant growth. J Exp Bot 64:433–443. https://doi.org/10.1093/jxb/ers330

Gao F, Chen J, Ma T, Li H, Wang N, Li Z, Zhang Z, Zhou Y (2014) The glutathione peroxidase gene family in *Thellungiella salsuginea*: genome-wide identification, classification, and gene and protein expression analysis under stress conditions. Int J Mol Sci 15(2):3319–3335. https://doi.org/10.3390/ijms15023319

Ghaderi N, Normohammadi S, Javadi T (2015) Morpho-physiological responses of strawberry (*Fragaria* × *Ananassa*) to exogenous salicylic acid application under drought stress. J Agric Sci Technol 17(1):167–178

Gholizadeh A, Kohnehrouz BB (2010) Activation of phenylalanine ammonia lyase as a key component of the antioxidative system of salt-challenged maize leaves. Braz J Plant Physiol 22(4):217–223

Gill SS, Tuteja N (2010) Reactive oxygen species and antioxidant machinery in abiotic stress tolerance in crop plants. Plant Physiol Biochem 48(12):909–930

Gondim FA, Gomes-Filho E, Costa JH, Alencar NLM, Prisco JT (2012) Catalase plays a key role in salt stress acclimation induced by hydrogen peroxide pretreatment in maize. Plant Physiol Biochem 56:62–71. https://doi.org/10.1016/j.plaphy.2012.04.012

Goodarzian-Ghahfarokhi M, Mansouri-Far C, Saeidi M, Abdoli M (2016) Different physiological and biochemical responses in maize hybrids subjected to drought stress at vegetative and reproductive stages. Acta Biol Szeged 60(1):27–37

Gou W, Zheng P, Tian L, Gao M, Zhang L, Akram NA, Ashraf M (2017) Exogenous application of urea and a urease inhibitor improves drought stress tolerance in maize (*Zea mays* L.). J Plant Res 130:599–609

Gunes A, Inal A, Alpaslam M, Erslan F, Bagsi EG, Cicek N (2007) Salicylic acid induced changes on some physiological parameters symptomatic for oxidative stress and mineral nutrition in maize (*Zea mays* L.) grown under salinity. J Plant Physiol 164:728–736. https://doi.org/10.1016/j.jplph.2005.12.009

Guo Z, Deshpande R, Paull TT (2010) ATM activation in the presence of oxidative stress. Cell Cycle 9(24):4805–4811

Han C, Liu Q, Yang Y (2009) Short-term effects of experimental warming and enhanced ultraviolet-B radiation on photosynthesis and antioxidant defense of *Picea asperata* seedlings. Plant Growth Regul 58(2):153–162

Hasanuzzaman M, Fujita M (2011) Selenium pretreatment upregulates the antioxidant defense and methylglyoxal detoxification system and confers enhanced tolerance to drought stress in rapeseed seedlings. Biol Trace Elem Res 143:1758–1776

Hasanuzzaman M, Alam MM, Rahman A, Hasanuzzaman M, Nahar K, Fujita M (2014) Exogenous proline and glycine betaine mediated upregulation of antioxidant defense and glyoxalase systems provides better protection against salt-induced oxidative stress in two rice (*Oryza sativa* L.) varieties. Biomed Res Int 2014:757219. https://doi.org/10.1155/2014/757219

Hasanuzzaman M, Nahar K, Hossain MS, Mahmud JA, Rahman A, Inafuku M, Oku H, Fujita M (2017) Coordinated actions of glyoxalase and antioxidant defense systems in conferring abiotic stress tolerance in plants. Int J Mol Sci 18(1):200. https://doi.org/10.3390/ijms18010200

Hasegawa PM, Bressan RA, Zhu JK, Bohnert HJ (2000) Plant cellular and molecular response to high salinity. Annu Rev Plant Physiol Plant Mol Biol 51:463–499

Hatch MD (1987) C4 photosynthesis: a unique blend of modified biochemistry, anatomy and ultrastructure. Biochim Biophys Acta Rev Biomembr 895(2):81–106

Hefny M, Abdel-Kader DZ (2009) Antioxidant-enzyme system as selection criteria for salt tolerance in forage sorghum genotypes (*Sorghum bicolor* L. Moench). In: Ashraf M, Ozturk M, Athar H (eds) Salinity and water stress. Tasks for vegetation sciences, vol 44. Springer, Dordrecht

Herbette S, de Labrouhe DT, Drevet JR, Roeckel-Drevet P (2011) Transgenic tomatoes showing higher glutathione peroxidase antioxidant activity are more resistant to an abiotic stress but more susceptible to biotic stresses. Plant Sci 180:548–553. https://doi.org/10.1016/j.plantsci.2010.12.002

Hernandez JA, Ferrer MA, Jiménez A, Barceló AR, Sevilla F (2001) Antioxidant systems and $O_2^{\bullet -}$/H_2O_2 production in the apoplast of pea leaves. Its relation with salt-induced necrotic lesions in minor veins. Plant Physiol 127(3):817–831

Hossain MA, Asada K (1984) Purification of dehydroascorbate reductase from spinach and its characterization as a thiol enzyme. Plant Cell Physiol 25(1):85–92

Hossain MA, Asada K (1985) Monodehydroascorbate reductase from cucumber is a flavin adenine dinucleotide enzyme. J Biol Chem 260(24):12920–12926

Huo Y, Wang M, Wei Y, Xia Z (2016) Overexpression of the maize *psbA* gene enhances drought tolerance through regulating antioxidant system, photosynthetic capability, and stress defense gene expression in tobacco. Front Plant Sci 6:1223. https://doi.org/10.3389/fpls.2015.01223

Hussein MM, Balbaa LK, Gaballah MS (2007) Salicylic acid and salinity effects on growth of maize plants. Res J Agric Biol Sci 3:321–328

IPCC [Intergovernmental Panel on Climate Change] (2014) Climate change 2014: impacts, adaptation, and vulnerability. Cambridge University Press, Cambridge. http://www.ipcc.ch/report/ar5/wg2/

Isendahl N, Schmidt G (2006) Drought in the Mediterranean: WWF policy proposals. Worldwide Fund for Nature. http://assets.panda.org/downloads/wwf_drought_med_report_2006.pdf

Jiang C, Zu C, Dianjun L, Zheng Q, Shen J, Wang H, Li D (2017) Effect of exogenous selenium supply on photosynthesis, Na^+ accumulation and antioxidative capacity of maize (*Zea mays* L.) under salinity stress. Sci Rep 7:42039. https://doi.org/10.1038/srep42039
Kanai R, Edwards GE, Sage RF, Monson RK (1999) The biochemistry of C_4 photosynthesis. In: Sage RF, Monson RK (eds) C_4 plant biology. Academic, San Diego, pp 49–87
Kaya C, Tuna AL, Dikilitas M, Cullu MA (2010) Responses of some enzymes and key growth parameters of salt-stressed maize plants to foliar and seed applications of kinetin and indole acetic acid. J Plant Nutr 33(3):405–422. https://doi.org/10.1080/01904160903470455
Kaya C, Ashraf M, Dikilitas M, Tuna AL (2013) Alleviation of salt stress induced adverse effects on maize plants by exogenous application of indoleacetic acid (IAA) and inorganic nutrients—a field trial. Aust J Crop Sci 7:249–254
Kaya C, Ashraf M, Sönmez O, Tuna AL, Aydemir S (2015) Exogenously applied nitric oxide confers tolerance to salinity-induced oxidative stress in two maize (*Zea mays* L.) cultivars differing in salinity tolerance. Turk J Agric For 39:909–919. https://doi.org/10.3906/tar-1411-26
Kingston-Smith AH, Foyer CH (2000) Over expression of Mn-superoxide dismutase in maize leaves leads to increased monodehydroascorbate reductase, dehydroreductase and glutathione reductase. J Expt Bot 51:1867–1877
Kobayashi K, Kumazawa Y, Miwa K, Yamanaka S (1996) ε-(γ-Glutamyl) lysine cross-links of spore coat proteins and transglutaminase activity in *Bacillus subtilis*. FEMS Microbiol Lett 144(2–3):157–160
Kocsy G, von Ballmoos P, Rüegsegger A, Szalai G, Galiba G, Brunold C (2001) Increasing the glutathione content in a chilling-sensitive maize genotype using safeners increased protection against chilling-induced injury. Plant Physiol 127(3):1147–1156
Koua D, Cerutti L, Falquet L, Sigrist CJ, Theiler G, Hulo N (2009) PeroxiBase: a database with new tools for peroxidase family classification. Nucleic Acids Res 37:261–266. https://doi.org/10.1093/nar/gkn680
Krieger-Liszkay A (2004) Singlet oxygen production in photosynthesis. J Exp Bot 56:337–346
Kumar S, Asif MH, Chakrabarty D, Tripathi RD, Dubey RS, Trivedi PK (2013) Expression of a rice Lambda class of glutathione *S*-transferase, *OsGSTL2*, in *Arabidopsis* provides tolerance to heavy metal and other abiotic stresses. J Hazard Mater 248–249:228–237
Kumar D, Al Hassan M, Naranjo MA, Agrawal V, Boscaiu M, Vicente O (2017) Effects of salinity and drought on growth, ionic relations, compatible solutes and activation of antioxidant systems in oleander (*Nerium oleander* L.). PLoS One 12(9):e0185017. https://doi.org/10.1371/journal.pone.0185017
Labudda M, Azam FMS (2014) Glutathione-dependent responses of plants to drought: a review. Acta Soc Bot Pol 83(1):3–12
Lacuesta M, Dever LV, Miñoz-Rueda A, Lea PJ (1997) A study of photorespiratory ammonia production in the C4 plant *Amaranthus edulis*, using mutants with altered photosynthetic capacities. Physiol Plant 99:447–455
Latef AAA, Tran S-LP (2016) Impacts of priming with silicon on the growth and tolerance of maize plants to alkaline stress. Front Plant Sci 7:243. https://doi.org/10.3389/fpls.2016.00243
Latif M, Akram NA, Ashraf M (2016) Regulation of some biochemical attributes in drought-stressed cauliflower (*Brassica oleracea* L.) by seed pre-treatment with ascorbic acid. J Hortic Sci Biotech 91:129–137. https://doi.org/10.1080/14620316.2015.1117226
Leterrier M, Francisco J, Corpas JB, Barroso LM, Sandalio LM, del Río LA (2005) Peroxisomal monodehydroascorbate reductase, genomic clone characterization and functional analysis under environmental stress conditions. Plant Physiol 138(4):2111–2123. https://doi.org/10.1104/pp.105.066225
Lisko KA, Aboobucker SI, Torres R, Lorence A (2014) Engineering elevated vitamin C in plants to improve their nutritional content, growth, and tolerance to abiotic stress. In: Phytochemicals, biosynthesis, function and application. Springer, Cham, pp 109–128
Lu P, Sang WG, Ma KP (2007) Activity of stress-related antioxidative enzymes in the invasive plant crofton weed (*Eupatorium adenophorum*). J Integr Plant Biol 49(11):1555–1564

Mahajan S, Tuteja N (2005) Cold, salinity and drought stresses: an overview. Arch Biochem Biophys 444:139–158

Maheshwari R, Dubey RS (2009) Nickel-induced oxidative stress and the role of antioxidant defense in rice seedlings. Plant Growth Regul 59(1):37–49

Manaa A, Mimouni H, Terras A, Chebil F, Wasti S, Gharbi E, Ahmed HB (2014) Superoxide dismutase isozyme activity and antioxidant responses of hydroponically cultured *Lepidium sativum* L. to NaCl stress. J Plant Interact 9(1):440–449. https://doi.org/10.1080/17429145.2013.850596

Marrs KA (1996) The functions and regulation of glutathione *S*-transferases in plants. Annu Rev Plant Physiol Plant Mol Biol 47:127–158. https://doi.org/10.1146/annurev.arplant.47.1.127. PMID: 15012285

Marrs KA, Walbot V (1997) Expression and RNA splicing of the maize glutathione *S*-transferase *Bronze2* gene is regulated by cadmium and other stresses. Plant Physiol 11(3):93–102

McGonigle B, Keeler SJ, Lau SM, Koeppe MK, O'Keefe DP (2000) A genomics approach to the comprehensive analysis of the glutathione *S*-transferase gene family in soybean and maize. Plant Physiol 124:1105–1120

Mir MA, John R, Alyemeni MN, Alam P, Ahmad P (2018) Jasmonic acid ameliorates alkaline stress by improving growth performance, ascorbate glutathione cycle and glyoxylase system in maize seedlings. Sci Rep 8(1):2831. https://doi.org/10.1038/s41598-018-21097-3

Mishra P, Bhoomika K, Dubey RS (2013) Differential responses of antioxidative defense system to prolonged salinity stress in salt-tolerant and salt-sensitive Indica rice (*Oryza sativa* L.) seedlings. Protoplasma 250(1):3–19

Mittler R (2002) Oxidative stress, antioxidants and stress tolerance. Trends Plant Sci 7:405–410

Mittler R (2006) Abiotic stress, the field environment and stress combination. Trends Plant Sci 11:15–19

Moller IM, Jensen PE, Hansson A (2007) Oxidative modifications to cellular components in plants. Annu Rev Plant Biol 58:459–481

Moussa HR, Abdel-Aziz SM (2008) Comparative response of drought tolerant and drought sensitive maize genotypes to water stress. Aust J Crop Sci 1(1):31–36

Mukhtar A, Akram NA, Aisha R, Shafiq S, Ashraf M (2016) Foliar applied ascorbic acid enhances antioxidative potential and drought tolerance in cauliflower (*Brassica oleracea* L. var. *botrytis*). Agrochimica 60:100–113

Mullineaux PM, Rausch T (2005) Glutathione, photosynthesis and the redox regulation of stress-responsive gene expression. Photosynth Res 86:459–474

Munns R (2002) Comparative physiology of salt and water stress. Plant cell & environment 25(2):239–50

Munns R (2005) Genes and salt tolerance: bringing them together. New Phytol 167:645–663. https://doi.org/10.1111/j.1469-8137.2005.01487.x

Munns R, Tester M (2008) Mechanisms of salinity tolerance. Ann Rev Plant Biol 59:651–681. https://doi.org/10.1146/annrev.arplant.59.032607.092911

Naeem M, Naeem MS, Ahmad R, Ahmad R, Ashraf MY, Ihsan MZ, Nawaz F, Athar HR, Ashraf M, Abbas HT, Abdullah M (2018) Improving drought tolerance in maize by foliar application of boron: water status, antioxidative defense and photosynthetic capacity. Arch Agron Soil Sci 64(5):626–639. https://doi.org/10.1080/03650340.2017.1370541

Nahar K, Hasanuzzaman M, Alam MM, Fujita M (2015) Glutathione-induced drought stress tolerance in mung bean: coordinated roles of the antioxidant defense and methylglyoxal detoxification systems. AoB Plants 7:plv069. https://doi.org/10.1093/aobpla/plv069

Nahar K, Hasanuzzaman M, Rahman A, Alam MM, Mahmud JA, Suzuki T, Fujita M (2016) Polyamines confer salt tolerance in mung bean (*Vigna radiata* L.) by reducing sodium uptake, improving nutrient homeostasis, antioxidant defense, and methylglyoxal detoxification systems. Front Plant Sci 7:1104. https://doi.org/10.3389/fpls.2016.01104

Navrot N, Collin V, Gualberto J, Gelhaye E, Hirasawa M, Rey P (2006) Plant glutathione peroxidases are functional peroxiredoxins distributed in several subcellular compartments and regulated during biotic and abiotic stresses. Plant Physiol 142:1364–1379. https://doi.org/10.1104/pp.106.089458

Navrot N, Rouhier N, Gelhaye E, Jaquot JP (2007) Reactive oxygen species generation and antioxidant systems in plant mitochondria. Physiol Plant 129:185–195

Naz H, Akram NA, Ashraf M (2016) Impact of ascorbic acid on growth and some physiological attributes of cucumber (*Cucumis sativus*) plants under water-deficit conditions. Pak J Bot 48:877–883

Noctor G, Foyer CH (1998) A re-evaluation of the ATP: NADPH budget during C_3 photosynthesis. A contribution from nitrate assimilation and its associated respiratory activity. J Exp Bot 49:1895–1908

Noctor G, Gomez L, Vanacker H, Foyer CH (2002) Interactions between biosynthesis, compartmentation, and transport in the control of glutathione homeostasis and signaling. J Exp Bot 53:1283–1304

Noctor G, Mhamdi A, Chaouch S, Han Y, Neukermans J, Marquez-Garcia B, Queval G, Foyer CH (2012) Glutathione in plants: an integrated overview. Plant Cell Environ 35:454–484

Noman A, Ali S, Naheed F, Ali Q, Farid M, Rizwan M, Irshad MK (2015) Foliar application of ascorbate enhances the physiological and biochemical attributes of maize (*Zea mays* L.) cultivars under drought stress. Arch Agron Soil Sci 61(12):1659–1672. https://doi.org/10.1080/03650340.2015.1028379

Omoto E, Taniguchi M, Miyake H (2012) Adaptation responses in C4 photosynthesis of maize under salinity. J Plant Physiol 169:469–477. https://doi.org/10.1016/j.jplph.2011.11.009

Ozyigit II, Filiz E, Vatansever R, Kurtoglu KY, Koc I, Öztürk MX, Anjum NA (2016) Identification and comparative analysis of H_2O_2-scavenging enzymes (ascorbate peroxidase and glutathione peroxidase) in selected plants employing bioinformatics approaches. Front Plant Sci 7:301. https://doi.org/10.3389/fpls.2016.00301

Pan Y, Wu LJ, Yu ZL (2006) Effect of salt and drought stress on antioxidant enzymes activities and SOD isoenzymes of liqurice (*Glycyrrhiza uralensis* Fisch). Plant Growth Regul 49:157–165. https://doi.org/10.1007/s10725-006-9101-y

Parida AK, Das AB (2005) Salt tolerance and salinity effects on plants: a review. Ecotoxicol Environ Saf 60(3):324–349

Peng CL, Ou ZY, Liu N, Lin GZ (2005) Response to high temperature in flag leaves of super high-yielding rice Pei'ai 64S/E32 and Liangyoupeijiu. Rice Sci 12(3):179–186

Puntarulo S, Sánchez RA, Boveris A (1988) Hydrogen peroxide metabolism in soybean embryonic axes at the onset of germination. Plant Physiol 86:626–630

Qin A, Shi Q, Yu X (2011) Ascorbic acid contents in transgenic potato plants overexpressing two dehydroascorbate reductase genes. Mol Biol Rep 38(3):1557–1566

Rajasekar M, Rabert GA, Manivannan P (2015) Triazole induced changes on biochemical and antioxidant metabolism of *Zea mays* L. (maize) under drought stress. J Plant Stress Physiol 1(1):35–42. https://doi.org/10.5455/jpsp.2015-08-024

Rao AR, Baskaran V, Sarada R, Ravishankar GA (2013) In vivo bioavailability and antioxidant activity of carotenoids from microalgal biomass—a repeated dose study. Food Res Int 54(1):711–717

Remme RN, Ivy NA, Mian MAK, Rohman MM (2013) Effect of salinity stress on glutathione *S*-transferases (GSTs) of maize. Iosr J Agril Vet Sci 4(1):42–52

Rios-Gonzalez K, Erdei L, Lips SH (2002) The activity of antioxidant enzymes in maize and sunflower seedlings as affected by salinity and different nitrogen sources. Plant Sci 162:923–930. https://doi.org/10.1016/S0168-9452(02)00040-7

Rohman MM, Suzuki T, Fujita M (2009) Identification of a glutathione *S*-transferase inhibitor in onion bulb (*Allium cepa* L.). Aust J Crop Sci 3(1):28–36

Rohman MM, Begum S, Akhi AH, Ahsan AFMS, Uddin MS, Amiruzzaman M, Banik BR (2015) Protective role of antioxidants in maize seedlings under saline stress: exogenous proline provided better tolerance than betaine. Bothalia J 45(4):17–35

Rohman MM, Talukder MZA, Hossain MG, Uddin MS, Amiruzzaman M, Biswas A, Ahsan AFMS, Chowdhury MAZ (2016a) Saline sensitivity leads to oxidative stress and increases the antioxidants in presence of proline and betaine in maize (*Zea mays* L.) inbred. Plant Omics 9(1):35–47

Rohman MM, Begum S, Talukder MZA, Akhi AH, Amiruzzaman M, Ahsan AFMS, Hossain Z (2016b) Drought sensitive maize inbred shows more oxidative damage and higher ROS scavenging enzymes, but not glyoxalases than a tolerant one at seedling stage. Plant Omics 9(4):220–232. https://doi.org/10.21475/poj.16.09.04.pne31

Rohman MM, Islam MR, Mahmuda BM, Begum S, Fakir OA, Amiruzzaman M (2018) Higher K^+/Na^+ and lower reactive oxygen species and lipid peroxidation are related to higher yield in maize under saline condition. Afr J Agric Res 13(5):239–247. https://doi.org/10.5897/AJAR2017.12878

Roxas VP, Lodhi SA, Garrett DK, Mahan JR, Allen RD (2000) Stress tolerance in transgenic tobacco seedlings that overexpress glutathione S-transferase/glutathione peroxidase. Plant Cell Physiol 41(11):1229–1234

Rubio MC, Bustos-Sanmamed P, Clemente MR, Becana M (2009) Effects of salt stress on the expression of antioxidant genes and proteins in the model legume *Lotus japonicus*. New Phytol 181(4):851–859

Saed-Moocheshi A, Shekoofa A, Sadeghi H, Pessarakli M (2014) Drought and salt stress mitigation by seed priming with KNO_3 and urea in various maize hybrids: an experimental approach based on enhancing antioxidant responses. J Plant Nutr 37(5):674–689. https://doi.org/10.1080/01904167.2013.868477

Sage RF (2004) The evolution of C_4 photosynthesis. New Phytol 161:341–370

Sánchez-Rodríguez E, del Mar Rubio-Wilhelmi M, Blasco B, Leyva R, Romero L, Ruiz JM (2012) Antioxidant response resides in the shoot in reciprocal grafts of drought-tolerant and drought-sensitive cultivars in tomato under water stress. Plant Sci 188:89–96

Sandhya VS, Ali SZ, Grover M, Reddy G, Venkateswarlu B (2010) Effect of plant growth promoting *Pseudomonas* spp. on compatible solutes, antioxidant status and plant growth of maize under drought stress. Plant Growth Regulation 62(1):21–30

Sandhya V, Shaik ZA, Minakshi G, Gopal R, Venkateswarlu B (2011) Drought-tolerant plant growth promoting *Bacillus* spp.: effect on growth, osmolytes, and antioxidant status of maize under drought stress. J Plant Interact 6(1):1742–9145

Saxena M, Deb Roy S, Singla-Pareek S-L, Sopory SK, Bhalla-Sarin N (2011) Overexpression of the glyoxalase II gene leads to enhanced salinity tolerance in *Brassica juncea*. Open Plant Sci 5:23–28

Sayfzadeh S, Habibi D, Taleghani FD, Kashani A, Vazan S, Qaen SHS, Khodaei AH, Khodaei MMA, Khodaei M (2011) Response of antioxidant enzyme activities and root yield in sugar beet to drought stress. Int J Agric Biol 13(3):357–362

Scandalias JG (1990) Response of plant antioxidant defence genes to environmental stress. Adv Genet 28:1–41

Scandalios JG (2005) Oxidative stress: molecular perception and transduction of signal triggering antioxidant gene defenses. Braz J Med Biol Res 38:995–1014

Serraj R, Sinclair TR (2002) Osmolyte accumulation: can it really help increase crop yield under drought conditions? Plant Cell Environ 25(2):333–341

Sgherri CLM, Maffei M, Navari-Izzo F (2000) Antioxidative enzymes in wheat subjected to increasing water deficit and rewatering. J Plant Physiol 157(3):273–279

Shan C, Liu H, Zhao D, Wang X (2014) Effects of exogenous hydrogen sulfide on the redox states of ascorbate and glutathione in maize leaves under salt stress. Biol Plant 58(1):169–173

Sharma P, Dubey RS (2005) Drought induces oxidative stress and enhances the activities of antioxidant enzymes in growing rice seedlings. Plant Growth Regul 46(3):209–221

Sharma P, Dubey RS (2007) Involvement of oxidative stress and role of antioxidative defense system in growing rice seedlings exposed to toxic concentrations of aluminum. Plant Cell Rep 26(11):2027–2038

Sharma P, Jha AB, Dubey RS, Pessarakli M (2012) Reactive oxygen species, oxidative damage, and antioxidative defense mechanism in plants under stressful conditions. J Bot 2012:217037. https://doi.org/10.1155/2012/217037

Sheikh FA, Dar ZA, Sofi PA, Lone AA (2017) Recent advances in breeding for abiotic stress (drought) tolerance in maize. Int J Curr Microbiol App Sci 6(4):2226–2243

Shin SY, Kim MH, Kim YH, Park HM, Yoo HS (2013) Co-expression of monodehydroascorbate reductase and dehydroascorbate reductase from *Brassica rapa* effectively confers tolerance to freezing-induced oxidative stress. Mol Cells 36:304–315. https://doi.org/10.1007/s10059-013-0071-4

Shiriga K, Sharma R, Kumar K, Yadav SK, Hossain F, Thirunavukkarasu N (2014) Expression pattern of superoxide dismutase under drought stress in maize. Int J Innov Res Sci Eng Technol 3(4):11333–11337

Singla-Pareek SL, Yadav SK, Pareek A, Reddy MK, Sopory SK (2006) Transgenic tobacco over expressing glyoxalase pathway enzymes grow and set viable seeds in zinc-spiked soils. Plant Physiol 140:613–623

Singla-Pareek SL, Yadav SK, Pareek A, Reddy MK, Sopory SK (2008) Enhancing salt tolerance in a crop plant by overexpression of glyoxalase II. Transgenic Res 17(2):171–180

Smirnoff N (2000) Ascorbic acid: metabolism and functions of a multifaceted molecule. Curr Opin Plant Biol 3:229–235

Stepien P, Klobus G (2005) Antioxidant defense in the leaves of C_3 and C_4 plants under salinity stress. Physiol Plant 125:31–40. https://doi.org/10.1111/j.1399-3054.2005.00534.x

Sudan J, Negi B, Arora S (2015) Oxidative stress induced expression of monodehydroascorbate reductase gene in *Eleusine coracana*. Physiol Mol Biol Plants 21(4):551–558

Sugimoto M, Oono Y, Gusev O, Matsumoto T, Yazawa T, Levinskikh MA (2014) Genome-wide expression analysis of reactive oxygen species gene network in Mizuna plants grown in long-term space flight. BMC Plant Biol 14:4. https://doi.org/10.1186/1471-2229-14-4

Szabados L, Savoure A (2010) Proline: a multifunctional amino acid. Trends Plant Sci 15(2):89–97

Szarka A, Horemans N, Kovacs Z, Grof P, Mayer M, Banhegyi G (2007) Dehydroascorbate reduction in plant mitochondria is coupled to the respiratory electron transfer chain. Physiol Plant 129:225–232

Takesawa T, Ito M, Kanzaki H, Kameya N, Nakamura I (2002) Over-expression of a glutathione *S*-transferase in transgenic rice enhances germination and growth at low temperature. Mol Breed 9:93–101

Tayefi-Nasrabadi H, Dehghan G, Daeihassani B, Movafegi A, Samadi A (2011) Some biochemical properties of guaiacol peroxidases as modified by salt stress in leaves of salt-tolerant and salt-sensitive safflower (*Carthamus tinctorius* L. cv.) cultivars. African Journal of Biotechnology 10(5):751–63

Tuna AL, Kaya C, Altunlu H, Ashraf M (2013) Mitigation effects of non-enzymatic antioxidants in maize (*Zea mays* L.) plants under salinity stress. Aust J Crop Sci 7(8):1181–1188

Uzilday B, Turkana I, Sekmena AH (2011) Comparison of ROS formation and antioxidant enzymes in *Cleome gynandra* (C_4) and *Cleome spinosa* (C_3) under drought stress. Plant Sci 182:59–70. https://doi.org/10.1016/j.plantsci.2011.03.015

Vadassery J, Tripathi S, Prasad R, Varma A, Oelmuller R (2009) Monodehydroascorbate reductase 2 and dehydroascorbate reductase 5 are crucial for a mutualistic interaction between *Piriformospora indica* and *Arabidopsis*. J Plant Physiol 166:1263–1274

Vardharajula S, Zulfikar Ali S, Grover M, Reddy G, Bandi V (2011) Drought-tolerant plant growth promoting *Bacillus* spp.: effect on growth, osmolytes, and antioxidant status of maize under drought stress. Journal of Plant Interactions 6(1):1-4

Valderrama R, Corpas FJ, Carreras A, Gómez-Rodríguez MV, Chaki M, Pedrajas JR, Fernandez-Ocana ANA, Del Río LA, Barroso JB (2006) The dehydrogenase-mediated recycling of NADPH is a key antioxidant system against salt-induced oxidative stress in olive plants. Plant Cell Environ 29(7):1449–1459

Vangronsveld J, Clijsters H (1994) Toxic effects of metals, in plants and the chemical elements. In: Farago ME (ed) Biochemistry, uptake, tolerance and toxicity. VCH, Weinheim, pp 150–177

Vitkauskaitė G, Venskaitytė L (2011) Differences between C_3 (*Hordeum vulgare* L.) and C_4 (*Panicum miliaceum* L.) plants with respect to their resistance to water deficit. Zemdirbyste-Agriculture 98(4):349–356

Viveros MFÁ, Inostroza-Blancheteau C, Timmermann T, González M, Arce-Johnson P (2013) Overexpression of *GlyI* and *GlyII* genes in transgenic tomato (*Solanum lycopersicum* Mill.) plants confers salt tolerance by decreasing oxidative stress. Mol Biol Rep 40(4):3281–3290

Volkov V (2015) Salinity tolerance in plants. Quantitative approach to ion transport starting from halophytes and stepping to genetic and protein engineering for manipulating ion fluxes. Front Plant Sci 6:873. https://doi.org/10.3389/fpls.2015.00873

Wang J, Zhang H, Allen RD (1999) Overexpression of an *Arabidopsis* peroxisomal ascorbate peroxidase gene in tobacco increases protection against oxidative stress. Plant Cell Physiol 40(7):725–732

Wang Y, Gu W, Meng Y, Xie T, Li L, Li J, Wei S (2017) γ-Aminobutyric acid imparts partial protection from salt stress injury to maize seedlings by improving photosynthesis and upregulating osmoprotectants and antioxidants. Sci Rep 7:43609. https://doi.org/10.1038/srep43609

Weisany W, Sohrabi Y, Heidari G, Siosemardeh A, Ghassemi-Golezani K (2012) Changes in antioxidant enzymes activity and plant performance by salinity stress and zinc application in soybean (*Glycine max* L.). Plant Omics 5(2):60–67

Xu J, Xing X-J, Tian Y-S, Peng R-H, Xue Y, Zhao W (2015) Transgenic *Arabidopsis* plants expressing tomato glutathione *S*-transferase showed enhanced resistance to salt and drought stress. PLoS One 10(9):e0136960. https://doi.org/10.1371/journal.pone.0136960

Yadav SK, Singla-Pareek SL, Ray M, Reddy MK, Sopory SK (2005a) Methylglyoxal levels in plants under salinity stress are dependent on glyoxalase I and glutathione. Biochem Biophys Res Commun 337(1):61–67

Yadav SK, Singla-Pareek SL, Reddy MK, Sopory SK, Sopory SK (2005b) Transgenic tobacco plants overexpressing glyoxalase enzymes resist an increase in methylglyoxal and maintain higher reduced glutathione levels under salinity stress. FEBS Lett 579(27):6265–6271

Yang XD, Li WJ, Liu JY (2005) Isolation and characterization of a novel PHGPxgene in *Raphanus sativus*. Biochim Biophys Acta 1728:199–205

Yang G, Wang Y, Xia D, Gao C, Wang C, Yang C (2014) Overexpression of a GST gene (*ThGSTZ1*) from *Tamarix hispida* improves drought and salinity tolerance by enhancing the ability to scavenge reactive oxygen species. Plant Cell Tissue Organ Cult 117:99–112. https://doi.org/10.1007/s11240-014-0424-5

Yang L, Fountain JC, Wang H, Ni X, Ji P, Lee RD, Kemerait RC, Scully BT, Guo B (2015) Stress sensitivity is associated with differential accumulation of reactive oxygen and nitrogen species in maize genotypes with contrasting levels of drought tolerance. Int J Mol Sci 16:24791–24819. https://doi.org/10.3390/ijms161024791

Ye J, Wang S, Deng X, Yin L, Xiong B, Wang X (2016) Melatonin increased maize (*Zea mays* L.) seedling drought tolerance by alleviating drought-induced photosynthetic inhibition and oxidative damage. Acta Physiol Plant 38:48. https://doi.org/10.1007/s11738-015-2045-y

Yoshida S, Tamaoki M, Shikano T (2006) Cytosolic dehydroascorbate reductase is important for ozone tolerance in *Arabidopsis thaliana*. Plant Cell Physiol 47(2):304–308

Yousuf PY, Hakeem KUL, Chandna R, Ahmad P (2012) Role of glutathione reductase in plant abiotic stress. Springer, New York, pp 149–158

Yu T, Li YS, Chen XF, Hu J, Chang X, Zhu YG (2003) Transgenic tobacco plants overexpressing cotton glutathione *S*-transferase (GST) show enhanced resistance to methyl viologen. J Plant Physiol 160(11):1305–1311

Zhang LX, Lai JH, Liang ZS, Ashraf M (2014) Interactive effects of sudden and gradual drought stress and foliar-applied glycinebetaine on growth, water relations, osmolytes accumulation and antioxidant defence system in two maize cultivars differing in drought tolerance. J Agron Crop Sci 200:425–433

Plants Behavior Under Soil Acidity Stress: Insight into Morphophysiological, Biochemical, and Molecular Responses

M. H. M. Borhannuddin Bhuyan, Mirza Hasanuzzaman, Kamrun Nahar, Jubayer Al Mahmud, Khursheda Parvin, Tasnim Farha Bhuiyan, and Masayuki Fujita

Introduction

Acidity stress (H^+ rhizotoxicity) is one of the major abiotic stresses that limits crop production and ranks second just after drought. Acid soils are widespread around the world covering 30–40% of arable land. From agricultural point of view, 76%

M. H. M. Borhannuddin Bhuyan
Laboratory of Plant Stress Responses, Department of Applied Biological, Sciences, Faculty of Agriculture, Kagawa University, Takamatsu, Kagawa, Japan

Bangladesh Agricultural Research Institute, Joydebpur, Gazipur, Bangladesh

Citrus Research Station, Bangladesh Agricultural Research Institute (BARI), Jaintiapur, Sylhet, Bangladesh

M. Hasanuzzaman (✉)
Department of Agronomy, Faculty of Agriculture, Sher-e-Bangla Agricultural University, Dhaka, Bangladesh
e-mail: mirzahasanuzzaman@sau.edu.bd

K. Nahar • T. F. Bhuiyan
Department of Agricultural Botany, Faculty of Agriculture, Sher-e-Bangla Agricultural University, Dhaka, Bangladesh

J. A. Mahmud
Department of Agroforestry and Environmental Science, Faculty of Agriculture, Sher-e-Bangla Agricultural University, Dhaka, Bangladesh

K. Parvin
Laboratory of Plant Stress Responses, Department of Applied Biological Sciences, Faculty of Agriculture, Kagawa University, Takamatsu, Kagawa, Japan

Department of Horticulture, Faculty of Agriculture, Sher-e-Bangla Agricultural University, Dhaka, Bangladesh

M. Fujita
Laboratory of Plant Stress Responses, Department of Applied Biological Sciences, Faculty of Agriculture, Kagawa University, Takamatsu, Kagawa, Japan

M. Hasanuzzaman et al. (eds.), *Plant Abiotic Stress Tolerance*,
https://doi.org/10.1007/978-3-030-06118-0_2

potentially arable land that we could use for crop production are acidic (Von Uexkůll and Mutert 1995). Acid soils occurred worldwide, especially in the humid tropics, subtropics, and temperate zones, characterized with infertility. Soil acidification is a normal natural process. According to Kochian et al. (2004), globally about 3 billion ha of land consisting acidic soil existed in 1970s, and by the next 20 years the amount of acidic land increased to 178 billion ha (Von Uexkůll and Mutert 1995). As plants are sessile, and cannot get away from acidity stress, their growth and development is hampered, where initial visual symptom showed as the reduction in root length (Kidd and Proctor 2001). Soil pH is a major, variable growth factor in natural and agricultural soils. The concentration of the H^+ ion in the soil represents its pH value. When the pH of surface soil reached in the point <5.0–5.5, the soil could be termed as acidic. This point is very critical for crop growth and development (Edmeades et al. 1995). But if pH further drops below 4.5, crop production becomes difficult. With the decline of the pH, the availability of plant nutrient within the soil decreases, plant cannot uptake sufficient amount of water, and suffers from nutrient deficiencies and toxicities. Soils become infertile due to calcium (Ca), magnesium (Mg), potassium (K), phosphorus (P), and molybdenum (Mo) deficiency. High concentration of H^+ ions present in acid soil is toxic to higher plants, which was not estimated for many decades (Kidd and Proctor 2001). A huge amount of H^+ ions exerts adverse effects on the soil nutrients availability, which increase with drop off of soil pH (Longnecker 1974; Menconi et al. 1995). This low pH increases the solubility of toxic metal/metalloid ions such as iron (Fe), zinc (Zn), copper (Cu), manganese (Mn), and aluminum (Al). Among these Al, the third most abundant element of earth crust is the most toxic. At neutral pH of the soil, Al goes for chemical reaction to form non-soluble compounds, but with the decrease of pH these compounds break and Al becomes available and creates toxicity (Sairam et al. 1998). Proton toxicity (acidity stress) and Al toxicity cannot be separated in field as Al is only soluble in acid solution. On the other hand, excessive H^+ compete with P, Mg, Ca, K, and Mo for absorption of plants and disturb uptake and translocation of these vital nutrients, ultimate plant growth reduction (Hernandez et al. 1993), thus accredited to significant crop losses.

Aluminum restricts nutrient scavenging ability of roots causing nutrient deficiencies and, also limits the penetration depth; as a result, root system developed poorly, and eventually reduced grain yields (Sairam 1994). In pH lower than 4.0 increased Al^{3+} concentrations provoke root hairs to disappear and curtail root growth (Hiscox and Isrealstam 1979). Besides different physiological aspects, such as, DNA synthesis, cell division, signal transduction pathways, plasma membrane integrity, and cytoskeleton structure are also severely hampered (Garcia-Oliveira et al. 2016). Moreover, enzymatic antioxidants showed impaired activities (Nahar et al. 2017) while, hydrolytic enzymes—phosphatases, glucosidase, and esterase were sharply obstructed at high Al^{3+} (Taranishi et al. 1974). Although Al toxicity is extensively investigated, the others are not widely studied yet. Besides the toxicities of certain minerals, at higher concentration, H^+ itself is toxic to the plants and linked with various morphophysiological and biochemical attributes of plants (Hinsinger et al.

2003; Felle et al. 2009). Spatial and temporal variation of H^+ within the soil may happen and plant roots may face disparity. On the other hand, with the change in soil water status soil H^+ concentration may be considerably modified (Misra and Tyler 1999). Again, plant itself is a major player for pH alteration in rhizosphere by root exudates and uptake of cations and anions. Rhizosphere pH may also be changed diurnally. Blossfeld and Gansert (2007) reported that upon light exposure, 0.5 unit increase in rhizosphere pH occurred within 1 h. Rhizosphere pH changed at the time of NH_4^+ and NO_3^- uptake by H^+ extrusion and cotransport (Crawford and Forde 2002). Alteration in the rhizosphere pH may not hamper the symplastic pH, but exhibited strong effect on apoplastic pH (Felle 1998; Gao et al. 2004). On the other hand, Ca^{2+} plays a vital role in plant cell growth (Bush 1995; Edel et al. 2017). Proton rhizotoxicity arrested root growth in various plant species (Kinraide and Parker 1987; Kinraide et al. 1994) and exerts its toxic effect by Ca^{2+} displacement (Kinraide et al. 1994; Koyama et al. 2001). Koyama et al. (1995) reported that the *Arabidopsis thaliana* root elongation is severely hampered in the low pH (4.5–4.8) growing media with low Ca^{2+} ionic strength, and root showed low viability, which indicates that the target site of H^+ is the Ca^{2+}-dependent physiological process of root growth. Similar results also reported in wheat (Kinraide et al. 1994), they also suggested that Ca^{2+} in the apoplast is the major targets of H^+ rhizotoxicity. Generally, the H^+ pump—H^+–ATPase, of the plasmalemma present in the root cells, works to keep the pH almost neutral in the cytoplasm. If the activity of the H^+ in the external growth medium exceeds the ability of the cell to maintain the cytoplasmic pH, the plant growth hampered (Shavrukov and Hirai 2016). However, the toxic effect of H^+ on the latter is still unclear. On the contrary, only few researches addressed the effects of external pH change on gene expression of plants.

Therefore, this chapter aims at reviewing the current knowledge on plants' responses to acidity stress or H^+ rhizotoxicity focusing on studies that analyzed the effects of H^+ toxicity and its relation to morphophysiological, biochemical, and molecular responses of crops. As there is a relationship between Al^{3+}, Mn^{2+} and Fe^{3+} toxicity with the acid soil as well as acidity stress, these toxicities were also addressed in necessary cases.

Soil pH and Acidic Soil

Soil pH has been used to define acid soils. The pH scale is the measuring unit for acidity vs. alkalinity of soils, which is a numeric system, ranging between 0 and 14 used to express acidity and alkalinity. The pH scale indicates the H^+ ion activity, which is expressed as the negative logarithm of the H^+ ion concentration in the soil solution. For instance, one-unit decrease in the pH (from 7.0 to 6.0) is corresponding to ten-fold increase in H^+ ion activity. The lower the pH, the more acidic the soil. Most of the soils range in pH from slightly <2.0 to slightly >11.0 (Soil Survey Division Staff 2017). The United States Department of Agriculture (USDA), Natural

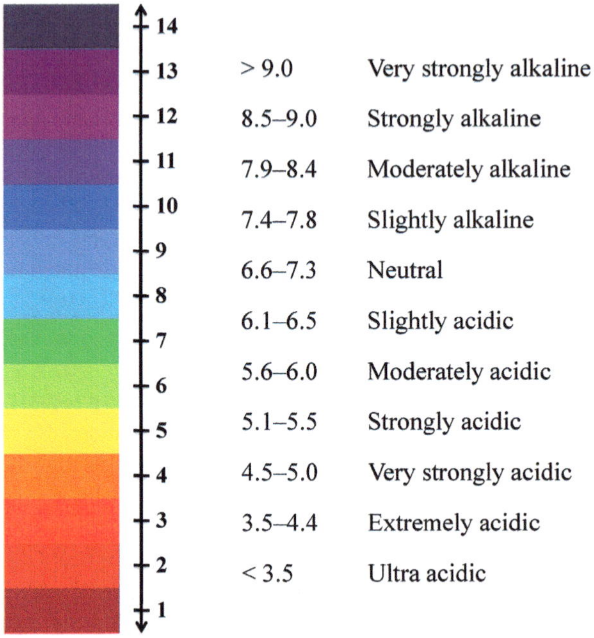

Fig. 1 The soil pH scale and soil classes based on pH and soil reaction. Adopted from Soil Survey Division Staff (2017)

Resources Conservation Service classified soils depending on their pH ranges into 11 classes (Fig. 1).

Both strongly alkaline and strongly acid conditions are generally detrimental to plant life. When the pH lay between 6.0 and 6.8, most of the plants grow best. Generally, acid soils can be termed as those soils having a pH (Potential of H^+) value below 7.0 on the pH scale, but from agricultural point of view acid soils are those that have a pH value of less than 5.5 around the year and associated with plant restricting conditions. Some plants can tolerate and grow in acidic condition thus termed as acid loving crops can grow between 4.5 and 5.5 pH values.

Causes

Soil acidity is a major ecological and economic concern. Usually, in acidic soils more often, when precipitation exceeds 450 mm $year^{-1}$ and about 30% (50 million sq. km) of the land turned into acidic. About 76% of potential arable land for agricultural production is suffering from acidity (Marschner 1991). Acid sulfate soils of coast may acidify when drained to produce crop or for urbanization, consequently affect aquatic life of the estuary, causing loss of fish, with successive rain after drought, when oxidation of soil S occurred (Goulding 2016, Table 1).

Soil acidity accelerated due to intensive agricultural practices and is particularly an important problem in sandy soil with low buffering capability, and relatively small clay particles become acidic due to their smaller reservoir of alkaline cations

Table 1 Processes that contribute to the increase hydrogen ions (H^+) levels in the soil, contributing to the soil acidity

H^+ ions source	Reaction
Parent material	Pyrite (iron/sulfur) $2FeS_2 + 6H_2O + 7O_2 = 4SO_4^{2-} + 8H^+ + 2Fe(OH)_2$
Environment	Carbon dioxide $H_2O + CO_2 = H_2CO_3$ $H_2CO_3 = H^+ + HCO_3^- = 2H^+ + CO_3$ Nitrogen $N_2 + O_2 = 2NO$ $2NO + O_2 = 2NO_2$ $3NO_2 + H_2O = 2HNO_3 + NO$ $HNO_3 = H^+ + NO3^-$ Sulfur (from burning fossil fuel) $S + O_2 = SO_2$ $2SO_2 + O_2 = 2SO_3$ $SO_3 + H_2O = H_2SO_4$ $H_2SO_4 = 2H^+ + SO_4^{2-}$
Fertilizer	Urea $(NH_2)_2CO + 2H_2O = (NH_4)_2CO_3$ $(NH_4)_2CO_3 + 2H^+ = 2NH_4 + CO_2 + H_2O$ $NH_4^+ + 2O_2 = NO_3^- + 2H^+ + H_2O$ Phosphate $Ca(H_2PO_4)_2 + H_2O = CaHPO_4 + H_3PO_4$ $H_3PO_4 = H^+ + H_2PO_4^- = 2H^+ + HPO_4^{2-} = 3H^+ + PO_4^{3-}$
Organic matter	Decomposition processes forming CO_2, NH_4^+, NO_3^-, $H_2PO_4^-$, SO_4^{2-}, H_2O (aerobic condition) CH_4, H_2S, NH_3 (anaerobic conditions)
Nutrient uptake by plant roots	H^+ ion extrusion in exchange for nutrient cations uptake
Excessive rain	Leaching of the cations from soil solution

Adopted from (Helyar and Porter 1989)

and higher leaching potential (Goulding 2016, Table 1). Although many soils are naturally acidic, agricultural practices and industrial processes promote soil acidification (Wenzl et al. 2003). Intensive crop production using nitrogenous fertilizers mainly urea and growing the same crop without rotation contribute to the increase of acidification. In such cases, H^+ is produced by natural process of ammonical nitrogen nitrification (Adams 1984, Table 1). Although less acidifying compared to ammonium, monocalcium phosphate [$Ca(H_2PO_4)_2$], one of the components of phosphate fertilizer, can also be an acidifying factor. It may react with water to form dicalcium phosphate ($CaHPO_4$) and phosphoric acid (H_3PO_4) that contributes in acidification process (Bouman et al. 1995, Table 1).

Acid rain is another problem, and in industrialized countries, especially Europe and the Northeastern United States, there is significant air pollution that produces acid rain—sulfuric and nitric acids clouds, which contains dissolved oxides of N and S, and can fall long far from source led to international disputes. In some areas, mining causes soil acidification (Wen et al. 2013, Table 1). The residues high in iron pyrite oxidized and after mixing with the run-off rain water go to the agricultural land and cause acidification (Table 1). Thus, exert ecological effects in different areas, decline in forest health and aquatic ecosystems.

Development of acid soils is often influenced by edaphic, climatic, and biological factors. For example, soils that develop from granite parent materials acidify at a faster rate than soils developed from calcareous parent materials. Sandy soils with relatively few clay particles acidify more rapidly due to their smaller reservoir of alkaline cations and higher leaching potential. High rainfall affects the rate of soil acidification depending on the rate of water percolation through the soil profile. Soil acidification is intensified by the removal of cations through the harvesting of crops, and organic matter decaying to form carbonic acid and other weak acids (Wen et al. 2013). Plant growth will contribute to acidification; a major nutrient uptake process is to exchange H^+ at the root surface for needed base cations (positively charged ions), such as Ca, Mg, and K. Leguminous plants are particularly acidifying because they take up more cations, in comparison to anions (negatively charged ions), than non-leguminous plants (Table 1). Many forms of organic matter can be acidifying, depending on the plant from which the organic matter is derived. Some plants contain significant quantities of organic acids. As their residues are decomposed, the organic acids naturally affect the soil pH. Other plants are acidifying simply because of the low concentrations of bases they contain. If the plant does not contain enough bases to satisfy microbial needs, the decomposition of the plant debris will not only give off carbon dioxide, but will also remove base nutrients, such as Ca and Mg from the soil (Table 1, Krug and Frink 1983; Edge et al. 1994; de Vries et al. 2014; Jin et al. 2013; Ramlall et al. 2015; Yu et al. 2017).

Worldwide Distribution

It is a cruel characteristic of some of the most populated areas in the world, particularly in the tropics and subtropics are the acid soils. They cover one-third of the world's total land area, including important areas of Africa, Asia, and Latin America, and 60% of the land that we could use for food crops. Today, about 30% of all arable land reaches toxic levels of acidity for crops.

There are several estimates about the extent of acid soils in the world. According to van Wambeke (1976), acid soils occupy 1455 million hectares (11%) of the world's land, while Haug and Foy (1984) estimate that 30–40% of the arable land in the world and up to 70% of the world's land. Von Uexküll and Mutert (1995) estimated that the global extent of acid soils (defined as soils with pH <5.5 in their surface layers) was 3.950 million hectares, or approximately 30% of the world's ice-free lands. This is in agreement with Eswaran et al. (1997), who estimate that around 26% of the land totally free of ice worldwide is suffering from soil acidity, and consequently unproductive.

Therefore, the total acid soils worldwide are estimated at 30% of the ice-free land and occur predominantly in two "belts" (Von Uexküll and Mutert 1995). The northern strip of acid soils is mainly forested (66.3%, or 2.21 million hectares), playing an important role in the ecology of the land; to be more specific, the northern belt in the cold humid temperate zone covering North America, South Asia, and Russia; and the southern belt in humid high rainfall tropical areas including South Africa, South America, Australia, and parts of New Zealand, while the southern strip of 17.7% (699 million hectares) is covered by tropical and subtropical savanna, prairie, and steppe. However, many soils in other areas are also acidic (Fig. 2). Only tropical acid soils comprise about two billion hectares, or 14% of the total area without ice in the world. About 40% of the world's arable soils (i.e., where crops could be grown) are acidic. Only 4.5% of acid soils are used for cultivation, and the remaining areas are not under agriculture yet (Von Uexküll and Mutert 1995) and could be used for plantations and pastures for animal production. Similar areas are found in Colombia, Venezuela, Central Africa, and Southeast Asia (Borlaug and Dowswell 1997). For example, acidity affects about 38% of farmland in Southeast Asia, 31% in Latin America, 20% in East Asia, 56% in Sub-Saharan Africa, and parts of North America (Wood et al. 2000; Hoekenga et al. 2006). In the USA, 1616 million hectares is affected, mostly in South America. In Australia and New Zealand, 239 million hectares of agricultural land is acidic (Von Uexküll and Mutert 1995). In China and India, 212 million hectares or 12% of agricultural land is classified as acidic (Edmonds 2012).

Nutrient Imbalance and Toxicity Due to Acidity

Acid soil toxicity—caused by H^+ rhizotoxicity is further accelerated by the combination of highly soluble toxic metal/metalloid elements like, Fe, Cu, Mn, Zn, and Al, together with essential nutrients deficiency like, P, Mg, Ca, K, and Na (Bian

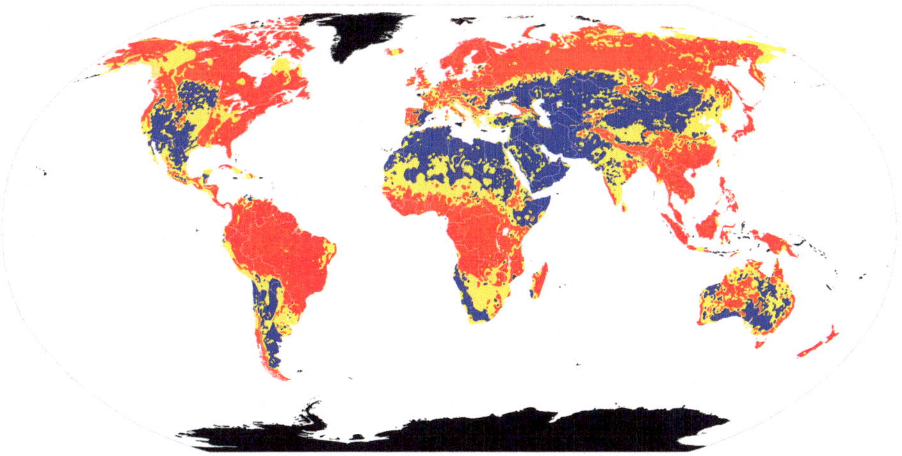

Fig. 2 Global distribution of acid, alkaline, and neutral soils (Wikipedia 2018)

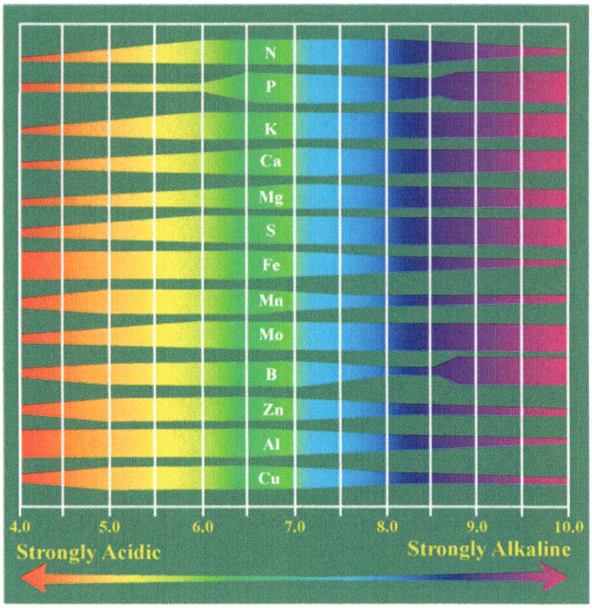

Fig. 3 Effect of soil pH on nutrient availability in plants

et al. 2013; Fig. 3). Low soil pH can therefore release excesses of Al, Fe, and Mn ion, which hamper crop productivity (Zeigler et al. 1995). High Al and Fe oxides and hydroxide at acidic soil pH fix P and create their unavailability to plants (Kisinyo et al. 2013). Up to date most of the attention on acidity stress directly or indirectly addressed to Al toxicity; contrary, very narrow attention was placed on Fe and Mn toxicities. Aluminum toxicity limits crop production by hampering their yield potential (Kochian et al. 2015; Rao et al. 2016). Aluminum toxicity results in root damage that hampers nutrient uptake, resulting in nutrient deficiency in crop plants (Steiner et al. 2012). Strong subsoil Al toxicity reduces plant rooting depth, hence increases susceptibility to drought, and decreases the use of subsoil nutrients (Ulrich et al. 1980). Moreover, Al toxicity can cause 25–80% yield losses in various crop plants.

Fe toxicity is associated with huge concentrations of reduced ferrous iron (Fe^{2+}) in the soil solution that occurs in flooded soils. Poorly aerated acid soils might be rich in iron content that is phytotoxic. Low pH-induced Fe toxicity is generally buffered through Al hydrolysis. In low soil pH, some of the anaerobic bacteria release high amounts of Fe^{2+}, which become toxic for plants. Acidic soil with Fe toxicity may results in yield reductions up to 100% in rice. High toxic levels of Fe, if accumulated in plants can cause lipid peroxidation, protein denaturation, and breakdown of nucleic acids. Furthermore, Fe toxicity damages the cell structures lead to stunted plant growth and leaf injury. Some tolerant genotypes can precipitate excess Fe on roots forming Fe plaque, which serve as a barrier against foreign Fe, and facilitates essential nutrients assimilation plants. (Ayeni et al. 2014; Sikirou et al. 2016).

Manganese toxicity is associated with Al and Fe hydrolysis, caused by soil acidity. Soil acidification accelerates solubility of Mn and increases its availability to toxic levels. In low soil pH, Mn toxicity is more prominent in susceptible plant species. Manganese toxicity appears as leaf chlorosis, which progresses to necrosis. Moreover, Mn toxicity disrupts chloroplast structure and functions, as a result reduced photosynthesis and transpiration hampered. Consequently, with the closure of stomata CO_2 fixation inhibited (Havlin et al. 2005; Hong et al. 2010). But up to now, there are a very limited number of reports regarding the Fe and Mn toxicity in plants. Hence, requires more investigations.

Biogeography of Plants in Relation to Acidic Soil

Acidity prevents crops from uptaking the balanced nutrients from the soil, thus, limiting yields. Among the abiotic stresses, its negative effect on global yield comes second immediately after drought and is particularly tough felt by marginal and small farmers, who cannot afford soil test or correct soil pH using calcium and magnesium-rich lime. Therefore, these farmers usually grow less profitable, acid-tolerant crops like millet, or suffer huge yield losses, when trying to grow popular high yielding crop varieties of wheat, rice, or maize. Many of the crop fields are kept vacant or cultivated once in rainy season.

Acid soils not only cause plant yield losses, but also effect plant distribution around the world. For example, barley—the fourth most important cereal crop in the world, is well known for its wide tolerance to abiotic stress, such as drought, alkaline conditions, cold, and heat (Forster et al. 2000). Due to its stress tolerance behavior, it is distributed all most all over the world and its growing area is extended from temperate to subtropical regions. However, as it cannot tolerate acidic soils, the barley production areas are distributed mainly on non-acid soil regions of the world.

Metal ion availability as well as toxicity depends on soil pH. Among the most studied metal toxicity in acidic soil–Al toxicity, which influence the distribution of some plant species. Clark and Ji (1995) found strong trends in distribution of some palm species, due to change in spatial soil pH and other chemical properties of the soil. Working on six alluvial terraces in Sumatra, Van Schaik and Mirmanto (1985) provide perhaps the clearest example of correlations between pH and plant patterns. They found that fruit production decreased, whereas forest stature and their life span increased with decrease in soil pH. Although they didn't study the species composition but mentioned that the plants of Dipterocarpaceae are more common on the higher terraces and mountain slopes with low soil pH, less fertile soil compared to the lower terraces, which clearly indicates strong relationships between soil pH and various morphophysiological plant processes with consequently change the species composition in an area.

In many regions of the world, this strong positive correlation is observed, between plant diversity and soil pH (Ewald 2003). Through a global-scale study, researchers have illustrated that the relationship between soil pH/plant diversity mainly depends on the regional primary or secondary evolutionary centers (Pärtel 2002). When the regional center for evolution of any species is located on low pH soils, its diversity relationship with soil pH is mostly negative. On the other hand, the diversity relationship between soil pH and plant is mostly positive, when its evolutionary center is located on high pH soils (Pärtel et al. 1996). As most of the soils at higher altitudes are acidic, there is little or no chance for high pH loving plants to grow on these mountain tops (Pärtel 2002). In these areas, species rarity is common due to lack of suitable habitants (Table 2).

Anthropogenic acidification has been considered to be a serious threat for many plants typical of intermediate pH soils (Bobbink et al. 2010). As the soil becomes more acidic due to anthropogenic activities, in course of time the plant productivity is reduced, and in some cases plants are unable to grow due to their critical pH limit. The activity of soil fauna and flora also decreases. Researchers suggested that the plants grown under a wide range of soil pH are distributed evenly around the globe, but plant species that require more restricted soil pH are somewhat threatened species and many of them are already extinct (Cleavitt 2001; Ingerpuu 2002). Therefore, we tried to review, collect, and compile the preferential pH range of some crop, ornamentals, and forest plants based on available literature, which is illustrated in Table 2.

Effects of Soil Acidity Stress on Plants

Acidic condition in plant growing medium disrupts the water uptake of plant (Kamaluddin and Zwiazek 2004; Long et al. 2017). Moreover, toxicities of H^+ with other metal/metalloids and a lack of essential nutrients cause severe yield reduction of crop species by hampering seed germination, vegetative and reproductive attributes of plants (Deska et al. 2011; Bahrami et al. 2012; Krstic et al. 2012). Proper pH level is imperative for germination process of any seed because germination initiating amylolytic enzymes require suitable pH level in germination medium (Lee et al. 1998). But acidic pH shrinks the germination process of seed; even stop at ultra-acidic condition (Deska et al. 2011; Kolodziejek and Patykowski 2015). Growth, developmental processes, and yield of plant are highly dependent on proper nutrient status of growing medium. Acidic condition demolishes the nutrient status by suppressing uptake of few important micro- and macronutrients (Kidd and Proctor 2001; Bian et al. 2013). Furthermore, elevated Al toxicity due to acidity hampers the root growth, which ultimately stunted plant establishment as well as reduce yield (Siecińska and Nosalewicz 2016) (Table 3).

Table 2 List of crops/plants according to their optimum pH requirements of soils

Name	Required pH	Name	Required pH
Field crops			
Oryza sativa L.	5.0–6.5	*Vigna mungo* (L.) Hepper	6.5–7.8
Triticum aestivum L.	6.3–7.0	*Lathyrus sativus* L.	6.0–7.8
Zea mays L. (Field corn)	5.5–6.7	*Vigna unguiculata* (L.) Walp.	5.5–6.5
Zea mays L. (Corn)	5.5–7.0	*Vigna radiata* (L.) R. Wilczek	6.2–7.2
Sorghum bicolor (L.) Moench	5.5–7.0	*Glycine max* (L.) Merr.	5.5–6.5
Medicago sativa L.	6.5–7.5	*Brassica napus* L.	6.0–7.5
Lens culinaris Medik.	5.5–7.0	*Arachis hypogaea* L.	5.0–6.5
Pisum sativum L.	6.0–7.5	*Helianthus annuus* L.	5.0–7.0
Cicer arietinum L.	6.0–7.0	*Brassica juncea* (L.) Czern.	5.0–8.0
Saccharum officinarum L.	5.0–6.5	*Beta vulgaris* L.	6.0–7.5
Vegetables, salads, and spices			
Solanum tuberosum L.	4.5–6.0	*Lepidium sativum* L.	6.0–7.0
Lycopersicon esculentum Mill.	5.5–7.5	*Cucumis sativus* L.	5.5–7.5
Solanum melongena L.	6.0–7.0	*Cucurbita maxima* L.	5.5–7.5
Ipomoea batatas (L.) Lam.	5.5–6.0	*Phaseolus vulgaris* L.	6.0–7.5
Capsicum annuum L.	5.5–7.0	*Asparagus officinalis* L.	6.0–8.0
Brassica oleracea L. var. Italica	6.0–7.0	*Apium graveolens* L.	6.0–7.0
Brassica oleracea L. var. Capitata	6.0–7.5	*Cichorium intybus* L.	5.0–6.5
Brassica oleracea L. var. Botrytis	5.5–7.5	*Lactuca sativa* L.	6.0–7.0
Brassica rapa L. var. Chinensis	6.0–7.5	*Eruca sativa* Mill.	6.0–6.8
Brassica oleracea L. var. Gemmifera	6.0–7.5	*Rheum rhabarbarum* L.	5.5–7.0
Brassica oleracea L. var. Acephala	6.0–7.5	*Spinacia oleracea* L.	6.0–7.5
Cynara cardunculus L.var. scolymus	6.5–7.5	*Colocasia esculenta* L.	5.5–6.5
Brassica oleracea L. var. Gongylodes	6.0–7.5	*Allium sativum* L.	5.5–7.5
Daucus carota subsp. Sativus (Hoffm.) Schübl. & G. Martens	5.5–7.5	*Armoracia rusticana* G. Gaertn., B. Mey. & Scherb.	6.0–7.0
Raphanus raphanistrum subsp. Sativus (L.) Domin	6.0–7.0	*Stevia rebaudiana* (Bertoni) Bertoni	6.7–7.2
Brassica rapa var. rapa L.	5.5–7.0	*Allium stipitatum* L.	5.5–7.0
Pastinaca sativa L.	5.5–7.5	*Allium ampeloprasum* L.	6.0–8.0
Nasturtium officinale Aiton.	5.0–8.0	*Petroselinum crispum* (Mill.) Fuss	5.0–7.0
Ocimum basilicum L.	5.5–6.5	*Mentha spicata* L.	5.5–7.5
Allium schoenoprasum L.	6.0–7.0	*Mentha × piperita* L.	6.0–7.5
Foeniculum vulgare Mill.	5.0–6.0	*Rosmarinus officinalis* L.	5.0–6.0
Allium sativum L.	5.5–7.5	*Allium cepa* L.	6.0–7.0
Zingiber officinale Roscoe	5.5–6.5	*Thymus vulgaris* L.	5.5–7.0
Origanum majorana L.	6.0–8.0		
Fruit crops			

(continued)

Table 2 (continued)

Name	Required pH	Name	Required pH
Citrullus lanatus (Thunb.) Matsum. & Nakai	5.5–6.5	*Artocarpus altilis* (Parkinson) Fosberg	5.0–6.0
Prunus domestica L.	6.0–8.8	*Litchi chinensis* Sonn.	5.5–6.5
Rubus occidentalis L.	5.5–7.0	*Musa* spp.	5.5–6.5
Rubus idaeus L.	6.0–7.5	*Mangifera indica* L.	5.5–6.8
Fragaria × *ananassa* (Weston) Duchesne ex Rozier	5.0–7.5	*Malus floribunda* Siebold ex Van Houtte	5.0–6.0
Rubus fruticosus L.	5.0–6.0	*Citrus* × *sinensis* (L.) Osbeck	5.8–6.5
Vaccinium caesariense Mack.	4.0–6.0	*Ficus carica* L.	6.0–6.8
Prunus cerasus L.	6.5–6.0	*Carica papaya* L.	5.8–6.5
Phoenix dactylifera L.	6.5–8.0	*Citrus* × *paradisi* Macfad.	6.0–6.8
Vitis vinifera L.	6.0–7.0	*Passiflora edulis* Sims.	5.0–6.0
Prunus persica (L.) Batsch	6.0–7.0	*Psidium guajava* L.	5.5–6.8
Pyrus spp.	6.0–7.5	*Ananas comosus* (L.) Merr.	4.7–5.5
Malus pumila Mill.	5.5–6.5	*Citrus japonica* Thunb.	6.0–6.8
Punica granatum L.	6.0–7.0	*Citrus maxima* Merr.	5.8–6.5
Persea Americana Mill.	6.2–6.5	*Citrus limon* (L.) Osbeck	6.0–6.8
Cocos nucifera L.	6.0–8.0	*Citrus tangerine* Tanaka	6.0–6.8
Citrus × *aurantiifolia* (Christm.) Swingle	6.0–6.8	*Macadamia integrifolia* Maiden & Betche	5.0–6.5
Grasses			
Digitaria eriantha Steud.	5.5–6.5	*Zoysia matrella* (L.) Merr.	6.0–6.5
Pennisetum clandestinum Hochst. ex Chiov.	5.5–6.5	*Eremochloa ophiuroides* (Munro) Hack.	6.0–6.5
Ophiopogon japonicus (L.f.) Ker Gawl.	6.0–6.8	*Stenotaphrum secundatum* (Walt.) Kuntze	6.0–6.5
Cynodon dactylon (L.) Pers.	6.0–6.5	*Miscanthus sinensis* Andersson	5.5–6.5
Agrostis canina L	6.0–6.5		
Flowers and ornamentals			
Trifolium repens L.	6.0–7.0	*Podospadix* spp.	5.5–6.5
Lotus corniculatus L.	5.5–6.0	*Abelmoschus moschatus* Medik.	5.8–6.5
Desmodium incanum DC.	5.5–6.5	*Ixora coccinea* L.	5.8–6.5
Saintpaulia shumensis B.L.Burtt	6.0–8.0	*Jacaranda* spp.	5.8–6.5
Aloe vera (L.) Burm.f.	6.5–7.0	*Bougainvillea* spp.	5.5–6.8
Amaryllis belladonna L.	4.5–6.0	*Lantana camara* L.	5.8–6.5
Araucaria araucana Pav.	4.5–6.0	*Camellia* spp.	4.5–5.5
Asparagus densiflorus (Kunth) Jessop	6.0–8.0	*Magnolia virginiana* L.	5.5–6.5
Rhododendron alabamense Rehder	4.5–6.0	*Dianthus caryophyllus* L.	6.0–6.5
Bambusa spp.	5.0–7.5	*Tagetes erecta* L.	5.8–6.5
Begonia oblique L.	6.0–8.0	*Chrysanthemum indicum* L.	6.0–6.5

(continued)

Table 2 (continued)

Name	Required pH	Name	Required pH
Opuntia spp. *Mammillaria* spp. *Melocactus* spp.	4.5–6.0	*Nerium oleander* L.	6.0–7.0
		Rosa spp.	5.8–6.5
		Salix babylonica L.	5.0–6.0
Camellia spp.	4.5–6.0	*Wisteria* spp.	6.5–8.0
Coffea arabica L.	4.5–6.0	*Taxus baccata* L.	5.5–7.5
Lilium longiflorum Thunb.	6.0–6.5	*Zinnia* spp.	5.5–7.5
Pteris spp.	4.5–6.0	*Tulipa* × *gesneriana* L.	6.0–7.0
Gardenia spp.	4.5–6.0	*Petunia* spp.	6.0–7.5
Geranium spp.	6.0–8.0	*Ipomoea nil* (L.) Roth	6.0–7.5
Hydrangea spp.	4.5–6.0	*Hyacinthus orientalis* L.	6.5–7.5
Orchis spp.	4.5–5.0	*Lavandula spica* L.	6.5–7.5
Oxalis spp.	6.0–8.0	*Dianthus caryophyllus* L.	6.0–7.5
Philodendron spp.	4.5–6.0	*Gladiolus* spp.	6.0–7.0
Euphorbia pulcherrima Willd. ex Klotzsch	6.0–6.5	*Chlorophytum comosum* (Thunb.) Jacques	6.1–6.5
Ficus elastica Roxb. ex Hornem.	6.0–6.5	*Iris germanica* L.	5.0–6.5
Hemerocallis lilioasphodelus L.	6.0–8.0	*Narcissus poeticus* L.	6.0–6.5
Delphinium spp.	6.0–7.5	*Cosmos* spp.	5.0–8.0
Myosotis sylvatica Ehrh.	6.0–7.0	*Alyssum* spp.	6.0–7.5
Digitalis spp.	6.0–7.5	*Calendula* spp.	5.5–7.0
Portulaca spp.	5.5–6.5	*Iberis* spp.	6.0–7.5
Philadelphus coronarius L.	6.0–6.8	*Dahlia pinnata* Cav.	6.0–7.5
Buxus spp.	6.0–6.8	*Dianthus barbatus* L.	6.0–7.5
Portulaca grandiflora Hook.	6.0–6.8	*Portulaca* spp.	5.5–7.5
Plectranthus amboinicus (Lour.) Spreng.	6.0–6.5	*Mertensia virginica* (L.) Pers. ex Link	6.5–7.5
Croton spp.	5.5–6.5	*Papaver somniferum* L.	6.0–7.5
Stephanotis thouarsii Brongn	6.0–6.8	*Chaenactis* spp.	6.0–7.5
Salvia officinalis L.	6.0–7.5	*Tropaeolum* spp.	5.5–7.5
Antirrhinum majus L.	5.5–7.0	*Viola tricolor*	5.5–7.0
Lathyrus odoratus L.	6.0–7.5	*Primula vulgaris* Huds.	5.5–6.5
Erysimum spp.	5.0–8.0	*Magnolia virginiana* L.	5.0–6.0
Hydrangea macrophylla (Thunb.) Ser. (blue)	4.0–5.0	*Hydrangea macrophylla* (Thunb.) Ser. (pink)	6.0–7.0
Syringa vulgaris Aiden C. Elharrar	6.0–7.5	*Forsythia suspensa* (Thunb.) Vahl	6.5–7.5
Forest plants			
Eucalyptus oblique L'Hér.	6.0–6.8	*Acer* spp.	6.0–7.5
Mesua ferrea L.	6.0–6.8	*Philadelphus* spp.	6.0–8.0
Thuja occidentalis L.	6.0–8.0	*Quercus* spp.	5.0–6.5
Fraxinus spp.	6.0–7.0	*Picea* spp.	4.5–5.5
Fagus spp.	5.0–6.5	*Gypsophila* spp.	6.0–7.5

(continued)

Table 2 (continued)

Name	Required pH	Name	Required pH
Betula spp.	5.0–6.5	*Juniperus* spp.	5.0–7.0
Strongylodon macrobotrys A.Gray	6.0–6.8	*Pithecellobium dulce* (Roxb.) Benth	6.0–6.8
Cornus spp.	6.0–7.0	*Buxus* spp.	6.5–7.5

Adopted from Uchida and Hue (2000)

Germination and Seedling Emergence

Seed germination is a standout among the most principal and fundamental stages in the developmental cycle of plants that decides the foundation and is profoundly responsive to its encompassing condition (Bahrami et al. 2012). Seeds only germinate under optimal environmental conditions and soil pH is one of the major environmental factors determining germination of seed as well as the further crop establishment. The acute effect of soil acidity on seed germination was studied previously by several researchers and the general conclusion from those examinations might be that an entirely acidic response applies a harmful impact on plants (Michaels 1910; Promsy 1911; Plate 1913). Moreover, the relationship of germination and acidity varies significantly with the seeds of distinctive plants and with the sort of acid utilized. It has been documented that less acidified condition is essential for seed germination (Marschner 1991) and the reason behind this phenomenon is the germination initiating amylolytic enzymes; those require proper pH of soil solution for germination (Lee et al. 1998). In addition, acidic pH exerts a direct impact on seeds in dissolving the seed coat and an indirect impact, which includes stimulating conditions for some fungi species development whose activity causes aperture of the seed coat (Vleeshouwers et al. 1995). Kolodziejek and Patykowski (2015) investigated the effects of acidic pH on *Rumex confertus* biennial species and found that seeds did not sprout at the point when pH was 4.0 with the highest germination at pH 5.6–6. Acidity stress in the growing medium decreases the seed germination both in the initial and final stages. Among the studied species, lowering of pH resulted in a significant germination inhibition of *Trifolium repens* seeds with the negatively affected green part of the seedlings. Decreasing pH even from 6.5 to 6.0, also brought about the reduced energy and capacity of germination as well as the seedling dry matter in *Medicago sativa* and *Festuca pratensis* (Deska et al. 2011). Long time ago, Singh et al. (1975) evaluated the response of four very important grass species *Iseilema anthephoroides, Sehima nervosum, Apluda mutica*, and *Dactyloctenium aegyptium* to acidic medium and revealed that at pH 2.0 no germination was observed in any of the species. With the course of time, it has been observed that the seed germination of *Betula luminifera* L. varied when subjected to different levels of pH stress. A complete inhibition of seed germination was observed under acidic condition of 3.0 and 3.5 range. With the expanding acidity level of ≥4.0, a significant and positive correlation among germination rate, germination

Table 3 Effects of acidity stress in different growth stages of plant life cycle

Plant species	Acidity level	Effect of acidity	References
Triticum aestivum L.	Al^{3+} toxicity, 30 mM/L of $AlCl_3$	Inhibited germination by 50% and poor seedling emergence	Zhang et al. (2010)
Pisum sativum L. (Rachana and Arkil)	0.2, 0.4, 0.6, and 0.8 g/kg of Al	Decreased seed germination and growth	Singh et al. (2011a, b)
Oryza sativa L. (Caiapó and IAC1289)	80, 160, 320 μM of Al	Decreased seed germination and complete inhibition of germination at 320 μM of Al	Marciano et al. (2010)
O. sativa cv. MR 84	pH 3.5	Reduced germination percentage, tiller no. and leaf no. by 4%, 14%, and 10%, respectively, as compared to control (5.6)	Zabawi et al. (2008)
Origanum compactum (Benth)	pH <3.5	Null seed germination were found under acidic pH (<3.5)	Laghmouchi et al. (2017)
Betula luminifera L.	pH <4.0	Completely inhibited germination at pH 3.0–3.5 Negatively affected germination rate, germination potential, germination index, and seed vitality index	Hai-yang et al. (2013)
Rumex confertus L.	pH 4.0	Non-sprouted seeds were found	Kolodziejek and Patykowski (2015)
Trifolium repens *Medicago sativa* *Festuca pratensis*	Lowering pH from 6.5 to 6.0	Hindered germination with damaged green part of seedlings Reduced energy and capacity of germination Decreased seedlings dry matter	Deska et al. (2011)
Iseilema anthephoroides *Sehima nervosum*, *Apluda mutica* *Dactyloctenium aegyptium*	pH 2.0	Complete inhibition of germination in all species	Singh et al. (1975)
Paulownia tomentosa L.	pH 1.5–7.0	None of the seed was germinated below pH 4.0 Significantly decreased seedling emergence	Turner et al. (1998)
O. sativa cv. Lalat	50 mM, $AlCl_3$	Altered cell cycles (mitotic and meiotic) Produced chromosomal abnormalities, binucleated, and multinucleated cells	Mohanty et al. (2004)
O. sativa	50 μM/L of Al	Severely suppressed root elongation	Zhao et al. (2013)

(continued)

Table 3 (continued)

Plant species	Acidity level	Effect of acidity	References
Glycine max Merrill	Al (800 mg/kg Al^{3+})	Decreased chl contents, photosynthetic rate, light, and water use efficiency Increased transpiration rate	Zhang et al. (2007)
Eucalyptus sp.	pH 3.0 and 4.0 with Al 4.4 mM	Reduced photosynthetic pigments. Altered leaf tissue structures and growth	Yang et al. (2015)
Zea mays L. *G. max*	pH 4.1	Decreased root growth, nutrient uptake, shoot biomass, and yield	Joris et al. (2013)
G. max	pH 4.0 and 6.5 and Al (150 μM)	Drastically reduced growth (shoot and root length and biomass), chl contents, and photosynthetic rate	Shamsi et al. (2008)
Z. mays	pH 4.5, 2.1 cmol Al kg^{-1}	Severely affected leaf size and appearance Decreased leaf area index, above-ground biomass, and root biomass by 60%, 30%, and 64%, respectively	Sierra et al. (2003)
Hordeum vulgare L., *Z. mays* *T. aestivum* *G. max*	pH < 5.5	Created nutritional imbalances Disturbed the uptake, transport, and utilization of Ca, Mg, P, K, and NH_4	Gupta et al. (2013)
Theobroma cacao L.	Al^{3+} (15, 30, 45, and 60 mg L^{-1})	Deformed leaf nuclear membrane cells, root epidermis cells, endodermis, and xylem parenchyma cells Disrupted plasma membrane and vacuole	de Almeida et al. (2015)
Phaseolus vulgaris L.	pH <5.0	Reduced plant height, leaf area index Extended days to maturity	Legesse et al. (2013)
Vicia faba L. (Dosha and NC 58)	pH 4.5 and 82 mM Al^{3+}	Reduced 3% and 40% of taproot length, respectively	Belachew and Stoddard (2017)
T. aestivum	pH 4.5	Restricted root development and yield due to water shortages in vegetative period	Caires et al. (2006)
O. sativa	Mn (500 μM)	Reduction of growth Deformation leaf structure	Lidon and Teixeira (2000)
H. vulgare	Mn (500 μM)	Appeared leaf chlorotic and necrotic lesions	Demirevska-Kepova et al. (2004)
Juncus effusus L.	Mn (500 μM)	Decreased plant height and biomass	Najeeb et al. (2009)

(continued)

Table 3 (continued)

Plant species	Acidity level	Effect of acidity	References
O. sativa	Mn (9.1 and 36.4 μM)	Curtailed amount of chl *a*, chl *b*, carotinoid, and reduction of net photosynthesis	Lidon et al. (2004)
Vicia faba L.	Mn (160 μM)	Reduced root and shoot development by 52% and 62.92%, respectively Chlorophyll content declined by maximum 42%	Arya and Roy (2011)
Saccharum officinarum L.	pH 3.74–4.84	Showed chlorosis in ratoon plantlet Hindered chl synthesis in their leaves	Huang et al. (2016)
Phragmites australis (Cav.) Trin ex. Steudel	Fe ($\geq$1 mg L^{-1})	Immediately decreased root growth, shoot growth Significantly inhibited plant growth	Batty and Younger (2003)
O. sativa cv. (BR IRGA 409; BR IRGA 412; BRA 041171 and BRA 041152)	Fe (4 mM)	Decreased the chl synthesis as well as photosynthesis	Pereira et al. (2013)
T. aestivum	Al (0.6 mM)	Inhibited cell division of the microspores Reproductive cytological alteration	Bakos et al. (2008)
Plantago almogravensis and *P. algarbiensis*	pH (4.0) Al (400 mM $AlCl_3$)	Altered development, chl contents, photosynthetic machinery, CHO synthesis, and accumulation	Martins et al. (2013a, b)
O. sativa cv. Pungsan (PS) and Geumgang (GG) and IR36	Al (150 μM)	Delayed flowering time Reduced grain yield	Kang et al. (2011)
Malus sylvestris Miller cv. Golden	pH (3.3 and 3.4)	Decreased pollen germination and pollen tube length	Munzuroglu et al. (2003)
Malus domestica Borkh	pH 3.0, 4.0, and 5.6	Negatively affected pollen grain viability, germination, pollen tube length	Bellani et al. (1997)
O. sativa cv. Lalat	Al (50 mM, $AlCl_3$)	Caused very high rate of pollen sterility of around 64%	Mohanty et al. (2004)
P. vulgaris	pH 2–4.5	Hindered buds and flowers developmental processes Produced abnormalities in reproductive processes at pH 2.0	Chehregani et al. (2006)
Pubescens ssp. tortuosa (Ledeb.) Nyman	pH 4.6 and 3.0	Nearly 50% reduction in the germ tube length and pollen germination	Neuvonen et al. (1991)

(continued)

Table 3 (continued)

Plant species	Acidity level	Effect of acidity	References
Mentha spicata L.	Mn (9–15.75 μM)	Decreased photosynthetic pigments during flowering stage Toxic Mn accumulation in both root and shoot	Asrar et al. (2005)
Lycopersicon esculentum Mill.	Al (500 μM as $Al_2(SO_4)_3{\cdot}18H_2O$)	Reduced pollen germination Decreased shoot and root growth of sporophyte	Searcy and Mulcahy (1990)
Prunus amygdalus Batsch and *Prunus domestica* L.	pH 2.8	Pollen germination and tube length were reduced by 93.43% and 93.41%; 100% and 99.63%, respectively	Nazmi et al. (2016)
O. sativa	pH 4.5 and 6.4	Resulted in sterility of spikelet	Fageria et al. (2004)
Camellia sinensis	pH 4.6 Al (0.2 mM)	Pollen growth was found to decrease with the increase of Al Severely repressed growth at pH 4.6	Konishi and Miyamoto (1983)
G. max	pH 2.8, 3.4, and 4.0	Reduced stalk no, pod hulls, no. of seeds per plant, no. of pods per plant, and no. of seeds per pod, when supplied with unfiltered ambient air	Troiano et al. (1983)
G. max	pH 5.3	Decreased dry weight of shoot, no. of pods per plant, and 100-grain weight	Fageria et al. (2013)
Vigna unguiculata L. (BRS Guariba)	pH 5.2 and 6.2	Pods no. per plant, seeds no. per pod, 100 seeds weight, grain yield were reduced significantly	Farias et al. (2016)
O. sativa	Low, 4.5 and high, 6.4	Grain yield and yield components were reduced significantly under low pH	Fageria et al. (2004)
O. sativa	Fe toxicity (pH 4.1, 4.5, and 4.7)	Severely reduced grain yield of by (16–78%) Damaged leaves (discoloration), decreased growth and tillering	Audebert and Fofana (2009)
Z. mays	Al (2.1 cmol kg^{-1}), pH 4.5	Grain yield reduced by 47% by affecting the leaf appearance, leaf area index, light harvest, and above-ground biomass	Sierra et al. (2003)

(continued)

Table 3 (continued)

Plant species	Acidity level	Effect of acidity	References
Z. mays	pH 4.2 or 4.5	Altered assimilate and nutrient partitioning Finally reduced the biomass production and yield	Sierra et al. (2006)
T. aestivum (tolerant and intolerant genotypes)	440 μM Al, 750 μM Mn, 625 μM Fe^{2+}, pH 4.2–4.8	A relative grain yield potential of 78% vs. 56% was found in tolerant to intolerant genotypes	Khabaz-Saberi et al. (2012)
Z. mays. *G. max*	pH 4.6	Severely affected grain yields	Caires et al. (2011)
G. max cv. Amsoy 71	pH 4.1, 3.3, and 2.7	Decreased seed yield by 10.7%, 16.8%, and 22.9%, respectively, compared to control (pH 5.6)	Evans et al. (1983)
G. max	pH 4.2	Negatively affected plant population, pods no. per plant, grains no. per pod, and 100 grains mass	Castro and Crusciol (2013)
P. vulgaris	Acrisols soil, pH 4.6–5.0	Pod no. per plant, seed no. per pod and seed weight decreased by 33, 82, and 93%, respectively	Buerkert et al. (1990)
T. aestivum Al-tolerant (ET8) and Al-sensitive (ES8)	pH 4.2	Higher shoot biomass and grain yield were obtained from ET8 as compared to ES8	Tang et al. (2001)
T. aestivum and *T. durum*	Lime (pH 5.3) and unlimed (4.5)	Spike numbers, above-ground biomass, and grain yield were more severely affected than plant height, spike length, and harvest index Limed soil plants produced higher yield	Bona et al. (1995)

potential, germination index, and seed vitality index were also observed (Hai-yang et al. 2013). Acidic pH 3.5, 4.0, 4.5, 5.0, and 5.6 (control) were imposed to *Oryza sativa* L. cv. MR 84 seeds on each day for 2 weeks. The result demonstrated that germination, tiller, and leaf number were lessened significantly at pH 3.5. Germination percentage, tiller number, and leaf number were reduced by about 4%, 14%, and 10%, respectively, as compared to control (pH 5.6) (Zabawi et al. 2008). As aluminum ion (Al^{3+}) is particularly very toxic and one of the major limiting factors in acidic soils, a study was conducted to investigate the effect of (Al^{3+}) on *Triticum aestivum* L. seed germination and seedling emergence. Germination as well as seedling emergence was hindered by high concentrations of 30 mM/L of $AlCl_3$. At this treatment, germination percentage was diminished by about 50% with the significant inhibition of seedling growth (Zhang et al. 2010). In a comparative study with tolerant (Rachana) and sensitive (Arkil) genotypes of *Pisum sativum* L., significantly decreased germination of seeds and growth of seedlings due to Al toxicity (0.2, 0.4, 0.6, and 0.8 g/kg of Al) were observed (Singh et al. 2011a). Suthar et al. (2009) conducted an experiment with *Solanum nigrum* L. and demonstrated that the germination percentage was lower under the acidic pH than the neutral pH condition. In addition, seed germination and seedling growth of *Paulownia tomentosa* were evaluated with various range of pH (1.5–7), where none of the seed was germinated below pH 4 with significantly decreased seedling emergence in soil pH 4.5 (Turner et al. 1998). Laghmouchi et al. (2017) carried out an experiment in vitro with *Origanum compactum* Benth medicinal plants and found null seed germination under acidic pH (<3.5). Maximum germination percentage (71%) was observed when provided with the optimum pH condition (7.0). Magidow et al. (2013) investigated the performance of swallowwort (*Vincetoxicum hirundinaria* Medik.), with low pH inceptisols and high pH alfisols. It was demonstrated that the final germination percentage was higher in high pH alfisols than in low pH inceptisols. Two *O. sativa* genotypes, Caiapó and IAC1289, were subjected to Al toxic acid soil. In Caiapó and IAC1289, decreased seed germination and complete inhibition of germination were found when exposed with 160, 320 μM and 80, 160 μM of Al, respectively. Mean germination time were also increased for both genotypes when exposed to Al toxicity (Marciano et al. 2010). At pH level 1.0, along with the extended average germination time, the germination rate, index, velocity, vigor index, shoot height, root length, shoot and root dry mass, and the stored substances transformation rate reduced significantly in *Zea mays* L. Effects of seed age and water solution pH on the germination, germination energy, root length, and stem height of *P. sativum* were investigated. It has been demonstrated that irrespective of the age of seeds, highest germination energy was recorded at neutral pH (7.0). In contrast, highest germination, root length, and stem height were observed at pH 6 and 5, respectively (Gordana et al. 2007).

Vegetative Stage

Soil acidity stress generally affects throughout the vegetative phage of plant life cycle by inhibiting the growth, visualizing damage symptoms, altering phonological events, hindering growth and developmental processes. A combination of factors limits plant growth togetherly results in soil acidic stress. In spite of the fact that a low pH itself is adverse and Al toxicity is one of the principal factors restricting plant growth in acidic soils. The primary indication of Al^{3+} stress in plants is the inhibition of root elongation as well as the final root growth via the Al association with the root apices cells (Siecińska and Nosalewicz 2016). Thus, immediately affected root growth is resulted from the quick inhibition of root tip meristem cell division by Al^{3+}, resulting in a stubby root tip, which impaired the water and nutrient uptake and ultimately making plant more susceptible to other stresses like drought (Samac and Tesfaye 2003; Jovanovic et al. 2006, 2007). However, the species of plants, particular condition of growth environments and the forms and concentrations of Al ion determines the degree of inhibition of root growth (Delhaize and Ryan 1995). Some other consequences of Al^{3+} toxicity includes hindered cell division and plant growth, destroyed plasma membrane, altered nutrient availability, certain metabolites and key enzymes as well as the susceptibility to several environmental stresses (Arunakumara et al. 2013). Aluminum stress (50 mM, $AlCl_3$) in *O. sativa* cv. Lalat resulted in cell cycles (mitotic and meiotic) alterations with varied chromosomal abnormalities and produced binucleated and multinucleated cells (Mohanty et al. 2004). More thickened and rigid cell wall along with the final reduction of the cell-wall extensibility required for normal cell expansion were also affected due to Al toxicity in *O. sativa* and *T. aestivum* (Yang et al. 2008). In acid soils, low P availability is also resulted from the formation of Al–P complexes, which act as a prime limiting factor for plant growth (Nian et al. 2009). Aluminum-induced Ca deficiencies also results in the young leaves rolling and curling and growing points collapse. Aluminum also inhibits plant development processes by extremely affecting DNA formation, blocking cell division, destroying cell wall, disrupting plasma membrane integrity, inhibiting signal transduction pathways and changing cytoskeleton (Bian et al. 2013). Additionally, soil acidity strongly affects plant growth by the deficiency of N in both tropical and temperate regions (Fageria and Baligar 2001). Soil acidity acts as a major limiting factor for nodulation and N fixation in legume plants. Especially plants more sensitive to acidic soils are those depending on the symbiotic N fixation for their growth because higher concentration of H^+, Al^{3+}, or Mn^{2+} and reduced concentration of Ca^{2+}, P, or Mo can restrict their nodule formation in acidic soils (Marschner 1991). Furthermore, the toxicity of Al presents in the acidic soil belated the vegetative growth of tropical legumes (Meda and Furlani 2005). Small, stunted, late matured, purpling of stem and leaf veins, yellowing and death of leaf apex (Bouma et al. 1981) and decreased opening of stomata, reduced photosynthesis, foliar necrosis, and chlorosis (Vitorello et al. 2005) are the consequences of the toxic Al. In most crop plants, μmol concentrations of Al are very much toxic. Despite being the most Al-tolerant one among the cereals, in *Oryza sativa* L. plants, 50 μM/L of Al can suppress root elongation (Zhao et al. 2013). Guo et al. (2012) studied the response of Al toxicity in two *O. sativa* varieties (Xiushui 132—Al tolerant) and (Yongyou 8—Al sensitive). Slight and severe growth inhibition and chlorophyll (chl) content reduction were observed in both sensitive and tolerant rice varieties, respectively. He et al. (2011) revealed that toxic level of Al and low P negatively affected the growth and photosynthesis of *Camellia oleifera* Abel. In two *Glycine max* (L.) Merr. varieties, low concentration of Al (200 mg/kg Al^{3+}) were found effective for growth whereas decreased chl content, photosynthetic rate, light and water use efficiency, and increased transpiration rate were evident under high concentration of Al (800 mg/kg Al^{3+}) (Zhang et al. 2007). Similarly, *Eucalyptus* trees subjected to acid soils (pH 3.0 and 4.0) as well as Al (4.4 mM) resulted in reduced photosynthetic parameters, altered leaf tissue structures, and overall limited growth (Yang et al. 2015). Depletion of Ca and other cation-based nutrition in acidified soil caused reduced regeneration, decreased canopy length with basal area growth in sugar maple tree (Sullivan et al. 2013). Joris et al. (2013) demonstrated that in *Z. mays* and *G. max*, acid loamy soil (pH 4.1) exhibited negative effects on root growth, nutrient uptake, shoot biomass, and yield which were further alleviated by the application of lime (0, 4, 8, and 12 Mg ha^{-1}). Together effects of Al and drought stress on *Phaseolus vulgaris* L. concluded both the damaged root and shoot system along with the impaired crop growth and yield (Yang et al. 2013). In a study with *G. max*, Shamsi et al. (2008) mentioned that a drastic reduction of growth (shoot and root length, and biomass), chl content, photosynthetic rate occurred under low pH (4.0 and 6.5) and Al (150 μM) treated soils. In *Z. mays*, the low pH, non-limed Al toxic (pH 4.5, 2.1 cmol Al kg^{-1}) soil resulted in the severely affected leaf size and appearance. Consequently, decreased leaf area index, aboveground biomass and root biomass by 60%, 30%, and 64%, respectively, were evident (Sierra et al. 2003). Aluminum exposure (pH below 5.5) created nutritional imbalances by disturbing the uptake, transport, and utilization of mainly Ca, Mg, P, K, and NH_4 in a number of crops namely, *Hordeum vulgare* L., *Z. mays*, *T. aestivum*, and *G. max* (Gupta et al. 2013). According to de Almeida et al. (2015), *Theobroma cacao* plant subjected to Al^{3+} toxicity (15, 30, 45, and 60 mg L^{-1}) showed a deformed leaf nuclear membrane cells, root epidermis cells, endodermis, and xylem parenchyma cells with disrupted plasma membrane and vacuole. An experiment with the two Al tolerant (CMS36, tolerant, spectral, moderately tolerant) and P-efficient *Z. mays* cultivars showed higher nutrient and assimilate partitioning in aboveground organs (Sierra et al. 2006). Martins et al. (2013a, b) observed significant changes due to low pH (4.0) and Al stress (400 mM $AlCl_3$) on the development, chl content, photosynthetic machinery, carbohydrate synthesis, and accumulation in *Plantago almogravensis* and *P. algarbiensis* where *P. almogravensis* showed more tolerance to acidity. Soil acidity has turned into a genuine risk in western Ethiopia for production of *P. vulgaris* by imposing negative effects on plant height, leaf area index, days to maturity, etc. (Legesse et al. 2013). Effects of soil acidity (pH 4.3–5.0, Al^{3+} conc. 4–16 mg kg^{-1}) on the development of primary roots, root hair, and rhizosheath in perennial grass showed noteworthy variations among the genotypes. Tall *Phalaris* cv. Sirosa were more sensitive to acidity showing poor root length and growth (Haling et al. 2010). Acidity (pH 4.5) and Al stress (82 mM/L Al^{3+}) were imposed to tolerant (Dosha and NC 58) and sensitive (Babylon) accessions of *Vicia faba* L. and showed 3% and 40% reduction of tap root length, respectively (Belachew and Stoddard 2017). Acidity stress (pH 4.5) restricted root development and yield of *T. aestivum* extremely, likely because of expanded water shortages in the vegetative stage. However, increased root development and grain yield were observed by 66% and 140%, respectively, after acidity amelioration by liming (Caires et al. 2006). In acid soils, the second most important concern is Mn toxicity that limits plant growth (Foy 1984). Mn toxicity primarily affects shoot growth and brings about stunted growth, chlorosis, and necrotic lesions in leaves (Kochian et al. 2004). Once Mn is highly accumulated in plants, it interferes with the metabolism and affects plant growth. Fernando et al. (2016) conducted a comparative study with both tolerant and sensitive genotypes of *Acer saccharum* L. to investigate the effects of excess Mn. It has been demonstrated that the relative growth rate in case of tolerant genotype was 98% as compared to the sensitive one, which was 71%. Enhanced leaf and canopy development were also evident in tolerant genotypes. Toxic level of Mn (500 μmol Mn L^{-1})-induced growth reduction and leaf structure deformation in a number of crop plants. Decreased shoot growth rate in *O. sativa* (Lidon and Teixeira 2000), dry weight in *Lolium perenne* L. (Mora et al. 2009), plant height and biomass in *Juncus effusus* L. (Najeeb et al. 2009) and leaf chlorotic and necrotic lesions in *H. vulgare* (Demirevska-Kepova et al. 2004), *L. perenne* (Rosas et al. 2007), *G. max* (Lavres et al. 2009), and *Brassica napus* L.(Moroni et al. 2003) were reported due to toxic amount of Mn in acidic soils. Excess Mn causes a reduction of photosynthesis and chl contents in crop plants. Curtailed amount of chl *a*, chl *b*, carotinoid, and reduction in net photosynthesis were observed in *O. sativa* subjected to excess amount of Mn (9.1 and 36.4 μM) (Lidon et al. 2004). Arya and Roy (2011) reported that increasing Mn concentration (160 μM) reduced the root and shoot development by 52% and 62.92%, respectively, along with the chl content declined by maximum 42% in *V. faba* plant. Acidic soil (pH 3.74–4.84) with excess Mn showed chlorosis in ratoon plantlet of *S. officinarum* due to excess Mn accumulation in the parent stalk with hindered chl synthesis in their leaves (Huang et al. 2016). Manganese treatments (3 and 6 mM $MnCl_2$) inhibited *O. sativa* growth by accumulating higher Mn in root, shoot, and damaging membrane. Manganese toxicity-induced oxidative stress through higher ROS accumulation was further balanced by the activation of antioxidant enzymes activities (Srivastava and Dubey 2011). To evaluate the effects of Fe on the *Phragmites australis* (Cav.) Trin ex. Steudel, a laboratory experiment was conducted. It was demonstrated that concentration above (1 mg L^{-1}) resulted in the immediate decrease in root growth, shoot growth, and ultimately significant inhibition of plant growth (Batty and Younger 2003). Excess Fe (4 mM) decreased the chl synthesis as well as photosynthesis in four *O. sativa* cultivars (BR IRGA 409; BR IRGA 412; BRA 041171 and BRA 041152) (Pereira et al. 2013). Besides the mineral toxicities, low pH in acidic soil can immediately inhibit root growth and thus, poor plant growth is triggered by the H^+ influx into the root tissues (Rangel et al. 2005).

Reproductive Stage

Acidity stress is very sensitive to reproductive processes of many plant species and ultimately results in poor plant production (Cox and Lins 1984). From the very ancient times, different research experiments have been showing the effect of acidity on reproductive processes. Acidity stress decreased the viability of pollen, which in turn reduces the number of seed available for germination. Acid deposition (pH 2.6–5.6) showed adverse effect on the pollen germination and pollen tube growth of *Picea glauca*. Pollen tube growth along with the germination was reduced significantly below pH 3.6 (Sidhu 1983). Aluminum stress inhibited the cell division of the microspores even at very low concentration of 0.6 mM, together with the reproductive cytological alteration in *T. aestivum* (Bakos et al. 2008). Kang et al. (2011) conducted an experiment with *O. sativa* employing two tolerant Pungsan and

Geumgang and one susceptible IR36 cultivars. Flowering time was delayed 27 days in susceptible IR36 along with the reduced grain yield due to Al (150 μM) stress. In another experiment, with the in vitro growth medium, at low pH (2.5 or 3.00), the conifer pollen germination percentage was reduced but the broad-leaved ones was completely inhibited (Paoletti 1991). Searcy and Mulcahy (1990) reported that Al toxicity imposed adverse effects on the gametophytes and sporophytes of four *Lycopersicon esculentum* Mill. cultivars. Al toxicity (500 μM, as $Al_2(SO_4)_3.18H_2O$) in sensitive cultivars significantly reduced the pollen germination as well as the shoot and root growth of sporophyte. But the shoot and root growth were varied in their response depending on the sensitivity of the cultivars. Pollen grain viability, germination, pollen tube length in *Malus domestica* Borkh was negatively affected by acidity (pH 3.0, 4.0, and 5.6) (Bellani et al. 1997). Similarly, at acidic pH (3.3 and 3.4) pollen germination and pollen tube length in *M. sylvestris* cv. Golden were decreased by 41.75% and 24.3%, respectively. Complete destruction of pollen tube and inhibition of germination were also evident below pH 3.1 and 3.0, respectively (Munzuroglu et al. 2003). Pollen germination and tube length have shown to decrease with the increasing level of acidity. In *Prunus amygdalus* Batsch and *P. domestica*, pollen germination and tube length were reduced by 93.43% and 93.41%; 100% and 99.63%, respectively, at pH 2.8 (Nazmi et al. 2016). Acidity effects on the pollination of *Oenothera parviflora* L. was examined in vitro. Low pH interferes in producing the viable pollen and inhibits the germination process of pollen. There was also a significant reduction in stigma receptivity at low pH (less than 3.6) condition (Cox 1983). Delhaize et al. (2009) reported that transgenic *H. vulgare* incorporated with wheat Al^{3+}-resistant gene (*TaALMT1*) showed more phosphorus use efficiency and produced greater than double the grain than the non-transformed sibling line. In *O. sativa*, low levels of soil acidity (4.5 and 6.4) gave rise to sterility of spikelet (Fageria et al. 2004). Kooistra (1967) verified that soil acidity caused the alteration of male to female flower ratio in *Cucumis sativus*. Decreased above-ground biomass and root biomass during flowering and grain yield were observed in soil acidity affected *Z. mays*. Severely reduced leaf area index, above-ground biomass and grain yield recorded were 60%, 30%, and 47%, respectively (Sierra et al. 2003). In *O. sativa* cv. Lalat, low dose of Al (50 mM, $AlCl_3$) caused very high rate of pollen sterility of around 64% (Mohanty et al. 2004). Egerton-Warbuton et al. (1993) conducted an experiment with *Eucalyptus calophylla* trees grown in two soil coal mine-site (pH 4.3) and forest-site (soil pH 5.3) types to investigate the effects of Al. It was revealed that all reproductive tissues (stigma, style, ovules) of mine-site flowers accumulated higher amount of Al than the forest-site flowers. Mine-site pistils also showed pollen tube abnormalities. Significant variations among the genotypes were observed in lime-treated and untreated acid soils. Pod length, pod number per plant, hundred seed weight, grain yield, and pod harvest index were higher in genotypes grown in lime-treated acidic soil (Legesse et al. 2013). Konishi and Miyamoto (1983) investigated the effects of Al and fluorine (F) on *Camellia sinensis* pollen tube growth. Pollen growth was found to decrease with the increase of Al and 0.2 mM Al was found effective in repressing the growth at pH 4.6. Chehregani et al. (2006) found that *P. vulgaris* subjected to acidity stress (pH 2–4.5) hindered buds and flowers developmental processes, where pH 2.0 was most destructive producing abnormalities in reproductive processes. Neuvonen et al. (1991) have observed nearly 50% reduction in the germ tube length and pollen germination in *B. pubescens* ssp. tortuosa (Ledeb.) Nyman, when subjected to acidity stress (pH 4.6 and 3.0). A greenhouse experiment was carried out with *Mentha spicata* L. to investigate the excess Mn (9–15.75 μM) effects at flowering stage. It was found that with the increase of the Mn concentrations, decreased photosynthetic pigments produced during flowering stage with the adverse consequences of the toxic Mn accumulation in both root and shoot (Asrar et al. 2005). Kapczyńska and Magdziarz (2015) investigated the effect of substrate pH upon flowering of *Mandevilla symphytocarpa*. They found positive correlation between low pH with flowering. Plants exposed to low pH (less than 5.0) produced more flowers and flower buds than those exposed to higher pH.

Yield Attributes and Yield

Plant germination, vegetative and reproductive attributes are affected by soil acidity stress, which consequently poses serious threat to crop production. Soluble forms of toxic metal/metalloid(s) and organic acids cause yield reduction under acidity. Aluminum toxicity is the major constraints for limiting crop productivity and yield in most acid soils (Krstic et al. 2012). The phytotoxic Al (150 μM) in acid soil resulted in the reduced flowering time, dry mass production, and grain yield (Kang et al. 2011). Exposure of *G. max* to acidity (pH 2.8, 3.4 and 4.0) along with ambient ozone resulted in a significant reduction of stalk number, pod hulls, number of seeds per plant, number of pods per plant, and number of seeds per pod (Troiano et al. 1983). In a field experiment with Brazilian oxisols, dry weight of shoot, number of pods per plant, and 100-grain weight of *G. max* were decreased by soil acidity (pH 5.3) and further increased when provided with lime and gave up to 90% economic yield (Fageria et al. 2013).

Negatively affected productivity of *Sesamum indicum* L. was documented due to soil acidity and S efficiency, which could be overcome by applying lime and S (Kumar et al. 2017). A study was conducted using the acidity-tolerant strains UFLA03-153, UFLA03-164, UFLA03-84, and INPA03-11B of *Vigna unguiculata* L. (BRS Guariba). Among the studied parameters, pods number per plant, seeds number per pod, 100 seeds weight, and grain yield were reduced by acidity (pH 5.2 and 6.2) (Farias et al. 2016). Responses of upland *O. sativa* to two levels of soil acidity (low, 4.5 and high, 6.4) were evaluated. Grain yield and yield components of *O. sativa* reduced significantly when subjected to low soil pH (4.5) compared to high pH (6.4).

Plant height, dry weight of shoot, number of panicle, harvest index, and 1000 grain weight showed negative correlations with yield under acidic pH (Fageria et al. 2004). Tang et al. (2002) conducted an experiment with two *T. aestivum* genotypes (Al-sensitive ES8 and Al-tolerant ET8) and demonstrated that the tolerant ET8 gave

51% higher yield than the sensitive ES8 one. Iron toxicity (pH 4.1, 4.5, and 4.7) severely reduced grain yield of rice (16–78%) by damaging leaves (discoloration), decreasing growth and tillering. Moreover, soil toxic amount of Fe (20–5000 mg kg^{-1}) have also showed 10–100% yield loss (Audebert and Fofana 2009). In some crop species, up to 30% yield loss have been recorded due to the toxic Fe-induced damaging of leaf tissues (Becker and Asch 2005).

Caires et al. (2008) investigated the effects of soil acidity and surface liming on the root growth and yield of *Z. mays*, *G. max*, and *T. aestivum*. Severely reduced root growth and yield were documented due to Al toxicity (pH 4.5) in all cases, which were ameliorated by lime application. In some field experiments carried out by The et al. (2001, 2006, 2012), tolerant maize cultivar (ATP-SR-Y) have shown to give 60% higher yield than sensitive cultivar (Tuxpeno-Sequia) against acidity. Because of the inhibited root growth and water, nutrients uptake in acid soils, yield potential of *Sorghum bicolor* was reduced significantly (Meda and Furlani 2005; Bernel and Clark 1998). In another report, it was observed that Al (2.1 cmol kg^{-1}) toxicity in the acidic soil (pH 4.5) reduced the *Z. mays* grain yield by 47% by affecting the leaf appearance, leaf area index, light harvest, and above-ground biomass (Sierra et al. 2003). Kariuki et al. (2007) reported that crop yield losses (grain and forage) can be overcome by using Al-tolerant winter wheat cultivars (Ok101, Ok102, 2137, 2174, Jagger, Jagalene, Custer, and AP502CL) in southern Great Plains. They also concluded cultivar differences in case of grain and forage yield was found more than 30% and 2137 cultivar showed the best result regarding yield. Subsoil acidity-induced yield reduction is a very common phenomenon and resulted from the consequences of disrupted water and nutrients uptake from soil (Tang et al. 2003).

Plants of *Z. mays* responded to the tropical acid soil (pH 4.2 or 4.5) through alteration of assimilate and nutrient partitioning, which finally reduced the biomass production and yield (Sierra et al. 2006). Acid soil (Oxisols or Ultisols) stress together with the soil salinity inhibited the N fixation and nodulation and final crop stand in *G. max* through disrupting the signal transduction processes (Miransari and Smith 2007). A long 17 years research result of Popescu (1998) demonstrated that 5.4–5.7 pH range is detrimental to and reduced bean yield compared to higher pH up to 7.0. Legesse et al. (2013) compared the performance of 25 genotypes of *P. vulgaris* in acid soil. Yield attributes differed among the genotypes due to the differences in the maturity, growth, and reproductive indices. Higher absolute and relative yield were obtained from Dimtu, new BILFA 58, Beshbesh, SER176, new BILFA51, and new BILFA genotypes. When investigating the comparative performance of tolerant and susceptible genotypes of *T. aestivum* in waterlogged acid soil (440 μM Al, 750 μM Mn, 625 μM Fe^{2+}, pH 4.2–4.8) a relative grain yield potential of 78% vs. 56% in tolerant to susceptible genotypes were found along with the reduced Al, Mn, and Fe toxicities (Khabaz-Saberi et al. 2012).

The increasing intensities of acidity caused the alteration of microbial activity in the soil and decreased the species richness and productivity (Chen et al. 2013a). In *Z. mays* and *G. max*, soil acidity (pH 4.6) affected grain yields were also revealed (Caires et al. 2011). Simulated acid rain-induced acidic pH 4.1, 3.3, and 2.7 decreased the *G. max* cv. Amsoy 71 seed yield by 10.7%, 16.8%, and 22.9%, respectively, compared to their control (pH 5.6). Whereas, reduced number of pods per plant resulted in decreased seed yield (Evans et al. 1983). Yield attributes, such as, plant population, pods number per plant, grains number per pod, and 100 grain mass of soybean suffered the main effects due to soil acidity (pH 4.2) and infertility. However, application of superficial liming and silicate under zero tillage system recovered the yield to some extent (Castro and Crusciol 2013). Soil acidity (Acrisols soil, pH 4.6–5.0) is the major factor that limits productivity of *P. vulgaris* L. in Southern Mexican State of Chiapas. Pod number per plant, seed number per pod and seed weight decreased by 33%, 82%, and 93%, respectively, due to severe acidity (Buerkert et al. 1990). Two near-isogenic *T. aestivum* Al-tolerant (ET8) and Al-sensitive (ES8) genotypes were subjected to sub-surface acidity (pH 4.2). One-third higher shoot biomass and grain yield were obtained from ET8 as compared to ES8 when provided with water (Tang et al. 2001). Bona et al. (1995) studied two common (*T. aestivum*) and two durum wheat (*T. durum*) cultivars against soil acidity with lime (pH 5.3) and unlimed (4.5) treatments. Spike numbers, above-ground biomass, and grain yield were more severely affected than plant height, spike length, and harvest index. Limed soil in all cases produced higher yield than unlimed soil and yield differences in limed and unlimed soil were significant only in *T. durum*.

Physiological Responses of Plants to Acidity Stress

Under acidity stress, plants showed varied reactions regarding physiological responses. At this point, plants tried to activate their defense mechanisms to reduce the detrimental effect and damages due to low pH of growing media. Many research reports illustrated the physiological response of plants regarding low pH stress, for example, membrane integrity, photosynthetic machinery, water relation and transpiration as well as oxidative damage and dysfunction of antioxidant system. Low pH caused membrane hyperpolarization, followed by depolarization hence increased electrolyte leakage (EL) in stressed plant in contrast to control plant. Similarly, malondialdehyde (MDA) content increased under acidic condition. It was observed that low pH increased membrane permeability and decreased the chl level. Low pH also decreased the water content in both root and shoot and decreased transpiration rate. Moreover, low pH induces greater H^+ concentration resulting oxidative stress through generation of ROS in cell and triggers cellular membrane damage in root due to higher H_2O_2 and MDA. Low pH like abiotic stresses also alter the activity of antioxidant enzymes, for instance, superoxide dismutase, (SOD), catalase (CAT), ascorbate peroxidase (APX), monodehydroascorbate reductase (MDHAR), dehydroascorbate reductase (DHAR), glutathione reductase (GR), glutathione peroxidase (GPX), glutathione *S*-transferase (GST) and peroxidases (POD), and non-enzymatic antioxidant components, for example, ascorbate (AsA) , glutathione (GSH) of cell (Hasanuzzaman et al. 2012). Therefore, in this section we reviewed different physiological responses of plants under low pH stress.

Membrane Integrity

Cell membrane integrity is very susceptible to any abiotic stress including acidity. Destruction of membrane integrity depends on intensity and duration of stress and plant type. Values of lipid peroxidation or EL considered as attributes of the cellular membrane integrity. Martins et al. (2013a, b) cultured shoots of *P. algarbiensis* and *P. almogravensis* for 7 days in acidic condition (pH of growing media was 4.0). The lipid peroxidation (expressed as the content of MDA) in *P. algarbiensis* shoots increased by 247% under acidic condition in contrast to control plants (pH of growing media was 5.75) whereas no significant change of MDA content was observed in *P. almogravensis* shoots. *Lotus corniculatus* plant grown under pH values close to medium acidic (at 5.5) caused membrane hyperpolarization, while acidification to pH 4.0 caused immediate depolarization (Pavlovkin et al. 2009). In *Cucurbita pepo* L., the exposure to acidity (low pH, 3.5) causes several physiological changes at shoot level (Rouphael et al. 2015). Electrolytic leakage increased by 78% in stressed plant in contrast to control plant (grown under pH 6.0). Similarly, Song et al. (2011) recorded that MDA content of *H. vulgare* seedling significantly increased under acidic condition (pH 4.5). In *Eucalyptus* leaves, it was observed that low pH increased membrane permeability (Yang et al. 2011). Zhang et al. (2015) checked the performance of two hybrid rice (*O. sativa*) cultivars (Yongyou 12, YY12, a japonica hybrid and Zhongzheyou 1, ZZY1, an indica hybrid) under acidity stress for 2 weeks. They recorded low pH stress (pH 3.5) prompted an enhancement of MDA in the roots of both rice varieties. In contrast to pH 5.5, the low pH considerably augmented the MDA level in YY12 and ZZY1 roots by 48% and 74%, respectively. Recently, Long et al. (2017) treated 4 weeks old *C. grandis* and *C. sinensis* seedlings with different level of acidic pH (2.5, 3, 4, 5, or 6) for 9 months. Root and leaf EL enhanced as the pH increased from 2.5 to 3, but EL remained comparatively steady under rising pH. Root and leaf EL was higher in *C. grandis* than in *C. sinensis* at pH 2.5, but it was alike between the citrus species tested at pH 3–6.

Photosynthetic Machinery

Extreme pH-induced ROS production might also be involved in the collapse of photosynthetic machinery. Inhibition of CO_2 assimilation and alteration of photosynthetic process including destruction of chl pigments is very common under acidity stress (Zhang et al. 2014). Yang et al. (2011) observed that a low pH decreased the chl level in *Eucalyptus* leaves. Saenen et al. (2014) carried out an experiment with *A. thaliana* under acidity stress (pH 4.5). They found considerable decrease of chl *a* and chl *b* content of plant at pH 4.5. Zhang et al. (2014) reported that acidic growing condition (pH 3.0) for *Juglans regia* L. decreased the leaf net photosynthetic rate, actual quantum yield of the photosystem II (PSII) electron transport.

Rouphael et al. (2015) grown zucchini squash (*C. pepo*) plants in nutrient solutions having different pH levels (Control, 6.0 and stressed/acidic, 3.5). Compared to control noteworthy reduction of SPAD index and leaf area was observed in response to low pH (pH 3.5). Thereafter, Yang et al. (2015) reported that low pH (3.0) decreased leaf photosynthesis and chl level in the four vegetatively propagated Eucalyptus clones (G9, G12, G3, and G4). Long et al. (2017) observed leaf CO_2 assimilation, stomatal conductance, and Rubisco activity did not change significantly as the pH decreased from 6 to 3, but they greatly decreased at pH 2.5. Moreover, Chl *a* and Chl *b* content considerably enhanced as the pH raised from 2.5 to 3, after which they remained unchanged or were only to some extent changed with increasing pH. But their concentrations did not differ significantly between the two citrus species at pH 3, 4, 5, and 6, but they were lower in *C. sinensis* leaves than in *C. grandis* leaves at pH 2.5. Similarly, Anugoolprasert et al. (2012) recorded that altering the pH between 5.7 and 3.6 did not decrease the chl content, photosynthetic rate, and stomatal conductance in sago palm leaves.

Water Relations and Transpiration

Water status and transpiration of plant altered due to abiotic stress including acidity. Martins et al. (2013a, b) observed that proline (Pro) content of *P. algarbiensis* shoots decreased by 28% under acidic stress condition (pH of growing media is 4.0, 7 days) in contrast to control (pH of growing media is 5.75), whereas in *P. almogravensis* shoots, Pro content increased by 34% in comparison with control plants. Proline is a low-molecular-weight amino acid, well known as osmoregulator and ROS scavenger. Elevated Pro-activity under the acidity stress is attributed for giving protection against oxidative injury including maintaining water status (Gill and Tuteja 2010; Cvikrová et al. 2013). Kamaluddin and Zwiazek (2004) observed that acidic condition rapidly reduced the water flow rate and the hydraulic conductivity of paper birch (*B. papyrifera*) roots. H^+-sensitive maize cultivar Adour 250 showed the reduction of whole-root water conductivity at pH 4.5. But it did not change in the H^+-tolerant maize cultivar BR 201 F (Gunsé et al. 1997). Yang et al. (2011) observed that a low pH decreased the water content in Eucalyptus roots, stems, and leaves. Recently, Long et al. (2017) reported that both pH 2.5 and 3 decreased the root RWC, while only pH 2.5 lowered the leaf RWC. Like water content, transpiration rate also hampered due to low pH. Acidity stress decreased transpiration rate in *Eucalyptus* (Yang et al. 2015), *Juglans regia* (Zhang et al. 2014), and sago palm (Anugoolprasert et al. 2012).

Oxidative Damages

Productions of ROS are very common in plant cells in times of different physiological and metabolic activities. Among these, singlet oxygen ($^1O_2$), superoxide anion ($O_2^{\bullet-}$), hydrogen peroxide (H_2O_2), and hydroxyl radical ($OH^\bullet$) are remarkable. Normally, this ROS are balanced by plant self-defense activities. Again, these ROS also acts as secondary messenger at very low concentration by associating with plant growth and developmental processes. But certain environmental stresses cause extra ROS production and goes beyond managing capacity. This excess ROS induces damage to cellular components and termed as oxidative stress. Oxidative damages including oxidation of lipid, proteins, and nucleic acid; enzymatic inhibition activities and programmed cell death (PCD) impose threat to plant survival resulting plant death (Hasanuzzaman et al. 2018). Under oxidative stress, many series events are happened where $O_2^{\bullet-}$ is produced first, turn to H_2O_2, subsequent lipid peroxidation and finally formation of MDA (Hasanuzzaman et al. 2018). Low pH induces greater H^+ concentration resulting oxidative stress through generation of ROS in cell (Shi et al. 2006). Acidic stress causes oxidative damage and indicated by higher MDA content which is confirmed by the report of Song et al. (2011) in barley. Accumulation of $O_2^{\bullet-}$ and H_2O_2 causes lipid peroxidation and indicated by higher MDA (Zhang et al. 2015). In *O. sativa*, Zhang et al. (2015) found that extremely low pH triggers cellular membrane damage in root due to higher H_2O_2. In *C. pepo*, acidic soil (pH 3.5) caused higher EL (Rouphael et al. (2015), while Shi et al. 2006 found elevated MDA in cucumber under pH 4.5. Thus, acidic stress causes oxidative damage to different kinds of plants, for example, Scots pine (*Pinus sylvestris* L.) (Ivanov et al. 2013), legume *Lotus corniculatus* (Pavlovkin et al. 2009; Pal'ove-Balang et al. 2012), and *Citrus* (Long et al. 2017). Model plant *Arabidopsis* also suffered from acidic conditions and showed excess H_2O_2 and MDA (Qiao et al. 2018). Similar result was also reported by Saenen et al. (2013, 2014) in *Arabidopsis* finding increased lipid peroxidation. Acidity-induced oxidative stress causes cell membrane damages as well as structural malformation which was found in tomato (Gabara et al. 2003). Chen et al. (2013b) conducted a study to evaluate the toxicity of acid rain-induced acidic condition (pH 3) on both sensitive (*Liquidambar formosana*) and tolerant (*Schima superb*) tree species. They found greater membrane lipid peroxidation through higher $O_2^{\bullet-}$ and H_2O_2 generation. But sensitive one showed severe injury compared to tolerant one (Table 4).

Antioxidative Defense

Plant itself is able to dismutase the excess ROS, when it faces stresses for better survival. Like other stresses low pH contributes to higher ROS due to H^+ toxicity. Naturally, plants have very efficient antioxidative capacity to detoxify surplus ROS including enzymatic (SOD, CAT, APX, MDHAR, DHAR, GR, GPX, GST, and

Table 4 Acidity-induced oxidative damage in different plant species

Crops	Acidity levels	Oxidative damages	References
Cucurbita pepo L. cv. Tempra	pH 3.5	Highest electrolyte leakage was found	Rouphael et al. (2015)
Cucumis sativus L. cv. Jinchun 5	pH 4.5	Increased MDA content significantly	Shi et al. (2006)
Arabidopsis thaliana	pH 3.0	Higher H_2O_2 content along with MDA Enhanced soluble protein significantly	Qiao et al. (2018)
Hordeum vulgare L.	pH 4.5	Higher level of MDA content was recorded $O_2^{\bullet-}$ and H_2O_2 highly accumulated confirmed by staining	Song et al. (2011)
Lycopersicon esculentum Mill.	pH 1.8	Malformed chloroplast and mitochondria at 13% and 95%, respectively	Gabara et al. (2003)
Pinus sylvestris L.	pH 4.5	Increased activity of SOD and CAT stated the higher accumulation of $O_2^{\bullet-}$ and H_2O_2	Ivanov et al. (2013)
Lotus corniculatus	pH 4.0	Induced rapid membrane depolarization in root Decreased diffusion potential	Pavlovkin et al. (2009)
Liquidambar formosana *Schima superba*	pH 3.0	Hastened $O_2^{\bullet-}$ and H_2O_2 production Provoked membrane lipid peroxidation Enhanced H_2O_2 production in *L. formosana* Increased MDA content significantly Detected severe injury by H_2O_2 by S-SiAR	Chen et al. (2013a, b)
A. thaliana	pH 4.5	Increased lipid peroxidation confirmed by thiobarbituric acid reactive substances (TBARS)	Saenen et al. (2014)
L. corniculatus	pH 4.0	Reduced membrane potential Suffered from membrane depolarization Severe damage to cytoplasm	Pal'ove-Balang et al. (2012)
Oryza sativa L.	pH 3.5	Increased H_2O_2 up to 47% and MDA by 74%	Zhang et al. (2015)
Citrus sinensis *Citrus grandis*	pH 2.5	Increased both leaf and root H_2O_2 Enhanced EL	Long et al. (2017)

POD) and non-enzymatic antioxidant (AsA, GSH) component (Hasanuzzaman et al. 2012). Higher activity of antioxidant components counteract oxidative damage and give better tolerance to H^+ toxicity through detoxification of ROS (Chen et al. 2013a, b).

Superoxide dismutase acts as first-line defense that dismutase $O_2^{\bullet-}$ by converting to H_2O_2 (Hasanuzzaman et al. 2012). Qiao et al. (2018) stated that very low pH (3.0) increased SOD activity above 50% along in *Arabidopsis* and gave protection. Similar result was also recorded from Song et al. (2011) in *H. vulgare*. Again, CAT activity is responsible for H_2O_2 detoxification. Increased activity of CAT was found in *H. vulgare* (Song et al. 2011). But several report mentioned lower CAT activity in different plant such as *L. esculentum* (Gabara et al. 2003), *Arabidopsis* (Qiao et al. 2018), *C. sativus* (Shi et al. 2006), *P. sylvestris* (Ivanov et al. 2013), and *O. sativa*

(Zhang et al. 2015). These results indicated that H_2O_2 scavenging capability in these crops is low through CAT, which again shows involvement of other antioxidants. Here AsA-GSH cycle also contributes actively in detoxification of H_2O_2. Hence, Zhang et al. (2015) found up-regulated APX activity due to low pH besides of lowered CAT activity and concluded APX abolish accumulated H_2O_2. In addition, both APX and AsA work together to detoxify H_2O_2. In *Arabidopsis*, acidic condition causes increased AsA concentration along with lowered CAT activity (Saenen et al. 2013, 2014). Chen et al. (2013a, b) reported increased GSH content under acidity stress. Similar result was also found from Qiao et al. (2018) in *Arabidopsis*. On the other hand, GSH also works by GPX and GST activity in scavenging H_2O_2 to H_2O. In *C. sativus*, increased GPX, DHAR, and GR activity was mentioned at low pH (Shi et al. (2006) and in *L. esculentum* increased GST activity was found (Gabara et al. 2003). Thus, enhanced activity of antioxidants stimulates antioxidant defense system in plant to scavenge ROS and bestow tolerance to low pH (Table 5).

Mechanisms of Acid Stress Tolerance in Plants

Low pH or acidity of growing media significantly affects the plant nutrients availability, increases toxic metals solubility, decreases soil microorganism activity, destroys root cells, and alters cation exchange (EC) capacity of rooting media. The toxic effects of acidity stress can be classified as either morphological or physiological, both of which lead to poor plant development, consequent yield loss as well as plant death (Martins et al. 2011; Zhang et al. 2015; Long et al. 2017).

Acidity stress is caused by a high concentration of H^+ ions in rooting media or soil, which has not been considered for many decades (Kinraide 1993), along with combinations of metal/metalloid toxicity, as well as lack of available essential nutrients (Foy et al. 1978). The large quantity of H^+ ions present in the acidic growing media adversely affects plant nutrient availability, which is negatively correlated with a decrease in soil pH (Kasai et al. 1992); contrary a positive correlation between low pH and metals (Fe, Cu, Mn, Zn, Al) solubility was reported, those intern become toxic for plants (Ginocchio et al. 2009). Excessive H^+ ions compete with other mineral elements (P, Mg, Ca, and Fe) for plants' uptake and alter their transportation as well as uptaking of other essential nutrients (Poschenrieder et al. 1995). Hence, many of the researchers emphasized H^+ rhizotoxicity (low pH toxicity) as the most important limiting factor that make the plants susceptible in acidic growing media. A high concentration of H^+ eventually leads to the gradual shutdown of PM H^+-ATPase activity in root cells and other enzymes throughout the entire plant, resulting in cessation of plant growth and stagnation (Dyhr-Jensen and Brix 1996).

Moreover, some researchers stated about the Al toxicity as the major limiting factor for plant growth in acidic condition as acidity stress (H^+ toxicity) and Al toxicity occurred simultaneously and difficult to separate since acidic growing media facilitate Al solubility. Kinraide (1993), when screening different *Holcus lanatus* L. and *B. pendula*, grasses, reported that H^+ toxicity is dominant over Al toxicity at a

Table 5 Acidity-induced up-regulation of antioxidant defense system in different plant species

Crops	Acidity levels	Antioxidant defense	References
Cucumis sativus L. cv. Jinchun 5	pH 4.5	Decreased CAT activity, while increased GPX, APX, DHAR, and GR activity	Shi et al. (2006)
Arabidopsis thaliana	pH 3.0	Increased SOD activity up to 58.51% where activity of CAT was reduced GSH and POD activity were higher NiR activity was decreased, while NR activity also showed significant variation	Qiao et al. (2018)
Hordeum vulgare L.	pH 4.5 and 5	Increased SOD, CAT, and POD	Song et al. (2011)
Lycopersicon esculentum Mill.	pH 1.8 (acid rain)	Increased SOD and CuZnSOD activity at 72 and 96 h of exposure Decreased GSH-Px and the activity of CAT after 24 h of treatment Exacerbated APX activity while found highest GST activity at 0.5, 48, and 72 h	Gabara et al. (2003)
Pinus sylvestris L.	pH 4.5	Stimulated the SOD activity at 2.6 times and lowered CAT activity about 2.9 times Lowered ascorbate peroxidase (APO) and guaiacol peroxidase (PO) activity	Ivanov et al. (2013)
Liquidambar formosana; Schima superba	Sulfuric-rich (S-SiAR), nitric-rich (N-SiAR), sulfate, and nitrate mixture (SN-SiAR)-induced pH 3.0 (acid rain)	Increased soluble protein content up to 53.4% Increased proline content up to 70.3% Decreased CAT activity by 75.72% and SOD activity by 60.69% in *L. formosana* Increased GSH content	Chen et al. (2013a, b)
A. thaliana	pH 4.5	Increased total AsA and decreased GSH concentration Enhanced SOD activity while CAT activity was decreased	Saenen et al. (2014)
Oryza sativa L.	pH 3.5	Decreased SOD and activity by 48% Enhanced APX activity by 74%	Zhang et al. (2015)

low Al concentration. While Kidd and Proctor (2001) concluded that plant species collected from acid organic soils were H^+ tolerant, whereas those collected from acidic mineral soils were Al^{3+} tolerant but not necessarily H^+ tolerant.

Plants species even strains within species vary significantly in terms of acid soil tolerance. Hence, it gives us the opportunity to breed acid-tolerant cultivars, which mainly depends on the proper understanding of the physiology, genetics, and gene regulation responsible for acid tolerance. Studies revealed that the acidity stress tolerance is governed by both internal and external mechanisms. For example, among the external mechanisms, organic acid exudation is very common in higher plants. But the matter of regret is that most of the acidity stress tolerance mechanisms only focused on the Al toxicity, and hence H^+ rhizotoxicity was overlooked for decades and there are a few available information on the H^+ toxicity tolerance. Thus, we have tried to mine the information and presented them in a row.

There are significant interspecific variations among plants in terms of critical pH ranges for optimal growth, development, and consequent tolerance to acidity or H^+ rhizotoxicity stress (Foy 1984; Ring et al. 1993; Nuruddin and Chang 1999; Tang et al. 2003). For instance, *Cistus salvifolius* L.—the Mediterranean shrub can adopt or tolerate extreme acid soils and grow normally (Bartoli et al. 2014), whereas most common crop species are more sensitive to acidity. Higher natural variations for the H^+ rhizotoxicity response and tolerance were observed among 260 accessions of *A. thaliana*, in terms of relative root growth compared with a non-acid medium along with a very high heritability (0.94), when the pH was dropped down by only 0.3 units (Ikka et al. 2007; Lefebvre et al. 2009). Genetic polymorphism was also found in many other experiments with various plants, for instance, in trees (Beech, Birch, Spruce), *P. vulgaris*, *O. sativa*, and even weed species in response to acidic stress in growing media (Murach and Ulrich 1988; Kidd and Proctor 2001; Rangel et al. 2005; Cha-Um et al. 2009; Martins et al. 2013a, b; Zhang et al. 2015). Actually, the genetic polymorphism for the high H^+ concentration tolerance within species can be wider compared to non-related species. For example, in legumes, sensitivity to low pH stress among two contrasting *P. vulgaris* genotypes was much higher than between selected cultivars of *P. sativum* and *G. max* (Lazof and Holland 1999). Slootmaker (1974) identified a contribution from the A and D genomes, which potentially carries one or more genes controlling tolerance to acidity in *Triticum* and *Aegilops*.

Most plants can manage in moderately low pH environments, but if the external environment is highly acidic the plants responded adversely. When the pH of external growth medium is decreased to 1 unit, the root cell cytoplasmic pH also dropped down by 0.1 units (Felle 1988; Wilkinson and Duncan 1989); as a result, the plasma membrane H^+–ATPase activity reduced (Dyhr-Jensen and Brix 1996), but vacuolar H^+–ATPase and vacuolar–PPase continue to force excessive H^+ from the cytoplasm to the vacuole, together with the pH-stat (Davies 1973, 1986; Sakano 1998), which also supports cellular pH homeostasis, but their efficiency to support cytoplasmic pH homeostasis reduced gradually, meanwhile external environment becoming more acidified (Shavrukov and Hirai 2016). Experiments with *Z. mays* (Yan et al. 1998) and *O. sativa* (Zhu et al. 2009) indicate that root H^+–ATPase activity increases under acidic growth conditions, which confers acidity tolerance. Consistent with this, a number of genes encoding ATPase subunits were up-regulated in *Arabidopsis* under acidity stress (Lager et al. 2010). It has been reported that root plasmalemma H^+–ATPases contribute significantly for the adaptation of *Z. mays* at low pH, which is the character to tolerate acidic growth environments (Yan et al. 1998).

Some transcription factors (TF) are also responsible for H^+ rhizotoxicity tolerance as well as regulate some downstream genes. The *STOP1* gene, encoding a Cys_2His_2 zinc-finger protein, was identified in *Arabidopsis* (Iuchi et al. 2007), which contributes to root growth at pH 4.3 in MS media compared with wild-type plants. This TF is only expressed when plants are exposed to low pH and regulate the Al-activated malate transporter *AtALMTl* (Iuchi et al. 2007). *STOP1* also regulates multiple downstream genes responsible for ion transport and pH regulation in *Arabidopsis* (Sawaki et al. 2009). Another TF, *STOP2* also partially able to restore root growth when *STOP2* mutant was grown in acidic growth solution, and this TF further activated the transcription of several genes involved in low pH and Al^{3+} stress tolerance, which are regulated by *STOP1*. Being localized in nucleus *STOP2* is less expressed than *STOP1*. On the other hand, *STOP2* could not regulate the expression of Al-induced *AtALMT1* expression, thus could not contribute in Al toxicity (Kobayashi et al. 2014). Orthologs of *STOP1* were also identified in *T. aestivum*, and three homologs of *TaSTOP1* are found on chromosomes 3AL, 3BL, and 3DL (Garcia-Oliveira et al. 2013), which are both genome and genotype specific in *T. aestivum* and might be responsive to acid growing media (Garcia-Oliveira et al. 2013). The large number of responsive TFs identified by Lager et al. (2010), and probable cell-type-specific responses as discovered for DREB family members (Iyer-Pascuzzi et al. 2011), implies that there is a complex process of transcriptional regulation of genes in response to acidity stress in *Arabidopsis*. The involvement of TFs other than *STOP*-like genes in the response of other plant species to acidity has not yet been reported.

As mentioned earlier, biochemical pH-stat regulates cellular pH, either producing or consuming protons during metabolic reactions with organic acids (Sakano 1998). Genes from biochemical pH-stat cooperatively regulate the cytoplasmic pH in conjunction with the proton pump H^+–ATPase and alternative oxidation (AOX) mechanism to reduce ROS in mitochondria (Sakano 1998). However, pH-stat genes are also strongly regulated by TFs like *STOP-1* (Sawaki et al. 2009).

Basically, the alternative oxidases (AOX) in plant cell mitochondria transfer electrons from the ubiquinone pool to oxygen without energy conservation. Evidence of alternate oxidation mechanism in mitochondria is involved in the tolerance of plants to Al stress, by means of organic acids, such as malate secretion that deactivate ROS (Kochian et al. 2004). But, it is still unclear how AOX is involved with respect to low pH stress alone and need further investigation. However, it has been reported that accumulation of ROS increases following acidification of the pH from 5.0 to 4.5 in growing media without the presence of toxic metal/metalloids in seedlings of *H. vulgare*, *P. sylvestris*, *O. sativa*, *C. sinensis*, and *C. grandis* (Song et al. 2011; Ivanov et al. 2013; Zhang et al. 2015; Long et al. 2017). The major enzymes catalyzing the reduction of hydrogen peroxide, detoxification of ROS, and protection of cells from ROS toxicity are SOD, POD, CAT, APX, MDHAR, DHAR, GR, GPX, and GST (Hasanuzzaman et al. 2017), are gradually but significantly regulated in plants under acidified liquid solution and medium, indicating the antioxidant defense that is directly involved in the response of plants to H^+ rhizotoxicity stress (Song et al. 2011; Ivanov et al. 2013; Martins et al. 2013a, b; Zhang et al. 2015). Similar results were found in *C. sativus* and *Arabidopsis* roots, in terms of the higher activities of other antioxidant enzymes, GPX and APX, in acidic growth solution (pH 4.5) compared with control pH of 6.5 (Saenen et al. 2013). At least 11 genes encoding toxic metal transport/detoxification proteins were identified by Lager et al. (2010). The expression of these genes was modulated by acidity stress only without the presence of any toxic metals, which indicates that these genes have multiple functions, with low pH being only one of the conditions implicated.

Another phenomenon named acid growth is one of the strategies to tolerate acidity is cell-wall extension, which is mediated by expansin genes that denature cell-wall proteins and increase cell-wall growth and extensibility, in low pH soils (Cosgrove 1999). Cell-wall acidification can promote expansin expression, inducing cell elongation and extension. However, it has been reported that the total length of coleoptiles of two cultivars of wheat was slightly increased, but not significantly, after the buffer pH was changed from 6.5 to 4.5 (Zhao et al. 2011).

Conclusion and Future Perspectives

With the increase of anthropogenic activities, the problem of soil acidification has been increasing. Due to soil acidity, a wide range of land area has become unsuitable for plant growth and productivity. Disruption of membrane integrity, photosynthetic machinery, water relations, transpiration, and oxidative damages are common direct detrimental effects of acidity stress. Making the soil nutrients unavailable, soil acidity creates nutrient deficiency stress, which then generates other damaging effects. Soil acidity is a major problem in different continents of the world. Cultural practices related to soil management are being executed to reclaim the soil fertility for normal plant growth. Due to pressure of increased population throughout the world, the crop intensification should be increased bringing the acidic land under cultivation. Use of lime is common practice to manage the acidic soil, which has limited effect and is considered as short term, and it may create adverse side effects. So, integrated management should be practiced combining agronomic, cultural, and biological practices and use of inorganic fertilizers. Although the standing crop in an acidic soil is brutally affected, progress on development of acidity-tolerant crop cultivar is not so forwarded. Even to understand the mechanism of damage caused by acidity; broad spectrum, depth and intensive research is needed. Moreover, exogenous use of phytoprotectants should be considered as a short gun approach to confer acidity stress tolerance in crop plants. Nonetheless, the anthropogenic reasons for increasing acidity stress should be minimized to prevent further increase of acidic land.

References

Adams F (1984) Crop response to lime in the southern United States. In: Adams F (ed) Soil acidity and liming, 2nd edn. American Society of Agronomy, Crop Science Society of America, and Soil Science Society of America, Madison, pp 211–265

Anugoolprasert O, Kinoshita S, Naito H, Shimizu M, Ehara H (2012) Effect of low pH on the growth, physiological characteristics and nutrient absorption of sago palm in a hydroponic system. Plant Prod Sci 15:125–131

Arunakumara KKIU, Walpola BC, Yoon M (2013) Aluminum toxicity and tolerance mechanism in cereals and legumes—a review. J Korean Soc Appl Biol Chem 56:1–9

Arya SK, Roy BK (2011) Manganese induced changes in growth, chlorophyll content and antioxidants activity in seedlings of broad bean (*Vicia faba* L.). J Environ Biol 32:707–711

Asrar Z, Khavari-nejad RA, Heidari H (2005) Excess manganese effects on pigments of *Mentha spicata* at flowering stage. Arch Agron Soil Sci 51:101–107

Audebert A, Fofana M (2009) Rice yield gap due to iron toxicity in West Africa. J Agron Crop Sci 195:66–76

Ayeni O, Kambizi L, Fatoki O, Olatunji O (2014) Risk assessment of wetland under aluminium and iron toxicities: a review. Aquat Ecosyst Health Manag 17:122–128

Bahrami H, Razmjoo J, Ostadi JA (2012) Effect of drought stress on germination and seedling growth of sesame cultivars (*Sesamum indicum L.*). Int J Agric Sci 2:423–428

Bakos F, Darkó É, Ascough G, Gáspár L, Ambrus H, Barnabás B (2008) A cytological study on aluminium-treated wheat anther cultures resulting in plants with increased Al tolerance. Plant Breed 127:235–240

Bartoli G, Bottega S, Forino LMC, Ciccarelli D, Spano C (2014) Plant adaptation to extreme environments: the example of *Cistus salviifolius* of an active geothermal alteration field. C R Biol 337:101–110

Batty LC, Younger PL (2003) Effects of external iron concentration upon seedling growth and uptake of Fe and phosphate by the common reed, *Phragmites australis* (Cav.) Trin ex. Steudel. Ann Bot 92:801–806

Becker M, Asch F (2005) Iron toxicity in rice-conditions and management concepts. J Plant Nutr Soil Sci 168:558–573

Belachew KY, Stoddard FL (2017) Screening of faba bean (*Vicia faba* L.) accessions to acidity and aluminium stresses. PeerJ 5:e2963. https://doi.org/10.7717/peerj.2963

Bellani LM, Rinallo C, Muccifora S, Gori P (1997) Effects of simulated acid rain on pollen physiology and ultrastructure in the apple. Environ Pollut 95:357–362

Bernel JH, Clark RB (1998) Growth traits among sorghum genotypes in response to Al^{3+}. J Plant Nutr 21:297–305

Bian M, Zhou M, Sun D, Li C (2013) Molecular approaches unravel the mechanism of acid soil tolerance in plants. Crop J 1:91–104

Blossfeld S, Gansert D (2007) A novel non-invasive optical method for quantitative visualization of pH dynamics in the rhizosphere of plants. Plant Cell Environ 30:176–186

Bobbink R, Hicks K, Galloway J, Spranger T, Alkemade R, Ashmore M, Bustamante M, Cinderby S, Davidson E, Dentener F, Emmett B (2010) Global assessment of nitrogen deposition effects on terrestrial plant diversity: a synthesis. Ecol Appl 20:30–59

Bona L, Baligar VC, Wright RJ (1995) Soil acidity effects on agribotanical traits of durum and common wheat. In: Date RA, Grundon NJ, Rayment GE, Probert ME (eds) Plant-soil interactions at low pH: principles and management. Developments in plant and soil sciences, vol 64. Springer, Dordrecht, pp 425–428

Borlaug NE, Dowswell CR (1997) The acid lands: one of agriculture's last frontiers. In: Moniz AC, Furlani AMC, Schaffert RE, Fageria NK, Rosolem CA, Cantarella H (eds) Plant-soil interactions at low pH: sustainable agriculture and forestry production. Sociedade Brasileira de CiEncia do Solo, Campinas, pp 5–15

Bouma D, Dowling EJ, David DJ (1981) Relations between plant aluminium content and the growth of lucerne and subterranean clover: their usefulness in the detection of aluminium toxicities. Aust J Exp Agric 21:311–317

Bouman OT, Curtin D, Campbell CA, Biederbeck VO, Ukrainetz H (1995) Soil acidification from long-term use of anhydrous ammonia and urea. Soil Sci Soc Am J 59:1488–1494

Buerkert AKG, de la Piedra CR, Munns DN (1990) Soil acidity and liming effects on stand, nodulation, and yield of common bean. Agron J 82:749–754

Bush DS (1995) Calcium regulation in plant cells and its role in signaling. Annu Rev Plant Physiol Plant Mol Biol 46:95–122

Caires EF, Corrêa JCL, Churka S, Barth G, Garbuio FJ (2006) Surface application of lime ameliorates subsoil acidity and improves root growth and yield of wheat in an acid soil under no-till system. Sci Agric (Piracicaba Braz) 63:502–509

Caires EF, Garbuio FJ, Churka S, Barth G, Corrêa JCL (2008) Effects of soil acidity amelioration by surface liming on no-till corn, soybean, and wheat root growth and yield. Eur J Agron 28:57–64

Caires EF, Joris HAW, Churka S (2011) Long-term effects of lime and gypsum additions on no-till corn and soybean yield and soil chemical properties in southern Brazil. Soil Use Manag 27:45–53

Castro GSA, Crusciol CAC (2013) Effects of superficial liming and silicate application on soil fertility and crop yield under rotation. Geoderma 195 & 196:234–242

Cha-Um S, Supaibulwattana K, Kirdmanee C (2009) Comparative effects of salt stress and extreme pH stress combined on glycinebetaine accumulation, photosynthetic abilities and growth characters of two rice genotypes. Rice Sci 16:274–282

Chehregani A, Malayeri BE, Kavianpour F, Yazdi HL (2006) Effect of acid rain on the development, structure and viability of pollen grains in bean plants (*Phaseolus vulgaris*). Pak J Biol Sci 9:1033–1036

Chen D, Lan Z, Bai X, Grace JB, Bai Y (2013a) Evidence that acidification-induced declines in plant diversity and productivity are mediated by changes in below-ground communities and soil properties in a semi-arid steppe. J Ecol 101:1322–1334

Chen J, Wang WH, Liu TW, Wu FH, Zheng HL (2013b) Photosynthetic and antioxidant responses of *Liquidambar formosana* and *Schima superba* seedlings to sulfuric-rich and nitric-rich simulated acid rain. Plant Physiol Biochem 64:41–51

Clark JS, Ji Y (1995) Fecundity and dispersal in plant populations: implications for structure and diversity. Am Nat 146:72–111

Cleavitt N (2001) Disentangling moss species limitations: the role of physiologically based substrate specificity for six species occurring on substrates with varying pH and percent organic matter. Bryologist 104:59–68

Cosgrove DJ (1999) Enzymes and other agents that enhance cell wall extensibility. Ann Rev Plant Physiol Plant Mol Biol 50:391–417

Cox RM (1983) Sensitivity of forest plant reproduction to long range transported air pollutants: *in vitro* sensitivity of pollen to simulated acid rain. New Phytol 95:269–276

Cox FR, Lins IDG (1984) A phosphorus soil test interpretation for corn grown on acid soils varying in crystalline clay content. Commun Soil Sci Plant Anal 15:1481–1491

Crawford NM, Forde BG (2002) Molecular and developmental biology of inorganic nitrogen nutrition. Arabidopsis Book 1:e0011. https://doi.org/10.1199/tab.0011

Cvikrová M, Gemperlová L, Martincová O, Vanková R (2013) Effect of drought and combined drought and heat stress on polyamine metabolism in proline-over-producing tobacco plants. Plant Physiol Biochem 73:7–15

de Almeida NM, de Almeida AF, Mangabeira PAO, Ahnert D, Reis GSM, de Castro AV (2015) Molecular, biochemical, morphological and ultrastructural responses of cacao seedlings to aluminum (Al^{3+}) toxicity. Acta Physiol Plant 37:1–17

de Vries W, Dobbertin M, Solberg S, Dobben H, Schaub M (2014) Impacts of acid deposition, ozone exposure and weather conditions on forest ecosystems in Europe: an overview. Plant Soil 380:1–45
Davies DD (1973) Control of and by pH. Symp Soc Exp Biol 27:513–529
Davies DD (1986) The fine control of cytosolic pH. Physiol Plant 67:702–706
Delhaize E, Ryan PR (1995) Aluminium toxicity and tolerance in plants. Plant Physiol 107:315–321
Delhaize E, Taylor P, Hocking PJ, Simpson RJ, Ryan PR, Richardson AE (2009) Transgenic barley *Hordeum vulgare* L. expressing the wheat aluminium resistance gene (*TaALMT1*) shows enhanced phosphorus nutrition and grain production when grown on an acid soil. Plant Biotechnol J 7:391–400
Demirevska-Kepova K, Simova-Stoilova L, Stoyanova Z, Holzer R, Feller U (2004) Biochemical changes in barley plants after excessive supply of copper and manganese. Environ Exp Bot 52:253–266
Deska J, Jankowski K, Bombik A, Jankowska J (2011) Effect of growing medium pH on germination and initial development of some grassland plants. Acta Sci Pol 10:45–56
Dyhr-Jensen K, Brix H (1996) Effects of pH on ammonium uptake by *Typha latifolia* L. Plant Cell Environ 19:1431–1436
Edel KH, Marchadier E, Brownlee C, Kudla J, Hetherington AM (2017) The evolution of calcium-based signalling in plants. Curr Biol 27:R667–R679
Edge CP, Bell SA, Ashenden TW (1994) Contrasting growth responses of herbaceous species to acidic fogs. Agric Ecosyst Environ 51:293–299
Edmeades DC, Blarney FPC, Farina MPW (1995) Techniques for assessing plant responses on acid soils. In: Date RA, Grundon NJ, Rayment GE, Probert ME (eds) Plant-soil interactions at low pH: principles and management. Developments in plant and soil sciences, vol 64. Springer, Dordrecht, pp 221–233
Edmonds RL (2012) Patterns of China's lost harmony: a survey of the country's environmental degradation and protection. Routledge, Newyork
Egerton-Warbuton LM, Griffin BJ, Lamont BB (1993) Pollen1pistil interactions in *Eucalyptus calophylla* provide no evidence of a selection mechanism for aluminium tolerance. Aust J Bot 41:541–552
Eswaran H, Almaraz R, van den Berg E, Reich P (1997) An assessment of the soil resources of Africa in relation to productivity. Geoderma 77:1–18
Evans LS, Lewin KF, Patti MJ, Cunningham EA (1983) Productivity of field-grown soybeans exposed to simulated acidic rain. New Phytol 93:377–388
Ewald J (2003) The calcareous riddle: why are there so many calciphilous species in the central European flora? Folia Geobot 38:357–366
Fageria NK, Baligar VC (2001) Improving nutrient use efficiency of annual crops in Brazilian acid soils for sustainable crop production. Commun Soil Sci Plant Anal 32:1303–1319
Fageria NK, Castro EM, Baliga VC (2004) Response of upland rice genotypes to soil acidity. In: Wilson MJ, He Z, Yang X (eds) The red soils of China: their nature, management and utilization. Springer, New York, pp 219–237
Fageria NK, Moreira A, Castro C, Moraes MF (2013) Optimal acidity indices for soybean production in Brazilian Oxisols. Commun Soil Sci Plant Anal 44:2941–2951
Farias TP, Trochmann A, Soares BL, Maringá FMS (2016) Rhizobia inoculation and liming increase cowpea productivity in Maranhão state. Acta Sci Agron 38:387–395
Felle HH (1988) Short term pH regulation in plants. Physiol Plant 74:583–591
Felle HH (1998) The apoplastic pH of the *Zea mays* root cortex as measured with pH-sensitive microelectrodes: aspects of regulation. J Exp Bot 49:987–995
Felle HH, Waller F, Molitor A, Kogel KH (2009) The mycorrhiza fungus *Piriformospora indica* induces fast root-surface pH signaling and primes systemic alkalinization of the leaf apoplast upon powdery mildew infection. Mol Plant-Microbe Interact 22:1179–1185

Fernando DR, Marshall AT, Lynch JP (2016) Foliar nutrient distribution patterns in sympatric maple species reflect contrasting sensitivity to excess manganese. PLoS ONE 11(7):e0157702. http://doi.org/10.1371/journal.pone.0157702

Forster BP, Ellis RP, Thomas WTB, Newton AC, Tuberosa R, This D, El-Enein RA, Bahri MH, Ben Salem M (2000) The development and application of molecular markers for abiotic stress tolerance in barley. J Exp Bot 51:19–27

Foy CD (1984) Physiological effects of hydrogen, aluminum, and manganese toxicities in acid soil. In: Adams F (ed) Soil acidity and liming, 2nd edn. American Society of Agronomy, Crop Science Society of America, and Soil Science Society of America, Madison, pp 57–97

Foy CD, Chaney RL, White MC (1978) The physiology of metal toxicity in plants. Annu Rev Plant Physiol 29:511–566

Gabara B, Sklodowska M, Wyrwicka A, Glinska S, Capinska M (2003) Changes in the ultrastructure of chloroplasts and mitochondria and antioxidant enzyme activity in *Lycopersicum esculentum* Mill. leaves sprayed with acid rain. Plant Sci 164:507–516

Gao D, Knight MR, Trewavas AJ, Sattelmacher B, Plieth C (2004) Self-reporting *Arabidopsis* expressing pH and [Ca^{2+}] indicators unveil ion dynamics in the cytoplasm and in the apoplast under abiotic stress. Plant Physiol 134:898–908

Garcia-Oliveira A, Benito C, Prieto P, Menezes RA, Rodrigues-Pousada C, Guedes-Pinto H, Martins-Lopes P (2013) Molecular characterization of TaSTOPI homoeologues and their response to aluminium and proton (H^+) toxicity in bread wheat (*Triticum aestivum* L.). BMC Plant Biol 13:134. https://doi.org/10.1186/1471-2229-13-134

Garcia-Oliveira AL, Chander S, Barcelo J, Poschenrieder C. (2016) Aluminium Stress in Crop Plants. In: Yadav P, Kumar S, Jain V (eds) Recent Advances in Plant Stress Physiology, Daya Publishing House, New Delhi, pp 237–263

Gill SS, Tuteja N (2010) Reactive oxygen species and antioxidant machinery in abiotic stress tolerance in crop plants. Plant Physiol Biochem 48:909–930

Ginocchio R, de la Fuente LM, Sanchez P, Bustamante E, Silva Y, Urrestarazu P, Rodriguez PH (2009) Soil acidification as a confounding factor on metal phytotoxicity in soils spiked with copper-rich mine wastes. Environ Toxicol Chem 28:2069–2081

Gordana B, Grljusic S, Rozman V, Lukic D, Lackovic R, Novoselovic D (2007) Seed age and pH of water solution effects on field pea (*Pisum sativum* L.) germination. Not Bot Hort Agrobot Cluj 35:20–26

Goulding KWT (2016) Soil acidification and the importance of liming agricultural soils with particular reference to the United Kingdom. Soil Use Manag 32:390–399

Gunsé B, Poschenrieder C, Barceló J (1997) Water transport properties of roots and root cortical cells in proton-and Al-stressed maize varieties. Plant Physiol 113:595–602

Guo TR, Yao PC, Zhang ZD, Wang JJ, Wang M (2012) Devolvement of antioxidative defense system in rice growing seedlings exposed to aluminum toxicity and phosphorus deficiency. Rice Sci 19:179–185

Gupta N, Gaurav SJ, Kumar A (2013) Molecular basis of aluminium toxicity in plants: a review. Am J Plant Sci 4:21–37

Hai-yang H, Chun-qin C, Qiang-qiang D, Zhi-bing W (2013) Effect of pH value on seed germination and seedling growth of *Betula luminifera*. J Southwest Forest Univ 5:006

Haling RE, Richardson AE, Culvenor RA, Lambers H, Simpson RJ (2010) Root morphology, root-hair development and rhizosheath formation on perennial grass seedlings is influenced by soil acidity. Plant Soil 335:457–468

Hasanuzzaman M, Hossain MA, Fujita M (2012) Exogenous selenium pretreatment protects rapeseed seedlings from cadmium-induced oxidative stress by upregulating antioxidant defense and methylglyoxal detoxification systems. Biol Trace Elem Res 149:248–261

Hasanuzzaman M, Nahar K, Hossain MS, Mahmud JA, Rahman A, Inafuku M, Oku H, Fujita M (2017) Coordinated actions of glyoxalase and antioxidant defense systems in conferring abiotic stress tolerance in plants. Int J Mol Sci 18:200. https://doi.org/10.3390/ijms18010200

Hasanuzzaman M, Nahar K, Alam MM, Bhuyan MHMB, Oku H, Fujita M (2018) Exogenous nitric oxide pretreatment protects *Brassica napus* L. seedlings from paraquat toxicity through the modulation of antioxidant defense and glyoxalase systems. Plant Physiol Biochem 126:173–186

Haug A, Foy CE (1984) Molecular aspects of aluminum toxicity. Crit Rev Plant Sci 1:345–373

Havlin JL, Beaton SL, Tisdale SL, Nelson WL (2005) In: Havlin J (ed) Soil fertility and fertilizers: an introduction to nutrient management, vol 515. Pearson Prentice Hall, Upper Saddle River, pp 97–141

He G, Zhang J, Hu Z, Wu J (2011) Effect of aluminum toxicity and phosphorus deficiency on the growth and photosynthesis of oil tea (*Camellia oleifera* Abel.) seedlings in acidic red soils. Acta Physiol Plant 33:1285–1292

Helyar KR, Porter WM (1989) Soil acidification, its measurement and the processes involved. In: Robson AD (ed) Soil acidity and plant growth. Academic Press, New York, pp 61–102

Hernandez A, Francisco J, Corpas FJ, Gomez GM, del Rio LA, Sevilla F (1993) Salt induced oxidative stresses mediated by activated oxygen species in pea leaf mitochondria. Physiol Plant 89:103–110

Hinsinger P, Plassard C, Tang C, Jaillard B (2003) Origins of root-mediated pH changes in the rhizosphere and their responses to environmental constraints: a review. Plant Soil 248:43–59

Hiscox JD, Isrealstam GF (1979) A method for the extraction of chlorophyll from leaf tissue without maceration. Can J Bot 57:1332–1334

Hoekenga OA, Maron LG, Pineros MA, Cancado GMA, Shaff J, Kobayashi Y, Ryan PR, Dong B, Delhaize E, Sasaki T, Matsumoto H, Yamamoto Y, Koyama H, Kochian LV (2006) AtALMT1, which encodes a malate transporter, is identified as one of several genes critical for aluminum tolerance in *Arabidopsis*. Proc Natl Acad Sci U S A 103:9738–9743

Hong E, Ketterings Q, Mcbride M (2010) Manganese. Agronomy fact sheet series–fact sheet 49. Nutrient Management Spear Program, Field Crop Extension, College of Agriculture and Life Sciences, Cornell University Cooperative Extension, Ithaca. http://nmsp.cals.cornell.edu. Accessed 30 Sept 2018

Huang YL, Yang S, Long GX, Zhao ZK, Li XF, Gu MH (2016) Manganese toxicity in sugarcane plantlets grown on acidic soils of southern China. PLoS One 11(3):e0148956. https://doi.org/10.1371/journal.pone.0148956

Ikka T, Kobayashi Y, luchi S, Sakurai N, Shibata D, Kobayashi M, Koyama H (2007) Natural variation of *Arabidopsis thaliana* reveals that aluminum resistance and proton resistance are controlled by different genetic factors. Theor Appl Genet 115:709–719

Ingerpuu N (2002) Bryophyte diversity and vascular plants. Tartu University Press, Tartu

Iuchi S, Koyama H, luchi A, Kobayashi Y, Kitabayashi S, Kobayashi Y, Ikka T, Hirayama T, Shinozaki K, Kobayashi M (2007) Zinc finger protein *STOP1* is critical for proton tolerance in Arabidopsis and coregulates a key gene in aluminum tolerance. Proc Natl Acad Sci U S A 104:9900–9905

Ivanov Y, Savochkin Y, Kuznetsov V (2013) Effect of mineral composition and medium pH on scots pine tolerance to toxic effect of zinc ions. Russ J Plant Physiol 60:260–269

Iyer-Pascuzzi AS, Jackson T, Cui H, Petricka JJ, Busch W, Tsukagoshi H, Benfey PN (2011) Cell identity regulators link development and stress responses in the *Arabidopsis* root. Dev Cell 21:770–782

Jin J, Jiang H, Zhang X, Wang Y, Song X (2013) Detecting the responses of Masson pine to acid stress using hyperspectral and multispectral remote sensing. Int J Remote Sens 34:7340–7355

Joris HAW, Caires EF, Bini AR, Scharr DA, Haliski A (2013) Effects of soil acidity and water stress on corn and soybean performance under a no-till system. Plant Soil 365:409–424

Jovanovic Z, Djalovic I, Komljenovic I, Kovacevic V, Cvijovic M (2006) Influences of liming on vertisol properties and yields of the field crops. Cereal Res Commun 34:517–520

Jovanovic Z, Djalovic I, Tolimir M, Cvijovic M (2007) Influence of growing sistem and NPK fertilization on maize yield on pseudogley of Central Serbia. Cereal Res Commun 35:1325–1329

Kalir A, Poljakoff-Mayber A (1981) Changes in activity of malate dehydrogenase, catalase, peroxidase and superoxide dimutase in the leaves of *Halimione portulacoides* (L.). Allen exposed to high sodium chloride concentrations. Ann Bot 47:75–85
Kamaluddin M, Zwiazek JJ (2004) Effects of root medium pH on water transport in paper birch (*Betula papyrifera*) seedlings in relation to root temperature and abscisic acid treatments. Tree Physiol 24:1173–1180. https://doi.org/10.1093/treephys/24.10.1173
Kang D, Seo Y, Futakuchi K, Vijarnsorn P, Ishii R (2011) Effect of aluminum toxicity on flowering time and grain yield on rice genotypes differing in al-tolerance. J Crop Sci Biotechnol 14:305–309
Kapczyńska A, Magdziarz K (2015) Influence of substrate pH on the growth and flowering of *Mandevilla* Lindl. Sundaville pretty red. Folia Hort 27:79–83
Kariuki SK, Zhang H, Schroder JL, Edwards J, Payton M, Carver BF, Raun WR, Krenzer EG (2007) Hard red winter wheat cultivar responses to a pH and aluminum concentration gradient. Agron J 99:88–98
Kasai M, Sasaki M, Yamamoto Y, Matsumoto H (1992) Aluminum stress increases K^+ efflux and activities of ATP- and PPj-dependent H^+ pumps of tonoplast-enriched membrane vesicles from barley roots. Plant Cell Physiol 33:1035–1039
Khabaz-Saberi H, Barker SJ, Rengel Z (2012) Tolerance to ion toxicities enhances wheat (*Triticum aestivum* L.) grain yield in waterlogged acidic soils. Plant Soil 354:371–381
Kidd PS, Proctor J (2001) Why plants grow poorly on very acid soils: are ecologists missing the obvious? J Exp Bot 52:791–799
Kinraide TB (1993) Aluminum enhancement of plant-growth in acid rooting media: a case of reciprocal alleviation of toxicity by 2 toxic cations. Physiol Plant 88:619–625
Kinraide TB, Parker DR (1987) Non-phytotoxicity of the aluminum sulfate ion, $AlSO_4^+$. Physiol Plant 71:207–212
Kinraide TB, Ryan PR, Kochian LV (1994) Al^{3+}-Ca^{2+} interactions in aluminium rhizotoxicity. II. Evaluating the Ca^{2+}-displacement hypothesis. Planta 192:104–109
Kisinyo PO, Othieno CO, Gudu SO, Okalebo JR, Opala PA, Maghanga JK, Ng'etich WK, Agalo JJ, Opile RW, Kisinyo JA, Ogola BO (2013) Phosphorus sorption and lime requirements of maize growing acids soil of Kenya. Sustain Agric Res 2:116–123
Kobayashi Y, Ohyama Y, Kobayashi Y, Ito H, Iuchi S, Fujita M, Zhao CR, Tanveer T, Ganesan M, Kobayashi M, Koyama H (2014) *STOP2* activates transcription of several genes for Al-and low pH-tolerance that are regulated by *STOP1* in *Arabidopsis*. Mol Plant 7:311–322
Kochian L, Hoekenga OA, Piñeros MA (2004) How do crop plants tolerate acid soils? Mechanisms of aluminum tolerance and phosphorous efficiency. Annu Rev Plant Biol 55:459–493
Kochian LV, Pineros MA, Liu J, Magalhaes JV (2015) Plant adaptation to acid soils: the molecular basis for crop aluminium resistance. Annu Rev Plant Physiol 66:571–598
Kolodziejek J, Patykowski J (2015) Effect of environmental factors on germination and emergence of invasive *Rumex confertus* in Central Europe. Sci World J 2015:170176. https://doi.org/10.1155/2015/170176
Konishi S, Miyamoto S (1983) Alleviation of aluminum stress and stimulation of tea pollen tube growth by fluorine. Plant Cell Physiol 24:857–862
Kooistra E (1967) Femaleness in breeding glasshouse cucumbers. Euphytica 16:1–17
Koyama H, Toda T, Yokota S, Zuraida D, Hara T (1995) Effects of aluminium and pH on root growth and cell viability in *Arabidopsis thaliana* strain Landsberg in hydroponic culture. Plant Cell Physiol 36:201–205
Koyama H, Toda T, Hara T (2001) Brief exposure to low-pH stress causes irreversible damage to the growing root in *Arabidopsis thaliana*: pectin–Ca interaction may play an important role in proton rhizotoxicity. J Exp Bot 52:361–368
Krstic D, Djalovic I, Nikezic D, Bjelic D (2012) Aluminium in acid soils: chemistry, toxicity and impact on maize plants. In: Aladjadjiyan A (ed) Food production—approaches, challenges and tasks. InTech, London. https://doi.org/10.5772/33077
Krug EC, Frink CR (1983) Acid rain on acid soil: a new perspective. Science 221:520–525

Kumar S, Meena RS, Yadav GS, Pandey A (2017) Response of sesame (*Sesamum indicum* L.) to sulphur and lime application under soil acidity. Int J Plant Soil Sci 14:1–9
Lager I, Andreasson O, Dunbar TL, Andreasson E, Escobar MA, Rasmusson AG (2010) Changes in external pH rapidly alter plant gene expression and modulate auxin and elicitor responses. Plant Cell Environ 33:1513–1528
Laghmouchi Y, Belmehdi O, Bouyahya A, Senhaji NS, Abrini J (2017) Effect of temperature, salt stress and pH on seed germination of medicinal plant *Origanum compactum*. Biocatal Agric Biotechnol 10:156–160
Lavres J, Malavolta E, Nogueira NL, Moraes MF, Rodrigues A, Lanzoni M, Pereira C (2009) Changes in anatomy and root cell ultrastructure of soybean genotypes under manganese stress. R Bras Ci Solo 33:395–403
Lazof DB, Holland MJ (1999) Evaluation of the aluminium-induced root growth inhibition in isolation from low pH effects in *Glycine max*, *Pisum sativum*, and *Phaseolus vulgaris*. Aust J Plant Physiol 26:147–157
Lee SS, Kim JH, Hong SB, Yun SH (1998) Effect of humidification and hardening treatment on seed germination of rice. Korean J Crop Sci 43:157–160
Lefebvre V, Kiani SP, Durand-Tardif M (2009) A focus on natural variation for abiotic constraints response in the model species *Arabidopsis thaliana*. Int J Mol Sci 10:3547–3582
Legesse H, Nigussie-Dechassa R, Gebeyehu S, Bultosa G, Mekbib F (2013) Response to soil acidity of common bean genotypes (*Phaseolus vulgaris* L.) under field conditions at Nedjo, Western Ethiopia. Sci Technol Arts Res J 2:3–15
Lidon FC, Teixeira MG (2000) Ricetolerance to excess Mn: implications in the chloroplast lamellae and synthesis of a novel Mnprotein. Plant Physiol Biochem 38:969–978
Lidon FC, Barreiro M, Ramalho J (2004) Manganese accumulation in rice: implications for photosynthetic functioning. J Plant Physiol 161:1235–1244
Long A, Zhang J, Yang L-T, Ye X, Lai N-W, Tan L-L, Lin D, Chen L-S (2017) Effects of low pH on photosynthesis, related physiological parameters, and nutrient profiles of citrus. Front Plant Sci 8:185. https://doi.org/10.3389/fpls.2017.00185
Longnecker DE (1974) The influence of high sodium upon fruiting and shedding boll characteristics, fiber properties and yields of two cotton species. Soil Sci 118:387–396
Magidow LC, Tommaso AD, Ketterings QM, Mohler CL, Milbrath LR (2013) Emergence and performance of two invasive swallowworts (*Vincetoxicum* spp.) in contrasting soil types and soil pH. Invasive Plant Sci Manag 6:281–291
Marciano DPRO, Ramos FT, Alvim MN, Magalhaes JR, França MGC (2010) Nitric oxide reduces the stress effects of aluminum on the process of germination and early root growth of rice. J Plant Nutr Soil Sci 173:885–891
Marschner H (1991) Mechanisms of adaptation of plants to acid soils. Plant Soil 134:1–20
Martins N, Gonçalves S, Palma T, Romano A (2011) The influence of low pH on in vitro growth and biochemical parameters of *Plantago almogravensis* and *P. algarbiensis*. Plant Cell Tissue Organ Cult 107:113–121
Martins N, Gonçalves S, Romano A (2013a) Metabolism and aluminum accumulation in *Plantago almogravensis* and *P. algarbiensis* in response to low pH and aluminum stress. Biol Plant 57:325–331
Martins N, Osório ML, Gonçalves S, Osório J, Palma T, Romano (2013b) A Physiological responses of *Plantago algarbiensis* and *P. almogravensis* shoots and plantlets to low pH and aluminum stress. Acta Physiol Plant 35:615–625
Meda AR, Furlani PR (2005) Tolerance to aluminum toxicity by tropical leguminous plants used as cover crops. Braz Arch Biol Technol 48:309–317
Menconi MCLM, Sgherri CLM, Pinzino C, Navari-Lzzo F (1995) Activated oxygen production and detoxification in wheat plants subjected to a water deficit programme. J Exp Bot 46:1123–1130
Michaels H (1910) Action of aqueous solutions of electrolytes on germination. Chem Abstr 4:1984–1985

MiransarI M, Smith DL (2007) Overcoming the stressful effects of salinity and acidity on soybean nodulation and yields using signal molecule genistein under field conditions. J Plant Nutr 30:1967–1992

Misra A, Tyler G (1999) Influence of soil moisture on soil solution chemistry and concentrations of minerals in the calcicoles *Phleum phleoides* and *Veronica spicata* grown on a limestone soil. Ann Bot 84:401–410

Mohanty S, Das AB, Das P, Mohanty P (2004) Effect of a low dose of aluminum on mitotic and meiotic activity, 4C DNA content, and pollen sterility in rice, *Oryza sativa* L. cv. Lalat. Ecotoxicol Environ Saf 59:70–75

Mora M, Rosas A, Ribera A, Rengel R (2009) Differential tolerance to Mn toxicity in perennial ryegrass genotypes: involvement of antioxidative enzymes and root exudation of carboxylates. Plant Soil 320:79–89

Moroni J, Scott B, Wratten N (2003) Differential tolerance of high manganese among rapeseed genotypes. Plant Soil 253:507–519

Munzuroglu O, Obek E, Geckil H (2003) Effects of simulated acid rain on the pollen germination and pollen tube growth of apple (*Malus sylvestris* Miller cv. Golden). Acta Biol Hung 54:95–103

Murach D, Ulrich B (1988) Destabilization of forest ecosystems by acid deposition. GeoJournal 17:253–259

Nahar K, Hasanuzzaman M, Suzuki T, Fujita M (2017) Polyamines-induced aluminum tolerance in mung bean: a study on antioxidant defense and methylglyoxal detoxification systems. Ecotoxicology 26:58–73

Najeeb U, Xu L, Shafaqat A, Jilani G, Gong HJ, Shen WQ, Zhou WJ (2009) Citric acid enhances the phytoextraction of manganese and plant growth by alleviating the ultrastructural damages in *Juncus effusus* L. J Hazard Mater 170:1156–1163

Nazmi G, Esma AYY, Aykut T (2016) Acidity effect in pollen germination and tube length of *Prunus amygdalus* Batsch and *Prunus domestica* L. J Appl Biol Sci 10:41–45

Neuvonen S, Nyyssönen T, Ranta H, Kiilunen S (1991) Simulated acid rain and the reproduction of mountain birch [*Betula pubescens* ssp. *tortuosa* (Ledeb.) Nyman]: a cautionary tale. New Phytol 118:111–117

Nian H, Yang C, Huang H, Hideaki M (2009) Effects on low pH and aluminum stresses on common beans (*Phaseolus vulgaris*) differing in low phosphorus and photoperiod responses. Front Biol 4:446–452

Nuruddin AA, Chang M (1999) Responses of herbaceous mimosa (*Mimosa strigillosa*), a new reclamation species to soil pH. Resour Conserv Recycl 27:287–298

Pal'ove-Balang P, Čiamporová M, Zelinová V, Pavlovkin J, Gurinová E, Mistrík I (2012) Cellular responses of two Latin-American cultivars of *Lotus corniculatus* to low pH and Al stress. Cent Eur J Biol 7:1046–1054

Paoletti E (1991) Effects of acidity and detergent on in vitro pollen germination and tube growth in forest tree species. Tree Physiol 10:357–366

Pärtel M (2002) Local plant diversity patterns and evolutionary history at the regional scale. Ecology 83:2361–2366

Pärtel M, Zobel M, Zobel K, van der Maarel E (1996) The species pool and its relation to species richness: evidence from Estonian plant communities. Oikos 75:111–117

Pavlovkin J, Pal'ove-Balang P, Kolarovič L, Zelinová V (2009) Growth and functional responses of different cultivars of *Lotus corniculatus* to aluminum and low pH stress. J Plant Physiol 166:1479–1487

Pereira EG, Oliva MA, Rosado-Souza L, Mendes GC, Colares DS, Stopato CH, Almeida AM (2013) Iron excess affects rice photosynthesis through stomatal and non-stomatal limitations. Plant Sci 201 & 202:81–92

Plate F (1913) Inhibition in the seed of *Avena sativa*. Chem Abstr 8:360

Popescu A (1998) Contributions and limitations to symbiotic nitrogen fixationin common bean (*Phaseolus vulgaris* L.) in Romania. Plant Soil 204:117–125

Poschenrieder C, Llugany M, Barcelo J (1995) Short-term effects of pH and aluminumon mineral-nutrition in maize varieties differing in proton and aluminum tolerance. J Plant Nutr 18:1495–1507
Promsy G (1911) Influence of acids on germination. Acad Sci 152:450–452
Qiao F, Zhang XM, Liu X, Chen J, Hu WJ, Liu TW, Liu JY, Zhu CQ, Ghoto K, Zhu XY, Zheng HL (2018) Elevated nitrogen metabolism and nitric oxide production are involved in *Arabidopsis* resistance to acid rain. Plant Physiol Biochem 127:238–247
Ramlall C, Varghese B, Ramdhani S, Pammenter NW, Bhatt A, Berjak P, Sershen (2015) Effects of simulated acid rain on germination, seedling growth and oxidative metabolism of recalcitrant-seeded *Trichilia dregeana* grown in its natural seed bank. Physiol Plant 153:149–160
Rangel AF, Mubin M, Rao IM, Horst WJ (2005) Proton toxicity interferes with the screening of common bean *Phaseolus vulgaris* genotypes for aluminium resistance in nutrient solution. J Plant Nutr Soil Sci 168:607–616
Rao IM, Miles JW, Beede SE, Horst WJ (2016) Root adaptations to soils with low fertility and aluminium toxicity. Ann Bot 118:593–605
Ring SM, Fisher RP, Poile GJ, Helyar KR, Conyers MK, Morris SG (1993) Screening species and cultivars for their tolerance to acidic soil conditions. Plant Soil 155:521–524
Rosas A, Rengel Z, Mora M (2007) Manganese supply and pH influence growth, carboxylate exudation and peroxidase activity of ryegrass and white clover. J Plant Nutr 30:253–270
Rouphael Y, Cardarelli M, Colla G (2015) Role of arbuscular mycorrhizal fungi in alleviating the adverse effects of acidity and aluminium toxicity in zucchini squash. Sci Hortic 188:97–105
Saenen E, Horemans N, Vanhoudt N, Vandenhove H, Biermans G, van Hees M, Wannijn J, Vangronsveld J, Cuypers A (2013) Effects of pH on uranium uptake and oxidative stress responses induced in *Arabidopsis thaliana*. Environ Toxicol Chem 32:2125–2133
Saenen E, Horemans N, Vanhoudt N, Vandenhove H, Biermans G, Hees MV, Wannijn J, Vangronsveld J, Cuypers A (2014) The pH strongly influences the uranium-induced effects on the photosynthetic apparatus of *Arabidopsis thaliana* plants. Plant Physiol Biochem 82:254–261
Sairam RK (1994) Effect of moisture stress on physiological activities of two contrasting wheat genotypes. Indian J Exp Biol 32:584–593
Sairam RK, Deshmukh PS, Saxena DC (1998) Role of antioxidant systems in wheat genotypes tolerance to water stress. Biol Plant 41:387–394
Sakano K (1998) Revision of biochemical pH-stat: involvement of alternative pathway metabolisms. Plant Cell Physiol 39:467–473
Samac DA, Tesfaye M (2003) Plant improvement for tolerance to aluminium in acid soils. Plant Cell Tissue Organ Cult 75:189–207
Sawaki Y, Iuchi S, Kobayashi Y, Kobayashi Y, Ikka T, Sakurai N, Fujita M, Shinozaki K, Shibata D, Kobayashi M, Koyama H (2009) STOP1 regulates multiple genes that protect *Arabidopsis* from proton and aluminum toxicities. Plant Physiol 150:281–294
Searcy KB, Mulcahy DL (1990) Comparison of the response to aluminum toxicity in gametophyte and sporophyte of four tomato (*Lycopersicon esculentum* Mill). Theor Appl Genet 80:289–295
Shamsi IH, Wei K, Zhang GP, Jilani GH, Hassan MJ (2008) Interactive effects of cadmium and aluminum on growth and antioxidative enzymes in soybean. Biol Plant 52:165–169
Shavrukov Y, Hirai Y (2016) Good and bad protons: genetic aspects of acidity stress responses in plants. J Exp Bot 67:15–30
Shi QH, Zhu ZJ, Juan LI, Qian QQ (2006) Combined effects of excess Mn and low pH on oxidative stress and antioxidant enzymes in cucumber roots. Agric Sci China 5:767–772
Sidhu SS (1983) Effects of simulated acid rain on pollen germination and pollen tube growth of white spruce (*Picea glauca*). Can J Bot 61:3095–3099
Siecińska J, Nosalewicz A (2016) Aluminium toxicity to plants as influenced by the properties of the root growth environment affected by other co-stressors. Rev Environ Contam Toxicol 243:1–26

Sierra J, Noël C, Dufour L, Ozier-Lafontaine H, Welcker C, Desfontaines L (2003) Mineral nutrition and growth of tropical maize as affected by soil acidity. Plant Soil 252:215–226
Sierra J, Ozier-Lafontaine H, Dufour L, Meunier A, Bonhomme R, Welcker C (2006) Nutrient and assimilate partitioning in two tropical maize cultivars in relation to their tolerance to soil acidity. Field Crops Res 95:234–249
Sikirou M, Saito K, Dramé KN, Saidou A, Dieng I, Ahanchédé A, Venuprasad R (2016) Soil-based screening for iron toxicity tolerance in rice using pots. Plant Prod Sci 19:489–496
Singh VP, Mall SL, Billore SK (1975) Effect of pH on germination of four common grass species of Ujjain (India). J Range Manag 6:497–498
Singh NB, Yadav K, Amist N (2011a) Phytotoxic effects of aluminum on growth and metabolism of *Pisum sativum* L. Int J Innov Biol Chem Sci 2:10–21
Singh VP, Tripathi DK, Kumar D, Chauhan DK (2011b) Influence of exogenous silicon addition on aluminium tolerance in rice seedlings. Biol Trace Elem Res 144:1260–1274
Slootmaker LAJ (1974) Tolerance to high soil acidity in wheat related species, rye and triticale. Euphytica 23:505–513
Soil Survey Division Staff (2017) Examination and description of soil profiles. In: Soil Survey Division Staff (eds) Soil survey manual. US Department of Agriculture Handbook 18Washington, DC pp 83–230
Song H, Xu X, Wang H, Tao Y (2011) Protein carbonylation in barley seedling roots caused by aluminum and proton toxicity is suppressed by salicylic acid. Russ J Plant Physiol 58:653–659
Srivastava S, Dubey RS (2011) Manganese-excess induces oxidative stress, lowers the pool of antioxidants and elevates activities of key antioxidative enzymes in rice seedlings. Plant Growth Regul 64:1–16
Steiner F, Zoz T, Junior ASP, Castagnara DD, Dranski JAL (2012) Effects of aluminium on plant growth and nutrient uptake in young physic nut plants. Semin Ciênc Agrar 33:1779–1788
Sullivan TJ, Lawrence GB, Bailey SW, McDonnell TC, Beier CM, Weathers KC, McPherson GT, Bishop DA (2013) Effects of acidic deposition and soil acidification on sugar maple trees in the adirondack mountains, New York. Environ Sci Technol 47:12687–12694
Suthar AC, Naik VR, Mulani RM (2009) Seed and seed germination in *Solanum nigrum* Linn. Am Eurasian J Agric Environ Sci 5:179–183
Tang C, Diatloff E, Rengel Z, McGann B (2001) Growth response to subsurface soil acidity of wheat genotypes differing in aluminium tolerance. Plant Soil 236:1–10
Tang C, Rengel Z, Abrecht D, Tennant D (2002) Aluminium-tolerant wheat uses more water and yields higher than aluminium-sensitive one on a sandy soil with subsurface acidity. Field Crop Res 78:93–103
Tang C, Asseng S, Diatloff E, Rengel Z (2003) Modelling yield losses of aluminium-resistant and aluminium-sensitive wheat due to subsurface soil acidity: effects of rainfall, liming and nitrogen application. Plant Soil 254:349–360
Taranishi Y, Tanaka A, Osumi N, Fukui S (1974) Catalase activity of hydrocarbon utilizing Candida yeast. Agric Biol Chem 38:1213–1216
The C, Calba H, Horst WJ, Zonkeng C (2001) Maize grain yield correlated responses to change in acid soil characteristics after 3 years of soil amendments. In: Seventh Eastern and Southern Africa regional maize conference, 11th to 15th February 2001, pp 222–227
The C, Calba H, Zonkeng C, Ngonkeu ELM, Adetimirin VO, Mafouasson HA, Meka SS, Horst WJ (2006) Responses of maize grain yield to changes in acid soil characteristics after soil amendments. Plant Soil 284:45–57
The C, Meka SS, Ngonkeu ELM, Bell JM, Mafouasson HA, Menkir A, Calba H, Zonkeng C, Atemkeng M, Horst WJ (2012) Maize grain yield responses to changes in acid soil characteristics with yearly leguminous crop rotation, fallow, slash, burn and liming practices. Int J Plant Soil Sci 1:1–15
Troiano J, Colavito L, Heller L, HcCune DC, Jacobson JS (1983) Effects of acidity of simulated rain and its joint action with ambient ozone on measures of biomass and yield in soybean. Environ Exp Bot 23:113–119

Turner GD, Lau RR, Young DR (1998) Effect of acidity on germination and seedling growth of *Paulownia tomentosa*. J Appl Ecol 25:561–567

Uchida R, Hue NV (2000) Soil acidity and liming. In: Silva JA, Uchida R (eds) Plant nutrient management in Hawaii's soils, approaches for tropical and subtropical agriculture. College of Tropical Agriculture and Human Resources, University of Hawaii, Manoa, pp 101–111

Ulrich B, Mayer R, Khanna PK (1980) Chemical changes due to acid precipitation in a loess-derived soil in Central Europe. Soil Sci 130:193–199

Van Schaik CP, Mirmanto E (1985) Spatial variation in the structure and litter fall of a Sumatran rain forest. Biotropica 17:196–205

Van Wambeke A (1976) Formation, distribution and consequences of acid soils in agricultural development. In: Wright MJ, Ferrari SA (eds) Proceedings of workshop on plant adaptation to mineral stress in problem soils. Special Publications Cornell University, Agricultural Experiment Station, Ithaca, pp 15–24

Vitorello VA, Capaldi FRC, Stefanuto VA (2005) Recent advances in aluminium toxicity and resistance in higher plants. Braz J Plant Physiol 17:129–143

Vleeshouwers LM, Bowmeester HJ, Karssen CM (1995) Redefining seed dormancy: an attempt to integrate physiology and ecology. J Ecol 83:1031–1037

Von Uexküll HR, Mutert E (1995) Global extent, development and economic impact of acid soils. Plant Soil 171:1–15

Wen XJ, Duan CQ, Zhang DC (2013) Effect of simulated acid rain on soil acidification and rare earth elements leaching loss in soils of rare earth mining area in southern Jiangxi Province of China. Environ Earth Sci 69:843–853

Wenzl P, Mancilla LI, Mayer JE, Albert R, Rao IM (2003) Simulating infertile acid soils with nutrient solutions. Soil Sci Soc Am J 67:1457–1469

Wikipedia (2018) Soil pH, from Wikipedia, the free encyclopedia. https://en.wikipedia.org/wiki/Soil_pH. Accessed 20 Sept 2018

Wilkinson RE, Duncan RR (1989) Sorghum seedling growth as influenced by H^+, Ca^{2+}, and Mn^{2+} concentrations. J Plant Nutr 12:1379–1394

Wood S, Sebastian K, Scherr SJ (2000) Pilot analysis of global ecosystems: agroecosystems. WRI and IFPRI, Washington

Yan F, Feuerle R, Schaffer S, Fortmeier H, Schubert S (1998) Adaptation of active proton pumping and plasmalemma ATPase activity of corn roots to low root medium pH. Plant Physiol 117:311–319

Yang JL, Li YY, Zhang YJ, Zhang SS, Wu YR, Wu P, Zheng SJ (2008) Cell Wall polysaccharides are specifically involved in the exclusion of aluminium from the rice root apex. Plant Physiol 146:602–611

Yang M, Huang SX, Fang SZ, Huang XL (2011) Response of seedling growth of four eucalyptus clones to acid and aluminum stress. Plant Nutr Fert Sci 17:195–201

Yang ZB, Rao IM, Horst WJ (2013) Interaction of aluminium and drought stress on root growth and crop yield on acid soils. Plant Soil 372:3–25

Yang M, Tan L, Xu Y, Zhao Y, Cheng F, Ye S, Jiang W (2015) Effect of low pH and aluminum toxicity on the photosynthetic characteristics of different fast growing eucalyptus vegetatively propagated clones. PLoS One 10:e0130963. https://doi.org/10.1371/journal.pone.0130963

Yu H, He N, Wang Q, Zhu J, Gao Y, Zhang Y, Jia Y, Yu G (2017) Development of atmospheric acid deposition in China from the 1990s to the 2010s. Environ Pollut 231:182–190

Zabawi AGM, Esa SM, Leong CP (2008) Effects of simulated acid rain on germination and growth of rice plant. J Trop Agric Food Sci 36:1–6

Zeigler RS, Pandey S, Miles J, Gourley LM, Sarkarung S (1995) Advances in the selection and breeding of acid-tolerant plants: rice, maize, sorghum and tropical forages. In: Date RA, Grundon NJ, Rayment GE, Probert ME (eds) Plant-soil interactions at low pH: principles and management. Developments in plant and soil sciences, vol 64. Springer, Dordrecht, pp 391–406

Zhang X, Liu P, Yang YS, Xu G (2007) Effect of Al in soil on photosynthesis and related morphological and physiological characteristics of two soybean genotypes. Bot Stud 48:435–444

Zhang H, Tan Z, Hu L, Wang S, Luo J, Jones RL (2010) Hydrogen sulfide alleviates aluminum toxicity in germinating wheat seedlings. J Integr Plant Biol 52:556–567

Zhang CP, Meng P, Li JZ, Wan XC (2014) Interactive effects of soil acidification and phosphorus deficiency on photosynthetic characteristics and growth in *Juglans regia* seedlings. China J Plant Ecol 38:1345–1355

Zhang Y-K, Zhu D-F, Zhang Y-P, Chen H-Z, Xiang J, Lin X-Q (2015) Low pH-induced changes of antioxidant enzyme and atpase activities in the roots of rice (*oryza sativa* l.) seedlings. PLoS One 10:e0116971. https://doi.org/10.1371/journal.pone.0116971

Zhao MR, Li F, Fang Y, Gao Q, Wang W (2011) Expansin-regulated cell elongation is involved in the drought tolerance in wheat. Protoplasma 248:313–323

Zhao XQ, Guo SW, Shinmachi F, Sunairi M, Noguchi A, Hasegawa I, Shen RF (2013) Aluminum tolerance in rice is antagonistic with nitrate preference and synergistic with ammonium preference. Ann Bot 111:69–77

Zhu Y, Di T, Xu G, Chen X, Zeng H, Yan F, Shen Q (2009) Adaptation of plasma membrane H^+-ATPase of rice roots to low pH as related to ammonium nutrition. Plant Cell Environ 32:1428–1440

Salinity: A Major Agricultural Problem—Causes, Impacts on Crop Productivity and Management Strategies

Abdul Majeed and Zahir Muhammad

Introduction

Global food security is greatly reliant on agricultural crops and their products which need substantial increases for maintenance of the gap between production and consumption. The significance of enhancing crop productivity is even more emergent in recent times because of the growing population of humankind which currently stands at 7.6 billion and is projected to exceed 9.7 billion by the year 2050 (FAOSTAT 2018). Undoubtedly, the increase in population will exert pressure for production of more crops and food resources which seems a challenging task for plant biologists. Concurrently, climate change and several biotic and abiotic stresses challenge the growth and production of agricultural crops. Among abiotic stresses, salinity is considered as one of the leading limiting factors responsible for growth and production decline of agricultural crops throughout the world principally in arid and semiarid regions (Kaashyap et al. 2018). About 20–33% of the cultivated and irrigated land throughout the world is affected by salinity, and the adverse effects are expected to reach to 50% in the year 2050 (Soda et al. 2016; Machado and Serralheiro 2017).

The problem of salinity emerges when ion concentration of different salts (predominantly NaCl) elevates in soils beyond threshold level required for normal germination, growth, and physiological activities of plants in root zones. Standard agricultural soils which promote the growth of cultivated crops have salinity level ~4 dS m^{-1} which is equivalent to 40 mM NaCl generally determined as the electrical conductivity of the saturated extract (Shrivastava and Kumar 2015). Ionic concentration exceeding 4 dS m^{-1} in rhizosphere result in stress conditions which challenge

A. Majeed (✉)
Department of Botany, Government Degree College Naguman,
Peshawar, Khyber Pakhtunkhwa, Pakistan

Z. Muhammad
Department of Botany, University of Peshawar, Peshawar, Khyber Pakhtunkhwa, Pakistan

M. Hasanuzzaman et al. (eds.), *Plant Abiotic Stress Tolerance*,
https://doi.org/10.1007/978-3-030-06118-0_3

the growth of crops in several ways. Leading causes of salinity may be the natural addition of salts from rocks and minerals through weathering process (primary salinity) or extensive human activities like irrigation, agricultural intensification, and deforestation (secondary salinity) (Athar and Ashraf 2009). In addition, high evaporation rate in tropics also seems to increase the level of salts in agricultural land. Salinity has drastic consequences on water availability in the soil for plants, the rate of transpiration and photosynthesis, stomatal opening and closures, and functional activities of plant roots (Khataar et al. 2018). The discrepancies triggered by salinity stress result in impaired growth, physiological functions, and low yield outputs of crops. Annual expenditures associated with crop losses as a result of salinity are documented as 27 Billion US dollars (Singh et al. 2016). Coupled with drought, salinity triggered crop losses range between 20 and 50% in documented studies (Shrivastava and Kumar 2015) while losses may even be much higher as a result of salinity stress in areas out of scientific investigations. From physiological aspects, almost all crops show sensitivity to the salinity stress albeit variation in responses to the imposed stress in different crops does exist based on their tolerance level. Major concerns about the adverse effects of growing salinity stress are associated with crops which have a substantial contribution to global food supply. Rice, maize, wheat, barley, sorghum, and potato are considered as key drivers in the fulfillment of food needs throughout the world, while their growth and productivity are considerably affected by salt stress in many parts.

To maintain a sustainable production of agricultural crops and to ensure future food sanctuary, efforts have been made over decades to address the adversities caused by salinity stress on crops. One of the basic approaches to prevent crop losses triggered by salinization is to induce salinity tolerance in salt-sensitive crops. Genomic and molecular approaches to understand the possibilities of salt resistance induction in crops have proven effective in further elaboration of making crops adaptable to saline conditions (Luo et al. 2017). In addition to molecular methods, breeding for salt tolerance, proper agronomic practices in irrigation, and search for cost-effective and feasible methods to reduce drastic effects of salinity on principal crops are crucially needed to be employed. This chapter presents an overview of the salinity imposed effects on crops, and employment of agronomic approaches, seed priming and plant growth promoting bacteria (PGPB) to induce tolerance in crops to salinity.

Causes of Salinity and General Effects

Root sphere is the zone of soil which provides water and minerals to plants in addition to harboring diverse beneficial microbial communities and plays a determining role in successful germination and the consequent establishment of growth and reproductive phases of plants. Imbalance in water contents and minerals in the rhizospheric soil can lead to deteriorated effects on plant growth and productive outputs. Soil salinity alters water and mineral contents and microbiota and results in poor germinability and growth of sensitive crops. Salinity is a physiological state of soil where the concentration of active ions of salts principally NaCl increases the threshold levels. It is established that salt concentration up to 4 dS m^{-1} (~40 mM) of NaCl in soils represent a suitable environment for most of the plant species; however, some plants known as halophytes can withstand salinity stress up to 100 mM (examples include *Vesicaria*, *Atriplex*, *Chenopodium* spp.) while most of the cultivated crops including cereals (known as glycophytes) show differential sensitivity to salinity stress and prefer to grow at $\leq$40 mM salt concentration (Rasool et al. 2013; Ismail and Horie 2017; Yang et al. 2018). Causes of salinity stress in the natural ecosystem and managed agriculture are both natural and anthropogenic. Naturally, the problem of salinity occurs when minerals and salty rocks are weathered and ion concentration accumulates in soil (Athar and Ashraf 2009). Transportation of salt contents to root growing zones from parent rocks (rich in salts) as a result of several types of weathering and from groundwater may result in an increased level of salt stress and could contribute to primary salinity (Daliakopoulos et al. 2016). Agricultural lands located near coastal regions are exposed to high salinity stress because of the presence of high concentrations of salts in seawater. In tropical regions, high temperature can stimulate evaporation rate and hence more salt ions in soils accumulate giving rise to salinity stress. Moreover, flooding and wind erosion may be regarded to some extent as natural causes of salinity due to the imbalanced flow of minerals and salts from one region to another. Shallow groundwater may also serve as a source of salinization when upward migration of salts occurs (Shrivastava and Kumar 2015). Land cover and climatic conditions are also thought to contribute to salinity (Fan et al. 2012). Anthropogenic salinity, representing a significant proportion of the overall global salinity, is caused by massive agricultural activities, poor irrigation, use of imbalanced fertilizers, soil degradation, and poor drainages (Shrivastava and Kumar 2015; Sandhu et al. 2017). Mining activities and the use of wastewater and industrial effluents enriched in diverse salt contents in addition to other hazardous materials may lead to agricultural salinity (Daliakopoulos et al. 2016). Deforestation is an imminent threat to changes in rainfall patterns and soil erosion which would lead to consequent salinity problems. Rhizospheric soils whether salinized by natural processes or human activities are generally classified on the basis of electrical conductivity (EC) as: (1) non-salinized (EC = 0–2 dS m^{-1}), (2) marginally salinized (EC = 2–4 dS m^{-1}), (3) moderately salinized (EC = 4–8 dS m^{-1}), (4) strongly salinized (EC = 8–16 dS m^{-1}), (5) rigorously salinized (EC =16–32 dS m^{-1}), and (6) exceptionally salinized (EC >32 dS m^{-1}) (Rasool et al. 2013). According to Munns and Gilliham (2015), the economic costs associated with salinity are variable in different regions which may range between 300 and 600 US dollars ton^{-1}. Daliakopoulos et al. (2016) stated that soil salinity may impart an economic burden of up to 600 million euros in European countries due to reduced crop yields and agricultural degradation. According to FAO and ITPS, African continent is the most salt-affected region where salinized area exceeds 122 million hectares followed by north and central Asia which represent almost 91.5 Mha of salt-affected land (Table 1).

Table 1 Salt- and sodic-affected soils in different continents of the world

Region	Salt-affected area (Mha)	Sodic-affected soils (Mha)	Total soil affected (salinity + sodicity) (Mha)
Africa	122.9	86.7	209.6
South Asia	82.3	1.8	84.1
North and Central Asia	91.5	120.2	211.7
Southeast Asia	20.0	–	20.0
South America	69.5	59.8	129.3
North America	6.2	9.6	15.8
Mexico/Central America	2.0	–	2.0
Australasia	17.6	340.0	357.0

Source: FAO and ITPS. 2015. Status of the World's Soil Resources (SWSR)—Main Report. Food and Agriculture Organization of the United Nations and Intergovernmental Technical Panel on Soils, Rome, Italy

Effects of Salinity on Crops

In general, plants are broadly categorized as glycophytes (sensitive to salinity) and halophytes (exhibit some degree of tolerance to salinity stress) on the basis of their tolerance in response to salinity stress. Halophytes are plants with typical examples *Vesicaria*, *Atriplex*, *Chenopodium* spp., and several others that can withstand high salt concentration while our major cultivated crops are glycophytes which show differential sensitivity to salinized soils (Borsani et al. 2003; Rasool et al. 2013; Yang et al. 2018). Differential responses of cultivated glycophytes depend on their growth phases, the soil characters, level of salinity, and agricultural practices (Daliakopoulos et al. 2016). Salinity stress primarily affects crops by causing water deficit condition in soils termed as "osmotic stress" while secondarily due to the accumulation of toxic ions which consequently lead to a poor or delayed germination and post-germination growth abnormalities (Läuchli and Grattan 2007; Munns and Tester 2008). The primary osmotic stress is considered as hastened in its effects to which plants show abrupt responses while the secondary effects of salinity exposures are relatively slower and the adverse effects appear at later stages when sufficiently larger amounts of sodium ions gather in plant tissues and correspond to photosynthetic abnormalities, damage to cells, and many metabolic malfunctions (Munns and Tester 2008; Parihar et al. 2015; Hanin et al. 2016). Tuteja (2007) asserted that when Na ion accumulation in plant tissues increases 100 mM, enzymes' functions, cell membrane structure, and cell division are severely influenced which result in a reduced growth. In leaves where transpiration proceeds, high level of ions may result in cell injuries and consequent retardation in growth (Parihar et al. 2015). Other drastic effects of salinity exposures on plants are misappropriated opening and closing of stomata, which lead to altered gaseous exchange, photosynthesis, and the rate of transpiration (Munns and Tester 2008; Hanin et al. 2016). Furthermore, it is strongly evident that higher concentration of salts' ions in soil negatively affects the uptake of other necessary ions which plants require for several metabolic and enzymatic activities (Hanin et al. 2016). The high buildup of toxic ions in leaves outside threshold levels causes eventual death of many cells which are actively involved in photosynthesis with a net result in limited photosynthetic activity (Läuchli and Grattan 2007). The altered ratio of potassium and sodium ions significantly influence the entry of former into plant tissues and intercellular vicinities which lead to drastic consequences on ionic balance inside plants (Tuteja 2007). These abnormalities in physiological and biochemical processes of exposed plants are generally associated with salt stress symptoms which may range from reduced and/or delayed germination, plant growth, reproductive parameters, and yield outputs. In sum, osmotic stress developed due to higher salts concentration, accumulation of Na^+ in plants, ion toxicity, altered ratio of Na^+ and K^+ results collectively in deficit water and nutrient uptake by germinating seeds and seedling roots, cell injuries, changes in vital enzymes and hormones, irregular stomatal opening and closing with disturbed gaseous exchange, rate of transpiration and photosynthesis; thus imposing stress environment for seed to properly germinate and for seedlings to properly grow. The overall result would be a reduction in growth and yields plants exposed to salinity.

In several studies, drastic consequences of salinity on germination, growth, yield, and physiological activities of major cultivated crops have been proven (Table 2). Cha-Um and Kirdmanee (2009) applied 100–400 mM NaCl stress to different genotypes of maize which responded to the imposed stress with reduced synthesis of chlorophylls while increased proline contents. Even lower salinity stress ranging between 3.5 and 8.5 dS m^{-1} caused a significant retardation in oil content and yield of safflower (Yeilaghi et al. 2012). In our previous study, we recorded a significant decline in germination and seedling length of wheat cultivars which were exposed to 8 dS m^{-1} NaCl stress (Muhammad et al. 2015). In okra, reduced germination and growth at 75 mM (Habib et al. 2016), while in chickpea arrested germination, growth, and biomass at 40 mM NaCl was observed (Atieno et al. 2017). In response to different level of salinity stress, reduced growth and biomass, leaf area, chlorophyll degradation, altered water status, defective stomatal functions, altered rate of transpiration and respiration, and imbalanced ion ratios in maize (Gul et al. 2017; Konuşkan et al. 2017), rice (Krishnamurthy et al. 2016); Islam et al. 2017; Shahzad et al. 2017; Radanielson et al. 2018), wheat (Gul et al. 2017; Fathi et al. 2017; Bajwa et al. 2018; Khataar et al. 2018), tomato (Rubio et al. 2017; Martinez et al. 2018), faba bean and common bean (Benidire et al. 2017; Hussein et al. 2017; Ahmad et al. 2018; Khataar et al. 2018), barley (Allel et al. 2018), tomato, cotton, and several other crops have been well documented (Meloni et al. 2003; Zhang et al. 2006; Sarabi et al. 2017; Sandhu et al. 2017; Martinez et al. 2018; Ahmadi et al. 2018).

Table 2 Effect of salinity on growth and physicochemical characters of different crops

Crop species	Salinity level/ concentration	Type of experiment	Crops' response	References
Corn (*Zea mays*)	100–400 mM	Laboratory assay	Decreased chlorophylls, increased proline	Cha-Um and Kirdmanee (2009)
Alfalfa (*Medicago sativa*)	30–150 mmol L^{-1}	Lab	Decreased K+, water content, stomatal conductance, chlorophyll pigments, transpiration, and photosynthesis	Li et al. (2010)
Wheat (*Triticum aestivum*)	150 mM	Pot experiment	Decreased CO_2 assimilation, rate of transpiration, dry biomass	Perveen et al. (2010)
Rice (*Oryza sativa*)	4–20 dS m^{-1}	Lab	Arrest in germination, growth of seedling, and biomass	Hakim et al. (2010)
Safflower (*Carthamus tinctorius*)	3.5–8.5 dS m^{-1}	Field trials	Reduced oil content and yield (7–29%)	Yeilaghi et al. (2012)
Tomato (*Solanum lycopersicum*)	120 mM	Growth chamber	Negative effects on root and shoot growth	Rivero et al. (2014)
Wheat (*T. aestivum*)	25–50 mM	Greenhouse	Reduction in growth and yield	Akhtar et al. (2015a, b)
Wheat (*T. aestivum*)	8 dS m^{-1}	Laboratory assay	Reduced germination and plumule length	Muhammad et al. (2015)
Okra (*Abelmoschus esculentus*)	75 mM	Pot experiment	Lower germination and plant growth	Habib et al. (2016)
Rice (*O. sativa*)	10 dS m^{-1}	Pot culture	44–97% reduction in grain yield	Krishnamurthy et al. (2016)
Rice (*O. sativa*)	4–8 dS m^{-1}	Pot culture	Retarded root growth, oxidative stress, and membrane damage	Islam et al. (2017)
Chickpea (*Cicer arietinum*)	40 mM NaCl	Greenhouse	15–32% reduction in plant growth, plant height, dry biomass, and yields	Atieno et al. (2017)
Maize (*Zea mays*)	100 mM	Growth chamber	Reduced fresh and dry biomass, chlorophyll degradation	Kaya et al. (2017)
Tomato (*S. lycopersicum*)	50–150 mM	Pot	Reduced leaf area, photosynthesis, and CO assimilation	Bacha et al. (2017)
Alfalfa (*M. sativa*)	18.3 and 24.5 dS m^{-1}	Greenhouse	Reduced biomass, shoot length, leaf area, stomata conductance	Sandhu et al. (2017)
Tomato (*S. lycopersicum*)	300 mM	Pot culture	Abnormal growth, gaseous exchange, fresh and dry weight	Rubio et al. (2017)

(continued)

Table 2 (continued)

Crop species	Salinity level/ concentration	Type of experiment	Crops' response	References
Wheat (*T. aestivum*)	100–200 mM	Pot	Decreased yield and growth	Ahanger and Agarwal (2017)
Maize (*Z. mays*)	40–80 mM	Pot	Altered biochemical parameters	Gul et al. (2017)
Rice (*O. sativa*)	120–250 mM	Lab assay	Growth discrepancy at high salinity stress	Shahzad et al. (2017)
Wheat (*T. aestivum*)	75–150 mM	Growth chamber	Reduced morphological traits and dry biomass	Fathi et al. (2017)
Melon (*Cucumis melo*)	30–90 mM	Greenhouse	Abnormalities in leaf water content, leaf area, K+ ions, chlorophylls	Sarabi et al. (2017)
Maize (*Z. mays*)	4–12 dS m^{-1}	Greenhouse	Reduction in emergence and early seedling growth by 57.9%	Konuşkan et al. (2017)
Faba bean (*Vicia faba*)	60–120 mM	Pot	Growth, biomass, nodulation capacity reduced	Benidire et al. (2017)
Faba bean (*V. faba*)	50–150 mM	Plastic trays	Biomass, photosynthesis, water content, stomatal conductance reduced	Hussein et al. (2017)
Faba bean (*V. faba*)	50–100 mM	Pot	Decrease in root and shoot length, dry biomass decreased by 67%	Ahmad et al. (2018)
Barley (*Hordeum vulgare*)	100–200 mm	Pot experiment	Growth and photosynthetic abnormality	Allel et al. (2018)
Tomato (*S. lycopersicum*)	75 mM	Growth chamber	45–60% reduction in biomass	Martinez et al. (2018)
Bean (*Phaseolus vulgaris*), wheat (*T. aestivum*)	2–20 dS m^{-1}	Closed growth chamber	Reduced shoot biomass and yield index at salinity $\geq$8 dS m^{-1}	Khataar et al. (2018)
Rice (*O. sativa*)	2–12 dS m^{-1}	Greenhouse and field	Up to 50 reduced growth, transpiration, photosynthesis	Radanielson et al. (2018)
Rapeseed (*Brassica napus*)	110–330 dS m^{-1}	Greenhouse	Disturbance in growth, respiration, photosynthesis	Ahmadi et al. (2018)
Wheat (*T. aestivum*)	4–10 dS m^{-1}	Pot	Growth, chlorophyll, and biochemical attributes drastically suffered	Bajwa et al. (2018)

Salinity Management Approaches

Modifications in Agricultural Practices

Rengasamy (2006) categorized three major types of salinity: (1) salinity associated with groundwater, (2) salinity associated with non-groundwater, and (3) salinity associated with irrigation. All the three types of salinity are manageable if preventive measures, depending on the types of salinity and location, are taken into account. Although it seems difficult to employ a particular method in agro farming systems for salinity management, a combination of several techniques would sufficiently help in reducing the drastic impacts of salinity on crops. As for natural factors responsible for salinity, man has little influence on bringing them under control. Since leading causes of anthropogenic salinity include land degradation, poor drainage, and use of substandard water and excessive irrigation, employment of appropriate practices will satisfactorily manage the salt menace. Foremost, increasing population pressure in urban areas stimulate migrations to rural areas—the main hub of agricultural activities—which consequently results in the utilization of agricultural land for dwelling purposes. In parts of fertile land destruction as a result of building constructions, the household outflow of water may cause disturbances in the quality of irrigated water. Thus, restoration and protection of fertile land would limit the potential secondary salinization. Rise in water table caused by irrigation practices, rainfall, and leaching of water which are perceived to be linked with the salinization process can be maintained by surface drainage. The exclusion of excess of water from rhizosphere is also an effective tool to maintain the permanent water table and avoid its rising (Tiwari and Goel 2017). Drainage can be achieved by making trenches in agricultural soils which are saturated with plenty of water or by installation of a drainage system although the latter one cannot be afforded by resource-poor farmers in many countries. Hanson and May (2004) stated that substrata drip irrigation could maintain water table in shallow water regions and correspond to better salinity management and improvement in crops in such adopted system. Ayars et al. (2006) proposed that subsurface drainage and management of water at specific depth in agricultural lands could significantly improve soil profile and salinization problems. They further suggested that proper design and post-installation maintenance of drainage system would be required for getting better results in arid soils with the saline profile. In a study conducted in Pakistan on the importance of different drainage projects in irrigated areas for salt, management revealed a 14–20% decrease in soil salinity (Azhar 2010). Valipour (2014), in a comprehensive review, presented the advantages of the drainage systems in reducing salinity problems in many countries where the problem prevails. He outlined that in Europe the drainage system was the most effective measure in controlling salinity stress; however, such systems were not recommended for areas where phosphorus deficiency remained a challenging problem. Hou et al. (2016) advocated the use of mulch-drip irrigation in soils enriched with salts to minimize salinity concentration and to promote plants' growth. Besides artificial drainage systems, cultivation of deep-rooted trees such as Eucalyptus sp. may prevent rise in the water table and consequent salinization. Canal system in different countries, particularly in Pakistan and other developing countries, is extensively used for irrigation purposes. Inappropriate designs of canal systems and lack of protective lining in most of the developing countries due to high costs result in water seepage. The seeped water move through different soil zones and lead to the solubility of minerals and their accumulation in the fertile agricultural land which consequently result in soil salinity and water logging.

Another leading cause of salinization throughout the world is excessive irrigation and the use of substandard water for the stated purpose. In areas with arid and semiarid conditions, irrigation processes facilitate the mobility of salts to non-saline areas where secondary salinization occurs and disturb the growth parameters of cultivated plants (Smedema and Shiati (2002). Rietz and Haynes (2003) debated that irrigation with poor water coupled with unmanaged drainage results in water level rise and accumulation of salts in the soil surface layer where plants are grown. The scarcity of fresh water for irrigation and other purposes in many parts of the world coerce farmers to use low standard water which mainly contains municipal wastes and salts, thereby reducing soil quality and elevating salinity and sodicity levels (Qadir and Oster 2004). It is the prime priorities of agricultural stakeholders to stimulate scientific practices in irrigation. This may include drip irrigation (Hou et al. 2016; Ortega-Reig et al. 2017), installation of tube wells (Memon et al. 2017), sprinkler irrigation (Rudrapur et al. 2017), and proper drainage by making drenches and installing water removal equipment. Although it seems costly particularly in developing countries, comparisons with crops losses and land degradation as a result of salinity and poor irrigation can highlight the potential benefits of sustainable irrigation methods for the long term. To minimize energy consumption costs incurred by tube wells irrigation, deployment of solar panels can be specifically useful. Cultivation of deep-rooted trees such as *Eucalyptus*, *Acacia*, and *Sesbania* species can efficiently manage water table and consequently avoid salinity problems. Deforestation and overgrazing are also some of the important indirect contributing factors towards salinization because of changes in climatic condition, altered rainfall, elevated evaporation, and flooding; thus, maximum vegetation and protection of forests can be helpful in long-term salinity management.

Seed Priming and Plant Growth Promoting Bacteria (Bio-Priming) for Salinity Management

Pre-germination soaking of seeds with different solutions (or pure water) for the different duration is termed as "seed priming" (Song et al. 2017). Solutions of different compounds, both of natural and synthetic origins, are usually used as pre-soaking materials to prepare seeds for better performance under stressful environment. Seeds may be primed either with water, or salts, hormones, and natural metabolic substances although different priming agents have variable effects on treated seeds in terms of their responses to exposed stress. Soaking duration with priming agents is also a necessary factor which enables seeds to architect stress-combating machinery. In practice, halo priming (treatment of seeds with different concentration of salts) and hydropriming (water treatment) seem economical because of low cost and easy availability of soaking gents (Jisha et al. 2013) although other priming compounds such as polyethylene glycol (PEG), auxins, gibberellins, cytokinins, salicylic acid, jasmonic acid, kinetin, urea, and several natural metabolic substances have been well proven in inducing defense and stress tolerance to subjected seeds (Islam et al. 2015; Salah et al. 2015; Savvides et al. 2016). Primary mechanism underlying the priming-induced stress tolerance in crops is the activation of enzymes, hormones, homeostatic regulation, synthesis of new messenger RNA, improvement in imbibition potential and dormancy, detoxification of reactive oxygen species, enhanced antioxidant systems, and induction of "pre-stress memory" (Bruce et al. 2007; Varier et al. 2010; Hussain et al. 2016) (Fig. 1). Primed seeds perform better than non-primed ones when sown either under normal conditions or in a stressed environment. Many studies suggest that priming cause better germination, seedling emergence, and vigor (Salah et al. 2015; Hussain et al. 2016). Ajouri et al. (2004) have documented that priming-induced salinity resistance in barley and better germination was achieved. Kaya et al. (2006) observed improved germination and growth of sunflower seedling in saline and drought conditions as a result of hydropriming. Under 100 mM NaCl stress, halopriming, and hydropriming invigorated wheat growth, yield, antioxidant activities, and ions regulation (Islam et al. 2015). Improved germination and reduced mean germination time of *Nigella sativa* were also recorded under 40 mM NaCl when seeds were treated with 1–2% NaCl and several other priming agents (Gholami et al. 2015). Khaliq et al. (2015) treated seeds of rice with 15–105 μmol L^{-1} concentrations of selenium and evaluated the primed seeds for germination, growth, and enzymatic activity under salinity stress. They observed the better performance of primed seed than control under stress conditions for all the studied parameters. Osmopriming of bread wheat seeds with $CaCl_2$ resulted in enhanced tolerance to salinity (100 mM) resulting in improved leaf area, Na/K ion ratio, and grain yield (Tabassum et al. 2017). Pretreatment of broccoli sprout with methyl jasmonate and KCl induced salt-stress tolerance in tested plants which resulted in growth and biochemical parameters elevation under salinity stress (Hassini et al. 2017). Recently, Bajwa et al. (2018) effectively induced salt tolerance in wheat by pre-soaking seeds with sorghum extracts and benzyl amino-purine. Germination, growth, and biochemical attributes exhibited improved performance in salinity imposed stress.

In addition to wide applicability of seed priming in agriculture, plant growth promoting bacteria (PGPB) have also a potential role in improving crops' stress responses. PGPB can manage drastic effects of salinity and other stress conditions on cultivated crops by several mechanisms. They may stimulate the production of specific proteins and osmoprotectants in crops exposed to stressful environment

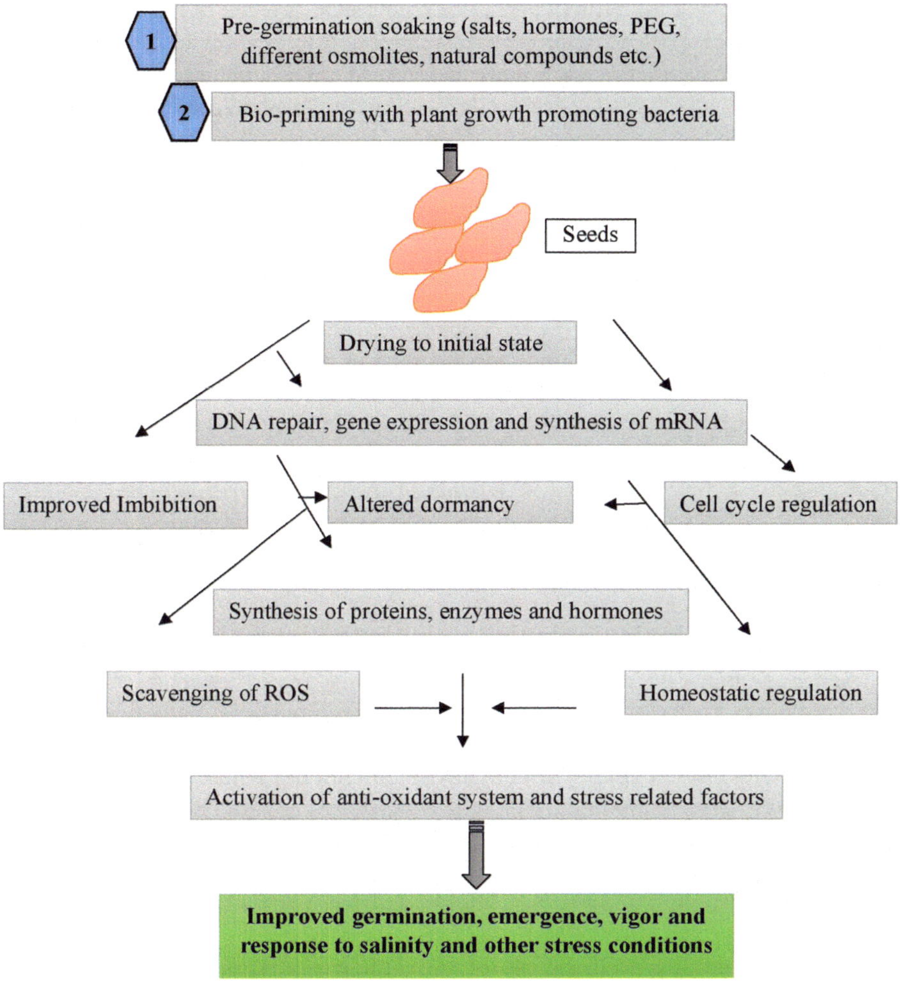

Fig. 1 Mechanism of stress tolerance induction in seeds by priming (after Varier et al. 2010 and Mahmood et al. 2016)

thereby reducing the adverse effects of stresses and protecting crops from stress injuries (Grover et al. 2011). Saleem et al. (2007) argued that one of the leading factors associated with stress is the elevated production of ethylene—a growth regulator produced indigenously by plants when they are challenged with stress conditions—is regulated by several strains of PGPB. It is established that many PGPB produce ACC deaminase and regulate the production and functions of cytokinins, antioxidants, and ethylene which not only help crops to take nutrient and water properly but also induce in them systemic resistance to salinity stress which imparts stimulatory effects on the growth and yield under such stresses (Yang et al. 2009). According to Glick (2012), plant growth suppression in fields is either caused by biotic (pathogenic interaction) or abiotic (salinity, intensive light, drought, temperature, etc.) stresses which are generally correlated with abnormal production of stress hormones, fluctuation in metabolic machinery. He provided evidence that many PGPB can manage the required production of ethylene, indole acetic acid (IAA), trehalose, cytokinin, and several other stress components. Shrivastava and Kumar (2015), in a comprehensive review, highlighted the role of PGPB in conferring salinity resistance to crops. They commented that PGPB induces systemic tolerance in crops besides their active role in the provision of nutrients, soil fertility, and disease suppression. Bharti et al. (2016) documented that *Dietzia natronolimnaea*, a plant growth promoting bacterium, was involved in regulating transcriptional factors which confer salt resistance to plants and protect them from salinity-induced injuries. Del Cerro et al. (2017) suggested that certain strains of *Rhizobium tropici* are involved in the production of nodulating factors under high salt concentration which promotes root nodulation in legume crops and helps in avoiding salinity stress. Some species of *Streptomyces* are recognized for their ability to produce 1-aminocyclopropane-1-carboxylate deaminase and promote the growth, ion uptake, and chlorophyll functions of rice under 150 mM NaCl stress (Jaemsaeng et al. 2018).

In the previous study, salinity tolerance and improved growth of tomato were recorded in response to PGPB (*Achromobacter piechaudii*) under 43 mM NaCl stress (Mayak et al. 2004). Tank and Saraf (2010) reported 50% increase in tomato growth and biomass under salt stress when plants were inoculated with different strains of *Pseudomonas* sp. under greenhouse conditions. Wheat seeds inoculated with *Arthrobacter* sp. exhibited efficient resistance to salt stress and yielded better than non-inoculated plants (Upadhyay and Singh 2015). Wheat inoculated with PGPB *Klebsiella* sp. (SBP8) and grown in 150–200 mM NaCl stress showed higher growth and chlorophyll activity than non-inoculated control (Singh et al. 2015). Edible peas grown under 70–130 mM NaCl stress after pretreatment with *Variovorax paradoxus* showed improved growth, photosynthetic performance and ion uptake (Wang et al. 2016). Bharti et al. (2016) reported that wheat inoculation with *Dietzia natronolimnaea* resulted in a significant elevation of plant height and dry biomass when salinity stress (150 mM NaCl) was imposed on test crop. Under similar salinity stress, *Streptomyces* sp. (strain GMKU 336) promoted plant growth, water status, chlorophyll and proline contents, and ion ratios of rice (Jaemsaeng et al. 2018). In other similar studies under high salt concentrations, *Burkholderia* and *Enterobacter* sp. stimulated the growth and physicochemical attributes of maize and wheat (Akhtar et al. 2015a, b), *Bacillus cereus* improved mung bean (Islam et al. 2016), and *Enterobacter* sp. promoted rice growth (Sarkar et al. 2018). These and dozens of other studies employing PGPB in salt stress environment indicate that several PGPB strains have a prospective role in inducing salt-stress tolerance to cultivated crops and enhancing their yield outputs.

Conclusions

Cultivated crops are challenged with several constraints among which salinity is an important one responsible for limited growth and crop production. Salinity stress imposes ion toxicity, water deficit conditions, imbalanced ion uptake, cellular damage, and degradation of chlorophylls, transpiration and respiration activities which definitely influence overall growth attributes of challenged plants. Both natural and man-triggered activities cause salinization in agriculture. Extensive irrigation, poor drainage, low-quality water for irrigation purposes are directly contributing to the problem of salinity. Employment of mechanisms and techniques to diminish the adverse effects of salinity and make crops adapted to withstand salinity would help in crop losses management. Thus, scientific approaches in agriculture such as sprinkler irrigation drench and pumping surface and groundwater, and lining of irrigating canals can significantly lower salinity levels in cultivated land. Seed priming with water, salts, hormones, and other chemicals induces stress adaptability in them and can prove effective tools to divert the adverse effects of salinity on crops. Biopriming with plant growth promoting bacteria has also been well recognized for their ability to induce salinity tolerance in crops. To make the maximum use of seed priming agents and PGPB for salt management in cultivated crops, commercial availability, economic accessibility, and safety issues concerned with priming chemicals and PGPB strains must be ensured by stakeholders. Since farmers represent a basic component agricultural system, their awareness about the significance of seed priming and PGPB in salinity can contribute to further expansion of these techniques in agricultural areas which are challenged with salinity.

References

Ahanger MA, Agarwal RM (2017) Salinity stress induced alterations in antioxidant metabolism and nitrogen assimilation in wheat (*Triticum aestivum* L) as influenced by potassium supplementation. Plant Physiol Biochem 115:449–460

Ahmad P, Alyemeni MN, Ahanger MA, Egamberdieva D, Wijaya L, Alam P (2018) Salicylic acid (SA) induced alterations in growth, biochemical attributes and antioxidant enzyme activity in Faba Bean (*Vicia faba* L.) seedlings under NaCl toxicity. Russian J Plant Physiol 65(1):104–114

Ahmadi FI, Karimi K, Struik PC (2018) Effect of exogenous application of methyl jasmonate on physiological and biochemical characteristics of *Brassica napus* L. cv. Talaye under salinity stress. S Afr J Bot 115:5–11

Ajouri A, Asgedom H, Becker M (2004) Seed priming enhances germination and seedling growth of barley under conditions of P and Zn deficiency. J Plant Nutr Soil Sci 167:630–636

Akhtar SS, Andersen MN, Liu F (2015a) Residual effects of biochar on improving growth, physiology and yield of wheat under salt stress. Agric Water Manag 158:61–68

Akhtar SS, Andersen MN, Naveed M, Zahir ZA, Liu F (2015b) Interactive effect of biochar and plant growth-promoting bacterial endophytes on ameliorating salinity stress in maize. Funct Plant Biol 42:770–781

Allel D, Ben-Amar A, Abdelly C (2018) Leaf photosynthesis, chlorophyll fluorescence and ion content of barley (*Hordeum vulgare*) in response to salinity. J Plant Nutr 41:497–508

Athar HR, Ashraf M (2009) Strategies for crop improvement against salinity and drought stress: an overview. In: Ashraf M, Ozturk M, Athar HR (eds) Salinity and water stress. Springer, New York, pp 1–16

Atieno J, Li Y, Langridge P, Dowling K, Brien C, Berger B, Sutton T (2017) Exploring genetic variation for salinity tolerance in chickpea using image-based phenotyping. Sci Rep 7(1):1300

Ayars E, Christe EW, Hornbuckle JW (2006) Controlled drainage for improved water management in arid regions irrigated agriculture. Agric Water Manag 86:128–139

Azhar AH (2010) Impact of subsurface drainage on soil salinity in Pakistan. J Anim Plant Sci 20:94–98

Bacha H, Tekaya M, Drine S, Guasmi F, Touil L, Enneb H, Ferchichi A (2017) Impact of salt stress on morpho-physiological and biochemical parameters of *Solanum lycopersicum* cv. Microtom leaves. S Afr J Bot 108:364–369

Bajwa AA, Farooq M, Nawaz A (2018) Seed priming with sorghum extracts and benzyl aminopurine improves the tolerance against salt stress in wheat (*Triticum aestivum* L.). Physiol Mol Biol Plants 24:239-249

Benidire L, Lahrouni M, El Khalloufi F, Göttfert M, Oufdou K (2017) Effects of rhizobium leguminosarum inoculation on growth, nitrogen uptake and mineral assimilation in *Vicia faba* plants under salinity stress. J Agric Sci Technol 19:889–901

Bharti N, Pandey SS, Barnawal D, Patel VK, Kalra A (2016) Plant growth promoting rhizobacteria *Dietzia natronolimnaea* modulates the expression of stress responsive genes providing protection of wheat from salinity stress. Sci Rep 6:34768

Borsani O, Valpuesta V, Botella MA (2003) Developing salt tolerant plants in a new century: a molecular biology approach. Plant Cell Tissue Organ Cult 73:101–115

Bruce TJ, Matthes MC, Napier JA, Pickett JA (2007) Stressful "memories" of plants: evidence and possible mechanisms. Plant Sci 173:603–608

Cha-Um S, Kirdmanee C (2009) Effect of salt stress on proline accumulation, photosynthetic ability and growth characters in two maize cultivars. Pak J Bot 41:87–98

Daliakopoulos IN, Tsanis IK, Koutroulis A, Kourgialas NN, Varouchakis AE, Karatzas GP, Ritsema CJ (2016) The threat of soil salinity: a European scale review. Sci Total Environ 573:727–739

Del Cerro P, Pérez-Montaño F, Gil-Serrano A, López-Baena FJ, Megías M, Hungria M, Ollero FJ (2017) The Rhizobium tropici CIAT 899 NodD2 protein regulates the production of Nod factors under salt stress in a flavonoid-independent manner. Sci Rep 7:46712

Fan X, Pedroli B, Liu G, Liu Q, Liu H, Shu L (2012) Soil salinity development in the yellow river delta in relation to groundwater dynamics. Land Degrad Dev 23:175–189

FAOSTAT (2018) Food and Agricultural Organization Statistical database, Crop production. http://faostat3.fao.org/download /Q/QC/ E (accessed on March 20, 2018)

Fathi A, Zahedi M, Torabian S, Khoshgoftar A (2017) Response of wheat genotypes to foliar spray of ZnO and Fe_2O_3 nanoparticles under salt stress. J Plant Nutr 40:1376–1385

Gholami M, Mokhtarian F, Baninasab B (2015) Seed halopriming improves the germination performance of black seed (*Nigella sativa*) under salinity stress conditions. J Crop Sci Biotechnol 18:21–26

Glick BR (2012) Plant growth-promoting bacteria: mechanisms and applications. Scientifica 2012:1

Grover M, Ali SZ, Sandhya V, Rasul A, Venkateswarlu B (2011) Role of microorganisms in adaptation of agriculture crops to abiotic stresses. World J Microbiol Biotechnol 27:1231–1240

Gul H, Kinza S, Shinwari ZK, Hamayun M (2017) Effect of selenium on the biochemistry of *Zea mays* under salt stress. Pak J Bot 49:25–32

Habib SH, Kausar H, Saud HM (2016) Plant growth-promoting rhizobacteria enhance salinity stress tolerance in okra through ROS-scavenging enzymes. Biomed Res Int 2016:6284547

Hakim MA, Juraimi AS, Begum M, Hanafi MM, Ismail MR, Selamat A (2010) Effect of salt stress on germination and early seedling growth of rice (*Oryza sativa* L.). Afr J Biotechnol 9:1911–1918

Hanin M, Ebel C, Ngom M, Laplaze L, Masmoudi K (2016) New insights on plant salt tolerance mechanisms and their potential use for breeding. Front Plant Sci 7:1787

Hanson B, May D (2004) Effect of subsurface drip irrigation on processing tomato yield, water table depth, soil salinity, and profitability. Agric Water Manag 68:1–17

Hassini I, Martinez-Ballesta MC, Boughanmi N, Moreno DA, Carvajal M (2017) Improvement of broccoli sprouts (*Brassica oleracea* L. var. italica) growth and quality by KCl seed priming and methyl jasmonate under salinity stress. Sci Hortic 226:141–151

Hou M, Zhu L, Jin Q (2016) Surface drainage and mulching drip-irrigated tomatoes reduces soil salinity and improves fruit yield. PLoS One 11:e0154799

Hussain S, Khan F, Cao W, Wu L, Geng M (2016) Seed priming alters the production and detoxification of reactive oxygen intermediates in rice seedlings grown under sub-optimal temperature and nutrient supply. Front Plant Sci 7:439

Hussein M, Embiale A, Husen A, Aref IM, Iqbal M (2017) Salinity-induced modulation of plant growth and photosynthetic parameters in faba bean (*Vicia faba*) cultivars. Pak J Bot 49:867–877

Islam F, Yasmeen T, Ali S, Ali B, Farooq MA, Gill RA (2015) Priming-induced antioxidative responses in two wheat cultivars under saline stress. Acta Physiol Plant 37:153

Islam F, Yasmeen T, Arif MS, Ali S, Ali B, Hameed S, Zhou W (2016) Plant growth promoting bacteria confer salt tolerance in *Vigna radiata* by up-regulating antioxidant defense and biological soil fertility. Plant Growth Regul 80:23–36

Islam F, Farooq MA, Gill RA, Wang J, Yang C, Ali B, Zhou W (2017) 2, 4-D attenuates salinity-induced toxicity by mediating anatomical changes, antioxidant capacity and cation transporters in the roots of rice cultivars. Sci Rep 7:10443

Ismail AM, Horie T (2017) Genomics, physiology, and molecular breeding approaches for improving salt tolerance. Annu Rev Plant Biol 68:405–434

Jaemsaeng R, Jantasuriyarat C, Thamchaipenet A (2018) Molecular interaction of 1-aminocyclopr opane-1-carboxylate deaminase (ACCD)-producing endophytic Streptomyces sp. GMKU 336 towards salt-stress resistance of *Oryza sativa* L. cv. KDML105. Sci Rep 8(1):1950

Jisha KC, Vijayakumari K, Puthur JT (2013) Seed priming for abiotic stress tolerance: an overview. Acta Physiol Plant 35:1381–1396

Kaashyap M, Ford R, Kudapa H, Jain M, Edwards D, Varshney R, Mantri N (2018) Differential regulation of genes involved in root morphogenesis and cell wall modification is associated with salinity tolerance in chickpea. Sci Rep 8:4855

Kaya MD, Okçu G, Atak M, Cıkılı Y, Kolsarıcı Ö (2006) Seed treatments to overcome salt and drought stress during germination in sunflower (*Helianthus annuus* L.). Eur J Agron 24:291–295

Kaya C, Ashraf M, Sonmez O, Polat T, Tuna AL (2017) The combined effects of nitric oxide and thiourea on plant growth and mineral nutrition of salt-stressed plants of two maize cultivars with differential salt tolerance. J Plant Nutr. https://doi.org/10.1080/01904167.2016.1270314

Khaliq A, Aslam F, Matloob A, Hussain S, Geng M, Wahid A, Rehman H (2015) Seed priming with selenium: consequences for emergence, seedling growth, and biochemical attributes of rice. Biol Trace Elem Res 166:236–244

Khataar M, Mohhamadi MH, Shabani F (2018) Soil salinity and matric potential interaction on water use, water use efficiency and yield response factor of bean and wheat. Sci Rep 8:2679

Konuşkan Ö, Gözübenli H, Atiş İ, Atak M (2017) Effects of salinity stress on emergence and seedling growth parameters of some maize genotypes (*Zea mays* L.). Turk J Agric Food Sci Technol 5:1668–1672

Krishnamurthy SL, Gautam RK, Sharma PC, Sharma DK (2016) Effect of different salt stresses on agro-morphological traits and utilisation of salt stress indices for reproductive stage salt tolerance in rice. Field Crops Res 190:26–33

Läuchli A, Grattan SR (2007) Plant growth and development under salinity stress. In: Jenks MA, Hasegawa PM, Mohan Jain S (eds) Advances in molecular breeding toward drought and salt tolerant crops. Springer, New York, pp 1–32

Li R, Shi F, Fukuda K, Yang Y (2010) Effects of salt and alkali stresses on germination, growth, photosynthesis and ion accumulation in alfalfa (*Medicago sativa* L.). Soil Sci Plant Nutr 56:725–733

Luo Y, Reid R, Freese D, Li C, Watkins J, Shi H, Song BH (2017) Salt tolerance response revealed by RNA-Seq in a diploid halophytic wild relative of sweet potato. Sci Rep 7:9624

Machado RMA, Serralheiro RP (2017) Soil salinity: effect on vegetable crop growth. Management practices to prevent and mitigate soil salinization. Horticulturae 3:30
Mahmood A, Turgay OC, Farooq M, Hayat R (2016) Seed biopriming with plant growth promoting rhizobacteria: a review. FEMS Microbiol Ecol 92(8):pii:fiw112
Martinez V, Nieves-Cordones M, Lopez-Delacalle M, Rodenas R, Mestre TC, Garcia-Sanchez F, Rivero RM (2018) Tolerance to stress combination in tomato plants: new insights in the protective role of melatonin. Molecules 23:535
Mayak S, Tirosh T, Glick BR (2004) Plant growth-promoting bacteria confer resistance in tomato plants to salt stress. Plant Physiol Biochem 42:565–572
Meloni DA, Oliva MA, Martinez CA, Cambraia J (2003) Photosynthesis and activity of superoxide dismutase, peroxidase and glutathione reductase in cotton under salt stress. Environ Exp Bot 49:69–76
Memon JA, Jogezai G, Hussain A, Alizai MQ, Baloch MA (2017) Rehabilitating traditional irrigation systems: assessing popular support for Karez rehabilitation in Balochistan, Pakistan. Hum Ecol 45:265–275
Muhammad Z, Hussain F, Rehmanullah, Majeed A (2015) Effect of halopriming on the induction of NaCl salt tolerance in different wheat genotypes. Pak J Bot 47: 1613-1620
Munns R, Tester M (2008) Mechanisms of Salinity Tolerance. Ann Rev Plant Biol 59:651-681
Munns R, Gilliham M (2015) Salinity tolerance of crops—what is the cost? New Phytol 208:668–673
Ortega-Reig M, Sanchis-Ibor C, Palau-Salvador G, García-Mollá M, Avellá-Reus L (2017) Institutional and management implications of drip irrigation introduction in collective irrigation systems in Spain. Agric Water Manag 187:164–172
Parihar P, Singh S, Singh R, Singh VP, Prasad SM (2015) Effect of salinity stress on plants and its tolerance strategies: a review. Environ Sci Pollut Res 22:4056–4075
Perveen S, Shahbaz M, Ashraf M (2010) Regulation in gas exchange and quantum yield of photosystem II (PSII) in salt-stressed and non-stressed wheat plants raised from seed treated with triacontanol. Pak J Bot 42:3073–3081
Qadir M, Oster JD (2004) Crop and irrigation management strategies for saline-sodic soils and waters aimed at environmentally sustainable agriculture. Sci Total Environ 323:1–19
Radanielson AM, Angeles O, Li T, Ismail AM, Gaydon DS (2018) Describing the physiological responses of different rice genotypes to salt stress using sigmoid and piecewise linear functions. Field Crops Res 220:46–56
Rasool S, Hameed A, Azooz MM, Siddiqi TO, Ahmad P (2013) Salt stress: causes, types and responses of plants. Ecophysiology and responses of plants under salt stress. Springer, New York, pp 1–24
Rengasamy P (2006) World salinization with emphasis on Australia. J Exp Bot 57:1017–1023
Rietz DN, Haynes RJ (2003) Effects of irrigation-induced salinity and sodicity on soil microbial activity. Soil Biol Biochem 35:845–854
Rivero RM, Mestre TC, Mittler RON, Rubio F, Garcia-Sanchez F, Martinez V (2014) The combined effect of salinity and heat reveals a specific physiological, biochemical and molecular response in tomato plants. Plant Cell Environ 37:1059–1073
Rubio MB, Hermosa R, Vicente R, Gómez-Acosta FA, Morcuende R, Monte E, Bettiol W (2017) The combination of *Trichoderma harzianum* and chemical fertilization leads to the deregulation of phytohormone networking, preventing the adaptive responses of tomato plants to salt stress. Front Plant Sci 8:294
Rudrapur S, Mundinamani SM, Kiresur VR, Hosamani SB, Manjunatha MV, Dolli SS (2017) Impact of sprinkler irrigation system on doubling Farmers' income and water use efficiency of tur in northern Karnataka. In: Agricultural Economics Research Review
Salah SM, Yajing G, Dongdong C, Jie L, Aamir N, Qijuan H, Jin H (2015) Seed priming with polyethylene glycol regulating the physiological and molecular mechanism in rice (*Oryza sativa* L.) under nano-ZnO stress. Sci Rep 5:14278
Saleem M, Arshad M, Hussain S, Bhatti AS (2007) Perspective of plant growth promoting rhizobacteria (PGPR) containing ACC deaminase in stress agriculture. J Ind Microbiol Biotechnol 34(10):635–648

Sandhu D, Cornacchione MV, Ferreira JF, Suarez DL (2017) Variable salinity responses of 12 alfalfa genotypes and comparative expression analyses of salt-response genes. Sci Rep 7:42958
Sarabi B, Bolandnazar S, Ghaderi N, Ghashghaie J (2017) Genotypic differences in physiological and biochemical responses to salinity stress in melon (*Cucumis melo* L.) plants: prospects for selection of salt tolerant landraces. Plant Physiol Biochem 119:294–311
Sarkar A, Ghosh PK, Pramanik K, Mitra S, Soren T, Pandey S, Maiti TK (2018) A halotolerant *Enterobacter* sp. displaying ACC deaminase activity promotes rice seedling growth under salt stress. Res Microbiol 169:20–32
Savvides A, Ali S, Tester M, Fotopoulos V (2016) Chemical Priming of Plants Against Multiple Abiotic Stresses: Mission Possible?. Trends Plant Sci 21:329–340
Shahzad R, Khan AL, Bilal S, Waqas M, Kang SM, Lee IJ (2017) Inoculation of abscisic acid-producing endophytic bacteria enhances salinity stress tolerance in *Oryza sativa*. Environ Exp Bot 136:68–77
Shrivastava P, Kumar R (2015) Soil salinity: a serious environmental issue and plant growth promoting bacteria as one of the tools for its alleviation. Saudi J Biol Sci 22:123–131
Singh RP, Jha P, Jha PN (2015) The plant-growth-promoting bacterium *Klebsiella* sp. SBP-8 confers induced systemic tolerance in wheat (*Triticum aestivum*) under salt stress. J Plant Physiol 184:57–67
Singh VK, Mishra A, Haque I, Jha B (2016) A novel transcription factor-like gene SbSDR1 acts as a molecular switch and confers salt and osmotic endurance to transgenic tobacco. Sci Rep 6:31686
Smedema LK, Shiati K (2002) Irrigation and salinity: a perspective review of the salinity hazards of irrigation development in the arid zone. Irrig Drain Syst 16:161–174
Soda N, Sharan A, Gupta BK, Singla-Pareek SL, Pareek A (2016) Evidence for nuclear interaction of a cytoskeleton protein (OsIFL) with metallothionein and its role in salinity stress tolerance. Sci Rep 6:34762
Song GC, Choi HK, Kim YS, Choi JS, Ryu CM (2017) Seed defense biopriming with bacterial cyclodipeptides triggers immunity in cucumber and pepper. Sci Rep 7:14209
Tabassum T, Farooq M, Ahmad R, Zohaib A, Wahid A (2017) Seed priming and transgenerational drought memory improves tolerance against salt stress in bread wheat. Plant Physiol Biochem 118:362–369
Tank N, Saraf M (2010) Salinity-resistant plant growth promoting rhizobacteria ameliorates sodium chloride stress on tomato plants. J Plant Interact 5:51–58
Tiwari P, Goel A (2017) An overview of impact of subsurface drainage project studies on salinity management in developing countries. Appl Water Sci 7:569–580
Tuteja N (2007) Mechanisms of high salinity tolerance in plants. In: Methods in Enzymology. Academic Press, pp 419–438
Upadhyay SK, Singh DP (2015) Effect of salt-tolerant plant growth-promoting rhizobacteria on wheat plants and soil health in a saline environment. Plant Biol 17:288–293
Valipour M (2014) Drainage, waterlogging, and salinity. Arch Agron Soil Sci 60:1625–1640
Varier A, Vari AK, Dadlani M (2010) The subcellular basis of seed priming. Curr Sci 99:450–456
Wang Q, Dodd IC, Belimov AA, Jiang F (2016) Rhizosphere bacteria containing 1-aminocyclop ropane-1-carboxylate deaminase increase growth and photosynthesis of pea plants under salt stress by limiting Na^+ accumulation. Func Plant Biol 43:161–172
Yang J, Kloepper JW, Ryu CM (2009) Rhizosphere bacteria help plants tolerate abiotic stress. Trend Plant Sci 14:1–4
Yang A, Akhtar SS, Iqbal S, Qi Z, Alandia G, Saddiq MS, Jacobsen SE (2018) Saponin seed priming improves salt tolerance in quinoa. J Agron Crop Sci 204:31–39
Yeilaghi H, Arzani A, Ghaderian M, Fotovat R, Feizi M, Pourdad SS (2012) Effect of salinity on seed oil content and fatty acid composition of safflower (*Carthamus tinctorius* L.) genotypes. Food Chem 130:618–625
Zhang ZJ, Li HZ, Zhou WJ, Takeuchi Y, Yoneyama K (2006) Effect of 5-aminolevulinic acid on development and salt tolerance of potato (*Solanum tuberosum* L.) microtubers in vitro. Plant Growth Regul 49:27–34

Salinity Stress Tolerance in Plants: Physiological, Molecular, and Biotechnological Approaches

Mojtaba Kordrostami and Babak Rabiei

Introduction: An Overview on Salinity

Plants are always exposed to the environmental changes. Due to the lack of mobility, they must be equipped with the mechanisms that respond appropriately to some of the acute changes that are called stresses (Rejeb et al. 2014). The stresses are divided into two distinct groups, biotic and abiotic. The biotic stresses include allelopathy, bacteria, competition, fungi, insects, weeds, viruses, and human activity (Pedrol et al. 2006). The abiotic stresses include increased salinity, drought, and temperature, decreased soil oxygen content, mineral nutrient deficiency, heavy metals, pollutants, and high UV radiation (Kanojia and Dijkwel 2018). A significant portion of available water resources in arid and semiarid regions, including underground saline water and run-off water results from the downstream drainage areas which should be managed and used appropriately (Misra 2014).

In arid and semiarid areas, water is the main limiting factor for agriculture. In these areas, due to the shortage of surface currents, there is a major pressure on groundwater. Failure to comply with technical points in the exploitation of underground water and overuse of them has caused the loss of groundwater in many areas (Calow et al. 2018). Plenty of lands around the world are affected by salinity and this issue is increasing. Salinity has affected more than 30% of irrigated land and, in total, 6% of the total world's land (Chaves et al. 2009; Parihar et al. 2015).

M. Kordrostami (✉)
Department of Plant Biotechnology, Faculty of Agricultural Sciences, University of Guilan, Rasht, Iran
e-mail: mkordrostami@webmail.guilan.ac.ir

B. Rabiei
Department of Agronomy and Plant Breeding, Faculty of Agricultural Sciences, University of Guilan, Rasht, Iran

M. Hasanuzzaman et al. (eds.), *Plant Abiotic Stress Tolerance*,
https://doi.org/10.1007/978-3-030-06118-0_4

Definition of Salinity

Salinity is an increase in the concentration of dissolved inorganic salts in the soil (soil solution) or water. The main cations include K^+, Mg^{2+}, Ca^{2+}, and Na^+, and the main anions are NO_3^-, HCO_3^-, SO_4^{2-}, Cl^-, and CO_3^{2-}. Other components in the highly saline soils and water are SiO_2, Al^{3+}, Sr^{2+}, B, Mo, and Ba^{2+} (Manchanda and Garg 2008). These compounds are reported individually in mg L^{-1} (ppm), mmol L^{-1}, or meq L^{-1}, while salinity is considered as the total concentration of total dissolved solids (TDS) in mg L^{-1} (Niu and Cabrera 2010). However, in the agricultural and horticultural section, the more common term and simpler measurement of total salinity is the determination of the electrical conductivity of water (EC_w) or the soil extract (EC_e) (Hardie and Doyle 2012). Water or solvent ability to conduct the electricity is directly related to the concentration of salts contained in it and units used for reporting EC (including mmhos cm^{-1} or dS m^{-1}) have the same unit value, while there is no exact relationship between TDS and EC. However, some relationships with approximate transformation are applicable, so that for water or solutions with EC up to 5 dS m^{-1}, the TDS value is obtained by multiplying EC (dS m^{-1}) by 640 and for solutions with a salinity greater than 5 dS m^{-1}, the TDS value is obtained by multiplying EC by 800 (Tanji 2002). In general, irrigation water is classified according to their salinity values as follows: water with electrical conductivity (EC_W) less than 0.7 dS m^{-1} or a salt concentration less than 500 mg L^{-1} is classified as drinking water and because it is non-saline, it is considered as high-quality water for irrigation. While water with EC_W between 0.7 and 2 dS m^{-1} or salt concentration between 500 and 1500 mg L^{-1} is considered as a little saline water for irrigation. And also, groundwater with EC_w between 2 and 10 dS m^{-1} or salt concentration between 1500 and 7000 mg L^{-1} is considered as fairly saline, the water with EC_W between 10 and 25 dS m^{-1} or salt concentration between 7000 and 25,000 mg L^{-1} is considered as very saline and the water with EC_w between 25 and 45 dS m^{-1} or salt concentration between 15,000 and 25,000 mg L^{-1} is considered as extremely saline. In comparison, sea water has EC_w more than 45 dS m^{-1} or salt concentration more than 45,000 mg L^{-1} (Rhoades and Chanduvi 1999).

On the other hand, the soils affected by the salt stress vary widely in terms of physical and chemical properties, and can be classified into the following groups (Hanin et al. 2016):

(a) Saline soils: pH is less than 8.5, ESP less than 15%, and EC more than 4 dS m^{-1}.
(b) Saline-sodic soils: pH is about 8.8, ESP is more than 15%, and the EC is more than 4 dS m^{-1}.
(c) Sodic soils: SAR is more than 15, pH higher than 8.5, ESP is more than 15%, and EC is less than 4 dS m^{-1}.

Types of Salinity and Its Causes

Primary or Natural Salinity

The accumulation of salts over the long periods through natural processes in the soil or underground waters causes initial salinity. Two natural processes created this event. The first factor is the weathering of native rocks containing soluble salts. The breakdown of rocks and the release of soluble salts of different types, mainly magnesium sodium, and calcium, and in lesser quantities, sulfates, and carbonates are caused by weathering processes. NaCl is the most abundant soluble salt. The second factor is the precipitation of ocean salt by rain and wind. "Cyclic salt" is the salt of the ocean, which is carried by wind and rain to other areas of the earth and is mainly sodium chloride (Munns 2002; Manchanda and Garg 2008; Munns and Tester 2008; Parihar et al. 2015).

Secondary Salinity or Salinity Created by Humans

Human activities cause the secondary salinity that change the soil's hydrological balance between primary water (rainfall or irrigation) and the transpiration (Dawood 2018). Clearing the land and replacing perennial crops with annual ones and irrigation schemes using salt-rich irrigation water or lack of adequate drainage are the most common causes of secondary salinity (Parihar et al. 2015). Irrigated systems are more willing to salinization, and now half of the irrigated crops in the world are affected by salinity or flooding due to the poor water or excessive drainage and rising water from underground aquifers (Hatfield et al. 2017). Irrigation programs account for about 15% the world's agricultural lands, but these lands have at least twice as much yield as drylands and produce about one-third of the world's food, and this fact highlights the importance of secondary salinity by humans and its impact on the food production (Scherr and McNeely 2008).

Problems Caused by Salinity in Plants

Most of the problems caused by salinity in plants are due to an increase in sodium chloride, which occurs in dry and coastal soils where their water sources have expanded. The high salinity of sodium chloride generates at least three specific problems in higher plants:

(a) **Osmotic effect**: The dissolved materials in the root developmental zone produce a negative osmotic potential, which reduces the water and soil potential (Mahajan and Tuteja 2005). When the water potential of the cells becomes less, the plant's ability to absorb the water decreases and this leads to a decrease in growth, which is called osmotic effect or physiologic dryness due to salinity (Kafi 2009).

(b) **Specific ion effect**: When the accumulation of SO_2, Na^+, or Cl^- concentrations in the cells reaches the threshold of damage, the specific ion effect occurs (Chartzoulakis 2011). At high concentrations, Na^+ can be replaced with Ca^{2+} in the plasma membrane of root hairs of cotton. As a result, a change in the permeability of the plasma membrane can occur, which can appear as K^+ leakage to the surrounding solution. Salinity causes inactivation of enzymes, cell death, and eventually whole plant death (HanumanthaRao et al. 2016).

(c) **Nutritional imbalance**: The nutritional balance of the plant depends on important elements such as zinc, potassium, iron, and calcium. The concentration of these elements in the plant is affected by sodium and calcium. Increasing sodium reduces other cations in the plant and disrupts the cationic balance of the plant. This increase also reduces the amount of calcium, magnesium, and potassium in the plant (Acosta-Motos et al. 2017). Therefore, the main reason for inhibiting growth (by sodium chloride) is the problems that arise in absorbing other minerals in competition with sodium (Gupta and Huang 2014).

In different references, the effects of salinity stress on plants are related to (Rasool et al. 2013):

1. Water stress or reduction of osmotic potential
2. Imbalance in nutrients
3. Salt stress or specific ionic effects
4. A combination of the above items

The above-mentioned factors have negative effects on the growth and development of plants at physiological, biochemical, and molecular levels (Pandey et al. 2017). In the review articles, the effects of salinity stress on plants such as effect on plant water status, leaf anatomy, proteins, photosynthetic pigments, lipid and carbohydrates metabolism, gas exchange, photosynthesis, chlorophyll fluorescence, antioxidant molecules and enzymes, and nitrogen metabolism have been reported (Munns and Tester 2008; Wani et al. 2013).

Effects of Salinity

The most important result of salinity stress is the apparent stop in plant growth (Negrão et al. 2017). The immediate response to salinity stress is the decrease in the rate of leaf area expansion, which results in stopping the increase in salt concentration in the plant. Salinity stress results in a significant reduction in the dry and fresh weight of leaves, stems, and roots (Qados 2011). Plant response to salinity is carried out in two steps over time: In the early stages of salinity stress (first phase), the decrease in growth is actually due to the water stress, which is due to the osmotic effects of salinity. Osmotic effects occur after salt increase outside the root environment and with increasing the levels of tolerance threshold (Munns 2002; Munns and Tester 2008). Reducing water absorption leads to a decrease in water pressure in the cell, resulting in a reduction in the size of the cell, which ultimately reduces the size of the plant (Acosta-Motos et al. 2017). The decrease in water entry to the plant due to its low osmotic potential and salinity stress leads to the stomatal closure in order to minimize the plant transpiration. The reduction in stomatal conductance reduces the amount of available carbon dioxide for carboxylation, reduces photosynthetic assimilation, and thus reduces the production of sufficient assimilates to grow (Geissler et al. 2009; Gupta and Huang 2014). With the continuation of salinity stress (Phase II), ionic effects arise from the accumulation of salts in the plant (Munns and Tester 2008). The entry of harmful salt into the plant causes toxicity in the leaves that are associated with yellowing and aging of them. The second stage occurs when salts accumulate in toxic levels in the old leaves, which is due to the lack of salt transfer for long periods of time to vacuole. This stage requires more time to develop than an osmotic phase. It has a negative effect on growth by reducing the supply of carbohydrates to the growing cells. This results in a further reduction in photosynthesis and ultimately a further decline in growth (Munns et al. 2006; Munns and Tester 2008).

Effect of Salinity on Plant Germination

The germination stage is very important because of the direct impact on the plant density (Hatfield and Prueger 2015). Ions in the soil or water can act as a stimulant or inhibitor in germination at this stage. Salinity can affect the seeds germination by reducing the osmotic potential of the growth medium, the toxicity of certain ions, such as sodium, chlorine, and reduced nutrient ions such as calcium and potassium (Afzali et al. 2011). These factors change the biochemical and physiological activity of the seed by preventing aerobic respiration or stimulating the catabolic stages. Plant tolerance to salinity is not a fixed feature and may vary in different stages of growth for different species (Shrivastava and Kumar 2015). According to the researches, by increasing the concentration of NaCl, the germination of many plants will decrease (Seffino 1998; Bae et al. 2006; Almodares et al. 2007).

Effects of Salinity on Plant Morphology

Salinity causes the stop in the vegetative growth of plants (Newell 2013). Munns (2002) generally proposed two phases of growth reduction in plants exposed to salinity. In the first phase, the decline in growth is very rapid and this is due to a kind of osmotic effect. In the second phase, the growth retardation is very slow and this is due to the accumulation of salt in the leaves, which leads to the toxicity of salts in the plants. In the case of salinity continuity, it may lead to the death of leaves, the reduction of the photosynthetic leaf area, the reduction of photosynthetic reserves in plants and, consequently, reduction of the yield. Colorless leaves and dry weight loss are the significant symptoms of salinity stress (Asghari and Ahmadvand 2018) because high concentrations of sodium in chromoplasts can stop photosynthesis. The effects of salinity appear more often in the old leaves. The lack of growth of the roots is another sign of salinity. Salinity also led to the suspension of the apical meristem, reduction in the diameter of the roots and their vascular tissue (Otitoloju 2016). One of the rapid effects of salinity stress in plants is the reduction of leaf area, which inhibits leaf growth and associated with an increase in salt concentration (Acosta-Motos et al. 2017). Salinity stress significantly reduces the fresh and dry weight of shoots, roots, and leaves (Chartzoulakis and Klapaki 2000). Salinity also reduces the intracellular space in the leaves (Chaves et al. 2009). In a research on rooted cuttings of kiwifruit under hydroponic conditions exposed to different levels of salinity (5, 10, 15, 20, and 30 mM sodium chloride) for 2 months, it was observed that leaf area and number of leaves, shoot growth, and plant dry weight decreased with increasing salt concentration (Chartzoulakis et al. 1995). In a study, Fisarakis et al. (2001) investigated the effect of different salinity concentrations (5, 25, 50 and 100 mM) on "Sultana" grapes cultivar in a 70-day period, they observed that shoot growth, leaf area, leaf number, and total dry weight were decreased at all salinity levels.

Effect of Salinity on Vegetative Growth

The most important effect of salinity is plant growth disturbance (Jouyban 2012). In spite of the study of the effect of salinity on the growth and metabolism of plants at the large level by researchers, there are two views in this regard. Some researchers have reported that salinity damage on the plant is due to the reduction of soil water potential due to the accumulation of salts (osmotic damage) and the formation of physiological dryness in the root environment, and another group considers the toxicity of ion as a cause of salinity damage (Gupta and Huang 2014). Munns (2002) states that salinity-resistant plants do not allow sodium and chloride to be transported to leaves and they have the ability to embed these ions in the vacuole to prevent accumulation in the cytoplasm or cell wall, thereby preventing ionic toxicity.

In the presence of high levels of salts in the plant environment, available water is reduced and on the one hand, it restricts the cell division and the volume of cells. In this way, leaf area expansion decreases and this decrease also affects the reduction in light absorption (Wang et al. 2001; Shrivastava and Kumar 2015). Salt damage to the leaves is one of the first signs of NaCl toxicity. In the salt-sensitive plants such as rice, even a small amount of foreign salt (50 mL of salt) can accumulate large quantities of salt (500 mmol) in the leaf apoplasts and cause damage to it. The reason for this is poor salt control in the transpiration route (Ahmad and Sharma 2008).

Effect of Salinity on Chlorophyll and Carotenoids

Leaf chlorophyll and carotenoid contents are generally decreased under salinity stress (Parida and Das 2005; Ashraf and Harris 2013). The oldest leaves begin to develop chlorosis, and they will fall with the prolonged stress. Sodium has been reported to cause chlorophyll degradation (Pinheiro et al. 2008; Li et al. 2011; Yang et al. 2011). Although salt stress reduces chlorophyll content, this reduction depends on the tolerance and resistance of the plant species (Ashraf and Harris 2013). As salt stress increases the chlorophyll content in the tolerant plants, while in the sensitive ones, it decreases chlorophyll content significantly (Khan et al. 2009; Akram and Ashraf 2011). Some researchers have reported the amount of chlorophyll as an indicator for salt tolerance, while in a number of studies, the correlation between salt tolerance and chlorophyll content was not observed (Ashraf and Harris 2013). Gomathi and Rakkiyapan (2011) reported that carotenoids were reduced by salinity stress, but resistant cultivars had more carotenoids than susceptible ones. In another study, the level of carotenoids was positively correlated with the salinity stress tolerance and was introduced as an index for assessing the salt tolerance (Ziaf et al. 2009).

Effect of Salinity on Photosynthesis

Environmental stresses degrade the photosynthetic pigments, resulting in reduced photo-reception efficiency of photosystems (PSI and PSII) which reduces photosynthesis (Geissler et al. 2009; Zhen-hua et al. 2012). Absorbed light by chlorophyll is consumed for photosynthesis, or converted to thermal energy or released as chlorophyll fluorescence. These three processes compete together so that the increase in one is accompanied by a decrease in other processes (Maxwell and Johnson 2000). Part of the light energy absorbed by chlorophyll which converts to chlorophyll fluorescence is not more than 1–2% of the total light absorbed by chlorophyll. But its measurement, especially in the stressed plants, provides valuable insights into the use of energy excited by photosystem II and other thylakoid membrane proteins. This method is also a fast and nondestructive method for the plant (Ashraf and Harris 2013).

Since chlorophyll a fluorescence analysis is an efficient, rapid, and sensitive method, it is widely used as a tool for nondestructive examination of plant photosynthesis, under different abiotic stresses like temperature, high light intensity, drought, salinity, nutrient deficiency, herbicide or toxicity of air pollutants, and heavy metals (Venkatesh et al. 2012). Fluorescence induction parameters, such as initial fluorescence (F_0), maximum fluorescence (F_m) and variable fluorescence (F_v), and especially their ratio, are commonly used to determine the number of metabolic disorders in the leaves of many plant species under different environmental stresses (Baker 2008; Kalaji et al. 2012). The maximum quantum efficiency of the photosystem II, which is shown by the F_v/F_m ratio, is an important parameter used to measure the linear transmission rate of an electron and is an index of total photosynthetic capacity (Balouchi 2010). F_v/F_m in healthy leaves under the stress conditions in many plants is about 0.8, and the decrease in this ratio indicates that part of the photosystem II reaction center is damaged and this event, called photoinhibition, is usually observed in plants under stress (Vaz and Sharma 2011).

There are many reports that salt stress has an adverse effect on plant growth and has attributed this phenomenon to reduced photosynthetic efficiency (Wu et al. 2010; Akram and Ashraf 2011; Silva et al. 2011). Still, its mechanisms are not completely clear, but because of the fact that photosystem II plays a key role in the plant's photosynthetic response to environmental conditions, the effect of salinity stress on photosystem II has been studied in a large number of studies (Ashraf and Harris 2013). Some studies have shown that salinity significantly reduces the activity of photosystem II (Akram and Ashraf 2011; Saleem et al. 2011), but in another study, such a significant decrease has not been observed (Abdeshahian et al. 2010). The effect of salinity stress on F_v/F_m depends on tolerance to salinity and varies between species or even among cultivars of a species. Some researchers report that in the early stages of salt stress, there is no major change in the maximum quantum yield because salinity initially decreases stomatal conductance, photosynthesis is reduced as a result, and therefore the photosystem II is not affected (Baker and Rosenqvist 2004; Morales et al. 2008).

Effect of Salinity on Nutrients

Several reports indicate that salinity reduces the absorption and accumulation of nutrients in plants. Nutrition disturbances may result in effects of salinity on the availability of nutrients, competitive absorption, transport, or distribution in the plant (Parihar et al. 2015). The lack of micronutrients under salt stress is very common due to high pH (Das 2014). Numerous studies have shown that salinity can reduce nitrogen accumulation in plants. Decreasing nitrogen absorption that occurs under salinity stress is due to the interaction between Na^+ and $NH4^+$ or between Cl^- and NO_3, which ultimately leads to reduced plant growth and yield. This decrease in NO_3 adsorption can be due to the antagonism effects of NO_3 with chlorine or due to reduced absorption of water under salt stress conditions (Parihar et al. 2015).

The availability of phosphorus also decreases in saline soils, which can be due to the effects of ionic strength that decreases PO_4^{3-} activity, or because of controlling the concentration of phosphate in the soil solution controlled by adsorption processes or the low solubility of calcium phosphate minerals. Hence, phosphate concentration in crops decreases with increasing salinity (Prapaga et al. 2015).

With increasing salinity stress, potassium, calcium, and nitrogen decrease in leaves (Tuna et al. 2007). High levels of sodium chloride in the root environment can reduce the absorption of nutrients, especially potassium and calcium, resulting in imbalances in potassium, calcium, and magnesium ions (Keutgen and Pawelzik 2009). In other words, rising salt concentrations (especially sodium chloride) may have antagonistic effects with other important ions to feed the plant (Tavakkoli et al. 2011) or change important cationic and anionic ratios such as Na^+/K^+, Na^+/Ca^{2+}, and Cl^-/NO_3^- (Pessarakli 2016). Recently, it has been reported that in response to the salinity of sodium chloride, the concentration of calcium and magnesium in all plant organs has been declined steadily (Hussin et al. 2013).

Due to competition between Na^+ and Cl^- with nutrients such as K^+, Ca^{2+}, and NO_3^-, salinity may cause nutrient imbalances or deficiencies (Jouyban 2012). Increasing salt absorption leads to the toxicity of certain ions, such as Na, Cl, or S, which reduces the absorption of essential nutrients such as P, K, N, and Ca (Jouyban 2012; Parihar et al. 2015). It has been reported that plant growth decline in high sodium concentrations is due to a lack of potassium and calcium uptake (Mousa et al. 2013; Juan et al. 2005). Potassium is known as a sodium ion exchangeable ion in the cell membrane system (Akram et al. 2009) since sodium excretion in the cortex is associated with potassium entry. The presence of large amounts of Na prevents the absorption of nutrients such as K by plant tissues, which leads to an increase in the Na^+/K^+ ratio (Benlloch et al. 1994). Munns (2002) stated that, under salt stress conditions, increasing sodium entry to the plant and its accumulation in the cytoplasmic sodium to replace potassium and produce ionic toxicity. Maintaining a high ratio of K^+/Na^+ in plant tissues is essential for salinity tolerance and the cell requires a high K^+/Na^+ ratio to perform its natural activities (Khan et al. 2009; López-Aguilar et al. 2012).

Effect of Salinity on Na^+ and K^+ Concentration in Plant and Its Relation with Salinity Tolerance

Many researchers have suggested that salinity increases sodium in the shoots and roots and reduces potassium in various plant organs. From the results of these studies, it seems that the basis of the tolerance of different cultivars is their ability to regulate sodium and chlorine adsorption so that the plant's ability to prevent sodium and chlorine adsorption is higher, its resistance will also be higher (D'Amelia et al. 2018). One of the effective mechanisms in salinity tolerance is the low Na/K ratio in root and shoot under salinity, which is obtained through the ability of seedlings to absorb potassium and prevent sodium entry to root (Kordrostami et al. 2016). So that, salinity-tolerant cultivars have different Na/K ratios in different organs and at different growth stages compared to salinity-sensitive cultivars (Sutcliffe and Baker 1981). The higher K/Na ratio increases plant salinity tolerance (Kordrostami et al. 2017). Salinity reduces the K/Na ratio in the plant, and there are significant differences between salinity-sensitive and salinity-tolerant cultivars for this trait (D'Amelia et al. 2018). In a study on rice in hydroponic conditions, Kordrostami et al. (2016) reported that under salinity conditions, the accumulation of sodium in seedlings of all genotypes doubled, and the accumulation of other elements, such as potassium, decreased significantly. However, as mentioned earlier, in this study, the tolerant cultivars accumulated less Na content in their shoots and roots.

Effect of Salinity on Osmolites and Osmoprotectants

During the stress period, plants need to keep down their internal water potential to maintain the turgor pressure of their cells and absorb water from the soil to continue their growth (Negrão et al. 2017). To do this, the plant needs to provide osmotic regulating agents either from a soil solution or by itself synthesizing these materials. To maintain ionic balance in vacuoles, plants accumulate low-molecular-weight compounds in the cytoplasm, which are called compatible solutes (Hayat et al. 2012). The reason for this naming is that they do not interfere with the normal biochemical reactions of the cell (Zhifang and Loescher 2003). Potassium ion (K^+) is one of the most important compatible osmolytes (Yokoi et al. 2002).

The most commonly used compatible osmolytes include simple sugars (mainly glucose and fructose), complex sugars (raffinose, trehalose, and fructan), alcoholic sugars (methylated inositol and glycerol), Quinine amino acid derivatives (pyrimidine, glycine betaine, proline betaine, proline, and beta-alanine betaine), and sulfonamide compounds (dimethylsulfonium propionate and choline sulfate) (Pessarakli 2016). In glycophyte plants, from different compatible osmolytes, sugars up to more than 50% provide a total osmotic potential in salt stress conditions. Despite a decrease in the rate of photosynthesis, there are several reports about the accumulation of soluble carbohydrates in plants in response to salinity or drought stress (Ahmad and Sharma 2008). Carbohydrates, such as sucrose, fructose, glucose, and fructan, accumulate under the salinity stress and play an important role in osmotic regulation, osmotic protection, carbon storage, and also as free radicals scavenging (Sami et al. 2016). It has also been reported that in plants under salinity stress, amino acids (such as glycine, arginine, alanine, serine, leucine, valine, proline, ornithine, and citrulline) and amides (such as asparagine and glutamine) accumulate (Ahmad et al. 2013).

Proline is the key amino acid in osmotic regulation. In addition, proline is considered as a source of carbon and nitrogen, as well as a free radical scavenger (Hayat et al. 2012). Proline is more effective in protecting against the stress compared with other commonly compatible osmolytes (especially sugars and alcoholic sugars) (Hossain et al. 2014). In plant species, there is a positive correlation between the osmotic potential of the leaf and the accumulation of alanine betaine, proline betaine, and glycine betaine. These organic compounds have osmoprotective effects in the cells. Among these ammonium compounds, glycine betaine is more likely to accumulate in the cells (Slama et al. 2015). Glycine betaine is predominantly present in chloroplasts and plays an important role in protecting thylakoid membranes in the chloroplast, thereby enhancing photosynthesis efficiency and maintaining the health of plasma membranes (Fariduddin et al. 2013).

Effect of Salinity on Proline Content in Plants

Proline accumulates in plants that are exposed to severe drought and salinity conditions. Possibly, the production of proline in the plant should be the result of a nonspecific reaction to the low water potential. Proline is probably involved in osmotic regulation and maintenance of the enzymatic activity of the plant under salinity stress. In addition to the osmotic adjustment, this amino acid plays a role in stabilizing cellular components, absorbing free radicals, and various processes against stresses (Kordrostami et al. 2017; Saleem et al. 2011). The effect of salinity on proline accumulation has been reported in many plant species such as cape gooseberry (Miranda et al. 2014), okra (Saleem et al. 2011), and rice (Kordrostami et al. 2017).

Effect of Salinity on Cell Membrane Stability

The stability of cell membranes under stress conditions is considered as the main cause of osmotic stress tolerance. As a result of the oxidation of fatty acids that occur following salinity, cell membranes are damaged. As a result of damage to the cell membrane, the permeability of the membrane is elevated and thus the electrolyte leakage from the cell causes the plant to wilt (Liang et al. 2003). Singh et al. (1992) reported a negative relationship between the percentage of damage to cell membrane with seed yield. The results of Kordrostami et al. (2017) research on rice under salt stress showed that the relative water content of the leaves and stomatal conductance of plants decreased with increasing salinity and electrolyte leakage increased significantly in the same situation. Tuna et al. (2008) reported an increase in the electrolyte leakage of leaves in corn under salt stress conditions.

Salinity Tolerance in Plants

Salinity tolerance in plants not only varies widely among different species but is also strongly influenced by the environmental conditions. Many factors related to plant, soil, water, and environment interact with each other and affect the salinity tolerance in plants. As a result, the plant's reaction to a specific salt concentration cannot be fully predicted.

Salinity tolerance in plants is usually investigated in one of the three ways (Gupta and Huang 2014):

- The plant's ability to survive in saline soils
- Growth or absolute yield of the plant
- Relative growth or relative yield of the plant in a saline soil and comparison with non-saline soil

The salinity factor affects the plant during all the growth stages. But in some cases, the sensitivity of a growth stage varies from one stage to another. Comparison of plant tolerance to salinity at seed germination stage with other growth stages would be difficult to consider different criteria for testing the plant response. Tolerance in the emergence stage is based on the survival of the plant, whereas this stage is determined on the basis of reduction in growth or yield (Gregorio et al. 1997).

Salt Tolerance Mechanisms

The salinity tolerance mechanisms of the plant are investigated at three levels of whole plant, cellular, and molecular levels. Particularly, the response at the whole plant is vital for some plants but is generally not used for all plants. It seems that cellular responses are conserved among many plants (Munns 2002).

Salinity Tolerance at the Whole Plant

The first line of defense against sodium entry into the plant is the plasma membrane of root cells, which has low sodium permeability in all studied species. On the other hand, root cells have a tendency to absorb potassium, which can accumulate unlike the concentration slope (Munns 2002). The presence of high concentrations of salt in the root zone results in the absorption of both salt and nutrients by the plant. After transferring water and solutes through the direction of the symplast to the center of the root, they are transmitted immediately to the elements of the vascular tube through the xylem parenchyma cells. The entrance to the symplast pathway is the main control point of salt entry to the plant (Maathuis et al. 2014). The accumulation of sodium in the root and preventing its transfer to the shoot is a mechanism that is found in some plants to cope with salinity (Acosta-Motos et al. 2017). Jeschke and Wolf (1993) found that Egyptian bean, castor, and barley could maintain their K levels in root cells to an acceptable level (130 mM sodium chloride) and transfer the contained potassium in the leaf for use in the roots and growing points of the shoots. Some salt-absorbing plants have special tubers at their leaf or stem that can repel large amounts of salt. Salt tubers are known in at least 11 plant families (10 families of dicotyledons and 1 family of monocotyledons). Many salinity-tolerant plants, by increasing the amount of water in their mesophilic cells, can tolerate the temporary increase in salt in apoplasts (Munns 2002).

Salinity Tolerance at Molecular Level

Salt tolerance is a complex and quantitative genetic phenomenon controlled by a large number of genes (Lang et al. 2017). The main difference in the relative tolerance of most plant species to salinity is in their membrane enzymes, which contribute to ion transfer in the cell membranes and tonoplast (Almeida et al. 2017). To transfer the ions throughout the membranes, the H^+-ATPase enzyme in the membrane and the tonoplast H^+-pyrophosphatase enzyme are required to produce hydrogen-dependent electrochemical differences. These antiporters transfer the excess salt from apoplast to cytoplasm and there transfer sodium to vacuole for the osmotic regulation (Segami et al. 2018). Studies have shown that salt regulates the expression of the H^+-ATPase gene in the plasma membrane of tobacco and *Atriplex nummularia*. In Atriplex, mRNA was induced by H^+-ATPase in growing roots and aged leaves by salt, but not in young leaves or stems. The mRNA response to salinity stress was a multigene or transcriptional effect that occurs at a specific point in the cell cycle (Niu et al. 1993). Research shows that three SOS proteins are the main component of the ion homeostasis signal control pathway (Ji et al. 2013). SOS1 has been identified as a plasma membrane Na^+/K^+ antiporter agent involved in Na^+ excretion, which plays a vital role in sodium excretion from root cells and sodium transfer at long distances from root to shoot. SOS2 is a serine-threonine protein kinase which has an autophosphorylation function. SOS3 is also a calcium-bound protein, which is believed to respond to salinity-induced calcium instability in the cytosol. SOS3 physically interacts with the SOS2 regulatory unit and stimulates its protein kinase activity by the calcium-dependent method. Then the SOS2/SOS3 complex is phosphorylated and activates SOS1, which results in the removal of sodium ions, and thus leads to the ion balance of the cell to the first state (Zhou et al. 2014). Also, the study of the *Arabidopsis* mutants has been found that the *AtHKT1* gene determines salinity tolerance and controls the sodium entry and potassium uptake with high affinity. Expression of the vacuolar H^+-ATPase gene (*Avp1*) in *Arabidopsis* stimulates Na^+ uptake into vacuoles and increases salinity tolerance (Jiang et al. 2013).

Salinity Tolerance at the Cellular Level

The high salinity stress disrupts the distribution of ions at both cellular and whole plant levels and the critical balance of water potential. In addition, long-term salinity stress may cause molecular degradation, growth stoppage, and even death. Salinity tolerance in many plants is achieved through three interconnected mechanisms (Shokri-Gharelo and Noparvar 2018):

1. Damage caused by injury should be prevented or reduced (detoxification): Extremely high salinity stress is responsible for destroying cellular structures, inhibiting enzymatic activity, absorbing nutrients, and photosynthetic functions. Plants react to the effects of toxic molecules and ions by synthesizing stress proteins or compatible osmolytes, which are involved in detoxification (Agrawal et al. 2015). Osmolytes, under stress conditions, preserve thylakoids and plasma membranes, and in transgenic plants, they also increase stress tolerance (Suprasanna et al. 2016). These osmolytes protect the cell from the reactive oxygen activity produced by ionic stress causing membrane degradation and ulti-

mately cell death, and which effectively activates enzymes that can tolerate the stress. Most progress in salinity tolerance is achieved through gene transfer technology to enhance the scavenging strategy (Ashraf and Akram 2009). Proline is known to regulate osmotic pressure in the cytoplasm, which is effective in storing energy and nitrogen during salinity stress. Increasing proline concentration in osmotic regulation of bacteria and plants is considered as the most abundant reaction that increases tolerance to water scarcity and salinity stress (Forlani et al. 2018). In plant tissues under normal conditions, ornithine is considered as a precursor of proline which δ-ornithine-aminotransferase (δ-OAT) converts it into glutamic-γ-semi-aldehyde (GSA). Then, GSA converted into pyrroline-5-carboxylate (P5C) and then it is reduced to proline (Kavi Kishor and Sreenivasulu 2014). Proline is also compatible with other cytoplasmic components and can easily convert to glutamate. These changes are very important because glutamate contributes to the production of other essential amino acids. Therefore, under salinity stress, proline in the plant acts both as nitrogen storage and as an osmotic pressure regulator (Acosta-Motos et al. 2017). When salt concentration increases, after several hours, due to the specific DNA activity, stress proteins are produced (Bohnert et al. 1995). In addition to regulation of plant growth and development (including seed dormancy, embryogenesis, leaf transpiration, and root and branch growth), abscisic acid (ABA) as the most important plant hormone, plays a vital role in plant response to abiotic stresses such as salinity and drought and low temperature (Vishwakarma et al. 2017). ABA increases the growth during plant exposure to non-lethal primary salinity and induces high salinity compatibility (Negrão et al. 2017). The plant's adaptation to salinity stress may be associated with the accumulation of osmotins, a kind of cationic protein, in vacuole, cytoplasm, plasma membrane, and vacuolar membrane components (Hanin et al. 2016). Synthesis of osmotins is induced by abscisic acid. Environmental stresses, especially drought stress and salinity, alter the activity and amount of enzymes involved in scavenging the oxygen radicals and their constant level of mRNA (Khan et al. 2015).

2. The homeostatic condition should be re-established in a new stressful environment (homeostasis): Another way to gain more salinity tolerance is to help the plant to create osmotic and ionic homeostasis again in a stressful environment. The final determinants of such mechanisms are various ion transporters that distribute toxic ions at both organs and cellular levels (Deinlein et al. 2014). The accumulation of high levels of sodium in the cytoplasm inhibits the activity of many enzymes. The important issue of salinity tolerance studies is to determine which transporters interfere with the entry of sodium into the cell (Assaha et al. 2017).
3. Growth, although at a slow rate, should be continued (growth regulator): Like the other abiotic stresses, salt stress inhibits plant growth at the final stage. Slower growth may be an adaptive property of plant to survive under stress conditions because it allows plants to store resources, repair damaged structures, and restart physiological functions. Some plants display a rapid response to mild stress and stop their growth. In other words, plants that do not respond well to stress condi-

tions may not be able to survive (Basu et al. 2016). Maintaining turgor pressure above the turgor pressure of the root environment requires energy that is associated with reduced growth. Most crop species, through the transfer of salt into the membrane of some of the organelles (vacuoles), are, to some extent, able prevent the absorption of salt into the transpiration (Acosta-Motos et al. 2017). The poor osmotic regulation results in the loss of cellular turgor and stomatal conductance, leading to decreased gas exchange and photosynthesis. The loss of turgor pressure can also have adverse effects on cell division and elongation (Munns 2002).

Screening Methods for Salinity Tolerance

Increasing salt tolerance in crops allows for the more efficient use of low-quality irrigation water. Improving salt tolerance of crop species requires access to new genetic and efficient variation (both natural and transgenic) for identifying salinity-tolerant resources. Probably, there is likely to be a wide range of genetic variations in salinity tolerance in existing international collections that have not been discovered or have not been used. It is difficult to screen a large number of varieties for tolerance to salinity in the field due to spatial heterogeneity, chemical and physical properties of the soil, and seasonal fluctuations in rainfall; therefore, most screening methods are performed under controlled environments (Turan et al. 2012).

Improved salinity tolerance requires an understanding of the physiology of plants against stress. Today, the merging of new molecular methods with plant physiological topics provides a new perspective on improving salinity tolerance which results in the improvement of the efficiency of important plants for human health and the preservation of the environment and sustainable agriculture. Plants are largely different in salinity tolerance, which shows different responses in growth. Also, tolerance to salinity in dicotyledon species is more than monocotyledon species (Munns and Tester 2008).

In recent decades, significant advances have been made in salinity tolerance of cultivars through conventional breeding and selection methods. However, many selection methods have been based on differences in agronomic characteristics. Common agricultural selection parameters for salinity tolerance include plant survival, relative growth retardation, yield, leaf area and damage, plant height, relative and growth rate. When screening is performed for components that comprise complex traits, physiological criteria are able to provide more information than agronomic parameters or visual assessment (Ashraf and Harris 2004).

Stress Tolerance Indices Based on Biomass

Selection and isolation of stress-tolerant genotypes are performed by direct (performance measurement) and indirect methods (based on morphological and physiological traits that correlate with stress tolerance) (Singh 2015). Ashraf and Wu (1994) stated that if the growth rate of a plant under salt stress conditions is measured relative to the growth under normal conditions, it could be a suitable standard for salinity tolerance, and in this case, the dry matter of the plant is a useful criterion for evaluation of salt tolerance. Dry matter aggregation is the best criterion that correlates with grain yield (Romero-Aranda et al. 2001). Selection based on the yield of cultivars in both stressed and non-stressed environments leads to the selection of high yield cultivars under stress conditions, since that the optimal alleles are selected under stress conditions, and the response to selection in non-stressed conditions is higher due to higher heritability in these conditions (Singh 2015). Also, selection based on the performance of a cultivar under stress conditions may be independent of its performance under favorable conditions. To determine the reaction of different cultivars in two stressed and non-stressed environments, Fernandez (1993) identified four types of reaction for genotypes:

- Genotypes that have high performance in both stress and normal environments (Group A)
- Genotypes that have only good performance under normal conditions (Group B)
- Genotypes that have only good performance under stress conditions (Group C)
- Genotypes with poor performance in both stressed and normal environments (Group D)

The average yield and yield stability in different environments are the main criteria for selection in the plant breeding programs. Many types of research have been performed to evaluate yield under stress conditions and selection of genotypes adapted to both stressed and normal environments is an important breeding goal. In this regard, various indices have been proposed for selection of superior cultivars for stress conditions (Fernandez 1993; Rosielle and Hamblin 1981). One of the mentioned indices is the stress susceptibility index (SSI), which was first introduced by Fischer and Maurer (1978). The lower the index, the more stable performance of a cultivar under stress conditions. The low amount of this index indicates low genotype sensitivity to stress. Indeed, the lower amount of the index, the genotype has a higher stress tolerance. Selection based on this index selects genotypes that have a high yield under stress conditions, but have low yields under normal conditions. As a result, this index cannot separate the genotypes of Group A and C from each other. In the SSI index, due to the use of stress intensity constant (SI), despite the higher efficiency in choosing superior genotypes, different results can be obtained in different years and environments (Anagholi et al. 2010).

Mean Productivity (MP) and Tolerance (TOL) indices are introduced to evaluate the stress tolerance of genotypes (Rosielle and Hamblin 1981). Based on TOL, which is calculated as the difference between the yield under stressed and non-stressed environments, the genotypes with lower numerical values of this index are more tolerant. Selection based on this index leads to the selection of genotypes which have a low yield under normal conditions but have a potentially high yield under the stress conditions. As a result, this index is not able to separate Group C from Group A. The MP index also selects genotypes that have high yields under normal conditions and have low yields under the stress conditions. Selection of superior genotypes in these indices is based on lower TOL and higher MP. They made the necessary condition for establishing a positive and logical relationship between the TOL and MP indices in this regard that the genetic variation in the stressed environment is greater than the normal environment.

Geometric Mean Productivity (GMP) and Stress Tolerance Index (STI) were used to identify high yielding genotypes in both stressed and non-stressed environments (Fernandez 1993). In his view, the appropriate selection criteria are those that identify Group A from other groups. The higher the amount of STI, the more tolerant of the stress and the higher yield potential in the genotype, and the higher the GMP value, the more tolerant to stress. GMP was more powerful than other MPs in differentiating the genotypes of group A from other groups and based on this, the STI index was also proposed based on GMP (Fernandez 1993). Indices that have a high correlation with performance in both stress and normal conditions are selected as the best indices. In the STI index, due to the use of a geometric mean constant, there is more efficiency in selecting superior genotypes, but due to the multiplication of numbers, a fixed geometric mean square may exist for pairings of numbers that are obviously different (Anagholi et al. 2010).

Stress Tolerance Based on Physiological Indices

In order to evaluate the variability among cultivars for salt tolerance, researchers have used various physio-biochemical properties as an appropriate selection index (Cuartero et al. 2006; Munns et al. 2006; Akram and Jamil 2007). For example, Essa and Al-Ani (2001) screened six soybean genotypes at different levels of salinity using leaf chlorophyll content as a selection index. Pakniyat and Armion (2007) evaluated 28 sugar beet cultivars (*Beta vulgaris* L.) at different levels of salinity (sodium chloride) using Na^+ in the aerial parts of the plant as a selection index. Akram and Jamil (2007) screened 34 canola (*Brassica napus* L.) cultivars using net photosynthesis as a physiological salinity index in 150 mM sodium chloride. Hanachi et al. (2014) introduced the photosystem II efficiency as a selection index in evaluating salinity tolerance of eggplant varieties (*Solanum melongena* L.).

Plant Breeding for Salinity Stress Tolerance

There are three methods for improving salinity tolerance (Gupta and Huang 2014). The first method is breeding for high performance under normal conditions. Since the maximum genetic potential of yield is expected to achieve in non-stress conditions, and there is a high correlation between yield in stressed and non-stressed conditions, high yielding genotypes in non-stressed conditions will have a relatively high yield in stressed conditions. This is the main philosophy of this method. However, the concept of occurrence of maximum genetic potentials in non-stressed conditions is discussed because the genotype × environment interaction can prevent the emergence of high yielding genotype to high yield under salinity stress conditions. Therefore, the second method, breeding for high performance under real salinity stress conditions, has been proposed. But the problem with this method is that the intensity of salinity stress from 1 year to another and, consequently, the environmental selection pressure on the breeding material varies greatly from one generation to the next. This issue, coupled with the low inheritance of performance (yield), causes the complexity and slowdown of the breeding programs (Kafi and Damghani 2001).

A third method, which can be a substitute for the two above methods, is the breeding of salinity tolerance in high yielding genotypes by introducing salinity-tolerant morphological and physiological mechanisms. But the transfer of salt tolerance to highly yielding genotypes is complicated because the physiological and genetic basis of adaptation to salinity stress is completely unknown. Conversely, the breeding of the yield potential of a tolerant genotype can be a more promising way, provided that genetic variation exists within such a genotype. To achieve salinity-tolerant and high yield genotypes, simultaneous selection can be used for yield in non-stressed environments and for yield stability under salinity stress conditions (Roy 2000).

Necessity to Improve Salinity Tolerance in Plants

Since most crops are not able to grow at high salt concentrations, only salt-tolerant plants can grow at high concentrations of sodium chloride (about 400 mmol). As a result, salinity is a threat to human food supply. Although there is already enough food for the world's population, more than 800 million people are affected by chronic malnutrition. Almost half of the world's land is desert or dry land. Production in these areas is only possible through irrigation. Irrigation of these lands will be accompanied by their salinity. The primary value of increased tolerance to crop salinity is irrigation stability. It seems that the change in plant salinity tolerance is an important aspect of plant breeding in the future to maintain food production (Xu 2010). Flowers and Flowers (2005) proposed five strategies for the production of salinity-tolerant plants.

- The cultivation of salt-tolerant plants as alternative products
- Use of hybridization between species for increasing salinity tolerance of crops
- Use of diversity among plants
- Genetic variation in plants using recurrent selection, mutation, or tissue culture
- Plant breeding for yield instead of breeding for tolerance

Considering the advances made in recent decades, breeding for increased tolerance through gene transfer and the production of transgenic plants can be added to this list.

Biotechnological Approaches for Salinity Tolerance

Biotechnology and genetic engineering are new knowledge whose first achievements are valued in an aura of fear and hope. Throughout history, many of the scientific phenomena in the initial stage have been faced with severe hesitation and resistance (Wani et al. 2015). Some biotechnological methods include:

(a) Inventing and developing the application of DNA and protein-based molecular markers, including the construction of genetic maps, study the genetic diversity of plants and select the appropriate parents using molecular markers (Kordrostami and Rahimi 2015).
(b) The production of transgenic crops using recombinant genes derived from very diverse living organisms (Berglund et al. 2017).

Genetic manipulation methods in crop plants have been used to identify salinity tolerance genes and their transfer (Roy et al. 2011). Transgenic plants have recently become available, but none have been tested on the farm, and the commercial use of these plants is still a long way (Key et al. 2008). Therefore, it is better to apply research in other fields such as selection based on molecular markers and their integration with breeding programs. Perhaps the most valuable outcome of the biotechnology program is to use molecular tools for the breeding programs (Kordrostami and Rahimi 2015). By discovering the ability to extract plant DNA and its cleavage by using the enzymes, sequencing the bases, the ability to understand DNA properties, the combination of DNA technology, and the progression of statistical methods, parts of the chromosome that controls quantitative traits were identified, which are called quantitative trait loci (QTLs) (Ghomi et al. 2013). In principle, using QTL, we can identify markers for plant selection. Several studies have shown that QTL depends on developmental stages. However, the success of any breeding program, especially for salinity tolerance, depends on the proper method of evaluation or screening (Abarshahr et al. 2011).

Breeding for salinity tolerance has been studied by many researchers (Ghomi et al. 2013; Kordrostami et al. 2016; Asghari and Ahmadvand 2018). The success achieved in the past has not been remarkable due to the complexity of work for improving salinity tolerance (Munns et al. 2006), lack of real sense of urgency to breed it, inadequate genetic diversity for salinity tolerance, the complexity of salinity × environmental factors interactions, and the lack of efficient selective techniques. At present, with the advances in germplasm modification, evaluation techniques, genetic inheritance, molecular markers, mapping, and software techniques, it has now facilitated improvements in salt tolerance and other abiotic stresses (Deshmukh et al. 2014). Salinity tolerance like other environmental stresses in the higher plants is a complex genetic and physiological trait. Most of the processes in a plant that are important in salinity tolerance have a few inheritances, show the continuous diversity, and are affected by environmental conditions (Kordrostami et al. 2016; Negrão et al. 2017). Effective genes in the transfer of sodium and potassium are different and their absorption is carried out in two different and independent ways (Blumwald 2000). Principally, sodium absorption in apoplasts and potassium transfer is a membrane-based process (Wakeel et al. 2011). Identifying tightly linked molecular markers with the target gene and mapping it on the chromosome is an important goal for cloning the genes and marker-assisted selection (MAS) (Kordrostami and Rahimi 2015). Mapping or tagging studies provide information on the number of genes controlling the trait and the location of these genes on the linkage map (Ghomi et al. 2013).

Conclusion

Salinity is an increase in the concentration of dissolved inorganic salts in the soil or water which affects the growth of plants in different stages (from germination to ripening) in varying degrees. The most important result of salinity stress is the apparent stop in plant growth. The immediate response to salinity stress is the decrease in the rate of leaf area expansion, which results in stopping the increase in salt concentration in the plant. Also, salinity stress has a significant role in reducing photosynthesis in plants. It has a negative effect on growth by limiting the supply of carbohydrates to the growing cells. This results in a further reduction in photosynthesis and ultimately a further decline in growth. Salt tolerance is a complex genetic and physiological trait that has little inheritance. Salinity tolerance in plants not only varies widely among different species but is also strongly influenced by the environmental conditions. Many factors related to plant, soil, water, and environment interact with each other and affect the salinity tolerance in plants. The salinity tolerance mechanisms of the plant are investigated at three levels of whole plant, cellular, and molecular levels. Particularly, the response at the whole plant is vital for some plants but is generally not used for all plants. It seems that cellular responses are conserved among many plants. There are three methods for improving salinity tolerance. The first method is breeding for high performance under normal conditions. The second method is breeding for high performance under real salinity stress conditions. A third method, which can be a substitute for the two above methods, is the breeding of salinity tolerance in high yielding genotypes by introducing salinity-tolerant morphological and physiological mechanisms. Considering the advances made in recent decades, breeding for increased tolerance through gene transfer and the production of transgenic plants can be added to this list. Perhaps the most valuable outcome of the biotechnology program is to use molecular tools for the breeding programs. Identifying tightly linked molecular markers with the target gene and mapping it on the chromosome is an important goal for cloning the genes and marker-assisted selection (MAS). Mapping or tagging studies provide information on the number of genes controlling the trait and the location of these genes on the linkage map.

References

Abarshahr M, Rabiei B, Lahigi HS (2011) Assessing genetic diversity of rice varieties under drought stress conditions. Not Sci Biol 3:114–123

Abdeshahian M, Nabipour M, Meskarbashee M (2010) Chlorophyll fluorescence as criterion for the diagnosis salt stress in wheat (*Triticum aestivum*) plants. Int J Chem Biol Eng 4:184–186

Acosta-Motos JR, Ortuño MF, Bernal-Vicente A, Diaz-Vivancos P, Sanchez-Blanco MJ, Hernandez JA (2017) Plant responses to salt stress: adaptive mechanisms. Agronomy 7:18. https://doi.org/10.3390/agronomy7010018

Afzali S, Shariatmadari H, Hajabbasi M (2011) Sodium chloride effects on seed germination, growth and ion concentration in chamomile (*Matricaria chamomilla*). Iran Agric Res 29:107–118

Agrawal C, Sen S, Chatterjee A, Rai S, Yadav S, Singh S, Rai L (2015) Signal perception and mechanism of salt toxicity/tolerance in photosynthetic organisms: cyanobacteria to plants. In: Bhumi NT, Maria M (eds) Stress responses in plants. Springer, Cham, pp 79–113

Ahmad P, Sharma S (2008) Salt stress and phyto-biochemical responses of plants. Plant Soil Environ 54:89–99

Ahmad P, Azooz M, Prasad M (2013) Salt stress in plants. Springer, Dordrecht

Akram MS, Ashraf M (2011) Exogenous application of potassium dihydrogen phosphate can alleviate the adverse effects of salt stress on sunflower. J Plant Nutr 34:1041–1057

Akram NA, Jamil A (2007) Appraisal of physiological and biochemical selection criteria for evaluation of salt tolerance in canola (*Brassica napus* L.). Pak J Bot 39:1593–1608

Akram MS, Ashraf M, Akram NA (2009) Effectiveness of potassium sulfate in mitigating salt-induced adverse effects on different physio-biochemical attributes in sunflower (*Helianthus annuus* L.). Flora-Morpho Dist Funct Ecolo Plant 204:471–483

Almeida DM, Oliveira MM, Saibo NJ (2017) Regulation of Na^+ and K^+ homeostasis in plants: towards improved salt stress tolerance in crop plants. Genet Mol Biol 40:326–345

Almodares A, Hadi M, Dosti B (2007) Effects of salt stress on germination percentage and seedling growth in sweet sorghum cultivars. J Biol Sci 7:1492–1495

Anagholi A, Tabatabaei S, Fouman A (2010) Evaluation of salinity tolerance of forage sorghum varieties with stress tolerance and susceptibility indices. Electron J Crop Prod 3:839–841

Asghari R, Ahmadvand R (2018) Salinity stress and its impact on morpho-physiological characteristics of Aloe Vera. Pertanika J Trop Agric Sci 41:411–422

Ashraf M, Akram NA (2009) Improving salinity tolerance of plants through conventional breeding and genetic engineering: an analytical comparison. Biotechnol Adv 27:744–752

Ashraf M, Harris P (2004) Potential biochemical indicators of salinity tolerance in plants. Plant Sci 166:3–16

Ashraf M, Harris P (2013) Photosynthesis under stressful environments: an overview. Photosynthetica 51:163–190

Ashraf M, Wu L (1994) Breeding for salinity tolerance in plants. Crit Rev Plant Sci 13:17–42

Assaha DV, Ueda A, Saneoka H, Al-Yahyai R, Yaish MW (2017) The role of Na^+ and K^+ transporters in salt stress adaptation in glycophytes. Front Physiol 8:509. https://doi.org/10.3389/fphys.2017.00509

Bae D, Yong K, Chun S (2006) Effect of salt (NaCl) stress on germination and early seedling growth of four vegetables species. J Cent Eur Agr 7:273–282

Baker NR (2008) Chlorophyll fluorescence: a probe of photosynthesis in vivo. Annu Rev Plant Biol 59:89–113

Baker NR, Rosenqvist E (2004) Applications of chlorophyll fluorescence can improve crop production strategies: an examination of future possibilities. J Exp Bot 55:1607–1621

Balouchi H (2010) Screening wheat parents of mapping population for heat and drought tolerance, detection of wheat genetic variation. Int J Biol Sci 6:56–66

Basu S, Ramegowda V, Kumar A, Pereira A (2016) Plant adaptation to drought stress. F1000Res 5:1–10
Benlloch M, Ojeda M, Ramos J, Rodriguez-Navarro A (1994) Salt sensitivity and low discrimination between potassium and sodium in bean plants. Plant and Soil 166:117–123
Berglund LE, Petersen TE, Fedosov SN, Nexo E, Laursen NB, Jensen EO (2017) Transgenic plants expressing cobalamin binding proteins. Google Patents
Blumwald E (2000) Sodium transport and salt tolerance in plants. Curr Opin Cell Biol 12:431–434
Bohnert HJ, Nelson DE, Jensen RG (1995) Adaptations to environmental stresses. Plant Cell 7:1099–1111
Calow R, MacDonald A, Le Sève MD (2018) The environmental dimensions of universal access to safe water. In: Oliver C, Tom S (eds) Equality in water and sanitation services. Routledge, London, pp 110–132
Chartzoulakis K (2011) The use of saline water for irrigation of olives: effects on growth, physiology, yield and oil quality. Acta Hortic 888:97–108
Chartzoulakis K, Klapaki G (2000) Response of two greenhouse pepper hybrids to NaCl salinity during different growth stages. Sci Hortic 86:247–260
Chartzoulakis K, Therios I, Misopolinos N, Noitsakis B (1995) Growth, ion content and photosynthetic performance of salt-stressed kiwifruit plants. Irrig Sci 16:23–28
Chaves MM, Flexas J, Pinheiro C (2009) Photosynthesis under drought and salt stress: regulation mechanisms from whole plant to cell. Ann Bot 103:551–560
Cuartero J, Bolarin M, Asins M, Moreno V (2006) Increasing salt tolerance in the tomato. J Exp Bot 57:1045–1058
D'Amelia L, Dell'Aversana E, Woodrow P, Ciarmiello LF, Carillo P (2018) Metabolomics for crop improvement against salinity stress. In: Vinay K, Shabir HW, Penna S, Lam-Son PT (eds) Salinity responses and tolerance in plants. Springer, Cham, pp 267–287
Das SK (2014) Role of micronutrient in rice cultivation and management strategy in organic agriculture—a reappraisal. Agric Sci 5:765–769
Dawood MG (2018) Stimulating plant tolerance against abiotic stress through seed priming. In: Amitava R, Harikesh BS (eds) Advances in seed priming. Springer, Singapore, pp 147–183
Deinlein U, Stephan AB, Horie T, Luo W, Xu G, Schroeder JI (2014) Plant salt-tolerance mechanisms. Trends Plant Sci 19:371–379
Deshmukh R et al (2014) Integrating omic approaches for abiotic stress tolerance in soybean. Front Plant Sci 5:244. https://doi.org/10.3389/fpls.2014.00244
Essa TA, Al-Ani DH (2001) Effect of salt stress on the performance of six soybean genotypes. Pak J Biol Sci 4:175–177
Fariduddin Q, Varshney P, Yusuf M, Ali A, Ahmad A (2013) Dissecting the role of glycine betaine in plants under abiotic stress. Plant Stress 7:8–18
Fernandez GC (1993) Effective selection criteria for assessing plant stress tolerance. In: Proceedings of the international symposium on "adaptation of vegetables and other food crops in temperature and water stress", 13-181, 992, 257, 270
Fisarakis I, Chartzoulakis K, Stavrakas D (2001) Response of Sultana vines (*V. vinifera* L.) on six rootstocks to NaCl salinity exposure and recovery. Agr Water Manage 51:13–27
Fischer R, Maurer R (1978) Drought resistance in spring wheat cultivars. I. Grain yield responses. Aust J Agr Res 29:897–912
Flowers T, Flowers S (2005) Why does salinity pose such a difficult problem for plant breeders? Agric Water Manag 78:15–24
Forlani G, Bertazzini M, Cagnano G (2018) Stress-driven increase in proline levels, and not proline levels themselves, correlates with the ability to withstand excess salt in a group of 17 Italian rice genotypes. Plant Biol. https://doi.org/10.1111/plb.12916
Geissler N, Hussin S, Koyro H-W (2009) Interactive effects of NaCl salinity and elevated atmospheric CO_2 concentration on growth, photosynthesis, water relations and chemical composition of the potential cash crop halophyte *Aster tripolium* L. Environ Exp Bot 65:220–231
Ghomi K, Rabiei B, Sabouri H, Sabouri A (2013) Mapping QTLs for traits related to salinity tolerance at seedling stage of rice (*Oryza sativa* L.): an agrigenomics study of an Iranian rice population. OMICS 17:242–251

Gomathi R, Rakkiyapan P (2011) Comparative lipid peroxidation, leaf membrane thermostability, and antioxidant system in four sugarcane genotypes differing in salt tolerance. Int J Plant Physiol Biochem 3:67–74

Gregorio GB, Senadhira D, Mendoza RD (1997) Screening rice for salinity tolerance. IRRI discussion paper series

Gupta B, Huang B (2014) Mechanism of salinity tolerance in plants: physiological, biochemical, and molecular characterization. Int J Genomics 2014:701596. https://doi.org/10.1155/2014/701596

Hanachi S, Van Labeke M-C, Mehouachi T (2014) Application of chlorophyll fluorescence to screen eggplant (*Solanum melongena* L.) cultivars for salt tolerance. Photosynthetica 52:57–62

Hanin M, Ebel C, Ngom M, Laplaze L, Masmoudi K (2016) New insights on plant salt tolerance mechanisms and their potential use for breeding. Front Plant Sci 7:1787. https://doi.org/10.3389/fpls.2016.01787

HanumanthaRao B, Nair RM, Nayyar H (2016) Salinity and high temperature tolerance in mungbean [*Vigna radiata* (L.) Wilczek] from a physiological perspective. Front Plant Sci 7:957. https://doi.org/10.3389/fpls.2016.00957

Hardie M, Doyle R (2012) Measuring soil salinity. In: Shabala S, Cuin T (eds) Plant salt tolerance. Methods in molecular biology (Methods and protocols). Humana Press, Totowa, pp 415–425

Hatfield JL, Prueger JH (2015) Temperature extremes: effect on plant growth and development. Weather Clim Extrem 10:4–10

Hatfield JL, Sauer TJ, Cruse RM (2017) Soil: the forgotten piece of the water, food, energy nexus. In: Donald LS (ed) Advances in agronomy. Academic Press, New York, pp 1–46

Hayat S, Hayat Q, Alyemeni MN, Wani AS, Pichtel J, Ahmad A (2012) Role of proline under changing environments: a review. Plant Signal Behav 7:1456–1466

Hossain MA, Hoque MA, Burritt DJ, Fujita M (2014) Proline protects plants against abiotic oxidative stress: biochemical and molecular mechanisms. In: Parvaiz A (ed) Oxidative damage to plants, antioxidant networks and signaling. Academic Press, New York, pp 477–522

Hussin S, Geissler N, Koyro H-W (2013) Effect of NaCl salinity on *Atriplex nummularia* (L.) with special emphasis on carbon and nitrogen metabolism. Acta Physiol Plant 35:1025–1038

Jeschke WD, Wolf O (1993) Importance of mineral nutrient cycling for salinity tolerance of plants. In: Lieth H, Al Masoom AA (eds) Towards the rational use of high salinity tolerant plants. Springer, Dordrecht, pp 265–277

Ji H, Pardo JM, Batelli G, Van Oosten MJ, Bressan RA, Li X (2013) The salt overly sensitive (SOS) pathway: established and emerging roles. Mol Plant 6:275–286

Jiang C, Belfield EJ, Cao Y, Smith JAC, Harberd NP (2013) An Arabidopsis soil-salinity–tolerance mutation confers ethylene-mediated enhancement of sodium/potassium homeostasis. Plant Cell 25(9):3535–3552. https://doi.org/10.1105/tpc.113.115659

Jouyban Z (2012) The effects of salt stress on plant growth. Tech J Eng Appl Sci 2:7–10

Juan M, Rivero RM, Romero L, Ruiz JM (2005) Evaluation of some nutritional and biochemical indicators in selecting salt-resistant tomato cultivars. Environ Exp Bot 54:193–201

Kafi M (2009) The effects of salinity and light on photosynthesis, respiration and chlorophyll fluorescence in salt-tolerant and salt-sensitive wheat (*Triticum aestivum* L.) cultivars. J Agric Sci Technol 11:535–547

Kafi M, Mahdavi Damghani A (2001) Mechanisms of environmental stress resistance in plants (in Persian). Ferdowsi University Press, Mashad.

Kalaji HM, Carpentier R, Allakhverdiev SI, Bosa K (2012) Fluorescence parameters as early indicators of light stress in barley. J Photochem Photobiol B 112:1–6

Kanojia A, Dijkwel PP (2018) Abiotic stress responses are governed by reactive oxygen species and age. Annu Plant Rev. 1:1–32. https://doi.org/10.1002/9781119312994.apr0611

Kavi Kishor PB, Sreenivasulu N (2014) Is proline accumulation per se correlated with stress tolerance or is proline homeostasis a more critical issue? Plant Cell Environ 37:300–311

Keutgen AJ, Pawelzik E (2009) Impacts of NaCl stress on plant growth and mineral nutrient assimilation in two cultivars of strawberry. Environ Exp Bot 65:170–176

Key S, Ma JK, Drake PM (2008) Genetically modified plants and human health. Proc R Soc Med 101:290–298

Khan M et al (2009) Role of proline, K/Na ratio and chlorophyll content in salt tolerance of wheat (*Triticum aestivum* L.). Pak J Bot 41:633–638

Khan MS, Ahmad D, Khan MA (2015) Utilization of genes encoding osmoprotectants in transgenic plants for enhanced abiotic stress tolerance. Electron J Biotechnol 18:257–266

Kordrostami M, Rahimi M (2015) Molecular markers in plants: concepts and applications. G3M 13:4024–4031

Kordrostami M, Rabiei B, Kumleh HH (2016) Association analysis, genetic diversity and haplotyping of rice plants under salt stress using SSR markers linked to *SalTol* and morpho-physiological characteristics. Plant Syst Evol 302:871–890

Kordrostami M, Rabiei B, Kumleh HH (2017) Biochemical, physiological and molecular evaluation of rice cultivars differing in salt tolerance at the seedling stage. Physiol Mol Biol Plants 23:529–544

Lang L et al (2017) Quantitative trait locus mapping of salt tolerance and identification of salt-tolerant genes in *Brassica napus* L. Front Plant Sci 8:1000. https://doi.org/10.3389/fpls.2017.01000

Li X-J, Yang M-F, Zhu Y, Liang Y, Shen S-H (2011) Proteomic analysis of salt stress responses in rice shoot. J Plant Biol 54:384. https://doi.org/10.1007/s12374-011-9173-8

Liang Y, Chen Q, Liu Q, Zhang W, Ding R (2003) Exogenous silicon (Si) increases antioxidant enzyme activity and reduces lipid peroxidation in roots of salt-stressed barley (*Hordeum vulgare* L.). J Plant Physiol 160:1157–1164

López-Aguilar R, Medina-Hernández D, Ascencio-Valle F, Arce-Montoya M, Larrinaga-Mayoral JA (2012) Differential responses of Chiltepin (*Capsicum annuum* var. glabriusculum) and Poblano (*Capsicum annuum* var. annuum) hot peppers to salinity at the plantlet stage. Afr J Biotechnol 11:2642–2653

Maathuis FJ, Ahmad I, Patishtan J (2014) Regulation of Na^+ fluxes in plants. Front Plant Sci 5:467. https://doi.org/10.3389/fpls.2014.00467

Mahajan S, Tuteja N (2005) Cold, salinity and drought stresses: an overview. Arch Biochem Biophys 444:139–158

Manchanda G, Garg N (2008) Salinity and its effects on the functional biology of legumes. Acta Physiol Plant 30:595–618

Maxwell K, Johnson GN (2000) Chlorophyll fluorescence—a practical guide. J Exp Bot 51:659–668

Miranda D, Fischer G, Mewis I, Rohn S, Ulrichs C (2014) Salinity effects on proline accumulation and total antioxidant activity in leaves of the cape gooseberry (*Physalis peruviana* L.). J Appl Bot Food Qual 87:67–73

Misra AK (2014) Climate change and challenges of water and food security. Int J Sustain Built Environ 3:153–165

Morales F, Abadía A, AbadÞa J (2008) Photoinhibition and photoprotection under nutrient deficiencies, drought and salinity. In: Demmig-Adams B, Adams WW, Mattoo AK (eds) Photoprotection, photoinhibition, gene regulation, and environment. Springer, Dordrecht, pp 65–85

Mousa MA, Al-Qurashi AD, Bakhashwain AA (2013) Response of tomato genotypes at early growing stages to irrigation water salinity. J Food Agr Environ 11:501–507

Munns R (2002) Comparative physiology of salt and water stress. Plant Cell Environ 25:239–250

Munns R, Tester M (2008) Mechanisms of salinity tolerance. Annu Rev Plant Biol 59:651–681

Munns R, James RA, Läuchli A (2006) Approaches to increasing the salt tolerance of wheat and other cereals. J Exp Bot 57:1025–1043

Negrão S, Schmöckel S, Tester M (2017) Evaluating physiological responses of plants to salinity stress. Ann Bot 119:1–11

Newell N (2013) Effects of soil salinity on plant growth plant physiology. Available at https://pages.stolaf.edu/wpcontent/uploads/sites/253/2015/03/Salinity-in-Plants.pdf (accessed January 2019).

Niu G, Cabrera RI (2010) Growth and physiological responses of landscape plants to saline water irrigation: a review. HortSci 45:1605–1609

Niu X, Narasimhan ML, Salzman RA, Bressan RA, Hasegawa PM (1993) NaCl regulation of plasma membrane H^+-ATPase gene expression in a glycophyte and a halophyte. Plant Physiol 103:713–718

Otitoloju K (2016) Salt-induced modifications in the vegetative anatomy of bottonweed and peruvian spikesedge. Int J Mar Sci 6:47. https://doi.org/10.5376/ijms.2016.06.0047

Pakniyat H, Armion M (2007) Sodium and proline accumulation as osmoregulators in tolerance of sugar beet genotypes to salinity. Pak J Biol Sci 10:4081–4086

Pandey P, Irulappan V, Bagavathiannan MV, Senthil-Kumar M (2017) Impact of combined abiotic and biotic stresses on plant growth and avenues for crop improvement by exploiting physio-morphological traits. Front Plant Sci 8:537. https://doi.org/10.3389/fpls.2017.00537

Parida AK, Das AB (2005) Salt tolerance and salinity effects on plants: a review. Ecotoxicol Environ Saf 60:324–349

Parihar P, Singh S, Singh R, Singh VP, Prasad SM (2015) Effect of salinity stress on plants and its tolerance strategies: a review. Environ Sci Pollut Res 22:4056–4075

Pedrol N, González L, Reigosa MJ (2006) Allelopathy and abiotic stress. In: Reigosa M, Pedrol N, González L (eds) Allelopathy. Springer, Dordrecht, pp 171–209

Pessarakli M (2016) Handbook of plant and crop stress, 3rd edn. CRC press, Boca Raton

Pinheiro HA et al (2008) Leaf gas exchange, chloroplastic pigments and dry matter accumulation in castor bean (*Ricinus communis* L) seedlings subjected to salt stress conditions. Ind Crop Prod 27:385–392

Prapaga K, Dasina S, Shanika W (2015) Effect of different salinity levels of a soil on nutrient availability of manure amended soil. In: 5th international symposium, IntSym 2015, SEUSL

Qados AMA (2011) Effect of salt stress on plant growth and metabolism of bean plant *Vicia faba* (L.). J Saudi Soc Agr Sci 10:7–15

Rasool S, Hameed A, Azooz M, Siddiqi T, Ahmad P (2013) Salt stress: causes, types and responses of plants. In: Ahmad P, Azooz M, Prasad M (eds) Ecophysiology and responses of plants under salt stress. Springer, New York, pp 1–24

Rejeb IB, Pastor V, Mauch-Mani B (2014) Plant responses to simultaneous biotic and abiotic stress: molecular mechanisms. Plants 3:458–475

Rhoades J, Chanduvi F (1999) Soil salinity assessment: methods and interpretation of electrical conductivity measurements. Food and Agriculture Organization of the United Nations, Rome

Romero-Aranda R, Soria T, Cuartero J (2001) Tomato plant-water uptake and plant-water relationships under saline growth conditions. Plant Sci 160:265–272

Rosielle A, Hamblin J (1981) Theoretical aspects of selection for yield in stress and non-stress environment 1. Crop Sci 21:943–946

Roy D (2000) Plant breeding: analysis and exploitation of variation. Alpha Science Int'l Ltd, Oxford

Roy B, Noren S, Mandal AB, Basu AK (2011) Genetic engineering for abiotic stress tolerance in agricultural crops. Biotechnology 10:1–22

Saleem A, Ashraf M, Akram N (2011) Salt (NaCl)-induced modulation in some key physio-biochemical attributes in okra (*Abelmoschus esculentus* L.). J Agron Crop Sci 197:202–213

Sami F, Yusuf M, Faizan M, Faraz A, Hayat S (2016) Role of sugars under abiotic stress. Plant Physiol Biochem 109:54–61

Scherr SJ, McNeely JA (2008) Biodiversity conservation and agricultural sustainability: towards a new paradigm of 'ecoagriculture' landscapes. Philos Trans R Soc Lond B Biol Sci 363:477–494

Seffino LG (1998) Salinity effects on the early development stages of *Panicum coloratum*: cultivar differences. Grass Forage Sci 53:270–278

Segami S, Asaoka M, Kinoshita S, Fukuda M, Nakanishi Y, Maeshima M (2018) Biochemical, structural, and physiological characteristics of vacuolar H^+-pyrophosphatase. Plant Cell Physiol 59:1300–1308

Shokri-Gharelo R, Noparvar PM (2018) Molecular response of canola to salt stress: insights on tolerance mechanisms. PeerJ 6:e4822

Shrivastava P, Kumar R (2015) Soil salinity: a serious environmental issue and plant growth promoting bacteria as one of the tools for its alleviation. Saudi J Biol Sci 22:123–131

Silva EN, Ribeiro RV, Ferreira-Silva SL, Viégas RA, Silveira JAG (2011) Salt stress induced damages on the photosynthesis of physic nut young plants. Sci Agric 68:62–68

Singh B (2015) Plant breeding: principles and methods. Kalyani Publishers, New Delhi

Singh M, Srivastava J, Kumar A (1992) Cell membrane stability in relation to drought tolerance in wheat genotypes. J Agron Crop Sci 168:186–190

Slama I, Abdelly C, Bouchereau A, Flowers T, Savoure A (2015) Diversity, distribution and roles of osmoprotective compounds accumulated in halophytes under abiotic stress. Ann Bot 115:433–447

Suprasanna P, Nikalje G, Rai A (2016) Osmolyte accumulation and implications in plant abiotic stress tolerance. In: Iqbal N, Nazar R, Khan NA (eds) Osmolytes and plants acclimation to changing environment: emerging omics technologies. Springer, New Delhi, pp 1–12

Sutcliffe JF, Baker DA (1981) Plants and mineral salts. Edward Arnold (Publishers) Ltd, London

Tanji KK (2002) Salinity in the soil environment. In: Läuchli A, Lüttge U (eds) Salinity: environment-plants-molecules. Springer, Dordrecht, pp 21–51

Tavakkoli E, Fatehi F, Coventry S, Rengasamy P, McDonald GK (2011) Additive effects of Na^+ and Cl^- ions on barley growth under salinity stress. J Exp Bot 62:2189–2203

Tuna AL, Kaya C, Ashraf M, Altunlu H, Yokas I, Yagmur B (2007) The effects of calcium sulphate on growth, membrane stability and nutrient uptake of tomato plants grown under salt stress. Environ Exp Bot 59:173–178

Tuna AL, Kaya C, Dikilitas M, Higgs D (2008) The combined effects of gibberellic acid and salinity on some antioxidant enzyme activities, plant growth parameters and nutritional status in maize plants. Environ Exp Bot 62:1–9

Turan S, Cornish K, Kumar S (2012) Salinity tolerance in plants: breeding and genetic engineering. Aust J Crop Sci 6:1337–1348

Vaz J, Sharma PK (2011) Relationship between xanthophyll cycle and non-photochemical quenching in rice (*Oryza sativa* L.) plants in response to light stress. Indian J Exp Biol 49:60–67

Venkatesh J, Upadhyaya CP, Yu J-W, Hemavathi A, Kim DH, Strasser RJ, Park SW (2012) Chlorophyll a fluorescence transient analysis of transgenic potato overexpressing D-galacturonic acid reductase gene for salinity stress tolerance. Hort Environ Biotechnol 53:320–328

Vishwakarma K et al (2017) Abscisic acid signaling and abiotic stress tolerance in plants: a review on current knowledge and future prospects. Front Plant Sci 8:161. https://doi.org/10.3389/fpls.2017.00161

Wakeel A, Farooq M, Qadir M, Schubert S (2011) Potassium substitution by sodium in plants. CRC Crit Rev Plant Sci 30:401–413

Wang D, Shannon M, Grieve C (2001) Salinity reduces radiation absorption and use efficiency in soybean. Field Crop Res 69:267–277

Wani AS, Ahmad A, Hayat S, Fariduddin Q (2013) Salt-induced modulation in growth, photosynthesis and antioxidant system in two varieties of *Brassica juncea*. Saudi J Biol Sci 20:183–193

Wani S, Gaur A, Shikari A, Iqbal A, Pandita D (2015) Transgenic rice: advancements and achievements. Adv Genet Eng 4:1–3

Wu X-X, Ding H-D, Chen J-L, Zhang H-J, Zhu W-M (2010) Attenuation of salt-induced changes in photosynthesis by exogenous nitric oxide in tomato (*Lycopersicon esculentum* Mill. L.) seedlings. Afr J Biotechnol 9:7837–7846

Xu Y (2010) Molecular plant breeding. Cabi, Wallingford

Yang J, Zheng W, Tian Y, Wu Y, Zhou D (2011) Effects of various mixed salt-alkaline stresses on growth, photosynthesis, and photosynthetic pigment concentrations of *Medicago ruthenica* seedlings. Photosynthetica 49:275–284

Yokoi S, Quintero FJ, Cubero B, Ruiz MT, Bressan RA, Hasegawa PM, Pardo JM (2002) Differential expression and function of *Arabidopsis thaliana* NHX Na^+/H^+ antiporters in the salt stress response. Plant J 30:529–539

Zhen-hua Z, Qiang L, Hai-xing S, Xiang-min R, Ismail AM (2012) Responses of different rice (*Oryza sativa* L.) genotypes to salt stress and relation to carbohydrate metabolism and chlorophyll content. Afr J Agr Res 7:19–27

Zhifang G, Loescher W (2003) Expression of a celery mannose 6-phosphate reductase in *Arabidopsis thaliana* enhances salt tolerance and induces biosynthesis of both mannitol and a glucosyl-mannitol dimer. Plant Cell Environ 26:275–283

Zhou H et al (2014) Inhibition of the *Arabidopsis* salt overly sensitive pathway by 14-3-3 proteins. Plant Cell 26(3):1166–1182. https://doi.org/10.1105/tpc.113.117069

Ziaf K, Amjad M, Pervez MA, Iqbal Q, Rajwana IA, Ayyub M (2009) Evaluation of different growth and physiological traits as indices of salt tolerance in hot pepper (*Capsicum annuum* L.). Pak J Bot 41:1797–1809

Water Deficit Stress Effects and Responses in Maize

Meeta Jain, Sunita Kataria, Mamta Hirve, and Rajkumar Prajapati

Introduction

Stress can be defined as an environmental factor capable of inducing a potentially injurious strain in living organisms. Plants are frequently exposed to biotic and abiotic stresses while growing in nature. Biotic stress involves damage to plants by other living organisms, such as bacteria, viruses, fungi, parasites, harmful insects, and weeds. Abiotic stress is a negative impact of non-living factors on plants in specific environment, such as light, temperature, water availability, nutrients, and soil structure. In relation to the water availability for plants, two terms—drought and water deficit stress—are often used in practice; they both are related to the availability of soil moisture to the plant. Thus, both may be called as "soil moisture stress" frequently faced by plants. This will vary according to the soil properties. Under drought situation, plants face inadequate water to drive their physiological functions, leading to leaf senescence and other changes, while "under irrigated" plants face water deficit stress where enough water is not available for absorption and transport of solutes from soil, leading to growth retardation. Among abiotic stresses, drought is one of the most adverse factors for plant growth and productivity (Noorka and Tabasum 2015).

Water requirement in crops varies in different tissues as well as across the growth stages from germination to maturity (Ihsan et al. 2016). Water deficit budgets lead to numerous physiological alterations, both in the long term and the short term. Long-term drought responses include altered root to shoot ratio (Blum and Arkin 1984) and reduced leaf area (Batanouny et al. 1991). Water deficit stress short-term responses include altered stomatal function (Stewart et al. 1995) and maintenance of cell turgor through *osmotic adjustment* (Turner et al. 1986). According to Kramer

M. Jain (✉) · S. Kataria · M. Hirve · R. Prajapati
School of Biochemistry, Devi Ahilya University, Indore, Madhya Pradesh, India

M. Hasanuzzaman et al. (eds.), *Plant Abiotic Stress Tolerance*,
https://doi.org/10.1007/978-3-030-06118-0_5

and Boyer (1995), plants respond to drought either by delaying dehydration where the plant maintains a relatively high plant water potential or by tolerating dehydration where the plant continues to function but at lower plant water potentials. Further, drought stress progressively reduces CO_2 assimilation rates due to decreased stomatal conductance. It affects leaf size, stems extension, and root proliferation; troubles plant water relations; and decreases water use efficiency. It disrupts photosynthetic pigments and reduces the gas exchange and the production of active oxygen species leading to decrease in plant growth and yield (Sairam et al. 1990; Datta et al. 2001; Keyvan 2010).

Plants respond, acclimate and withstand to drought stress by altering several morphological, physiological, and biochemical responses, such as changes in plant structure, growth rate, tissue osmotic potential, and antioxidant defenses (Duan et al. 2007; Lichtfouse et al. 2009). Accumulation of free proline by plant tissue under water deficit conditions is an adaptive response. It can be due to increased protein degradation or enhanced proline synthesis. During water deprivation or extreme salinity, proline serves as a compatible solute that adjusts the osmotic potential in the cytoplasm (Caballero et al. 2005). The root is the first organ to be exposed to water deficit. Leaf growth is very sensitive to water stress, and may be inhibited by a slight reduction of water potential in the tissue (Hsiau and Xu 2000). Therefore, under the conditions of water scarcity, osmotic adjustment in the root occurs prior to that in the leaf so that turgor pressure required for continued root growth and absorption of water and nutrients could be facilitated. Thus, osmotic adjustment in the root is likely to delay the onset of water deficit in the shoots, which otherwise could reduce the stomatal conductance and photosynthetic activity. There are different types of plant reactions to this stress: (1) stress escape, (2) stress avoidance, and (3) stress tolerance. Plant drought tolerance involves changes at whole-plant functions. Hence, it is necessary to determine the most suitable conditions in which to observe the type of response that is better in order to improve plant performance.

Globally maize (*Zea mays* L.) is the third most important crop, after wheat and rice. It is produced on nearly 100 million hectares in developing countries, with almost 70% of the total maize production in the developing world coming from low and lower middle income countries (FAOSTAT 2012). Maize is a versatile crop grown over a wide range of agro-climatic zones. In fact, the suitability of maize to diverse environments is unmatched by any other crop. It is grown from below sea level to altitudes higher than 3000 m, and in areas with 250 mm to more than 5000 mm of rainfall per year and with a growing cycle ranging from 3 to 10 months. Maize is a direct staple food for millions of individuals and, through indirect consumption as a feed crop, is an essential component of global food security. In maize, water requirement is low at early growth stages, maximum at reproductive growth stages and again lowers down during terminal growth stages. Leaf rolling has been commonly observed in corn fields especially during severe soil compaction. Thus, pollination and fertilization period at the time of drought stress will result in yield loss. About 2 weeks before silk emergence, corn enters the period of grain yield determination. Corn is the most sensitive to drought stress during this period. Continued wilting of the plant at this stage can decrease yield 3–4% per day. Inadequate plant water can also delay silk elongation and silks that do emerge may become non-receptive to pollen. Obviously this can result in poor pollination. During the silking and pollen shed period, severe stress may reduce yield up to 8% per day. Water stress during grain filling reduces yield 2.0–6.0% with each day of stress. Abortion of kernels during the first 2 weeks following pollination is common during drought. Kernels can also abort during blister and milk stages if there is severe drought stress. Once kernels have reached the dough stage of development, yield losses will occur mainly from reduced kernel test weight. Drought stress during dough and dent stages can lead to premature black layer formation in the kernels. Grain filling and soft dough formation are most sensitive to water deficiency, whereas pre-tasseling and physiological maturity are relatively insensitive to water deficiency (Grant et al. 1989; Ahmed-Amal and Mekki 2005). Effects of drought on maize at different growth stages and organizational levels have been presented in Fig. 1.

Several techniques have been used to simulate drought; one of them which is widely used is to employ metabolically inactive compounds, such as polyethylene glycol (PEG) to study the effects of water stress in different groups of plants (Ashraf et al. 1996; Kauser et al. 2006; Shamim et al. 2014). In this chapter, a brief attempt

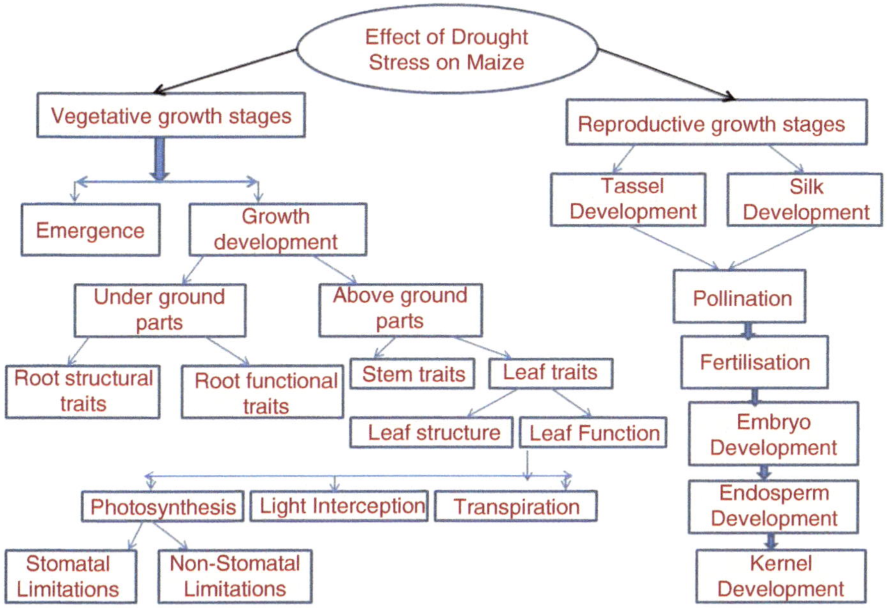

Fig. 1 Effects of drought stress on vegetative and reproductive growth stages of maize

has been made to summarize the effects of water deficit stress on maize plant and the adaptive responses by the crop.

Water Deficit Stress Effects in Maize

Water Deficit Stress and Germination

Seed germination is regarded as the first stage of plant's life cycle exposed to environmental stress and a successful establishment of seedling is required to obtain the plants with optimal ability of growth and development. Seed vigor of crops can be a key to increase crop yields per unit area because it can improve crop resilience against climate change effects and biotic impediments to crop yields. The very first step in the seed germination is imbibition, i.e., absorption of water by the dry seed. After absorbing water, the seeds swell. It is the chemical composition of seed which decides the amount of water absorbed. Proteins, mucilage, and pectin are more hydrophilic colloid and absorb more water than starch (Rahmani 2006). Water is the main factor stimulating germination and access to water by osmotic and matric potential (suction) is reduced. Crop establishment is accomplished up to development of seventh or eighth leaf. These early growth stages are critically affected by drought stress. Proper seed germination is dependent on availability of appropriate moisture contents for metabolic activation to breakdown the dormancy or to convert stored food into consumable form. Crop survival, growth, and development are determined by efficacy of seedling establishment (Hadas 2004). Drought stress reduces the germination potential of maize seeds by reducing their viability. Poor maize seed germination is directly associated with poor post-germination performance (Radić et al. 2007). Severity of drought stress is found to be directly linked with poor imbibition, germination and seedling establishment in maize (Achakzai 2009). Germination index has been found to be reduced by water deficiency (Almansouri et al. 2001). Germination velocity index (GVI) which measures the *speed* of seedling *germination* is corroborated with seed strength and always GVI was greater for maize hybrids than landraces due to hybrid vigor (Mabhaudhi 2009). Maize grain size is greater than other cereals like wheat, rice, and barley, therefore water requirement is greater for maintenance of osmotic potential and conversion of stored food into consumable form for proper germination (Gharoobi et al. 2012). Seed vigor is considered as important parameter in maize breeding which is badly reduced by drought stress (Khodarahmpour 2011). After germination, water deficiency significantly reduced the plumule and radicle growth which resulted in unusual seedling growth (Gharoobi et al. 2012). Hydropriming, magnetopriming, and osmopriming of maize seed result in improved seed germination by regulation of enzymatic activity to break the dormancy which clearly highlights the importance of water availability for exploitation of full germination potential (Janmohammadi et al. 2008; Vaishnav and Jain 2015; Kataria et al. 2017). Root and shoot elongations are parameter of seedling growth and these are subjected to reduction by drought stress. At seedling stage in maize, reduction in shoot elongation is more than root elongation under drought stress (Khodarahmpour 2011; Jain et al. 2010). Rate and degree of seedling establishment of maize are critical factors for determination of time of physiological maturity and grain yield (Rauf et al. 2007).

Water Deficit Stress and Root Characteristics

Roots are vital parts of the plants because these are the primary detectors or sensors of drought stress. Root length, root volume, root density, and number of roots are the characteristic structural traits which are disturbed under drought stress, as a result of which whole aerial plant parts are affected. Spatial water uptake and temporal water uptake are functional characteristics of roots. Root system of maize is comprised of axillary and lateral roots. Axillary roots are further comprised of primary, seminal, nodal, or crown roots (Cahn et al. 1989). Primary and seminal roots are collectively known as embryonic roots. Seminal roots are permanent and have functional role in growth and development of plant (Navara et al. 1994). Root density, volume, and number of roots are reduced under mild and severe drought stress (Nejad et al. 2010). Roots of maize plant become elongated under mild drought stress to explore more soil foils for more water uptake, whereas under severe drought stress root length is reduced. Further, it has been confirmed that reduced crown root number improves water acquisition under water deficit stress in maize genotypes from drying soil (Gao and Lynch 2016). More decrease in root biomass than the decrease in shoot biomass leading to an increased root:shoot ratio with increase in water deficit stress at all growth stages in maize has been reported by Benjamin et al. (2014).

Water Deficit Stress and Leaf Properties

At the onset of water stress, inhibition of cell growth leads to reduction in leaf development. Lower leaf surface causes less water uptake from the soil and transpiration is also reduced. Restrictions on the leaf surface could be considered as the first line of defense against water deficit stress (Kafi and Damghany Mahdavi 1999). After encountering with the deficiency of water, old leaves start falling. The process of shedding leaves during water stress is largely the result of increased synthesis and sensitivity to the ABA hormone in plants (Kabiri 2010). Leaves in maize are ranged from 8 to 20 and these are present alternatively on nodes. Both structural and functional components of leaf are affected by water deficit stress. Leaf size and number of leaves are the structural components and photosynthesis, transpiration, and light interception are the functional traits of leaf. Turgor pressure, light interception, and flux assimilation are determinant of leaf elongation (Rucker et al. 1995). Under drought stress turgor of leaves is reduced and leaves are curled or folded (Du Plessis 2003). Due to leaf folding, the leaf area is decreased and so the light interception, which results into reduced photosynthetic activity. Leaf area and photosynthesis are directly related with each other (Stoskopf 1981). Under the conditions of drought stress, cell division and cell elongation are decreased which results into reduction of leaf area. Reduction in leaf area under drought stress conditions is employed as adaptive strategy by maize plants. Considering this fact, leaf area index has been taken as an important consideration for maize breeding against drought stress (Hajibabaee et al. 2012). Plant water requirement is reduced by reducing the leaf area and probability of plant survival is increased under limited water availability (Belaygue et al. 1996) but chlorophyll contents, chloroplast contents, and photosynthetic activity are reduced which reduced the grain yield (Flagella et al. 2002; Goksoy et al. 2004).

Water Deficit Stress and Water Relations

Water relations of the plants can be measured by various characteristics, such as relative water content (RWC), leaf water potential, stomatal resistance, rate of transpiration, leaf temperature, and canopy temperature. Water potential (Y_w) varies greatly, depending on the type of plant and on environmental conditions. The influence of drought stress on internal water status and the final grain yield of three Egyptian corn (*Zea mays L.*) genotypes GIZA2, TWC310, and TWC320 at different developmental stages were investigated by Atteya (2003). Plants grown in pots were subjected to four levels of water stress at vegetative and tassel emergence stages. Exposure of plants to drought led to noticeable decrease in leaf water potential, RWC, and osmotic potential (Y_p). Maintenance of high RWC has been considered to be a drought-resistance rather than drought-escape mechanism, and it is a consequence of adaptive characteristics, such as osmotic adjustment and/or bulk modulus of elasticity (Grashoff and Ververke 1991). Exposure of wheat and rice plants to drought stress has been found to substantially decrease the leaf water potential, relative water content, and transpiration rate, with a concomitant increase in leaf temperature (Siddique et al. 2001). On the other hand, marginal reduction in RWC due to supply of sorbitol to etiolated maize leaf segments during greening has been reported (Tiwary and Jain 2016). In another study, lower leaf water potential, turgor potential and stomatal conductance was recorded by Craufurad et al. 2000 in groundnut on imposing moisture stress, where stomatal conductance was more strongly affected than the leaf water status. Under drought stress, relative water content and water potential is reduced; consequently, leaf temperature is increased due to reduced transpiration cooling (Siddique et al. 2001). The transpiration ratio is the ratio of the mass of water transpired to the mass of dry matter produced; its value for crops is in the range of 200–1000 (i.e., crop plants transpire 200–1000 kg of water for every kg of dry matter produced). For maize, its value is 388, and for soybean and wheat it is 704 and 613, respectively. This shows that maize is relatively efficient water user crop (Jensen 1973). However, regardless of being efficient water user maize is badly affected by drought stress. Reduced transpiration levels of Si-fed maize and rice primarily due to the lower transpiration through stomatal pores have been reported (Agarie et al. 1998; Gao et al. 2004, 2006), which mainly ascribed to the turgor loss of guard cells originating from Si deposition and changing of the physical and mechanical properties of their cell walls (Ueno and Agarie 2005; Savvas and Ntatsi 2015).

Water Deficit Stress and Stomatal Closure

Stomatal closure is the first response of virtually all plants to acute water deficit to prevent the transpirational water loss (Mansfield and Atkinson 1990). It can be considered as the first line of defense against desiccation, since it is much quicker than changes in roots growth, leaf area, chloroplast ultrastructure, and pigment proteins. This may result in response to either a decrease in leaf turgor and/or water potential (Ludlow and Muchow 1990) or to a low-humidity atmosphere (Maroco et al. 1997). Under mild drought stress, stomatal closure has protective role in saving the water loss and increasing water use efficiency; however, under severe drought stress it becomes dangerous (Chaves et al. 2009). Stomatal conductance, usually measured in mmol m^{-2} s^{-1}, is the measure of the rate of passage of CO_2 entering, or water vapor exiting through the stomata of a leaf. Therefore, the regulation of leaf stomatal conductance (*g*) is a key phenomenon in plants as it is vital for both prevention of desiccation and CO_2 acquisition. Thus, the decrease in photosynthetic rate under stressful conditions such as drought and salinity is normally attributed to a suppression in the mesophyll conductance and the stomatal closure at moderate and severe stress (Chaves et al. 2009; Sage and Zhu 2011). Passive and active stomata closures occur under normal conditions and stress prevalence, respectively (Fig. 2). Reactive oxygen species (ROS), electron acceptors and electron carriers have potential role in regulation of stomatal conductance (Chaves et al. 2009). As the rate of transpiration decreases, the amount of heat that can be dissipated increases (Yokota et al. 2002). Various studies have shown that stomatal responses are often more closely linked to soil moisture content than to leaf water status. This is due to stomatal response to chemical signals, e.g., abscisic acid (ABA), produced by dehydrating roots (Turner et al. 2001). Environmental conditions that enhance the rate of transpiration also increase the pH of leaf sap, which can promote ABA accumulation and concomitantly diminish stomatal conductance. Closure of stomata after 24 h PEG treatment has been reported in leaves of two maize cultivars 704 and 301 (Mohammadkhani and Heidari 2008).

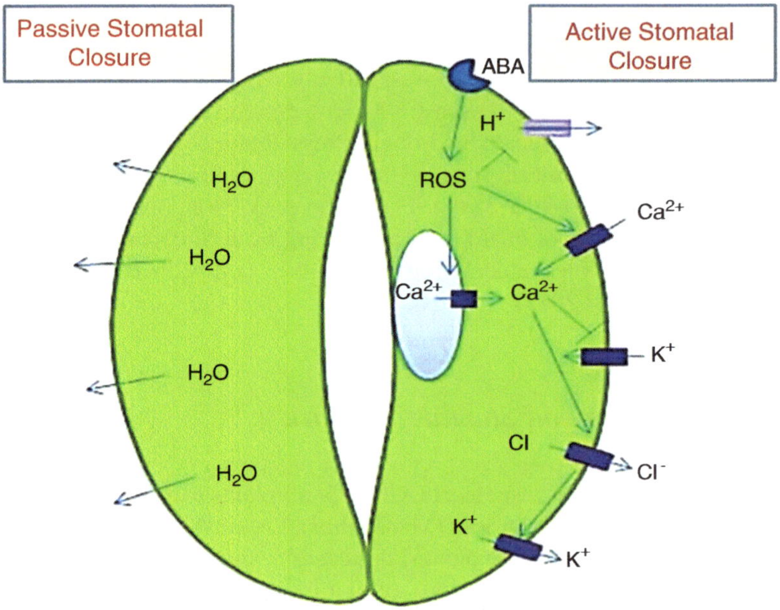

Fig. 2 Passive and active stomatal closure. Passive stomatal closure occurs under normal conditions and active stomatal closure occurs under drought stress (Arve et al. 2011)

Water Deficit Stress and Hormone Homeostasis

Water deficit stress is known to alter hormone homeostasis in plants (Pospíšilová 2003; Peleg et al. 2011). Hormonal changes further modulate the gene expression essential for their acclimation to the stress conditions. One of the early responses to drought is the accumulation of ABA, which causes stomatal closure, which reduces water loss through transpiration and results into preventing cell expansion, and limiting CO_2 fixation (Wilkinson and Davies 2002). Furthermore, several researches postulated jasmonic acid (JA) as a possible signaling molecule mediating the response to water stress in plants (Creelman and Mullet 1995; Fujita et al. 2006). These studies confirmed that there is a temporary accumulation of JA followed by progressive accumulation of ABA under drought conditions. This trend indicated that both ABA and JA share various transcription factors associated with responses to abiotic stress. Shinozaki and Yamaguchi-Shinozaki (2007) suggest that there is a cross-talk between ABA and JA in the signaling cascade that is triggered in water-stressed plants. Cytokinins (CKs), which participate in several aspects of plant development, such as seed germination, vascular development, meristem function, stimulation of photosynthesis and of sink strength, and counteract leaf senescence, are often considered as antagonists to ABA (Hare et al. 1999). In general, application of CKs to the leaf epidermis reverses ABA-induced stomatal closure and maintains normal transpiration rates in many plants (Pospíšilová 2003; Ha et al. 2016).

In maize, this phenomenon has been reported to occur in both young and old leaves (Blackman and Davies 1985). Increased cytokinin concentration in the xylem sap promotes stomatal opening directly and affects the sensitivity of stomata towards abscisic acid (Wilkinson and Davies 2002). Ethylene production under abiotic stresses, such as waterlogging and drought, can reduce photosynthesis (Rajala and Peltonen-Sainio 2001), inhibit root growth, reduce shoot/leaf expansion (Sharp and LeNoble 2002; Pierik et al. 2007) and decrease grain yield by grain abortion, among other actions (Wilkinson and Davies 2010).

Respiration

A considerable amount of energy is spent to cope up with drought; hence, drought tolerance is a cost-intensive phenomenon. The carbohydrate fraction that is lost during respiration determines the overall metabolic efficiency of the plant (Davidson et al. 2000). As compared to other physiological processes, studies on effects of water deficit stress on plant respiration are very rare. Increasing concentration of PEG has been found to decrease oxygen consumption percentage in both roots and leaves of two maize cultivars 704 and 301 (Mohammadkhani and Heidari 2008). Further, The decrease of oxygen consumption percentage in 704 plants was higher than 301 plants. Therefore, water stress has a higher effect in 704 plant's respiration than 301 plants. Oxygen consumption percentage in high water stress in roots was lower than leaves and roots were more sensitive than leaves. Drought probably acted directly on roots, because the roots were immerged in PEG solutions and water stress in roots was higher than shoots, whereas leaves could reduce oxygen consumption by stomatal regulations.

Water Deficit Stress and Photosynthesis

Water deficit results into progressive suppression of photosynthesis by disrupting all major components including the thylakoid electron transport, the carbon reduction cycle, and the stomatal control of the CO_2 supply, together with an increased accumulation of sugars (Farooq et al. 2009; Carmo-Silva and Salvucci 2012). Both stomatal and non-stomatal limiting factors in maize plant reduce the process of photosynthesis. The stomatal closure is triggered by reduced leaf turgor and root originated signals along with lower plant water status. The reduced water potential in the roots transduces the signals for stomatal closure and thus results into decreased CO_2 diffusion in the leaves affecting ribulose 1,5-bisphosphate carboxylase/oxygenase (RuBisCO) (Flexas et al. 2007). Thus, reduced CO_2 diffusion is considered as main reason for decline of photosynthesis. Stomatal closure has protective role in saving the water loss and increasing water use efficiency under mild drought stress but under severe drought stress stomatal closure becomes inevitable evil (Chaves et al. 2009). Leaf structural characters and biochemical parameters are components of non-stomatal inhibition of photosynthesis. According to Von Caemmerer (2000) and Ghannoum (2009), carboxylation is changed by RuBisCO, PEPC (phosphoenolpyruvate carboxylase) and regeneration of PEP (phosphoenolpyruvate). Besides its negative effects on dark reactions of photosynthesis, drought also disrupts the cyclic and non-cyclic types of electron transport during the light reaction of photosynthesis (Obidiegwu et al. 2015). Further, chlorophyll contents are reduced due to drought either by activation of cellular protein degradation or by limited nitrate synthesis (Ghannoum 2009; Obidiegwu et al. 2015; Becker and Fock 1986). It is reported that in C_4 plants, like maize, intercellular spaces and chloroplast positions are misplaced by drought stress; resultantly CO_2 diffusion and light penetration are disturbed followed by decreased photosynthetic activity (Flexas et al. 2004). Under drought stress, photorespiration and Mahler's reaction act as alternative electron sinks (Ghannoum 2009). Mahler's reaction develops oxidative stress under drought stress as it is involved in generation of reactive oxygen species. As a result of a direct reduction reaction in photosystem-I, oxygen molecule is converted into superoxide (Haupt-Herting and Fock 2002). Carboxylation activity of RuBisCO and regeneration of RuBP and ATP are reduced by inhibited CO_2 concentration in the leaves under drought stress (Tezara et al. 1999). Leaf biochemistry, membrane permeability (aquaporin activity), leaf shrinkage, alterations in intercellular spaces, intercellular structure, internal diffusion, and internal conductance are altered under drought stress which results in reduction of CO_2 diffusion through mesophyll (Chaves et al. 2009; Lawlor and Cornic 2002).

Water Deficit Stress and Oxidative Damage

Exposure of plants to environmental stresses quite often leads to the generation of reactive oxygen species (ROS), including superoxide radicals ($O_2^{\bullet -}$), hydroxyl radicals ($^{\bullet}OH$), hydrogen peroxide (H_2O_2), alkoxy radicals ($RO^{\bullet}$), and singlet oxygen ($^1O_2$) (Munné-Bosch and Penuelas 2003). Many cell compartments produce ROS, such as chloroplasts, mitochondria, and peroxisomes; of these, chloroplasts are a potentially important source because excited pigments in thylakoid membranes may interact with O_2 to form strong oxidants (Niyogi 1999; Reddy et al. 2004). Production of ROS in peroxisomes during abiotic stress is mainly the outcome of enhanced photorespiration resulting in the production of H_2O_2 by glycolate oxidase (Baishnab and Ralf 2012; Kerchev et al. 2016). These reactive species may react with proteins, lipids, and DNA, causing oxidative damage and impairing the normal functions of cells (Foyer and Fletcher 2001). Several reports have shown the deleterious effects of ROS, whose production is stimulated under water stress (Blokhina et al. 2003; Yang et al. 2014).

Plants possess an efficient antioxidant (enzymatic and non-enzymatic) defense system to cope with ROS-induced oxidative stress (Anjum et al. 2011a, b; Ashraf et al. 2015). Both enzymatic, i.e., ascorbate peroxidase (APX), superoxide dismutase (SOD), peroxidase (POD), catalase (CAT), dihydro ascorbate reductase (DHAR), and mono dehydro ascorbate reductase (MDHAR), and non-enzymatic, i.e., dihydro ascorbate (DHA), ascorbate (AsA), and glutathione (GSH), antioxidants diminish the oxidative damage caused by stressful conditions. In plants subjected to a long-term drought stress, the contribution of both enzymatic and non-enzymatic antioxidants may ensure the stress tolerance (Sharma et al. 2012). It has been reported that in drought tolerance of maize there is a direct or indirect contribution of these antioxidants. For instance, Adebayo and Menkir (2015) stated that sustained yields in maize under drought stress were directly related to better antioxidant activities. Farooq et al. (2009) also concluded that improved activities/levels of enzymatic and non-enzymatic antioxidants scavenge ROS and hence, may enhance the drought tolerance. Upregulation of the activities of anti-oxidative defense systems in three different maize hybrids (Dong Dan 80, Wan Dan 13, and Run Nong 35) due to drought stress has been reported (Anjum et al. 2017). On the other hand, under severe drought stress antioxidant enzyme activities declined significantly ($P < 0.01$) in later stages, during the tasseling, blister, and milk stages (Bai et al. 2006). The activity of ROS scavenging enzymes, catalase, superoxide dismutase, ascorbate peroxidase, guaiacol peroxidase (Gu-POX), and glutathione reductase (GR) has been found to enhance with the supply of sorbitol in maize leaves (Tiwary et al. 2016). Further, protective effect of methane against osmotic stress by modulating sugar and ascorbate has been demonstrated in maize by Han et al. (2017).

Water Deficit Stress and Protein Response

A common response to abiotic stress is the accumulation of proteins in plants, such as heat shock proteins, signal transduction proteins, enzymes involved in primary and secondary metabolic processes, chaperones, and dehydrins (Lee et al. 2005; Sun et al. 2016). These proteins confer direct and indirect protection to plants through the synthesis of osmo-protectants, detoxifying enzymes and to encode regulatory proteins, such as protein kinases, transcription factors (TFs), and phosphatases (Krasensky and Jonak 2012). Late embryogenesis abundant (LEA) proteins have also been found to accumulate in high levels as a common mechanism against various abiotic stress conditions, like heat, cold, drought, salinity, osmotic stress, leaf desiccation, and seedling dehydration (Asghar et al. 1994; Brini et al. 2011; Hanin et al. 2011; Graether and Boddington 2014). Dehydrins are unstructured proteins that constitute the biochemical group of LEA proteins—group 2 proteins (Asghar et al. 1994; Brini et al. 2011; Hanin et al. 2011) and are thought to play a significant role in cellular protection against dehydration (Hanin et al. 2011), leading to cell collapse. Dehydrins fill into spaces, accumulate and bind water which helps in maintaining cell volume during dehydration (Hanin et al. 2011). Although their specific functions are uncertain, it is believed that dehydrins contribute significantly to plant survival (Lee et al. 2005). They are localized in various parts of the plants including cytosol, plasma membrane (Koag et al. 2003), mitochondria, protein bodies, nucleus, chloroplast, and vacuoles (Hara et al. 2005; Koag et al. 2009). The difference in expression and tissue location suggest that individual members of the DHN multigene family have somewhat distinct biological functions (Koag et al. 2009). The expression of dehydrin-like proteins in the roots and leaves of maize varieties 704 and 301 has been reported by Mohammadkhani and Heidari (2008); however, there was no relationship between protein changes and drought tolerance. Osmotin is also reported as accumulated protein under water stress in several plant species such as tobacco, triplex, tomato, and maize (Ramagopal 1993). Proteomic analysis under drought treatments has been shown to report the significant alteration in expression of 70 and 36 proteins in B73 and Lo964 maize inbred lines, respectively. Further, in the sensitive genotype B73, the numbers and levels of differentially expressed proteins were generally higher, implying an improved sensitivity to drought given the function of the observed differentially expressed proteins, such as redox homeostasis, cell rescue/defense, hormone regulation, and protein biosynthesis and degradation (Yang et al. 2014).

Water Deficit Stress and Growth and Development

For establishment of normal plant structure, proper growth and development of crop plants is important that carry out all physiological and metabolic processes and give potential yield. Drought stress critically hindered the growth and development of maize. Growth is described as increase in size of plant which is directly associated with increase in number of cells and cell size. Meristematic tissues are involved in active elongation of plant by active cell division. Further, several component parameters comprised growth and development and these parameters are estimated by different traits like plant height, leaf area, structural and functional characters of root, plant fresh weight, plant dry weight, plant biomass, and stem diameter. Drought stress has been observed to reduce under plant height, stem diameter, plant biomass, and leaf area (Khan et al. 2001; Zhao et al. 2006). Because of reduction in water potential of cells, cell division and cell size are reduced which further causes stunting of plant growth (Nonami 1998). Drought stress increases the leaf to stem ratio which is indication of high level of growth retardation in stems than leaves (Hajibabaee et al. 2012). Reduced water potential in roots interrupts the optimal water supply to the elongating cells and resultantly cell elongation is reduced. Light interception is reduced after reduction of leaf area. Less interception of solar radiations causes the reduction in biomass production (Delfine et al. 2001). Besides light interception, stomatal activity is also responsible for lower biomass production (Delfine et al. 2001; Medrano et al. 2002). Rise in leaf temperature under drought stress inhibits the enzymatic activity and reduces photosynthesis (Chaves et al. 2002). Photosynthetic machinery is inactivated by increase in leaf temperature above threshold temperature which is 30 °C (Crafts-Brandner and Salvucci 2002).

Drought has inhibitory effects on maize life cycle; particularly reproductive growth phase is most vulnerable to drought stress. It has been reported that during reproductive growth stage under drought stress, the most probable reason for more susceptibility of maize plant is translocation of photosynthetic assimilates to the reproductive parts rather than roots for their extensive elongation (Setter et al. 2001; Taiz and Zeiger 2006). Sequential effects of drought stress on reproductive growth stages of maize are depicted in Fig. 1. Pollen and silk development, pollination, embryo development, endosperm development, and kernel development are the different component phases of reproductive growth stage which are severely threatened by drought stress.

Adaptive Responses Against Water Deficit Stress

Depending on the particular drought circumstances usually plants have several strategies to overcome from the water deficit stress (Tardieu 2012). On the other hand, there is water conservation strategies planned to decrease cumulative transpiration to arrange the prolonged water accessibility for the plants through slow stomatal conductance and reducing leaf evaporative area, like slow leaf growth rate and early leaf senescence. Under severe drought conditions, these strategies can have advantageous effects; however since they reduce the capability for photosynthesis and consequently decrease the biomass accumulation, they possibly will impose high yield penalty under mild to moderate water deficit. On the other hand, under mild to moderate water deficits strategies that support high stomatal conductance and continuance of high photosynthesis are more beneficial, because during and after the stress they promote sustaining growth capacity. Furthermore, persistent high stomatal conductance helps in reduction of leaf temperature and reduced the harmful effects of heat stress that generally occurs along with drought (Tardieu 2012; Lopes et al. 2011).

In surviving abiotic conditions, stressed plants induce the production of osmoprotectants. These protectants include both primary and secondary metabolites. Primary metabolites are plant chemicals that are generally responsible for core housekeeping functions, such as energy production, regulation of essential metabolites and molecules. They include proline, glycine betaine, carbohydrates, and soluble sugars which tend to protect cellular structures by regulating osmotic activity (Nilsen and Orcutt 1996; Bita and Gerats 2013). The secondary metabolites are plant chemicals that are produced in specific metabolic pathways and are necessary for growth or sometimes required under specified conditions. They are phenolics, anthocyanins, flavonoids, and plants steroids which play a significant role in tolerance against heat stress. Glycine betaine is an amino acid derivative found in a wide range of crops and accumulates in some plants under various environmental stress conditions including drought (Zhang et al. 2012). Glycine betaine is accumulated as a compatible solute in maize and sugarcane, while others such as rice, mustard,

Arabidopsis, and tobacco do not naturally produce glycine betaine under stress conditions.

Understanding these mechanisms by which plants survive stress condition serves as pre-requisite in developing strategies to improve their tolerance to stress. Plants in surviving stress conditions activate various defense mechanisms of the plant. Plant response to stress varies with the level of stress condition it suffers, thus metabolites in plants also vary with the level of stress they are subjected to (Barchet et al. 2014). Therefore, maize in responding to stress conditions undergoes changes which are observed at various levels of plant growth (Barchet et al. 2014; Bowne et al. 2012).

Drought Tolerance Mechanisms

In terms of annual metric tons, maize is the cereal crop among highest worldwide production (FAO. 2011) and the majority of its cultivation is rain-fed, with inadequate potential for mitigation of water deficit stress. Hence, it is essential to generate drought-tolerant varieties either by conventional breeding or by genetic engineering. During flowering time maize is more vulnerable to drought, with the most severe reductions in yield taking place in the 3-week period bracketing male (anthesis) and female (silking) flowering events (Hall et al. 1982). Usually, maize plants exhibit protandry (i.e., anticipated anthesis with respect to silking), and the main effect of water deficits that occurs immediately before anthesis is to enhance the anthesis-silking interval (ASI), with the simultaneous negative effects on ovary pollination (Hall et al. 1982; Bolanos and Edmeades 1993). In addition, when water deficits persist during silking, significant abortion of fertilized ovaries takes place due to reduced assimilate availability for successful kernel set (Anderson et al. 2004). In this species, the mobilization of recently fixed carbon is an important determinant of plant and ear growth rates, because reserves stored before silking do not contribute to alleviate a reduction in current assimilate availability (Bruce et al. 2002) and to lessen kernel abortion (Schussler and Westgate 1994). Thus, maintenance of functional source leaves during drought episodes seems crucial for minimizing the negative effects of this constraint on final grain yield and global productivity.

Quantitative Trait Loci (QTL)

To advance our insight into crop development, the understanding of maize response to water stress requires a complete assessment of changes in gene expression induced by stress. To identify the loci for stress tolerance in plants, the Quantitative trait loci (QTL) had been utilized as a potent and equitable approach (Ismail et al. 1999; Tuberosa et al. 2002). However, the loci identified by QTL account for only a portion of the genetic variances in adaptation between genotypes, and delineating a QTL to a single gene using genetic approaches is time-consuming and technically demanding (Fridman et al. 2000, 2004). As an alternative, microarray technology is a useful tool for analyzing genome-wide gene expression (Schena et al. 1995). Recently, this technology has been applied to analyze maize gene expression profiles under various stress conditions (Andjelkovic and Thompson 2006; Zheng et al. 2004; Yu and Setter 2003). Under drought stress, cDNA microarrays were used to monitor the gene expression in the leaves and roots of maize seedlings (Zheng et al. 2004; Yu and Setter 2003; Jia et al. 2006). Wang et al. (2003) reported that the transcriptome in maize roots quickly altered at the 3-h time point under salt stress. How the reproductive organs act in response to stress conditions is another part of gene expression analysis of maize (Andjelkovic and Thompson 2006; Yu and Setter 2003; Zinselmeier et al. 2002). Under water deficit stress, placenta/pedicel and endosperm differed considerably in their transcriptional responses at 9 days after pollination. Of stress response genes, 89% were upregulated in placenta/pedicel and 82% were downregulated in endosperm (Yu and Setter 2003). Zinselmeier et al. (2002) used oligonucleotide microarray containing 1502 genes to study the genes expression in response to water deficit stress at 4 days after silking and 8 days after pollination in maize ear and kernel. These authors showed that 17 genes (10 upregulated, 7 downregulated) were affected by stress among these tissues and confirmed that as compared to other tissues gene expression in the pedicel were more responsive (Andjelkovic and Thompson 2006). When the maize plants experience drought stress, two peaks of sensitivity have been identified which may reduce the crop yield (Saini 1997). The first peak is centered on the period from meiosis to tetrad breakup in anthers, and second peak of sensitivity takes place during anthesis and initial stages of grain development. Meiotic-stage stress changes the successive development of the male gametophyte and even causes pollen sterility (Downey 1969). When the accessible soil moisture was inadequate, then the significant yield reductions were observed during male gametes development (Denmead and Show 1960).

Use of Potassium (K) for Drought Tolerance

Addition of drought stress with nutrient deficiency and toxicity is responsible for severe losses of crop production in the entire globe. Therefore, mineral nutrients in plant body can enhance its tolerance to drought stress. Among these nutrients, K is important element to develop tolerance in the plant body. The low quantity of K in the plant body decreases the photosynthetic carbon metabolism and also the consumption of fixed carbon resources (Cakmak and Engels 1999; Mengel and Kirkby 2001); as a result of this, huge deposition of carbohydrates take place in the source leaves. Consequently, excess of nonutilized light energy and photoelectron are there in the plant bodies which create photo oxidative damage. Abortion of kernels at the top of cob and smaller grain size may be resulted due to the deficiency of potassium (Bly et al. 2002). Silicon (Si) has been known to augment plant defense against biotic and abiotic pressures. In this context, Si has been proved as an efficient strategy to boost tolerance of maize plants to drought/water deficit stress (Zargar and Agnihotri 2013; Amin et al. 2014).

Conclusions

Global warming is changing the climatic conditions all over the world; consequently, it generates abnormal environmental conditions, one of them is water deficit stress. Drought affects more or less every plant development processes, at a cellular level from membrane conformation, organization of chloroplast and enzyme activity, reduction in growth and yield at the whole plant level. Drought stress stimulates stomatal closure, reduces transpiration and photosynthetic rates, and leads to earlier crop maturity and poor productivity. Plants have various signaling pathways to minimize the harmful effects of water stress, which respond by altering their growth pattern, upregulation of antioxidants, accumulation of compatible solutes and by producing the stress proteins and chaperones.

Maize (*Zea mays* L.) is well recognized as one of the most important cereals worldwide. Normally, it needs 500–800 mm of water during its life cycle (80–110 days) however; occurrence of drought stress during maize growth period may hamper the various physiological processes. Given the current trends towards higher global temperatures, the development of crop varieties with enhanced tolerance to water deficit stress and higher water use efficiency has become a high priority for plant breeders and genetic engineers in order to meet future food demands.

References

Achakzai AKK (2009) Effect of water stress on imbibition, germination and seedling growth of maize. Sarhad J Agric 25:165–172

Adebayo MA, Menkir A (2015) Combining ability of adapted and exotic drought-tolerant maize inbred lines under full irrigation and rainfed conditions in Nigeria. J Crop Improv 29:117–130

Agarie S, Uchida H, Qgata W, Kubota F, Kaufman PB (1998) Effects of silicon on transpiration and leaf conductance in rice plants (*Oryza sativa* L). Jpn J Crop Sci 1:89–95

Ahmed-Amal O, Mekki BB (2005) Yield and yield components of two maize hybrids as influenced by water deficit during different growth stages. Egypt J Appl Sci 20:64–79

Almansouri M, Kinet JM, Lutts S (2001) Effect of salt and osmotic stresses on germination in durum wheat (*Triticum durum* Desf.). Plant Soil 231:243–254

Amin M, Ahmad R, Basra SMA, Murtaza G (2014) Silicon induced improvement in morpho-physiological traits of maize (*Zea Mays* L.) under water deficit. Pak J Agric Sci 51:187–196

Anderson SR, Lauer MJ, Schoper J, Shibles R (2004) Pollination timing effects on kernel set and silk receptivity in four maize hybrids. Crop Sci 44:464–473

Andjelkovic V, Thompson R (2006) Changes in gene expression in maize kernel in response to water and salt stress. Plant Cell Rep 25:71–79

Anjum SA, Wang LC, Farooq M, Hussain M, Xue LL, Zou CM (2011a) Brassinolide application improves the drought tolerance in maize through modulation of enzymatic antioxidants and leaf gas exchange. J Agron Crop Sci 197:177–185

Anjum SA, Wang LC, Farooq M, Xue LL, Ali S (2011b) Fulvic acid application improves the maize performance under well-watered and drought conditions. J Agron Crop Sci 197:409–417

Anjum SA, Ashraf U, Tanveer M, Khan I, Hussain S, Shahzad B, Zohaib A, Abbas F, Saleem MF, Ali I, Wang LC (2017) Drought induced changes in growth, osmolyte accumulation and antioxidant metabolism of three maize hybrids. Front Plant Sci 8:69

Asghar R, Fenton RD, Demason DA, Close TJ (1994) Nuclear and cytoplasmic localization of maize embryo and aleurone dehydrin. Protoplasma 177:87–94

Arve LE, Torre S, Olsen JE, Tanino KK (2011) Stomatal Responses to Drought Stress and Air Humidity In book: Abiotic Stress in Plants - Mechanisms and Adaptations, Intech, pp 267–280.

Ashraf MY, Naqvi MH, Khan AH (1996) Evaluation of four screening techniques for drought tolerance in wheat (*Triticum aestivum* L.). Acta Agron Hung 44:213–220

Ashraf U, Kanu AS, Mo Z, Hussain S, Anjum SA, Khan I et al (2015) Lead toxicity in rice: effects, mechanisms, and mitigation strategies—a mini reviews. Environ Sci Pollut Res 22:18318–18332

Atteya AM (2003) Alteration of water relations and yield of corn genotypes in response to drought stress. Bulg J Plant Physiol 29:63–76

Bai LP, Sui FG, Ge TD, Sun ZH, Lu YY, Zhou G (2006) Effect of soil drought stress on leaf water status, membrane permeability and enzymatic antioxidant system of maize. Pedosphere 16:326–332

Baishnab CT, Ralf O (2012) Reactive oxygen species generation and signaling in plants. Plant Signal Behav 7:1621–1633

Barchet GLH, Dauwe R, Guy RD, Schroeder WR, Soolanayakanahally RY, Campbell MM, Mansfield SD (2014) Investigating the drought-stress response of hybrid poplar genotypes by metabolite profiling. Tree Physiol 34:1203–1219

Batanouny KH, Hussein MM, Abo El Kheir MSA (1991) Response of *Zea mays* to temporal variation of irrigation and salinity under farm conditions in the Nile Delta, Egypt. In: Proc. inter. conf. plant growth, drought and salinity in the Arab region, pp 189–204

Becker TW, Fock HP (1986) The activity of nitrate reductase and poll sizes of some amino acids and some sugars in water stressed maize leaves. Photosynth Res 8: 267–274

Belaygue C, Wery J, Cowan AA, Tardieu F (1996) Contribution of leaf expansion, rate of leaf appearance, and stolon branching to growth of plant leaf area under water deficit in white clover. Crop Sci 36:1240–1246

Benjamin JG, Nielsen DC, Vigil MF, Mikha MM, Calderon F (2014) Water deficit stress effects on corn (Zea mays L.) root: shoot ratio. Open J Soil Sci 4:151–160

Bita CE, Gerats T (2013) Plant tolerance to high temperature in a changing environment: scientific fundamentals and production of heat stress-tolerant crops. Front Plant Sci 4:273

Blackman PG, Davies WJ (1985) Cytokinins, abscisic acid and the control of plant water balance. Acta Hortic 171:255–262

Blokhina O, Virolainen E, Fagerstedt KV (2003) Antioxidants, oxidative damage and oxygen deprivation stress: a review. Ann Bot 91:179–194

Blum A, Arkin B (1984) Sorghum root growth and water use as affected by water supply and growth duration. Field Crops Res 9:131–142

Bly A, Gelderman RH, Gerwing J, Murrell TS (2002) Symptoms associated with potassium deficiency in corn. Better Crops Plant Food 86:12–15

Bolanos J, Edmeades GO (1993) Eight cycles of selection for drought tolerance in tropical maize. I. Responses in grain yield, biomass, and radiation utilization. Field Crops Res 31:233–252

Bowne JB, Erwin TA, Juttner J, Schnurbusch T, Langridge P, Bacic A, Roessner U (2012) Drought responses of leaf tissues from wheat cultivars of differing drought tolerance at the metabolite level. Mol Plant 5:418–429

Brini F, Yamamoto A, Jlaiel L, Takeda S, Hobo T, Dinh HQ, Hanin M (2011) Pleiotropic effects of the wheat dehydrin DHN-5 on stress responses in Arabidopsis. Plant Cell Physiol 52:676–688

Bruce WB, Edmeades GO, Barker TC (2002) Molecular and physiological approaches to maize improvement for drought tolerance. J Exp Bot 53:13–25

Caballero JI, Verduzco CV, Galan J, Jimenez ESD (2005) Proline accumulation as a symptom of drought stress in maize: a tissue differentiation requirement. J Exp Bot 39:889–897

Cahn MD, Zobel RW, Bouldin DR (1989) Relationship between root elongation rate and diameter and duration of growth of lateral roots of maize. Plant Soil 119:271–279

Cakmak I, Engels C (1999) Role of mineral nutrients in photosynthesis and yield formation. In: Rengel Z (ed) Mineral nutrition of crops: mechanisms and implications. The Haworth Press, New York, pp 141–168

Carmo-Silva AE, Salvucci ME (2012) The temperature response of CO_2 assimilation, photochemical activities and Rubisco activation in *Camelina sativa*, a potential bioenergy crop with limited capacity for acclimation to heat stress. Planta 236:1433–1445

Chaves MM, Pereira JS, Maroco J, Rodrigues ML, Ricardo CPP, Osório ML, Carvalho I, Faria T, Pinheiro C (2002) How plants cope with water stress in the field? Photosynthesis and growth. Ann Bot 89:907–916

Chaves MM, Flexas J, Pinheiro C (2009) Photosynthesis under drought and salt stress: regulation mechanisms from whole plant to cell. Ann Bot 103:551–560

Crafts-Brandner SJ, Salvucci ME (2002) Sensitivity of photosynthesis in C_4 plant, maize, to heat stress. Plant Physiol 129:1773–1780

Craufurad PQ, Wheeler TR, Ellis RH, Summerfield RJ, Prasad PVV (2000) Escape and tolerance to high temperature at flowering in groundnut. J Agric Sci 135:371–378

Creelman RA, Mullet JE (1995) Jasmonic acid distribution and action in plants: regulation during development and response to biotic and abiotic stress. Proc Natl Acad Sci U S A 92:4114–4119

Datta JK, Mondal T, Banerjee A, Mondal NK (2001) Assessment of drought tolerance of selected wheat cultivars under laboratory condition. J Agric Technol 7:1383–1393

Davidson EA, Verchot LV, Cattanio JH, Ackerman IL, Carvalho HM (2000) Effects of soil water content on soil respiration in forests and cattle pastures of eastern Amazonian. Biogeochemistry 48:53–69

Delfine S, Loreto F, Alvino A (2001) Drought-stress effects on physiology, growth and biomass production of rainfed and irrigated Bell Pepper plants in the Mediterranean region. J Am Soc Hortic Sci 126:297–304

Denmead OT, Show RH (1960) The effects of soil moisture stress at different stages of growth on the development and yield of corn. Agron J 52:272–274

Downey LA (1969) Crop density and water use studies with irrigated maize (*Zea mays* L) on Sodic clay soils. MSc thesis, University of Sydney, Sydney

Du Plessis J (2003) Maize production. Directorate agricultural information services, department of agriculture in cooperation with ARC-Grain Crops Institute

Duan B, Yang Y, Lu Y, Korpelainen H, Berninger F, Li C (2007) Interactions between drought stress, ABA and genotypes in *Picea asperata*. J Exp Bot 58:3025–3036

FAO (2011) FAOSTAT online database. http://faostat.fao.org/. Accessed Dec 2011

FAOSTAT (2012.) http://faostat.fao.org/site/567/desktopdefault.aspx#ancor

Farooq M, Wahid A, Kobayashi N, Fujita D, Basra SMA (2009) Plant drought stress: effects, mechanisms and management. Agron Sustain Dev 29:185–212

Flagella Z, Rotunno T, Tarantino E, Di Caterina R, De Caro A (2002) Changes in seed yield and oil fatty acid composition of high oleic sunflower (*Helianthus annuus* L.) hybrids in relation to the sowing date and the water regime. Eur J Agron 17:221–230

Flexas I, Bota J, Cifre J, Escalona JM, Galmes J, Gulias J, Left tK, Martinez-Canellas SF, Moreno MT, Ribas-Carbo M, Riera D, Samplo B, Medrano H (2004) Understanding down-regulation of photosynthesis under water stress: future prospects and searching for physiological tools for irrigation management. Ann Appl Biol 144:273–223

Flexas J, Diaz-espejo A, Galmés J, Kaldenhoff R, Medrano H, Ribas-Carbo M (2007) Rapid variations of mesophyll conductance in response to changes in CO_2 concentration around leaves. Plant Cell Environ 30:1284–1298

Foyer CH, Fletcher JM (2001) Plant antioxidants: colour me healthy. Biologist 48:115–120

Fridman E, Pleban T, Zamir D (2000) A recombination hotspot delimits a wild-species quantitative trait locus for tomato sugar content to 484 bp within an invertase gene. Proc Natl Acad Sci U S A 97:4718–4723

Fridman E, Carrari F, Liu YS, Fernie AR, Zamir D (2004) Zooming in on a quantitative trait for tomato yield using interspecific introgressions. Science 305:1786–1789

Fujita M, Fujita Y, Noutoshi Y, Takahashi F, Narusaka Y, Yamaguchi-Shinozaki K, Shinozaki K (2006) Crosstalk between abiotic and biotic stress responses: a current view from the points of convergence in the stress signaling networks. Curr Opin Plant Biol 9:436–442

Gao Y, Lynch JP (2016) Reduced crown root number improves water acquisition under water deficit stress in maize *(Zea mays L.)*. J Exp Bot 67:4545–4557

Gao XP, Wang XF, Lu YF, Zhang LY, Shen YY, Liang Z, Zhang DP (2004) Jasmonic acid is involved in the water-stress-induced betaine accumulation in pear leaves. Plant Cell Environ 27:497–507

Gao X, Zou C, Wang L, Zhang F (2006) Silicon decreases transpiration rate and conductance from stomata of maize plants. J Plant Nutr 29:1637–1647

Ghannoum O (2009) C4 photosynthesis and water stress. Ann Bot 103:635–644

Gharoobi B, Ghorbani M, Nezhad MG (2012) Effects of different levels of osmotic potential on germination percentage and germination rate of barley, corn and canola. Iran J Plant Physiol 2:413–417

Goksoy AT, Demir AO, Turan ZM, Dagustu N (2004) Responses of sunflower to full and limited irrigation at different growth stages. Field Crops Res 87:167–178

Graether SP, Boddington KF (2014) Disorder and function: a review of the dehydrin protein family. Front Plant Sci 5:576

Grant RF, Jakson BS, Kinuy JR, Arkin F (1989) Water deficit timing effects on yield components in maize. Agron J 81:61–65

Grashoff C, Ververke DR (1991) Effect of pattern of water supply on *Vicia faba* L. 3. Plant water relations, expansive growth and stomatal reactions. Neth J Agric Sci 39:247–262

Ha Y, Shang Y, Hee Nam K (2016) Brassinosteroids modulate ABA-induced stomatal closure in *Arabidopsis*. J Exp Bot 67:6297–6308

Hadas A (2004) Seedbed preparation: the soil physical environment of germinating seeds. In: Benech-Arnold RL, Sanchez RA (eds) Handbook of seed physiology: applications to agriculture. Food Product Press, New York, p 480

Hajibabaee M, Azizi F, Zargari K (2012) Effect of drought stress on some morphological, physiological and agronomic traits in various foliage corn hybrids. Am Eurasian J Agric Environ Sci 12:890–896

Hall AJ, Villela F, Trapani N, Chimenti C (1982) The effects of water stress and genotype on the dynamics of pollen-shedding and silking in maize. Field Crops Res 5:349–363

Han B, Duan X, Wang Y, Zhu K, Zhang J, Wang R, Hu H, Qi F, Pan J, Yan Y, Shen W (2017) Methane protects against polyethylene glycol-induced osmotic stress in maize by improving sugar and ascorbic acid metabolism. Sci Rep 7:46185

Hanin M, Brini F, Ebel C, Toda Y, Takeda S, Masmoudi K (2011) Plant dehydrins and stress tolerance: versatile proteins for complex mechanisms. Plant Signal Behav 6:1503–1509

Hara M, Fujinaga M, Kuboi T (2005) Metal binding by citrus dehydrin with histidine rich domains. J Exp Bot 56:2695–2703

Hare PD, Cress WA, van Staden J (1999) Proline synthesis and degradation: a model system for elucidating stress-related signal transduction. J Exp Bot 50:413–434

Haupt-Herting S, Fock HP (2002) Oxygen exchange in relation to carbon assimilation in water-stressed leaves during photosynthesis. Ann Bot 89:851–859

Hsiau TC, Xu LK (2000) Sensitivity of growth of roots versus leaves to water stress: biophysical analysis and relation to water transport. J Exp Bot 51:1595–1616

Ihsan MZ, El-Nakhlawy FS, Ismail SM, Fahad S, Daur I (2016) Phenological development and growth studies as affected by drought and late season high temperature stress under arid environment. Front Plant Sci 7:795

Ismail A, Hall A, Close T (1999) Allelic variation of a dehydrin gene co segregates with chilling tolerance during seedling emergence. Proc Natl Acad Sci U S A 96:9

Jain M, Tiwary S, Gadre R (2010) Sorbitol-induced changes in various growth and biochemical parameters in maize. Plant Soil Environ 56:263–267

Janmohammadi M, Dezfuli PM, Sharifzadeh F (2008) Seed invigoration techniques to improve germination and early growth of inbred line of maize under salinity and drought stress. Gen Appl Plant Physiol 34:215–226

Jensen ME (1973) Consumptive use of water and irrigation water requirements. American Society of Civil Engineers, New York

Jia J, Fu J, Zheng J, Zhou X, Huai J, Wang J, Wang M, Zhang Y, Chen X, Zhang J, Zhao J, Su Z, Lv Y, Wang G (2006) Annotation and expression profile analysis of 2073 full-length cDNAs from stress-induced maize (*Zea mays* L) seedlings. Plant J 48:710–727

Kabiri R (2010) Effect of Salicylic acid to reduce the oxidative stress caused by drought in the hydroponic cultivation of *Nigella sativa* (*Nigella sativa*). MA thesis, Kerman University Shahid Bahonar

Kafi M, Damghany Mahdavi A (1999) Mechanism of resistance of plants to environmental stresses. University of Mashhad

Kataria S, Baghel L, Guruprasad KN (2017) Pre-treatment of seeds with static magnetic field improves germination and early growth characteristics under salt stress in maize and soybean. Biocatal Agric Biotechnol 10:83–90

Kauser R, Athar HUR, Ashraf N (2006) Chlorophyll fluorescence: a potential indicator for rapid assessment of water stress tolerance in canola (*Brassica napus* L.). Pak J Bot 38:1501–1509

Kerchev PI, Waszczak C, Lewandowska A et al (2016) Lack of glycolate oxidase-1, but not glycolate oxidase-2, attenuates the photorespiratory phenotype of catalase-2 deficient Arabidopsis. Plant Physiol 171:1704–1719

Keyvan S (2010) The effects of drought stress on yield, relative water content, proline, soluble carbohydrates and chlorophyll of bread wheat cultivars. J Anim Plant Sci 8:1051–1060

Khan MB, Hussain N, Iqbal M (2001) Effect of water stress on growth and yield components of maize variety YHS 202. J Res (Sci) 12:15–18

Khodarahmpour Z (2011) Effect of drought stress induced by polyethylene glycol (PEG) on germination indices in corn (*Zea mays* L.) hybrids. Afr J Biotechnol 10:18222–18227

Koag MC, Fenton RD, Wilkens S, Close TJ (2003) The binding of maize DHN1 to lipid vesicles. Gain of structure and lipid specificity. Plant Physiol 131:309–316

Koag MC, Wilkens S, Fenton RD, Resnik J, Vo E, Close TJ (2009) The K segment of maize DHN1 mediates binding to anionic phospholipid vesicles and concomitant structural changes. Plant Physiol 150:1503–1514

Kramer P, Boyer S (1995) Water relations plants and soils. Academic, New York

Krasensky J, Jonak C (2012) Drought, salt, and temperature stress-induced metabolic rearrangements and regulatory networks. J Exp Bot 63:1593–1608

Lambers H, Atkin OK, Scheureater I (1996) Respiratory patterns in roots in relation to their function. In: Waisel Y (ed) Plant roots. The hidden half. Marcel Dekker, New York

Lawlor DW, Cornic G (2002) Photosynthetic carbon assimilation and associated metabolism in relation to water deficits in higher plants. Plant Cell Environ 25:275–294

Lee SC, Lee MY, Kim SJ, Jun SH, An G, Kim SR (2005) Characterization of an abiotic stress-inducible dehydrin gene, OsDhn1, in rice (*Oryza sativa* L.). Mol Cell 19:212–218

Lichtfouse E, Navarrete M, Debaeke P, Souchère V, Alberola C, Menassieu J (2009) Agronomy for sustainable agriculture: a review. Agron Sustain Dev 29:1–6

Lopes MS, Araus JL, van Heerden PDR, Foyer CH (2011) Enhancing drought tolerance in C_4 crops. J Exp Bot 62:3135–3153

Ludlow MM, Muchow RC (1990) A critical evolution of traits for improving crop yields in water-limited environments. Adv Agron 43:107–153

Mabhaudhi T (2009) Responses of maize (*Zea mays* L.) landraces to water stress compared with commercial hybrids. MSc thesis, University of KwaZulu-Natal, Pietermaritzburg, South Africa

Mansfield TA, Atkinson CJ (1990) Stomatal behavior in water-stressed plants. In: Stress responses in plants: adaptation and acclimation mechanisms. Wiley-Liss Inc., pp 241–264

Maroco JP, Pereira JS, Chaves MM (1997) Stomatal responses to leaf-to-air vapor pressure deficit in Sahelian species. Aust J Plant Physiol 24:381–387

Medrano H, Escalona JM, Bota J, Gulias J, Flexas J (2002) Regulation of photosynthesis of C3 plants in response to progressive drought. Stomatal conductance as a reference parameter. Ann Bot 89:895–905

Mengel K, Kirkby EA (2001) Principles of plant nutrition, 5th edn. Kluwer Academic Publishers, Dordrecht, p 849

Mohammadkhani N, Heidari R (2008) Effects of drought stress on soluble proteins in two maize varieties. Turk J Biol 32:23–30

Munné-Bosch S, Penuelas J (2003) Photo and antioxidative protection, and a role for salicylic acid during drought and recovery in fieldgrown *Phillyrea angustifolia* plants. Planta 217:758–766

Navara L, Jesko T, Duchoslav S (1994) Participation of seminal roots in water uptake by maize root system. Biologia (Bratislava) 49:91–95

Nejad SK, Bakhshande A, Nasab SB, Payande K (2010) Effect of drought stress on corn root growth. Rep Opin 2:1–7

Nilsen ET, Orcutt DM (1996) Physiology of plants under stress: abiotic factors. Wiley, New York

Niyogi KK (1999) Photoprotection revisited: genetic and molecular approaches. Annu Rev Plant Physiol 50:333–359

Nonami H (1998) Plant water relations and control of cell elongation at low water potentials. J Plant Res 111:373–382

Noorka IR, Tabasum S (2015) Dose-response behaviour of water scarcity towards genetical and morphological traits in spring wheat (*Tricticum aestivum* L.). Pak J Bot 47:1225–1230

Obidiegwu JE, Bryan GJ, Jones HG, Prashar A (2015) Coping with drought: stress and adaptive responses in potato and perspectives for improvement. Front Plant Sci 6:542

Peleg Z, Reguera M, Tumimbang E, Walia H, Blumwald E (2011) Cytokinin-mediated source/sink modifications improve drought tolerance and increase grain yield in rice under water-stress. Plant Biotechnol J 9:747–758

Pierik R, Sasidharan R, Voesenek LACJ (2007) Growth control by ethylene: adjusting phenotypes to the environment. J Plant Growth Regul 26:188–200

Pospíšilová J (2003) Participation of phytohormones in the stomatal regulation of gas exchange during water stress. Biol Plant 46:491–506

Radić V, Beatović D, Jelena M (2007) Salt tolerance of corn genotypes (*Zea mays* L.) during germination and later growth. J Agric Sci 52:115–120

Rahmani N (2006) Effect of irrigation and nitrogen application on the quantity and quality of medicinal plant Marigold (*Calendula Officinalis* L.). Master Thesis Agronomy, University of Takestan Branch. Responses of plants to drought stress. J Afr Agric Res 6:2026–2032

Rajala A, Peltonen-Sainio P (2001) Plant growth regulator effects on spring cereal root and shoot growth. Agron J 9:936–943

Ramagopal S (1993) Advances in understanding the molecular biology of drought and salinity tolerance in plants—the first decade. In: Lodha ML, Mehta SL, Ramagopal S, Srivastava GP (eds) Adv Pl Biotech Biochem. Indian Society of Agriculture Biochemists, Kanpur, pp 39–48

Rauf M, Munir M, Hassan M, Ahmad M, Afzal M (2007) Performance of wheat genotypes under osmotic stress at germination and early seedling growth stage. Afr J Biotechnol 6:971–975

Reddy AR, Chaitanya KV, Vivekanandan M (2004) Drought-induced responses of photosynthesis and antioxidant metabolism in higher plants. J Plant Physiol 161:1189–1202

Rucker KS, Kvien CK, Holbrook CC, Hook JE (1995) Identification of peanut genotypes with improved drought avoidance traits. Peanut Sci 24:14–18

Sage RF, Zhu XL (2011) Exploiting the engine of C_4 photosynthesis. J Exp Bot 62:2989–3000

Saini HS (1997) Effects of water stress on male gametophyte development in plants. Sex Plant Reprod 10:67–73

Sairam RK, Deshmukh PS, Shukla DS, Ram S (1990) Metabolic activity and grain yield under moisture stress in wheat genotypes. Indian J Plant Physiol 33:226–231

Savvas D, Ntatsi G (2015) Biostimulant activity of silicon in horticulture. Sci Hortic 196:66–81

Schena M, Shalon D, Davis RW, Brown PO (1995) Quantitative monitoring of gene expression patterns with a complimentary DNA microarray. Science 270:467–470
Schussler JR, Westgate ME (1994) Increasing assimilate reserve does not prevent kernel abortion at low water potential in maize. Crop Sci 35:1074–1080
Setter TL, Flannigan BA, Melkonian J (2001) Loss of kernel set due to water deficit and shade in maize: carbohydrate supplies, abscisic acid, and cytokinins. Crop Sci 41:1530–1540
Shamim F, Saqlan SM, Athar HUR, Waheed A (2014) Screening and selection of tomato genotypes/cultivars for drought tolerance using multivariate analysis. Pak J Bot 46:1165–1178
Sharma P, Jha AB, Dubey RS, Pessarakli M (2012) Reactive oxygen species, oxidative damage and antioxidative defense mechanism in plants under stressful conditions. J Bot:217037
Sharp RE, LeNoble ME (2002) ABA, ethylene and the control of shoot and root growth under water stress. J Exp Bot 53:33–37
Shinozaki K, Yamaguchi-Shinozaki K (2007) Gene networks involved in drought stress response and tolerance. J Exp Bot 58:221–227
Siddique MRB, Hamid A, Islam MS (2001) Drought stress effects on water relations of wheat. Bot Bull Acad Sin 41:35–39
Stewart JD, Zine El-Abidine A, Bernier PY (1995) Stomatal and mesophyll limitations of photosynthesis in black spruce seedlings during multiple cycles of drought. Tree Physiol 15:57–64
Stoskopf CN (1981) Understanding crop production. Reston Publishing Company Inc., Reston, pp 91–99
Sun CX, Li MQ, Gao XX, Liu LN, Wu XF, Zhou JH (2016) Metabolic response of maize plants to multi-factorial abiotic stresses. Plant Biol 18:120–129
Taiz L, Zeiger E (2006) Plant physiology, 4th edn. Sinauer Associates Inc., Sunderland
Tardieu F (2012) Any trait or trait-related allele can confer drought tolerance: just design the right drought scenario. J Exp Bot 63:25–31
Tezara W, Mitchell VJ, Driscoll SP, Lawlor DW (1999) Water stress inhibits plant photosynthesis by decreasing coupling factor and ATP. Nature 401:914–917
Tiwary S, Jain M (2016) Impact of osmotic stress on biochemical and physiological parameters in *Zea mays* L. cv. Ganga safed-2 genotype. Maize Genom Genet 7:1–14
Tiwary S, Gadre R, Jain M (2016) Osmotic stress induced changes in chlorophyll biosynthesis and antioxidative system in light grown maize leaves. J Bio Innov 5:123–137
Tuberosa R, Sanguineti MC, Landi P, Giuliani MM, Salvi S, Conti S (2002) Identification of QTLs for root characteristics in maize grown in hydroponics and analysis of their overlap with QTLs for grain yield in the field at two water regimes. Plant Mol Biol 48:697–712
Turner NC, Toole JCO, Cruz RT, Yambao EB, Ahmad S, Narnuco OS, Dingkhum M (1986) Response of seven diverse rice cultivars to water deficits. Osmotic adjustment, leaf elasticity, leaf expansion, leaf death, stomatal conductance and photosynthesis. Field Crops Res 13:273–286
Turner NC, Wright GC, Siddique KHM (2001) Adaptation of grain legumes (pulses) to water-limited environments. Adv Agron 71:123–231
Ueno O, Agarie S (2005) Silica deposition in cell walls of the stomatal apparatus of rice leaves. Plant Prod Sci 8:71–73
Vaishnav J, Jain M (2015) Influence of halopriming and hydropriming on seed germination and growth characteristics of *Zea mays* L. cv. GSF-2 under salt stress. Res J Chem Environ 19:1–6
Von Caemmerer S (2000) Biochemical models of leaf photosynthesis. CSIRO Publishing, Collingwood, p 2
Wang H, Miyazak S, Kawai K, Deyholos M, Galbraith DW, Bohnert HJ (2003) Temporal progression of gene expression responses to salt shock in maize roots. Plant Mol Biol 52:873–891
Wilkinson S, Davies WJ (2002) ABA-based chemical signalling: the coordination of responses to stress in plants. Plant Cell Environ 25:195–210
Wilkinson S, Davies WJ (2010) Drought, ozone, ABA and ethylene: new insights from cell to plant to community. Plant Cell Environ 33:510–525
Yang PM, Huang QC, Qin GY, Zhao SP, Zhou JG (2014) Different drought-stress responses in photosynthesis and reactive oxygen metabolism between autotetraploid and diploid rice. Photosynthetica 52:193–202

Yokota A, Kawasaki S, Iwano M, Nakamura C, Miyake C, Akashi K (2002) Citrulline and DRIP-1 protein (ArgE Homologue) in drought tolerance of wild watermelon. Ann Bot 89:825–832
Yu LX, Setter TL (2003) Comparative transcriptional profiling of placenta and endosperm in developing maize kernels in response to water deficit. Plant Physiol 131:568–582
Zargar SM, Agnihotri A (2013) Impact of silicon on various agro-morphological and physiological parameters in maize and revealing its role in enhancing water stress tolerance. Emir J Food Agric 25:138–141
Zhang XQ, Huang GQ, Bian XM, Zhao QG (2012) Effects of nitrogen fertilization and root separation on the plant growth and grain yield of maize and its rhizosphere microorganisms. J Appl Ecol 23:3369–3376
Zhao TJ, Sun S, Liu Y, Liu JM, Liu Q, Yan YB, Zhou HM (2006) Regulating the drought-responsive element (DRE)-mediated signaling pathway by synergic functions of trans-active and transinactive DRE binding factors in *Brassica napus*. J Biol Chem 281:10752–10759
Zheng J, Zhao J, Tao Y, Wang J, Liu Y, Fu J, Jin Y, Gao P, Zhang J, Bai Y, Wang G (2004) Isolation and analysis of water stress induced genes in maize seedlings by subtractive PCR and cDNA macroarray. Plant Mol Biol 55:807–823
Zinselmeier C, Sun Y, Helentjaris T, Beatty M, Yang S, Smith H, Habben J (2002) The use of gene expression profiling to dissect the stress sensitivity of reproductive development in maize. Field Crop Res 75:111–121

Temperature Extremes: Impact on Rice Growth and Development

Sadam Hussain, Abdul Khaliq, Basharat Ali, Hafiz Athar Hussain, Tauqeer Qadir, and Saddam Hussain

Introduction

Rice is an important cereal crop, feeding approximately four billion people worldwide. It is the second most cereal consumed after wheat and fulfills 80% calories requirements for more than half of the world's population (FAO 2008). The demands for rice production are increasing due to continuous population growth. The rate of population growth is higher in rice consuming regions such as Asia, and Africa (Dawe 2008). Rice can be cultivated at 50°N in China to 35°S in South Wales in Australia and Uruguay with more than 3000 mm rainfall, but also in desert regions with less than 50 mm rainfall during the growing season (Krishnan et al. 2011). The production of rice is susceptible in the tropical, subtropical, and temperate areas due to continuous events of high (heat stress) and low (cold stress) temperature. In addition, the IPCC (2014) also predicted that extreme temperature stress would be increasing in coming years.

Plant growth and development is dependent on surrounding temperature and every species showed different optimal temperatures for growth, development, and reproduction (Hatfield et al. 2008, Hatfield and Prueger 2011). The optimal temperature for rice cultivation is between 25 and 35 °C, and rice growth is favored in an area with moderate temperature (Reyes et al. 2003). Temperature beyond optimum is detrimental for rice and negatively affects growth, development and ultimately reduces the grain yield (Fahad et al. 2015).

S. Hussain · A. Khaliq · B. Ali · T. Qadir · S. Hussain (✉)
Department of Agronomy, University of Agriculture, Faisalabad, Punjab, Pakistan
e-mail: shussain@uaf.edu.pk

H. A. Hussain
College of Agronomy and Biotechnology, Southwest University, Chongqing, China

Institute of Environment and Sustainable Development in Agriculture, Chinese Academy of Agricultural Sciences, Beijing, China

M. Hasanuzzaman et al. (eds.), *Plant Abiotic Stress Tolerance*,
https://doi.org/10.1007/978-3-030-06118-0_6

In tropical and subtropical climates, high temperature is a major limiting factor for rice production (Wahid et al. 2007). The impact of high temperature stress mainly depends on duration, intensity, and timing of stress; however, it is more harmful during reproductive stage (Cao et al. 2008; Tenorio et al. 2013). High temperature affects the reproductive and developmental stages of rice by decreasing plant height and root elongation, causing poor anther dehiscence and spikelet sterility, and hindering the process of pollination (Jagadish et al. 2010). The predicted mean annual temperature increase is 0.7–0.9 °C per decade in Southeast Asia, which associated to 4.8 °C by 2100 (IPCC 2014; Asian Development Bank (ADB) 2009; Burhan et al. 2017). Peng et al. (2004) and Fahad et al. (2016) explain that every 1 °C increase in temperature may decrease the grain yield by 10% in rice.

The occurrence of cold stress is common in many Asian countries (Koike et al. 1990; Zhang et al. 2014a, b). In other regions, such as East and West Africa, Europe, South America, and the United States, rice crops must be cold tolerant as the frequent incidence of cold stress threatens its productivity. The southernmost state of Brazil (Rio Grande do Sul) produces more than 60% of the total rice production in South America and is prone to low temperatures. This region (South America), along with Uruguay and Argentina, is prevalent for the cultivation of indica rice cultivars. Rice, being a crop of tropical to subtropical climate, is very sensitive to cold stress than other cereals (Hussain et al. 2016a, b). The physiological causes of yield losses in rice from cold stress vary between vegetative and reproductive stage. During vegetative growth, low temperature reduces seedling vigor (Ali et al. 2006), decreases the number of tillers (Shimono et al. 2002), enhances mortality rate (Farrell et al. 2006a, b; Baruah et al. 2009; Fujino et al. 2004a, b) and extends the growth period of rice (Alvarado and Hernaiz 2007). At reproductive stage, cold stress affects the sink formation (Kuroki et al. 2007; Oliver et al. 2007), causes panicle sterility, hampers grain formation, affects the grain quality and ultimately reduces yield (Shimono et al. 2002; Jena et al. 2012).

Recent studies on the impacts of heat (Cao et al. 2008; Tenorio et al. 2013) and cold stresses (Jena et al. 2012; Zhang et al. 2014b) mainly focused on reproductive processes and grain yield. These reviews demonstrated the limited facts about the consequence of temperature stress (cold and heat) on reproductive stage. In this chapter, the effects of heat and cold stress on different stages of rice are discussed, and strategies and opportunities for improving the rice tolerance to heat and cold stress are described.

Heat Stress in Rice

Vegetative Growth

Among different climatic factors, increasing temperature adversely affects the crop growth and productivity (Southworth et al. 2000). Temperature is a main driving force for development and normal growth of rice (Kropff et al. 1995). Optimum temperature ranges vary between 27 and 32 °C (Yin et al. 1996). Heat stresses can

Table 1 Effect of high temperature on growth and development of rice

Reference	Study traits	Growth stage	Effects
Soleymani and Shahrajabian (2012)	Temperature stress at germination stage	Germination	Heat stress reduces the coleoptile and radicle length, radicle weight, germination rate
Prasad et al. (2006a)	Heat stress and number of leaves	–	Heat stress can result in significant increases in leaf number
Ashraf and Hafeez (2004); Rodríguez et al. (2005)	Heat stress and photosynthesis	–	During photosynthesis, heat stress altered the thylakoid structure, swelling of grana, and loss of grana staking
Morales et al. (2003)	Activity of photosystem II, under heat stress	–	Heat stress reduces the activity of photosystem (PS II) during photosynthesis
Hurkman et al. (2009)	Photosynthesis rate of different cultivars under heat stress	–	High temperature (33 °C) for 5 days decreased the photosynthesis rate by 15% and 16% in the variety T219 and Shuanggui 1, respectively
Mohammed and Tarpley (2009)	Respiration under heat stress	–	Under high temperature stress, a negative association present between rice yield and leaf respiration
Cao et al. (2008)	Capacity of pollen production under heat stress	Pollen development	Heat stress possesses the negative effects on pollen production and poor development of microspores
Jagadish et al. (2007)	Heat stress at anthesis stage	Anthesis	At anthesis, temperature above 35 °C for 5 days results in complete failure of seed production
Prasad et al. (2006)	Pollination process under heat stress	Pollination and fertilization	High temperature significantly reduces the pollen production and the shedding of pollen grain during the flowering
Tashiro and Wardlaw (1991)	Effects of heat stress during grain development period	Grain development	High temperature (>25 °C) shortens the grain filling period and reduces average grain weight

influence crop performance at any stage of development, but the seed growth, emergence and developmental phase is the most sensitive in this regard (Table 1). Frequent episodes of heat stress also reduce the coleoptile and radicle length, radicle weight, germination rate (Soleymani and Shahrajabian 2012), seed vigor (Grass and Burris 1995), and dry mass (Wahid et al. 2007). While seed germination stage plays a vital role for sustainable cropping, and considered as the most sensitive stage to heat stress (Spiertz et al. 2006; Dias et al. 2011). These intraspecific and interspecific variations at various growth stages (Ashraf 2004) may increase early germination and faster emergence to cope with antagonistic germination conditions (McDonald 2000). In rice, panicle differentiation occurs between 18 and 30 °C. Panicle number will increase if air temperature is below 20 °C (Yamamoto et al. 1985). After tillering stage, heat stress in rice decreases the number of panicles. Under heat stress, brassinolide (a growth regulator hormone) plays an important role in protection of rice seedling from high temperature stress (Cao and Zhao 2008). After emergence stage, root structure in new seedlings showed higher weight than shoot, at that time, soil temperature also impacts their growth. Hence, it is essential to investigate the specific seed germination traits and seedling growth when influenced by high air and soil temperature.

At high temperature, leaf-elongation rate increases but the total duration for elongation decreases. Heat stress also results in premature senescence of leaves, which cause the earlier cessation of grain filling in temperate rice cultivar (Kim et al. 2011). Cell growth and division are two major primary processes, which are involved in plant growth under heat stress and also stimulate the cell elongation and division. Baker and Allen (1993) reported that the biomass of rice plant decreased by 16% when temperature increased from 25 to 27 °C. Biomass production increased by 13–16% when temperature increased from 25 to 28 °C (Oh-e et al. 2007), but no difference in biomass production was observed when temperature increased from 25 to 31 °C (Kim et al. 1996). In summary, heat stress affects almost all the developmental stages of rice, from emergence to harvesting. During the early growth stage, temperature stress reduces the coleoptile length and radicle length, radicle weight, germination rate, seed vigor, and dry mass. After tillering stage, heat stress in rice may decrease the number of panicles.

Photosynthesis and Respiration

Among different physiological process, photosynthesis is one of the most heat sensitive processes in plants (Crafts-Brandner and Salvucci 2002; Biswal et al. 2011). Heat stress has a great influence on the photosynthesis capacity in C3 plants (Yang et al. 2006). In chloroplast, the photochemical reaction in thylakoid lamellae and carbon metabolism of stroma are considered as the primary injury sites under high temperature (Wang et al. 2009; Marchand et al. 2005). Heat stress altered the thylakoid structure, swelling of grana, and loss of grana staking (Rodríguez et al. 2005; Ashraf and Hafeez 2004). Heat stress also reduces the activity of photosystem II (PS II) (Morales et al. 2003) and reduces the aggregate of photosynthetic pigments (Marchand et al. 2005). Closure of stomata under heat stress is a major reason for impaired photosynthesis (Ashraf and Hafeez 2004). In rice plant, high temperature (33 °C) for 5 days decreased the photosynthesis rate by 15% and 16% in the variety T219 and Shuanggui 1, respectively (Hurkman et al. 2009). This reduction is mainly due to the reduction of soluble protein and rubisco binding protein under heat stress (Sumesh et al. 2008). Heat stress also affects the sucrose and starch synthesis (Rodríguez et al. 2005; Djanaguiraman et al. 2009), reduces the leaf water potential, leaf area, and premature leaf senescence which negatively reduce the total photosynthesis (Greer and Weedon 2012; Young et al. 2004).

Plant growth and development is highly dependent on respiration. Higher dark respiration rate may be associated with increased amount of substrate for respiration during the ripening period. Morita et al. (2004) concluded from their studies that increased respiration loss results in decreased average grain weight. Under high temperature stress, a negative association exists between rice yield and leaf respiration (Mohammed and Tarpley 2009). Although, the effects of heat stress on photosynthesis and respiration have been studied in recent years; however, a deep understanding of its impact on photosynthesis and respiration during different stages remains elusive.

Reproductive Stage

Reproductive phase spans initiation of panicle through to physiological grain maturity. This section summarizes the recent knowledge about the heat stress on flower initiation, pollen development, flowering and anthesis, and grain yield (Table 1).

Flower Initiation and Development

In rice, booting and heading stages are sensitive stages to heat stress. During the heading stage, heat stress restricts the swelling of pollen grains. While heat stress during the anthesis stage can decline the floral production due to embryo absorption (Matsui et al. 2000), deformed floral organs also appear along with a reduction in size and number (Cao et al. 2008).

Pollen Development

In rice, heat stress reduces the pollen viability and its germination (Jagadish et al. 2010) and also disturbs a number of tapetum functions which required for pollen germination and anther dehiscence on stigma in rice plant (Endo et al. 2009). Heat stress may result in loss of pollen viability and spikelet sterility (Das et al. 2014). Heat stress has a negative impact on pollen production and poor development of microspores (Cao et al. 2008). Reproductive structure such as pores in anther, stigma length, pollen number, and protein expression in anther also altered under heat stress (Jagadish et al. 2010). Before flowering, heat stress reduced the pollen viability and spikelet fertility and during the landing of pollen grain, water content in pollen is modified under environment fluctuations (Das et al. 2014). The water contents of pollen are essential for production and dispersion of pollen grains.

Anthesis

At anthesis stage, heat stress reduced the anther dehiscence in rice, and an episode of high temperature (>39 °C) caused anther dehiscence one day before flowering (Matsui and Omasa 2002). Anthesis is critical phase under heat stress (Matsui and Omasa 2002), and heat stress at anthesis reduces the anther dehiscence as well as affects the pollen germination, pollen tube growth, and pollination (Prasad et al. 2006). At anthesis, temperature above 35 °C for 5 days results in complete failure of seed production (Jagadish et al. 2007). Under extreme capricious climate, plants adopt a number of avoidance mechanisms to cope with heat stress by making canopy cool through effective transpiration. Under heat stress, spikelet sterility declined with early morning flowering by completing flower opening before temperatures reached 35 °C, a general threshold value for spikelet sterility (Jagadish et al. 2008).

Pollination and Fertilization

Pollination is a temperature-sensitive process (Matsui et al. 2000), and factors contributing to pollination such as pollen viability, reception, and its germination play a vital role in crop production. High temperature adversely affects pre-zygotic and post-zygotic reproductive development more than vegetative development, and post-fertilization development is less sensitive than pre-fertilization development (Cao et al. 2008). High temperature stress significantly reduces the pollen production and the shedding of pollen grain during the flowering (Prasad et al. 2006), also restricts the swelling of pollen grain and reduces the pollen grain release. Furthermore, after anthesis, heat stress results in poor pollen germination and inhibits the pollen tube growth (Tang et al. 2008). In summary, high temperature stress at flower initiation leads to poor anther dehiscence and results in spikelet sterility and during anthesis may reduce the pollination and formation of pollen grain. In rice, heat stress delayed the flowering, and affects pre-fertilization development more than post development. During flowering, heat stress reduces the pollen production and severe loss in grain yield.

Grain Development and Yield Formation

During the grain filling, fluctuation in ambient temperature disturbs the proportion of amylose to amylopectin in the endosperm (Ahmed et al. 2008) and starch accumulation (Umemoto et al. 1995; Ito et al. 2009), which are the major components of grain and major contributors in the formation of its weight. High temperature influences the grain filling by disturbing the enzyme activity, which plays vital role in starch synthesis (Oh-e et al. 2007). At ripening, heat stress results in chalky grains, which are loosely packed, and round shaped (Mitsui et al. 2013). During grain development, high temperature influences the starch composition and altered the fine amylopectin structure, which is necessary for chalkiness of rice grain (Inouchi et al. 2000). In rice, heat stress reduces the grain filling period and 1 °C rise in ambient temperature (25 °C), decreased the development period up to 3 days, which decreases the fraction of mature grain and reduces average grain weight (Tashiro and Wardlaw 1991). During the primary maturation stage, heat stress boosted the buildup of all types of storage protein, while the accumulation of prolamin protein decreased at maturity (Lin et al. 2010). Less prolamin in chalky grain rice, representing an association between prolamin and chalky structure in rice grain, and that high temperature stress affect the expression of storage protein during grain filling phase (Lin et al. 2010). At flowering and grain filling, heat stress decreases the grain yield. High temperature in the night reduces the grain weight, as compared with control (22/22 °C) and day-time temperature (34/22 °C) (Morita et al. 2005). In rice, heat stress reduces the translocation of assimilates to spikelets (Cao et al. 2008), results in low amylose content and reduces the milling quality (Zhang et al. 2014a). In summary, heat stress at grain filling reduces the starch synthesis in grain, resulting in less starch accumulation during ripening stage of rice. At ripening, heat stress results in chalky grain with an abnormal shape, reduces grain development phase and decreases the grain weight and yield of rice.

Cold Stress in Rice

Vegetative Growth

Rice is a cold sensitive crop (Nakagahra et al. 1997; Table 2) and 15 million hectares of rice crop planted in the world exposed to low temperature. At low temperature, rice plants demonstrate a wide range of damage, depending on the stage of development and duration of exposure (Kim et al. 2012). Germination, booting, flowering, and filling stages are comparatively more sensitive to cold stress (Satake 1976). Germination ability is important for better stand establishment, and several traits such as root length, shoot length, and seed vigor are associated with germination ability. Optimum temperature for early seedling and germination ranges between 25 and 35 °C (Fujino et al. 2004a, b). Cold stress during germination stage results in poor germination and early seedling rot, reduces the seedling or complete death of seedling (Han et al. 2002; Lou et al. 2007; Hussain et al. 2018) and results in poor seedling vigor (Reyes et al. 2003).

Cold stress can equally injure or damage both the vegetative and generative organs in rice, and low temperature at vegetative stage reduces the seedling vigor and growth (Ali et al. 2006), reduces number of seedling and tillering (Shimono et al. 2002), increases plant mortality (Farrell et al. 2006a, b; Baruah et al. 2009; Fujino et al. 2004a, b) and extends growth period (Ghadirnezhad and Fallah 2014).

Table 2 Effect of cold stress on growth and development of rice

Reference	Study traits	Growth stage	Effects
Han et al. (2002); Lou et al. (2007)	Cold stress during germination and seedling establishment	Seedling establishment	Cold stress during germination stage results in poor germination and early seedling rot, reduces the seedling or complete death of seedling
Shimono et al. (2002)	Cold stress during tillering stage	Tillering	Cold stress reduces the tillering
Janda et al. (1994); Van Heerden et al. (2002); Li et al. (2004)	Cold stress on the photosynthesis	–	Cold stress substantially reduces the photosynthesis rate
Kanneganti and Gupta (2008); Kim et al. (2009); Sharma et al. (2005)	Photosynthesis under the cold stress	–	In rice, cold stress affects the florescence and chlorophyll content, thus inhibiting the photosynthesis
Shrestha et al. (2013)	Cold stress during the flowering stage	Flowering	Cold temperature stress, with non-inductive photoperiod, results in subsequent flowering and delay in panicle initiations
Mamun et al. (2006)	Cold stress during the meiotic stage	Gametophytes development	During the meiotic stage, cold stress declines the anther respiration, results in sucrose accumulation in anther, changes the composition if amino acids, causes protein degradation and increases the asparagine concentrations
Saito et al. (2001); Oliver et al. (2007)	Cold stress effect on reproductive stage	Reproductive	Cold stress declines the grain filling period, which leads to small grain size
Oliver et al. (2007)	Cold stress during grain development stage	Reproductive stage	During the grain development, cold stress results in partial and late maturation of grain

Photosynthesis and Respiration

Photosynthesis is a key phenomenon, which significantly contributes to the plant growth and development. In cold sensitive plants, rate of photosynthesis is highly dependent on the temperature conditions, and cold stress substantially reduces the photosynthesis rate (Janda et al. 1994; Van Heerden et al. 2002; Li et al. 2004; Fig. 1), by increasing the concentration of soluble sugars and decreasing the starch content in all the organs (Jouve et al. 1993). Chlorophyll content and fluorescence are two main photosynthesis properties which are negatively influenced by cold stress. In rice, cold stress affects the florescence and chlorophyll content, and ultimately reduces the photosynthesis rate (Kanneganti and Gupta 2008; Kim et al.

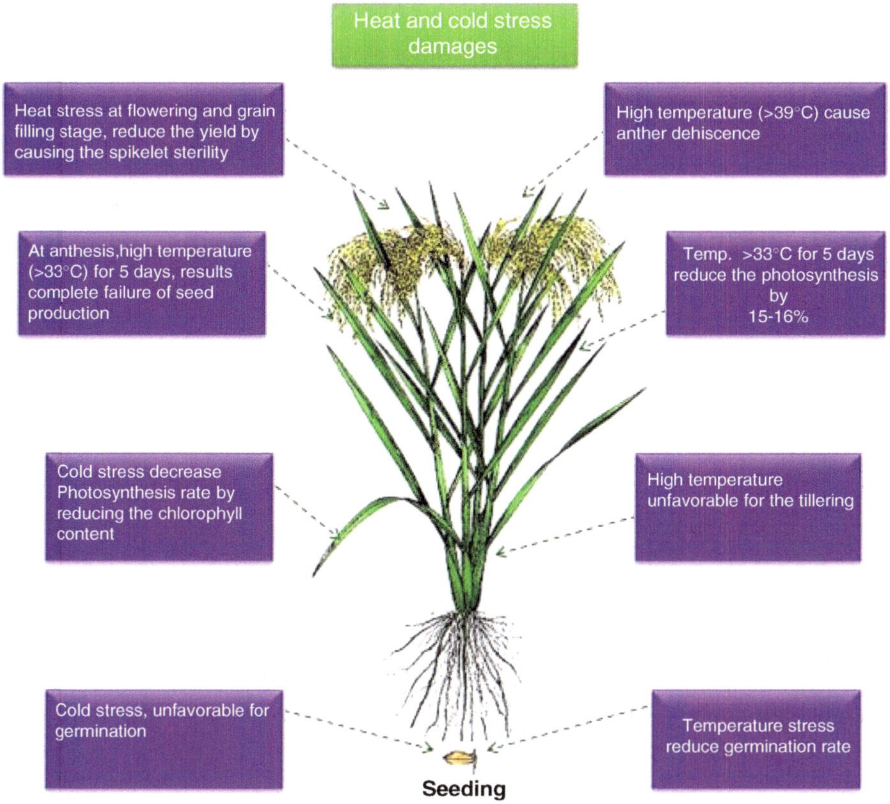

Fig. 1 Effect of temperature stresses on different growth stages of rice

2009; Sharma et al. 2005). Cold stress also inhibits the chlorophyll synthesis, increases the membrane permeability (Zeng et al. 2000), damages the chloroplast (Chen et al. 1997), and finally decreases the photosynthetic activity. In chloroplast stroma, cold stress lowered the activity of dark reaction (Weeden and Buchanan 1983), reduced the light dependent reaction activity in thylakoid membrane (Li et al. 1990) and decreased the carbon dioxide (CO_2) assimilation activity. Cold stress also reduces the Hill reaction activity and inhibited the electron transmission activity (Kaniuga et al. 1979).

In chilling sensitive crops, low temperature changed the respiratory rate and resulted in destruction of mitochondrial structure and also reduced the activity of some enzymes by reducing the kinetic energy (Lyons et al. 1979; Munro et al. 2004). At early growth stages, cold stress reduces the cytochrome path of electron transport and enhances alternative respiratory pathways (Prasad et al. 1994; Ribascarbo et al. 2000). In short, cold stress in rice reduces the photosynthesis activity by reducing the chlorophyll content and inflorescence. Moreover, further investigations are required to investigate whether cold stress played a direct role in the photosynthetic rate and chlorophyll inflorescence parameters.

Reproductive Stage

Flower Initiation and Development

Rice is more likely to suffer from cold stress (Table 2) especially during floral development (Lu et al. 1999; Kuroki et al. 2007). In rice, panicle initiation is strongly influenced by thermal conditions prior to reproductive stage, as rice requires to accumulate the specific degree days (heat sum) before the initiation of the reproductive stage (Shrestha et al. 2013). Thus, cold temperature stress, with non-inductive photoperiod, results in subsequent flowering and delay in panicle initiations (Shrestha et al. 2013). At booting, low temperature stress inhibits the growth of pollen, which affects spikelet fertility in rice.

Gametophytes Development

In higher plants, haploid gametophytes are produced from the diploid cell through process of meiosis. Under cold stress, sensitivity of male organ increased after the onset of meiosis (Oliver et al. 2005). Low temperature causes the protein denaturation in anthers and results in substantial pollen infertility (Oliver et al. 2005). During the meiotic stage, cold stress declined the anther respiration, resulted in sucrose accumulation in anther, changes the composition if amino acids, causes protein degradation and increased the asparagine concentrations (Mamun et al. 2006). In rice plants, cold stress interferes with the formation of microspore wall (Mamun et al. 2006). A callose wall is formed before meiosis, which surrounds the microspore mother cells. This wall (callose) is critical in microspore wall formation (Mamun et al. 2006; Arshad et al. 2017). Low temperature results in early termination of callose wall, which hides segregation of microspores (involved in late meiosis), resulting in poor wall growth in consequence microspores and primary growth of microspore is very sensitive to low temperature stress (Mamun et al. 2006).

Until now, only a few research studies focused on the effect of low temperature stress at panicle initiation of female organs in rice. Since the macrospore and its early development constitute at least 50% of the reproductive process, so it is suggested that future research efforts are made on investigating the effects of low temperature stress on female flower development.

Pollen Development

In rice, low temperature affects both male and female reproductive developmental organs; however, the male part is more sensitive to cold stress (Koike et al. 1990). In temperate areas, cold stress affected the young microspores during the pollen growth (Andaya and Mackill 2003). Cold stress causes pollen sterility in rice due to disruption in the meiotic phase of pollen development, limits the starch accumulation in pollen, restricts pollen development and permanently disrupts the pollen (Oliver et al. 2005).

Anthesis

In rice, cold stress during anthesis causes the spikelet sterility and probably partial panicle exertion and spikelet absorption (Koike et al. 1990; Suh et al. 2010). Takeuchi et al. (2001) reported that floral structures are more tolerant to low temperature stress in rice during the first 3 days of anthesis. But, after 5 days, a steady decline occurs in spikelet fertility leading to spikelet sterility.

Pollination and Fertilization

Cold stress affects the flower initiation, meiosis during booting and results in ovule sterility and pollen infertility. During the meiotic stage, cold stress disrupts the anther respiration, reduces proline level, altered the compositions of amino acids and increased the asparagine levels. Early development of microspores is very sensitive to low temperature. Cold stress causes early breakdown of starch, poor development of pollen tube and results in early anthesis, which leads to the spikelet sterility in rice.

Grain Development and Yield Formation

Inferior spikelet and delay in flowering under cold stress lead to poor grain development and low grain yield. Poor grain development is attributed to a limited carbohydrates supply but this may not be an only single reason, since vital enzymes are involved in carbon metabolism. In rice, increase or decrease in amylose concentration is dependent on particular cultivar, but production of amylose is higher at low temperature compared with high temperature (Ahmed et al. 2008). At 15–20 °C, low temperature increased the amylose concentration in grain of several japonica rice cultivars (Hirano and Sano 1998). During the grain development, cold stress results in partial and late maturation of grain (Oliver et al. 2007) and grain development is regulated through source–sink relationship, which is adversely affected by temperature. Under cold stress, grain filling period and rate are declined, which lead to small grain size (Saito et al. 2001; Oliver et al. 2007). However, grain development in rice is also regulated by different plant hormones such as ABA and cytokinins (CKTs) (Oliver et al. 2005, 2007). Due to increased concentration of ABA in leaves, stomatal closure occurs, which restricts the photosynthesis due to low carbon dioxide (CO_2) within cells. In summary, amylose concentration in grain increases under cold stress conditions. Cold stress reduces the duration and rate of grain filling and results in the production of shriveled grains. Grain development in rice is mainly regulated by different plant hormones such as ABA and CTKs. ABA concentration increased in plants under cold stress, and is played as a regulator in response to stress.

Strategies to Improve Tolerance Against Heat and Cold Stresses

Fortunately, the rice species have a broad range of adaptation to temperature stress, and tolerant ecotypes are available for breeding purposes. Breeding and adaptation to cold stress are so far more successful than heat stress; however, an appropriate selection method is necessary for assessing the high temperature stress tolerance in segregating populations using controlled temperature conditions. During the different stages of rice, breeding practices have been effective in development of heat tolerance; both tolerance and avoidance are the suitable traits for initiating the breeding programs.

In rice, different crop management strategies can improve the resistance against cold and heat stresses. In tropical low-altitude systems, increasing the deepness of water up to panicle initiation is the most effective practice used for minimizing the losses due to cold stress (Singh et al. 2005). Exogenous application of growth regulating compounds or seed treatment may help plant withstand the temperature stress (Fahad et al. 2016; Wang et al. 2016). Chemical substances such as jasmonic acid, salicylic acid, and abscisic acid are found to enhance the tolerance against heat stress. Jasmonic acid which is a natural growth regulator is very helpful against stress through expression of relative genes (Farooq et al. 2009). Non-hormonal growth regulators such as paklobutrazol, triazoles, unikonazol, and chlorocholin-chloride have been used for improving the chilling tolerance of cultivated plants (Anderson et al. 1994; Lurie et al. 1994; Feng et al. 2003). Sodium benzoate, glutathione, tyrone, and formate slow down the degradation of unsaturated fatty acids and reduce the chilling damage in chilling sensitive plants (Xu et al. 2000).

Balanced use of crop nutrients is another strategy to avoid rice from succumbing to thermal stress. During changing weather conditions, less than optimal amount of nitrogen is a good practice to increase rice tolerance against cold stress. Changing the planting time is another strategy in some rice grown areas. Late planting escapes the heat stress and early planting reduces the cold stress affect (Singh et al. 2005). All these practices may be successful in prevailing climatic conditions of a region; still development of new cultivar in collaboration with biologists, agronomist, and breeders may overcome these problems more efficiently.

Conclusions and Future Studies

In rice, substantial yield loss occurs due to temperature stresses (Tables 1 and 2). Heat stress affects almost all the stages of rice, from emergence to harvesting. The seed growth, emergence, and development are severely affected by heat stress. In rice, heat stress reduces the coleoptile and radicle length, radicle weight, germination rate, leaf-elongation duration, and biomass. At reproductive stage, heat stress causes anther indehiscence and reduces the pollen dispersal. Heat stress delays flowering and affects post-fertilization development. It may increase the rate of grain filling but it shortens the period of grain filling and leads to substantial loss in yield. Heat stress at ripening stage also results in chalky grains, which are round and irregular shaped.

Cold stress also hampers the various growth and reproductive processes in rice, and it delays panicle initiation, affects anther respiration, enhances sucrose accumulation and triggers protein denaturation. Low temperature causes the pollen sterility and impairs the pollination due to inhibited anther dehiscence which ultimately reduce the final grain yield in rice. Cold stress also reduces the period for grain filling which reduces the grain size and final productivity.

In rice, high temperature stress negatively affects the mitosis; its role to encourage pollen abortion at mitosis through its effects on the tapetum is not well understood and needs to be investigated further. For improving the rice production, the response and tolerance mechanisms need to be explored at the molecular level and functional genomics approach should also be used for the understanding of molecular basis regarding rice response to thermal stress tolerance. Different ecophysiological and genomics investigations may help to understand the interaction between temperature stress and genotypes. New rice varieties for heat and cold tolerance at the different stages are needed to reduce the yield losses. Despite progress in temperature stress tolerance, heat and cold stresses still reduce rice yield, mainly where indica rice cultivars are grown. Fortunately, difference rice cultivars have widespread adaptation to low temperature stress and cold-tolerant ecotypes are presented for breeding purposes. An inclusive approach is required to identify the basic mechanisms for cold tolerance. Different crop management strategies such as balanced use of crop nutrients and changing the planting time should be studied more in the future for enhancing the rice tolerance against temperature stresses.

References

Ahmed N, Maekawa M, Tetlow IJ (2008) Effect of low temperature on grain filling, amylose content and activity of starch biosynthesis enzymes in endosperm of basmati rice. Aust J Agric Res 59:599–604

Ali MG, Naylor REL, Matthews S (2006) Distinguishing the effects of genotype and seed physiological age on low temperature tolerance of rice (*Oryza sativa* L.). Exp Agric 42(3):337–349

Alvarado R, Hernaiz S (2007) Antecedentes generales sobreel arroz en Chile. En: Alvarado R (ed), Arroz. Manejo tecnológico. Boletín INIA N° 162. Instituto de Investigaciones Agropecuarias INIA, Centro Regional de Investigación Quilamapu, Chillán, 179 p

Andaya VC, Mackill DJ (2003) QTLs conferring cold tolerance at the booting stage of rice using recombinant inbred lines from a japonica x indica cross. Theor Appl Genet 106:1084–1090

Anderson MD, Prasad TK, Martin BA, Stewart CR (1994) Differential gene expression in chilling-acclimated maize seedlings and evidence for the involvement of abscisic acid in chilling tolerance. Plant Physiol 105(1):331–9

Arshad MS, Farooq M, Asch F, Krishna JS, Prasad PV, Siddique KH (2017) Thermal stress impacts reproductive development and grain yield in rice. Plant Physiol Bioch 115:57–72

Ashraf M (2004) Some important physiological selection criteria for salt tolerance in plants. Flora 199:361–376
Ashraf M, Hafeez M (2004) Thermotolerance of pearl millet and maize at early growth stages: growth and nutrient relations. Biol Plant 48:81–86
Asian Development Bank (ADB) (2009) The economics of climate change in Southeast Asia: a regional review. Manila, The Philippines
Baker JT, Allen LHJ (1993) Effects of CO2 and temperature on rice: a summary of five growing seasons. J Agric Meteorol 48:575–582
Baruah AR, Ishigo-Oka N, Adachi M, Oguma Y, Tokizono Y, Onishi K, Sano Y (2009) Cold tolerance at the early growth stage in wild and cultivated rice. Euphytica 165(3):459–470
Biswal B, Joshi PN, Raval MK, Biswal UC (2011) Photosynthesis, a global sensor of environmental stress in green plants: stress signalling and adaptation. Curr Sci 101:47–56
Burhan A, Rasul G, Qadir T, Hussain S, Saqib M, Bukhari SAA (2017) Environmental policies to protect pollinators: attributes and actions needed to avert climate borne crisis of oil seed agriculture in Pakistan. AIMS Agric Food 2(3):233–250
Cao YY, Zhao H (2008) Protective roles of brassinolide on rice seedlings under high temperature stress. Rice Sci 15:63–68
Cao YY, Duan H, Yang LN, Wang ZQ, Zhuo SC, Yang JC (2008) Effective heat stress during meiosis on grain yield of rice cultivars differing in heat tolerance and its physiological mechanism. Acta Agron Sin 34:2134–2142
Chen SN, Zou XJ, Liang B (1997) Electron microscope observation on membrane system of leaf cells of some varieties of rice seedlings with different cold-resistance. Plant Physiol Commun 33(3):191–194. In Chinese
Crafts-Brandner SJ, Salvucci ME (2002) Sensitivity of photosynthesis in a C4 plant, maize, to heat stress. Plant Physiol 129:1773–1780
Das S, Krishnan P, Nayak M, Ramakrishnan B (2014) High temperature stress effects on pollens of rice (Oryza sativa L.) genotypes. Environ Exp Bot 101:36–46
Dawe D (2008) Agricultural research, poverty alleviation, and key trends in Asia's rice economy. In: Charting new pathways to C4 rice. pp 37–53
Dias PMB, Brunel-Muguet S, Durr C, Huguet T, Demilly D, Wagner MH, Teulat-Merah B (2011) QTL analysis of seed germination and pre-emergence growth at extreme temperature in Medicago truncatula. Theor Appl Genet 122:429–444
Djanaguiraman M, Sheeba JA, Devi DD, Bangarusamy U (2009) Cotton leaf senescence can be delayed by nitrophenolate spray through enhanced antioxidant defense system. J Agron Crop Sci 195:213–224
Endo M, Tsuchiya T, Hamada K, Kawamura S, Yano K, Ohshima M, Higashitani A, Watanabe M, Kawagishi-Kobayashi M (2009) High temperatures cause male sterility in rice plants with transcriptional alterations during pollen development. Plant Cell Physiol 50:1911–1922
Fahad S, Hussain S, Bano A, Saud S, Hassan S, Shan D, Khan FA, Khan F, Chen Y, Wu C, Tabassum MA (2015) Potential role of phytohormones and plant growth-promoting rhizobacteria in abiotic stresses: consequences for changing environment. Environ Sci Pollut Res 22(7):4907–4921
Fahad S, Hussain S, Saud S, Hassan S, Ihsan Z, Shah AN, Wu C, Yousaf M, Nasim W, Alharby H, Alghabari F, Huang J (2016) Exogenously applied plant growth regulators enhance the morpho-physiological growth and yield of rice under high temperature. Front Plant Sci 7:1250
FAO (2008) Food and agriculture statistics global outlook. Food and Agricultural Organization of the United Nations. http://faostat.fao.org/Portals/_Faostat/documents/pdf/world.pdf
Farooq M, Basra SM, Wahid A, Khaliq A, Kobayashi N (2009) Rice seed invigoration: a review. In: Organic farming, pest control and remediation of soil pollutants. Springer Dordrecht, pp 137–175
Farrell TC, Fox KM, Williams RL, Fukai S (2006a) Genotypic variation for cold tolerance during reproductive development in rice: screening with cold air and cold water. Field Crops Res 98:178–194. https://doi.org/10.1016/j.fcr.2006.01.003

Farrell TC, Fox KM, Williams RL, Fukai S, Lewin LG (2006b) Minimising cold damage during reproductive development among temperate rice genotypes. II. Genotypic variation and flowering traits related to cold tolerance screening. Aust J Agric Res 57:89–100
Feng Z, Guo A, Feng Z (2003) Amelioration of chilling stress by triadimefon in cucumber seedlings. Plant Growth Regul 39(3):277–83
Fujino K, Sekiguchi H, Sato T, Kiuchi H, Nonoue Y, Takeuchi Y, Ando T, Lin S, Yano M (2004a) Mapping of quantitative trait loci controlling low-temperature germinability in rice (Oryza sativa L.). Theor Appl Genet 108:794–799
Fujino K, Sekiguchi H, Sato T et al (2004b) Mapping of quantitative trait loci controlling low-temperature germinability in rice (Oryza sativa L.). Theor Appl Genet 108(5):794–799
Ghadirnezhad R, Fallah A (2014) Temperature effect on yield and yield components of different rice cultivars in flowering stage. Int J Agron. Article ID 846707, 4 p. https://doi.org/10.1155/2014/846707
Grass L, Burris JS (1995) Effect of heat stress during seeds development and maturation on wheat (Triticum durum) seed quality 1 Seed germination and seedling vigors. Can J Plant Sci 75:821–829
Greer DH, Weedon MM (2012) Modelling photosynthetic responses to temperature of grapevine (Vitisvinifera cv. Semillon) leaves on vines grown in a hot climate. Plant Cell Environ 35:1050–1064
Han LZ, Kou HJ, Piao ZZ (2002) Status and prospects of genetic and QTLs analysis for cold tolerance in rice. Chin J Rice Sci 16(2):193–198
Hatfield JL, Prueger JH (2011) Agroecology: implications for plant response to climate change. In: Yadav SS, Redden RJ, Hatfield JL, Lotze-Campen H, Hall AE (eds). Wiley-Blackwell, West Sussex, pp 27–43
Hatfield JL, Boote KJ, Fay P, Hahn L, Izaurralde RC, Kimball BA, Mader T, Morgan J, Ort D, Polley W, Thomson A, Wolfe D (2008) Agriculture. In: The effects of climate change on agriculture, land resources, water resources, and biodiversity in the United States
Hirano HY, Sano Y (1998) Enhancement of Wx gene expression and the accumulation of amylose in the response to cool temperatures during seed development in rice. Plant Cell Physiol 39:807–812
Hurkman WJ, Vensel WH, Tanaka CK, Whitehand L, Altenbach SB (2009) Effect of high temperature on albumin and globulin accumulation in the endosperm proteome of the developing wheat grain. J Cereal Sci 49:12–23
Hussain S, Khan F, Hussain HA, Nie L (2016a) Physiological and biochemical mechanisms of seed priming-induced chilling tolerance in rice cultivars. Front Plant Sci 7:116
Hussain S, Khan F, Cao W, Wu L, Geng M (2016b) Seed priming alters the production and detoxification of reactive oxygen intermediates in rice seedlings grown under sub-optimal temperature and nutrient supply. Front Plant Sci 7:439
Hussain HA, Hussain S, Khaliq A, Ashraf U, Anjum SA, Men S, Wang L (2018) Chilling and drought stresses in crop plants: implications, cross talk, and potential management opportunities. Front Plant Sci 9:393
Inouchi N, Ando H, Asaoka M, Okuno K, Fuwa H (2000) The effect of environmental temperature on distribution of unit chains of rice amylopectin. Starch 52:8–12
IPCC Climate Change (2014) Mitigation of climate change working group III contribution to the fifth assessment report of the intergovernmental panel on climate change. Cambridge University Press
Ito S, Hara T, Kawanami Y, Watanabe T, Thiraporn K, Ohtake N, Sueyoshi K, Mitsui T, Fukuyama T, Takahashi Y, Sato T, Sato A, Ohyama T (2009) Carbon and nitrogen transport during grain filling in rice under high-temperature conditions. J Agron Crop Sci 195:368–376
Jagadish SVK, Craufurd PQ, Wheeler TR (2007) High temperature stress and spikelet fertility in rice. J Exp Bot 58:1627–1635
Jagadish SVK, Craufurd PQ, Wheeler TR (2008) Phenotyping parents of mapping populations of rice (Oryza sativa L.) for heat tolerance during anthesis. Crop Sci 48:1140–1146

Jagadish SVK, Muthurajan R, Oane R, Wheeler TR, Heuer S, Bennet J, Craufurd PQ (2010) Physiological and proteomic approaches to address heat tolerance during anthesis in rice (Oryza sativa L.). J Exp Bot 61:143–156

Janda T, Szalai G, Kissimon J, Paldi E, Marton C, Szigeti Z (1994) Role of irradiance in the chilling injury of young maize plants studied by chlorophyll fluorescence induction measurements. Photosynthetica 30:293–299

Jena KK, Kim SM, Suh JP, Yang CI, Kim YJ (2012) Identification of cold-tolerant breeding lines by quantitative trait loci associated with cold tolerance in rice. Crop Sci 51(2):517–523

Jouve L, Engelmann F, Noirot M, Charrier A (1993) Evaluation of biochemical markers (sugar, proline, malonedialdehyde and ethylene) for cold sensitivity in microcuttings of two coffee species. Plant Sci 91:109–116

Kaniuga Z, Zabek J, Sochanoeicz B (1979) Photosynthetic apparatus in chilling-sensitive plants. Contribution of loosely bound manganese to the mechanism of reversible inactivation of Hill reaction activity following cold and dark storage and illumination of leaves. Planta 144:490–502

Kanneganti V, Gupta AK (2008) Overexpression of OsiSAP8, a member of stress associated protein (SAP) gene family of rice confers tolerance to salt, drought and cold stress in transgenic tobacco and rice. Plant Mol Biol 66(5):445–462

Kim HY, Horie T, Nakagawa H, Wada K (1996) Effects of elevated CO2 concentration and high temperature on growth and yield of rice, In: the effect on development, dry matter production and some growth characteristics. Jpn J Crop Sci 65:634–643

Kim SJ, Lee SC, Hong SK, An K, An G, Kim SR (2009) Ectopic expression of a cold responsive OsAsr1 cDNA gives enhanced cold tolerance in transgenic rice plants. Mol Cells 27(4):449–458

Kim SH, Choi H-S, Cho Y-C, Kim SR (2011) Cold-responsive regulation of a flower-preferential class III peroxidase gene, OsPOX1, in Rice (Oryza sativa L.). J Plant Biol 55(2):123–131

Kim SI, Kim DM, Tai TH (2012) Evaluation of rice seedling tolerance to constant and intermittent low temperature stress. Rice Sci 19(4):295–308

Koike S, Takeda K, Satake T (1990) Comparison of cold tolerance between isogenic pairs differing in grain size. Rice Genet Newslett 7:133–134

Krishnan P, Ramakrishnan B, Raja Reddy K, Reddy VR (2011) High temperature effects on rice growth, yield, and grain quality. In: Sparks DL (ed) Advances in agronomy, vol 111. Academic, Burlington, pp 87–206

Kropff MJ, Mathews RB, Van Laar HH, Ten Berge HFM (1995) The rice model Oryza 1 and its testing. In: Mathews RB, Kropff MJ, Bachelet D, van Laar HH (eds) Modeling the impact of climate change on rice production in Asia. CABI & IRRI, Wallingford, Oxon, and Los Banos, pp 27–50

Kuroki M, Saito K, Matsuba S, Yokogami N, Shimizu H, Ando I, Sato Y (2007) A quantitative trait locus for cold tolerance at the booting stage on rice chromosome 8. Theor Appl Genet 115:593–600

Li P, Liu HX, Wang YR, Liang CY, Guo ZY (1990) Effect of chilling on photosynthesis in flag leaves at primary heading stage of F1 hybrid rice cultivars and their parental lines (three lines). Acta Bot Sin 32(6):456–464. In Chinese

Li XG, Wang XM, Meng QW, Zou Q (2004) Factors limiting photosynthetic recovery in sweet pepper leaves after short-term chilling stress under low irradiance. Photosynthetica 42(2):257–262

Lin CJ, Li CY, Lin SK, Yang FH, Huang JJ, Liu YH, Lur HS (2010) Influence of high temperature during grain filling on the accumulation of storage proteins and grain quality in rice (Oryza sativa L.). J Agric Food Chem 58:10545–10552

Lou QJ, Chen L, Sun ZX, Xing YZ, Li J, Xu XY, Mei HW, Luo LJ (2007) A major QTL associated with cold tolerance at seedling stage in rice (*Oryza sativa* L.). Euphytica 158:87–94

Lu C, Zou J, Ikehash H (1999) Spikelet fertility affected by low temperature in indica-japonica hybrids of rice (*Oryza sativa* L.). Jpn J Trop Agric 43:254–259

Lurie S, Ronen R, Meier S (1994) Determining chilling injury induction in green peppers using nondestructive pulse amplitude modulated (PAM) fluorometry. J Am Soc Hortic Sci 119(1):59–62

Lyons JM, Raison JK, Steponkus PL (1979) The plant membrane in response to low temperature: an overview. In: Low temperature stress in crop plants: the role of the membrane. New York, pp 1–24
Mamun EA, Alfred S, Cantrill LC, Overall RL, Sutton BG (2006) Effects of chilling on male gametophyte development in rice. Cell Biol Int 30:583–591
Marchand FL, Mertens S, Kockelbergh F, Beyens L, Nijs I (2005) Performance of high arctic tundra plants improved during but deteriorated after exposure to a simulated extreme temperature event. Glob Chang Biol 11:2078–2089
Matsui T, Omasa K (2002) Rice (*Oryza sativa* L.) cultivars tolerant to high temperature at flowering: anther characteristics. Ann Bot 89:683–687
Matsui T, Omasa K, Horie T (2000) High temperature at flowering inhibits swelling of pollen grains, a driving force for thecae dehiscence in rice (*Oryza sativa* L.). Plant Prod Sci 3:430–434
McDonald MB (2000) Seed priming. In: Black M, Bewley JD (eds) Seeds technology and its biological basis. Sheffield Academic Press Ltd, Sheffield, pp 287–325
Mitsui T, Shiraya T, Kaneko K, Wada K (2013) Proteomics of rice grain under high temperature stress. Front Plant Sci 4:1–5
Mohammed AR, Tarpley L (2009) Impact of high nighttime temperature on respiration, membrane stability, antioxidant capacity, and yield of rice plants. Crop Sci 49:313–325
Morales D, Rodriguez P, Dellamico J, Nicolas E, Torrecillas A, Sanchez-Blanco MJ (2003) High-temperature preconditioning and thermal shock imposition affects water relations, gas exchange and root hydraulic conductivity in tomato. Biol Plant 47:203–208
Morita S, Shiratsuchi H, Takanashi JI, Fujita K (2004) Effect of high temperature on grain ripening in rice plants: analysis of the effects of high night and high day temperatures applied to the panicle and other parts of the plant. Jpn J Crop Sci 73:77–83
Morita S, Yonemaru JI, Takanashi JI (2005) Grain growth and endosperm cell size under high night temperatures in rice (*Oryza sativa* L.). Ann Bot 95:695–701
Munro KD, Hodges DM, DeLong JM, Forney CF, Kristie DN (2004) Low temperature effects on ubiquinone content, respiration rates and lipid peroxidation levels of etiolated seedlings of two differentially chilling-sensitive species. Physiol Plant 121(3):488–497
Nakagahra M, Okuno K, Vaughan D (1997) Rice genetic resources: history, conservation, investigative characterization and use in Japan. Plant Mol Biol 35:69–77
Oh-e I, Saitoh K, Kuroda T (2007) Effects of high temperature on growth, yield and dry-matter production of rice grown in the paddy field. Plant Prod Sci 10:412–422
Oliver SN, Van Dongen JT, Alfred SC, Mamun EA, Zhao X, Saini HS, Fernandes SF, Blanchard CL, Sutton BG, Geigenberger P, Dennis ES, Dolferus R (2005) Cold-induced repression of the rice anther-specific cell wall invertase gene OSINV4 is correlated with sucrose accumulation and pollen sterility. Plant Cell Environ 28:1534–1551
Oliver SN, Dennis ES, Dolferus R (2007) ABA regulates apoplastic sugar transport and is a potential signal for cold-induced pollen sterility in rice. Plant Cell Physiol 48:1319–1330
Peng SB, Huang JL, Sheehy JE, Laza RC, Visperas RM, Zhong X, Centeno GS, Khush GS, Cassman KG (2004) Rice yield decline with higher night temperature from global warming. Proc Natl Acad Sci U S A 101:9971–9975
Prasad TK, Anderson MD, Martin BA, Stewart CR (1994) Evidence for chilling-induced oxidative stress in maize seedlings and a regulatory role for hydrogen peroxide. Plant Cell 6:65–74
Prasad PVV, Boote KJ, Allen LH Jr (2006a) Adverse high temperature effects on pollen viability, seed-set, seed yield, and harvest index of grain-sorghum [Sorghum bicolor (L.) Moench] are more severe at elevated carbon dioxide due to higher tissue temperatures. Agric For Meteorol 139:237–251
Prasad PVV, Boote KJ, Allen LH Jr, Sheehy JE, Thomas JMG (2006b) Species, ecotype and cultivar differences in spikelet fertility and harvest index of rice in response to high temperature stress. Field Crops Res 95:398–411

Reyes BG, Myers SJ, McGrath JM (2003) Differential induction of glyoxylate cycle enzymes by stress as a marker for seedling vigor in sugar beet (Beta vulgaris). Mol Genet Genomics 269:692–698

Ribascarbo M, Aroca R, Gonzalez-Meler MA, Irigoyen JJ, Sanchezdiaz M (2000) The electron partitioning between the cytochrome and alternative respiratory pathways during chilling recovery in 2 cultivars of maize differing in chilling sensitivity. Plant Physiol 122(1):199–204

Rodríguez M, Canales E, Borrás-Hidalgo O (2005) Molecular aspects of abiotic stress in plants. Biotecnol Apl 22:1–10

Saito K, Miura K, Nagano K, Hayano-Saito Y, Araki H, Kato A (2001) Identification of two closely linked quantitative trait loci for cold tolerance on chromosome 4 of rice and their association with anther length. Theor Appl Genet 103(6–7):862–8

Satake T (1976) Determination of the most sensitive stage to sterile-type cool injury in rice plants. Res Bull Hokkaido Natl Agric Exp Stn 113:1–33

Sharma P, Sharma N, Deswal R (2005) The molecular biology of the low temperature response in plants. Bioessays 27(10):1048–1059

Shimono H, Hasegawa T, Iwama K (2002) Response of growth and grain yield in paddy rice to cool water at different growth stages. Field Crop Res 73:67–79

Shrestha S, Asch F, Brueck H, Giese M, Dusserre J, Ramanantsoanirina A (2013) Phenological responses of upland rice grown along an altitudinal gradient. Environ Exp Bot 89:1–10

Singh RP, Brennan JP, Farrell T, Williams R, Reinke R, Lewin L, Mullen J (2005) Economic analysis of breeding for improved cold tolerance in rice in Australia. Australas Agribus Rev 13:1–9

Soleymani A, Shahrajabian MH (2012) Study of cold stress on the germination and seedling stage and determination of recovery in rice varieties. Int J Biol 4(4):23. https://doi.org/10.5539/ijb.v4n4p23

Southworth J, Randolph J, Harbeck M, Doering O, Pfeifer R, Rao D, Johnston J (2000) Consequences of future climate change and changing climate variability on maize yields in the Midwestern United States. Agric Ecosyst Environ 82:139–158

Spiertz JHR, Hamer RJ, Xu H, Primo-Martin C, Don C, Van der Putten PE (2006) Heat stress in wheat (Triticum aestivum L.): effect on grain growth and quality traits. Eur J Agron 25:89–95

Suh JP, Jeung JU, Lee JI, Choi YH, Yea JD, Virk PS, Mackill DJ, Jena KK (2010) Identification and analysis of QTLs controlling cold tolerance at the reproductive stage and validation of effective QTLs in cold-tolerant genotypes of rice (Oryza sativa L.). Theor Appl Genet 120:985–995

Sumesh KV, Sharma-Natu P, Ghildiyal MC (2008) Starch synthase activity and heat shock protein in relation to thermal tolerance of developing wheat grains. Biol Plant 52:749–753

Takeuchi Y, Hayasaka H, Chiba B, Tanaka I, Shimano T, Yamagishi M, Nagano K, Sasakiand T, Yano M (2001) Mapping quantitative trait loci controlling cool-temperature tolerance at booting stage in temperate japonica rice. Breed Sci 51:191–197

Tang RS, Zheng JC, Jin ZQ, Zhang DD, Huang YH, Chen LG (2008) Possible correlation between high temperature-induced floret sterility and endogenous levels of IAA, GAs and ABA in rice (Oryza sativa L.). Plant Growth Regul 54:37–43

Tashiro T, Wardlaw IF (1991) Effect of high temperature on kernel dimensions and the type and occurrence of kernel damage in rice. Aust J Agric Res 42:485–496

Tenorio FA, Ye C, Redoña E, Sierra S, Laza M, Argayoso MA (2013) Screening rice genetic resources for heat tolerance. SABRAO J Breed Genet 45:371–381

Umemoto T, Nakamura Y, Ishikura N (1995) Activity of starch synthase and the amylose content in rice endosperm. Phytochemistry 40:1613–1616

Van Heerden PDR, Kruger GHJ, Loveland JE, Parry MAJ, Foyer CH (2002) Dark chilling imposes metabolic restrictions on photosynthesis in soybean. Plant Cell Environ 26(2):323–337

Wahid A, Gelani S, Ashraf M, Foolad MR (2007) Heat tolerance in plants: an overview. Environ Exp Bot 61:199–223

Wang JZ, Cui LJ, Wang Y, Li JL (2009) Growth, lipid peroxidation and photosynthesis in two tall fescue cultivars differing in heat tolerance. Biol Plant 53:247–242

Wang W, Chen Q, Hussain S, Mei J, Dong H, Peng S, Huang J, Cui K, Nie L (2016) Pre-sowing seed treatments in direct-seeded early rice: consequences for emergence, seedling growth and associated metabolic events under chilling stress. Sci Rep 6:19637

Weeden NF, Buchanan BB (1983) Leaf cytosolic fructose-1, 6-bisphosphatase-A potential targets in low temperature stress. Plant Physiol 72:259–261

Xu CC, Li L, Kuang T (2000) Photoprotection in chilling-sensitive and-resistant plants illuminated at a chilling temperature: role of the xanthophyll cycle in the protection against lumen acidification. Funct Plant Biol 27(7):669–75

Yamamoto Y, Tamori T, Kawaguchi S (1985) Relations between weather and growth of rice plant. I. Effects of air-temperature on the growth of rice plant in the first half stage. Bull Toyama Agric Exp Stn 16:20–26

Yang X, Chen X, Ge Q, Li B, Tong Y, Zhang A, Li Z, Kuang T, Lu C (2006) Tolerance of photosynthesis to photoinhibition, high temperature and drought stress in flag leaves of wheat: a comparison between a hybridization line and its parents grown under field conditions. Plant Sci 171:389–397

Yin X, Kroff MJ, Goudriann J (1996) Differential effects of day and night temperature on development to flowering in rice. Ann Bot 77:203–213

Young LW, Wilen RW, Bonham-Smith PC (2004) High temperature stress of Brassica napus during flowering reduces micro- and megagametophyte fertility, induces fruit abortion, and disrupts seed production. J Exp Bot 55:485–495

Zeng NY, He JX, Zhao W, Liang HG (2000) Changes in components of pigments and proteins in rice photosynthetic membrane during chilling stress. Acta Botan Boreali-Occiden Sin 20(1):8–14. In Chinese

Zhang H, Duan L, Dai JS, Zhang CQ, Li J, Gu MH, Liu QQ, Zhu Y (2014a) Major QTLs reduce the deleterious effects of high temperature on rice amylose content by increasing splicing efficiency of Wx pre-mRNA. Theor Appl Genet 127:273–282

Zhang Q, Chen Q, Wang S, Hong Y, Wang Z (2014b) Rice and cold stress, in rice: methods for its evaluation and summary of cold tolerance-related quantitative trait loci. Rice 7:1–12

Submergence Stress in Rice: Physiological Disorders, Tolerance Mechanisms, and Management

Umer Mahmood, Hafiz Athar Hussain, Sadam Hussain, Umair Ashraf, Abdul Khaliq, and Saddam Hussain

Introduction

Rice (*Oryza sativa*) is an important cereal crop which feeds about one-half of the world population (Liu et al. 2015). It is majorly produced and consumed in Asia and has a wide range of adaptation, i.e., from lowland to upland, rain-fed to deep water and wetland areas (Sarkar et al. 2006). The traditional method of raising rice is transplantation of rice nursery into a puddled field with continuous flooding. This is one of the key unique and distinguished characteristics of the rice which enables it to withstand in excess water supply. Submergence is a condition where the whole plant is immersed in water. It deprives the plants for free atmospheric oxygen and reduces the rates of photosynthesis and respiration (Bailey-Serres et al. 2010). It usually has harmful effects on crop plants; however, rice has capability to withstand such conditions to some extent (Fig. 1).

Based on its duration and depth, submergence is divided into two classes, i.e., flash flooding and deep water flooding. In flash flooding, the depth of water is low

U. Mahmood
Department of Agronomy, University of Agriculture, Faisalabad, Punjab, Pakistan

College of Agronomy and Biotechnology, Southwest University, Chongqing, China

H. A. Hussain (✉)
College of Agronomy and Biotechnology, Southwest University, Chongqing, China

Institute of Environment and Sustainable Development in Agriculture, Chinese Academy of Agricultural Sciences, Beijing, China

S. Hussain · A. Khaliq · S. Hussain (✉)
Department of Agronomy, University of Agriculture, Faisalabad, Punjab, Pakistan
e-mail: shussain@uaf.edu.pk

U. Ashraf
Department of Botany, University of Education (Lahore), Faisalabad, Punjab, Pakistan

Department of Agronomy, University of Agriculture, Faisalabad, Punjab, Pakistan

M. Hasanuzzaman et al. (eds.), *Plant Abiotic Stress Tolerance*,
https://doi.org/10.1007/978-3-030-06118-0_7

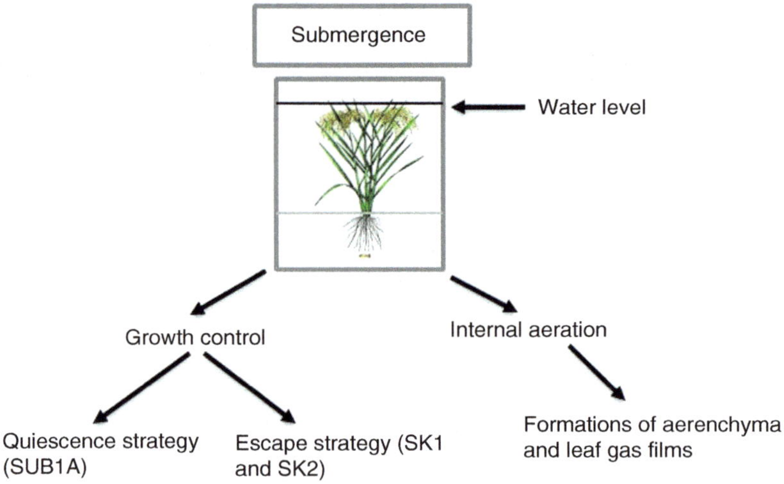

Fig. 1 Strategies of adaptation to excess water stress in the form of submerge in rice plants

and ponding conditions persist for few days and/or weeks as a consequence of heavy rainfall (Bailey-Serres et al. 2010; Jackson and Ram 2003). On the contrary, in deep water flooding, several meters water depth prolonged for several months (Hattori et al. 2011). During submergence, the seed germination and seedling growth of rice seeds are desirable characters for good stand establishment (Miro and Ismail 2013). Submergence often results in low light interception, reduced gaseous exchange, chlorosis, cell wall damage, and higher rate of pest attack (Panda et al. 2006). Hussain et al. (2016b) reported that submergence has deleterious effects on rice and reduced the seed germination and growth. Extended periods of complete submergence for a period of 7 days may cause ultimate death of rice plants (Xu et al. 2006; Bailey-Serres et al. 2010). During submerged conditions, the water fills the air spaces within the soil and make it fully saturated which may result in deficiency of carbon dioxide (CO_2), low light intensity, and decreased photosynthesis in rice (Pedersen et al. 2013). Moreover, decline in photosynthetic pigments, inhibited growth, alterations in enzyme activities, lipid peroxidation, and membrane injury are the added consequences of submergence (Jackson and Ram 2003; Nishiuchi et al. 2012).

Submergence immensely reduces the gaseous exchange (O_2 and CO_2) due to the low diffusion rate (Kende et al. 1998; Gibbs and Greenway 2003). Development of aerenchyma cells allow the plants to survive under hypoxic conditions by supplying oxygen to the submerged plant parts (Parlanti et al. 2011; Steffens et al. 2011). In the radial zone of the roots, radial oxygen loss (ROL) might be an adaptive strategy of plants to withstand in wetlands and/or submerged conditions (Visser et al. 2000; McDonald et al. 2002). Moreover, the epidermis and/or upper cuticle layer of submerged plant leaves are developed in such a way that the resistance against gaseous exchange is minimum (Mommer and Visser 2005; Mommer et al. 2007).

Hence, submergence negatively affects the plant growth and productivity by disturbing source–sink relationships and resource acquisition strategies. The submergence also results in oxidative stress in plants by changing various physio-biochemical processes. This chapter discusses the physio-biochemical, anatomical, cellular, and molecular responses of plants to submergence stress and highlights the possible tolerance mechanisms of rice plants under submergence. The possible management strategies for augmenting submergence tolerance in rice are also proposed.

Physiological Effects of Submergence

Effects on Photosynthesis and Respiration

Under normal conditions, stomata are open and CO_2 easily enters without any restriction. On the contrary, under submerged conditions, surrounding water molecules restrict the CO_2 entry into the plants (Mommer and Visser 2005). In a liquid medium, diffusion of CO_2 is restricted by different cell tissues and organelles, i.e., cell wall, chloroplast membrane, stroma, and cytoplasm (Tholen et al. 2012), which results in reduced photosynthesis due to the low availability of CO_2 to the chloroplast (Giuliani et al. 2013). In submerged plants, the rate of photosynthesis directly relates to the availability of CO_2 concentration and light (Winkel et al. 2013). Low light availability under submerged conditions not only limits the photosynthesis but also lowers the CO_2 concentration (Maberly and Madsen 1998; Sand-Jensen and Frost-Christensen 1999; Fig. 2). Under submergence, terrestrial plants adjust

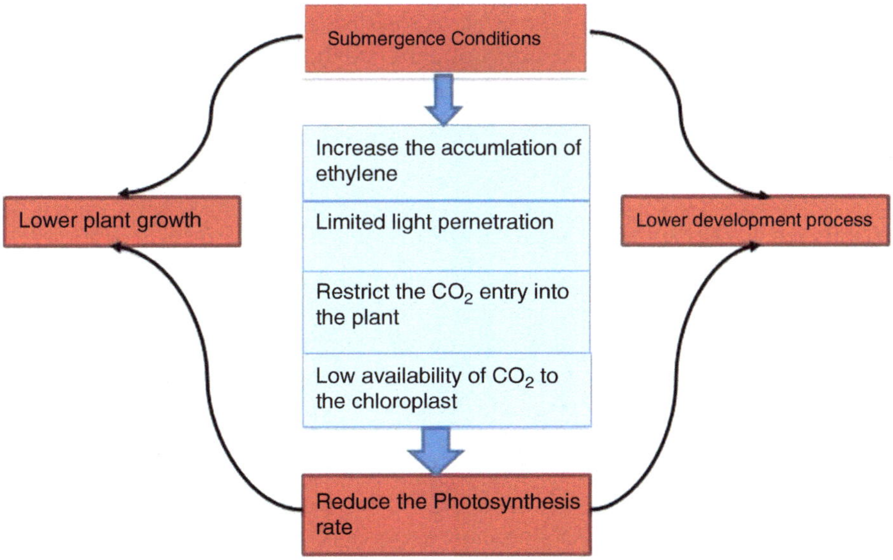

Fig. 2 Physiological effects of submergence

themselves by modifying the leaves (thin leaves), readjust their chloroplast content near to the epidermis and reduce their cuticle to promote the gaseous diffusion in water. Such modifications help chloroplast to get sufficient CO_2 from the atmosphere (Mommer and Visser 2005; Mommer et al. 2005). Submergence not only disturbs the gaseous exchange but also induces ethylene accumulation in plants (Setter et al. 1988; Mommer and Visser 2005; Voesenek et al. 2006).

Submergence interrupts all oxygen dependent mechanisms inside the plant cells by shifting aerobic respiration to anaerobic respiration (Ayi et al. 2016). The level of respiration is directly related to temperature, e.g., increase in temperature causes increase in the respiration resulting in reduced oxygen inside the cells (Colmer et al. 2011). Submerged rice plant faces energy crisis due to the low level of O_2 (anoxic condition) which limits the respiration, and ultimately causes plant death (Crawford and Braendle 1996). In submerged plants, reduced photosynthesis and enhanced consumption of photo-assimilates in anaerobic respiration (fermentation) also lower the carbohydrates reserves, and such imbalance and over-consumption of sugars during submergence may lead to complete failure of plant growth or even plant death (Bailey-Serres and Voesenek 2008; Colmer and Voesenek 2009).

Effects on Chlorophyll Contents

In addition to decreased rates of photosynthesis, chlorophyll contents are also reduced under submergence that may cause leaf senescence (Ella et al. 2003a, b). Submergence stress enhances levels of ethylene production and accumulation which causes the degradation of chlorophyll and reduced carbon fixation (Das et al. 2000). Ethylene production enhances the gene expression and activates chlorophyllase enzyme which triggers the breakdown of chlorophyll (Singh et al. 2014a, b), and consequently reduces carbon dioxide (CO_2) fixation under submergence (Sarkar et al. 2001; Ella et al. 2003a, b).

Effects on Carbohydrate Consumption

The restricted gas diffusion in submerged conditions alters the plant metabolism (Ito et al. 1999) which on the other hand depends upon the oxygen availability. In submerged plants, restricted amount of O_2 lowers the metabolic performance and survival rate of plants (Vartapetian and Jackson 1997). Low level of O_2 inside the plant cells triggers the utilization of stored carbohydrates, which may cause the death of cell (Bailey-Serres and Chang 2005). Reduced chlorophyll contents in submerged plants generally cause inhibition of C-fixation which depletes carbohydrates reserves in plants (Das et al. 2009). Submergence accelerates the utilization of non-structural carbohydrates (NSC) and triggers protein hydrolysis (Vartapetian 2005).

Submergence Tolerance in Rice

Quiescence Strategy in "Flash Flooding"

Rice plants have developed some adaptive strategies to survive in submerged conditions. For example, Nishiuchi et al. (2012) reported that during submergence, many rice varieties extend their leaves above the water surface to fetch the atmospheric oxygen. However, in some varieties, more energy consumption for leaf elongation makes the plants unable to recover after the submergence (Jackson and Ram 2003). Jackson and Ram (2003) studied the genetic basis of tolerance against flash flooding, quantitative trait loci (QTL) present on the chromosome number 9 known as *submergence-1 (SUB1)* locus was found to be responsible for the submergence tolerance (Fukao et al. 2006).

Tolerant cultivars are also characterized by ethylene production, where ethylene response factor (ERF) is coded by three similar genes viz. *Sub1A, B, and C* (Hattori et al. 2011). During submergence, ERF limits the shoot elongation in rice (Saika et al. 2007; Fukao and Bailey-Serres 2008a, b). The gene *SUB1A* is missing in non-tolerant cultivars, and introduction of this gene could induce tolerance against submergence and is also responsible for the decreased growth and energy usage under submergence stress (Hattori et al. 2011).

During the period of flash flooding, plants conserve energy by suppressing the expression of genes involving in the coding of sucrose synthase and α-amylase which are involved in sucrose and starch metabolism, respectively. This energy is used in various physiological functions after submergence (Fukao et al. 2006).

Besides energy preservation, *SUB1A* gene also responsible for the positive regulation of alcohol fermentation thus helps the plant to survive under flash flooding (Fukao et al. 2006). In submergence, alcohol dehydrogenase activity increases, and triggers the synthesis of ethanol and NAD^+ regeneration (Kato-Noguchi et al. 2010). Fukao et al. (2008) stated that tolerant cultivars reduce their growth during submergence due to the low level of gibberellic acid (GA). GA is a growth hormone involved in cell division (Kende et al. 1998), and its synthesis reduced by the negative regulation of *SLENDER RICE-1* (*SLR1*) and *SLR1* like-1 (*SLRL1*) genes to decrease the leaf elongation during submergence, *SUB1A* is the enhancer of both these genes (Fukao and Bailey-Serres 2008a, b). Abscisic acid (ABA) synthesis is promoted by the *SUB1A* and also enhances the expressions of genes which are involved to detoxify the reactive oxygen species (ROS), thus lowering its accumulation (Nishiuchi et al. 2012). *SUB1* gene hinders the ethylene mediated elongation under submergence (Singh et al. 2014a, b). It also accelerates the activity of *SLR1* and *SLR2* (repressors of GA) and ultimately restricts the shoot elongation due to low level of GA (Fukao and Bailey-Serres 2008a, b; Bailey-Serres et al. 2010).

Escape Strategy in Deep Water Flooding

The duration of this type of flooding is long as compared to the flash flooding. Anoxic conditions reduce plant energy, and thus plants must have to escape such conditions for survival (Nishiuchi et al. 2012). The escape mechanism involves the elongation of internodal distance. Some rice cultivars are capable of increasing their height at the rate of 25 cm per 24 h (Kende et al. 1998). Ethylene is held responsible as an initiation factor for internodal elongation (Yin et al. 2017). This helps the plants to extend their leaves over the surface of water for gaseous exchange, thus increasing the rates of photosynthesis and respiration (Bailey-Serres and Voesenek 2008). During deep water flooding, biosynthesis of ethylene is activated (Nishiuchi et al. 2012).

Ethylene accumulation is related to the increased concentrations of GA and decreased concentration of abscisic acid (ABA). The increased GA:ABA ratio thus facilitates elongation (Kende et al. 1998; Sauter 2000). Ethylene production is also responsible for the production of aerenchyma in plant tissues (Steffens et al. 2011). Leaf expansion is promoted due to loosening of cell wall (Lee and Kende 2001) which results in the aerenchyma formation and triggers ethylene accumulation (Steffens et al. 2011). Cellulose microfibrils change their alignment in the internodal regions during elongation (Sauter et al. 1993). The expression of two genes SNORKEL1 (SK1) and SNORKEL2 (SK2) enables the deep water rice cultivars to elongate their internodes (Hattori et al. 2009). Hattori et al. (2009) reported that *SK* genes, located on chromosome number 12, positively regulate the ethylene response factor (ERF), a transcriptional factor involved in the elongation of internodes. Ethylene also induces the formation of adventitious roots, which are formed by the injury of epidermal cells (Steffens and Sauter 2005).

Gaseous Exchange in Submerged Rice Plants

Absorption of oxygen in submerged environment decreases by 10,000 times as compared to air (Colmer 2003a, b). Diffusion of CO_2 in the aqueous medium is also decreased by four times as compared with diffusion in the air (Nobel 2009). Gaseous exchange is dramatically reduced by submergence whereby the exchange of O_2 and CO_2 between plant and their environment is inhibited (Kende et al. 1998; Gibbs and Greenway 2003). Due to inhibited gaseous exchange, concentration of O_2 is reduced which negatively affects the rate of photosynthesis (Fukao and Bailey-Serres 2008b). Exchange of ethylene is also reduced along with oxygen and carbon dioxide under submerged conditions (Jackson 2008). Increasing levels of ethylene activate the formation of adventitious roots and aerenchyma, carbohydrate usage, and elongation of shoots (Fukao et al. 2006). Formation of aerenchyma facilitates the transportation of O_2 from shoots to roots (Jackson and Armstrong 1999).

Formation of Barrier to Radial Oxygen Loss (ROL)

Diffusion of oxygen from shoot to root tip is increased by the formation of a barrier. This barrier hinders the intake of toxic metal ions, CO_2, and ethylene into the roots (Nishiuchi et al. 2012). In many wetland species, with the development of aerenchyma, they also have a barrier, i.e., ROL mostly in the basal zone of rice roots to reduce the oxygen loss (Colmer 2003a). This barrier is formed by the deposition of lignin and suberin, which forms the woody texture exterior to basal side of roots (Soukup et al. 2007). This physical barrier to ROL, despite of enabling the plant to survive under excess water stress, also lowers the nutrient and water uptake (Polthanee and Changdee 2008; Nishiuchi et al. 2012; Sauter 2013).

Formation of Aerenchyma

Formation of aerenchyma plays an important role in the supply of oxygen during submerged conditions from shoot to terminal portion of the roots and aeration of gases like methane and carbon dioxide (Colmer 2003a; Evans 2003). Ethylene accumulation in submerged plants is responsible for the formation of aerenchyma (Zhou et al. 2002; Geisler-Lee et al. 2010; Lenochova et al. 2009). Ethylene triggers the formation of aerenchyma in maize and rice (Drew et al. 2000; Shiono et al. 2008). On the contrary, Visser and Bögemann (2006) concluded that ethylene is not always responsible for the aerenchyma formation, as in the case of *Juncus effusus*, where the aerenchyma formation is not altered by the application of gaseous ethylene action inhibitor 1-methylcyclopropene (1-MCP). Rice roots in oxygen deficit conditions can store oxygen in the aerenchyma during the transportation of water (Ranathunge et al. 2011). The aerenchyma is of two types, i.e., (1) schizogenous and (2) lysigenous. The schizogenous aerenchyma is formed by changes in growth patterns during development and dissociation of neighboring cells (Evans 2003). The number of these tissues increases in hypoxic conditions (Visser et al. 2003). Lysigenous aerenchyma is formed by the death of adjacent cells, and these spaces are filled with gas (Evans 2003), e.g., in rice (Nishiuchi et al. 2012). The formation of lysigenous aerenchyma is a highly controlled process and all the protective layers and vascular tissues remain undamaged during their formation (Yamauchi et al. 2011). Generally, formation of aerenchyma takes 1–3 days after the onset of hypoxic conditions (Malik et al. 2003; Haque et al. 2010; Rajhi et al. 2011). In the formation of this type of aerenchyma, degradation enzymes (cellulases and pectinases) play an important role (Nishiuchi et al. 2012). Action of cellulase is enhanced by the accumulation of ethylene and other chemicals which elevate Ca^{2+} level inside the cell; on the contrary, the performance of cellulose is reduced by the increase in the level of K252a (inhibitor of protein kinases) and the inhibitors of Ca^{2+} (He et al. 1996; Drew et al. 2000).

Formation of Adventitious Roots

Formation of adventitious roots ensures rapid supply of nutrients and water to the growing shoots (Fukao and Bailey-Serres 2008a, b). During submergence, epidermal tissues get damaged and initiate the growth of adventitious roots (Mergemann and Sauter 2000). Nodal adventitious root formation is governed by the signal transduction pathway whereby, submergence stimulates the ethylene accumulation resulting in formation of root primordia (Lorbiecke and Sauter 1999; Steffens and Rasmussen 2016). Accumulation of ethylene is mediated by transcription factors ADVENTITIOUS ROOTLESS1 (CRL1) and CROWN ROOTLESS5 (CRL5) for the formation of root primordia (Inukai et al. 2005; Liu et al. 2005).

Ethylene besides playing an important role in the formation of adventitious roots also helps the plants to withstand in low oxygen environment (Steffens and Sauter 2005). Rich et al. (2012) reported that in submerged plants, the number of aerenchyma in adventitious roots is greater than leaves. Owing to this adaptation, adventitious roots consume less amount of O_2 for respiration as compared to leaves (Ayi et al. 2016).

Formation of Leaf Gas Films

During submergence, leaf surface of rice becomes hydrophobic and gas films are formed which help the plant to exchange gases while remaining underwater (Colmer and Pedersen 2008; Pedersen et al. 2009). Increase in gaseous exchange by the formation of leaf gas films enhances the aeration inside the submerged plants (Winkel et al. 2011); as a result, CO_2 uptake and photosynthesis in submergence increases at day time and internal aeration increases at night (Colmer et al. 2011). It is hypothesized that in submerged leaves, gas films facilitate the stomata to avoid cuticle resistance and uptake of O_2 (Verboven et al. 2014).

Antioxidant Defense Mechanism Against Reactive Oxygen Species (ROS) in Submerged Rice

Biotic and abiotic stresses trigger the production of reactive oxygen species (ROS) (Xu et al. 2015; Lin et al. 2016). Low oxygen levels (anaerobic conditions in submergence) also lead to the formation of ROS, because oxygen acts as a final electron acceptor in ETC at complex-iv and dumps electrons in a very safe manner by making water. Increasing levels of ROS such as superoxide, hydrogen peroxide, and hydroxyl radical act as signal for leaf senescence (Van Breusegem and Dat 2006; Fig. 3). Santosa et al. (2007) have reported that rice genotypes containing SUB1 show less destruction due to the activity of free radicals. Here are a number of

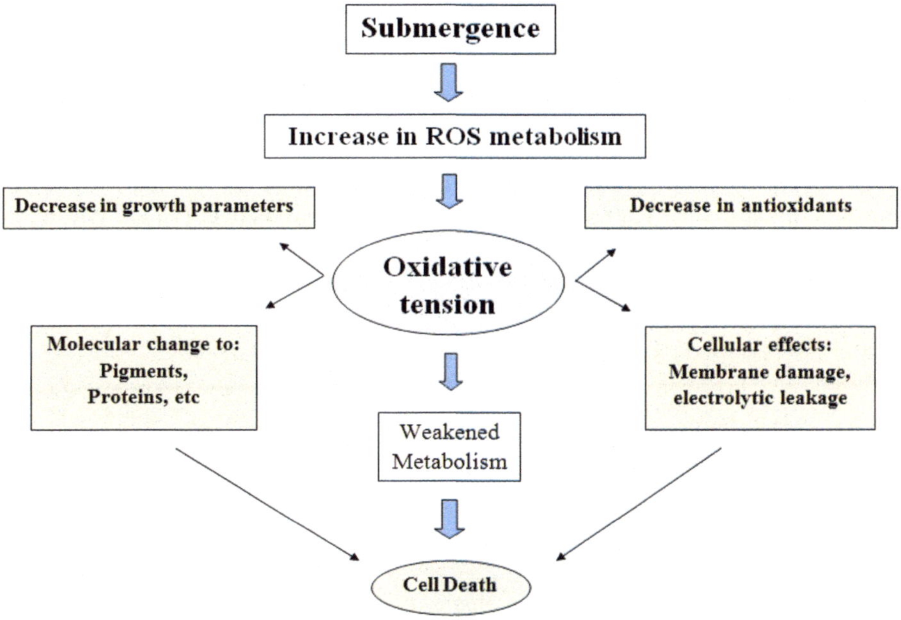

Fig. 3 Submergence increases the ROS production in cells. Prolonged conditions ultimately cause cell death (Adopted from Upadhyay 2018)

mechanisms explaining the production of ROS in cell membrane, mitochondria, chloroplast, endoplasmic reticulum, and peroxisomes which depends on the types of tissue and cell, and causes cell damage by oxidation (Ribeiro et al. 2017).

ROS are synthesized in various plant organelles; however, their production is more in those organelles (peroxisome, mitochondria, and chloroplast) where the continuous flow of electrons generates greater metabolic activity (Corpas et al. 2001; Dat et al. 2000). ROS are charged particles (except H_2O_2) and unable to pass through the membranes, therefore endogenous antioxidant defense is more essential than overall synthesis of antioxidative enzymes (Upadhyay 2018).

Plant cells have enzymatic as well as non-enzymatic mechanisms to retain ROS at lower levels (Ribeiro et al. 2017). The agents playing an important role in the detoxification of ROS include ascorbate peroxidase (APX), glutathione peroxidase (GPX), and superoxide dismutase (SOD) (Del Río 2015). SOD reacts with superoxide radicals and form oxygen (O_2) and hydrogen peroxide (H_2O_2). Later the H_2O_2 is detoxified by the ascorbate peroxidase and catalase (CAT) (Srivalli et al. 2003). Antioxidant defensive mechanism plays very important role in the detoxification of ROS and reduces the negative effects (Gill and Tuteja 2010).

Re-aeration or De-submergence

De-submergence has deleterious effects on plant due to higher levels of oxygen, and causes oxidative damage to rice seedlings (Ushimaru et al. 1999). Tolerant cultivars on de-submergence show increase in fresh biomass due to conservation of carbohydrate metabolism during submergence. On the contrary, the reserves in non-tolerant cultivars are hydrolyzed and thus unable to recover (Upadhyay et al. 2009; Kang et al. 2009). Cultivars having *SUB1* gene are capable of good recovery phase during the onset of de-submergence (Singh et al. 2014a, b). Re-aeration responses also confer tolerance and protect the leaf from dehydration (Setter et al. 2010).

Management Strategies

Seed Priming

In rice, different crop management strategies can improve the resistance under submerged conditions. Seed priming is used to improve tolerance in various plant species against abiotic stresses including heavy metal, salinity, drought, chilling, and submergence stress (Jisha et al. 2013; Paparella et al. 2015; Hussain et al. 2016a, b). Seed priming has positive effects on growth and metabolic activities, under stressed as well as normal conditions (Farooq et al. 2009; Khaliq et al. 2015; Zheng et al. 2016). It ensures uniform and timely germination of rice seedlings under normal and stressed conditions (Khaliq et al. 2015; Hussain et al. 2016a, b).

Seed priming enhances the tolerance by increasing the activity of antioxidant defense mechanisms, carbohydrate metabolism, and seedling vigor (Ella et al. 2011). Hussain et al. (2016a, b) found that priming with salicylic acid (SA) and selenium (Se) improved the seed germination and other morphological traits under submergence stress. Efficient use of nutrients and other inputs will enhance the productivity of transgenic rice cultivars, e.g., Khao Hlan On, Chiknal, Tilakachari, and Sirambe Putih in submerged areas of Myanmar, Bangladesh, India, and Indonesia, respectively (Vergara et al. 2014). Exogenous potassium (K) application could improve plant growth, chlorophyll contents and photosynthetic capacities as well as nutrients uptake in plants under submerged conditions (Ashraf et al. 2011). Furthermore, application of K fertilizer alone or in combination with phosphorus could maintain internal gas diffusion and energy levels required for normal growth and developmental processes under submergence conditions in rice (Singh et al. 2014a, b).

Molecular Approach to Develop Submergence Tolerant Cultivars

Traditional breeding approaches that comprise bulk and pedigree selection based on morphological markers have been successfully used in rice breeding for submergence stress tolerance. Transgenesis of SUB1 gene enhances the tolerance against submergence in those rice cultivars which are already good in agronomic traits (Siangliw et al. 2003; Xu et al. 2006; Neeraja et al. 2007; Septiningsih et al. 2009, 2013; Singh et al. 2010; Thomson et al. 2010; Manzanilla et al. 2011; Mackill et al. 2012; Collard et al. 2013). Introduction of *SUB1* gene amplifies the yield by 2–5 times, following complete submergence for 12–17 days (Iftekharuddaula et al. 2011; Dar et al. 2013). Rice cultivars containing *SUB1* gene complete their life cycle earlier and give more yield as compared to the non-*SUB1* cultivars facing submerged conditions (Singh et al. 2009; Manzanilla et al. 2011). In future, molecular breeding approaches are required for developing superior submergence tolerant varieties.

Conclusions

Poor early growth and substantial yield loss occur in rice due to submergence stress. In rice, submergence impedes various growth phases. Submergence not only disrupts the gaseous exchange but also affects the photosynthetic rate due to increase in accumulation of ethylene and lower level of CO_2. Different crop management strategies such as seed priming and efficient nutrients use could improve the resistance in rice against submergence stress. Submergence interrupts all oxygen dependent mechanisms inside the plant cells by shifting aerobic respiration to anaerobic respiration. Breeding and molecular approaches can be employed to develop resistant rice cultivars against submerged conditions. Previously, several genes (e.g., *SLR1, SK1,* and *SUB1*) have been identified to be responsible for submergence resistance/tolerance; however, further attention is requiring from researchers and breeders.

References

Ashraf MA, Ahmad MS, Ashraf M, Al-Qurainy F, Ashraf MY (2011) Alleviation of waterlogging stress in upland cotton (*Gossypium hirsutum* L.) by exogenous application of potassium in soil and as a foliar spray. Crop Pasture Sci 62:25–38

Ayi Q, Zeng B, Liu J, Li S, van Bodegom PM, Cornelissen JH (2016) Oxygen absorption by adventitious roots promotes the survival of completely submerged terrestrial plants. Ann Bot 118:675–683

Bailey-Serres J, Chang R (2005) Sensing and signaling in response to oxygen deprivation in plants and other organisms. Ann Bot 96:507–518

Bailey-Serres J, Voesenek LACJ (2008) Flooding stress: acclimations and genetic diversity. Annu Rev Plant Biol 59:313–339

Bailey-Serres J, Fukao T, Ronald P, Ismail A, Heuer S, Mackill D (2010) Submergence tolerant rice: SUB1's journey from landrace to modern cultivar. Rice 3:138–147

Collard BCY, Septiningsih EM, Das SR, Carandang JJ, Pamplona AM, Sanchez DL, Kato Y, Ye G, Reddy JN, Singh US, Iftekharuddaula KM, Venuprasad R, Vera-Cruz CN, Mackill DJ, Ismail AM (2013) Developing new flood-tolerant varieties at the International Rice Research Institute (IRRI). SABRAO J Breed Genet 45:42–56

Colmer TD (2003a) Long-distance transport of gases in plants: a perspective on internal aeration and radial oxygen loss from roots. Plant Cell Environ 26:17–36

Colmer TD (2003b) Aerenchyma and an inducible barrier to radial oxygen loss facilitate root aeration in upland, paddy and deep-water rice (*Oryza sativa* L.). Ann Bot 91:301–309

Colmer TD, Pedersen O (2008) Under water photosynthesis and respiration in leaves of submerged wetland plants: gas films improve CO_2 and O_2 exchange. New Phytol 177:918–926

Colmer TD, Voesenek LACJ (2009) Flooding tolerance: suites of plant traits in variable environments. Funct Plant Biol 36:665–681

Colmer TD, Winkel A, Pedersen O (2011) A perspective on underwater photosynthesis in submerged terrestrial wetland plants. AoB Plants 2011:plr030. https://doi.org/10.1093/aobpla/plr030

Corpas FJ, Barroso JB, del Rio LA (2001) Peroxisomes as a source of reactive oxygen species and nitric oxide signal molecules in plant cells. Trends Plant Sci 6:145–150

Crawford RMM, Braendle R (1996) Oxygen deprivation stress in a changing environment. J Exp Bot 47:145–159

Dar MH, Janvry AD, Emerick K, Raitzer D, Sadoulet E (2013) Flood tolerant rice reduces yield variability and raises expected yield, differentially benefitting socially disadvantaged groups. Sci Rep 3:1–8

Das A, Nanda BB, Sarkar RK, Lodh SB (2000) Effect of complete submergence on the activity of starch phosphorylase enzyme in rice (Oryza sativa L.) leaves. J Plant Biochem Biot 9(1):41–3

Das KK, Panda D, Sarkar RK, Reddy JN, Ismail AM (2009) Submergence tolerance in relation to variable floodwater conditions in rice. Environ Exp Bot 66(3):425–434. https://doi.org/10.1016/j.envexpbot.2009.02.015

Dat S, Vandenabeele E, Vranová M, Van Montagu M, Inzé D, Van Breusegem F (2000) Dual action of the active oxygen species during plant stress responses. Cell Mol Life Sci 57:779–795

Del Río LA (2015) ROS and RNS in plant physiology: an overview. J Exp Bot 66:2827–2837

Drew MC, He CJ, Morgan PW (2000) Programmed cell death and aerenchyma formation in roots. Trends Plant Sci 5:123–127

Ella ES, Kawano N, Yamauchi Y, Tanaka K, Ismail AM (2003a) Blocking ethylene perception during submergence reduced chlorophyll degradation and improved seedling survival in rice. Funct Plant Biol 30:813–819

Ella ES, Kawano N, Ito O (2003b) Importance of active oxygen scavenging system in the recovery of rice seedlings after submergence. Plant Sci 165:85–93

Ella ES, Dionisio-Sese ML, Ismail AM (2011) Seed pre-treatment in rice reduces damage, enhances carbohydrate mobilization and improves emergence and seedling establishment under flooded conditions. AoB Plants 2011:plr007. https://doi.org/10.1093/aobpla/plr007

Evans DE (2003) Aerenchyma formation. New Phytol 161:35–49

Farooq M, Basra SMA, Wahid A, Khaliq A, Kobayashi N (2009) Rice seed invigoration: a review. In: Lichtfouse E (ed) Organic farming, pest control and remediation of soil pollutants. Springer, New York, pp 137–175

Fukao T, Harris T, Bailey-Serres J (2008) Evolutionary analysis of the Sub1 gene cluster that confers submergence tolerance to domesticated rice. Ann Bot 103(2):143–50

Fukao T, Bailey-Serres J (2008a) Submergence tolerance conferred by Sub1A is mediated by SLR1 and SLRL1 restriction of gibberellin responses in rice. Proc Natl Acad Sci U S A 105:16814–16819

Fukao T, Bailey-Serres J (2008b) Ethylene—a key regulator of submergence responses in rice. Plant Sci 175:43–51

Fukao T, Xu K, Ronald PC, Bailey-Serres J (2006) A variable cluster of ethylene response factor-like genes regulates metabolic and developmental acclimation responses to submergence in rice. Plant Cell 18:2021–2034

Geisler-Lee J, Caldwell C, Gallie DR (2010) Expression of the ethylene biosynthetic machinery in maize roots is regulated in response to hypoxia. J Exp Bot 61:857–871

Gibbs J, Greenway H (2003) Mechanisms of anoxia tolerance in plants. I. Growth, survival and anaerobic catabolism. Funct Plant Biol 30:1–47

Gill SS, Tuteja N (2010) Reactive oxygen species and antioxidant machinery in abiotic stress tolerance in crop plants. Plant Physiol Biochem 48:909–930

Giuliani R, Koteyeva N, Voznesenskaya E, Evans M, Cousins A, Edwards G (2013) Coordination of leaf photosynthesis, transpiration and structural traits in Rice and wild relatives (Genus Oryza). Plant Physiol 162:1632–1651

Haque ME, Abe F, Kawaguchi K (2010) Formation and extension of lysigenous aerenchyma in seminal root cortex of spring wheat (*Triticum aestivum* cv. Bobwhite line SH 98 26) seedlings under different strengths of waterlogging. Plant Root 4:31–39

Hattori Y, Nagai K, Furukawa S, Song XJ, Kawano R, Sakakibara H, Wu J, Matsuoka T, Yoshimura A, Kitano H, Matsuoka M, Mori H, Ashikari M (2009) The ethylene response factors SNORKEL1 and SNORKEL2 allow rice to adapt to deep water. Nature 460:1026–1030

Hattori Y, Nagai K, Ashikari M (2011) Rice growth adapting to deep water. Curr Opin Plant Biol 14:100–105

He CJ, Morgan PW, Drew MC (1996) Transduction of an ethylene signal is required for cell death and lysis in the root cortex of maize during aerenchyma formation induced by hypoxia. Plant Physiol 112:463–472

Hussain S, Khan F, Hussain HA, Nie L (2016a) Physiological and biochemical mechanisms of seed priming-induced chilling tolerance in rice cultivars. Front Plant Sci 7:116. https://doi.org/10.3389/fpls.2016.00116

Hussain S, Yin H, Peng S, Khan FA, Khan F, Sameeullah M, Hussain HA, Huang J, Cui K, Nie L (2016b) Comparative transcriptional profiling of primed and non-primed rice seedlings under submergence stress. Front Plant Sci 7:1125

Iftekharuddaula K, Newaz M, Salam M, Ahmed H, Mahbub M, Septiningsih E, Collard B, Sanchez D, Pamplona A, Mackill D (2011) Rapid and high-precision marker assisted backcrossing to introgress SUB1 QTL into BR11, the rainfed lowland rice mega variety of Bangladesh. Euphytica 178:83–97

Inukai Y, Sakamoto T, Ueguchi-Tanaka M, Shibata Y, Gomi K, Umemura I, Hasegawa Y, Ashikari M, Kitano H, Matsuoka M (2005) Crown rootless1, which is essential for crown root formation in rice, is a target of an AUXIN RESPONSE FACTOR in auxin signaling. Plant Cell 17:1387–1396

Ito O, Ella E, Kawano N (1999) Physiological basis of submergence tolerance in rainfed lowland rice ecosystem. Field Crops Res 64:75–90

Jackson MB (2008) Ethylene-promoted elongation: an adaptation to submergence stress. Ann Bot 101:229–248

Jackson MB, Armstrong W (1999) Formation of aerenchyma and the processes of plant ventilation in relation to soil flooding and submergence. Plant Biol 1:274–287

Jackson MB, Ram PC (2003) Physiological and molecular basis of susceptibility and tolerance of rice plants to complete submergence. Ann Bot 91:227–241

Jisha KC, Vijaya kumari K, Puthur JT (2013) Seed priming for abiotic stress tolerance: an overview. Acta Physiol Plant 35:1381–1396. https://doi.org/10.1007/s11738-012-1186-5

Kang YY, Guo SR, Li J, Duan JJ (2009) Effect of root applied 2, 4-epibrassinolide on carbohydrate status and fermentative enzyme activities in cucumber (*Cucumis sativus* L.) seedlings under hypoxia. Plant Growth Regul 57:259–269

Kato-Noguchi H, Yasuda Y, Sasaki R (2010) Soluble sugar availability of aerobically germinated barley, oat and rice coleoptiles in anoxia. J Plant Physiol 167:1571–1576

Kende H, van der Knaap E, Cho HT (1998) Deep water rice: a model plant to study stem elongation. Plant Physiol 118:1105–1110

Khaliq A, Aslam F, Matloob A, Hussain S, Geng M, Wahid A, ur Rehman H (2015) Seed priming with selenium: consequences for emergence, seedling growth, and biochemical attributes of rice. Biol Trace Elem Res 166:236–244. https://doi.org/10.1007/s12011-015-0260-4

Lee Y, Kende H (2001) Expression of β-expansins is correlated with internodal elongation in deep water rice. Plant Physiol 127:645–654

Lenochova Z, Soukup A, Votrubova O (2009) Aerenchyma formation in maize roots. Biol Plant 53:263–270

Lin KC, Jwo WS, Chandrika NNP, Wu TM, Lai MH, Wang CS, Hong CY (2016) A rice mutant defective in antioxidant-defense system and sodium homeostasis possesses increased sensitivity to salt stress. Biol Plant 60:86–94

Liu H, Wang S, Yu X, Yu J, He X, Zhang S, Shou H, Wu P (2005) ARL1, a LOB-domain protein required for adventitious root formation in rice. Plant J 43:47–56

Liu E, Liu X, Zeng S, Zhao K, Zhu C, Liu Y, Breria MC, Zhang B, Hong D (2015) Time-course association mapping of the grain-filling rate in rice (Oryza sativa L.). PLoS One 10(3):e0119959.

Lorbiecke R, Sauter M (1999) Adventitious root growth and cell-cycle induction in deep water rice. Plant Physiol 119:21–29

Maberly SC, Madsen TV (1998) Affinity for CO_2 in relation to the ability of freshwater macrophytes to use HCO3. Funct Ecol 12:99–106

Mackill DJ, Ismail AM, Singh US, Labios RV, Paris TR (2012) Development and rapid adoption of submergence-tolerant (Sub1) rice varieties. Adv Agron 115:299–352

Malik AI, Colmer TD, Lambers H, Schortemeyer LM (2003) Aerenchyma formation and radial O_2 loss along adventitious roots of wheat with only the apical root portion exposed to O_2 deficiency. Plant Cell Environ 26:1713–1722

Manzanilla DO, Paris TR, Vergara GV, Ismail AM, Pandey S, Labios RV, Tatlonghari GT, Acda RD, Chi TTN, Duoangsila K, Siliphouthone I, Manikmas MOA, Mackill DJ (2011) Submergence risk and farmers' preference: implications for breeding Sub1 rice in Southeast Asia. Agric Syst 104:335–347

McDonald MP, Galwey NW, Colmer TD (2002) Similarity and diversity in adventitious root anatomy as related to root aeration among a range of wetland and dryland grass species. Plant Cell Environ 25:441–451

Mergemann H, Sauter M (2000) Ethylene induces epidermal cell death at the site of adventitious root emergence in rice. Plant Physiol 124:609–614

Miro B, Ismail AM (2013) Tolerance of anaerobic conditions caused by flooding during germination and early growth with in rice (*Oryza sativa* L.). Front Plant Sci 4:269. https://doi.org/10.3389/fpls.2013.00269

Mommer L, Visser EJW (2005) Under water photosynthesis in flooded terrestrial plants: a matter of leaf plasticity. Ann Bot 96:581–589

Mommer L, Pons TL, Wolters-Arts M, Venema JH, Visser EJW (2005) Submergence-induced morphological, anatomical, and biochemical responses in a terrestrial species affect gas diffusion resistance and photosynthetic performance. Plant Physiol 139:497–508

Mommer L, Wolters-Arts M, Andersen C, Visser EJW, Pedersen O (2007) Submergence-induced leaf acclimation in terrestrial species varying in flooding tolerance. New Phytol 176:337–345

Neeraja CN, Maghirang-Rodriguez R, Pamplona A, Heuer S, Collard BCY, Septiningsih EM, Vergara G, Sanchez DLM, Xu K, Ismail AM, Mackill DJ (2007) A marker-assisted backcross approach for developing submergence-tolerant rice cultivars. Theor Appl Genet 115:767–776

Nishiuchi S, Yamauchi T, Takahashi H, Kotula L, Nakazono M (2012) Mechanisms for coping with submergence and waterlogging in rice. Rice 5:11–14

Nobel PS (2009) Physicochemical and environmental plant physiology, 4th edn. Academic Press, San Diego

Panda D, Rao DN, Sharma SG, Strasser RJ, Sarkar RK (2006) Submergence effects on rice genotypes during seedling stage: probing of submergence driven changes of photosystem 2 by chlorophyll a fluorescence induction O-J-I-P transient. Photosynthetica 44:69–75

Paparella S, Araújo SS, Rossi G, Wijayasinghe M, Carbonera D, Balestrazzi A (2015) Seed priming: state of the art and new perspectives. Plant Cell Rep 34:1281–1293. https://doi.org/10.1007/s00299-015-1784-y

Parlanti S, Kudahettige NP, Lombardi L, Mensuali-Sodi A, Alpi A, Perata P, Pucciariello C (2011) Distinct mechanisms for aerenchyma formation in leaf sheaths of rice genotypes displaying a quiescence or escape strategy for flooding tolerance. Ann Bot 107:1335–1343

Pedersen O, Rich SM, Colmer TD (2009) Surviving floods: leaf gas films improve O2 and CO2 exchange, root aeration, and growth of completely submerged rice. Plant J 58:147–156

Pedersen O, Colmer TD, Sand-Jensen K (2013) Underwater photosynthesis of submerged plants recent advances and methods. Front Plant Sci 4:140

Polthanee A, Changdee T (2008) Influence of adventitious root removing and timing of fertilizer application in flooded soil on growth, yield and N, P, K uptake of kenaf (*Hibiscus cannabinus* L.) under greenhouse and field conditions. Asian J Plant Sci 7:352–359

Rajhi I, Yamauchi T, Takahashi H, Nishiuchi S, Shiono K, Watanabe R, Mliki A, Nagamura Y, Tsutsumi N, Nishizawa NK, Nakazono M (2011) Identification of genes expressed in maize root cortical cells during lysigenous aerenchyma formation using laser microdissection and microarray analyses. New Phytol 190:351–368

Ranathunge K, Lin J, Steudle E, Schreiber L (2011) Stagnant deoxygenated growth enhances root suberization and lignifications, but differentially affects water and NaCl permeabilities in rice (*Oryza sativa* L.) roots. Plant Cell Environ 34:1223–1240

Ribeiro CW, Korbes AP, Garighan JA, Jardim-Messeder D, Carvalho FE, Sousa RH, Caverzan A, Teixeira FK, Silveira JA, Margis-Pinheiro M (2017) Rice peroxisomal ascorbate peroxidase knockdown affects ROS signaling and triggers early leaf senescence. Plant Sci 263:55–65

Rich SM, Ludwig M, Colmer TD (2012) Aquatic adventitious root development in partially and completely submerged wetland plants *Cotula coronopifolia* and *Meionectes brownii*. Ann Bot 110:405–414

Saika H, Matsumura H, Takano T, Tsutsumi N, Nakazono M (2007) A point mutation of Adh1 gene is involved in the repression of coleoptile elongation under submergence in rice. Breed Sci 56:69–74

Sand-Jensen K, Frost-Christensen H (1999) Plant growth and photosynthesis in the transition zone between land and stream. Aquat Bot 63:23–35

Santosa IE, Ram PC, Boamfa EI, Laarhoven LJJ, Reuss J, Jackson MB, Harren FJM (2007) Patterns of peroxidative ethane emission from submerged rice seedlings indicate that damage from reactive oxygen species takes place during submergence and is not necessarily a post-anoxic phenomenon. Planta 226:193–202

Sarkar RK, Das S, Ravi I (2001) Changes in certain antioxidative enzymes and growth parameters as a result of complete submergence and subsequent re-aeration of rice cultivars differing in submergence tolerance. J Agron Crop Sci 187:69–74

Sarkar RK, Reddy JN, Sharma SG, Ismail AM (2006) Physiological basis of submergence tolerance in rice and implications for crop improvement. Curr Sci 10:899–906

Sauter M (2000) Rice in deep water: "how to take heed against a sea of troubles". Naturwissensehaften 87:289–303

Sauter M (2013) Root responses to flooding. Curr Opin Plant Biol 16:282–286

Sauter M, Seagull RW, Kende H (1993) Internodal elongation and orientation of cellulose microfibrils and microtubules in deep water rice. Planta 190:354–362

Septiningsih EM, Pamplona AM, Sanchez DL, Neeraja CN, Vergara GV, Heuer S, Ismail AM, Mackill DJ (2009) Development of submergence-tolerant rice cultivars: the Sub1 locus and beyond. Ann Bot 103:151–160

Septiningsih EM, Collard BCY, Heuer S, Bailey-Serres J, Ismail AM, Mackill DJ (2013) Applying genomics tools for breeding submergence tolerance in rice. In: Varshney RK, Tuberosa R (eds) Translational genomics for crop breeding: abiotic stress, yield and quality. Wiley, Chichester, pp 9–30

Setter TL, Kupkanchanakul T, Kupkanchanakul K, Bhekasut P, Wiengweera A, Greenway H (1988). Environmental factors in deepwater rice areas in Thailand: oxygen, carbon dioxide, and ethylene. In: International Deepwater Rice Workshop, Bangkok (Thailand), 26–30 Oct 1987. IRRI

Setter TL, Bhekasut P, Greenway H (2010) Desiccation of leaves after de-submergence is one cause for intolerance to complete submergence of the rice cultivar IR42. Funct Plant Biol 37:1096–1104

Shiono K, Takahashi H, Colmer TD, Nakazono M (2008) Role of ethylene in acclimations to promote oxygen transport in roots of plants in waterlogged soils. Plant Sci 175:52–58

Siangliw M, Toojinda T, Tragoonrung S, Vanavichit A (2003) Thai Jasmine rice carrying QTLCh9 (Sub1 QTL) is submergence tolerant. Ann Bot 91:225–261

Singh S, Mackill DJ, Ismail AM (2009) Responses of SUB1 rice introgression lines to submergence in the field: yield and grain quality. Field Crops Res 113:12–23

Singh N, Dang T, Vergara G, Pandey D, Sanchez D, Neeraja C, Septiningsih E, Mendioro M, Tecson Mendoza R, Ismail A, Mackill D, Heuer S (2010) Molecular marker survey and expression analysis of rice submergence-tolerance genes SUB1A and SUB1C. Theor Appl Genet 121:1441–1453

Singh S, Mackill DJ, Ismail AM (2014a) Physiological basis of tolerance to complete submergence in rice involves genetic factors in addition to the SUB1 gene. AoB Plants 6:plu060. https://doi.org/10.1093/aobpla/plu060

Singh AK, Singh P, Singh VN, Khan AH, Singh S, Srivastava AK, Singh US, Ismail AM, Haefele SM (2014b) The beneficial effects of applying potassium alone or with phosphorus during nursery management in enhancing the survival and yield of the rice variety Swarna-Sub1 in a flood-prone ecosystem. Better Crops 38:3–11

Soukup A, Armstrong W, Schreiber L, Franke R, Votrubová O (2007) Apoplastic barriers to radial oxygen loss and solute penetration: a chemical and functional comparison of the exodermis of two wetland species, *Phragmites australis* and *Glyceria maxima*. New Phytol 173:264–278

Srivalli B, Sharma G, Khanna-Chopra R (2003) Antioxidative defense system in an upland rice cultivar subjected to increasing intensity of water stress followed by recovery. Physiol Plantarum 119(4):503–12

Steffens B, Rasmussen A (2016) The physiology of adventitious roots. Plant Physiol 170:603–617

Steffens B, Sauter M (2005) Epidermal cell death in rice (*Oryza sativa* L.) is regulated by ethylene, gibberellin and abscisic acid. Plant Physiol 139:713–721

Steffens B, Wang J, Sauter M (2006) Interactions between ethylene, gibberellin and abscisic acid regulate emergence and growth rate of adventitious roots in deepwater rice. Planta 223:604–612

Steffens B, Geske T, Sauter M (2011) Aerenchyma formation in the rice stem and its promotion by H_2O_2. New Phytol 190:369–378

Tholen D, Ethier G, Genty B, Pepin S, Zhu XG (2012) Variable mesophyll conductance revisited: theoretical background and experimental implications. Plant Cell Environ 35:2087–2103

Thomson MJ, Ismail AM, McCouch SR, Mackill MJ (2010) Marker assisted breeding. In: Pareek A, Sopory SK, Bohnert HJ, Govindjee (eds) Abiotic stress adaptation in plants: physiological, physiological, molecular and genomic foundation. Photosynthetica. Springer, New York, pp 474–474

Upadhyay RK (2018) Oxidative injury and its detoxification in rice plants after submergence stress. Proc Natl Acad Sci India Sect B Biol Sci 88(1):15–21. https://doi.org/10.1007/s40011-016-0724-0

Upadhyay RK, Panda SK, Dutta BK (2009) Growth, chlorophyll and electrical conductivity responses of rice cultivars to different levels of submergence and post submergence stress. J Phytology 1:325–432

Ushimaru T, Kanemotsu S, Shibasaka M, Tsuji H (1999) Effect of hypoxia on the antioxidative enzymes in aerobically grownrice (*Oryza sativa* L.) seedlings. Physiol Plant 107:181–187
Van Breusegem F, Dat JF (2006) Reactive oxygen species in plant cell death. Plant Physiol 141:384–390. https://doi.org/10.1104/pp.106.078295.384
Vartapetian BB (2005) Plant anaerobiosis stress as novel trend in ecological physiology, biochemistry, and molecular biology. 1. Establishment of a new scientific discipline. Russ J Plant Physiol 52:826–844
Vartapetian BB, Jackson MB (1997) Plant adaptations to anaerobic stress. Ann Bot 79:3–20
Verboven P, Pedersen O, Ho QT, Nicolai BM, Colmer TD (2014) The mechanism of improved aeration due to gas films on leaves of submerged rice. Plant Cell Environ 37:2433–2452
Vergara GV, Nugraha Y, Esguerra MQ, Mackill DJ, Ismail AM (2014) Variation in tolerance of rice to long-term stagnant flooding that submerges most of the shoot will aid in breeding tolerant cultivars. AoB Plants 6:plu055. https://doi.org/10.1093/aobpla/plu055
Visser EJW, Bögemann GM (2006) Aerenchyma formation in the wetland plant Juncus effusus is independent of ethylene. New Phytol 171:305–314
Visser EJW, Colmer TD, Blom CWPM, Voesenek LACJ (2000) Changes in growth, porosity, and radial oxygen loss from adventitious roots of selected mono- and dicotyledonous wetland species with contrasting types of aerenchyma. Plant Cell Environ 23:1237–1245
Visser EJW, Voesenek L, Vartapetian BB, Jackson MB (2003) Flooding and plant growth. Ann Bot 91:107–109
Voesenek LACJ, Colmer TD, Pierik R, Millenaar FF, Peeters AJM (2006) How plants cope with complete submergence. New Phytol 170:213–226
Winkel A, Colmer TD, Pedersen O (2011) Leaf gas films of *Spartina anglica* enhance rhizome and root oxygen during tidal submergence. Plant Cell Environ 34:2083–2092
Winkel A, Colmer TD, Ismail AM, Pedersen O (2013) Internal aeration of paddy field rice (*Oryza sativa*) during complete submergence—importance of light and floodwater O_2. New Phytol 197:1193–1203. https://doi.org/10.1111/nph.12048
Xu K, Xu X, Fukao T, Canlas P, Maghirang RR, Heuer S, Ismail AM, Bailey-Serres J, Ronald PC, Mackill DJ (2006) Sub1A is an ethylene-response-factor-like gene that confers submergence tolerance to rice. Nature 442:705–708
Xu K, Chun S, Li T, Ma X, Liang X, Ding X, Liu H, Luo L (2015) OsGRAS23, a rice GRAS transcription factor gene, is involved in drought stress response through regulating expression of stress responsive genes. BMC Plant Biol 15:141
Yamauchi T, Rajhi I, Nakazono M (2011) Lysigenous aerenchyma formation in maize root is confined to cortical cells by regulation of genes related to generation and scavenging of reactive oxygen species. Plant Signal Behav 6:759–761
Yin C-C, Zhao H, Ma B, Chen S-Y, Zhang J-S (2017) Diverse roles of ethylene in regulating agronomic traits in Rice. Front Plant Sci 8:1676
Zheng M, Tao Y, Hussain S, Jiang Q, Peng S, Huang J, Cui K, Nie L (2016) Seed priming in dry direct-seeded rice: consequences for emergence, seedling growth and associated metabolic events under drought stress. Plant Growth Regul 78:167–178. https://doi.org/10.1007/s10725-015-0083-5
Zhou Z, de Almeida Engler J, Rouan D, Michiels F, Van Montagu M, Van Der Straeten D (2002) Tissue localization of a submergence-induced 1-aminocyclopropane-1-carboxylic acid synthase in rice. Plant Physiol 129:72–84

Oxidative Stress and Antioxidant Defense Mechanisms in Plants Under Salt Stress

Riaz Ahmad, Sajjad Hussain, Muhammad Akbar Anjum, Muhammad Fasih Khalid, Muhammad Saqib, Iqra Zakir, Ahmad Hassan, Shah Fahad, and Shakeel Ahmad

Introduction

The world population is expanding rapidly and may grow from 6 billion to 9.3 billion by the year 2050. Meanwhile, the crop yield is reducing quickly because of negative effects of several environmental stresses. Consequently, there is an urgent need to produce stress-tolerant germplasms to survive the coming difficulties in food security. Among different environmental stresses, abiotic stress is known as the major source of stress, reducing the average production or yield of all field crops by approximately 50% and, ultimately, causing losses worth US$100 million every year (Mahajan et al. 2008). Plants have vulnerability or resistance to abiotic stresses having numerous genes activate and interact with other stress indicator. Salinization is the most severe environmental stress factor among all abiotic stresses, damaging at least 20% of the crop production of arable land all over the

R. Ahmad · M. A. Anjum · M. F. Khalid · M. Saqib · A. Hassan
Department of Horticulture, Bahauddin Zakariya University, Multan, Pakistan

S. Hussain (✉)
Department of Horticulture, Faculty of Agricultural Sciences and Technology, Bahauddin Zakariya University, Multan, Punjab, Pakistan
e-mail: sajjad.hussain@bzu.edu.pk

I. Zakir
Department of Agronomy, Bahauddin Zakariya University, Multan, Pakistan

Department of Agriculture, The University of Swabi, Swabi, Khyber Pakhtunkhwa, Pakistan

S. Fahad
Department of Agriculture, The University of Swabi, Swabi, Khyber Pakhtunkhwa, Pakistan

S. Ahmad
Department of Agronomy, Faculty of Agricultural Sciences and Technology, Bahauddin Zakariya University, Multan, Punjab, Pakistan

M. Hasanuzzaman et al. (eds.), *Plant Abiotic Stress Tolerance*,
https://doi.org/10.1007/978-3-030-06118-0_8

globe (Hussain et al. 2017). Additionally, high salinity levels on irrigated land are estimated to have devastating effects and are predicted to cause approximately 50% losses of arable land by the middle of the twenty-first century. Worldwide, the total land area with high salinity is more than 9 billion hectares, which poses a serious danger for agriculture (Mahajan et al. 2008). Moreover, each year, an additional 2 million hectares (approximately 1% of the global agricultural land) becomes contaminated by salinity. When roots are exposed to salts in the rhizosphere, they are clearly affected by salinity stress because the roots are major organs in plants (Cosme et al. 2016). Multiple factors—including insufficient irrigation, high levels of evaporation, frequent irrigation, and contact with seawater—result in soil salinity.

The amount of salt dissolved in a body of water is its salinity. Four main causes of soil salinity are listed below:

- The amount of salt that plants uptake from the groundwater by capillary action through their roots
- Salt accumulated when plants are irrigated with salty water
- Windblown salt accumulation
- Salt accumulation from rainfall

Only small amounts of salts are obtained from rainwater. However, such salts may build up to poisonous levels over time. Therefore, three major causes of soil salinity have been listed below (Rengasamy 2006):

- Groundwater-associated salinity
- Non-groundwater-associated salinity from parent rock or from rainwater
- Windblown salt accumulation

Anthropogenic activities are the main cause of salinization. Soil salinity is the main cause of shrinkage of limited and valuable agricultural land, which also decreases crop efficiency and productivity all over the world. Salinity levels have been increasing gradually. Salt accumulation in the soil is the major cause of water stress and creates nutrient-deficient conditions for plants, which are unfavorable for plant growth and development. Salt accumulation reduces water uptake, as well as nutrient uptake, through the roots. This affects plant growth and ultimately can cause the death of the plant. Alteration in redox homeostasis is a primary stress that is increased through salinity. Salinization causes differences in electron flow from central transport chains in organelles to oxygen-reduction pathways that are prominent in overproduction of reactive oxygen species (ROS), which are involved in oxidation of requisite biomolecules, i.e., lipids, proteins, nucleic acids, and carbohydrates. Therefore, fluctuations in their properties and functions will eventually lead to metabolic and physiological problems. Overproduction of ROS is the cause of imbalances in homeostasis at the cellular level, as well as at the subcellular level, and may ultimately cause cell death (Ivanova et al. 2016). It is necessary to reduce toxic ROS, as well as combating oxidative damage. Plants have an antioxidant defense mechanism, containing several enzymatic and nonenzymatic constituents that stabilize the negative effects of ROS by changing them into less toxic constituents. Therefore, it is considered that increased levels of antioxidants usually help to increase plant tolerance (Lei et al. 2015). Salinity stress increases a variety of different reactions in plants, causing morphological, physiological, biochemical, and molecular changes.

Many crop species can be categorized into two types—glycophytes and halophytes—based on salinity levels. Glycophytes include those crop species that are sensitive to salinization, while halophytes are those crop species that are salt tolerant. Most glycophytes may not be cultivable in the presence of 100 mM of NaCl, while halophytes may be cultivable in the presence of more than 250 mM of NaCl. Those plants that are salt susceptible can reduce their intake of salts and try to maintain an osmotic balance by production of beneficial solutes (e.g., proline, glycine betaine, and numerous sugars), while salt-tolerant crops may have the competence to isolate and accumulate salts in their cell vacuoles. Hence, accumulation of salts in the cytosol is inhibited by maintenance of high cytosolic K^+ and Na^+ ratios in plant cells. Competition of sodium (Na^+) ions and chloride (Cl^+) ions with other several nutrients—such as potassium (K^+) ions, calcium (Ca^{2+}) ions, and nitrates (NO^{3-})—may decrease plant growth by disturbing the accessibility of nutrients, hindering passage of all nutrients, and splitting various nutrients. Therefore, high levels of salts are a major cause of nutrient imbalances. Chlorophyll content, as well as photosynthetic proficiency, is reduced in numerous crops (e.g., peanut and pumpkin) as a result of high salinity (Kurum et al. 2013; Sarkar et al. 2014). When salinity stress occurs in plants, NaCl ions are accumulated in the roots of salt-tolerant and salt-sensitive crop plants. Negrão et al. (2017) reported that NaCl ions are accumulated in young leaves and in older leaves of salt-sensitive cultivars and salt-tolerant cultivars, respectively. The buildup of NaCl ions in the roots causes nutrient differences in the root tissues as a result of a reduced ratio of KCl to NaCl, due to increased salt levels. Nutritional imbalance is resulting in reduced growth, as well as dry matter accumulation in the roots. Such consequences have been observed in maize crops, which are salt sensitive (Hussain et al. 2017). Understanding of molecular approaches is necessary for addressing high levels of salinity stress. Molecular work is successively evolving salinity-tolerant cultivars and is considered a crucial factor in resolving current problems with salinity-related reductions in crop yields all over the world.

It is necessary to reduce salinization problems for higher crop production all over the world. One possible solution is reclamation of wasteland, but this is beyond the economic means of poor farmers. Another promising solution is development of salt-tolerant cultivars (Miranda et al. 2018). Variations in salt sensitivity among various genotypes of a species provide a basis for screening salt-sensitive and salt-tolerant cultivars, which may be used for further experimentation in the future (Hussain et al. 2017). Salt stress increases changes in cellular mechanism by amending expression of both specific and housekeeping genes. These may affect a great number of cellular proteins. Proteomics is considered an excellent tool to address environmental pressures, molecular influences, stress-influenced variations, and genotypic inconsistency (Miranda et al. 2018). In the current scenario, it is of vital necessity to study oxidative stress and antioxidant defense mechanisms of plants during salinity stress.

Physiological and Biochemical Mechanisms

Plants produce several physical, biological, and biochemical systems to persist in soils under high salinity. The principal mechanisms they use are listed below:

- Ion homeostasis, as well as compartmentalization in different cells and tissues
- Uptake and transport of nutrient ions
- Induction of osmoprotectants and beneficial solutes
- Triggering and production of antioxidant enzymes, as well as antioxidant compounds
- Production of polyamines and spermine
- Production of nitric oxide
- Hormonal variations

Adverse Effects of Salinity Stress

High-salinity stress has various major adverse affects on plants, such as those listed below:

- Plant growth and development are reduced by high levels of salinity (Lei et al. 2015). High salt levels are a major cause of drought, as well as ion toxicity. For that reason, a high level of salinity can frequently affect all characteristics of a plant's functioning and its metabolism or structure. Hyperionic and hyperosmotic stresses are caused by high salinity, which can ultimately lead to plant death.
- Seed germination and plant growth and development in all agricultural crops are affected by exposure with salt. During the early phases, salt stress mostly reduces the germination rate by altering structural measures, as well as decreasing seedling emergence. Seedling exposure to salt stress at the time of transplantation decreases the stem growth, as well as the dry biomass (Lee et al. 2013).
- When plants are very susceptible to salt stress, the opposing effects of salinity on the seed germination percentage, as well as during the time of seedling growth in all such agricultural crops, are very serious (Läuchli and Grattan 2007).
- The basic features of a high level of salinity and drought stress may interact with one another as a result of high salt accumulation in lands resulting in reduction of water in topsoil. Therefore, it makes it progressively harder for plants to obtain water and essential nutrients.
- Ion-specific stress causes variations in the ratio of K^+ to N^+ in conditions of high salinity. The outside Na^+ ions may harmfully influence intracellular K^+ influx.
- The accumulation or buildup of Na^+ and Cl^- concentrations is due to high salinity in the cytosol, which may eventually damage plant cells. Na^+ ions can disrupt cell membrane integrity and then attract Cl^- ions toward the gradient.

- Sodium (Na^+) ion concentrations of more than 100 mM are very dangerous for cell structures during abortion and may not only decrease the movement of several vital enzymes, cell division, and cell enlargement but also cause membrane disorder and osmotic imbalance, which finally may arrest plant growth.
- Higher sodium ion concentrations may also cause a decrease in the rate of photosynthesis, as well as causing production of ROS.
- Potassium (K^+) ions are also necessary constituents for plant growth. Variations in K^+ ions due to high-salinity stress may change the osmotic stability and the opening and closing of stomata, as well as changing the utility of certain enzymes.
- High salinization may damage cells in emerging leaves, resulting in growth reduction. Moreover, such salt-specific effects of salinity may produce a harmful effect of salts within plant cells. Salts may accumulate within older leaves, as well as causing the death of leaves that are essential for the persistence of the plant (Lee et al. 2013).

Oxidative Stress Under Salinity

Roots are considered the first major organs that are directly exposed to salinity stress, resulting in a greater decrease in growth than that caused by salinity stress in all other plant tissues, such as shoots. An increased level of NaCl ions in roots has been reported to be accompanied by increased oxidative stress constraints such as electrolyte leakage (EL), malondialdehyde (MDA), and hydrogen peroxide (H_2O_2), and reduced glutathione (GSH) and ascorbate (ASC) redox levels in plant root cells (Ghosh et al. 2015). A very high degree of salinization has been described as causing oxidative stress in several tissues and cells of plants. Because of high-salinity stress, the levels of ROS are increased in various plant cells and tissues as a result of irregularities in the electron transport chain, as well as accumulation of photoreducing influence (Sharma et al. 2012). Excess electrochemical energy may be degenerated by the Mehler reaction, which is a major cause of ROS and H_2O_2 production. Membrane injury is revealed by prominent EL and MDA levels.

Under high-salinity conditions, it is possible for oxidative defense mechanisms to be triggered by enzymatic and nonenzymatic reactions. These are well known as collective reactions during salinity in several tissues and also include organ-specific responses. Roots and mature leaves respond with increased levels of polyphenols, which are nonenzymatic components of antioxidants, with a large buildup of NaCl in such tissues or organs. Because of high salinity, tocopherols may not build up in the root cells, because of the specificity of tocopherols in scavenging singlet oxygen radicals in photosystem II. Correspondingly, it has been shown that plants under high salinity may build up large amounts of polyphenols. Furthermore, it has been confirmed that more polyphenols are produced in young leaves than in mature leaves (Samira et al. 2015). Use of triazole under high salinity has been studied for protection of tocopherols against oxidative stress in tomato leaves, as well as in wheat crops (Tanou et al. 2009). Additionally, Hussain et al. (2017) showed that as a result of large amounts of salt, ASC and tocopherols are decreased in de-etiolated rice leaves; such contradictory observations may be due to specificity of cells or tissues. Similarly, as a result of high salinity, GSH levels can be increased at the organ level, and this may be root specific. This condition may improve the demand for and absorption of sulfur, in times of stress, for production of various components of antioxidants, such as GSH. Furthermore, it has been shown that roots produce stress hormones such as abscisic acid (ABA), which is well known as a stress signal in plant shoots. Levels of this hormone may be increased by several environmental stresses, and a sulfur source is also required. Gallardo et al. (2014) identified that salt-stressed plants contain upregulated sulfate transporter genes, such as *AtSULTR3*, as well as huge amount of sulfate. Consequently, it has been observed that the amount of sulfate influences the levels of ABA and GSH in various plant cells because of high salt concentrations (Cao et al. 2014). For reducing ROS, the ASC-GSH cycle plays an important role in increasing ASC and GSH content, the ASC and GSH contents and significantly reduced the ASC and GSH redox status. It has been confirmed that salt-resistant cultivars produce higher amount of transcriptional activities of ASC-GSH cycle, but susceptible cultivars do not (Mahajan and Sanejouand 2015).

Generally, antioxidant defense mechanisms are very common in all plants and their tissues, including in different developmental growth stages. There is a possibility that an alternative antioxidant defense system will be found in all plant cells or even tissues. Every organ such as the root or shoot may use a more effective mechanism in harmony with many other metabolic pathways within that particular organ to survive under salinity, which may result in oxidative stress. Interpreting such detailed substitutes can aid our understanding of effective and helpful metabolic systems used by many cells or tissues to cope in particular stress circumstances (Gallardo et al. 2014). For example, Giza119 is an engineered maize variety, which has higher levels of key enzymes involved in the ASC-GSH cycle in its roots (Sytykiewicz 2016). This variety produces greater amounts of tocopherols in its shoots or even in whole plants, with condensed transport of NaCl from the root to the shoot. During the seedling stage of plant life, this may be very helpful for management of salinity stress (Sytykiewicz 2016).

Production and Scavenging of ROS

Generally, ROS production is the most important concern arising from aerobic respiration in all plants. ROS include numerous free radicals (e.g., superoxide anions and hydroxyl radicals), while the nonradicals include H_2O_2, as well as singlet oxygen. A stepwise decrease in molecular oxygen (O_2) through excess energy release and electron transfer processes indicates production of too much ROS. In numerous plants, ROS is continuously produced through obligatory outflow of electrons into O_2 from electron transfer processes in the chloroplasts, mitochondria, and plasma membranes as a derivative of numerous respiration paths limited to many cellular components. Abiotic or environmental stresses (e.g., chilling, metal toxicity, drought, and salinity), as well as pathogen attack, mainly increase ROS production in several plants because of interference with cellular homeostasis. In extremely high concentrations, ROS are known to be very detrimental to several microorganisms. A cell is considered to be under "oxidative stress" conditions when the ROS level goes beyond the level that is manageable by the antioxidant defense system. Different environmental stress conditions produce higher concentrations of ROS, which pose a serious risk to plant organs through initiation of lipid peroxidation, as well as corrosion of proteins. Such abiotic stresses may also damage DNA, RNA, proteins, and various enzymes, and initiation of programmed cell death pathways, and even organ death, may occur.

Regardless of their potentially harmful actions, ROS play roles as secondary messengers for the selection of cellular mechanisms involved in resistance against ecological stresses. ROS can operate as destructive or signaling molecules, depending upon the balance between ROS generation and ROS-scavenging activities. Because ROS fulfill several functions, it is compulsory for organs to regulate ROS levels very actively to avoid oxidative harm while not eliminating them completely. Reclamation of higher ROS is attained through an effectual antioxidative defense system containing enzymatic and nonenzymatic antioxidants. Several enzymatic and nonenzymatic antioxidants are produced during salinity stresses within the plant cell. Many studies have been carried out in several plants regarding the function and induction of numerous enzymes involved in the defense mechanisms that cope the stress in plants produced by different abiotic stresses. Regarding these environmental conditions, conservation of antioxidant capacity may reduce toxic ROS and plays an important role in enhancing stress tolerance mechanisms in plants. Significant advancements have been used to improve oxidative stress tolerance in numerous plants by developing transgenic plants having higher amount of different antioxidants. Immediate manifestation of numerous antioxidant enzymes has been shown to be most helpful and more powerful than single or double enzyme expression. Therefore, such mechanisms are most useful for evaluation and characterization of transgenic plants with improved resistance against different environmental stresses. Additionally, antioxidant defense systems play signifcant roles in tolerance of plants under stresses.

Production Sites and Effects of ROS

ROS comprise several free radicals, reactive fragments, and ions, which mostly result from O_2. Furthermore, it has been estimated that approximately 1% of the molecular oxygen expended by many plants is used to generate ROS in numerous subcellular fragments. So, ROS are mostly acknowledged as having different characters, being both harmful and beneficial, depending on their absorption in plants. Serious damaging effects are caused when ROS concentrations are high. In moderate and lower concentrations, however, they act as secondary messengers in intracellular signaling, facilitating numerous reactions of plant cells, tissues, or organs.

ROS and Oxidative Loss of Biomolecules

Generation and destruction of ROS need to be precisely balanced to reduce oxidative stress. A cell is considered to be in a state of oxidative stress when the amount of ROS causes enhancement in the defense mechanism in the plant. The balance between generation and scavenging of ROS is disturbed by various stressful circumstances, such as high salinity, drought effects, high light conditions, toxicity of different metals, and various pathogens (Han et al. 2009). Lipids, proteins, and DNA are some of the major biomolecules in plants that may be lost when high ROS concentrations are produced as a consequence of large amounts of salt being absorbed through the roots. These responses may possibly interfere with essential membrane characteristics such as fluidity and ion transportation, hindering enzyme action, protein cross-linkage, and protein synthesis, and causing DNA loss, which may eventually lead toward tissue or even whole-plant death (Tanou et al. 2009).

Lipids

Lipid peroxidation mostly affects cellular functioning and is found in both cellular and organellar membranes when ROS concentrations go above the threshold level. Sharma and Dubey (2005) reported that DNA and proteins can be damaged when lipid peroxidation magnifies oxidative stress, with lipid synthesis resulting in radicals that may react with themselves. Furthermore, lipid peroxidation has been extensively applied for an indication of ROS facilitating cell membrane injuries in different stressful circumstances. Mishra et al. (2011) reported that deprivation or destruction of numerous lipids have been described in cultivated crop plants because of several environmental stresses. Additionally, there is proliferation of lipid peroxidation due to various stresses, which is equivalent to increased ROS generation. MDA is the final product during peroxidation of unsaturated fatty acids in phospholipids and is also liable for cell membrane destruction. ROS attack two specific locations: unsaturated or double bonds between two carbon atoms in phospholipid molecules, and ester bonds between glycerol and fatty acid molecules. During the production of ROS, polyunsaturated fatty acids are also found in membrane phospholipids. The reaction involved in peroxidation of numerous polyunsaturated fatty acids with a single OH group is known as a part of a cyclic chain reaction. Generally, the complete mechanism of lipid peroxidation involves different successive stages such as introduction or initiation, development, and closure or ending. The preliminary stage of lipid peroxidation involves stimulation of O_2, as a level controlling factor. Miller et al. (2008) confirmed that $O_2^{\bullet-}$ and $^{\bullet}OH$ groups may react with methylene groups of polyunsaturated fatty acids, increasing lipid peroxy radicals, as well as hydro peroxides.

Proteins

The effect of ROS on proteins might possibly be the source of changes in protein behavior; some effects are direct, while others may be indirect. Total modification contains variation of a proteins movement by carbonylation, nitrosylation, and development of disulfide bonds, as well as glutathionylation. Yamauchi et al. (2008) found that proteins may remain amended ultimately though conjugation using breakdown products of fatty acid peroxidation. At the time of unnecessary ROS production, site-specific amino acid alterations, fragmentation of peptide chains, accumulation of cross-linked reaction products, and different absorbing mechanisms, as well as enlarged exposure of proteins to proteolysis, occur.

Plant cells damaged by oxidative stress normally require better attentions of carbonylated proteins, which is a broadly used indicator of protein oxidation (Møller and Kristensen 2004). Enriched alteration of various proteins is described in all plants in conditions of environmental stress (Maheshwari and Dubey 2009). The amino acids present in a peptide differ in their vulnerability to attack by ROS. Thiol groups and sulfur-containing amino acids are more vulnerable to damage by ROS (Tanou et al. 2009). Numerous metals—such as Cd, Pb, and Hg—exposed to cause the reduction of protein-bound thiol groups. Oxygen can also remain added to methionine to form methionine sulfoxide. Tyrosine is freely cross-linked to bityrosine products in the presence of ROS (Tanou et al. 2009).

Enzyme inactivation is due to irreversible oxidation of iron–sulfur through $O_2^{\bullet-}$ divalent cations which drag through metal (Fe) on the binding site on proteins. Then the metal (Fe) reacts in a Fenton reaction to form $^{\bullet}OH$, which quickly oxidizes an amino acid remaining at or adjacent to the cation-binding site of the proteins. Reacted proteins support by means of different sources by proteolytic assimilation. Cabiscol et al. (2000) recommended that oxidation might influenced the ubiquitination mechanism and proteasomes degradation. Pea leaf rough extracts with cumulative H_2O_2 concentrations from Cd-treated plants and peroxisomes disinfected from pea leaves displayed proliferations in carbonyl content. Oxidized proteins were more competently degraded, and the proteolytic activity was increased by 20% as a result of the metal treatment (Cabiscol et al. 2000). Different scientists explored that, under stress condition more crosslinked occur it will not only protects protein from damaging; nevertheless, they also may prevent proteases from damaging other oxidized proteins (Grune et al. 1997; Cabiscol et al. 2000).

DNA

ROS are known as a main cause of DNA loss. ROS may be a source of oxidative harm to the cell nucleus, mitochondria, and chloroplast DNA (Imlay and Linn 1988). DNA is well known as the genetic material of cells, and attacks on DNA result in alterations in proteins, and can cause breakdowns or thorough inactivation of the proteins. Oxidative occurrence to DNA can cause deoxyribose oxidation—with damage, insertion, or deletion of various nucleotides—and a diversity of alterations in the organic foundations of the nucleotides, including DNA–protein crosslinks. Furthermore, alterations in the nucleotides of a single constituent may cause divergence with the nucleotides in the other strand, resulting in mutations. Meriga et al. (2004) reported that high levels of DNA deprivation have been detected in several plants as manifestations of abiotic stresses such as salinity and metal toxicity. Oxidative stress on DNA bases usually includes ˙OH additions to double bonds, whereas sugar damages mostly due to hydrogen produced from deoxyribose. The hydroxyl radical is known to react with all purine and pyrimidine bases, as well as the deoxyribose backbone (Evans et al. 2004). ROS-induced DNA harm can cause different mutagenic changes. Mutations arising from alterations at G:C sites, in particular, show oxidative effects on DNA caused by ROS. ROS produced reactive yields which damages the bases of DNA (Evans et al. 2004).

ROS effects on DNA sugars result in single-strand disruptions (Oleinick et al. 1987). DNA protein crosslinks cannot be freely recovered and might be dead when transcrpiton is done. As compared with nuclear DNA, the mitochondria and chloroplasts are very vulnerable to oxidative DNA damage because of the absence of defending proteins and histones, and ROS generating sites (Richter 1992). Although repair of damaged DNA may be possible, extreme alterations initiated by ROS result in permanent damage to the DNA, with possibly detrimental consequences for the cell (Yilancioglu et al. 2014).

The Antioxidant Defense System

When plants are exposed to salt stress conditions, ROS are produced and damage cellular activity. To overcome these ROS, the cellular defense mechanisms of the plant are activated and protect the cells. The cellular defense systems protect the plant in three different ways: (1) protection, (2) interception, and (3) restoration. Maintenance of sufficient antioxidant levels helps to conserve compounds, as well as enzymes. Long-term and short-term alterations in these functions are new areas of interest in research. Synthetic antioxidants are being used in simulations of numerous biological approaches in plants. Enzymatic antioxidants (superoxide dismutase (SOD), peroxide (POD), catalase (CAT), ascorbate peroxidase (APX), and glutathione reductase (GR)) and nonenzymatic antioxidants (proline, glycine betaine, sugars, polyols, and phenols) produced in plant cells are the main constituents of the antioxidant defense systems and limit salt-induced damage.

Components of the Nonenzymatic Antioxidative Defense System

Components of the nonenzymatic antioxidant defense system are involved in different metabolic processes in the cell and act as cofactors for different enzymes that directly affect plant growth and development, from cell multiplication to cell death. Plants with reduced nonenzymatic antioxidant levels have been shown to be hypersensitive to environmental stress (Pinto and Gara 2004). There are different components of the nonenzymatic antioxidant defense system, such as tocopherols, carotenoids, and phenolic compounds. Tocopherols are a group of lipophilic antioxidants involved in scavenging of oxygen free radicals. Tocopherols inhibit the chain proliferation phase in lipid oxidation, which sorts it as an active free radical. An increase in tocopherol has been shown to increase tolerance of salinity in diverse plant species (Bafeel and Ibrahim 2008). It has been established that metabolic causes of tocopherol biosynthesis affect endogenous ascorbate and GSH pools in leaves. Furthermore, the apparent levels of genes encrypting enzymes involved in the Halliwell–Asada cycle (ascorbate peroxidase (APX), dehydroascorbate reductase (DHAR), and monodehydroascorbate reductase (MDHAR)) were upregulated. Mutants of *Arabidopsis thaliana* with transfer DNA (T-DNA) insertions in the tocopherol biosynthesis genes tocopherol cyclase (vte1) and γ-tocopherol methyltransferase (vte4) exhibited higher concentrations of protein carbonyl groups and glutathione disulfide (GSSG) in comparison with the wild type, representing development of oxidative stress. Transgenic rice plants with OsVTE1 RNA interference (OsVTE1-RNAi) were highly sensitive to salt stress, whereas transgenic plants overexpressing OsVTE1 (OsVTE1-OX) exhibited higher tolerance of salt stress. The OsVTE1-OX plants also generated less H_2O_2 than the control plants.

Carotenoids are also part of the group of lipophilic antioxidants and have the ability to detoxify several forms of ROS. Carotenoids are present in several plants, including microorganisms. Carotenoids also assist signaling molecules that affect plant development and biotic or abiotic stress reactions (Li et al. 2010). Gomathi and Rakkiyapan (2011) observed that high carotenoid content favored excellent variation of sugarcane plants under saline conditions.

Phenolic compounds (e.g., flavonoids, tannins, hydroxycinnamate esters, and lignin) are considered diverse secondary metabolites with antioxidant properties. Phenolic compounds, especially flavonoids and phenylpropanoids, are oxidized by peroxidase and act in H_2O_2 scavenging and the phenolic/ascorbic acid (AsA)/POD system. There is a certain indication of stimulation of phenolic metabolism in plants as a reaction to various stresses. Janas et al. (2009) detected that ROS might provide a collective signal for acclimatization to Cu^{2+} stress and might cause accumulation of total phenolic compounds in dark-grown lentil roots. Also, transgenic potato plants with increased concentrations of flavonoids were shown to exhibit greater antioxidant capacity.

Components of the Enzymatic Antioxidative Defense System

The enzymatic antioxidant defense mechanisms comprise different enzymes (e.g., SOD, CAT, ascorbate, and GSH). These enzymes operate in different subcellular compartments and respond in concert when cells are exposed to oxidative stress.

Superoxide dismutase (SOD) plays a main role in tolerance against oxidative stress in all aerobic organisms. SOD activity showed visible affect to numerous environmental stresses. Increased activity of SOD is frequently linked to better plant tolerance of environmental stresses. It has been recommended that SOD be used as an indirect selection measure for screening salinity-resistant plant materials. Overproduction of SOD has been described as resulting in enhanced oxidative stress tolerance in plants (Gupta et al. 1993).

Among antioxidant enzymes, catalase (CAT) was the first enzyme to be exposed and described. CAT scavenges H_2O_2 production in cell organelles during photorespiratory oxidation, β-oxidation of fatty acids, and several other enzyme activities (e.g., xanthine oxidase (XOD) attachment to SOD) (Corpas et al. 2008). H_2O_2 is degraded by CAT in an energy-effective way. Various environmental stresses are major causes of either enhancement or depletion of CAT activity, depending on the stress intensity, duration, and type (Moussa and Aziz 2008).

Ascorbic acid (AsA) is a very rich and low molecular weight antioxidant that plays a major part in protection against oxidative stress produced by increased levels of ROS. AsA is known as the most powerful and effective antioxidant because of its capability to donate electrons in a number of enzymatic and nonenzymatic reactions. AsA has been shown to play an important role in numerous physiological processes in plants such as growth, differentiation, and metabolism. Apoplastic AsA is assumed to characterize the first line of defense, besides potentially damaging external oxidants (Barnes et al. 2002). AsA defends major macromolecules from oxidative damage. In normal physiological conditions, AsA frequently occurs in a condensed state in chloroplasts, where it also acts as a cofactor for violaxanthin deepoxidase, consequently supporting tolerance of excess excitation energy (Smirnoff 2000). AsA plays an important role in removal of H_2O_2 through the AsA-GSH cycle.

Glutathione (GSH) is one of the fundamental low molecular weight nonprotein thiols and plays an important role in intracellular protection against ROS-induced oxidative loss. Because of its high reducing power, GSH plays an important role in different biological practices such as cell growth, cell division, regulation of sulfate transportation, signal transduction, conjugation of metabolites, enzymatic regulation, synthesis of proteins and nucleic acids, synthesis of phytochelatins for metal chelation, reclamation of xenobiotics, and expression of stress reactive genes (Foyer and Noctor 2003). GSH acts as an antioxidant in numerous ways. It can react chemically with $O_2^{\bullet-}$, $^{\bullet}OH$, and H_2O_2, and therefore functions as a free radical scavenger. GSH may protect macromolecules such as proteins, lipids, and DNA. When apple trees were exposed to progressive salinity, the primary reaction was slight oxidation of the GSH pool, followed by enhanced GSH applications.

Conclusion and Future Prospects

It is necessary to characterize and evaluate salt-tolerant cultivars with precise and site-specific construction techniques that will tolerate high production and high throughput in salt-affected zones. Advances in the fields of genetics, genomics, and biotechnology, combined with conservative breeding programs, have a great prospective capability to develop transgenic cultivars for excellent yield performance under different environmental or abiotic stress circumstances. Furthermore, exogenous use of several osmoprotectants, plant hormones (natural or synthetic), nutrient balancing, and seed-priming practices may be efficient and useful for low-cost cultivar development in saline zones.

Salt stress is one of the main threats to global food security. Little research is available on the development of salt-resistant germplasms. Salinity produces osmotic stress, as well as ionic toxicity, in plants, with opposing effects on seed germination and plant life-spans. Salt stress disrupts the functions of cytosolic enzymes, which may cause nutritional imbalances, as well as oxidative destruction. All of these effects severely reduce crop growth and ultimately crop yield. Growth of salt-resistant plants by conservative breeding, along with genetic engineering, has shown that conventional breeding can be made much more effective to resolves such problems. However, in future different techniques such as RNA interference, transposon insertional knockouts are useful for stress-resistant genes. Increased understanding of signaling systems in plants is expected to result in production of salt-resistant cultivars in the future. Application of molecular markers for mass screening and breeding, marker-assisted selection, quantitative trait locus (QTL) studies, different exogenous growth hormones and osmoprotectants for growing seeds or plants, cellular-based stress signaling, and ion homeostasis tools to produce a useful model of an entire plant could be helpful for development of salt-resistant cultivars.

References

Bafeel SO, Ibrahim MM (2008) Antioxidants and accumulation of α-tocopherol induce chilling tolerance in *Medicago sativa*. Int J Agric Biol 10:593–598

Barnes JD, Zheng Y, Lyons TM (2002) Plant resistance to ozone: the role of ascorbate. In: Omasa K, Saji H, Youssefian S, Kondo N (eds) Air pollution and plant biotechnology. Springer, Tokyo, pp 235–254

Cabiscol E, Piulats P, Echave E, Herrero E, Ros J (2000) Oxidative stress promotes specific protein damage in *Saccharomyces cerevisiae*. J Biol Chem 275:27393–27398

Cao MJ, Wang Z, Zhao Q, Mao JL, Speiser A, Wirtz M, Hell R, Zhu JK, Xiang CB (2014) Sulfate availability affects ABA levels and germination response to ABA and salt stress in *Arabidopsis thaliana*. Plant J 77:604–615

Corpas FJ, Chaki M, Fernandez-Ocana A, Valderrama R, Palma JM, Carreras A, Begara-Morales JC, Airaki M, del Río LA, Barroso JB (2008) Metabolism of reactive nitrogen species in pea plants under abiotic stress conditions. Plant Cell Physiol 49:1711–1722

Cosme M, Lu J, Erb M, Stout MJ, Franken P, Wurst S (2016) A fungal endophyte helps plants to tolerate root herbivory through changes in gibberellin and jasmonate signaling. New Phytol 211:1065–1076

Evans MD, Dizdaroglu M, Cooke MS (2004) Oxidative DNA damage and disease: induction, repair and significance. Mutat Res 567:1–61

Foyer CH, Noctor G (2003) Redox sensing and signaling associated with reactive oxygen in chloroplasts, peroxisomes and mitochondria. Physiol Plant 119:355–364

Gallardo K, Courty PE, Signor LC, Wipf D, Vernoud V (2014) Sulfate transporters in the plant's response to drought and salinity: regulation and possible functions. Front Plant Sci 5:580

Ghosh S, Mitra S, Paul A (2015) Physiochemical studies of sodium chloride on mungbean (*Vigna radiata* L. Wilczek) and its possible recovery with spermine and gibberellic acid. Sci World J 2015:858016
Gomathi R, Rakkiyapan P (2011) Comparative lipid peroxidation, leaf membrane thermostability, and antioxidant system in four sugarcane genotypes differing in salt tolerance. Int J Plant Physiol Biochem 3:67–74
Grune T, Reinheckel T, Davies KJA (1997) Degradation of oxidized proteins in mammalian cells. FASEB J 11:526–534
Gupta AS, Heinen JL, Holaday AS, Burke JJ, Allen RD (1993) Increased resistance to oxidative stress in transgenic plants that overexpress chloroplastic Cu/Zn superoxide dismutase. Proc Natl Acad Sci U S A 90:1629–1633
Han C, Liu Q, Yang Y (2009) Short-term effects of experimental warming and enhanced ultraviolet-B radiation on photosynthesis and antioxidant defense of *Picea asperata* seedlings. Plant Growth Regul 58:153–162
Hussain M, Ahmad S, Hussain S, Lal R, Allah SU, Nawaz A (2017) Rice in saline soils: physiology, biochemistry, genetics, and management. Adv Agron 20:1
Imlay JA, Linn S (1988) DNA damage and oxygen radical toxicity. Science 240:1302–1309
Ivanova D, Zhelev Z, Aoki I, Bakalova R, Higashi T (2016) Overproduction of reactive oxygen species-obligatory or not for induction of apoptosis by anticancer drugs. Chin J Cancer Res 28:383
Janas KM, Amarowicz R, Zielin J, Tomaszewska A, Kosin SA, Posmyk MM (2009) Induction of phenolic compounds in two dark-grown lentil cultivars with different tolerance to copper ions. Acta Physiol Plant 31:587–595
Kurum R, Ulukapi K, Aydinsakir K, Onus AN (2013) The influence of salinity on seedling growth of some pumpkin varieties used as rootstock. Not Bot Horti Agrobot Cluj Napoca 41:219
Läuchli A, Grattan SR (2007) Plant growth and development under salinity stress. In: Jenks MA, Hasegawa PM, Jain SM (eds) Advances in molecular breeding toward drought and salt tolerant crops. Springer, Dordrecht, pp 1–32
Lee MH, Cho EJ, Wi SG, Bae H, Kim JE, Cho JY, Lee S, Kim JH, Chung BY (2013) Divergences in morphological changes and antioxidant responses in salt-tolerant and salt-sensitive rice seedlings after salt stress. Plant Physiol Biochem 70:325–335
Lei XG, Zhu JH, Cheng WH, Bao Y, Ho YS, Reddi AR, Holmgren A, Arnér ES (2015) Paradoxical roles of antioxidant enzymes: basic mechanisms and health implications. Physiol Rev 96:307–364
Li C, Lv J, Zhao X, Ai X, Zhu X, Wang M, Zhao S, Xia G (2010) TaCHP-a wheat zinc finger protein gene downregulated by ABA and salinity stress plays a positive role in stress tolerance. Plant Physiol pp110
Mahajan S, Sanejouand YH (2015) On the relationship between low-frequency normal modes and the large-scale conformational changes of proteins. Arch Biochem Biophys 567:59–65
Mahajan S, Pandey GK, Tuteja N (2008) Calcium-and salt-stress signaling in plants: shedding light on SOS pathway. Arch Biochem Biophys 471:146–158
Maheshwari R, Dubey RS (2009) Nickel-induced oxidative stress and the role of antioxidant defence in rice seedlings. Plant Growth Regul 59:37–49
Meriga B, Reddy BK, Rao KR, Reddy LA, Kishor PBK (2004) Aluminium-induced production of oxygen radicals, lipid peroxidation and DNA damage in seedlings of rice (*Oryza sativa*). J Plant Physiol 161:63–68
Miller G, Shulaev V, Mittler R (2008) Reactive oxygen signaling and abiotic stress. Physiol Plant 133:481–489
Miranda AJ, Achalandabaso YA, Aguirresarobe A, Canto DA, López PU (2018) Similarities and differences between the responses to osmotic and ionic stress in quinoa from a water use perspective. Agric Water Manag 203:344–352
Mishra S, Jha AB, Dubey RS (2011) Arsenite treatment induces oxidative stress, upregulates antioxidant system, and causes phytochelatin synthesis in rice seedlings. Protoplasma 248:565–577

Møller IM, Kristensen BK (2004) Protein oxidation in plant mitochondria as a stress indicator. Photochem Photobiol Sci 3:730–735
Moussa R, Aziz SM (2008) Comparative response of drought tolerant and drought sensitive maize genotypes to water stress. Aust J Crop Sci 1:31–36
Negrão S, Schmöckel SM, Tester M (2017) Evaluating physiological responses of plants to salinity stress. Ann Bot 119:1–11
Oleinick NL, Chiu SM, Ramakrishnan N, Xue LY (1987) The formation, identification, and significance of DNA-protein cross-links in mammalian cells. Br J Cancer 8:135–140
Pinto MCD, Gara LD (2004) Changes in the ascorbate metabolism of apoplastic and symplastic spaces are associated with cell differentiation. J Exp Bot 55:2559–2569
Rengasamy P (2006) World salinization with emphasis on Australia. J Exp Bot 57:1017–1023
Richter C (1992) Reactive oxygen and DNA damage in mitochondria. Mutat Res 275:249–255
Samira M, Hichem H, Boughalleb F, Mounir D (2015) Effect of light-salinity interaction on photosystem activity of excised corn leaves. Afr Crop Sci J 23:343–354
Sarkar T, Thankappan R, Kumar A, Mishra GP, Dobaria JR (2014) Heterologous expression of the AtDREB1A gene in transgenic peanut-conferred tolerance to drought and salinity stresses. PLoS One 9:e110507
Sharma P, Dubey RS (2005) Drought induces oxidative stress and enhances the activities of antioxidant enzymes in growing rice seedlings. Plant Growth Regul 46:209–221
Sharma P, Jha AB, Dubey RS, Pessarakli M (2012) Reactive oxygen species, oxidative damage, and antioxidative defense mechanism in plants under stressful conditions. J Bot 2012:1–26
Smirnoff N (2000) Ascorbic acid: metabolism and functions of a multi-faceted molecule. Curr Opin Plant Biol 3:229–235
Sytykiewicz H (2016) Expression patterns of genes involved in ascorbate--glutathione cycle in aphid-infested maize (*Zea mays* L.) seedlings. Int J Mol Sci 17:268
Tanou G, Molassiotis A, Diamantidis G (2009) Induction of reactive oxygen species and necrotic death-like destruction in strawberry leaves by salinity. Environ Exp Bot 65:270–281
Yamauchi Y, Furutera A, Seki K, Toyoda Y, Tanaka K, Sugimoto Y (2008) Malondialdehyde generated from peroxidized linolenic acid causes protein modification in heat-stressed plants. Plant Physiol Biochem 46:786–793
Yilancioglu K, Cokol M, Pastirmaci I, Erman B, Cetiner S (2014) Oxidative stress is a mediator for increased lipid accumulation in a newly isolated *Dunaliella salina* strain. PLoS One 9:e91957

Oxidative Stress and Antioxidant Defense in Plants Under Drought Conditions

Sajjad Hussain, Muhammad Junaid Rao, Muhammad Akbar Anjum, Shaghef Ejaz, Iqra Zakir, Muhammad Arif Ali, Niaz Ahmad, and Shakeel Ahmad

Introduction

Globally, climate change causes various environmental stresses, which pose threats to agricultural crop production. In natural environments, plants are exposed to different biotic and abiotic stresses, which have severe impacts on plant growth and yield (Mittler 2002; Suzuki et al. 2014; Zandalinas et al. 2017). Among the different environmental stresses (such as drought, heat, irregular temperatures, and high salinity), drought is one of the major abiotic factors throughout the world, as it directly affect on plant growth, thereby threatening the economy by limiting crop production (Martinez et al. 2018; Zhou et al. 2007; Jacobsen et al. 2012).

S. Hussain (✉)
Department of Horticulture, Faculty of Agricultural Sciences and Technology, Bahauddin Zakariya University, Multan, Punjab, Pakistan
e-mail: sajjad.hussain@bzu.edu.pk

M. J. Rao
Key Laboratory of Horticultural Plant Biology (Ministry of Education), Huazhong Agricultural University, Wuhan, People's Republic of China

College of Horticulture and Forestry Sciences, Huazhong Agricultural University, Wuhan, People's Republic of China

M. A. Anjum · S. Ejaz
Department of Horticulture, Bahauddin Zakariya University, Multan, Pakistan

I. Zakir
Department of Agronomy, Bahauddin Zakariya University, Multan, Pakistan

M. A. Ali · N. Ahmad
Department of Soil Science, Bahauddin Zakariya University, Multan, Pakistan

S. Ahmad
Department of Agronomy, Faculty of Agricultural Sciences and Technology, Bahauddin Zakariya University, Multan, Punjab, Pakistan

M. Hasanuzzaman et al. (eds.), *Plant Abiotic Stress Tolerance*,
https://doi.org/10.1007/978-3-030-06118-0_9

Drought leads to molecular, physiological, and biochemical responses, of which photosynthesis is the prime physiological target (Zinta et al. 2014; Chaves et al. 2009). Usually, drought stress decreases the crop yield, alters chlorophyll components, hinders the photosynthetic process (Muller et al. 2011), and changes the activity of key enzymes involved in carbon metabolism and antioxidant processes (Devi et al. 2012; Kaur et al. 2007). It also inhibits the enzymes involved in the Calvin cycle and causes a decline in photochemical activities (Dias and Brüggemann 2010). Drought directly decreases the photosynthetic abilities of plants and reduces their growth by creating an imbalance between production and scavenging of reactive oxygen species (ROS), which ultimately facilitates high accumulation of ROS in cells and induces damage to the nucleus, membrane lipids, proteins, and other components of cells, leading to cell death (Kabiri et al. 2014).

The responses of plants to drought depend on the intensity and duration of the stress, but also on the genetic characteristics of the individual species. If drought stress is sustained for a long period, it will indeed cause oxidative injury due to excessive increments in ROS (Hussein and Safinaz 2013; Miller et al. 2010). Under challenging environmental conditions, plants trigger molecular, biochemical, and physiological responses to acclimatize and persist against abnormal conditions. Unusual abiotic situations force plants to make changes in their metabolism to minimize the harmful effects of the environmental stress. Study of a single abiotic aspect does not reflect the plant's response to stress, because in natural environments, plants have to face more than one abiotic stress such as heat, water deficit, and irregular temperatures (Mittler and Blumwald 2010; Suzuki et al. 2014; De Boeck et al. 2015; Hu et al. 2015; Liu et al. 2015; Zhang et al. 2015). Plants undergo changes in their gene expression patterns to fight against multiple abiotic stresses (such as heat, water deficit, temperature, etc.), which empower the plants to react in such a way that helps them to normalize their functioning under extreme environmental conditions (Mittler 2006). These types of alterations in gene expression patterns force plants to undergo changes at the metabolic level to deal with unfavorable environmental circumstances. This kind of alteration adopted by plants varies among species and according to the type and intensity of the stress.

To expose the mechanism of drought stress and to detect drought-related genes in plants, a wide range of studies have been conducted on *Arabidopsis* (as a model plant). Enormous success has been achieved in understanding the mechanisms of drought tolerance and in determining drought-associated genes (Seki et al. 2002). Recognition of drought tolerance genes is a major step toward breeding resistant cultivars through breeding programs or via the latest genome-editing technologies.

Oxidative Stress Under Drought Conditions

Drought stress induces a series of metabolic alterations, morphological variations, and physiological changes in plants, which disrupt the normal homeostasis of the plants. Water deficit conditions impose oxidative stress on plants, which leads to production of various ROS such as hydrogen peroxide (H_2O_2), hydroxyl radicals

Table 1 Reactive oxygen species (*ROS*) types, sites of production, reactivity, and scavenging systems in plants cells [Ref. in text]

ROS	ROS symbol (+/−)	Sites of production	Reactivity	Scavenging systems	
				Enzymatic	Nonenzymatic
Hydrogen peroxide	H_2O_2, nonradical	Peroxisomes, chloroplasts, mitochondria, cytosol, apoplast	Reacts with proteins by acting on cysteine and methionine residues; reacts with heme proteins; reacts with DNA	APX, CAT, GPX, PER, PRX, GSH	Ascorbate
Hydroxyl radical	$OH^{\bullet}$, radical	Iron, H_2O_2 (Fenton reaction)	Extremely reactive with all biomolecules, including DNA, RNA, lipids, and proteins		Flavonoids, proline, ascorbate, sugars
Singlet oxygen	$^1O_2$, nonradical	Membranes, chloroplasts, nucleus	Oxidizes lipids, proteins (Trp, His, Tyr, Met, and Cys residues), and G residues of DNA		Carotenoids, α-tocopherol
Superoxide	$O_2^{\bullet-}$, radical	Apoplast (RBOHs), chloroplasts, mitochondria, peroxisomes, electron transfer chains	Reacts with Fe–S proteins and is dismutated into H_2O_2	SOD	Flavonoids, ascorbate

APX ascorbate peroxidase, *CAT* catalase, *GPX* glutathione, *PER* peroxidase, *GSH* glutathione, *PER* peroxidase, *PRX* peroxiredoxin, *RBOH* respiratory burst oxidase homologue

($OH^{\bullet}$), superoxide (O_2^-), and singlet oxygen ($^1O_2$). In plants, O_2^- and H_2O_2 govern different reactions at cellular level, such as the Fenton reaction (an iron-catalyzed reaction) and different enzyme reactions involving xanthine oxidase, peroxidases (PODs), reduced nicotinamide adenine dinucleotide phosphate hydrogen (NADPH) oxidase, and lipoxygenases, as shown in Table 1. The major cellular components that are more prone to damage by these radicals are nucleic acid, proteins, carbohydrates, lipids (through peroxidation of unsaturated fatty acids in the membrane), and enzymes (through denaturation) (Mattos and Moretti 2015).

Generally, plants produce ROS in an aerobic environment. Any variation in the external environment causes variations at the metabolic level, which trigger ROS production in plant cells (Suzuki et al. 2013). High production of ROS causes inequality in the electron transport process in the mitochondria and chloroplasts. Under water deficit conditions, approximately 70% of total H_2O_2 is produced through photorespiration, which is the key factor in ROS production. Drought stress causes stomatal closure, which prompts ROS accumulation. ROS such as hydrogen peroxide are well known for causing cellular damage, toxicity, and inhibition of photosynthesis under stress conditions (Choudhury et al. 2017), whereas in normal conditions, these molecule mediate the signaling transduction response in plant cells (Mittler 2017). However, enormous increments in ROS and scavenging are key processes that mediate more efficient consumption of ROS as a primary plant defense strategy to fight against numerous stress conditions (biotic or abiotic). ROS increments reveal a negative impact on normal plant growth and homeostasis, which restricts cellular processes. On the other hand, at low levels, ROS function as signal transduction molecules, which help plants to maintain their normal cellular processes (Nath et al. 2016).

Production and Scavenging of ROS

ROS production is stimulated in a highly systematic manner in defined cellular organelles of the plant cell, governed by ROS signaling. This may be initiated by the enzyme responsible for ROS production in the apoplast and plasma membrane, NADPH oxidase [in plants, this is termed a respiratory burst oxidase homologue (RBOH)] (Sumimoto 2008; Suzuki et al. 2011; Laurindo et al. 2014; Sirokmany et al. 2016). This enzyme family may be located in the endoplasmic reticulum, mitochondria, vacuole, or nucleus, and they are regulated through various phosphorylation/dephosphorylation and calcium-dependent reactions (Suzuki et al. 2011; Laurindo et al. 2014). Moreover, in the case of RBOHs, the signaling of ROS is initiated through PODs in the apoplast, along with ROS accumulation in various cellular organelles such as the nucleus, mitochondria, chloroplasts, and peroxisomes (Mittler et al. 2004, 2011; Halliwell and Gutteridge 2007; Konig et al. 2012; Foyer and Noctor 2013; Vaahtera et al. 2014; Mignolet-Spruyt et al. 2016). Therefore, an appropriate balance is maintained between metabolic ROS production, ROS detoxification, and ROS production for signaling; the rates of ROS reactivity and diffusion; and ROS perception in the various cellular organelles (ROS network) of the plant cell (Mittler et al. 2004).

To retain the normal balance of ROS levels under drought stress, plant tissues produce various enzymes that are scavengers of ROS (Mattos and Moretti 2015). Increased tolerance of abiotic stress in plants is thought to be associated with rapid detoxification of excessive ROS (Martinez et al. 2018). These ROS-dependent signals/reactions regulate the broad reaction of the plant cell to cope with the specific circumstances.

Under abiotic stresses, ROS has some beneficiary functions until a normal ROS level is reached, such as stimulation of plant acclimatization pathways through induction of signal transduction reactions (Foyer and Noctor 2013; Vaahtera et al. 2014; Mignolet-Spruyt et al. 2016; Mittler 2017). During drought stress, ROS-scavenging systems are highly sensitive and are also unable to initiate systemic signaling (Davletova et al. 2005; Suzuki et al. 2013; Mittler et al. 2004; Suzuki et al. 2011). Plants stimulate different pathways, such as the Asada–Foyer–Halliwell pathway, which consume energy in the form of NADPH to detoxify ROS toxicity;

once the energy (NADPH) is consumed, these pathways stop and the plant is unable to protect itself from ROS toxicity (Mittler et al. 2004).

Typically, ROS seem to be beneficial, provided that the cell reserves have sufficient energy to detoxify the ROS during drought stress (Choudhury et al. 2017). ROS are mostly tolerable for cells, as they support cellular sustainability and development. It is only in the event of oxidative stress that they consequently initiate programmed cell death pathways such as cell necrosis or ferroptosis (Berghe et al. 2014; Xie et al. 2016; Conrad et al. 2016). Thus, plants that can maintain their metabolism under drought stress are able to withstand environmental stress and mount an appropriate acclimatization response.

Antioxidant Defense System

Plants initiate various mechanisms to maintain normal homeostasis of cells, such as enzymatic and nonenzymatic scavenging systems to protect cells from oxidative damage. These enzymatic and nonenzymatic scavenging systems are mediated by plants to detoxify the detrimental effects of drought stress, as shown in Tables 2 and 3.

Table 2 Enzymatic antioxidant defense systems used by plants to enhance drought tolerance under stress conditions [See text for references]

Enzymatic scavenging system	Enzyme symbol; EC	Reacts with
Superoxide dismutase	SOD; EC 1.15.1.1	$O_2^{\bullet-}$
Peroxidase	POD; EC 1.11.1.x	ROS (detoxification)
Catalase	CAT; EC 1.11.1.6	H_2O_2
Glutathione reductase	GR; EC 1.6.4.2	GSH cycle
Glutathione *S*-transferases	GST; EC 2.5.1.18	Hydroxyperoxide
Glutathione peroxidase	GPX; EC 1.11.1.9	H_2O_2
Ascorbate peroxidase	APX; EC 1.11.1.11	H_2O_2

EC International Union of Biochemistry and Molecular Biology (IUBMB) Enzyme Commission number, *GSH* glutathione, *ROS* reactive oxygen species

Table 3 Nonenzymatic antioxidant defense systems used by plants to enhance drought tolerance under stress conditions [See text for references]

Nonenzymatic scavenging system	Class of compound	Reaction with ROS
Proline	Osmolyte	Helps to preserve drought tolerance
Glycine betaine	Osmolyte	Helps to preserve drought tolerance
α-Tocopherols	Vitamin E	Helps to maintain cell homeostasis
Ascorbic acid	Vitamin C	Helps to maintain cell homeostasis
Carotenoids	Tetraterpenoid	Photoprotective role in ROS scavenging
Flavonoids	Secondary metabolite	Neutralizes free radicals
Glutathione	Glutathione	Protects the photosynthetic apparatus from ROS

ROS reactive oxygen species

Enzymatic ROS-Scavenging System

Superoxide Dismutase

The superoxide dismutase (SOD) metalloenzyme is localized in almost all aerobic organisms and is the foremost antioxidant enzyme in all subcellular sections that are prone to ROS-assisted oxidative stress. Abiotic stresses trigger the production of ROS; here, SOD reacts as a frontline defense system, enhancing the plant's tolerance of the harmful effects of ROS by catalyzing the $O_2^{\bullet-}$ into H_2O_2 and O_2 as shown in Table 2 (Mittler 2002).

Peroxidase

In addition to RBOHs, peroxidase (POD) is involved in mediating ROS signaling in the apoplast, along with ROS accumulation in various cellular organelles such as the nucleus, mitochondria, chloroplasts, and peroxisomes (Mittler et al. 2004, 2011; Halliwell and Gutteridge 2007; Konig et al. 2012; Foyer and Noctor 2013; Vaahtera et al. 2014; Mignolet-Spruyt et al. 2016). High accumulation of ROS occurs during photorespiration and photosynthesis in peroxisomes and chloroplasts, along with production of ROS in the apoplast/cell wall (Suzuki et al. 2011), prompting the plant to stimulate its enzymatic ROS-scavenging system, including POD, ascorbate peroxidase (APX), catalase (CAT), and SOD (Suzuki et al. 2011).

Catalase

Catalase (CAT) enzymes contain tetrameric heme, giving them the tendency to react with H_2O_2 and convert it into H_2O (water) and O_2 (molecular oxygen). CATs are important enzymes for detoxification of ROS under stress conditions (Garg and Manchanda 2009) because CAT has the highest runover rates of all enzymes; in 1 min, one CAT molecule can transform about 6 million H_2O_2 molecules into H_2O (water) and O_2 (molecular oxygen) as shown in Table 2. CAT enzymes have been studied widely because of their dynamic role in elimination of H_2O_2 molecules formed in peroxisomes through oxidases involved in purine catabolism, photorespiration, and β-oxidation of fatty acids (Polidoros and Scandalios 1999).

Glutathione Reductase

Glutathione reductase (GR), a flavor protein, is a diverse enzyme class found in both eukaryotes and prokaryotes (Romero-Puertas et al. 2006). GR is an important enzyme in the ascorbic acid–glutathione cycle, which helps in detoxifying the detrimental effects of ROS. This enzyme is found in chloroplasts, and a small amount is localized in the cytosol (Edwards et al. 1990; Creissen et al. 1994).

Glutathione S*-Transferase*

In plants, glutathione *S*-transferase (GST) is a large and complex enzyme class with diverse functions in plants, such as hormone homeostasis, hydroxyperoxide detoxification, herbicide detoxification, vacuolar sequestration of anthocyanin, tyrosine metabolism, plant responses to biotic and abiotic stresses, and regulation of apoptosis; additionally, it catalyzes the conjugation of electrophilic xenobiotic substrates through tripeptide glutathione (Dixon et al. 2010).

Glutathione Peroxidase

Glutathione peroxidase (GPX) is a large family of diverse enzymes. GPX uses glutathione for H_2O_2 reduction and also aids in reduction of lipid and organic hydroperoxides, thus protecting cells from oxidative damage (Noctor et al. 2002). In *Arabidopsis*, about seven GPX-correlated protein families have been identified in the mitochondria, endoplasmic reticulum, cytosol, and chloroplasts (Millar et al. 2003).

Ascorbate Peroxidase

Ascorbate peroxidase (APX) plays a critical role in ROS scavenging and cell protection in algae, *Euglena*, higher plants, and some other organisms. APX is involved in H_2O_2 scavenging in the ascorbate–glutathione cycle (which uses ASH as an electron donor) and the water–water cycle (Table 1). The APX family contains five different types of isoform, including glyoxysome membrane (gmAPX), thylakoid (tAPX), cytosolic (cAPX), mitochondria isoforms (mAPX), and chloroplast stromal soluble (sAPX) forms (Noctor and Foyer 1999).

Nonenzymatic ROS-Scavenging System

Proline

Proline is a well-known osmolyte, which is now regarded as a strong inhibitor of programmed cell death and a potential antioxidant. Thus, proline is considered a nonenzymatic antioxidant required by animals, plants, and microbes to minimize the adversarial effects of ROS (Chen and Dickman 2005). In plants, L-proline is catalyzed from L-glutamic acid through D1-pyrroline-5-carboxylate synthetase (Verbruggen and Hermans 2008).

Glycine Betaine

Glycine betaine is one of the main organic osmolytes produced by plants under abiotic stress conditions such as irregular temperatures, drought, heavy metals, salinity, and ultraviolet radiation. Accumulation of glycine betaine enhances plant tolerance of abnormal environmental conditions through mediation of osmotic adjustment, membrane integrity, and enzyme regulation (Ashraf and Foolad 2007).

α-Tocopherol (Vitamin E)

Tocopherols are potent lipid radicals and ROS scavengers (Hollander-Czytko et al. 2005). They are major lipid-soluble antioxidants in biomembranes and act in a dual role (antioxidant as well as nonantioxidant). In plants, tocopherols are abundant in the thylakoid membrane of chloroplasts, acting as antioxidants in maintenance of membrane stability and scavenging or quenching ROS such as $^1O_2$. Four different isomers of tocopherols have been found in plants, of which α-tocopherol is prominent because it has three methyl groups in its molecular structure, which makes it a strong antioxidant (Kamal-Eldin and Appelqvist 1996).

Ascorbic Acid (Vitamin C)

Ascorbic acid is a water-soluble, effective, and potent antioxidant found abundantly in meristems and in photosynthetic cells. It acts as a defender by reducing the damage caused by ROS (Smirnoff 2005; Athar et al. 2008). Ascorbic acid concentrations are highest in the developed chloroplasts of mature leaves with abundant chlorophyll (Smirnoff 2000), such as in stromal compartments (with a concentration of 30 mM) as shown in Table 3. About 30–40% of the total ascorbic acid is present in chloroplasts (Foyer and Noctor 2005).

Carotenoids

Carotenoids are pigments found abundantly in all photosynthetic organisms. Those such as zeaxanthin and β-carotene play a photoprotective role by scavenging ROS or dispersing excess energy in the form of heat and suppressing lipid peroxidation.

In nature, more then 600 carotenoids are present. Carotenoids have three major roles: absorbing light wavelengths between 400 and 550 nm (Sieferman-Harms 1987), protecting the photosynthetic apparatus from free radicals produced during photosynthesis (Collins 2001), and protecting and stabilizing photosystem I and the thylakoid membrane as shown in Table 3 (Niyogi et al. 2001; Gill et al. 2009).

Flavonoids

Flavonoids are distributed throughout the plant kingdom. They are usually found in floral parts, pollens, and leaves. Most flavonoids are accumulated as glycosides in plant vacuoles, and they are also found in the aerial parts of plants and as exudates on leaves. In plant cells they are present in a 1-mM concentration and play major roles in pigmentation of seeds, flowers, and fruit; defense against pathogens; protection from harmful ultraviolet light; and germination of pollen and plant fertility (Olsen et al. 2010). Flavonoids include well-known antioxidants such as α-tocopherol and ascorbic acid, which are vital for plants in unfavorable environmental conditions and assist in neutralization of free radicals and scavenging of ROS before they can do damage (Løvdal et al. 2010).

Glutathione

Glutathione (GSH) is a key metabolite, which plays a central role in scavenging of ROS to avoid oxidative damage in almost all cellular compartments such as the endoplasmic reticulum, apoplast, peroxisomes, mitochondria, vacuoles, chloroplasts, and cytosol (Mittler and Zilinskas 1992; Jimenez et al. 1998). GSH is abundant in all plant tissues and is important for protection of the photosynthetic apparatus against ROS-induced oxidative damage (Creissen et al. 1999). Moreover, GSH is vital for regulating various physiological processes such as signal transduction, sulfate transport, detoxification of xenobiotics, and conjugation of metabolites (Xiang et al. 2001).

Conclusion

Plant growth and development are influenced by different kinds of environmental stresses during the plant's life cycle. Among different environmental factors, drought is one of the most injurious abiotic factors, which limits crop production by targeting photosynthesis and various cellular functions. Under normal water conditions, reactive oxygen species (ROS) have various important functions in maintaining and regulating the normal functioning of the cells, but when drought stress conditions trigger production of ROS, this disrupts the dynamic balance between ROS production and ROS-scavenging systems. Accumulation of ROS also depends on the intensity and duration of water stress, and it varies among species. There are two categories of ROS: radical ROS, which include hydroxyl radicals ($OH^\bullet$) and superoxide (O_2^-); and nonradical ROS, which include hydrogen peroxide (H_2O_2) and singlet oxygen ($^1O_2$). In plants, O_2^- and H_2O_2 are involved in numerous cellular reactions, such as the Fenton reaction and different enzymatic reactions involving xanthine oxidase, peroxidases, NADPH oxidase, and lipoxygenases. The major cellular components that are more prone to damage by these radicals include nucleic acid, proteins, carbohydrates, lipids, and enzymes (through denaturation).

A plant species that has a better inherited genetic response, allowing it to rapidly deploy its antioxidant enzymatic and nonenzymatic defense system, can tolerate drought better than a plant species with a poor antioxidant defense system. Enzyme and protein–encoding drought-specific genes have the ability to enhance drought tolerance. Different crops have already been genetically engineered for drought tolerance, such as wheat, sugarcane, and soybean. These two enzymatic and genetic engineering strategies are unique and vital tools, which can be used to help alleviate the world's future problems related to energy, food, and environment stresses.

References

Ashraf M, Foolad MR (2007) Roles of glycine betaine and proline in improving plant abiotic stress resistance. Environ Exp Bot 59:206–216

Athar HR, Khan A, Ashraf M (2008) Exogenously applied ascorbic acid alleviates salt-induced oxidative stress in wheat. Environ Exp Bot 63:224–231

Berghe VT, Linkermann A, Jouanlanhouet S (2014) Regulated necrosis: the expanding network of non-apoptotic cell death pathways. Nat Rev Mol Cell Biol 15:135–147

Chaves MM, Flexas J, Pinheiro C (2009) Photosynthesis under drought and salt stress: regulation mechanisms from whole plant to cell. Ann Bot 103:551–560

Chen C, Dickman MB (2005) Proline suppresses apoptosis in the fungal pathogen *Colletotrichum trifolii*. Proc Natl Acad Sci 102:3459–3464

Choudhury FK, Rivero RM, Blumwald E, Mittler R (2017) Reactive oxygen species, abiotic stress and stress combination. Plant J 90:856–867

Collins A (2001) Carotenoids and genomic stability. Mutat Res 475:1–28

Conrad M, Angeli JPF, Vandenabeele P, Stockwell BR (2016) Regulated necrosis: disease relevance and therapeutic opportunities. Nat Rev Drug Discov 15:348–366

Creissen G, Firmin J, Fryer M, Kular B, Leyland N, Reynolds H, Pastori G, Wellburn F, Baker N, Wellburn A, Mullineaux P (1999) Elevated glutathione biosynthetic capacity in the chloroplasts of transgenic tobacco plants paradoxically causes increased oxidative stress. Plant Cell 11:1277–1291

Creissen GP, Broadbent P, Kular B, Reynolds H, Wellburn AR, Mullineaux PM (1994) Manipulation of glutathione reductase in transgenic plants: implications for plant responses to environmental stress. Proc R Soc Edinb Biol Sci 102:167–175

De Boeck HJ, Bassin S, Verlinden M, Zeiter M, Hiltbrunner E (2015) Simulated heat waves affected alpine grassland only in combination with drought. New Phytol 209:531–541

Davletova S, Rizhsky L, Liang H, Shengqiang Z, Oliver DJ, Coutu J, Shulaev V, Schlauch K, Mittler R (2005) Cytosolic ascorbate peroxidase 1 is a central component of the reactive oxygen gene network of *Arabidopsis*. Plant Cell 17:268–281

Devi R, Kaur N, Gupta AK (2012) Potential of antioxidant enzymes in depicting drought tolerance of wheat (*Triticum aestivum* L.). Indian J Biochem Biophys 49:257–265
Dias MC, Brüggemann W (2010) Limitations of photosynthesis in *Phaseolus vulgaris* under drought stress: gas exchange, chlorophyll fluorescence and Calvin cycle enzymes. Photosynthetica 48:96–102
Dixon DP, Skipsey M, Edwards R (2010) Roles for glutathione transferases in plant secondary metabolism. Phytochemistry 71:338–350
Edwards EA, Rawsthorne S, Mullineaux PM (1990) Subcellular distribution of multiple forms of glutathione reductase in leaves of pea (*Pisum sativum* L.). Planta 180:278–284
Foyer CH, Noctor G (2005) Redox homeostis and antioxidant signaling: a metabolic interface between stress perception and physiological responses. Plant Cell 17:1866–1875
Foyer CH, Noctor G (2013) Redox signaling in plants. Antioxid Redox Signal 18:2087–2090
Garg N, Manchanda G (2009) ROS generation in plants: boon or bane? Plant Biosys 143:8–96
Gill SS, Khan NA, Anjum NA, Tuteja N (2009) Amelioration of cadmium stress in crop plants by nutrients management: morphological, physiological and biochemical aspects. Plant Stress 5:1–23
Halliwell B, Gutteridge JMC (eds) (2007) Free radicals in biology and medicine, 5th edn. Oxford University Press, Oxford
Hollander-Czytko H, Grabowski J, Sandorf I, Weckermann K, Weiler EW (2005) Tocopherol content and activities of tyrosine aminotransferase and cystine lyase in *Arabidopsis* under stress conditions. J Plant Physiol 162:767–770
Hu X, Wu L, Zhao F, Zhang D, Li N, Zhu G, Li C, Wang W (2015) Phosphoproteomic analysis of the response of maize leaves to drought, heat and their combination stress. Front Plant Sci 6:1–21
Hussein MM, Safinaz SZ (2013) Influence of water stress on photosynthetic pigments of some fenugreek varieties. J Appl Sci Res 9:5238–5245
Jacobsen SE, Jensen CR, Liu F (2012) Improving crop production in the arid Mediterranean climate. Field Crops Res 128:34–47
Jimenez A, Hernandez JA, Pastori G, del Rio LA, Sevilla F (1998) Role of the ascorbate--glutathione cycle of mitochondria and peroxisomes in the senescence of pea leaves. Plant Physiol 118:1327–1335
Kabiri R, Nasibi F, Farahbakhsh H (2014) Effect of exogenous salicylic acid on some physiological parameters and alleviation of drought stress in *Nigella sativa* plant under hydroponic culture. Plant Prot Sci 50:43–51
Kamal-Eldin A, Appelqvist LA (1996) The chemistry and antioxidant properties of tocopherols and tocotrienols. Lipids 31:671–701
Kaur K, Gupta AK, Kaur N (2007) Effect of water deficit on carbohydrate status and enzymes of carbohydrate metabolism in seedlings of wheat cultivars. Indian J Biochem Biophys 44:223–230
Konig J, Karl-Josef MM, Dietz DJ (2012) Mechanisms and dynamics in the thiol/disulfide redox regulatory network: transmitters, sensors and targets. Curr Opin Plant Biol 15:261–268
Laurindo FR, Araujo TL, Abrahão TB (2014) Nox NADPH oxidases and the endoplasmic reticulum. Antioxid Redox Signal 20:2755–2775
Liu Z, Xin M, Qin J, Peng H, Ni Z, Yao Y, Sun Q (2015) Temporal transcriptome profiling reveals expression partitioning of homeologous genes contributing to heat and drought acclimation in wheat (*Triticum aestivum* L.). BMC Plant Biol 15:152. https://doi.org/10.1186/s12870-015-0511-8
Løvdal T, Olsen KM, Slimestad R, Verheul M, Lillo C (2010) Synergetic effects of nitrogen depletion, temperature, and light on the content of phenolic compounds and gene expression in leaves of tomato. Phytochemistry 71:605–613
Martinez V, Nieves-Cordones M, Lopez-Delacalle M, Rodenas R, Mestre TC, Garcia-Sanchez F, Rubio F, Nortes PA, Mittler R, Rivero RM (2018) Tolerance to stress combination in tomato plants: new insights in the protective role of melatonin. Molecules 23:535. https://doi.org/10.3390/molecules23030535

Mattos LM, Moretti CL (2015) Oxidative stress in plants under drought conditions and the role of different enzymes. Enzyme Eng 5:1

Mignolet-Spruyt L, Xu E, Idänheimo N, Hoeberichts FA, Mühlenbock P, Brosché M, Van Breusegem F, Kangasjärvi J (2016) Spreading the news: subcellular and organellar reactive oxygen species production and signalling. J Exp Bot 67:3831–3844

Millar AH, Mittova V, Kiddle G, Heazlewood JL, Bartoli CG, Theodoulou FL, Foyer CH (2003) Control of ascorbate synthesis by respiration and its implication for stress responses. Plant Physiol 133:443–447

Miller G, Suzuki N, Ciftci-Yilmaz S, Mittler R (2010) Reactive oxygen species homeostasis and signalling during drought and salinity stresses. Plant Cell Environ 33:453–467

Mittler R (2002) Oxidative stress, antioxidants and stress tolerance. Trends Plant Sci 7:405–410

Mittler R (2006) Abiotic stress, the field environment and stress combination. Trends Plant Sci 11:15–19

Mittler R (2017) ROS are good. Trends Plant Sci 22:11–19

Mittler R, Blumwald E (2010) Genetic engineering for modern agriculture: challenges and perspectives. Annu Rev Plant Biol 61:443–462

Mittler R, Vanderauwera S, Suzuki N, Miller G, Tognetti VB, Vandepoele K, Gollery M, Shulaev V, Van Breusegem F (2011) ROS signaling: the new wave? Trends Plant Sci 16:300–309

Mittler R, Zilinskas BA (1992) Molecular cloning and characterization of a gene encoding pea cytosolic ascorbate peroxidase. J Biol Chem 267:21802–21807

Mittler R, Vanderauwera S, Gollery M, Van Breusegem F (2004) Reactive oxygen gene network of plants. Trends Plant Sci 9:490–498

Muller B, Pantin F, Génard M, Turc O, Freixes S, Piques M, Gibon Y (2011) Water deficits uncouple growth from photosynthesis, increase C content, and modify the relationships between C and growth in sink organs. J Exp Bot 62:1715–1729

Nath M, Bhatt D, Prasad R, Gill SS, Anjum NA and Tuteja T (2016) Reactive Oxygen Species Generation-Scavenging and Signaling during Plant-Arbuscular Mycorrhizal and Piriformospora indica Interaction under stress condition. Front. Plant Sci. 7:1574

Niyogi KK, Shih C, Chow WS, Pogson BJ, DellaPenna D, Bjorkman O (2001) Photoprotection in a zeaxanthin-and lutein-deficient double mutant of *Arabidopsis*. Photosynth Res 67:139–145

Noctor G, Foyer CH (1999) A re-evaluation of the ATP: NADPH budget during C3 photosynthesis. A contribution from nitrate assimilation and its associated respiratory activity? J Exp Bot 49:1895–1908

Noctor G, Gomez L, Vanacker H, Foyer CH (2002) Interactions between biosynthesis, compartmentation, and transport in the control of glutathione homeostasis and signaling. J Exp Bot 53:1283–1304

Olsen KM, Hehn A, Jugde H, Slimestad R, Larbat R, Bourgaud F, Lillo C (2010) Identification and characterisation of CYP75A31, a new flavonoid 3050-hydroxylase, isolated from *Solanum lycopersicum*. BMC Plant Biol 3:10–21

Polidoros NA, Scandalios JG (1999) Role of hydrogen peroxide and different classes of antioxidants in the regulation of catalase and glutathione *S*-transferase gene expression in maize (*Zea mays* L.). Physiol Plant 106:112–120

Romero-Puertas MC, Corpas FJ, Sandalio LM, Leterrier M, Serrano RM, del Rio LA, Palma JM (2006) Glutathione reductase from pea leaves: response to abiotic stress and characterization of the peroxisomal isozyme. New Phytol 170:43–52

Seki M, Narusaka M, Ishida J, Nanjo T, Fujita M, Oono Y, Kamiya A, Nakajima M, Enju A, Sakurai T (2002) Monitoring the expression profiles of 7000 *Arabidopsis* genes under drought, cold and high-salinity stresses using a full-length cDNA microarray. Plant J 31:279–292

Sieferman-Harms D (1987) The light harvesting function of carotenoids in photosynthetic membrane. Plant Physiol 69:561–568

Sirokmany G, Donkó Á, Geiszt M (2016) Nox/Duox family of NADPH oxidases: lessons from knockout mouse models. Trends Pharmacol Sci 37:318–327

Smirnoff N (2000) Ascorbic acid: metabolism and functions of a multifacetted molecule. Curr Opin Plant Biol 3:229–235

Smirnoff N (2005) Ascorbate, tocopherol and carotenoids: metabolism, pathway engineering and functions. In: Smirnoff N (ed) Antioxidants and reactive oxygen species in plants. Blackwell, Oxford, pp 53–86

Sumimoto H (2008) Structure, regulation and evolution of Nox family NADPH oxidases that produce reactive oxygen species. FEBS J 275:3249–3277

Suzuki N, Miller G, Morales J, Shulaev V, Torres MA, Mittler R (2011) Respiratory burst oxidases: the engines of ROS signaling. Curr Opin Plant Biol 14:691–669

Suzuki N, Miller G, Salazar C (2013) Temporal--spatial interaction between reactive oxygen species and abscisic acid regulates rapid systemic acclimation in plants. Plant Cell 25:3553–3569

Suzuki N, Rivero RM, Shulaev V, Blumwald E, Mittler R (2014) Abiotic and biotic stress combinations. New Phytol 203:32–43

Vaahtera L, Brosché M, Wrzaczek M, Kangasjärvi J (2014) Specificity in ROS signaling and transcript signatures. Antioxid Redox Signal 21:1422–1441

Verbruggen N, Hermans C (2008) Proline accumulation in plants: a review. Amino Acids 35:753–759

Xiang C, Werner BL, Christensen EM, Oliver DJ (2001) The biological functions of glutathione revisited in *Arabidopsis* transgenic plants with altered glutathione levels. Plant Physiol 126:564–574

Xie Y, Hou W, Song X, Yu Y, Huang J, Sun X, Kang R, Tang D (2016) Ferroptosis: process and function. Cell Death Differ 23:369–379

Zandalinas SI, Balfagón D, Arbona V, Gómez-Cadenas A (2017) Modulation of antioxidant defense system is associated with combined drought and heat stress tolerance in citrus. Front Plant Sci 8:953–963

Zhang YP, E ZG, Jiang H, Wang L, Zhou J, Zhu DF (2015) A comparative study of stress-related gene expression under single stress and intercross stress in rice. Genet Mol Res 14:3702–3717

Zhou Y, Lam HM, Zhang J (2007) Inhibition of photosynthesis and energy dissipation induced by water and high light stresses in rice. J Exp Bot 58:1207–1217

Zinta G, Abd Elgawad H, Domagalska MA, Vergauwen L, Knapen D, Nijs I, Janssens IA, Beemster GTS, Asard H (2014) Physiological, biochemical, and genome-wide transcriptional analysis reveals that elevated CO_2 mitigates the impact of combined heat wave and drought stress in *Arabidopsis thaliana* at multiple organizational levels. Glob Chang Biol 20:3670–3685

Reactive Oxygen Species Metabolism and Antioxidant Defense in Plants Under Metal/Metalloid Stress

Jubayer Al Mahmud, M. H. M. Borhannuddin Bhuyan, Taufika Islam Anee, Kamrun Nahar, Masayuki Fujita, and Mirza Hasanuzzaman

Introduction

Rapid industrial revolution and urbanization in modern world release considerable amount of toxic metal/metalloid to the environment which is mostly responsible for soil, water, and air pollution (Hasanuzzaman and Fujita 2012; Zhang and Gao 2015; Rizwan et al. 2016). Recently, toxic metals/metalloids are considered major environmental contaminants for world agricultural production. There are numerous source of toxic metal/metalloid which disturbs the growth of plant, e.g., ore mining and refining, use of agrochemicals (fertilizer and pesticide), metal industries,

J. Al Mahmud
Department of Agroforestry and Environmental Science, Faculty of Agriculture, Sher-e-Bangla Agricultural University, Dhaka, Bangladesh

M. H. M. Borhannuddin Bhuyan
Laboratory of Plant Stress Responses, Department of Applied Biological Sciences, Faculty of Agriculture, Kagawa University, Takamatsu, Kagawa, Japan

Bangladesh Agricultural Research Institute, Joydebpur, Gazipur, Bangladesh

Citrus Research Station, Bangladesh Agricultural Research Institute (BARI), Jaintiapur, Sylhet, Bangladesh

T. I. Anee · M. Hasanuzzaman (✉)
Department of Agronomy, Faculty of Agriculture, Sher-e-Bangla Agricultural University, Dhaka, Bangladesh
e-mail: mirzahasanuzzaman@sau.edu.bd

K. Nahar
Department of Agricultural Botany, Faculty of Agriculture, Sher-e-Bangla Agricultural University, Dhaka, Bangladesh

M. Fujita
Laboratory of Plant Stress Responses, Department of Applied Biological Sciences, Faculty of Agriculture, Kagawa University, Takamatsu, Kagawa, Japan

M. Hasanuzzaman et al. (eds.), *Plant Abiotic Stress Tolerance*,
https://doi.org/10.1007/978-3-030-06118-0_10

burning of fossil fuels, vehicular exhaust emissions, power stations, solid wastes disposal (including sewage sludge), irrigation of soil with wastewater, and contiguous industrial activities (Yadav 2010; Hasanuzzaman and Fujita 2012; Rehman et al. 2017). Metal/metalloid pollution in the crop growing area increases their accumulation in plant as well as facilitates entry of them into the human food cycle. Moreover, it harshly alters the usual growth, physiology, and biochemistry of plant (Sarwar et al. 2017; Mahmud et al. 2018). So, toxic metal/metalloid has become an ever increasing concern and scientists are giving attention and showing huge research interest in recent years.

Some metals, e.g., zinc (Zn), iron (Fe), boron (B), copper (Cu), manganese (Mn), and nickel (Ni), are essential element for plant growth up to certain concentration that is critically involved in wide range of physiological and biochemical processes in plants for the activation of different metal-dependent enzymes and proteins. However, at excess concentrations, they can become injurious to plants. Metals like cadmium (Cd), lead (Pb), mercury (Hg), chromium (Cr), aluminum (Al) and metalloid like arsenic (As) are nonessential and potentially highly toxic for any living organisms including plants (Flora et al. 2008; Tangahu et al. 2011; Hasanuzzaman and Fujita 2012; Zhou et al. 2014; Emamverdian et al. 2015). They inhibit seed germination, reduce plant growth, alter plant metabolism, decrease nutrient uptake, and inhibit photosynthesis by negatively affecting chlorophyll (chl) formation and chloroplast ultrastructure (Sharma and Dietz 2009; Hasanuzzaman and Fujita 2012; Hossain et al. 2012). So, in a whole they are able to create nutritional disorder and oxidative stress for plant, which jointly hinder the key mechanisms of plant growth and development.

Toxic metal/metalloid is extremely reactive and creates toxicity to plant cells in many ways depending on their types, nature, and concentration. At the very beginning toxic metal/metalloids overgenerate toxic reactive oxygen species (ROS) (Hasanuzzaman and Fujita 2012; Panda et al. 2016; Mahmud et al. 2018). Moreover, metal/metalloid toxicity alters a wide range of important functions of plant in physiological and molecular level. They inactivate or downregulate different enzymes, denature protein and destroy the integrity of cell membrane which lastly hinder plant metabolism, photosynthesis, respiration along with production of toxic ROS or free radicals. Cellular ultrastructures like lipids, proteins, and nucleic acids are oxidized by the elevated level of ROS (Sigfridsson et al. 2004; Hasanuzzaman and Fujita 2012). In addition, methylglyoxal (MG), a cytotoxic reactive oxidative compound, impulsively formed under a variety of abiotic stresses including heavy metal/metalloid toxicity, which damages cellular ultrastructure and causes mutation and even cell death (Yadav et al. 2008; Hasanuzzaman et al. 2017a). But plants have antioxidant defense and glyoxalase systems to scavenge overgenerated ROS and MG. Plant cells contain non-enzymatic components (ascorbic acid, AsA; glutathione, GSH; phenolic compounds; alkaloids; α-tocopherol and non-protein amino acids) and enzymatic components (superoxide dismutase, SOD; catalase, CAT; ascorbate peroxidase, APX; glutathione reductase, GR; monodehydroascorbate reductase, MDHAR; dehydroascorbate reductase, DHAR; glutathione peroxidase, GPX; and glutathione *S*-transferase, GST) within the antioxidant defense system

(Gill and Tuteja 2010; Hasanuzzaman and Fujita 2012). Glyoxalase or MG detoxification system established with two efficient vital enzymes, glyoxalase I (Gly I) and glyoxalase II (Gly II) (Yadav et al. 2008; Hasanuzzaman et al. 2017a). Beside these two system different osmolytes (proline, Pro; glycine betaine, GB; trehalose, Tre; etc.), chelating agents (phytochelatins, PCs; metallothioneins, MTs; etc.) were synthesized in plant cell to work against stress (Cobbett and Goldsbrough 2002; Singh et al. 2015). So, effective function of antioxidant defense and glyoxalase systems against ROS and MG, improvement of osmolytes synthesis and production of different chelating agent under stress condition determines the tolerance capability of plants. However, the efficiency of this system varies greatly with plant genotypes and stress intensity. With the increase of metal/metalloid concentration, plants lose surviving capacity. So, enhancement of abovementioned molecular approaches is one of the major strategies to cope up against metal/metalloid stress. In recent times, using exogenous phytoprotectants such as plant hormones, organic acids, signaling molecules, and trace elements is now common in research and is anticipated to metal/metalloid stress tolerance. In this chapter we primarily focused on metal/metalloid-induced oxidative stress in plants and different molecular approaches in increasing antioxidant defense.

Reactive Oxygen Species Generation Under Metal/Metalloid Stress

Production of ROS is an inevitable phenomenon in higher plants due to diverse cellular metabolic activities (Hossain et al. 2012). But under different abiotic stress including metal/metalloid toxicity, ROS production increased dramatically and destroys the balance of it with antioxidant enzymes (Hasanuzzaman and Fujita 2012). Under normal conditions, ROS play numerous indispensable functions in cell for regulating the expression of different genes. Moreover, ROS control numerous processes like the cell cycle, plant growth, abiotic stress responses, systemic signaling, programmed cell death, pathogen defense and development. The imbalance activation or reduction of oxygen at the time of different metabolic activities (photosynthesis, respiration, etc.) under stress condition is accountable for overgeneration of ROS in different important parts of plant cell including plastids, peroxisomes, mitochondria, cytosol, and apoplast. Among the cellular parts, chl is the principal site for ROS production (Hasanuzzaman et al. 2012b; Tripathy and Oelmüller 2012; Hossain et al. 2012). Inequity between ROS production and safe detoxification produce oxidative stress and the accumulating ROS are detrimental for the plants. Conversely, certain level of ROS can act as signaling molecule in response to various stresses (Tripathy and Oelmüller 2012). Two categories of toxic metal are available in the earth including redox active (e.g., Fe, Cu, Cr, Co) and non-redox active (e.g., Cd, Pb, Zn, Ni, Al), where redox active metal can generate ROS directly through Haber–Weiss and Fenton reactions. Rest metals/metalloids (Cd,

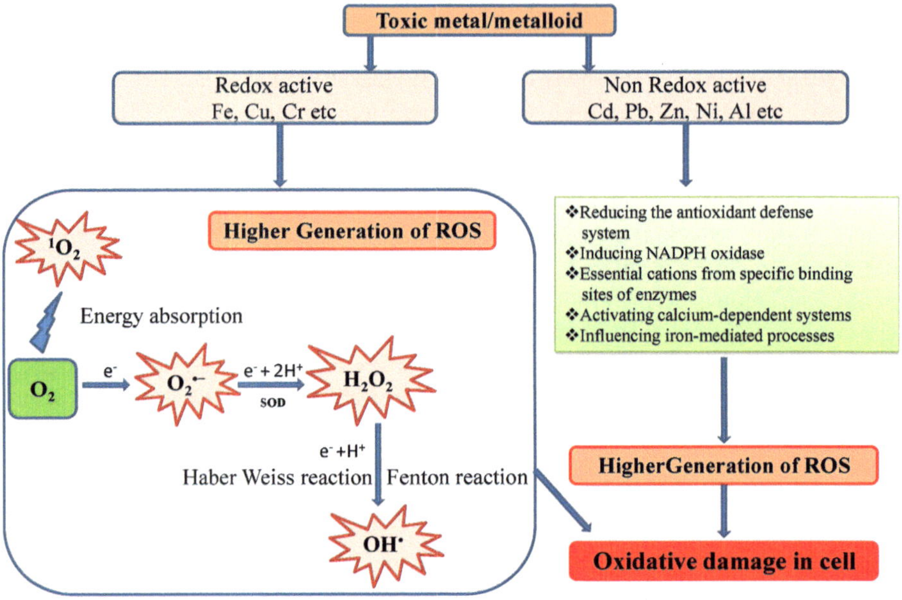

Fig. 1 Metal/metalloid-induced ROS generation in plants. ROS, $^1O_2$, $O_2^{\bullet-}$, H_2O_2, and $OH^{\bullet}$ indicate reactive oxygen species, singlet oxygen, superoxide anion, hydrogen peroxide, and hydroxyl radical

Pb, Zn, Ni, Al, etc.—without redox capacity) enhance ROS production by reducing the antioxidant glutathione pool, inducing NADPH oxidase, displacing essential cations from specific binding sites of enzymes, activating calcium-dependent systems and influencing iron-mediated processes (Fig. 1; Chou et al. 2011; Hasanuzzaman and Fujita 2012; Panda et al. 2016; Mahmud et al. 2018). Reactive oxygen species or free radicals principally include singlet oxygen, $^1O_2$; superoxide anion, $O_2^{\bullet-}$; hydrogen peroxide, H_2O_2; and hydroxyl radical, $OH^{\bullet}$ (Sigfridsson et al. 2004; Hasanuzzaman and Fujita 2012). High amounts of $O_2^{\bullet-}$ and H_2O_2 were produced in chl through the Mehler reaction, fundamentally during the reduced rate of photosynthetic carbon fixation, a typical situation during abiotic stresses (Takahashi and Murata 2008).

Productions of excess ROS under different metal/metalloids stress were recorded in a diverse range of studies. Reactive oxygen species (H_2O_2 content) of rapeseed seedlings increased by 37 and 60% exposed to 0.5 and 1.0 mM $CdCl_2$ for 2 days, respectively (Hasanuzzaman et al. 2012b). Shah et al. (2001) recorded elevated level of superoxide anion generation of two rice cultivars (Ratna and Jaya) which were grown in sand cultures for a period of 5–20 days in the presence of 100 and 500 mM $Cd(NO_3)_2$ in the medium. Mahmud et al. (2017a) histochemically stained Cd-induced ROS accumulation in the leaves of three *Brassica* species and found dark blue spots of $O_2^{\bullet-}$ and brown spots of H_2O_2 increasing with the increasing dose of Cd (Fig. 2). Compared with control, Cd stress increased H_2O_2 and $O_2^{\bullet-}$

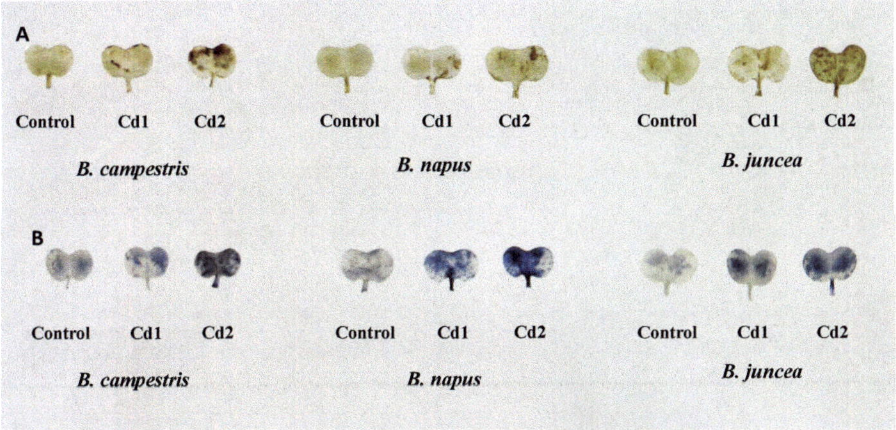

Fig. 2 Histochemical detection of H_2O_2 (**a**) and $O_2^{\bullet-}$ (**b**) of different *Brassica* species under Cd stress. Cd1 and Cd2 indicate 0.25 mM $CdCl_2$ and 0.5 mM $CdCl_2$, respectively (Source: Mahmud et al. 2017a)

generation in the leaves of mustard (Mahmud et al. 2018), mung bean (Nahar et al. 2016a, b), and rice (Rahman et al. 2016) seedlings. Similarly As stress for 5 days in 14-day-old rice seedlings increased H_2O_2 content (Rahman et al. 2015). *Luffa acutangula* was exposed to two levels of As stress (5 and 50 μM) which resulted with dose-dependent increase of ROS ($O_2^{\bullet-}$ and H_2O_2) generation (Singh et al. 2013). Mahmud et al. (2017b, c) carried out two separate experiments under Cr stress with mustard plant and observed higher generation of H_2O_2 and $O_2^{\bullet-}$ in the leaves. Lower ATP production, higher lipid peroxidation, membrane impairment, and DNA damage due to high generation of ROS under Pb stress were observed in different plants (Ali et al. 2014; Shahid et al. 2015; Ashraf and Tang 2017). Similar with other metal excess, Zn (Wu et al. 2015a, b; Hamed et al. 2017a; Pramanick et al. 2017), Cu (Dong et al. 2014), Ni (Rajpoot et al. 2015), Mn (Srivastava and Dubey 2011), Al (Nahar et al. 2017), B (Soylemezoglu et al. 2009), etc. also enhance ROS generation in different plants.

Metal/Metalloid-Induced Oxidative Stress in Plants

Metals/metalloids stresses are considered as imperative environmental hazards to plant because besides initiating nutrient imbalance they create serious oxidative stress in cellular level (Gjorgieva et al. 2013; Štolfa et al. 2015). Principally both redox active and non-redox active metal/metalloids enhance ROS production by Haber–Weiss and Fenton reactions or reducing the antioxidant glutathione pool, inducing NADPH oxidase, displacing essential cations from specific binding sites of enzymes, activating calcium-dependent systems and influencing iron-mediated

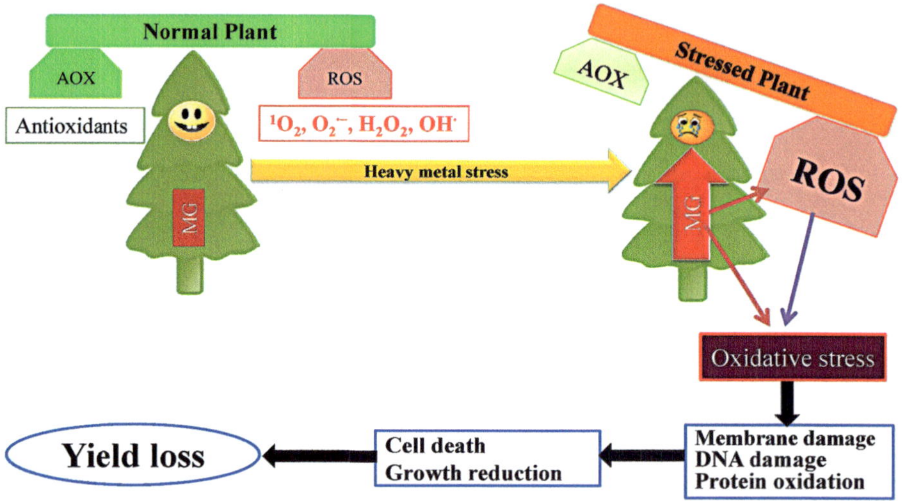

Fig. 3 Metal/metalloid-induced oxidative stress in plants. ROS, MG, $^1O_2$, $O_2^{\bullet-}$ H_2O_2, and $OH^{\bullet}$ indicate reactive oxygen species, methylglyoxal, singlet oxygen, superoxide anion, hydrogen peroxide, and hydroxyl radical

processes (Chou et al. 2011; Hasanuzzaman and Fujita 2012; Panda et al. 2016; Mahmud et al. 2018). Production of toxic MG under metal stress also recorded in many plants which directly or indirectly involved with oxidative stress (Nahar et al. 2016a, b; Hasanuzzaman et al. 2017a, b, c; Mahmud et al. 2017a, b, 2018). Overgenerated ROS and MG destroy the balance of them with antioxidative enzymes in cellular level which causes a wide range of detrimental physiological responses in plants including peroxidation of lipids, proteins, DNA, etc. and consequently leads to cell death (Hasanuzzaman and Fujita 2012; Sharma et al. 2012; Hasanuzzaman et al. 2017c). There is enough evidence that exposure of plants to excess concentrations of metals/metalloids face serious oxidative stress as they destroy the balance of antioxidant defense and ROS production in cell. In this section, we discussed different metals/metalloids-induced oxidative stress in plants (Fig. 3).

Cadmium

Cadmium has been assumed as a serious environmental contaminant in terms of damage to plant growth and human health. It accumulates in plants excessively from growing media/soil and subsequently transport into human food cycle. Increased Cd uptake in plants interrupts metabolism of plant and obstructs crop growth and development (Mahmud et al. 2017a, 2018). As a redox inactive metal

Cd is unable to develop oxidative stress directly through Fenton or Haber–Weiss reaction, but enhance production of ROS by weakening antioxidant defense system (Srivastava et al. 2004; Benavides et al. 2005; Gill and Tuteja 2010; Mahmud et al. 2017a) and hindering transfer of electron at photosystem II (Sigfridsson et al. 2004; Mahmud et al. 2017a). Being a nonessential inactive redox metal, Cd is not involved in generating oxidative stress through the Fenton or Haber–Weiss reaction. However, Cd can block the photosynthetic electron transport chain, which is the main reason for increased ROS production. Cadmium hinders the photoactivation of photosystem II (PS II) by inhibiting electron transfer and leads to the generation of ROS (Sigfridsson et al. 2004). Cadmium impedes with the cellular structures and functions of several important molecules with marked endpoints such as cell proliferation and cell death (Cuypers et al. 2010). Chou et al. (2011) observed that Cd is involved with the striking increase of lipid peroxidation and H_2O_2 rice seedlings. Considerable amount of electrolyte leakage was also observed under Cd stress (Agami and Mohamed 2013). Nahar et al. (2016a) carried out an experiment with mung bean plant and observed that Cd stress increased the oxidative stress markers ($O_2^{\bullet-}$ generation, H_2O_2 and MDA contents) in a concentration-dependent manner. In their study lipoxygenase activity also increased under Cd stress, which may finally increase lipid peroxidation. Higher MG content also recorded under Cd stress. Similar findings were observed in previous experiments on Cd stress in different plants (Szollosi et al. 2009; Zheng et al. 2010; Nahar et al. 2016b). Oxidative stress also increased in rice seedling along with growth reduction under Cd stress (0.25 mM and 0.5 mM Cd) for 3 days (Rahman et al. 2016). They demonstrated MDA content increased by 83 and 136%, H_2O_2 content increased by 46 and 84%, LOX activity increased by 60 and 85% under 0.25 mM and 0.5 mM $CdCl_2$, respectively. They also observed considerable increase of MG content under both levels of stress. In 2017, Mahmud et al. (2017a) checked the relative tolerance of three *Brassica* species (*B. napus*, *B. campestris*, and *B. juncea*) seedlings exposed to two concentration of Cd stress (0.25 and 0.5 mM $CdCl_2$) for 3 days. They recorded Cd stress decreased seedlings biomass, leaf relative water content (RWC), and chlorophyll (chl) content, whereas proline (Pro), MDA, and H_2O_2 content, lipoxygenase (LOX) activity and MG content increased in all species in dose-dependent manner. Malondialdehyde content increased by 74, 57, and 55% under 0.25 mM $CdCl_2$, and 114, 94, and 85% under 0.5 mM $CdCl_2$ in *B. campestris*, *B. napus*, and *B. juncea*, respectively, compared with the control plants. Similarly, Wu et al. (2017) demonstrated a study with Chinese cabbage under Cd stress and found overproduction of MDA and H_2O_2. Very recently, Mahmud et al. (2018) treated 12-day-old *B. juncea* seedlings with Cd (0.5 mM Cd and 1.0 mM $CdCl_2$) in a semi-hydroponic medium for 3 days. They found increasing Cd concentrations in growing media led to reduced growth, biomass, water status and chl content of seedlings by increasing oxidative damage. Malondialdehyde content; H_2O_2 level; superoxide, $O_2^{\bullet-}$ generation; LOX activity; and MG content increased substantially under Cd stress, whereas activity of different major enzymes of antioxidant defense and glyoxalase system downregulated resulting oxidative stress.

Lead

Lead is a heavy metal well-known for its eco-toxicological properties like long-term persistence and non-biodegradability (Ashraf et al. 2015; Ashraf and Tang 2017). It has a broad range of harmful effects to both plants and animals affecting their morphological, physiological, and biochemical properties. Many studies demonstrated the negative outcome of plants, e.g., reduced germination, impaired plant growth, root growth, seedling development, transpiration, reduced water and protein content, and chlorophyll synthesis upon exposure to Pb toxicity (Flora et al. 2012; Kumar et al. 2012a, 2017). It also results in lower ATP production, higher lipid peroxidation, membrane impairment, and DNA damage by producing ROS in a higher amount (Ali et al. 2014; Ashraf and Tang 2017). However, the toxic properties of Pb are depended on the applied chemical form and duration of exposure (Shahid et al. 2015). *Vicia faba* is considered to be a metal sensitive plant and hence Shahid et al. (2014a, b) used this plant to evaluate the role of different chelating agents on the Pb toxicity to plants. They used 5 μM Pb as lead nitrate for different durations (4, 8, 12, and 24 h) and observed that thiobarbituric acid reactive substances (TBARS) content was slightly higher at 4 h of Pb exposure while significantly increased by 115, 74, and 62% at 8, 12, and 24 h of Pb stress, respectively, compared to the control plants. Unlike TBARS content H_2O_2 production was unaffected at 4 h of Pb exposure but significantly rose after 8 h (61%) compared to control. But it was then declined to basal level at 12 h of same stress conditions and again enhanced by 80% after 24 h of Pb stress treatment compared to non-stressed control treatment (Shahid et al. 2014a). Two bursts of lipid peroxidation and H_2O_2 induction at 1 and 12 h of same level of Pb stress in same plants were reported by Shahid et al. (2015), while after 4, 8, and 24 h of Pb exposure the oxidative stress levels were also evident. They explained in their findings that Pb induced lipid peroxidation and H_2O_2 induction occurs rapidly and simultaneously during first 24 h of Pb induction but these are not continuous process and occur in steps which might be due to the activation of different antioxidant enzymes involved in ROS scavenging (Shahid et al. 2015). Such increase in lipid peroxidation has also been reported in *Triticum aestivum* L. seedlings exposed to 3 mM Pb stress for 7 days (Lamhamdi et al. 2016) and in *Juncus effusus* L. treated with 0.5 and 1.0 mM of Pb for 15 days (Najeeb et al. 2017). In *Medicago sativa* plants, Pb (10 and 100 μM) stress for 2 and 7 days showed time- and dose-dependent accumulation of Pb (Hattab et al. 2016). Lead accumulation induced lipid peroxidation in both roots and shoots in a time- and dose-dependent manner having stronger phytotoxic effects in roots than in shoots as the translocation rate from root to shoot was less (Hattab et al. 2016). Another medicinal herb *Acalypha indica* L. was found to accumulate more Pb in roots (121.6 mg g^{-1} DW) than in shoots (17.5 mg g^{-1} DW) when exposed to Pb. This accumulation of lead resulted in about 49% higher MDA content in stressed plants compared to untreated control at various regimes during 5, 8, and 12 days of Pb exposure in both roots and shoots (Venkatachalam et al. 2017). Ashraf and Tang (2017) tested two rice cultivars under four different Pb levels viz., control (0), low (400), medium (800), and high (1200) mg $kg^{-1}$of soil. They observed increased lipid peroxidation, H_2O_2 content and electrolyte leakage under Pb stress while photosynthetic pigments production was reduced. Lead toxicity also variably affected protein and soluble sugar contents in both rice cultivars.

Arsenic

Arsenic (As) is one of the most toxic metalloids which has no known function as nutrient (Farooq et al. 2016). It is considered to be the 20th most ubiquitous element in the environment having more than 200 As-containing minerals (Zhao et al. 2010; Hoang et al. 2010). Uptake of As in different plant parts and further accumulation, translocation, and biomagnification not only affect the plant growth and productivity but also increase threat to human (Kumar et al. 2015). Chlorosis, inhibited growth and water uptake, disrupted photosynthesis, transpiration and respiration, and interference with other metabolic activities are some of the results observed in plants exposed to As stress (Stoeva et al. 2005; Verbruggen et al. 2009). Like other metal stress As also promotes ROS production which induces oxidative stress and also exhibits generation of another toxic compound MG (Hasanuzzaman and Fujita 2013; Rahman et al. 2015). Both ROS and MG cause serious damage to protein, lipid, DNA, and other biomolecules, if there is no protective mechanism involved (Yadav et al. 2005; Rahman et al. 2015). Arsenic induced oxidative stress and the consequences have been demonstrated by many researchers. Hasanuzzaman and Fujita (2013) tested *T. aestivum* L. plants under As stress (0.25 and 0.5 mM) and observed that the contents of MDA and H_2O_2 were markedly increased in a dose-dependent manner in the leaves of stressed plants. Rice (*O. sativa*) field is often flooded during the growing period and As is the most commonly occurring and mobile element in soil pore water. So, rice has been considered as a plant material in most of the experiments regarding As toxicity study. An interactive study was conducted by Ren et al. (2014) with As and mercury (Hg) where rice plants were exposed to different concentrations of arsenite (AsIII) and both shoots and roots were observed to show higher levels of lipid peroxidation and As accumulation in a dose-dependent manner (under As stress only). Rice seedlings grown under 0.5 and 1.0 mM of As stress treatments resulted in 66 and 176% higher MDA, 69 and 89% higher H_2O_2, respectively, compared to the non-stressed control plants (Rahman et al. 2015). Enhancement of toxic MG content was also recorded in stressed leaves at both levels of As stress (Rahman et al. 2015). Accumulation of As in both shoots and roots of rice plants and subsequent acceleration of lipid peroxidation (denoted by higher MDA production) were also reported by other researchers (Singh et al. 2016; Moulick et al. 2016, Nath et al. 2014). Vegetable like *Luffa acutangula* was also experimented under two levels of As stress (5 and 50 μM) which resulted with dose-dependent increase of superoxide radical ($O_2^{\bullet-}$), MDA and H_2O_2 production (Singh et al. 2013). Under 5 and 50 μM As stress, generation of $O_2^{\bullet-}$ was raised by 163 and 296%, H_2O_2 by 192 and 317%, and MDA content by 178 and 334%, respectively, compared to their respective control.

Zinc

Being an essential micronutrient, zinc has significant roles in various enzymatic reactions involved in the metabolism of protein, carbohydrate, nucleic acid, and lipid (Broadley et al. 2007; Feigl et al. 2015). But, when it is present in soil at a concentration above the optimum level (300 mg kg^{-1}), it becomes toxic to plant growth (Marschner 1995; Anwaar et al. 2015). Inhibited seed germination, root development, plant growth, impaired membrane integrity, leaf chlorosis, altered cell division, nutrient imbalance, hampered photosynthesis, ultimate cell death, etc. are some effects of Zn toxicity documented so far by different researchers (Wang et al. 2009; Michael and Krishnaswamy 2011; Anwaar et al. 2015; Feigl et al. 2015; He et al. 2016). Like other metals, excess amount of Zn promotes ROS generation which induces lipid peroxidation, protein oxidation, and DNA and membrane damage (Wu et al. 2015a, b; Hamed et al. 2017a; Pramanick et al. 2017). Zinc induces this oxidative stress not by the performance of univalent oxido-reduction reaction itself but by the inactivation of crucial enzymes restricting relevant functional groups that bind to oxygen, nitrogen, or sulfur atoms (Morina et al. 2010; Pramanick et al. 2017). Oxidative stress was induced by the accelerated production of H_2O_2 and $O_2^{\bullet-}$ radicals in rice seedlings upon exposure to 100 μM Zn for 12 h (Chen et al. 2017). While reduced production of $O_2^{\bullet-}$ radicals was reported in wheat roots exposed to 3 mM Zn stress for 6 days (Duan et al. 2015). However in case of MDA and H_2O_2 contents, the values were higher compared to control which denotes the Zn-induced oxidative stress (Duan et al. 2015). Another Poaceae crop maize was earlier tested under Zn stress (250 and 1000 mg kg^{-1} soil, 7 weeks) and a dose-dependent increase in MDA content by 36 and 80%, respectively, compared to the treatments without Zn was observed (Cui and Zhao 2011). Higher MDA, H_2O_2 contents and electrolyte leakage (%) were also observed in both roots and shoots of maize plants after 10, 20, and 30 days of 50 μM Zn stress (Islam et al. 2014). Zinc stress (10 mM) prevailing for 7 days increased the rate of $O_2^{\bullet-}$ and MDA production in leaves of *Solanum melongena* plants (He et al. 2016) along with the higher content of H_2O_2 and electrolyte leakage (Wu et al. 2015a, b). Similarly, cotton plants showed increasing MDA, H_2O_2 and electrolyte leakage with increasing Zn concentrations under 25 and 50 μM of Zn stress conditions for 60 days (Anwaar et al. 2015). Mukhopadhyay et al. (2013) experimented tea (*Camellia sinensis*) under Zn deficient, optimum, and excess levels, and observed that both deficient and excess amounts of Zn resulted in higher levels of MDA, H_2O_2, $O_2^{\bullet-}$ and electrolyte leakage compared to the optimum condition. Similar results were observed in *Raphanus sativus* L. plants (under 5 mM Zn stress) (Ramakrishna and Rao 2015). Crops like *Morus alba* (Tewari et al. 2008), *Solanum lycopersicum* (Cherif et al. 2011), *Vigna Mungo* (Gupta et al. 2011), and *Phaseolus vulgaris* (Michael and Krishnaswamy 2011) were also investigated earlier which evidenced the Zn-induced oxidative stress. Seeds of a medicinal herb *Plantago ovata* were grown on agar-sucrose medium and exposed to different Zn toxicity levels to study the changes in expression levels of *MT2* genes. Results revealed that there is a correlation between the enhanced expressions of *MT2* gene and increased total antioxidant activity under

Zn-induced oxidative stress (Pramanick et al. 2017). Algae were also subjected to Zn stress where Zn-induced development of oxidative stress was demonstrated (Hamed et al. 2017a).

Manganese

Manganese, a micronutrient for plant, has role in enzyme activation of several vital metabolic pathways linked to carbohydrates production through photosynthesis, respiration, and protein synthesis (Boojar and Goodarzi 2008). Manganese deficiency altered the electron supply chain of photosystem II, resulted in upsetting of the water photolysis, which resulted in reduced photosynthesis (Fernando and Lynch 2015). Contrarily, excess Mn can be phytotoxic when concentrations are high in foliage (González and Lynch 1999). In acidic soil condition (pH below 5.5), Mn is readily uptake by the plant roots, hence become phytotoxic, and interfere with a range of metabolic processes and reduce plant growth (Lidon and Teixeira 2000; Hauck et al. 2003). Photo-oxidation is the principal consequence of Mn toxicity in plants (González et al. 1998). Photo-oxidative stress due to Mn toxicity includes alteration of electron transport chain, inactivation of Rubisco and disrupting energy dissipation. It also increase ROS generation, intervene in metal uptake, alter translocation and metabolism of the antioxidant enzymes (Houtz et al. 1988) and damage lipids and proteins by reaction with sulfhydryl groups (Van Asshe and Clijsters 1990). A sharp photosynthetic decline is caused in excess of Mn due to hindered Chl biosynthesis (Hauck et al. 2002). Leaf chlorosis, correspondent to lack of Chl *a* and *b*, attributed to higher accumulation of Mn in leaves, damaging photosynthetic machinery (Millaleo et al. 2013, Weng et al. 2013, Rojas-Lillo et al. 2014). Thus, Mn toxicity possibly targets the chloroplast at the very beginning (Lidon et al. 2004; Chen et al. 2015), attributed to overproduction of hazardous ROS, such as $O_2^{\bullet-}$, H_2O_2, $^1O_2$, and $OH^{\bullet}$ ahead of scavenging capability of plants causing oxidative stress (Faize et al. 2011; Fischel et al. 2015). Like other metal/metalloid stress plant cells overproduce $O_2^{\bullet-}$ and H_2O_2. Rice seedlings treated with 6 mM Mn^{2+} for 20 days resulted in increased $O_2^{\bullet-}$ level (197% and 199% in roots and shoots, respectively) in comparison to controls (Srivastava and Dubey 2011). These ROS overloads are the main causal entity for damaging membrane lipids, proteins, enzymes, nucleic acids, etc., damaging metabolic process and in most cases causing programmed cell death (Fernando et al. 2013). Furthermore, lower oxygen availability caused less reduction of $NADP^+$ because of lesser efficiency of electron transport chain, which in turn dropped off CO_2 assimilation rate (Foyer and Shigeoka 2011; Gururani et al. 2015). Rice somewhat can tolerate Mn, by oxidizing toxic Mn^{2+} into less toxic Mn^{4+} (Wang et al. 2002). Lipid peroxidation—a useful indicator of oxidative damage (Verma and Dubey 2003), but increased TBARS level (22% in root and 63% shoot compared to control) were observed in rice seedlings with excess Mn in growing media suggested the Mn persuade oxidative damage of cell macro-molecules (Srivastava and Dubey 2011).

Copper

Copper (Cu) is one of the essential micronutrients for plants which plays crucial roles in various metabolic activities including photosynthesis, respiration, protein metabolism, cell wall formation, electron transport, nitrogen fixation, ethylene sensing, oxidative stress protection, etc. (Kong et al. 2015; Mei et al. 2015). But, even a slightly higher concentration of this transitional metal than the optimum range can cause severe toxic effects on plants' morpho-physiological and biochemical processes (Thounaojam et al. 2013; Dong et al. 2014; Mei et al. 2015; Jadid et al. 2017). Copper stress interferes with the Fenton's reaction and hereby generates ROS in a higher amount leading to oxidative stress (Moenne et al. 2016; Hamed et al. 2017b). Copper stress induced oxidative damages were previously recorded by many researchers in different plant species (Mostofa and Fujita 2013; Mostofa et al. 2014, 2015; Sáez et al. 2015 and references therein). When rice plants exposed to two concentrations of Cu (200 and 500 μM) for 24 h and 7 days durations, a higher amount of MDA and H_2O_2 productions was observed (Thounaojam et al. 2013), which was also true for 75 and 150 μM Cu for 48 h (Mostofa and Fujita 2013), 100 μM Cu for 48 h (Mostofa et al. 2014), and 100 μM Cu for 4 and 7 days (Mostofa et al. 2015) durations. Cu-induced overproduction of ROS like $O_2^{\bullet-}$ and H_2O_2 was also visualized in rice leaves which was done by histochemical detection method (Mostofa and Fujita 2013; Mostofa et al. 2015). Mostofa et al. (2015) also observed that the toxic MG production increases with the increasing duration of stress exposure. Habiba et al. (2015) experimented with *B. napus* seedlings under Cu stress conditions (50 and 100 μM Cu, 8 weeks) and observed significant increases in MDA content, H_2O_2 concentrations and electrolyte leakage with reducing photosynthetic pigments and protein contents. Similarly, in ryegrass (*Lolium perenne* L.), MDA and H_2O_2 production and $O_2^{\bullet-}$ generation rate were significantly higher under Cu stress (200 μM) conditions for 14 days (Dong et al. 2014). Three cotton genotypes were exposed to 100 μM of Cu stress for 4 days and all of those resulted in increased MDA and H_2O_2 contents in both leaves and roots (Mei et al. 2015). *Capsicum frutescens* (Jadid et al. 2017) and two medicinal plants *Cassia angustifolia* (Nanda and Agrawal 2016) and *Medicago lupulina* (Kong et al. 2015) were tested under different levels of Cu stress and increased lipid peroxidation was reported in all cases. Experiments with different forms of algae were also done to study the oxidative stress responses under Cu-induced stress condition where similar trends of increase in lipid peroxidation were observed (Sáez et al. 2015; Hamed et al. 2017b). But, when Moso bamboo (*Phyllostachys pubescens*) plants were kept for 15 days under different concentrations of Cu (5, 25, and 100 μM), a non-significant increase in MDA content was recorded (Chen et al. 2015).

Nickel

Nickel is another plant essential micronutrient which renders beneficial effects on plant growth at a lower concentration (0.01–5.00 mg kg^{-1}) but creates structural, metabolic, and physiological constraints when present in excess amount (Molas

2001; Theriault et al. 2016). Nickel plays an important role in nitrogen metabolism as it is the key component of urease enzyme and also exerts plant disease resistance (Yan et al. 2008; Dubey and Pandey 2011; Rajpoot et al. 2015). However, higher concentration of nickel negatively affects seed germination, plant growth, photosynthesis, mineral nutrition, water relations, sugar transport, and root development, beside causing wilting, necrosis, and chlorosis symptoms (Martínez-Ruiz and Martínez-Jerónimo 2016; Soltani et al. 2016; Rizwan et al. 2017). At cellular level, overproduction of ROS and induction of lipid peroxidation, protein damage, DNA damage followed by modulation of antioxidant defense system have been documented as the results of Ni toxicity in various crop species (Ghasemi et al. 2013; Khaliq et al. 2015; Soltani et al. 2016; Farid et al. 2017; Rizwan et al. 2017). Overproduction of $O_2^{\bullet-}$ and $OH^{\bullet}$ radicals, higher H_2O_2 and TBARS contents were observed in *O. sativa* seedlings under Ni stress (200 μM) for both 4 and 8 days durations (Rajpoot et al. 2015). For longer duration (14 days), even with lower concentrations of Ni (10, 50, and 100 μM) *O. sativa* shoots exhibited higher MDA content with higher Ni doses (Rizwan et al. 2017). Farid et al. (2017) reported that electrolyte leakage, MDA and H_2O_2 contents of *B. napus* plants increased in a dose-dependent (10, 100, and 500 μM Ni) manner. Such dose-dependent increase of TBARS and H_2O_2 under Ni stress had also been reported in barley (Kumar et al. 2012b). For 50 and 100 μM Ni exposure, cotton plants showed higher MDA and H_2O_2 contents, and electrolyte leakage compared to the non-stressed control plants (Khaliq et al. 2015). Earlier, Dubey and Pandey (2011) experimented with *V. mungo* plants under different concentrations of Ni (20, 40, 60, 80, and 100 μM) and recorded a very consistent dose-dependent increase in MDA content. Ni stress induced enhanced lipid peroxidation was also reported in wheat (Gajewska et al. 2013), *Eruca sativa* (Kamran et al. 2016), *Melissa officinalis* (Soltani et al. 2016), and *Pistia stratiotes* (Singh and Pandey 2011).

Chromium

Chromium is one of the most abundant, highly toxic and soluble environmental pollutant that harmfully affects plant growth and development through inducing oxidative stress (Homa et al. 2016; Gorny et al. 2016; Mahmud et al. 2017b, c). Like other redox active heavy metals, Cr is able to cause oxidative damage in different plants by reducing CO_2 fixation and photosynthetic activities through hindering electron transport and inactivating the enzymes of Calvin cycle (Shanker et al. 2005; Sundaramoorthy et al. 2010; Zhang et al. 2010). Chromium increased ROS generation, which completely distorted the chloroplastic membrane with harsh change in thylakoids resulting in oxidative stress and causing chlorosis (Choudhury and Panda 2004; Panda and Choudhury 2005). Chromium-induced lipid peroxidation due to ROS generation destroys the integrity and function of cell membranes as well as results in cell death (Panda and Choudhury 2005). Gill et al. (2015) checked Cr-induced physio-chemical and ultrastructural changes in four cultivars of *B. napus* L. viz. ZS 758, Zheda 619, ZY 50, and Zheda 622. They observed ROS (H_2O_2

and $O_2^{•-}$) generation and MDA content were linearly enhanced in both the leaves and roots of all cultivars due to Cr stress. As a result, net photosynthetic rate, stomatal conductance, intercellular CO_2 concentration, and transpiration rate were hampered negatively in all cultivars. Similar upsetting in physiological activities owing to oxidative stress was observed in other plants by different group of scientists (Choudhary et al. 2012; Kováčik et al. 2013; Ali et al. 2015a, b). The damage enhanced in dose-dependent manner. Mahmud et al. (2017b) carried out an experiment to observe the performance of *B. juncea* L. under two levels of Cr stress (0.15 mM and 0.3 mM K_2CrO_4, 5 days). They recorded overgeneration of ROS in the leaves of *B. juncea* L. under both levels of Cr stress. As a result, LOX activity and MDA content drastically enhanced. Moreover, MG content also increased. As a result, plant suffered severe oxidative stress and growth reduction occurred. In the same time, Mahmud et al. (2017c) checked the performance of Indian mustard seedlings under Cr stress (0.15 mM and 0.3 mM K_2CrO_4, 5 days) in another experiment. The TBARS content was increased by 30 and 65% under 0.15 mM and 0.3 mM K_2CrO_4 stress, respectively, in comparison with control plants. Similarly compared with control plants H_2O_2 content increased by 24 and 46% and LOX activity increased by 68 and 101% under 0.15 mM and 0.3 mM K_2CrO_4 stress, respectively. Furthermore, in contrast to control plants MG content increased by 47 and 82% under 0.15 mM and 0.3 mM K_2CrO_4 stress, respectively. All are responsible for oxidative damage as well as reduction of growth and biomass.

Aluminum

Aluminum is the metal that found most abundantly in the earth crust; as a result, it is one of the major limiting factor for crop productivity in 30% of the world's acidic soils (Kochian et al. 2015). In cellular level, Al formed aluminum superoxide (AlO_2^{2+}) a semi-reduced radical ion that is more powerful oxidant than $O_2^{•-}$ and boosts ROS production, which create oxidative damage in cellular level (Sun et al. 2014; de Sousa et al. 2016). In the meantime, these overproduced ROS initiate lipid peroxidation and formed reactive carbonyl species (RCS), which are also toxic for the plant cells (Mano 2012; Biswas and Mano 2015) and believed to be the most injurious at the time of Al toxicity (Yin et al. 2010). Therefore, both upstream ROS and downstream RCS induced cell injury in plants during Al toxicity. Aluminum inhibits cell elongation and division in root tip, which is the first Al toxicity symptoms, inhibits root growth and showed stunted appearance; as a result, plant cannot uptake water and essential nutrients (Nahar et al. 2017; Kochian et al. 2015). It also decreases Chl content as well as the photosynthetic rate, which inhibit the growth and development of plants (Ali et al. 2008). As Al is very reactive, and it can bind to many binding sites of cells, for instance cell wall, plasma membrane surface, the cytoskeleton, and nucleus, these are the target of Al-induced injury (Panda et al. 2009). Aluminum alters the cation exchange capacity, thus hampers the nutrient uptake (Matsumoto 2000) and can also trigger PCD (Panda et al. 2009). Like other metals, Al induces oxidative stress in plant cells. The damage due to Al in cells may occur in different ways, for example Al induced radical chain reactions by Fe that enhances peroxidation of lipids (Nahar et al. 2017). Several reports on Al toxicity suggested that Al induced ROS overproduction such as $^1O_2$, $O_2^{\bullet -}$, H_2O_2, and $OH^{\bullet}$ that cause lipid peroxidation (Nahar et al. 2017; Ali 2017; Chowra et al. 2017; Pontigo et al. 2017; Zhao et al. 2017; Yin et al. 2017; Ramírez-Duarte et al. 2017). In Arabidopsis roots, various RCS are present, but after Al treatment the content of the most of the RCS boosted up to 163% causing oxidative stress (Yin et al. 2017). Rice seedlings exposed to Al (100 μM) showed higher lipid peroxidation, resulting in loss of membrane integrity (Awasthi et al. 2017). In rice, upon exposure to Al growing seedlings absorbed Al and translocation took place in shoots. But prolonged exposure caused lipid peroxidation, and furthermore caused DNA damage (Meriga et al. 2004). They also reported that Al accumulation boosted oxygen radicals that induce membrane lipid peroxidation, thus increased MDA levels were observed (Meriga et al. 2004). Research reports suggested that the Al tolerant maize line usually used avoidation technique to reduce oxidative damage due to Al. Therefore, 2–2.5 times lower amount of Al was found in tolerant line. This line also showed higher Ca^{2+}, Mg^{2+}, and K^+ concentration in roots with threefold increase in Pro content. In contrast, sensitive lines showed higher Al accumulation in roots, lower Pro content, and higher MDA content (Giannakoula et al. 2008). Achary et al. (2008) evaluated Al for induction of oxidative stress and DNA damage in the growing roots of *Allium cepa* L. and found overgeneration of ROS, thus caused lipid peroxidation, protein oxidation, and DNA damage in a dose-dependent manner. Therefore, these data provide some information about the Al-induced oxidative damage in different crops but further investigation is needed to clarify the reasons for the oxidative damages in genetic level and possible mechanism that is used by the Al tolerant variety to reduce the oxidative damages.

Others

Boron

Among the other metalloids, B is an essential element that is required for normal growth, development, and reproduction of higher plants, but its deficiency and toxicity threshold is very narrow (Singh et al. 2010). Some soils are naturally rich in B, as well sometimes ground water may contain high amount of B, or people may use excess B fertilizer to the soil or irrigate the B rich water to soil, which causes a considerably higher concentration of B in crop fields leading to toxicity (Aftab et al. 2012). Research findings suggested that B rich soils are the cause of poor plant growth and development as well as reducing yield (Papadakis et al. 2004). The typical symptom exhibited by the plants exposed to B toxicity is the loss of vigor, delayed growth, leaf burn, and small number and size of fruits (Nable et al. 1997). Regardless of these, B toxicity creates excess generation and accumulation of ROS, had reported by various researchers in several crops in apple rootstock (Molassiotis et al. 2006), wheat (Gunes et al. 2007), barley (Inal et al. 2009), tomato (Cervilla et al. 2007), mung bean (Yusuf et al. 2011), and sweet basil (Landi et al. 2013). Furthermore, these overgenerated ROS have reported to induce peroxidation of membrane lipids and toxic H_2O_2 accumulation (Zhang et al. 2017; Soylemezoglu et al. 2009). Boron toxicity is common in arid and semiarid regions, responsible for significant growth and yields loss (Nable et al. 1997; Eser and Aydemir 2016). Basically, B is relatively nonreactive but forms strong complexes with other metabolites having more than one OH group (Reid 2010). Unlike other plant nutrients, B exists as boric acid—an uncharged molecule, at normal physiological pH and can readily enter through the membrane layers (Eser and Aydemir 2016; Dordas et al. 2000), and exerts its toxic effects when accumulated in a large amount in plant cell (Tombuloglu et al. 2012; Kaya and Ashraf 2015). Grape vine rootstock exposed to B toxicity was found with higher stomatal resistance, MDA, H_2O_2 and Pro concentrations, which clearly indicates the oxidative damage due to B (Soylemezoglu et al. 2009). Eraslan et al. (2007) conducted experiment on lettuce to show the effects of B toxicity and finally concluded that B toxicity induced oxidative stress, which resulted in lipid peroxidation and losses of membrane integrity.

Iron

Another well-known plant nutrient Fe is an essential metal at lower concentration, but may exert toxic effects on plants when remain in abundant in soil. This problem is common in low land rice soils around the world, where Fe exists as water-soluble Fe^{2+} ions. Although, Fe plays vital role in important biological processes like Chl biosynthesis, chloroplast development as well as photosynthesis (Olaleye et al. 2009). The Fe (II)/Fe (III) redox couple takes part in plant growth by modulating the enzyme catalyzed redox reaction (Gill and Tuteja 2010). Furthermore, a high Fe^{2+}concentration may boost several redox reactions by donating electron (Sahrawat 2003). But crops grown in low land especially rice may suffer iron toxicity, which is a major cause of low yield in many countries of Asia, South America, West and Central Africa (Sahrawat 2004) and lowered as much as 30–60% yield (Majerus et al. 2007). Toxicity at early vegetative stage hampers the growth and complete yield loss (Becker and Asch 2005). On the other hand, Fe^{2+} irons in excess boost up the production of ROS ($O_2^{\bullet-}$, H_2O_2, $^1O_2$, HO_2, $OH^{\bullet}$, OH^-, and RO), which are highly toxic and reactive and cause protein oxidation, lipid peroxidation, carbohydrate and DNA damage leading to cell death (Baruah and Bharali 2015). In medicinal plant *Bacopa monnieri* L. Fe toxicity leads to overproduction of lipid peroxidation product MDA in leaves and shoots (Sinh and Saxena 2006).

Mercury

Nowadays, Hg has gaining attention to researchers due to high toxicity and distribution (Regier et al. 2013). It has no known physiological role and neither acts as plant nutrient, and the fact to be feared is that it is not metabolized (Israr et al. 2006). The predominant form of Hg in soil and water is Hg^{2+}, which is highly soluble and can readily accumulate in higher plants (Elbaz et al. 2010). On the other hand, Hg is highly toxic to plants as it can react with sulfhydryl groups (Zhou et al. 2007). This metal can also disrupt the membrane integrity by lipid peroxidation and can cause DNA damage (Malar et al. 2015). The most important feature of oxidative stress triggered by Hg is overproduction of ROS (Cargnelutti et al. 2006; Deng et al. 2013). The accumulation of Hg increased in a concentration- and duration-dependent manner, and was positively correlated with the leaf damage. Oxidative stress after Hg exposure was evidenced in *Lemna minor* by a significant decrease in photosynthetic pigments and an increase in MDA. Exposure to Hg also caused the overaccumulation of Pro and soluble sugars (Zhang et al. 2017).

Selenium

Another important commonly found metalloid Se is emerged as a phytoprotectant and signaling molecule in plant science (Hasanuzzaman et al. 2011), but it has potential role in creating health hazard for plants when applied in a large amount. It is accumulated in soil extensively from fertilizer or other anthropogenic activities which can cause phytotoxic effects in plants (Chen et al. 2014; El-Ramady et al. 2014; Wu et al. 2015a, b). Very few researchers have investigated the effect of toxic Se. Lipid peroxidation was the main response observed at 0.05 mM sodium selenite (Gomes-Junior et al. 2007). Experiment with rice reported that Se caused phytotoxic effects on plants by inducing chlorosis and exacerbating oxidative stress by overproduction of H_2O_2 and $O_2^{\bullet-}$ causing lipid peroxidation measured by increased MDA content (Mostofa et al. 2017).

Molecular Approaches in Enhancing Antioxidant Defense Under Metal/Metalloids Toxicity

Improving Osmolyte Synthesis

To prevent the adverse effects of various environmental stresses including metal and metalloids stress, plants exhibit a range of adaptive strategies mutually at the morphological and cellular levels. Under metal/metalloid toxicity conditions, to cope with the metal-induced osmotic, ionic as well as oxidative stresses, plants produce and build up osmolyte or organic compatible solutes or osmoprotectants that balance water status, take part in the cellular energy transfer, stabilize membranes and proteins, scavenge ROS, and chelate heavy metal as well as lowering metal uptake (Ashraf and Foolad 2007; Gill and Tuteja 2010; Kaur and Asthir 2015). Osmoprotectants are small, highly soluble, uncharged, and nontoxic organic molecules, including α-amino acids (proline, ectoine), ammonium compounds (glycine betaine, β-alanine betaine, dimethylsulfoniopropionate, choline), polyols, sugars (trehalose), sugar alcohols (sorbitol, mannitol), etc. Thus, better stress adaptability and tolerance can be observed in the plants with enhanced biosynthesis of osmoprotectants or over-expressed osmoprotectant biosynthesis pathway genes (Singh et al. 2015). Therefore, scientists are working to enhance the metal stress tolerance in plants by improving the osmolyte synthesis in cellular level by exogenous application of phytoprotectants, together with tinkering in the genetic level. In this part several of these research findings are highlighted, where phytoprotectants favored in osmolyte biosynthesis as well as noteworthy protection against adverse effects of metal/metalloid stresses. Among the osmoprotectants or compatible solutes Pro is the most common, and accumulated when plants face various abiotic stress including metal/metalloid stress (Asgher et al. 2013; Kaur et al. 2011; Min et al. 2013), thus, suit cellular atmosphere favorable for phytochelation and metal sequestration, hence cells get protection against ROS damage (Nahar et al. 2016b; Mahmud et al. 2017a, b, c).

Proline is also an important parameter in stress marker to determine the metal toxicity among plants (Sharma et al. 2016). When toxic metals start to rise around the root zone, plants increase Pro accumulation. Patel et al. (2016) found up to 112% more Pro compared with control as the tannery sludge (TS) concentration increased in the growing media. Quantity of Pro in Cd-affected mung bean seedlings also increased by 53% compared with control (Nahar et al. 2016b) putting enormous pressure on the plant's survival.

Toxic effect of the metals/metalloids can be avoided in a significant extent by exogenous application of polyamines (PAs) such as di-amineputrescine (Put), tri-aminespermidine (Spd), and tetra-aminespermine (Spm). Nahar et al. (2016b) found up to 31% increase in Pro content after application of Put and/or sodium nitroprusside (SNP) in Cd-affected mung bean seedling, which reveals the influence of PAs in osmolyte biosynthesis. Exogenous Put increases Pro biosynthesis in *Populus cathayana* up to 60% and influences in the sex expression of male and female under Cu stress (Chen et al. 2013). Amooaghaie et al. (2017) found a noticeable stimulatory effect of the Pro accumulation in *Sesamum indicum* roots under Pb stress, while treating with SNP. Furthermore, a number of studies recommended that plant hormones can increase Pro synthesis when applied from exogenous source. For example, brassinosteroids and indole acetic acid can induce Pro biosynthesis in abiotic stress conditions, thus conferring metal/metalloid stress (Kaya et al. 2010). The application of 1 mM epibrassinolide (EBL) with Cr stress in rice seedlings showed considerable increase of Pro level (7.25 mmole g^{-1} FW) as compared to Cr-alone treated seedlings (4.76 mmole g^{-1}FW) eliminate the Cr stress positively (Sharma et al. 2016). Spraying EBL on stressed or stress free chickpea plants also had additive effect compared with respective controls at 45-day-old plant. The highest Pro content was obtained from combined stress and EBL spray by 52% (Wani et al. 2017). In addition, EBL supports in the transcription and subsequent translation of specific genes that improve the plant's tolerance at metal/metalloid stress (Hayat et al. 2014). Reports also suggested that jasmonic acid (JA) mediates accumulation of osmolytes in Cu stress and diminishes oxidative stress.

Proline accumulation increased up to 154% in 1 nM JA treated *Cajanus cajan* seedling compared to control (Poonam et al. 2013). Applying JA is known to biosynthesize betaine—another important osmolyte, under abiotic stress conditions (Gao et al. 2004). Moreover, kinetin is reported to alleviate toxic effects of B in wheat seedlings by increasing Pro biosynthesis up to 90% compared with corresponding control (Eser and Aydemir 2016). Cellular free Pro content was enhanced by 590, 612, and 622%, respectively, in soybean (*Glycine max* L.) when exogenous EBL or Pro applied to As toxicity as compared to the control (Chandrakar et al. 2017). This enhanced Pro level was the probable result of over-expressed activity of P5CS, the key enzyme responsible in Pro biosynthetic path (Singh et al. 2015; Reddy et al. 2015).

Organic acids and amino acids applied exogenously can promote osmolyte biosynthesis and can confer tolerance of plants at metal stress. For example, salicylic acid (SA) has linked to the alleviation of heavy metal-induced growth inhibition in *Cucumis melo* L. by endorsing Pro accumulation, antioxidant defense, and photosynthesis (Zhang et al. 2015). Proline was reported to employ against metal/metalloid toxicity in some plants to induce cellular Pro biosynthesis. Supplemental Pro supply by irrigation water to young date palm plants exposed to Cd showed increased free cellular Pro content by 38% giving osmotic adjustment (Zouari et al. 2016). In another experiment they found Pro increased by 55% and 102% in leaves and roots respectively under Cd stress in *Olea europaea* L. cv Chemlali. They suggested that Pro is a metabolite for conferring Cd stress (Zouari et al. 2016). The amounts of Pro increased significantly in *Hyoscyamus niger* plant under Ni stress treated with arginine (Arg), where maximum accumulation of Pro was observed at 50 μM Ni stress treated with 20 μM Arg (Nasibi et al. 2013). Proline as a protectant amino acid was also reported in Indian mustard (Asgher et al. 2013) and wheat (Khan et al. 2015). Moreover, it was reported that the exogenous application of ALA noticeably improved the Pro content in *Brassica napus* leaves under variable Cd stress (Ali et al. 2015a, b). Heat shock protein may also contribute in osmolyte accumulation. In duckweed (*Lemna minor*), Pro level increased by 88% higher than that of control (Zhang et al. 2017).

Plant nutrients and trace elements are often used to eliminate metal/metalloid stress by enhancing osmolyte synthesis and accumulation. Mangrove plants *Avicennia marina* and *Kandelia obovata* were found with enhanced Pro content when supplemented with P at Cd stress owing to the optimistic association among P and water (Dai et al. 2017). Application of Si in As toxicity significantly increased the cysteine and Pro content by 30% and 38%, respectively, over the value of respective control (Pandey et al. 2016). Proline accumulation was found in *B. juncea* leaves in response to Cu stress supplemented with Se and EBL, and 87.3% over the non-treated control plants (Yusuf et al. 2016). Moreover, Se increased Pro synthesis in *B. napus* plants exposed to metal/metalloids stress, and enhanced the antioxidant defense (Hasanuzzaman et al. 2012a).

Phytochelatins and Metallothioneins Synthesis

Similar with essential nutrients plants naturally uptake metal/metalloid from contaminated growing media by their roots. But, plants have potential to minimize toxicity at cellular level as they are able to sequestrate metal/metalloid into cell vacuole as an immobile form through a variety of chelating agent (Nahar et al. 2016a, b; Mahmud et al. 2017b, c, 2018). So, chelation and sequestration of metal/metalloid by different ligands are the key mechanisms of plants to work against stress. Phytochelatins (PCs) and metallothioneins (MTs) are common and best characterized chelating agent or ligands in plant which are involved in metal/metalloid detoxification under stress condition (Cobbett and Goldsbrough 2002). Phytochelatin is an oligomer of the γ-glutamyl (Glu)-cysteinyl (Cys) unit and enzymatically synthesized polypeptides which produced from GSH with the help of the enzyme phytochelatin synthase (Vatamaniuk et al. 2000; Zagorchev et al. 2013; Mahmud et al. 2018). Like PCs, MTs also Cys-rich compound but it is gene-encoded polypeptide (Cobbett and Goldsbrough 2002). Rate of PCs synthesis is plant-specific and/or metal/metalloid-specific. Among the different metal/metalloid, Cd, Hg, As, and Fe confirmed higher synthesis of PCs in plant cell. On the other hand, Cu and Ni are moderate inducers while Pb and Zn are weak inducers of PCs (Anjum et al. 2015). Beside functioned as a metal detoxifier through chelating, MTs are involved in metal homeostasis and oxidative stress protection (Hossain et al. 2012). Therefore, high PCs and MTs synthesis under stress condition is considered as essential intrinsic criteria for decreasing metal/metalloid toxicity or increasing tolerance of plants. Plant scientists are trying to increase the metal/metalloid stress tolerance in plants by enhancing the cellular PCs and MTs content by applying different phytoprotectant exogenously. Recently, Nahar et al. (2016a) exposed mung bean plant to two levels of Cd stress (1 mM and 1.5 mM) and observed PC content increased by 190 and 267%, respectively, compared to control. A further increase in PC content was recorded when Cd-stressed mung bean seedlings were pretreated with exogenous spermine. Correspondingly, Nahar et al. (2016b) found similar result in other experiment. They found Cd stress increased PC content in mung bean plant which further increased by different polyamines. Similarly Mahmud et al. (2018) raised mustard seedlings to Cd stress and found that PC content increased in a dose-dependent manner. They also exogenously applied citric acid to Cd-affected seedlings and observed that PC content further increased. Metal/metalloid toxicity activates the enzyme phytochelatin synthase which contributes to enhance synthesis of PCs and finally makes a HM-PCs complex. In contrast, cellular MTs are remarkably diverse and it is subdivided into three types based on the arrangement of the Cys residue (Zhou et al. 2006). The Cys-Cys, Cys-X-Cys, and Cys-X-X-Cys motifs (X indicates any type of amino acid) are characteristic and invariant for MTs. The association or allocation of cysteine residues bestows different MT iso-forms and their ability to bind and sequester different metal ions for detoxification and homeostasis. Metallothioneins biosynthesis is regulated at the transcriptional level and is induced by several factors, including hormones, cytotoxic agents, and toxic metals, such as

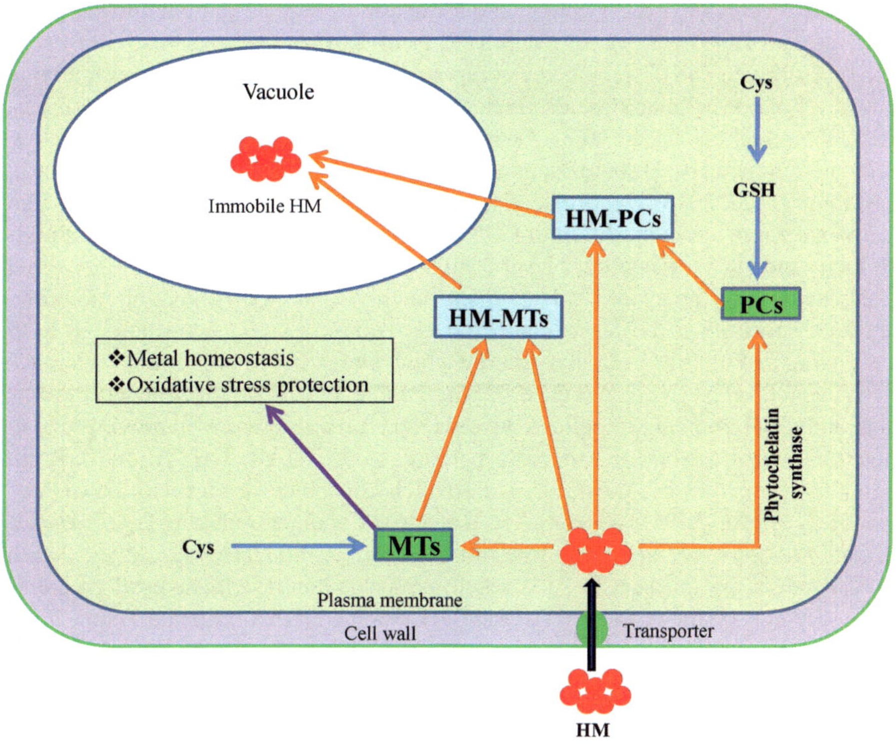

Fig. 4 Phytochelatins and MTs synthesis and their function in metal/metalloid chelation and sequestration. PCs, MTs, Cys, GSH, and HM indicate phytochelatins, metallothioneins, cysteine, glutathione, and heavy metal, respectively

Cd, Zn, Hg, Cu, Au, Ag, Co, Ni, etc. (Kagi 1991, Yang et al. 2005; Hossain et al. 2012). Ahn et al. (2012) carried out an experiment and found that three *Brassica rapa* MT genes (BrMT1, BrMT2, and BrMT3) are differentially regulated under various heavy metal stresses (Fig. 4).

Antioxidant Defense

Plants have an inherent well-established apparatus termed as antioxidant defense system which regulates cellular ROS levels according to the requirements at a particular occasion. Antioxidant defense system is composed of non-enzymatic components (AsA, GSH, phenolic compounds, alkaloids, α-tocopherol, non-protein amino acids, etc.) and enzymatic components (SOD, CAT, APX, GR, MDHAR, DHAR, GPX, GST, etc.) which helps to develop stress tolerance (Gill and Tuteja 2010; Hasanuzzaman and Fujita 2012). Under the any abiotic stress condition,

AsA-GSH cycle performs a fundamental function in regulating cellular redox balance through ROS scavenging (Mahmood et al. 2010; Hasanuzzaman and Fujita 2012). Ascorbate and GSH, the key compounds of AsA-GSH pool, are available in almost all major cellular structure including cytoplasm, chloroplast, apoplast, peroxisome, mitochondria, etc. Both AsA and GSH are very efficient to scavenge H_2O_2 (Gill and Tuteja 2010; Hasanuzzaman and Fujita 2012; Hasanuzzaman et al. 2017b, c; Mahmud et al. 2018). Ascorbate, a water-soluble non-enzymatic antioxidant, can easily make reaction with different ROS including $O_2^{\bullet -}$ and $OH^{\bullet}$ and able to diminish their contents in the cellular level (Gill and Tuteja 2010). Similar to AsA, GSH is also a non-enzymatic antioxidant which functioned as stress signaling molecule, acts as a substrate of GPX, regulates the glyoxalase system, contributes in ROS scavenging and inhibits protein oxidation which altogether improve stress tolerance mechanism of plant (Noctor et al. 2002; Mahmud et al. 2017c). So, enhancement of AsA and GSH contents in cellular level is very urgent task for improving metal/metalloid stress tolerance. Ascorbate peroxidase, MDHAR, DHAR, and GR are major four enzymes of AsA-GSH cycle which contribute in regeneration of AsA and GSH. Where, APX utilizes AsA as a electron donor for reducing H_2O_2 to water (Hasanuzzaman et al. 2017b). Furthermore, MDHAR and DHAR are engaged with AsA regeneration in the cell. As a result, AsA content at cellular level predominantly depends on the activity of APX, MDHAR, and DHAR (Gill and Tuteja 2010; Hasanuzzaman and Fujita 2012; Mahmud et al. 2017a, b, c). On the other hand, renaissance of GSH greatly depends on the activity of GR (Gill and Tuteja 2010). Superoxide dismutase is considered a key enzyme in regulating intracellular concentrations of ROS because the first step in ROS detoxification starts with dismutation of $O_2^{\bullet -}$ to H_2O_2, which is catalyzed by SOD (Gill and Tuteja 2010; Hasanuzzaman and Fujita 2012). Accordingly, CAT eliminates H_2O_2 by converting it to O_2 and H_2O (Miller et al. 2008). Glutathione peroxidase plays an important role in protecting cell components from oxidative damage by scavenging peroxides and different electrophiles (Gill and Tuteja 2010). On the other hand, GST is a multidimensional enzyme which works to improve stress tolerance and is very familiar for their function in enzymatic detoxification of xenobiotics. It performs through catalyzing the conjugation of GSH with electrophilic. Sometimes GST catabolized the toxic substances or sequestered into the cell vacuole (Dixon and Edwards 2010; Hossain et al. 2012).

A lot of studies described the roles of antioxidant in conferring metal/metalloid stress tolerance. Here we discussed some of them. Mahmud et al. (2017a) checked relative tolerance of three *B. species* under Cd stress (two levels; 0.25 mM and 0.5 mM $CdCl_2$) and found AsA content reduction was lower and GSH content enhancement was higher in *B. juncea* compared with other species. Monodehydroascorbate reductase, GR, and SOD activities increased significantly in *B. juncea* under Cd stress compared with the other species. Catalase activity did not decrease in *B. juncea* due to Cd stress, compared with the other species. Activity of DHAR decreased with both levels of Cd stress in all species except for *B. juncea* under 0.25 mM $CdCl_2$ stress. These results prove that activity of antioxidants depends on plant type and stress intensity. Similar diverse upregulation and down-regulation of antioxidant enzymes under Cd stress were observed in mung bean (Nahar et al. 2016a, b), rice (Rahman et al. 2016), rapeseed (Hasanuzzaman et al. 2012b), and mustard (Mahmud et al. 2018) seedlings. Hasanuzzaman et al. (2018) carried out an experiment with wheat seedlings under Pb stress and found Pb treatment disrupted antioxidant enzyme activities and status of endogenous ascorbate and GSH pool. They found APX activity increased but MDHAR and DHAR activity decreased under Pb stress. On the other hand, GR activity increased under mild stress but decreased under severe stress. Under Cr stress AsA content decreased and GSH content increased in Indian mustard seedlings (Mahmud et al. 2017b, c). They also recorded APX and GR activity increased but MDHAR and DHAR activity decreased under Pb stress. Similar trend was observed in rice seedlings under As stress except MDHAR. The activity of MDHAR was increased under As stress (Rahman et al. 2015). Noteworthy decrease in antioxidants pool along with elevated enzymatic antioxidant activity was observed in rice seedling treating with both 3 mM and 6 mM $MnCl_2$. With the increase in Mn concentration and exposure length, AsA and GSH along with their redox ratios declined. Plants treated with 6 mM $MnCl_2$ for 20 days exhibited 48 and 65% reduction in roots and shoots, respectively, whereas DHA content declined by 27 and 32%. Like AsA the level of GSH also declined by the toxic effect of Mn, while GSSG displayed increased level in mild stress and declined at severe stress. The ratio GSH/GSSG was also declined steadily in the seedlings with severe stress treatments, with redox states alteration (Srivastava and Dubey 2011). Responses of Mn toxicity in soybean plants include upregulation of different antioxidative enzymes such as CAT, POD, and SOD (Santos et al. 2017). In rice, upon exposure to Al growing seedlings absorbed Al and translocation took place in shoots. But prolonged exposure caused lipid peroxidation and altered the activities of SOD and POD, furthermore caused DNA damage (Meriga et al. 2004). So, it is very clear that under stress condition some of the antioxidants upregulated and want to give tolerance to plant. But with the increase of stress intensity, antioxidant defense system becomes inefficient. As a result, scientists are using different exogenous protectants (organic acid, osmolytes, etc.) to increase activity of antioxidant enzymes as well as stress tolerance.

Conclusions and Future Perspectives

Metal/metalloid pollution has become an indispensable incidence to the modern world due to urbanization and industrialization. This is resulting in the pollution of environment— soil, air, and water, ultimately leading to hazardous condition for both flora and fauna. Metals like Cd, Pb, Hg, Cr, Al and metalloid like As are potentially toxic to plants, while some essential elements (e.g., Zn, Fe, B, Cu, Mn, Ni) may also become injurious when present in excess amount. These elements not only hamper plant physiology but also result in oxidative stress in plants. Another important fact is, metal stress induced toxic MG production has also been recorded in many plants which is involved with oxidative stress, directly or indirectly (Nahar et al. 2016a, b; Hasanuzzaman et al. 2017b; Mahmud et al. 2017a, b, 2018). This overproduction of ROS and MG destroys the balance of them with antioxidative enzymes at cellular level which causes multiple negative responses in plant physiology including lipid peroxidation, protein peroxidation, DNA degradation, etc. or even cell death (Hasanuzzaman and Fujita 2012; Sharma et al. 2012; Hasanuzzaman et al. 2017b). Therefore, it is of prime need to understand the physiological and molecular mechanism of metal stress and identify the ways to develop tolerance within the plants essential for agricultural survival. However, rational knowledge of crop physiology and modulated crop management procedures can make it possible to lessen the losses caused by metal/metalloid stress. Researchers of divergent fields have so far experimented with a number of crop varieties and metal forms to understand and disclose the possible mechanisms of inducing tolerance against metal/metalloid stress. Among the molecular approaches endogenous synthesis of osmolytes, PCs and MTs, and antioxidant defense enzymes is of utmost importance. Application of different types of exogenous protectants (e.g., organic acids, osmolytes, etc.) may play a crucial role in this case. In addition, modified agronomic practices can also be a promising way to minimize the uptake or to diminish the effect of metal/metalloid on plants. On the other hand, geneticists can also play a vital role by producing transgenic plants related to antioxidant which induces tolerance to metal/metalloid stress. Improving genetic potential in plants by tailoring the genes responsible for the upregulation of antioxidant defense components can be a very efficient way of phytoremediation of metals/metalloids which would be a sustainable green technology.

References

Achary VM, Jena S, Panda KK, Panda BB (2008) Aluminium induced oxidative stress and DNA damage in root cells of *Allium cepa* L. Ecotoxicol Environ Saf 70:300–310

Aftab T, Khan MMA, Naeem M, Idrees M, da Silva JAT, Ram M (2012) Exogenous nitric oxide donor protects *Artemisia annua* from oxidative stress generated by boron and aluminium toxicity. Ecotoxicol Environ Saf 80:60–68

Agami RA, Mohamed GF (2013) Exogenous treatment with indole-3-acetic acid and salicylic acid alleviates cadmium toxicity in wheat seedlings. Ecotoxicol Environ Saf 94:164–171

Ahn YO, Kim SH, Lee J, Kim HR, Lee H-S, Kwak S-S (2012) Three *Brassica rapa* metallothionein genes are differentially regulated under various stress conditions. Mol Biol Rep 39:2059–2067

Ali B (2017) Salicylic acid induced antioxidant system enhances the tolerence to aluminium in mung bean (*Vigna radiata* L. Wilczek) plants. Indian J Plant Physiol 22:178–189

Ali B, Hasan SA, Hayat S, Hayat Q, Yadav S, Fariduddin Q, Ahmad A (2008) A role for brassinosteroids in the amelioration of aluminum stress through antioxidant system in mung bean (*Vigna radiate* L. Wilczek). Environ Exp Bot 62:153–159

Ali B, Xu X, Gill RA, Yang S, Ali S, Tahir M, Zhou W (2014) Promotive role of 5-aminolevulinic acid on mineral nutrients and antioxidative defense system under lead toxicity in *Brassica napus*. Ind Crop Prod 52:617–626

Ali S, Bharwana SA, Rizwan M, Farid M, Kanwal S, Ali Q, Ibrahim M, Gill RA, Khan MD (2015a) Fulvic acid mediates chromium (Cr) tolerance in wheat (*Triticum aestivum* L.) through lowering of Cr uptake and improved antioxidant defense system. Environ Sci Pollut Res 22:10601–10609. https://doi.org/10.1007/s11356-015-4271-7

Ali B, Gill RA, Yang S, Gill MB, Farooq MA, Liu D, Daud MK, Ali S, Zhou W (2015b) Regulation of cadmium-induced proteomic and metabolic changes by 5-aminolevulinic acid in leaves of *Brassica napus* L. PLoS One 10:e0123328. https://doi.org/10.1371/journal.pone.0123328

Amooaghaie R, Zangene-Madar F, Enteshari S (2017) Role of two-sided crosstalk between NO and H_2S on improvement of mineral homeostasis and antioxidative defense in *Sesamum indicum* under lead stress. Ecotoxicol Environ Saf 139:210–218

Anjum NA, Hasanuzzaman M, Hossain MA, Thangavel P, Roychoudhury A, Gill SS, Rodrigo MAM, Adam V, Fujita M, Kizek R, Duarte AC, Pereira E, Ahmad I (2015) Jacks of metal/metalloid chelation trade in plants—an overview. Front Plant Sci 6:192. https://doi.org/10.3389/fpls.2015.00192

Anwaar SA, Ali S, Ali S, Ishaque W, Farid M, Farooq MA, Najeeb U, Abbas F, Sharif M (2015) Silicon (Si) alleviates cotton (*Gossypium hirsutum* L.) from zinc (Zn) toxicity stress by limiting Zn uptake and oxidative damage. Environ Sci Pollut Res 22(5):3441–3450. https://doi.org/10.1007/s11356-014-3938-9

Asgher M, Khan MIR, Iqbal N, Masood A, Khan NA (2013) Cadmium tolerance in mustard cultivars: dependence on proline accumulation and nitrogen assimilation. J Funct Environ Bot 3:30–42

Ashraf M, Foolad M (2007) Roles of glycine betaine and proline in improving plant abiotic stress resistance. Environ Exp Bot 59:206–216

Ashraf U, Tang X (2017) Yield and quality responses, plant metabolism and metal distribution pattern in aromatic rice under lead (Pb) toxicity. Chemosphere 176:141–155

Ashraf U, Kanu AS, Mo Z, Hussain S, Anjum SA, Khan I, Abbas RN, Tang X (2015) Lead toxicity in rice; effects, mechanisms and mitigation strategies—a mini review. Environ Sci Pollut Res 22:18318–18332

Awasthi JP, Saha B, Regon P, Sahoo S, Chowra U, Pradhan A, Roy A, Panda SK (2017) Morpho-physiological analysis of tolerance to aluminum toxicity in rice varieties of North East India. PLoS One 12:e0176357. https://doi.org/10.1371/journal.pone.0176357

Baruah KK, Bharali A (2015) Physiological basis of iron toxicity and its management in crops: recent advances in crop physiology. Daya Publishing House, New Delhi

Becker M, Asch F (2005) Iron toxicity in rice—conditions and management concepts. J Plant Nutr Soil Sci 168:558–573

Benavides MP, Gallego SM, Tomaro ML (2005) Cadmium toxicity in plants. Braz J Plant Physiol 17:21–34. https://doi.org/10.1590/S1677-04202005000100003

Biswas S, Mano J (2015) Lipid peroxide-derived short-chain carbonyls mediate hydrogen peroxide-induced and salt-induced programmed cell death in plants. Plant Physiol 168:885–898

Boojar MM, Goodarzi F (2008) Comparative evaluation of oxidative stress status and manganese availability in plants growing on manganese mine. Ecotoxicol Environ Saf 71:692–699

Broadley MR, White PJ, Hammond JP, Zelko I, Lux A (2007) Zinc in plants. New Phytol 173:677–702

Cargnelutti D, Tabaldi LA, Spanevello RM, Jucoski GO, Battisti V, Redin M, Linares CEB, Dressler VL, Flores ÉMM, Nicoloso FT, Morsch VM, Schetinger MRC (2006) Mercury toxicity induces oxidative stress in growing cucumber seedlings. Chemosphere 65:999–1006

Cervilla LM, Blasco B, Ríos JJ, Romero L, Ruiz JM (2007) Oxidative stress and antioxidants in tomato (*Solanum lycopersicum*) plants subjected to boron toxicity. Ann Bot 100:747–756

Chandrakar V, Yadu B, Meena RK, Dubey A, Keshavkant S (2017) Arsenic-induced genotoxic responses and their amelioration by diphenylene iodonium, 24-epibrassinolide and proline in *Glycine max* L. Plant Physiol Biochem 112:74–86. https://doi.org/10.1016/j.plaphy.2016.12.023

Chen L, Wang L, Chen F, Korpelainen H, Li C (2013) The effects of exogenous putrescine on sex-specific responses of *Populus cathayana* to copper stress. Ecotoxicol Environ Saf 97:94–102

Chen Y, Mo HZ, Hu LB, Li YQ, Chen J, Yang LF (2014) The endogenous nitric oxide mediates selenium-induced phytotoxicity by promoting ROS generation in *Brassica rapa*. PLoS One 9:e110901

Chen J, Shaf M, Li S, Wang Y, Wu J, Ye Z, Peng D, Yan W, Liu D (2015) Copper induced oxidative stresses, antioxidant responses and phytoremediation potential of Moso bamboo (*Phyllostachys pubescens*). Sci Rep 5:13554. https://doi.org/10.1038/srep13554

Chen Q, Zhang X, Liu Y, Wei J, Shen W, Shen Z, Cui J (2017) Hemin-mediated alleviation of zinc, lead and chromium toxicity is associated with elevated photosynthesis, antioxidative capacity; suppressed metal uptake and oxidative stress in rice seedlings. Plant Growth Regul 81:253–264. https://doi.org/10.1007/s10725-016-0202-y

Cherif J, Mediouni C, Ammar WB, Jemal F (2011) Interactions of zinc and cadmium toxicity in their effects on growth and in antioxidative systems in tomato plants (*Solanum lycopersicum*). J Environ Sci 23:837–844

Chou T-S, Chao Y-Y, Huang W-D, Hong C-Y, Kao CH (2011) Metal/metalloid-induced oxidative stress in plants. J Plant Physiol 168:1021–1030

Choudhary SP, Kanwar M, Bhardwaj R, Yu J-Q, Tran L-SP (2012) Chromium stress mitigation by polyamine-brassinosteroid application involves phytohormonal and physiological strategies in *Raphanus sativus* L. PLoS One 7:e33210. https://doi.org/10.1371/journal.pone.0033210

Choudhury S, Panda SK (2004) Toxic effects, oxidative stress and ultrastructural changes in moss *Taxithelium nepalense* (Schwaegr.) Broth. under chromium and lead phytotoxicity. Water Air Soil Pollut 167:73–90. https://doi.org/10.1007/s11270-005-8682-9

Chowra U, Yanase E, Koyama H, Panda SK (2017) Aluminium-induced excessive ROS causes cellular damage and metabolic shifts in black gram *Vigna mungo* (L.) Hepper. Protoplasma 254:293–302

Cobbett C, Goldsbrough P (2002) Phytochelatins and metallothioneins: roles in heavy metal detoxification and homeostasis. Annu Rev Plant Biol 53:159–182

Cui Y, Zhao N (2011) Oxidative stress and change in plant metabolism of maize (*Zea mays* L.) growing in contaminated soil with elemental sulfur and toxic effect of zinc. Plant Soil Environ 57:34–39

Cuypers A, Plusquin M, Remans T, Jozefczak M, Keunen E, Gielen H, Opdenakker K, Nair AR, Munters E, Artois TJ, Nawrot T, Vangronsveld J, Smeets K (2010) Cadmium stress: an oxidative challenge. Biometals 23:927–940. https://doi.org/10.1007/s10534-010-9329-x

Dai M, Lu H, Liu W, Jia H, Hong H, Liu J, Yan C (2017) Phosphorus mediation of cadmium stress in two mangrove seedlings *Avicennia marina* and *Kandelia obovata* differing in cadmium accumulation. Ecotoxicol Environ Saf 139:272–279. https://doi.org/10.1016/j.ecoenv.2017.01.017

de Sousa A, AbdElgawad H, Han A, Teixeira J, Matos M, Fidalgo F (2016) Oxidative metabolism of rye (*Secale cereale* L.) after short term exposure to aluminum, uncovering the glutathione-ascorbate redox network. Front Plant Sci 7:685. https://doi.org/10.3389/fpls.2016.00685

Deng C, Zhang D, Pan X, Chang F, Wang S (2013) Toxic effects of mercury on PSI and PSII activities, membrane potential and transthylakoid proton gradient in *Microsorium pteropus*. J Photochem Photobiol B 127:1–7

Dixon DP, Edwards R (2010) Glutathione transferases. In: The Arabidopsis book, vol 8. The American Society of Plant Biologists, Austin

Dong Y, Xu L, Wang Q, Fan Z, Kong J, Bai X (2014) Effects of exogenous nitric oxide on photosynthesis, antioxidative ability, and mineral element contents of perennial ryegrass under copper stress. J Plant Interact 9:402–411

Dordas C, Chrispeels MJ, Brown PH (2000) Permeability and channel-mediated transport of boric acid across membrane vesicles isolated from squash roots. Plant Physiol 124:1349–1362

Duan X, Li X, Ding F, Zhao J, Guo A, Zhang L, Yao J, Yang Y (2015) Interaction of nitric oxide and reactive oxygen species and associated regulation of root growth in wheat seedlings under zinc stress. Ecotoxicol Environ Saf 113:95–102

Dubey D, Pandey A (2011) Effect of nickel (Ni) on chlorophyll, lipid peroxidation and antioxidant enzymes activities in black gram (*Vigna mungo*) leaves. Int J Sci Nat 2:395–401

Elbaz A, Wei YY, Meng Q, Zheng Q, Yang ZM (2010) Mercury-induced oxidative stress and impact on antioxidant enzymes in *Chlamydomonas reinhardtii*. Ecotoxicology 19:1285–1293

El-Ramady HR, Domokos-Szabolcsy E, Abdalla NA, Alshaal TA, Shalaby TA, Sztrik A, Prokisch J, Fari M (2014) Selenium and nano-selenium in agroecosystems. Environ Chem Lett 12:495–510
Emamverdian A, Ding Y, Mokhberdoran F, Xie Y (2015) Heavy metal stress and some mechanisms of plant defense response. Sci World J 2015:756120. https://doi.org/10.1155/2015/756120
Eraslan F, Inal A, Savasturk O, Gunes A (2007) Changes in antioxidative system and membrane damage of lettuce in response to salinity and boron toxicity. Sci Hortic 114:5–10
Eser A, Aydemir T (2016) The effect of kinetin on wheat seedlings exposed to boron. Plant Physiol Biochem 108:158–164
Faize M, Burqos L, Piqueras A, Nicolas E, Barba-Espin G, Clement-Moreno MJ, Alcobendas R, Artlip T, Hernandez JA (2011) Involvement of cytosolic ascorbate peroxidase and Cu/Zn-superoxide dismutase for improved tolerance against drought stress. J Exp Bot 62:2599–2613
Farid M, Ali S, Rizwan M, Saeed R, Tauqeer HM, Sallah-Ud-Din R, Azam A, Raza N (2017) Microwave irradiation and citric acid assisted seed germination and phytoextraction of nickel (Ni) by *Brassica napus* L.: morpho-physiological and biochemical alterations under Ni stress. Environ Sci Pollut Res 24(26):21050–21064. https://doi.org/10.1007/s11356-017-9751-5
Farooq MA, Islam F, Ali B, Najeeb U, Mao B, Gill RA, Yan G, Siddique KHM, Zhou W (2016) Arsenic toxicity in plants: cellular and molecular mechanisms of its transport and metabolism. Environ Exp Bot 132:42–52
Feigl G, Lehotai N, Molnár Á, Ördög A, Rodríguez-Ruiz M, Palma JM, Corpas FJ, Erdei L, Kolbert Z (2015) Zinc induces distinct changes in the metabolism of reactive oxygen and nitrogen species (ROS and RNS) in the roots of two *Brassica* species with different sensitivity to zinc stress. Ann Bot 116:613–625
Fernando DR, Lynch JP (2015) Manganese phytotoxicity: new light on an old problem. Ann Bot 116:313–319
Fernando DR, Marshall AT, Forster PI, Hoebee SE, Siegele R (2013) Multiple metal accumulation within a manganese-specific genus. Am J Bot 100:690–700
Fischel JS, Fischel MH, Sparks DL (2015) Advances in understanding reactivity of manganese oxides with arsenic and chromium in environmental systems. In: Feng X, Li W, Zhu M, Sparks DL (eds) Advances in the environmental biogeochemistry of manganese oxides. American Chemical Society, pp 1–27
Flora SJS, Mittal M, Mehta A (2008) Heavy metal induced oxidative stress & its possible reversal by chelation therapy. Indian J Med Res 128:501–523
Flora G, Gupta D, Tiwari A (2012) Toxicity of lead: a review with recent updates. Interdiscip Toxicol 5(2):47–58
Foyer CH, Shigeoka S (2011) Understanding oxidative stress and antioxidant functions to enhance photosynthesis. Plant Physiol 155:93–100
Gajewska E, Drobik D, Wielanek M, Nalewajko JS, Gocławski J, Mazur J, Skłodowska M (2013) Alleviation of nickel toxicity in wheat (*Triticum aestivum* L.) seedlings by selenium supplementation. Biol Lett 50:63–76
Gao X, Wang X, Lu Y, Zhang L, Shen Y, Liang Z, Zhang D (2004) Jasmonic acid is involved in the water stress induced betaine accumulation in pear leaves. Plant Cell Environ 27:497–507. https://doi.org/10.1111/j.1365-3040.2004.01167.x
Ghasemi F, Heidari R, Jameii R, Purakbar L (2013) Responses of growth and antioxidative enzymes to various concentrations of nickel in *Zea mays* leaves and roots. Rom J Biol Plant Biol 58:37–49
Giannakoula A, Moustakas M, Mylona P, Papadakis I, Yupsanis T (2008) Aluminium tolerance in maize is correlated with increased levels of mineral nutrients, carbohydrates and proline and decreased levels of lipid peroxidation and Al accumulation. J Plant Physiol 165:385–396
Gill SS, Tuteja N (2010) Reactive oxygen species and antioxidant machinery in abiotic stress tolerance in crop plants. Plant Physiol Biochem 48:909–930
Gill RA, Zang L, Ali B, Farooq MA, Cui P, Yang S, Ali S, Zhou W (2015) Chromium-induced physio-chemical and ultrastructural changes in four cultivars of *Brassica napus* L. Chemosphere 120:154–164. https://doi.org/10.1016/j.chemosphere.2014.06.029

Gjorgieva D, Panovska TK, Ruskovska T, Bačeva T, Stafilov T (2013) Mineral nutrient imbalance, total antioxidants level and DNA damage in common bean (*Phaseolus vulgaris* L.) exposed to heavy metals. Physiol Mol Biol Plants 19:499–507

Gomes-Junior RA, Gratão PL, Gaziola SA, Mazzafera P, Lea PJ, Azevedo RA (2007) Selenium-induced oxidative stress in coffee cell suspension cultures. Funct Plant Biol 34:449–456

González A, Lynch J (1999) Tolerance of tropical common bean genotypes to manganese toxicity: performance under different growing conditions. J Plant Nutr 22:511–525

González A, Steffen KL, Lynch JP (1998) Light and excess manganese. Implications for oxidative stress in common bean. Plant Physiol 118:493–504

Gorny J, Billon G, Noiriel C, Dumoulin D, Lesven L, Madé B (2016) Chromium behavior in aquatic environments: a review. Environ Rev 24:503–516. https://doi.org/10.1139/er-2016-0012

Gunes A, Inal A, Bagci EG, Coban S, Pilbeam DJ (2007) Silicon mediates changes to some physiological and enzymatic parameters symptomatic for oxidative stress in spinach (*Spinacia oleracea* L.) grown under B toxicity. Sci Hortic 113:113–119

Gupta B, Pathak GC, Pandey N (2011) Induction of oxidative stress and antioxidant responses in *Vigna mungo* by zinc stress. Russ J Plant Physiol 58:85–91

Gururani MA, Venkatesh J, Tran LSP (2015) Regulation of photosynthesis during abiotic stress induced photoinhibition. Mol Plant 8:1304–1320

Habiba U, Ali S, Farid M, Shakoor MB, Rizwan M, Ibrahim M, Abbasi GH, Hayat T, Ali B (2015) EDTA enhanced plant growth, antioxidant defense system, and phytoextraction of copper by *Brassica napus* L. Environ Sci Pollut Res 22(2):1534–1544. https://doi.org/10.1007/s11356-014-3431-5

Hamed SM, Zinta G, Klöck G, Asard H, Selim S, AbdElgawad H (2017a) Zinc-induced differential oxidative stress and antioxidant responses in *Chlorella sorokiniana* and *Scenedesmus acuminatus*. Ecotoxicol Environ Saf 140:256–263

Hamed SM, Selim S, Klöck G, AbdElgawad H (2017b) Sensitivity of two green microalgae to copper stress: growth, oxidative and antioxidants analyses. Ecotoxicol Environ Saf 144:19–25

Hasanuzzaman M, Fujita M (2012) Heavy metals in the environment: current status, toxic effects on plants and possible phytoremediation. In: Anjum NA, Pereira MA, Ahmad I, Duarte AC, Umar S, Khan NA (eds) Phytotechnologies: remediation of environmental contaminants. CRC Press, Boca Raton, pp 7–73

Hasanuzzaman M, Fujita M (2013) Exogenous sodium nitroprusside alleviates arsenic-induced oxidative stress in wheat (*Triticum aestivum* L.) seedlings by enhancing antioxidant defense and glyoxalase system. Ecotoxicology 23:584–596

Hasanuzzaman M, Anwar Hossain M, Masayuki F (2011) Selenium-induced up regulation of the antioxidant defense and methylglyoxal detoxification system reduces salinity-induced damage in rapeseed seedlings. Biol Trace Elem Res 143:1704–1721

Hasanuzzaman M, Hossain MA, Fujita M (2012a) Exogenous selenium pretreatment protects rapeseed seedlings from cadmium-induced oxidative stress by upregulating antioxidant defense and methylglyoxal detoxification systems. Biol Trace Elem Res 149:248–261

Hasanuzzaman M, Hossain MA, Teixeira da Silva JA, Fujita M (2012b) Plant responses and tolerance to abiotic oxidative stress: antioxidant defense is a key factor. In: Bandi V, Shanker AK, Shanker C, Mandapaka M (eds) Crop stress and its management: perspectives and strategies. Springer, Berlin, pp 261–316

Hasanuzzaman M, Nahar K, Hossain MS, Mahmud JA, Rahman A, Inafuku M, Oku H, Fujita M (2017a) Coordinated actions of glyoxalase and antioxidant defense systems in conferring abiotic stress tolerance in plants. Int J Mol Sci 18:200. https://doi.org/10.3390/ijms18010200

Hasanuzzaman M, Nahar K, Gill SS, Alharby HF, Razafindrabe BH, Fujita M (2017b) Hydrogen peroxide pretreatment mitigates cadmium-induced oxidative stress in *Brassica napus* L.: an intrinsic study on antioxidant defense and glyoxalase systems. Front Plant Sci 8:115. https://doi.org/10.3389/fpls.2017.00115

Hasanuzzaman M, Nahar K, Anee TI, Fujita M (2017c) Exogenous silicon attenuates cadmium-induced oxidative stress in *Brassica napus* L. by modulating AsA-GSH pathway and glyoxalase system. Front Plant Sci 8:1061. https://doi.org/10.3389/fpls.2017.01061

Hasanuzzaman M, Nahar K, Rahman A, Mahmud JA, Alharby HF, Fujita M (2018) Exogenous glutathione attenuates lead-induced oxidative stress in wheat by improving antioxidant defense and physiological mechanisms. J Plant Interact 13:203–212

Hattab S, Hattab S, Flores-Casseres ML, Boussetta H, Doumas P, Hernandez LE, Banni M (2016) Characterisation of lead-induced stress molecular biomarkers in *Medicago sativa* plants. Environ Exp Bot 123:1–12

Hauck M, Hesse V, Runge M (2002) Correlations between the Mn/Ca ratio in stem flow and epiphytic lichen abundance in a dieback-affected spruce forest of the Harz Mountains Flora. Funct Ecol Plants 197:361–369

Hauck M, Paul A, Gross S, Raubuch M (2003) Manganese toxicity in epiphytic lichens: chlorophyll degradation and interaction with iron and phosphorus. Environ Exp Bot 49:181–191

Hayat S, Khalique G, Wani AS, Alyemeni MN, Ahmad A (2014) Protection of growth in response to 28-homobrassinolide under the stress of cadmium and salinity in wheat. Int J Biol Macromol 64:130–136

He J, Wang Y, Ding H, Ge C (2016) Epibrassinolide confers zinc stress tolerance by regulating antioxidant enzyme responses, osmolytes, and hormonal balance in *Solanum melongena* seedlings. Braz J Bot 39(1):295–303. https://doi.org/10.1007/s40415-015-0210-6

Hoang TH, Ju-Yong K, Sunbaek B, Kyoung-Woong K (2010) Source and fate of As in the environment. Geosyst Eng 13:35–42

Homa D, Haile E, Washe AP (2016) Determination of spatial chromium contamination of the environment around industrial zones. Int J Anal Chem 7214932. https://doi.org/10.1155/2016/7214932

Hossain MA, Piyatida P, da Silva JAT, Fujita M (2012) Molecular mechanism of heavy metal toxicity and tolerance in plants: central role of glutathione in detoxification of reactive oxygen species and methylglyoxal and in heavy metal chelation. J Bot 2012:872875. https://doi.org/10.1155/2012/872875

Houtz RL, Nable RO, Cheniae GM (1988) Evidence for effects on the in vivo activity of ribulose-biphosphate carboxylase/oxygenase during development of Mn toxicity in tobacco. Plant Physiol 86:1143–1149

Inal A, Pilbeam DJ, Gunes A (2009) Silicon increases tolerance to boron toxicity and reduces oxidative damage in barley. J Plant Nutr 32:112–128

Islam F, Yasmeen T, Riaz M, Arif MS, Ali S, Raza SH (2014) *Proteus mirabilis* alleviates zinc toxicity by preventing oxidative stress in maize (*Zea mays*) plants. Ecotoxicol Environ Saf 110:143–152

Israr M, Sahi S, Datta R, Sarkar D (2006) Bioaccumulation and physiological effects of mercury in *Sesbania drummondii*. Chemosphere 65:591–598

Jadid N, Maziyah R, Nurcahyani DD, Mubarokah NR (2017) Growth and physiological responses of some *Capsicum frutescens* varieties to copper stress. AIP Conf Proc 1854:020018. https://doi.org/10.1063/1.4985409

Kagi JHR (1991) Overview of metallothionein. Methods Enzymol 205:613–626

Kamran MA, Musstjab SA, Eqani AS, Bibi S, Xu R-K, Amna, Monis MFH, Katsoyiannis A, Bokhari H, Chaudhary HJ (2016) Bioaccumulation of nickel by *E. sativa* and role of plant growth promoting rhizobacteria (PGPRs) under nickel stress. Ecotoxicol Environ Saf 26:256–263

Kaur G, Asthir B (2015) Proline: a key player in plant abiotic stress tolerance. Biol Plant 59:609–619

Kaur G, Kumar S, Thakur P, Malik JA, Bhandhari K, Sharma KD, Nayyar H (2011) Involvement of proline in response of chickpea (*Cicer arietinum* L.) to chilling stress at reproductive stage. Sci Hortic 128:174–118

Kaya C, Ashraf M (2015) Exogenous application of nitric oxide promotes growth and oxidative defense system in highly boron stressed tomato plants bearing fruit. Sci Hortic 185:43–47

Kaya C, Tuna AL, Okant AM (2010) Effect of foliar applied kinetin and indole acetic acid on maize plants grown under saline conditions. Turk J Agric For 34:529–538

Khaliq A, Ali S, Hameed A, Farooq MA, Farid M, Shakoor MB, Mahmood K, Ishaque W, Rizwan M (2015) Silicon alleviates nickel toxicity in cotton seedlings through enhancing growth, photosynthesis and suppressing Ni uptake and oxidative stress Silicon alleviates nickel toxicity in cotton. Arch Agron Soil Sci. https://doi.org/10.1080/03650340.2015.1073263

Khan MIR, Nazir F, Asgher M, Per TS, Khan NA (2015) Selenium and sulfur influence ethylene formation and alleviate cadmium-induced oxidative stress by improving proline and glutathione production in wheat. J Plant Physiol 173:9–18

Kochian L, Piñeros MA, Liu J, Magalhaes JV (2015) Plant adaptation to acid soils: the molecular basis for crop aluminum resistance. Annu Rev Plant Biol 66:571–598

Kong Z, Glick BR, Duan J, Ding S, Tian J, McConkey BJ, Wei G (2015) Effects of 1-aminocyclopropane-1-carboxylate (ACC) deaminase-overproducing Sinorhizobium meliloti on plant growth and copper tolerance of *Medicago lupulina*. Plant Soil 391(1–2):383–398

Kováčik J, Babula P, Klejdus B, Hedbavny J (2013) Chromium uptake and consequences for metabolism and oxidative stress in chamomile plants. J Agric Food Chem 61:7864–7873. https://doi.org/10.1021/jf401575a

Kumar A, Prasad MNV, Sytar O (2012a) Lead toxicity, defense strategies and associated indicative biomarkers in *Talinum triangulare* grown hydroponically. Chemosphere 89:1056–1065

Kumar H, Sharma D, Kumar V (2012b) Nickel-induced oxidative stress and role of antioxidant defence in Barley roots and leaves. Int J Environ Biol 2:121–128

Kumar S, Dubey RS, Tripathi RD, Chakrabarty D, Trivedi PK (2015) Omics and biotechnology of arsenic stress and detoxification in plants: current updates and prospective. Environ Int 74:221–230

Kumar B, Smita K, Flores LC (2017) Plant mediated detoxification of mercury and lead. Arab J Chem 10:S2335–S2342

Lamhamdi M, Lafont R, Rharrabe K, Sayah F, Aarab A, Bakrim A (2016) 20-Hydroxyecdysone protects wheat seedlings (*Triticum aestivum* L.) against lead stress. Plant Physiol Biochem 98:64–71

Landi M, Pardossi A, Remorini D, Guidi L (2013) Antioxidant and photosynthetic response of a purple-leaved and a green-leaved cultivar of sweet basil (*Ocimum basilicum*) to boron excess. Environ Exp Bot 85:64–75

Lidon FC, Teixeira MG (2000) Rice tolerance to excess Mn: implications in the chloroplast lamellae and synthesis of a novel Mn protein. Plant Physiol Biochem 38:969–978

Lidon FC, Barreiro M, Ramalho J (2004) Manganese accumulation in rice: implications for photosynthetic functioning. J Plant Physiol 161:1235–1244

Mahmood Q, Ahmad R, Kwak SS, Rashid A, Anjum NA (2010) Ascorbate and glutathione: protectors of plants in oxidative stress. In: Anjum NA, Chan MT, Umar S (eds) Ascorbate glutathione pathway and stress tolerance in plants. Springer, Dordrecht, pp 209–229. https://doi.org/10.1007/978-90-481-9404-9_7

Mahmud JA, Hasanuzzaman M, Nahar K, Rahman A, Fujita M (2017a) Relative tolerance of different species of Brassica to cadmium toxicity: coordinated role of antioxidant defense and glyoxalase systems. Plant Omics 10:107–117. https://doi.org/10.21475/poj.10.02.17.pne409

Mahmud JA, Hasanuzzaman M, Nahar K, Rahman A, Hossain MS, Fujita M (2017b) γ-aminobutyric acid (GABA) confers chromium stress tolerance in *Brassica juncea* L. by modulating the antioxidant defense and glyoxalase systems. Ecotoxicology 26:675–690. https://doi.org/10.1007/s10646-017-1800-9

Mahmud JA, Hasanuzzaman M, Nahar K, Rahman A, Hossain SM, Fujita M (2017c) Maleic acid assisted improvement of metal chelation and antioxidant metabolism confers chromium tolerance in *Brassica juncea* L. Ecotoxicol Environ Saf 144:216–226. https://doi.org/10.1016/j.ecoenv.2017.06.010

Mahmud JA, Hasanuzzaman M, Nahar K, Bhuyan MHMB, Fujita M (2018) Insights into citric acid-induced cadmium tolerance and phytoremediation in *Brassica juncea* L.: coordinated functions of metal chelation, antioxidant defense and glyoxalase systems. Ecotoxicol Environ Saf 147:990–1001

Majerus V, Bertin P, Swenden V, Fortemps A, Lobréaux S, Lutts S (2007) Organ-dependent responses of the African rice to short-term iron toxicity: ferritin regulation and antioxidative responses. Biol Plant 51:303–312

Malar S, Sahi SV, Favas PJC, Venkatachalam P (2015) Mercury heavy-metalinduced physiochemical changes and genotoxic alterations in water hyacinths [*Eichhornia crassipes* (Mart.)]. Environ Sci Pollut Res 22:4597–4608

Mano J (2012) Reactive carbonyl species, their production from lipid peroxides, action in environmental stress, and the detoxification mechanism. Plant Physiol Biochem 59:90–97

Marschner H (1995) Mineral nutrition of higher plants. Academic, London, pp 411–430

Martínez-Ruiz EB, Martínez-Jerónimo F (2016) How do toxic metals affect harmful cyanobacteria? An integrative study with a toxigenic strain of *Microcystis aeruginosa* exposed to nickel stress. Ecotoxicol Environ Saf 133:36–46

Matsumoto H (2000) Cell biology of aluminum toxicity and tolerance in higher plants. Int Rev Cytol 200:1–46

Mei L, Daud MK, Ullah N, Ali S, Khan M, Malik Z, Zhu SJ (2015) Pretreatment with salicylic acid and ascorbic acid significantly mitigate oxidative stress induced by copper in cotton genotypes. Environ Sci Pollut Res 22(13):9922–9931. https://doi.org/10.1007/s11356-015-4075-9

Meriga B, Reddy BK, Rao KR, Reddy LA, Kishor PB (2004) Aluminium-induced production of oxygen radicals, lipid peroxidation and DNA damage in seedlings of rice (*Oryza sativa*). J Plant Physiol 161:63–68

Michael PI, Krishnaswamy M (2011) The effect of zinc stress combined with high irradiance stress on membrane damage and antioxidative response in bean seedling. Environ Exp Bot 74:171–177

Millaleo R, Reyes-Diaz M, Alberdi M, Ivanov AG, Krol M, Huner NP (2013) Excess manganese differentially inhibits photosystem I versus II in *Arabidopsis thaliana*. J Exp Bot 64:343–354

Miller G, Shulaev V, Mitter R (2008) Reactive oxygen signaling and abiotic stress. Physiol Plant 133:481–489. https://doi.org/10.1111/j.1399-3054.2008.01090.x

Min S, Wen-jing X, Xiang-yong P, Fan-hua K (2013) Effects of exogenous proline on the growth of wheat seedlings under cadmium stress. Chin J Appl Ecol 24:129–134

Moenne A, Gonzáleza A, Sáez CA (2016) Mechanisms of metal tolerance in marine macroalgae, with emphasis on copper tolerance in Chlorophyta and Rhodophyta. Aquat Toxicol 176:30–37

Molas J (2001) Comparison of nickel toxicity and resistance strategies of cabbage plants grown in soil with addition of inorganic and organic Ni (II) complexes. Dev Plant Soil Sci 92:464–465

Molassiotis A, Sotiropoulos T, Tanou G, Diamantidis G, Therios I (2006) Boron-induced oxidative damage and antioxidant and nucleolytic responses in shoot tips culture of the apple rootstock EM 9 (*Malus domestica* Borkh). Environ Exp Bot 56:54–62

Morina F, Jovanovic L, Mojovic M, Vidovic M, Pankovic D, Veljovic Jovanovic S (2010) Zinc-induced oxidative stress in *Verbascum thapsus* is caused by an accumulation of reactive oxygen species and quinhydrone in the cell wall. Physiol Plant 140(3):209–224

Mostofa MG, Fujita M (2013) Salicylic acid alleviates copper toxicity in rice (*Oryza sativa* L.) seedlings by up-regulating antioxidative and glyoxalase systems. Ecotoxicology 22(6):959–973. https://doi.org/10.1007/s10646-013-1073-x

Mostofa MG, Seraj ZI, Fujita M (2014) Exogenous sodium nitroprusside and glutathione alleviate copper toxicity by reducing copper uptake and oxidative damage in rice (*Oryza sativa* L.) seedlings. Protoplasma 251(6):1373–1386. https://doi.org/10.1007/s00709-014-0639-7

Mostofa MG, Hossain MA, Fujita M, Tran L-SP (2015) Physiological and biochemical mechanisms associated with trehalose-induced copper-stress tolerance in rice. Sci Rep 5:11433. https://doi.org/10.1038/srep11433

Mostofa MG, Hossain MA, Siddiqui MN, Fujita M, Tran LSP (2017) Phenotypical, physiological and biochemical analyses provide insight into selenium-induced phytotoxicity in rice plants. Chemosphere 178:212–223

Moulick D, Ghosh D, Santra SC (2016) Evaluation of effectiveness of seed priming with selenium in rice during germination under arsenic stress. Plant Physiol Biochem 109:571–578

Mukhopadhyay M, Das A, Subba P, Bantawal P, Sarkar B, Ghosh P, Mondal TK (2013) Structural, physiological, and biochemical profiling of tea plants under zinc stress. Biol Plant 57:474–480. https://doi.org/10.1007/s10535-012-0300-2

Nable RO, Banuelos GS, Paull JG (1997) Boron toxicity. Plant Soil 193:181–198

Nahar K, Hasanuzzaman M, Alam MM, Rahman A, Suzuki T, Fujita M (2016a) Polyamine and nitric oxide crosstalk: antagonistic effects on cadmium toxicity in mung bean plants through upregulating the metal detoxification, antioxidant defense, and methylglyoxal detoxification systems. Ecotoxicol Environ Saf 126:245–255. https://doi.org/10.1016/j.ecoenv.2015.12.026

Nahar K, Rahman M, Hasanuzzaman M, Alam MM, Rahman A, Suzuki T, Fujita M (2016b) Physiological and biochemical mechanisms of spermine-induced cadmium stress tolerance in mung bean (*Vigna radiata* L.) seedlings. Environ Sci Pollut Res 23:21206–21218. https://doi.org/10.1007/s11356-016-7295-8

Nahar K, Hasanuzzaman M, Suzuki T, Fujita M (2017) Polyamines-induced aluminum tolerance in mung bean: a study on antioxidant defense and methylglyoxal detoxification systems. Ecotoxicology 26:58–73

Najeeb U, Ahmad W, Zia MH, Malik Z, Zhou W (2017) Enhancing the lead phytostabilization in wetland plant *Juncus effusus* L. through somaclonal manipulation and EDTA enrichment. Arab J Chem 10:3310–3317

Nanda R, Agrawal V (2016) Elucidation of zinc and copper induced oxidative stress, DNA damage and activation of defence system during seed germination in *Cassia angustifolia* Vahl. Environ Exp Bot 125:31–41. https://doi.org/10.1016/j.envexpbot.2016.02.001

Nasibi F, Heidari T, Asrar Z, Mansoori H (2013) Effect of arginine pre-treatment on nickel accumulation and alleviation of the oxidative stress in *Hyoscyamus niger*. J Soil Sci Plant Nutr 13:680–689

Nath S, Panda P, Mishra S (2014) Arsenic stress in rice: redox consequences and regulation by iron. Plant Physiol Biochem 80:203–210

Noctor G, Gomez L, Vanacker H, Foyer CH (2002) Interactions between biosynthesis, compartmentation and transport in the control of glutathione homeostasis and signalling. J Exp Bot 53:1283–1304. https://doi.org/10.1093/jexbot/53.372.1283

Olaleye AO, Ogunkunle AO, Singh BN, Akinbola GE, Tabi FO, Fayinminu OO, Iji ME (2009) Ratios of nutrients in lowland rice grown on two iron toxic soils in Nigeria. J Plant Nutr 32:1–17

Panda SK, Choudhury S (2005) Chromium stress in plants. Braz J Plant Physiol 17:95–102

Panda SK, Baluška F, Matsumoto H (2009) Aluminum stress signaling in plants. Plant Signal Behav 4:592–597

Panda SK, Choudhury S, Patra HK (2016) Heavy metal induced oxidative stress in plants: physiological and molecular perspectives. In: Tuteja N, Gill SS (eds) Abiotic stress response in plants. Wiley Online Library. https://doi.org/10.1002/9783527694570.ch11

Pandey C, Khan E, Panthri M, Tripathi RD, Gupta M (2016) Impact of silicon on Indian mustard (*Brassica juncea* L.) root traits by regulating growth parameters, cellular antioxidants and stress modulators under arsenic stress. Plant Physiol Biochem 104:216–225. https://doi.org/10.1016/j.plaphy.2016.03.032

Papadakis IE, Dimassi KN, Bosabadilis AM, Therios IN, Pataks A, Giannakoula A (2004) Boron toxicity in 'Clementine' mandarin plants grafted on two rootstocks. Plant Sci 166:539–547

Patel A, Pandey V, Patra DD (2016) Metal absorption properties of *Mentha spicata* grown under tannery sludge amended soil-its effect on antioxidant system and oil quality. Chemosphere 147:67–73

Pontigo S, Godoy K, Jiménez H, Gutiérrez-Moraga A, Mora ML, Cartes P (2017) Silicon-mediated alleviation of aluminum toxicity by modulation of Al/Si uptake and antioxidant performance in ryegrass plants. Front Plant Sci 8:642

Poonam S, Kaur H, Geetika S (2013) Effect of jasmonic acid on photosynthetic pigments and stress markers in *Cajanus cajan* (L.) Millsp. seedlings under copper stress. Am J Plant Sci 4:817

Pramanick P, Chakraborty A, Raychaudhuri SS (2017) Phenotypic and biochemical alterations in relation to *MT2* gene expression in *Plantago ovata* forsk under zinc stress. Biometals 30(2):171–184. https://doi.org/10.1007/s10534-017-9990-4

Rahman A, Mostofa MG, Alam MM, Nahar K, Hasanuzzaman M, Fujita M (2015) Calcium mitigates arsenic toxicity in rice seedlings by reducing arsenic uptake and modulating the antioxidant defense and glyoxalase systems and stress markers. BioMed Res Int 2015:340812. https://doi.org/10.1155/2015/340812

Rahman A, Mostofa MG, Nahar K, Hasanuzzaman M, Fujita M (2016) Exogenous calcium alleviates cadmium-induced oxidative stress in rice (*Oryza sativa* L.) seedlings by regulating the antioxidant defense and glyoxalase systems. Braz J Bot 39:393–407. https://doi.org/10.1007/s40415-015-0240-0

Rajpoot R, Rani A, Srivastava RK, Pandey P, Dubey RS (2015) *Terminalia arjuna* bark extract alleviates nickel toxicity by suppressing its uptake and modulating antioxidative defence in rice seedlings. Protoplasma 253(6):1449–1462. https://doi.org/10.1007/s00709-015-0899-x

Ramakrishna B, Rao SSR (2015) Foliar application of brassinosteroids alleviates adverse effects of zinc toxicity in radish (*Raphanus sativus* L.) plants. Protoplasma 252(2):665–677. https://doi.org/10.1007/s00709-014-0714-0

Ramírez-Duarte WF, Kurobe T, Teh SJ (2017) Impairment of antioxidant mechanisms in Japanese Medaka (*Oryzias latipes*) by acute exposure to aluminum. Comp Biochem Physiol Toxicol Pharmacol 198:37–44

Reddy PS, Jogeswar G, Rasineni GK, Maheswari M, Reddy AR, Varshney RK, Kishor PK (2015) Proline over-accumulation alleviates salt stress and protects photosynthetic and antioxidant enzyme activities in transgenic sorghum [*Sorghum bicolor* (L.) Moench]. Plant Physiol Biochem 94:104–113

Regier N, Larras F, Bravo AG, Ungureanu VG, Amouroux D, Cosio C (2013) Mercury bioaccumulation in the aquatic plant Elodea nuttallii in the field and in microcosm: accumulation in shoots from the water might involve copper transporters. Chemosphere 90:595–602

Rehman MZU, Rizwan M, Ali S, Ok YS, Ishaque W, Saifullah, Nawaz MF, Akmal F, Waqar M (2017) Remediation of heavy metal contaminated soils by using *Solanum nigrum*: a review. Ecotoxicol Environ Saf 143:236–248

Reid R (2010) Can we really increase yields by making crop plants tolerant to boron toxicity? Plant Sci 178:9–11

Ren J-H, Sun H-J, Wang S-F, Luo J, Ma LQ (2014) Interactive effects of mercury and arsenic on their uptake, speciation and toxicity in rice seedling. Chemosphere 117:737–744

Rizwan M, Ali S, Abbas T, Rehman MZ, Hannan F, Keller C, Al-Wabel MI, Ok YS (2016) Cadmium minimization in wheat: a critical review. Ecotoxicol Environ Saf 130:43–53

Rizwan M, Imtiaz M, Dai Z, Mehmood S, Adeel M, Liu J, Tu S (2017) Nickel stressed responses of rice in Ni subcellular distribution, antioxidant production, and osmolyte accumulation. Environ Sci Pollut Res 24(25):20587–20598. https://doi.org/10.1007/s11356-017-9665-2

Rojas-Lillo Y, Alberdi M, Acevedo P, Inostroza-Blancheteau C, Rengel Z, de la Luz Mora M, Reyes-Díaz M (2014) Manganese toxicity and UV-B radiation differentially influence the physiology and biochemistry of highbush blueberry (*Vaccinium corymbosum*) cultivars. Funct Plant Biol 41:156–167

Sáez CA, Roncarati F, Moenne A, Moody AJ, Brown MT (2015) Copper-induced intra-specific oxidative damage and antioxidant responses in strains of the brown alga *Ectocarpus siliculosus* with different pollution histories Claudio. Aquat Toxicol 159:81–89

Sahrawat KL (2003) Iron toxicity in wetland rice: occurrence and management through integration of genetic tolerance with plant nutrition. J Indian Soc Soil Sci 51:409–417

Sahrawat KL (2004) Iron toxicity in wetland rice and the role of other nutrients. J Plant Nutr 27:1471–1504

Santos E, Santini JM, Paixao A, Júnior E, Lavres J, Campos M, Reis A (2017) Physiological highlights of manganese toxicity symptoms in soybean plants: Mn toxicity responses. Plant Physiol Biochem 113:6–19

Sarwar N, Imran M, Shaheen MR, Ishaque W, Kamran MA, Matloob A, Rehim A, Hussain S (2017) Phytoremediation strategies for soils contaminated with heavy metal: modifications and future perspectives. Chemosphere 171:710–721

Shah K, Kumar RG, Verma S, Dubey RS (2001) Effect of cadmium on lipid peroxidation, superoxide anion generation and activities of antioxidant enzymes in growing rice seedlings. Plant Sci 161:1135–1144

Shahid M, Dumat C, Pourrut B, Sabir M, Pinelli E (2014a) Assessing the effect of metal speciation on lead toxicity to *Vicia faba* pigment contents. J Geochem Explor 144:290–297

Shahid M, Pinelli E, Pourrut B, Dumat C (2014b) Effect of organic ligands on lead-induced oxidative damage and enhanced antioxidant defense in the leaves of *Vicia faba* plants. J Geochem Explor 144:282–289

Shahid M, Dumat C, Pourrut B, Abbas G, Shahid N, Pinelli E (2015) Role of metal speciation in lead induced oxidative stress to *Vicia faba* roots. Russ J Plant Physiol 6:448–454

Shanker AK, Cervantes C, Loza-Tavera H, Avudainayagam S (2005) Chromium toxicity in plants. Environ Int 31:739–753. https://doi.org/10.1016/j.envint.2005.02.003

Sharma SS, Dietz KJ (2009) The relationship between metal toxicity and cellular redox imbalance. Trends Plant Sci 14:43–50

Sharma P, Jha AB, Dubey RS, Pessarakli M (2012) Reactive oxygen species, oxidative damage, and antioxidative defense mechanism in plants under stressful conditions. J Bot 2012:217037. https://doi.org/10.1155/2012/217037

Sharma P, Kumar A, Bhardwaj R (2016) Plant steroidal hormone epibrassinolide regulate—heavy metal stress tolerance in *Oryza sativa* L. by modulating antioxidant defense expression. Environ Exp Bot 122:1–9

Sigfridsson KGV, Bernát G, Mamedov F, Styring S (2004) Molecular interference of Cd^{2+} with photosystem II. Biochim Biophys Acta 1659:19–31

Singh K, Pandey SN (2011) Effect of nickel-stresses on uptake, pigments and antioxidative responses of water lettuce, *Pistia stratiotes* L. J Environ Biol 32:391–394

Singh AL, Jat RS, Chaudhari V, Bariya H, Sharma SJ (2010) Toxicities and tolerance of mineral elements boron, cobalt, molybdenum and nickel in crop plants. Plant Stress 4:31–56

Singh VP, Srivastava PK, Prasad SM (2013) Nitric oxide alleviates arsenic-induced toxic effects in ridged Luffa seedlings. Plant Physiol Biochem 71:155–163

Singh M, Singh VP, Dubey G, Prasad SM (2015) Exogenous proline application ameliorates toxic effects of arsenate in *Solanum melongena* L. seedlings. Ecotoxicol Environ Saf 117:164–173

Singh N, Marwa N, Mishra SK, Mishra J, Verma PC, Rathaur S, Singh N (2016) *Brevundimonas diminuta* mediated alleviation of arsenic toxicity and plant growth promotion in *Oryza sativa* L. Ecotoxicol Environ Saf 125:25–34

Sinh S, Saxena R (2006) Effect of iron on lipid peroxidation, and enzymatic and non-enzymatic antioxidants and bacoside-A content in medicinal plant *Bacopa monnieri* L. Chemosphere 62:1340–1350

Soltani E, Radjabian T, Abrishamchi P, Talei D (2016) Physiological and biochemical responses of *Melissa officinalis* L. to nickel stress and the protective role of salicylic acid. Arch Agron Soil Sci. https://doi.org/10.1080/03650340.2016.1207241

Soylemezoglu G, Demir K, Inal A, Gunes A (2009) Effect of silicon on antioxidant and stomatal response of two grapevine (*Vitis vinifera* L.) rootstocks grown in boron toxic, saline and boron toxic-saline soil. Sci Hortic 123:240–246

Srivastava S, Dubey RS (2011) Manganese-excess induces oxidative stress, lowers the pool of antioxidants and elevates activities of key antioxidative enzymes in rice seedlings. Plant Growth Regul 64:1–16

Srivastava S, Tripathi RD, Dwivedi UN (2004) Synthesis of phytochelatins and modulation of antioxidants in response to cadmium stress in *Cuscuta reflexa*—an angiospermic parasite. J Plant Physiol 161:665–674

Stoeva N, Berova M, Zlatev ZL (2005) Effect of arsenic on some physiological parameters in bean plants. Biol Plant 49:293–296

Štolfa I, Pfeiffer TŽ, Špoljarić D, Teklić T, Lončarić Z (2015) Heavy metal-induced oxidative stress in plants: response of the antioxidative system. In: Gupta D, Palma J, Corpas F (eds) Reactive oxygen species and oxidative damage in plants under stress. Springer, Cham. https://doi.org/10.1007/978-3-319-20421-5_6

Sun CL, Lu LL, Liu LJ, Liu WJ, Yu Y, Liu XX, Hu Y, Jin CW, Lin XY (2014) Nitrate reductase-mediated early nitric oxide burst alleviates oxidative damage induced by aluminum through enhancement of antioxidant defenses in roots of wheat (*Triticum aestivum*). New Phytol 201:1240–1250

Sundaramoorthy P, Chidambaram A, Ganesh KS, Unnikannan P, Baskaran L (2010) Chromium stress in paddy: (i) Nutrient status of paddy under chromium stress; (ii) Phytoremediation of chromium by aquatic and terrestrial weeds. C R Biol 333:597–607. https://doi.org/10.1016/j.crvi.2010.03.002

Szollosi R, Varga IS, Erdei L, Mihalik E (2009) Cadmium-induced oxidative stress and antioxidative mechanisms in germinating Indian mustard (*Brassica juncea* L.) seeds. Ecotoxicol Environ Saf 72:1337–1342

Takahashi S, Murata N (2008) How do environmental stresses accelerate photoinhibition? Trends Plant Sci 13:178–182

Tangahu BV, Abdullah SRS, Basri H, Idris M, Anuar N, Mukhlisin M (2011) A review on heavy metals (As, Pb, and Hg) uptake by plants through phytoremediation. Int J Chem Eng 2011:939161

Tewari RK, Kumar P, Sharma PN (2008) Morphology and physiology of zinc-stressed mulberry plants. J Plant Nutr Soil Sci 171:286–294

Theriault G, Michael P, Nkongolo K (2016) Comprehensive transcriptome analysis of response to nickel stress in White Birch (*Betula papyrifera*). PLoS One 11:e0153762. https://doi.org/10.1371/journal.pone.0153762

Thounaojam TC, Panda P, Choudhury S, Patra HK, Panda SK (2013) Zinc ameliorates copper-induced oxidative stress in developing rice (*Oryza sativa* L.) seedlings. Protoplasma 251(1):61–69

Tombuloglu H, Semizoglu N, Sakcali S, Kekec G (2012) Boron induced expression of some stress-related genes in tomato. Chemosphere 86:433–438

Tripathy BC, Oelmüller R (2012) Reactive oxygen species generation and signaling in plants. Plant Signal Behav 7:1621–1633. https://doi.org/10.4161/psb.22455

Van Asshe F, Clijsters H (1990) Effects of metals on enzyme activity in plant. Plant Cell Environ 13:195–206

Vatamaniuk OK, Mari S, Lu Y, Rea PA (2000) Mechanism of heavy metal ion activation of phytochelatin (PC) synthase. J Biol Chem 275:31451–31459. https://doi.org/10.1074/jbc.M002997200

Venkatachalam P, Jayalakshmi N, Geetha N, Sahi SV, Sharma NC, Rene ER, Sarkar SK, Favas PJC (2017) Accumulation efficiency, genotoxicity and antioxidant defense mechanisms in medicinal plant *Acalypha indica* L. under lead stress. Chemosphere 171:544–553

Verbruggen N, Hermans C, Schat H (2009) Mechanisms to cope with arsenic or cadmium excess in plants. Curr Opin Plant Biol 12:364–372

Verma S, Dubey RS (2003) Lead toxicity induces lipid peroxidation and alters the activities of antioxidant enzymes in growing rice plants. Plant Sci 164:645–655

Wang YX, Wu P, Wu YR, Yan XL (2002) Molecular marker analysis of manganese toxicity tolerance in rice under greenhouse conditions. Plant Soil 238:227–233

Wang C, Zhang SH, Wang PF, Hou J, Zhang WJ, Li W (2009) The effect of excess Zn on mineral nutrition and antioxidative response in rapeseed seedlings. Chemosphere 75:1468–1476

Wani AS, Tahir I, Ahmad SS, Dar RA, Nisar S (2017) Efficacy of 24-epibrassinolide in improving the nitrogen metabolism and antioxidant system in chickpea cultivars under cadmium and/or NaCl stress. Sci Hortic 225:48–55. https://doi.org/10.1016/j.scienta.2017.06.063

Weng XY, Zhao LL, Zheng CJ, Zhu JW (2013) Characteristics of the hyperaccumulator plant *Phytolacca acinosa* (Phytolaccaceae) in response to excess manganese. J Plant Nutr 36:1355–1365

Wu ZL, Banuelos GS, Lin ZQ, Liu Y, Yuan LX, Yin XB, Li M (2015a) Biofortification and phytoremediation of selenium in China. Front Plant Sci 6:136. https://doi.org/10.3389/fpls.2015.00136
Wu X, He J, Ding H, Zhu Z, Chen J, Xu S, Zha D (2015b) Modulation of zinc-induced oxidative damage in *Solanum melongena* by 6-benzylaminopurine involves ascorbate–glutathione cycle metabolism. Environ Exp Bot 116:1–11
Wu Z, Liu S, Zhao J, Wang F, Du Y, Zou S, Li H, Wen D, Huang Y (2017) Comparative responses to silicon and selenium in relation to antioxidant enzyme system and the glutathione-ascorbate cycle in flowering Chinese cabbage (*Brassica campestris* L. ssp. *chinensis* var. *utilis*) under cadmium stress. Environ Exp Bot 133:1–11
Yadav SK (2010) Heavy metals toxicity in plants: an overview on the role of glutathione and phytochelatins in heavy metal stress tolerance of plants. S Afr J Bot 76:167–179
Yadav SK, Singla-Pareek SL, Reddy MK, Sopory SK (2005) Methylglyoxal detoxifcation by glyoxalase system: a survival strategy during environmental stresses. Physiol Mol Biol Plants 11:1–11
Yadav SK, Singla-Pareek SL, Sopory SK (2008) An overview on the role of methylglyoxal and glyoxalases in plants. Drug Metabol Drug Interact 23(1–2):51–68
Yan R, Gao S, Yang W, Cao M, Wang S, Chen F (2008) Nickel toxicity induced antioxidant enzyme and phenylalanine ammonia-lyase activities in *Jatropha curcas* L. cotyledons. Plant Soil Environ 54(7):294–300
Yang XE, Jin XF, Feng Y, Islam E (2005) Molecular mechanisms and genetic basis of heavy metal tolerance/hyperaccumulation in plants. J Integr Plant Biol 47:1025–1035
Yin L, Mano J, Wang S, Tsuji W, Tanaka K (2010) The involvement of lipid peroxide-derived aldehydes in aluminum toxicity of tobacco roots. Plant Physiol 152:1406–1417
Yin L, Mano JI, Tanak K, Wang S, Zhang M, Deng X, Zhang S (2017) High level of reduced glutathione contributes to detoxification of lipid peroxide derived reactive carbonyl species in transgenic Arabidopsis overexpressing glutathione reductase under aluminum stress. Physiol Plant 161:211–223
Yusuf M, Fariduddin Q, Ahmad A (2011) 28-Homobrassinolide mitigates boron induced toxicity through enhanced antioxidant system in *Vigna radiata* plants. Chemosphere 85:1574–1584
Yusuf M, Khan TA, Fariduddin Q (2016) Interaction of epibrassinolide and selenium ameliorates the excess copper in *Brassica juncea* through altered proline metabolism and antioxidants. Ecotoxicol Environ Saf 129:25–34. https://doi.org/10.1016/j.ecoenv.2016.03.001
Zagorchev L, Seal CE, Kranner I, Odjakova M (2013) A central role for thiols in plant tolerance to abiotic stress. Int J Mol Sci 14:7405–7432. https://doi.org/10.3390/ijms14047405
Zhang J, Gao X (2015) Heavy metals in surface sediments of the intertidal Laizhou Bay, Bohai Sea, China: distributions, sources and contamination assessment. Mar Pollut Bull 98:320–327
Zhang H, Hu LY, Li P, Hu KD, Jiang CX (2010) Hydrogen sulfide alleviated chromium toxicity in wheat. Biol Plant 54:743–747
Zhang Y, Xu S, Yang S, Chen Y (2015) Salicylic acid alleviates cadmium-induced inhibition of growth and photosynthesis through upregulating antioxidant defense system in two melon cultivars (*Cucumis melo* L.). Protoplasma 252:911–924. https://doi.org/10.1007/s00709-014-0732-y
Zhang T, Lu Q, Su C, Yang Y, Hu D, Xu Q (2017) Mercury induced oxidative stress, DNA damage, and activation of antioxidative system and Hsp70 induction in duckweed (*Lemna minor*). Ecotoxicol Environ Saf 143:46–56
Zhao FJ, McGrath SP, Meharg AA (2010) Arsenic as a food chain contaminant: mechanisms of plant uptake and metabolism and mitigation strategies. Annu Rev Plant Biol 61:535–559
Zhao X, Chen Q, Wang Y, Shen Z, Shen W, Xu X (2017) Hydrogen-rich water induces aluminum tolerance in maize seedlings by enhancing antioxidant capacities and nutrient homeostasis. Ecotoxicol Environ Saf 144:369–379
Zheng G, Lv HP, Gao S, Wang SR (2010) Effects of cadmium on growth and antioxidant responses in *Glycyrrhiza uralensis* seedlings. Plant Soil Environ 56:508–515
Zhou G, Xu Y, Li J, Yang L, Liu J-Y (2006) Molecular analyses of the metallothionein gene family in rice (*Oryza sativa* L.). J Biochem Mol Biol 39:595–606

Zhou ZS, Huang SQ, Guo K, Mehta SK, Zhang PC, Yang ZM (2007) Metabolic adaptations to mercury-induced oxidative stress in roots of *Medicago sativa* L. J Inorg Biochem 101:1–9

Zhou B, Yao W, Wang S, Wang X, Jiang T (2014) The metallothionein gene, TaMT3, from *Tamarix androssowii* confers Cd^{2+} tolerance in Tobacco. Int J Mol Sci 15:10398–10409

Zouari M, Ahmed CB, Elloumi N, Bellassoued K, Delmail D, Labrousse P, Rouina BB (2016) Impact of proline application on cadmium accumulation, mineral nutrition and enzymatic antioxidant defense system of *Olea europaea* L. cv Chemlali exposed to cadmium stress. Ecotoxicol Environ Saf 128:195–205. https://doi.org/10.1016/j.ecoenv.2016.02.024

Reactive Oxygen Species Signaling in Plants

Muhammad Nafees, Shah Fahad, Adnan Noor Shah, Muhammad Adnan Bukhari, Maryam, Ishtiaq Ahmed, Shakeel Ahmad, and Sajjad Hussain

Introduction

The presence of reactive oxygen species (ROS) in plants is an inevitable consequence of their aerobic metabolism (Halliwell and Gutteridge 1989; Halliwell 2006). They are produced during aerobic metabolism, which is limited to cellular organelles with a powerful electron flow. The list of such organelles includes the

M. Nafees (✉) · I. Ahmed
Department of Horticultural Sciences, University College of Agriculture and Environmental Sciences, The Islamia University of Bahawalpur, Bahawalpur, Punjab, Pakistan
e-mail: muhammad.nafees@iub.edu.pk

S. Fahad (✉)
Department of Agriculture, The University of Swabi, Swabi, Khyber Pakhtunkhwa, Pakistan
e-mail: shahfahad@uoswabi.eedu.pk

A. N. Shah
Department of Agronomy, Faculty of Agriculture, Gomal University, Dera Ismail Khan, Khyber Pakhtunkhwa, Pakistan

M. A. Bukhari
Department of Agronomy, University College of Agriculture and Environmental Sciences, The Islamia University of Bahawalpur, Bahawalpur, Punjab, Pakistan

Maryam
Department of Botany, Government Sadiq College Women University, Bahawalpur, Punjab, Pakistan

S. Ahmad
Department of Agronomy, Faculty of Agricultural Sciences and Technology, Bahauddin Zakariya University, Multan, Punjab, Pakistan

S. Hussain (✉)
Department of Horticulture, Faculty of Agricultural Sciences and Technology, Bahauddin Zakariya University, Multan, Punjab, Pakistan
e-mail: sajjad.hussain@bzu.edu.pk

M. Hasanuzzaman et al. (eds.), *Plant Abiotic Stress Tolerance*,
https://doi.org/10.1007/978-3-030-06118-0_11

mitochondria, peroxisomes, and chloroplasts. ROS comprise singlet oxygen ($^1O_2$), superoxide radical (O_2^-), hydroxyl radical ($HO^•$), hydrogen peroxide (H_2O_2), etc. In plants, various physiological responses have been studied besides cellular structural changes and degradation of enzymes, nucleic acids, proteins, etc. It has been assumed that under abiotic stress, ROS production could be a crucial indication of phytotoxicity (Choudhury et al. 2013).

ROS are oxygen-containing molecules exhibiting higher chemical reactivity than O_2. Their ability to cause cellular damage has been mitigated to some extent through evolutionary pressure for plants to expand their range of enzymatic and nonenzymatic ROS scavengers. Rapid changes in the compartmental redox balance and ROS homeostasis are among the initial symptoms that follow fluctuations in environmental conditions (Waszczak et al. 2018). Plants detect these parameter changes and exploit them as signals in multiple processes that serve to regulate metabolism or physiology at the tissue level, at the whole-plant level, or in specific subcellular compartments.

ROS are constantly formed as derivatives of various metabolic pathways, which are confined to a small area in cellular compartments (Foyer and Harbinson 1994). These molecules are scavenged by antioxidative defense components located in particular sections of the cell, under physiological steady-state conditions (Alscher et al. 1997). Recent research has revealed that ROS work as major signals because they show reactions to a wide range of biomolecules, result in irreversible damage, and may be directed toward death and necrosis (Rebeiz et al. 1988; Girotti 2001).

On the other hand, it has also been observed that ROS can control various biological processes and can manipulate signal transduction pathways and gene expression, thereby signifying cellular strategies for ROS utilization (Dalton et al. 1999). ROS can operate as signaling molecules, since they are very small and can disperse over small distances. Among the different ROS members, only H_2O_2 has the ability to diffuse through plant membranes and therefore manipulate function directly in cell-to-cell signaling. The composition and accessibility of the antioxidant system are determining factors for concentration and longevity of ROS. $HO^•$ is the most reactive form of ROS, with an estimated lifetime measured in nanoseconds, while that of $^1O_2$ is measured in microseconds.

The most stable ROS—H_2O_2 and O_2^-—have considerably longer lifetimes (measured in milliseconds to seconds), which depend upon the presence and activity of dedicated ROS scavengers (Mhamdi et al. 2010). The concept of oxidative stress has changed during the past few decades. The belief that ROS are destructive substances that indiscriminately oxidize various molecules and structures has been superseded by the concept of ROS signaling (Foyer and Noctor 2005). According to our current understanding, effective antioxidative systems in symplastic compartments keep ROS concentrations low even in the presence of increased ROS production rates (Foyer and Noctor 2016; Noctor et al. 2016). In this chapter, we discuss recent progress in the understanding of the functions of ROS. Their roles in cellular responses, cellular retrograde signaling, stomatal closures, and defense against pathogens in the presence of biotic and abiotic stress are the focuses of this chapter.

Main Cellular Basis of the Effects of Reactive Oxygen Species

For effectiveness, a signaling molecule needs to be formed rapidly and effectively on demand for induction of distinct effects within the cell, and when it is no longer required, it should be removed promptly (Neill et al. 2003). The onset of stress results in rapid formation of ROS. These are so reactive that they even show reactions with membrane carbohydrates, proteins, lipids, and DNA. H_2O_2 is able to pass through biological membranes, using aquaporins (Bienert et al. 2007; Dynowski et al. 2008; Mubarakshina et al. 2010; Borisova et al. 2012), and is directed at systemic responses.

There is genetic confirmation that ROS can also operate as signaling molecules for regulation of diverse plant functions (Foyer and Noctor 2005). The primary ROS (i.e., O_2^- and H_2O_2) in plants can also be considered as a main ROS because in plants it can a vital functions as a secondary envoys by monitoring miscellaneous functions of growth and development (Foreman et al. 2003). ROS generated in cellular organelles (e.g., chloroplasts and mitochondria) can cause changes in the nucleus, but the complete mechanism of signaling is still not fully understood (Apel and Hirt 2004). Considerable research has achieved acceptance of ROS signaling mechanisms in plants, and now it is beyond doubt that ROS are indeed major signaling molecules in various processes in plants (Pitzschke et al. 2006). For example, production of H_2O_2 is activated for the duration of biotic and abiotic stresses. During abiotic stress the ROS production triggered by cytosolic membrane–bound reduced nicotinamide adenine dinucleotide phosphate (NADPH) oxidase is considered a signal (Laloi et al. 2004).

As expression of several genes is influenced by ROS, this suggests that ROS work as biological signals in stress regulation (Laloi et al. 2004; Neill et al. 2002a). Laloi et al. (2004) confirmed that ROS interaction with targeted molecules is very selective, as the ROS concentration is amplified and alters gene expression. Additionally, it has been explained that oxidation of signaling pathways induces changes at the gene level that result in the creation and activation of transcription factors that are probably redox sensitive (Neill et al. 2002a). Research has revealed that in *Arabidopsis*, H_2O_2 can activate two stress mitogen-activated protein kinases (MAPKs)—AtMPK3 and AtMPK6—and strongly stimulate nucleotide diphosphate kinase 2 (AtNDPK2) (Kovtun et al. 2000). Because they have controlled antioxidant systems, plants are tolerant of H_2O_2, as it helps in elimination and upholding the steady redox state (Foyer and Noctor 2000, 2003). Key roles of glutathione and ascorbate in plant redox nodding have been described in numerous reports, and the roles of these species in plant signaling have been confirmed in previous reports (Horling et al. 2003; Foyer and Noctor 2000, 2003). In the redox state the chloroplasts of plants have been found to respond to redox-signaling gene expression. Chloroplast protein expression is affected by changes in the chloroplast redox state. In chloroplasts, glutathione, ascorbate, plastoquinone (PQ), and ROS work as key signaling factors, along with ferredoxin and thioredoxin (Pfannschmidt et al. 1999).

Photosystem II (PSII) is associated with the manufacture of $^1O_2$, while production of $O_2^{\bullet-}$ is associated equally with photosystem I (PSI) and PSII (Asada 2006; Møller and Sweetlove 2010). Flu mutant studies of *Arabidopsis* have revealed that $^1O_2$ signaling is related to planned cell death (PCD) acquire precise characteristic for gene induction, in contrast to other ROS (Pitzschke et al. 2006; Triantaphylidès and Havaux 2009). ROS are excessively produced in peroxisomes through various biochemical reactions, as these are major sites. In C_3 plants, during photosynthesis, peroxisomes produce prominent quantities of H_2O_2, and antioxidant activity is greatly elevated in them (Foyer and Noctor 2000). The enzymes that are involved include ascorbate peroxidase (APX), catalase (CAT), and others that are associated with the ascorbate–glutathione system (Jimenez et al. 1997; Foyer and Noctor 2003).

These enzymes are requisites for H_2O_2scavenging. During photorespiration, CAT activity decreases, which leads to oxidized glutathione accumulation (Foyer and Noctor 2003). It has also been shown that glutathione and ascorbate accumulation can stable CAT scarcity, and the glycollate oxidase effect is responsible for transferring the indication from chloroplasts to peroxisomes (Robson and Vanlerberghe 2002). Such measures have been studied under drought conditions and high temperature stress in plants. Direct correlations have been reported in cellular redox homeostasis and mitochondrial redox conditions, but the scavenging activity of ROS in the mitochondria is lesser than that in peroxisomes and chloroplasts, so the stability of the self-redox state depends on the total cellular redox status (Foyer and Noctor 2003).

ROS production in chloroplasts and peroxisomes far exceeds that in the mitochondria. Despite this, the mitochondria are rich in oxidized protein. Oxidized protein exists in high concentrations because of its vulnerability to ROS, and these oxidized proteins are important components of mitochondrial electron transport complexes I and III (Møller and Sweetlove 2010). Production of ROS in the mitochondria is primarily due to the existence of alternative oxidase (AOX) (Robson and Vanlerberghe 2002; Vanlerberghe et al. 2002). However, PCD in plants is controlled by higher mitochondrial ROS generation (Tiwari et al. 2002). Numerous phytocompounds—including abscisic acid (ABA), jasmonic acid (JA), salicylic acid (SA), and phytohormones—standardize the defensive reaction in plants under abiotic stress. However, ROS with these phytocompounds have not been yet implicated. ABA is broadly associated with abiotic stress and is involved in directing growth and development. Phytocompounds such as SA, ethylene, and JA (but not ABA) have noteworthy functions in biotic stress, and sometime ABA behaves as a negative regulator of disease resistance (Fujita et al. 2006).

Cellular Retrograde Signaling in Plants

The nucleus of the cell controls the expression of genes in organelles. Retrograde signaling is organ signaling regulated by gene expression in the nucleus. Retrograde signaling synchronizes gene communication for development and metabolism between the organelles and the nucleus, and afterward it adjusts retrograde progression (Suzuki et al. 2012). ROS are produced in the mitochondria and chloroplasts in adverse stress conditions, and the metabolism and redox index in organelles are significant sources of backward signals that play a prospective function in stress acclimatization in plants (Woodson and Chory 2008).

It has been claimed that enough time has been spend on studying chloroplast retrograde signaling. There are multiple pathways for chloroplast–nucleus retrograde signaling. Mg–protoporphyrin IX (Mg-PPIX) has been studied predominantly (Nott et al. 2006) but is still not properly understood. Under developed chloroplast in mutants of *Arabidopsis* and carotenoid biosynthesis inhibitors norflurazon showed that the nucleus receives signals from chloroplasts that change nuclear gene expression (Suzuki et al. 2012; Nott et al. 2006), but this is generally dependent on the existence of AB14 in the nucleus and GUN1 in the chloroplasts.

Moreover, gene expression of *Arabidopsis* is altered with accumulation of a methyl ester of Mg-PPIX (Nott et al. 2006). Studies of Mg-PPIX in *Arabidopsis* have revealed that stress responses are related to approximately 35% of identified proteins (Suzuki et al. 2012). These include peroxidases (ATP15, PER22, and ATP3) and glutathione *S*-transferases (AtGST10 and AtGSTF3), which play a considerable role in deprivation of the methyl ester of Mg-PPIX (Kindgren et al. 2011). The protein phosphorylation process is involved in chloroplast–nucleus signaling mediated by ROS (Nott et al. 2006). Microarrays have been used to study the association of $^1O_2$ in chloroplast backward signaling in mutants of *Arabidopsis*, exposing various sets of alleles that are stimulated by $^1O_2$ (Suzuki et al. 2012; Gadjev et al. 2006).

Mitochondrial retrograde signaling is still not apparently implicited compared to chloroplast. In mitochondrial ROS signaling, alternative oxidase 1 (AOX1) expression is used. The nucleus encodes the AOX1, which plays a major role as a key marker for mitochondrial reversing signaling (Suzuki et al. 2012). In mitochondrial signaling, the involvement of protein has still not been evidenced. In *Arabidopsis* mitochondria, the mutants scarce in backward signaling not capable to persuade luciferase activity motivated by AOX1 in reply as treatment of antimycin-A (Zarkovic et al. 2005; Rhoads and Subbaiah 2007). So, evidently, it is clear that though a significant advance has been achieved in comprehension of ROS backward signaling in plants, the exact regulation percentage is still not fully understood (Asada 2006).

Effects of Reactive Oxygen Species Signaling in Plants

Reactive oxygen species are an indispensable part of aerobic life (Mittler et al. 2004). ROS can react with a large number of biomolecules, causing irreversible damage and leading to necrosis and death (Rebeiz et al. 1988; Girotti 2001). However, ROS can also manipulate gene expression and signaling pathways, suggesting that cells have evolved a strategy to use ROS for signaling to control various biological processes (Dalton et al. 1999).

ROS such as H_2O_2, $OH^•$, and $O_2^{•-}$ either are produced by redox (oxidation–reduction) reactions or are active derivatives of O_2. Among these ROS, only H_2O_2

Table 1 Various sites of reactive oxygen species (*ROS*) production

ROS	Main sites of production	References
H_2O_2	Chloroplast, peroxisome, plasma membrane, apoplast, mitochondria, endoplasmic reticulum	Halliwell and Gutteridge (1989), Cona et al. (2006), Rasmusson et al. (2008), Khan et al. (2012)
$OH^{\bullet}$	Chloroplast, mitochondria	Elstner (1991), Turrens (2003), Murphy (2009)
$O_2^{\bullet -}$	Chloroplast (electron transport chain), mitochondria (respiratory chain), plasma membrane	Elstner (1991), Turrens (2003), Apel and Hirt (2004)
$^1O_2$	Chloroplast, plasma membrane, mitochondria	Halliwell and Gutteridge (1989), Foyer and Harbinson (1994)

can cross the plant membrane, so they can directly act on the cell signal. These ROS are permanently produced in different organelles, are highly reactive and toxic, and can be oxidized to damage proteins, nucleic acids, and lipids (Suzuki et al. 2012; Singh et al. 2016) (Table 1). Rather than focusing on their toxic nature, recent research has concentrated on their signaling roles in several key physiological processes in plants (Baxter et al. 2014). In the development process, the involvement of ROS as a signaling molecule suggests that plants may have developed the toxicity of highly tolerant ROS in their evolution (Bhattacharjee 2014; Mattila et al. 2015). ROS are amazing, multipurpose molecular species that are constantly produced in plants as an inevitable result of oxygen-metabolizing redox cascades.

Effects of Reactive Oxygen Species on Stomata

Stomata are well known for arbitration of photosynthetic CO_2 exchange and efficient utilization of water produced by transpiration-driven sap ascent (Song et al. 2014). The signaling mechanisms and networks involved in stomatal activities are of great interest. Each network has unique receptors and early signal elements, but they also have a common component, such as the plasma membrane anion channel and the potassium channel, with solute flux in the actual stomatal movement promoting the inflow of water.

Plant hormone ABA—which is synthesized in buds, roots, and especially in seeds, veins, and guard cells—plays an important role in various physiological processes, such as stomatal function development and regulation, to deal with abiotic stress (Boursiac et al. 2013). Shinozaki and Yamaguchi-Shinozaki (2007) have reported that in the case of high salinity and water stress, ABA begins to accumulate in plant cells, and its accumulation directs the change of gene expression and stomatal closure, then reduces transpiration and water loss.

Gas exchange decreases as a result of stomatal closure, thereby leading to a reduction of photosynthetic activity (Song et al. 2014; Boursiac et al. 2013). Under stress conditions, the concentration of ABA increases with the release of a conjugate form or enhancement of biosynthesis, and the degradation rate decreases. These steps occur in the affected cells or in adjacent cells, resulting in the absorption of ABA by pressureless cells. Therefore, ABA can be directly involved in stomatal closure. For example, to contend with increased ROS, the activity of superoxide dismutase (SOD) may increase, along with those of APX and CAT, and proteins such as dehydrins may be produced to mitigate the effects of cell dehydration (Neill et al. 2008).

Effects of Reactive Oxygen Species on Pathogens

When a cell is exposed to a pathogen, this results in production of ROS at an early stage. Exposure to different types of pathogens leads to production of O_2^- or its disproportionation product, H_2O_2, in the apoplast (Doke 1983; Grant et al. 2000a, b). Abdollahi and Ghahremani observed the function of chloroplasts in the synergy between *Erwinia amylovora* and a recipient by using uracil as a chloroplast electron transport chain (ETC) inhibitor. As a result of the presence of uracil, ROS production was reduced at the time of the pathogen–host interaction, and ROS production is related to the release of necrosis in all species (Abdollahi and Ghahremani 2011). Research by Liu et al. showed that activation of the salicylic acid–induced protein kinase/neutrotrophin 4/wound-induced protein kinase (SIPK/Ntf4/WIPK) cascade by disease-causing agent activity enhances the production of ROS in plastids, and mainly in chloroplasts, and this plays a key role in the signaling for and/or execution of hypersensitive response (HR) cell death in plants. They finally deduced that chloroplast rupture occurs before an NADPH oxidase burst and ROS generated by the mitochondria might play a crucial role in the acceleration of the cell death process (Liu et al. 2007). According to pharmacological studies, various portions of the overall ROS formation due to infection appear to be affected by a lot processes. The active role of NADPH oxidase has been shown to be predominant in most cases (Torres and Dangl 2005). ROS formation due to the host–pathogen interaction is affected by NADPH oxidases and cell wall peroxides (Grant et al. 2000a, b).

Redox Oxygen Species Under Biotic Stress

Various biotic and abiotic stresses occur in the plant life cycle in a natural environment. Plants have evolved complex mechanisms to identify various stresses and mount their best responses to these conditions; thus, plants have evolved and developed mechanisms to maintain the balance between energy production and utilization (Suzuki et al. 2012). It is not easy to attribute a change in the physiology of the metabolism of a crop to a specific stress, because various complex stress factors affect the plants simultaneously, with interconnections existing between different and opposite signaling response pathways for security against insects and pathogens, depending on the specific stress conditions (Kusnieczyk et al. 2007).

Biotic stresses include damage due to living organisms (bacteria, viruses, fungi, harmful insects, and weeds) (Lata et al. 2018), causing diseases in plants. Wilting in plants, root rot, leaf spots, and seed spoilage are caused by microorganisms. However, plants produce various natural products of a toxic nature to protect themselves against microbial pathogens, and these products are also important for communication between plants and other organisms (Schafer and Wink 2009). Del-Rio et al. (2002) proposed a protoplastic source centered upon chloroplastic and mitochondrial ROS generation systems. Some biotic stresses lead to synthesis of superoxide on plasma membranes from NADPH oxidase (Lamb and Dixon 1997), similar to synthesis of reduced nicotinamide adenine dinucleotide (NADH)–dependent superoxide (Auh and Murphy 1995), which is dismutated to H_2O by superoxide dismutase; moreover, hydrogen peroxide originates from a cell wall in a three-component system requiring ion fluxes and leading to extracellular alkalinization (Bolwell et al. 1999). Some insects act as vectors for viruses and bacteria, and most of them cause severe physiological damage in plants. Disturbance of plants through allelopathy results in severe damage to native plants. Microorganisms living in healthy plant tissues without causing any apparent damage to the host plants are known as endophytes (Bacon and White 2000), and some fungi and bacteria are known to live their whole lives inside the plant body (Wilson 1995). The term "endophyte" was first used by De Bary in 1866. In 1846, Leveille recognized fungi in wheat leaves and named them endophytic fungi (Riesen and Close 1987). Backman and Sikora (2008) divided endophytes into the following three types based on the nature of their pathogenicity:

1. Pathogens of another host that are not pathogenic in an endophytic relationship
2. Nonpathogenic microbes
3. Nonpathogenic pathogens that are capable of colonization through selection methods

The presence of redox oxygen species is inevitable in organisms leading an aerobic existence; however, plants have adopted successful defenses against the lethal effects of ROS during the course of evolution. ROS are chemical compounds produced in all living organisms in response to any type of stress (Halliwell and Gutteridge 1989; and Halliwell 2006). Plants produce an extensive series of physiological responses with changes in cellular assembly and degradation of nucleic acids, proteins, enzymes, antioxidants, etc. (Halliwell and Gutteridge 1989). Biosynthesis of ROS is considered phytotoxicity, and the mechanism of ROS production has been widely studied in plant species under various abiotic stresses (Choudhury and Panda 2004). High production of ROS is a distinctive property of living plants, exerting severe impacts on plant metabolism that cause negative impacts on plant growth and development. In higher plants, productivity under abiotic stresses (drought, salinity, heat stress, heavy metals, etc.) results in high production of redox oxygen species, which are prominent in a wide range of biological and chemical fluctuations, such as degradation of antioxidants, lipid peroxidation, and gene mutations (Choudhury et al. 2013). Production of H_2O_2 is activated by cytosolic membrane–bound NADPH oxidase for plant signaling under biotic and abiotic stresses (Laloi et al. 2004), with changes in gene expression, signifying that ROS acts as a natural signal in regulating various environmental stresses (Neill et al. 2002b; Laloi et al. 2004).

Plant growth and development are adversely affected by high activity of ROS in cells during abiotic stress. During high stress, ROS are produced in the mitochondria and these ROS play a role in direct signaling that play a possible role in acclimatization to stress in plants (Suzuki et al. 2012; Woodson and Chory 2008). Mg–protoporphyrin IX is the best-studied chloroplast–nucleus retrograde signaling pathway in higher plants (Nott et al. 2006); however, the chloroplast reverse-signaling mechanism is still under study. Chloroplasts send signals to the nucleus that alter nuclear gene expression, as shown in studies of carotenoid biosynthesis inhibitors of *Arabidopsis* (Nott et al. 2006; Suzuki et al. 2012).

In plant organelles, metabolic processes are responsive to alterations in environmental conditions, and metabolic variations can lead to oxidative stress in cells by enhancing the production and accumulation of ROS, which retards metabolic activities and causes oxidation of components of the cell (Suzuki et al. 2012). Changes in the energy balance and carbon metabolism in response to stress have been revealed in both mitochondria and chloroplasts (Takahashi and Murata 2008), and to avoid uncontrolled production of ROS and oxidative destruction, significant metabolic coordination is needed to maintain the flow of energy through these organelles in all possible growing conditions. During stress situations, ROS metabolism in the mitochondria and chloroplasts and the organelle redox state are initiators for retrograde (organelle-to-nucleus) signals, which play a key role in the adaptation of plants (Woodson and Chory 2008). With elevated temperatures or high light intensities, active transport or leakage and diffusion of ROS from the mitochondria or chloroplasts can cause destruction of normal metabolism and ultimately cell death (Suzuki et al. 2012).

At the early stage of plant exposure to salinity, nutrient deficiency occurs and regular patterns of growth and development are lost (Munns and Tester 2008). Drought inhibits photosynthesis, reduces membrane integrity, and increases ROS generation in the body (Greenberg et al. 2008). Xu et al. (2016) showed that prolonged water stress reduces leaf size, root growth, leaf water potential, and stomatal opening; delays flowering and fruiting; and finally limits the plant's growth and productivity.

Heavy metals hamper photosynthesis, respiration, nutrient uptake, and nitrogen and protein metabolism in all types of plants (Zhang et al. 2009). High accumulation of Cd in the growth medium results in oxidative stress through excessive production of ROS, which damage the plant's antioxidant system (Rahman et al. 2016). Because of their function as cell signaling molecules and their innate reactivity, ROS are of significant importance, with one key target of ROS signals being amino acids (Dautreaux and Toledano 2007). Accumulation of ROS is naturally associated with generation of the plant signaling molecule nitric oxide (Neill et al. 2002b).

Antioxidant enzymes such as APX, CAT, and others determine the lifetime of ROS (Munné-Bosch et al. 2013). ROS signal mediation due to hormones, pathogens, and atmospheric pollutants can be controlled by use of ascorbate (Pignocchi et al. 2003).

Conclusion

Aerobic life makes the existence of reactive oxygen species (ROS) inevitable. During evolution, plants have developed the ability to detect harmful effects of ROS and use ROS in different biological processes. In the presence of abiotic stress, the concentration of ROS in cells is high, which limits the growth and development of plants. Plants deploy complex antioxidant defense mechanisms that limit the production of ROS and remove them from the cellular environment. In this chapter the important role of reactive oxygen–mediated gene expression in regulating the development and survival of plants has been described. Although much progress has been made in our understanding of the role of ROS in plants, it is far from clear that reactive oxygen species play a key role in stress regulation and metabolism.

References

Abdollahi H, Ghahremani Z (2011) The role of chloroplasts in the interaction between *Erwinia amylovora* and host plants. Acta Hortic 896:215–221

Alscher RG, Donahue JH, Cramer CL (1997) Reactive oxygen species and antioxidants: relationships in green cells. Physiol Plant 100:224–233

Apel K, Hirt H (2004) Reactive oxygen species: metabolism, oxidative stress, and signal transduction. Annu Rev Plant Biol 55:373–399

Asada K (2006) Production and scavenging of reactive oxygen species in chloroplasts and their functions. Plant Physiol 141:391–406

Auh CK, Murphy TM (1995) Plasma membrane redox enzyme is involved in the synthesis of O_2 and H_2O_2 by *Phytophthora* elicitor-simulated rose cells. Plant Physiol 107:1241–1247

Backman PA, Sikora RA (2008) Endophytes: an emerging tool for biological control. Biol Control 46:1–3

Bacon CW, White J (2000) Microbial endophytes. Marcel Dekker, CRC, Boca Raton

Baxter A, Mittler R, Suzuki N (2014) ROS as key players in plant stress signalling. J Exp Bot 65:1229–1240

Bhattacharjee S (2014) Membrane lipid peroxidation and its conflict of interest: the two faces of oxidative stress. Curr Sci 107:1811–1823

Bienert GP, Moller AL, Kristiansen KA, Schulz A, Moller IM, Schjoerring JK et al (2007) Specific aquaporins facilitate the diffusion of hydrogen peroxide across membranes. J Biol Chem 282:1183–1192

Bolwell GP, Blee KA, Butt VS, Davies DR, Gardner SL, Gerrish C, Minibayeva F, Rowntree EG, Wojtaszek P (1999) Recent advances in understanding the origin of the apoplastic oxidative burst in plant cells. Free Radic Res 31:137–145

Borisova MM, Kozuleva MA, Rudenko NN, Naydov IA, Klenina IB, Ivanov BN (2012) Photosynthetic electron flow to oxygen and diffusion of hydrogen peroxide through the chloroplast envelope via aquaporins. Biochim Biophys Acta 1817:1314–1321

Boursiac Y, Léran S, Corratgé-Faillie C, Gojon A, Krouk G, Lacombe B (2013) ABA transport and transporters. Trends Plant Sci 18:325–333
Choudhury S, Panda SK (2004) Toxic effects, oxidative stress and ultrastructural changes in moss Taxithelium nepalense (schwaegr) Broth. under chromium and lead phytotoxicity.Water Air Pollution (submitted)
Choudhury S, Panda P, Sahoo L, Panda SK (2013) Reactive oxygen species signaling in plants under abiotic stress. Plant Signal Behav 8:23681–23686
Cona A, Rea G, Angelini R, Federico R, Tavladoraki P (2006) Functions of amine oxidases in plant development and defence. Trends Plant Sci 11:80–88
Dalton TP, Shertzer HG, Puga A (1999) Regulation of gene expression by reactive oxygen. Annu Rev Pharmacol Toxicol 39:67–101
Dautreaux B, Toledano MB (2007) ROS as signalling molecules: mechanisms that generate specificity in ROS homeostasis. Nat Rev Mol Cell Biol 8:813–824
Del-Rio LA, Corpas FJ, Sandalio LM, Palma JM, Gomez M, Barroso JB (2002) Reactive oxygen species, antioxidant systems and nitric oxide in peroxisome. J Exp Bot 53:1255–1272
Doke N (1983) Generation of superoxide anion by potato tuber protoplasts during the hypersensitive response to hyphal wall components of *Phytophthora infestans* and specific inhibition of the reaction by suppressors of hypersensitivity. Physiol Plant Pathol 23:359–367
Dynowski M, Schaaf G, Loque D, Moran O, Ludewig U (2008) Plant plasma membrane water channels conduct the signaling molecule H_2O_2. Biochem J 414:53–61
Elstner EF (1991) Mechanisms of oxygen activation in different compartments of plant cells, In: Pell E, Steffen KL (eds) Active oxygen/oxidative stress and plant metabolism. American Society of Plant Physiologists, Rockville, pp 13–25
Foreman J, Demidchik V, Bothwell JH, Mylona P, Miedema H, Torres MA, Linstead P, Costa S, Brownlee C, Jones JD, et al. (2003) Reactive oxygen species produced by NADPH oxidase regulate plant cell growth. Nature 422:442–446
Foyer CH, Harbinson JC (1994) Oxygen metabolism and the regulation of photosynthetic electron transport. In: Foyer CH, Mullineaux PM (eds) Causes of photooxidative stress and amelioration of defense systems in plants. CRC, Boca Raton, pp 1–42
Foyer CH, Noctor G (2000) Oxygen processing in photosynthesis: regulation and signalling. New Phytol 146:234–240
Foyer CH, Noctor G (2003) Redox sensing and signalling associated with reactive oxygen in chloroplast, peroxisomes and mitochondria. Physiol Plant 119:355–364
Foyer CH, Noctor G (2005) Oxidant and antioxidant signaling in plants: a re-evaluation of the concept of oxidative stress in a physiological context. Plant Cell Environ 28:1056–1071
Foyer CH, Noctor G (2016) Stress-related redox signaling: what's in pROSpect? Plant Cell Environ 39:951–964
Fujita M, Fujita Y, Noutoshi Y, Takahashi F, Narusaka Y, Yamaguchi-Shinozaki K et al (2006) Crosstalk between abiotic and biotic stress responses: a current view from the points of convergence in the stress signaling networks. Curr Opin Plant Biol 9:436–442
Gadjev I, Vanderauwera S, Gechev TS, Laloi C, Minkov IN, Shulaev V et al (2006) Transcriptomic footprints disclose specificity of reactive oxygen species signaling in *Arabidopsis*. Plant Physiol 141:436–445
Girotti AW (2001) Photosensitized oxidation of membrane lipids: reaction pathways, cytotoxic effects and cytoprotective mechanisms. J Photochem Photobiol B 63:103–113
Grant M, Brown I, Adams S, Knight M, Ainslie A, Mansfield J (2000a) The RPM1 plant disease resistance gene facilitates a rapid and sustained increase in cytosolic calcium that is necessary for the oxidative burst and hypersensitive cell death. Plant J 23:441–450
Grant JJ, Yun BW, Loake GJ (2000b) Oxidative burst and cognate redox signalling reported by luciferase imaging: identification of a signal network that functions independently of ethylene, SA and Me-JA but is dependent on MAPKK activity. Plant J 24:569–582
Greenberg BM, Huang XD, Gerwing P, Yu XM, Chang P, Wu SS, Gerhardt K, Nykamp J (2008) Phytoremediation of salt impacted soils: greenhouse and the field trials of plant growth promoting rhizobacteria (PGPR) to improve plant growth and salt phytoaccumulation. In: Proceedings

of the 33rd AMOP Technical Seminar on Environmental Contamination and Response, Halifax, 8–10 June 2010, pp 627–637
Halliwell B (2006) Reactive species and antioxidants. Redox biology is a fundamental theme of aerobic life. Plant Physiol 141:312–322
Halliwell B, Gutteridge JMC (1989) Free radicals in biology and medicine. Clarendon, Oxford
Horling F, Lamkemeyer P, König J, Finkemeier I, Kandlbinder A, Baier M et al (2003) Divergent light-, ascorbate-, and oxidative stress-dependent regulation of expression of the peroxiredoxin gene family in *Arabidopsis*. Plant Physiol 131:317–325
Jimenez A, Hernandez JA, Del Rio LA, Sevilla F (1997) Evidence for the presence of ascorbate glutathione cycle in mitochondria and peroxisomes of pea leaves. Plant Physiol 114:275–284
Khan BR, Adham AR, Zolman BK (2012) Peroxisomal Acyl-CoA oxidase 4 activity differs between *Arabidopsis* accessions. Plant Mol Biol 78:45–58
Kindgren P, Eriksson MJ, Benedict C, Mohapatra A, Gough SP, Hansson M et al (2011) A novel proteomic approach reveals a role for Mg--protoporphyrin IX in response to oxidative stress. Physiol Plant 141:310–320
Kovtun Y, Chiu WL, Tena G, Sheen J (2000) Functional analysis of oxidative stress--activated mitogen-activated protein kinase cascade in plants. Proc Natl Acad Sci U S A 97:2940–2945
Kusnieczyk A, Winge P, Midelfart H, Armbruster WS, Rossiter JT, Bones AM (2007) Transcriptional responses of *Arabidopsis thaliana* ecotypes with different glucosinolate profiles after attack by polyphagous *Myzus persicae* and oligophagous *Brevicoryne brassicae*. J Exp Bot 58:2537–2552
Laloi C, Apel K, Danon A (2004) Reactive oxygen signalling: the latest news. Curr Opin Plant Biol 7:323–328
Lamb C, Dixon RA (1997) The oxidative burst in plant disease resistance. Annu Rev Plant Physiol Plant Mol Biol 48:251–275
Lata R, Chowdhury S, Gond SK, White JF (2018) Induction of abiotic stress tolerance in plants by endophytic microbes. Lett Appl Microbiol 66:268–276
Liu Y, Ren D, Pike S, Pallardy S, Gassmann W, Zhang S (2007) Chloroplast-generated reactive oxygen species are involved in hypersensitive response-like cell death mediated by a mitogen-activated protein kinase cascade. Plant J 51:941–954
Mattila H, Khorobrykh S, Havurinne V, Tyystjärvi E (2015) Reactive oxygen species: reactions and detection from photosynthetic tissues. J Photochem Photobiol B 152(Pt B):176–214
Mhamdi A, Queval G, Chaouch S,Vanderauwera S,Van Breusegem F, Noctor G (2010) Catalase function in plants: a focus on *Arabidopsis* mutants as stress-mimic models. J Exp Bot 61:4197–4220. https://doi.org/10.1093/jxb/erq282
Mittler R, Vanderauwera S, Martin G, Frank VB (2004) Reactive oxygen gene network of plants. TRENDS in Plant Science. https://doi.org/10.1016/j.tplants.2004.08.009
Møller IM, Sweetlove LJ (2010) ROS signalling—specificity is required. Trends Plant Sci 15:370–340
Mubarakshina MM, Ivanov BN, Naydov IA, Hillier W, Badger MR, Krieger-Liszkay A (2010) Production and diffusion of chloroplastic H_2O_2 and its implication to signaling. J Exp Bot 61:3577–3587
Munné-Bosch S, Queval G, Foyer CH (2013) The impact of global change factors on redox signaling underpinning stress tolerance. Plant Physiol 161:5–19
Munns R, Tester M (2008) Mechanisms of salinity tolerance. Annu Rev Plant Biol 59:651–681
Murphy MP (2009) How mitochondria produce reactive oxygen species. Biochem J 417:1–13
Neill S, Desikan R, Hancock J (2002a) Hydrogen peroxide signalling. Curr Opin Plant Biol 5: 388–395
Neill SJ, Desikan R, Clarke A, Hurst RD, Hancock JT (2002b) Hydrogen peroxide and nitric oxide as signalling molecules in plants. J Exp Bot 53:1237–1247
Neill SJ, Desikan R, Hancock JT (2003) Nitric oxide signaling in plants. New Phytol 159:11–35
Neill S, Barros R, Bright J, Desikan R, Hancock J, Harrison J, Morris P, Ribeiro D, Wilson I (2008) Nitric oxide, stomatal closure, and abiotic stress. J Exp Bot 59:165–176

Noctor G, Mhamdi A, Foyer CH (2016) Oxidative stress and antioxidative systems: recipes for successful data collection and interpretation. Plant, Cell Environ 39(5):1140–1160. ISSN 0140-7791

Nott A, Jung HS, Koussevitzky S, Chory J (2006) Plastid-to-nucleus retrograde signaling. Annu Rev Plant Biol 57:739–759

Pfannschmidt T, Nilsson A, Allen JF (1999) Photosynthetic control of chloroplast gene expression. Nature 406:731–734

Pignocchi C, Fletcher JM, Wilkinson JE, Barnes JD, Foyer CH (2003) The function of ascorbate oxidase in tobacco. Plant Physiol 132:1631–1641

Pitzschke A, Forzani C, Hirt H (2006) Reactive oxygen species signaling in plants. Antioxid Redox Signal 8:1757–1764

Rahman A, Mostofa MG, Nahar K, Hasanuzzaman M, Fujita M (2016) Exogenous calcium alleviates cadmium-induced oxidative stress in rice (*Oryza sativa* L.) seedlings by regulating the antioxidant defense and glyoxalase systems. Braz J Bot 39:393–407

Rasmusson AG, Geisler DA, Møller IM (2008) The multiplicity of dehydrogenases in the electron transport chain of plant mitochondria. Mitochondrion 8:47–60

Rebeiz CA, Montazer-Zouhoor A, Mayasich JM, Tripathy BC, Wu SM, Rebeiz C (1988) Photodynamic herbicides. Recent developments and molecular basis of selectivity. CRC Crit Rev Plant Sci 6:385–436

Rhoads DM, Subbaiah CC (2007) Mitochondrial retrograde regulation in plants. Mitochondrion 7:177–194

Riesen TK, Close RC (1987) Endophytic fungi in propiconazole-treated and untreated barley leaves. Mycologia 79:546–552

Robson CA, Vanlerberghe GC (2002) Transgenic plant cells lacking mitochondrial alternative oxidase have increased susceptibility to mitochondria-dependent and -independent pathways of programmed cell death. Plant Physiol 129:1908–1920

Schafer H, Wink M (2009) Medicinally important secondary metabolites in recombinant microorganisms or plants: progress in alkaloid biosynthesis. Biotechnol J 4:1684–1703

Shinozaki K, Yamaguchi-Shinozaki K (2007) Gene networks involved in drought stress response and tolerance. J Exp Bot 58:221–227

Singh R, Singh S, Parihar P, Mishra RK, Tripathi DK, Singh VP, Chauhan DK, Prasad SM (2016) Reactive oxygen species (ROS): beneficial companions of plants' developmental processes. Front Plant Sci 7:1299. https://doi.org/10.3389/fpls.2016.01299

Song Y, Miao Y, Song CP (2014) Behind the scenes: the roles of reactive oxygen species response of plants to abiotic stress. Plant Cell Environ 35:259–270

Suzuki N, Koussevitzky S, Mittler R, Miller G (2012) ROS and redox signalling in the response of plants to abiotic stress. Plant Cell Environ 35:259–270

Takahashi S, Murata N (2008) How do environmental stresses accelerate photoinhibition? Trends Plant Sci 13:178–182

Tiwari BS, Belenghi B, Levine A (2002) Oxidative stress increased respiration and generation of reactive oxygen species, resulting in ATP depletion, opening of mitochondrial permeability transition, and programmed cell death. Plant Physiol 128:127–181

Torres MA, Dangl JL (2005) Functions of the respiratory burst oxidase in biotic interactions, abiotic stress and development. Curr Opin Plant Biol 8:397–403

Triantaphylidès C, Havaux M (2009) Singlet oxygen in plants: production, detoxification and signaling. Trends Plant Sci 14:219–228

Turrens JF (2003) Mitochondrial formation of reactive oxygen species. J Physiol 552:335–344

Vanlerberghe GC, Robson CA, Yip JY (2002) Induction of mitochondrial alternative oxidase in response to a cell signal pathway down-regulating the cytochrome pathway prevents programmed cell death. Plant Physiol 129:1829–1842

Waszczak C, Melanie C, Jaakko K (2018) Reactive oxygen species in plant signaling. Annu Rev Plant Biol. https://doi.org/10.1146/annurev-arplant-042817-040322

Wilson D (1995) Endophyte: the evolution of a term, and clarification of its use and definition. Oikos 73:274–276

Woodson JD, Chory J (2008) Coordination of gene expression between organellar and nuclear genomes. Nat Rev Genet 9:383–395

Xu Z, Jiang Y, Jia B, Zhou G (2016) Elevated-CO_2 response of stomata and its dependence on environmental factors. Front Plant Sci 7:657. https://doi.org/10.3389/fpls.2016.00657

Zarkovic J, Anderson SL, Rhoads DM (2005) A reporter gene system used to study developmental expression of alternative oxidase and isolate mitochondrial retrograde regulation mutants in *Arabidopsis*. Plant Mol Biol 57:871–888

Zhang F, Zhang H, Wang G, Xu L, Shen Z (2009) Cadmium-induced accumulation of hydrogen peroxide in the leaf apoplast of *Phaseolus aureus* and *Vicia sativa* and the roles of different antioxidant enzymes. J Hazard Mater 168:76–84

Role of Selective Exogenous Elicitors in Plant Responses to Abiotic Stress Tolerance

Bilal Ahmad, Abbu Zaid, Yawar Sadiq, Shafia Bashir, and Shabir H. Wani

Introduction

In view of the unsettling increase in the human population coupled with climate change and global warming, considerable enhancement in agricultural productivity is the need of the hour. However, a variety of environmental constraints like biotic and abiotic pressures act as confronting factors to the crop productivity (Wani and Sah 2014; Rejeb et al. 2014). Abiotic stress is currently being realized as one of the most ubiquitous and potential threats to human existence resulting in the overwhelming consequences to our health as well as agricultural systems (Pereira 2016; Roberts and Mattoo 2018). Decreased crop yield and soil fertility together with the accumulation of the harmful elements in the food chains are questioning human survival, an annoying gift of abiotic stress to humanity, leading to severe health as well as ecological consequences (McLaughlin et al. 1999; Verstraeten et al. 2008). In order to meet the feeding requirements of the world population, 70% increase in the agricultural productivity has to be achieved to feed a supplementary 2.3 billion

Bilal Ahmad and Abbu Zaid contributed equally to this work.

B. Ahmad · A. Zaid (✉) · Y. Sadiq
Plant Physiology and Biochemistry Division, Department of Botany,
Aligarh Muslim University, Aligarh, India
e-mail: azaid@myamu.ac.in

S. Bashir
Department of Botany, Punjab Agricultural University, Ludhiana, Punjab, India

S. H. Wani
Mountain Research Centre for Field Crops, Sher-e-Kashmir University of Agricultural Sciences and Technology of Kashmir, Anantnag, Jammu and Kashmir, India

M. Hasanuzzaman et al. (eds.), *Plant Abiotic Stress Tolerance*,
https://doi.org/10.1007/978-3-030-06118-0_12

population by 2050 (Tilman et al. 2011). Thus, in order to meet the desired goal, stress tolerance in plants is of chief importance. However, stress tolerance mechanisms as well as responses in plants are far more complex as compared to animals (Qin et al. 2011). This complexity of the stress tolerance traits in plants has forced us to look for alternative mechanistic approaches as plant-breeding methods have proven unsuccessful in this regard. Thus, potent and novel approaches are the demand of the hour and plant biologists are leaving no stone unturned to meet the desired aim. Among the various approaches, engineering of phytohormones has proven a very effective method of choice for the production of climate-resilient crops with prominent yields (Wani et al. 2016, 2018).

In order to respond to the wide range of internal and external stimuli, plants need to regulate different aspects of growth and development. A diverse and discrete assemblage of signaling compounds, known as phytohormones, often found in minute quantities in the cells of the plants, come to their rescue in mediating such responses playing fundamental roles in promoting plant acclimatization to constantly changing environments. Such acclimatization in plants brought about by the phytohormones-mediated regulation of different aspects of growth and development, nutrient allocation, and source/sink transitions has been well established (Fahad et al. 2015; Ciura and Kruk 2018). Although responses of plants to abiotic stresses are determined by a range of factors, phytohormones have been observed to be the most imperative endogenous compounds responsible for modulating diverse physiological and molecular responses, a decisive prerequisite for survival of plants as sessile life forms (Fahad et al. 2015; Bücker-Neto et al. 2017; Pál et al. 2018a). In the recent past, exogenous application of plant growth regulators (PGRs) has proven quite successful in alleviating different types of abiotic stresses in plants. These signaling compounds either act at the site of biosynthesis or are transported elsewhere to act there (Peleg and Blumwald 2011). Exogenously applied elicitors are of imperative significance in augmenting plant stress tolerance mechanisms in plants and include PGRs or their derived products like (abscisic acid, ethylene, 5-aminolevulinic acid, salicylic acid, brassinosteroids, jasmonates, nitric oxide, etc.), degraded and non-degraded polysaccharides (carrageenan, chitosan, sodium alginate) (Hayat et al. 2012; Ahammed et al. 2015; Wani et al. 2016; Ahmad et al. 2017a, b, 2018; Akram and Ashraf 2013; Singh et al. 2017a, b), strigolactones (Banerjee and Roychoudhury 2018; Mostofa et al. 2018), hydrogen sulfide (Zhu et al. 2018; Banerjee et al. 2018), polyamines (Paul and Roychoudhury 2017; Mustafavi et al. 2016), melatonin and serotonin (Zhang et al. 2018; Mukherjee 2018; Ke et al. 2018), non-enzymatic antioxidant like ascorbic acid (Alamri et al. 2018), glucose (Sami and Hayat 2018), vitamin derivative (Rasheed et al. 2018). In this chapter, we will discuss the role of selective exogenous phytohormones abscisic acid (ABA), salicylic acid (SA), methyl jasmonate (MeJA), hydrogen peroxide (H_2O_2), and nitric oxide (NO) in abiotic stress tolerance.

Role of Exogenous Elicitors in Conferring Abiotic Stress Tolerance in Plants

Abscisic Acid

Abscisic acid (ABA), an isoprenoid phytohormone synthesized via 2-*C*-methyl-D-erythritol-4-phosphate (MEP) pathway in plastids, is one of the exceptionally studied signaling molecules playing an exceptionally distinct role in plant abiotic stress tolerance, hence also called as stress hormone in plants (Dar et al. 2017; Fernando and Schroeder 2016). In addition to its distinct role in different physiological processes, its function as a fundamental messenger in the adaptive response of plants to diverse abiotic stresses in stress tolerance mechanism has attracted substantial attention (Sreenivasulu et al. 2010; Peleg and Blumwald 2011). Endogenous levels of ABA increase swiftly in response to environmental perturbations and modify gene expression by the stimulation of specific signal transduction pathways (Sreenivasulu et al. 2010; O'Brien and Benková 2013). ABA has been recognized to act as an endogenous signal enabling plants to overcome the detrimental environmental conditions and has also been reported to govern transcriptional regulation of up to 10% of the imperative protein-encoding genes involved in diverse physiological processes (Nemhauser et al. 2006; Keskin et al. 2010). Acting as a signal under water scarcity, ABA enables the plants to transduce signal to the shoot eventually activating antitranspirant activity and helping the plant to save water particularly by reduced leaf expansion and closure of stomata (Wilkinson et al. 2012). Besides other architectural modifications, ABA is efficiently implicated in vigorous root growth under drought stress and nitrogen-deficient conditions (Giuliani et al. 2005; Zhang et al. 2007a, b). Expression of various stress-responsive genes has been attributed to ABA in addition to those involved in the synthesis of dehydrins and LEA proteins (Verslues et al. 2006; Sreenivasulu et al. 2012). Synthesis of antioxidant enzymes and osmoprotectants which confer drought tolerance and the up-regulation of processes involved in the maintenance of cell turgor have been endorsed to ABA (Chaves et al. 2003). ABA concentration has been reported to increase proportionately in the plants exposed to salt stress indicating its possible role in overcoming salt stress (Zhang et al. 2006). In an experiment involving *Landoltia punctata* plants, Liu et al. (2018) tested the efficacy of ABA application on carbohydrate metabolism enzyme activity, endogenous phytohormone content, photosynthetic pigments, starch content, and fresh and dry weights and found that ABA supplementation regulated the activity of key starch metabolism enzymes, and hormone content to increase biomass and starch accumulation.

Salicylic Acid

Salicylic acid (SA) (ortho-hydroxy benzoic acid), an imperative endogenous plant growth regulator of phenolic nature ubiquitously distributed in the plant kingdom, regulates various aspects of growth and development in plants, chiefly those implicated in response to biotic and abiotic stress (Miura and Tada 2014; Asgher et al. 2015; Hasanuzzaman et al. 2017). As a naturally occurring endogenous signaling compound, it helps the plant to establish and modulate its defense in response to a wide variety of stresses including the metal/metalloid stress in different crop plants (Hayat et al. 2010; Patel and Hemantaranjan 2012; Mostofa and Fujita 2013; Singh et al. 2015; Ahmad et al. 2018). Furthermore, it plays a vital role in regulating various physiological processes in plants, like uptake of nutrients, transpiration, stomatal closure, chlorophyll biosynthesis, photosynthesis and synthesis of protein, and antioxidant regulation (Shakirova and Sakhabutdinova 2003; Khan et al. 2015; Fardus et al. 2018). In addition, increased rate of transpiration, quantum yield, water use efficiency, internal CO_2 concentration, sugar translocation, stomatal conductance and net photosynthetic rate, mineral nutrient concentrations, and prevention of degradation of CO_2 fixing enzymes (RuBisCO) have been endorsed to the exogenous SA application (Fariduddin et al. 2003; Khan et al. 2003; Gondor et al. 2016; Garg and Bharti 2018). Moreover, exogenous sourcing of SA impedes ethylene biosynthesis, enriches photosynthetic pigments, triggers the induction of flavonoid biosynthesis pathways, arouses photosynthetic machinery, and ameliorates the detrimental effects produced by heavy metal stress in plants (Zhao et al. 1995; Zhang and Chen 2011; Gondor et al. 2016). In a more recent experiment, Soliman et al. (2018) reported the preventive role of acetylsalicylic acid in *Phaseolus vulgaris* plants to chilling stress conditions in which acetylsalicylic acid improved photosynthesis, antioxidants, and growth and induced cold stress resistance by up-regulating the expression of cold-responsive genes *CBF3* and *COR47*, respectively.

Jasmonic Acid

In plant kingdom, jasmonic acid (JA) and its methyl ester, methyl jasmonate (MeJA), are cosmopolitan in distribution (Per et al. 2018; Zaid and Mohammad 2018). JA as well as its methyl ester methyl jasmonate (MeJA) are known to influence diverse processes in plants thereby acting as natural plant growth regulators (Ueda and Saniewski 2006; Norastehnia et al. 2007; Ahmad et al. 2016; Mir et al. 2018). The initial isolation of JA was reported from the culture filtrate of the fungus *Lasiodiplodia theobromae* (Aldridge et al. 1971). Reports regarding the role of JAs in protection of plants against abiotic stresses are contradictory as MeJA has been generally observed to inhibit photosynthetic activity, stomatal opening, and plant growth (Anjum et al. 2011; Yan et al. 2015). However, reports have also revealed that MeJA at stumpy concentrations augments growth and enhances plant abiotic stress tolerance, thereby acting as stress modulators (Walia et al. 2007; Keramat et al. 2009). JAs have been well documented for their vital roles in plant responses to wide range of abiotic stresses including drought (Brossa et al. 2011), salt (Dong et al. 2013; Qiu et al. 2014; Zhao et al. 2014), heavy metals (Maksymiec et al. 2005; Ahmad et al. 2017a, b), and heat stress (Clarke et al. 2009). Application of MeJA modulates photosynthesis and antioxidant system considerably by regulating the expression at the gene level (Maserti et al. 2011).

Nitric Oxide

Nitric oxide (NO), a major signaling molecule involved in various signal transduction cascades, is beneficial to plants exposed to different abiotic stresses and acts diversely as a free radical, released from numerous noxious by-products of oxidative metabolism and also aids in plant sustenance. NO is exceedingly diffusible ubiquitous bioactive endogenous signaling molecule which gained substantial engrossment in last decade or two due to its widespread participation in mediating different plant physiological processes. A research outbreak during the last two decades in plant NO biology has shown the association of this key signaling molecule with germination, growth, pollen growth, photosynthesis, leaf senescence and reorientation, floral regulation, hypocotyl growth, pathogen defense, and root organogenesis (Mur et al. 2013; Beligni and Lamattina 2000; Šírová et al. 2011; He et al. 2004; Jasid et al. 2009; Neill et al. 2008; Pagnussat et al. 2003; Correa-Aragunde et al. 2004). NO also finds association with apoptosis and biosynthesis of phytoalexins in plants (Misra et al. 2011). The beneficial effects in plants have been documented at lower concentrations while higher doses repress growth. Exogenously application of sodium nitropruside (SNP) in *Citrus aurantium* has been reported to prevent carbonylation of prevents thereby helping to overcome the damaging effects of salinity (Beligni and Lamattina 2000). During heat stress, NO acts as an antioxidant, limiting the excessive production of ROS (Suzuki and Mittler 2006). Exogenous SNP application and other NO donors protect plant leaves by reducing oxidative damage through increased activity of antioxidant enzymes like CAT, SOD, and POX under heat stress (Song et al. 2006). Exogenous SNP application promisingly ameliorates the decline in chlorophyll content and net photosynthetic rate under osmotic stress (Tan et al. 2008).

Hydrogen Peroxide

The growth and development of plant species has apprehensively regulated by minute concentrations of hydrogen peroxide (H_2O_2) which is regarded as a versatile second messenger in phytohormone signaling biomolecule for myriad of processes under stressful environments (Wojtyla et al. 2016; Maruta et al. 2012; Khan et al. 2018). H_2O_2 controls various functions in plants owing to its diverse properties like small size, high diffusibility, and as one of the main components of reactive oxygen species (Leshem et al. 1998; Si et al. 2018). In cells, the source of H_2O_2 is NADPH oxidase (respiratory burst oxidase homolog, RBOH), whose production mediates initial response to various environmental stresses (Gilroy et al. 2014; Dietz et al. 2016; Neill et al. 2002a, b; Gaupels et al. 2016). Various studies have established the role of external application of H_2O_2 in controlling plant metabolism under stress conditions by stimulating the activation and expression of genes related to stress tolerance for physiological adjustment and their capacity to endure continued subjection to environmental change (Saxena et al. 2016; Wang et al. 2018a, b; Iqbal et al. 2018). Several complex signaling pathways and defense responses triggered by H_2O_2 have been studied in detail nevertheless; there is great scope of future research which could clarify the underlying mechanisms and signaling pathways. H_2O_2 affects growth, development, photosynthesis, antioxidant, and osmoprotectant systems. In a field experiment, Farouk and Qados (2018) assess the role of different (0, 2, and 4%) doses of H_2O_2 on growth, anatomical characters, yield, and some biochemical aspects and of pea plant. Results indicate that H_2O_2 caused a significant increment in growth, photosynthetic attributes, pigment contents, yield as well as the quality of pea plants. In quinoa plants, Iqbal et al. (2018) applied two modes of H_2O_2, 80 mM through seed priming and 15 mM as foliar supplementation under drought stress conditions. 80 mM H_2O_2 improved emergence attributes of plants while 15 mM H_2O_2 caused a significant increment in stomatal conductance, transpiration rate, net photosynthetic rate, chlorophyll content, proline, and total soluble sugar contents. Both H_2O_2 application modes reduced endogenous ABA content and also enhanced the activities of enzymatic antioxidant batteries, including superoxide dismutase, catalase, peroxidase, and ascorbate peroxidase. In a recent study, Bagheri et al. (2019) working on pistachio seedlings observed that H_2O_2 improved salt tolerance through coordination with antioxidant systems.

Signaling Cross Talk Between Phytohormones in Abiotic Stress Tolerance

Phytohormones play an important role in regulating diverse metabolic and physiological processes in plants by their synergistic and antagonistic interactions among themselves or with other signaling agents like transcription factors, secondary messengers, and in order to maintain a balance between defense mechanisms and plant growth and development, their signaling pathways are either interconnected or biosynthesis or responses are modulated by each (Kazan 2015; Hu et al. 2017; Asgher et al. 2018). Nevertheless, unraveling molecular mechanisms on these aspects to mechanistically understand how distinct signaling cascades interplay with each other to fine tune growth, development, and physiology of plants under biotic and abiotic environmental cues. Moreover, the interplay between various phytohormones can program new machinery at genetic level which could affect defense mechanisms, ameliorate the impacts of environmental pressures thereby enhance plant abiotic stress tolerance.

ABA has diverse functions during the growth and development of plants together with its role during various abiotic stresses like drought and salinity stresses (Zhang et al. 2006). On the other hand, NO has its hand in varied plant metabolic and physiological processes and studies reveal that numerous interactions (synergistic and antagonistic) between NO and ABA (Freschi 2013; Simontacchi et al. 2013). Studies show an interdependence of NO and ABA; NO generation is essential for ABA-induced stomatal closure and ABA regulates biosynthesis of NO inside the guard cells (Bright et al. 2006; Neill et al. 2002a, b). The aforesaid fact can be corroborated with studies regarding *Vicia faba* (Garcia-Mata and Lamattina 2002) and *Pisum sativum* (Garcia-Mata and Lamattina 2003). Garcia-Mata et al. (2003) reported that NO maintain Ca^{2+} ion concentration in guard cell (*Vicia faba*) by regulating K^+ channels resulting in stomatal closure. Defective ABA-induced stomatal closure in *Arabidopsis* NR-deficient mutant (nia1::Ds) that have no NR1-associated NO generation, as well as by the cPTIO, an NO scavenger supports that NO is essential for ABA-regulated signaling pathway. Since *Arabidopsis* mutant, *rcn1* (insensitive to ABA and methyl jasmonate) also did not accumulate NO, thus conclusively NO regulates downstream signaling of ABA as well as methyl jasmonate (Garcia-Mata and Lamattina 2003; Saito et al. 2009; Hancock et al. 2011). Considering the earlier reports NO synthase is a key enzyme involved in NO generation. Out of the two isoforms of NR (*NR1*; *NR2*), NR1 is involved in ABA-mediated signaling and nitrite-NO oxido-reductase (Ni-NOR) helps in nitrate assimilation but recent reports suggest that NOA1 is the enzyme involved in regulation of NO production. Reports support that NOA (mitochondria) in *Triticum aestivum* is up-regulated by ABA as well as by NaCl treatment, thus giving a further support to ABA-associated NO generation (Hao et al. 2010; Moreau et al. 2010; Stohr and Stremlau 2006). A similar interface has been observed in *Arabidopsis thaliana* between salicylic acid and *NOA1*-mediated NO generation regulated by ABA (Sun et al. 2010).

Under stressful conditions such as water deficit and UV-B radiation, both NO and ABA extensively cross talk, leading to different plant adaptive responses viz., stomatal closure and antioxidant defenses (Neill et al. 2008; Tossi et al. 2009). A significant enhancement in the concentration of ABA, NO, and H_2O_2 was observed in the leaves of *Zea mays* after UV-B irradiation. A mutant, *viviparous 14 (vp14)*, defective in ABA synthesis shows higher UV-B damage, shows ameliorative response after the application of ABA, and reinstates NO and ABA accumulation (Tossi et al. 2009; Wilkinson and Davies 2010). The effect is brought about by the ABA-induced H_2O_2 generation that escalates the NO accumulation in the leaves due to the activity of mitogen-activated protein kinase and up-regulation of antioxidant enzymes activity (Zhang et al. 2007a, b). Nevertheless, H_2O_2 and NO levels in the leaves of *Nicotiana tabacum* are enhanced by the overexpression of *SgNCED1* (9-*cis* epoxycarotenoid dioxygenase) gene by increased ABA levels that leads to the up-regulation of the expression and activity of various antioxidant enzymes (Zhang et al. 2009). Under water and nutrient-limited conditions, bromeliads ensure their survival from crassulacean acid metabolism which is under the control of NO and ABA interplay in these plants (Freschi et al. 2010; Mioto and Mercier 2013). Tominaga et al. (2010) has reported that *enf1*(enhanced nitrogen fixation 1) mutant of *Lotus japonicus* have higher rates of nitrogen fixation and number of root nodules with lower ABA sensitivity and reduced NO production in nodules. However, in *Lactuca sativa*, NO invigorates seed germination with a gradual increment in ABA catabolism and NO accumulation (Hancock et al. 2011). S-nitrosylation promotes degradation of a transcription factor *ABI5* that ameliorate seed germination in *Arabidopsis* (Albertos et al. 2015) although the ABA receptors inactivation via tyrosine nitration adjusts cellular responsiveness to ABA after the formation of NO and ROS in the cell (Castillo et al. 2015). Some of the representative examples of interaction between various exogenous elicitors are listed in Table 1.

Conclusion Remarks and Future Perspectives

In view of the collected literature above, it is clear that plant stress signaling is governed by the complex interactions and cross talks between various phytohormones. Phytohormones undergo synergistic and antagonistic cross talks during stress signaling in a tissue-specific manner. Abiotic stress causes oxidative stress in plants by the overproduction of reactive oxygen species (ROS) (Fig. 1). Phytohormones retained the oxidative state either directly quenching the ROS or stimulating the antioxidant batteries indirectly. Phytohormones also modify the expression of genes responsible for undesired, negative effects on plants' metabolism. A schematic overview of phytohormones-mediated abiotic stress tolerance is given in Fig. 1. Thus, there is an urgent need to exploit the knowledge of stress-induced promoters in abiotic stress tolerance for genetic modifications. In addition to traditional lab-based experiments, the engineering effect of phytohormones should be tested under open field conditions in a changing and natural environment.

Table 1 Representative studies of interaction between various elicitors and in diverse plant species under optimal and stress environments

Elicitor 1	Elicitor 2	Condition	Plant species	Parameter studied	Reference
Nitric oxide	Methyl jasmonate	Optimal	*Zea mays*	Increased activities of ascorbate peroxidase (APX), glutathione reductase (GR), monodehydroascorbate reductase (MDHAR), L-galactono-1,4-lactone dehydrogenase (GalLDH), dehydroascorbate, (DHAR), Gamma-glutamylcysteine synthetase (γ-ECS) enzymes. Results indicate that jasmonic acid-induced NO activated the phosphorylation level of *MEK1/2* which brings the up-regulation of ascorbate and glutathione metabolism	Shan and Sun (2018)
Salicylic acid	Methyl jasmonate	Chilling stress	*Citrus limon*	Interaction of SA and MeJ increased the synthesis of total phenolics and phenylalanine ammonia lyase, decreasing the activities of polyphenol oxidase and peroxidase, reduced chilling-induced membrane permeability and membrane lipid peroxidation, while decreased total amount of sugar and amylase activity	Siboza et al. (2014)
Salicylic acid	Hydrogen peroxide	Freezing stress	*Triticum aestivum*	Less cell membrane damage, maintained maximum photosystem II quantum yield, electron transport rates, up-regulation of antioxidant ability and expression of cold-responsive genes. They concluded that H_2O_2 and ABA signaling cross talk mediate SA-induced freezing tolerance in wheat	Wang et al. (2018a, b)
Salicylic acid	Nitric oxide	Water deficit	*Zea mays*	Increased activities of Rubisco and Rubisco activase enzymes. Moreover, higher transcription rates of *Rbc* L, *ZmRCAα*, and *ZmRCAβ* mRNA. The study highlights the interactive role of SA and NO in regulating photosynthesis	Shao et al. (2018)
Salicylic acid	28-homobrassinolide	Salt stress	*Brassica juncea*	Increased SPAD chlorophyll, growth attributes, photosynthetic traits, carbonic anhydrase, nitrate reductase, quantum yield of photosystem II. Salt stress was completely ameliorated by the combined supplementation of two hormones	Hayat et al. (2012)

(continued)

Table 1 (continued)

Elicitor 1	Elicitor 2	Condition	Plant species	Parameter studied	Reference
Salicylic acid	Nitric oxide	Metalloid phytotoxicity	*Oryza sativa*	Signaling of nitric oxide and salicylic acid showed a protective role against arsenite stress. Both decreased arsenic translocation and accumulation in the roots as well as shoots. Their study demonstrated that there is close relationship between nitric oxide and salicylic acid that work in integrated manner to modulate the signaling response during metalloid stress in plants	Singh et al. (2017a, b)
Salicylic acid	Nitric oxide	Heat stress	*Lablab purpureus*	Both substantially modulated the growth and biophysical process under stress environment, remarkably regulated antioxidant enzyme activities. Both modulated mRNA transcription level of certain enzymes, improvement in plant water relation and recovered photosynthesis under heat stress. Their results indicated that both regulated DNA methylation pattern under heat stress	Rai et al. (2018)
Jasmonic acid	Salicylic acid	Wounding and fungal stress	*Noccaea praecox*	Modulation in endogenous JA and SA levels, failure in SA signaling increase susceptibility to stress conditions	Llugany et al. (2013)
Salicylic acid	Nitric oxide	Pathogen stress	*Arabidopsis thaliana*	Both SA and NO trigger activation of plant defense responses after pathogen attack and both work synergistically to activate the defense signaling by targeting the same effector proteins and/or their genes	Klessig et al. (2000)
Polyamines	Abscisic acid	Osmotic stress	*Triticum aestivum*	Abscisic acid increased the putrescine, but decreased the spermidine contents. Abscisic acid and osmotic stress altered metabolism of proline. The study unravels the connection between abscisic acid signaling and polyamine metabolism during osmotic stress conditions in plants	Pál et al. (2018b)

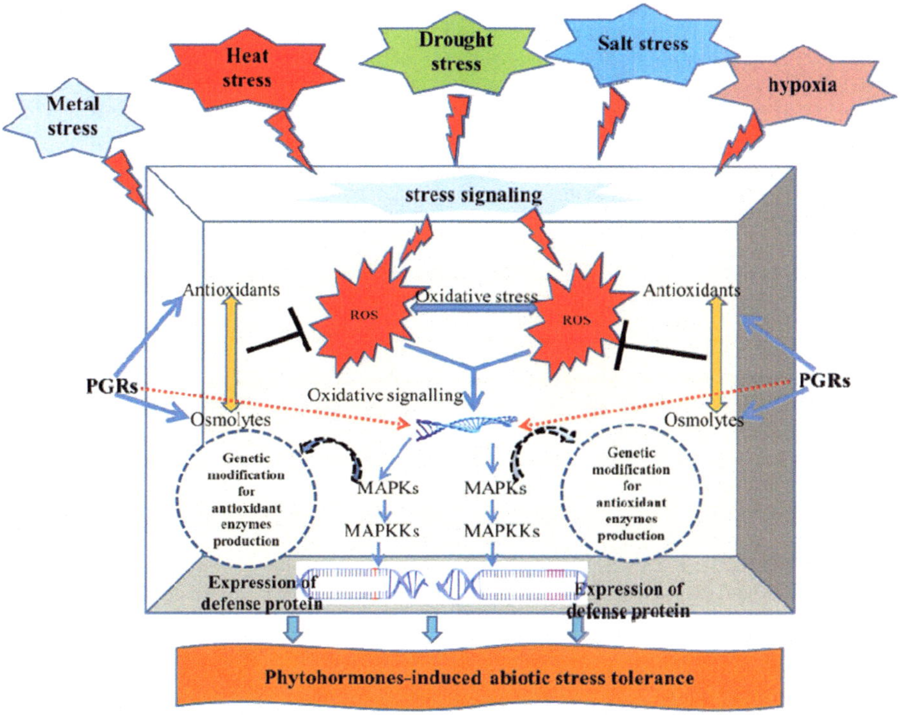

Fig. 1 Showing involvement of different exogenous elicitors in inducing tolerance against different kinds of abiotic stresses. As represented in figure, abiotic stress causes oxidative stress by orchestrating the biosynthesis of reactive oxygen species (ROS) whose low concentration promotes oxidative signaling. Plant growth regulators (PGRs) in turn alleviate the abiotic stress-induced oxidative stress by directly decreasing the overproduction of reactive oxygen species through the activation of the antioxidant defense system which comprises antioxidant proteins and osmolytes or indirectly modulating the signal transduction cascades which initiate stress responses. Various mitogen-activated protein kinases (MAPKs) and mitogen-activated protein kinase kinases (MAPKKs) induce the expression of defense-related genes. Pointed arrows show promotive while blunt arrows show inhibitory interactions, respectively

References

Ahammed G, Xia XJ, Li X, Shi K, Yu JQ, Zhou YH (2015) Role of brassinosteroid in plant adaptation to abiotic stresses and its interplay with other hormones. Curr Protein Pept Sci 16(5):462–473

Ahmad P, Rasool S, Gul A, Sheikh SA, Akram NA, Ashraf M, Gucel S (2016) Jasmonates: multifunctional roles in stress tolerance. Front Plant Sci 7:813

Ahmad B, Khan MM, Jaleel H, Sadiq Y, Shabbir A, Uddin M (2017a) Exogenously sourced γ-irradiated chitosan-mediated regulation of growth, physiology, quality attributes and yield in Mentha piperita L. Turk J Biol 41(2):388–401

Ahmad P, Alyemeni MN, Wijaya L, Alam P, Ahanger MA, Alamri SA (2017b) Jasmonic acid alleviates negative impacts of cadmium stress by modifying osmolytes and antioxidants in faba bean (*Vicia faba* L.). Arch Agron Soil Sci 63:1–11

Ahmad B, Jaleel H, Sadiq Y, Khan MMA, Shabbir A (2018) Response of exogenous salicylic acid on cadmium induced photosynthetic damage, antioxidant metabolism and essential oil production in peppermint. J Plant Growth Regul 86(2):273–286. https://doi.org/10.1007/s10725-018-0427-z

Akram NA, Ashraf M (2013) Regulation in plant stress tolerance by a potential plant growth regulator, 5-aminolevulinic acid. J Plant Growth Regul 32(3):663–679

Alamri SA, Siddiqui MH, Al-Khaishany MY, Nasir Khan M, Ali HM, Alaraidh IA, Mateen M (2018) Ascorbic acid improves the tolerance of wheat plants to lead toxicity. J Plant Interact 13(1):409–419

Albertos P, Romero-Puertas MC, Tatematsu K, Mateos I, Sanchez-Vicente I, Nambara E, Lorenzo O (2015) S-nitrosylation triggers ABI5 degradation to promote seed germination and seedling growth. Nat Commun 6:8669

Aldridge DC, Galt S, Giles D, Turner WB (1971) Metabolites of *Lasiodiplodia theobromae*. J Chem Soc C Org 1623–1627

Anjum SA, Xie XY, Farooq M, Wang LC, Xue LI, Shahbaz M (2011) Effect of exogenous methyl jasmonate on growth, gas exchange and chlorophyll contents of soybean subjected to drought. Afr J Biotechnol 24:9647–9656

Asgher M, Khan M, Anjum NA, Khan NA (2015) Minimising toxicity of cadmium in plants-role of plant growth regulators. Protoplasma 252:399–413

Asgher M, Khan MIR, Anjum NA, Verma S, Vyas D, Per TS, Khan NA (2018) Ethylene and polyamines in counteracting heavy metal phytotoxicity: a crosstalk perspective. J Plant Growth Regul 37(4):1050–1065

Bagheri M, Gholami M, Baninasab B (2019) Hydrogen peroxide-induced salt tolerance in relation to antioxidant systems in pistachio seedlings. Sci Hortic 243:207–213

Banerjee A, Roychoudhury A (2018) Strigolactones: multi-level regulation of biosynthesis and diverse responses in plant abiotic stresses. Acta Physiol Plant 40(5):86

Banerjee A, Tripathi DK, Roychoudhury A (2018) Hydrogen sulphide trapeze: environmental stress amelioration and phytohormone crosstalk. Plant Physiol Biochem 132:46–53. https://doi.org/10.1016/j.plaphy.2018.08.028

Beligni MV, Lamattina L (2000) Nitric oxide stimulates seed germination and de-etiolation, and inhibits hypocotyl elongation, three light-inducible responses in plants. Planta 210(2):215–221

Bright J, Desikan R, Hancock JT, Weir IS, Neill SJ (2006) ABA-induced NO generation and stomatal closure in Arabidopsis are dependent on H2O2 synthesis. Plant J 45(1):113–122

Brossa R, López-Carbonell M, Jubany-Marí T, Alegre L (2011) Interplay between abscisic acid and jasmonic acid and its role in water-oxidative stress in wild-type, ABA-deficient, JA-deficient, and ascorbate-deficient *Arabidopsis* plants. J Plant Growth Regul 30:322–333

Bücker-Neto L, Paiva ALS, Machado RD (2017) Interactions between plant hormones and heavy metals responses. Genet Mol Biol 40:373–386

Castillo MC, Lozano-Juste J, Gonzalez-Guzman M, Rodriguez L, Rodriguez PL, Leon J (2015) Inactivation of PYR/PYL/RCAR ABA receptors by tyrosine nitration may enable rapid inhibition of ABA signaling by nitric oxide in plants. Sci Signal 8:ra89

Chaves MM, Maroco JP, Pereira JS (2003) Understanding plant responses to drought—from genes to the whole plant. Funct Plant Biol 30(3):239–264

Ciura J, Kruk J (2018) Phytohormones as targets for improving plant productivity and stress tolerance. J Plant Physiol 229:32–40

Clarke SM, Cristescu SM, Miersch O, Harren FJ, Wasternack C, Mur LA (2009) Jasmonates act with salicylic acid to confer basal thermotolerance in *Arabidopsis thaliana*. New Phytol 182:175–187

Correa-Aragunde N, Graziano M, Lamattina L (2004) Nitric oxide plays a central role in determining lateral root development in tomato. Planta 218(6):900–905

Dar NA, Amin I, Wani W, Wani SA, Shikari AB, Wani SH, Masoodi KZ (2017) Abscisic acid: a key regulator of abiotic stress tolerance in plants. Plant Gene 11:106–111

Dietz KJ, Mittler R, Noctor G (2016) Recent progress in understanding the role of reactive oxygen species in plant cell signaling. Plant Physiol 171:1535–1539

Dong W, Wang M, Xu F, Quan T, Peng K, Xiao L, Xia G (2013) Wheat oxophytodienoate reductase gene TaOPR1 confers salinity tolerance via enhancement of abscisic acid signaling and reactive oxygen species scavenging. Plant Physiol 161:1217–1228

Fahad S, Hussain S, Bano A, Saud S, Hassan S, Shan D, Tabassum MA (2015) Potential role of phytohormones and plant growth-promoting rhizobacteria in abiotic stresses: consequences for changing environment. Environ Sci Pollut Res 22:4907–4921

Fardus J, Matin MA, Hasanuzzaman M, Hossain MA, Hasanuzzaman M (2018) Salicylic acid-induced improvement in germination and growth parameters of wheat under salinity stress. J Anim Plant Sci 28:197–207

Fariduddin Q, Hayat S, Ahmad A (2003) Salicylic acid influences net photosynthetic rate, carboxylation efficiency, nitrate reductase activity, and seed yield in *Brassica juncea*. Photosynthetica 41:281–284

Farouk S, Qados AMA (2018) Enhancing seed quality and productivity as well as physio-anatomical responses of pea plants by folic acid and/or hydrogen peroxide application. Sci Hortic 240:29–37

Fernando VD, Schroeder DF (2016) Role of ABA in *Arabidopsis* salt, drought, and desiccation tolerance. In: Abiotic and biotic stress in plants-recent advances and future perspectives. InTech

Freschi L (2013) Nitric oxide and phytohormone interactions: current status and perspectives. Front Plant Sci 4:398

Freschi L, Rodrigues MA, Domingues DS, Purgatto E, Van Sluys MA, Magalhaes JR, …, Mercier H (2010) Nitric oxide mediates the hormonal control of Crassulacean acid metabolism expression in young pineapple plants. Plant Physiol 152(4):1971–1985

Garcia-Mata C, Lamattina L (2002) Nitric oxide and abscisic acid cross talk in guard cells. Plant Physiol 128:790–792

Garcia-Mata C, Lamattina L (2003) Abscisic acid, nitric oxide and stomatal closure—is nitrate reductase one of the missing links? Trends Plant Sci 8:20–26

Garcia-Mata C, Gay R, Sokolovski S, Hills A, Lamattina L, Blatt MR (2003) Nitric oxide regulates K^+ and Cl^- channels in guard cells through a subset of abscisic acid-evoked signaling pathways. Proc Natl Acad Sci U S A 100:11116–11121

Garg N, Bharti A (2018) Salicylic acid improves arbuscular mycorrhizal symbiosis, and chickpea growth and yield by modulating carbohydrate metabolism under salt stress. Mycorrhiza 28(8):727–746

Gaupels F, Furch ACU, Zimmermann MR, Chen F, Kaever V, Buhtz A (2016) Systemic induction of NO-, Redox-, and cGMP signaling in the pumpkin extrafascicular phloem upon local leaf wounding. Front Plant Sci 7:154

Gilroy S, Suzuki N, Miller G, Choi WG, Toyota M, Devireddy AR (2014) A tidal wave of signals: calcium and ROS at the forefront of rapid systemic signaling. Trends Plant Sci 19:623–630

Giuliani S, Sanguineti MC, Tuberosa R, Bellotti M, Salvi S, Landi P (2005) Root-ABA1, a major constitutive QTL, affects maize root architecture and leaf ABA concentration at different water regimes. J Exp Bot 56:3061–3070

Gondor OK, Janda T, Soós V, Pál M, Majláth I, Adak MK, Balázs E, Szalai G (2016) Salicylic acid induction of flavonoid biosynthesis pathways in wheat varies by treatment. Front Plant Sci 7:1447

Hancock JT, Neill SJ, Wilson ID (2011) Nitric oxide and ABA in control of plant function. Plant Sci 181:555–559

Hao L, Yu C, Li B, Wang D (2010) Molecular cloning and preliminary analysis 1204 of TaNOA in common wheat. Sheng Wu Gong Cheng Xue Bao 26:48–56

Hasanuzzaman M, Nahar K, Bhuiyan TF, Anee TI, Inafuku M, Oku H, Fujita M (2017) Salicylic acid: an all-rounder in regulating abiotic stress responses in plants. In: Phytohormones-signaling mechanisms and crosstalk in plant development and stress responses. InTech

Hayat Q, Hayat S, Irfan M, Ahmad A (2010) Effect of exogenous salicylic acid under changing environment: a review. Environ Exp Bot 68:14–25

Hayat S, Maheshwari P, Wani AS, Irfan M, Alyemeni MN, Ahmad A (2012) Comparative effect of 28 homobrassinolide and salicylic acid in the amelioration of NaCl stress in *Brassica juncea* L. Plant Physiol Biochem 53:61–68

He Y, Tang RH, Hao Y, Stevens RD, Cook CW, Ahn SM, Fiorani F (2004) Nitric oxide represses the *Arabidopsis* floral transition. Science 305(5692):1968–1971

Hu Y, Jiang Y, Han X, Wang H, Pan J, Yu D (2017) Jasmonate regulates leaf senescence and tolerance to cold stress: crosstalk with other phytohormones. J Exp Bot 68:1361–1369

Iqbal H, Yaning C, Waqas M, Rehman H, Shareef M, Iqbal S (2018) Hydrogen peroxide application improves quinoa performance by affecting physiological and biochemical mechanisms under water deficit conditions. J Agron Crop Sci 204(6):541–553. https://doi.org/10.1111/jac.12284

Jasid S, Galatro A, Villordo JJ, Puntarulo S, Simontacchi M (2009) Role of nitric oxide in soybean cotyledon senescence. Plant Sci 176(5):662–668

Kazan K (2015) Diverse roles of jasmonates and ethylene in abiotic stress tolerance. Trends Plant Sci 20:219–229

Ke Q, Ye J, Wang B, Ren J, Yin L, Deng X, Wang S (2018) Melatonin mitigates salt stress in wheat seedlings by modulating polyamine metabolism. Front Plant Sci 9:914

Keramat B, Kalantari KM, Arvin MJ (2009) Effects of methyl jasmonate in regulating cadmium induced oxidative stress in soybean plant (*Glycine max* L.). Afr J Microbiol Res 31:240–244

Keskin BC, Yuksel B, Memon AR, Topal-Sarıkaya A (2010) Abscisic acid regulated gene expression in bread wheat ('*Triticum aestivum*' L.). Aust J Crop Sci 4(8):617

Khan W, Prithviraj B, Smith DL (2003) Photosynthetic responses of corn and soybean to foliar application of salicylates. J Plant Physiol 160:485–492

Khan MIR, Fatma M, Per TS, Anjum NA, Khan NA (2015) Salicylic acid-induced abiotic stress tolerance and underlying mechanisms in plants. Front Plant Sci 6:462

Khan TA, Yusuf M, Fariduddin Q (2018) Hydrogen peroxide in regulation of plant metabolism: signalling and its effect under abiotic stress. Photosynthetica 56(4):1237–1248

Klessig DF, Durner J, Noad R, Navarre DA, Wendehenne D, Kumar D, Trifa Y (2000) Nitric oxide and salicylic acid signaling in plant defense. Proc Natl Acad Sci 97:8849–8855

Leshem Y, Wills RBH, Ku V (1998) Evidence for the function of the free radical gas-nitric oxide (NO) as an endogenous maturation and senescence regulating factor in higher plants. Plant Physiol Biochem 36:825–833

Liu Y, Chen X, Wang X, Fang Y, Huang M, Guo L, Zhao H (2018) Improving biomass and starch accumulation of bioenergy crop duckweed (*Landoltia punctata*) by abscisic acid application. Sci Rep 8:9544

Llugany M, Martin SR, Barceló J, Poschenrieder C (2013) Endogenous jasmonic and salicylic acids levels in the Cd-hyperaccumulator *Noccaea (Thlaspi) praecox* exposed to fungal infection and/or mechanical stress. Plant Cell Rep 32:1243–1249

Maksymiec W, Wianowska D, Dawidowicz AL, Radkiewicz S, Mardarowicz M, Krupa Z (2005) The level of jasmonic acid in *Arabidopsis thaliana* and *Phaseolus coccineus* plants under heavy metal stress. J Plant Physiol 162:1338–1346

Maruta T, Noshi M, Tanouchi A, Tamoi M, Yabuta Y, Yoshimura K (2012) H_2O_2-triggered retrograde signaling from chloroplasts to nucleus plays specific role in response to stress. J Biol Chem 287:11717–11729

Maserti BE, Del Carratore R, Della Croce CM, Podda A, Migheli Q, Froelicher Y, Luro F, Morillon R, Ollitrault P, Talon M, Rossignol M (2011) Comparative analysis of proteome changes induced by the two spotted spider mite *Tetranychus urticae* and methyl jasmonate in citrus leaves. J Plant Physiol 168:392–402

McLaughlin MJ, Parker DR, Clarke JM (1999) Metals and micronutrients—food safety issues. Field Crop Res 60:143–163

Mioto PT, Mercier H (2013) Abscisic acid and nitric oxide signaling in two different portions of detached leaves of Guzmania monostachia with CAM up-regulated by drought. J Plant Physiol 170:996–1002

Mir MA, Sirhindi G, Alyemeni MN, Alam P, Ahmad P (2018) Jasmonic acid improves growth performance of soybean under nickel toxicity by regulating nickel uptake, redox balance, and oxidative stress metabolism. J Plant Growth Regul 37(4):1195–1209

Misra AN, Misra M, Singh R (2011) Nitric oxide: a ubiquitous signaling molecule with diverse role in plants. African J Plant Sci 5:57–74

Miura K, Tada Y (2014) Regulation of water, salinity, and cold stress responses by salicylic acid. Front Plant Sci 5:4

Moreau M, Lindermayr C, Durner J, Klessig DF (2010) NO synthesis and signaling in plants where do we stand? Physiol Plant 138:372–383

Mostofa MG, Fujita M (2013) Salicylic acid alleviates copper toxicity in rice (*Oryza sativa* L.) seedlings by up-regulating antioxidative and glyoxalase systems. Ecotoxicology 22:959–973

Mostofa MG, Li W, Nguyen KH, Fujita M, Tran LSP (2018) Strigolactones in plant adaptation to abiotic stresses: an emerging avenue of plant research. Plant Cell Environ 41(10):2227–2243. https://doi.org/10.1111/pce.13364

Mukherjee S (2018) Novel perspectives on the molecular crosstalk mechanisms of serotonin and melatonin in plants. Plant Physiol Biochem 132:33–45. https://doi.org/10.1016/j.plaphy.2018.08.031

Mur LA, Prats E, Pierre S, Hall MA, Hebelstrup KH (2013) Integrating nitric oxide into salicylic acid and jasmonic acid/ethylene plant defense pathways. Front Plant Sci 4:215

Mustafavi SH, Shekari F, Hatami Maleki H (2016) Influence of exogenous polyamines on antioxidant defence and essential oil production in valerian (*Valeriana officinalis* L.) plants under drought stress. Acta Agric Slov 107:81–91

Neill S, Desikan R, Hancock JT (2002a) Hydrogen peroxide signalling. Curr Opin Plant Biol 5:388–395

Neill SJ, Desikan R, Clarke A, Hancock JT (2002b) Nitric oxide is a novel component of abscisic acid signaling in stomatal guard cells. Plant Physiol 128:13–16

Neill S, Barros R, Bright J, Desikan R, Hancock J, Harrison J, Wilson I (2008) Nitric oxide, stomatal closure, and abiotic stress. J Exp Bot 59:165–176

Nemhauser JL, Hong F, Chory J (2006) Different plant hormones regulate similar processes through largely nonoverlapping transcriptional responses. Cell 126:467–475

Norastehnia A, Sajedi RH, Nojavan-Asghari M (2007) Inhibitory effects of methyl jasmonate on seed germination in maize (Zea mays): effect on a-amylase activity and ethylene production. Gen Appl Plant Physiol 33:13–23

O'Brien JA, Benková E (2013) Cytokinin cross-talking during biotic and abiotic stress responses. Front Plant Sci 4:451

Pagnussat GC, Lanteri ML, Lamattina L (2003) Nitric oxide and cyclic GMP are messengers in the indole acetic acid-induced adventitious rooting process. Plant Physiol 132:1241–1248

Pál M, Janda T, Szalai G (2018a) Interactions between plant hormones and thiol-related heavy metal chelators. Plant Growth Regul 85(2):173–185

Pál M, Tajti J, Szalai G, Peeva V, Végh B, Janda T (2018b) Interaction of polyamines, abscisic acid and proline under osmotic stress in the leaves of wheat plants. Sci Rep 8:12839

Patel PK, Hemantaranjan A (2012) Salicylic acid induced alteration in dry matter partitioning, antioxidant defence system and yield in chickpea (*Cicer arietinum* L.) under drought stress. Asian J Crop Sci 4:86–102

Paul S, Roychoudhury A (2017) Seed priming with spermine and spermidine regulates the expression of diverse groups of abiotic stress-responsive genes during salinity stress in the seedlings of indica rice varieties. Plant Gene 11:124–132

Peleg Z, Blumwald E (2011) Hormone balance and abiotic stress tolerance in crop plants. Curr Opin Plant Biol 14:290–295

Per TS, Khan MIR, Anjum NA, Masood A, Hussain SJ, Khan NA (2018) Jasmonates in plants under abiotic stresses: Crosstalk with other phytohormones matters. Environ Exp Bot 145:104–120
Pereira A (2016) Plant abiotic stress challenges from the changing environment. Front Plant Sci 7:1123
Qin F, Shinozaki K, Yamaguchi-Shinozaki K (2011) Achievements and challenges in understanding plant abiotic stress responses and tolerance. Plant Cell Physiol 52:1569–1582
Qiu Z, Guo J, Zhu A, Zhang L, Zhang M (2014) Exogenous jasmonic acid can enhance tolerance of wheat seedlings to salt stress. Ecotoxicol Environ Saf 104:202–208
Rai KK, Rai N, Rai SP (2018) Salicylic acid and nitric oxide alleviate high temperature induced oxidative damage in *Lablab purpureus* L. plants by regulating bio-physical processes and DNA methylation. Plant Physiol Biochem 128:72–88
Rasheed R, Ashraf MA, Kamran S, Iqbal M, Hussain I (2018) Menadione sodium bisulphite mediated growth, secondary metabolism, nutrient uptake and oxidative defense in okra (*Abelmoschus esculentus* Moench) under cadmium stress. J Hazard Mater 360:604–614. https://doi.org/10.1016/j.jhazmat.2018.08.043
Rejeb IB, Pastor V, Mauch-Mani B (2014) Plant responses to simultaneous biotic and abiotic stress: molecular mechanisms. Plants 3(4):458–475
Roberts DP, Mattoo AK (2018) Sustainable agriculture—enhancing environmental benefits, food nutritional quality and building crop resilience to abiotic and biotic stresses. Agriculture 8(1):8
Saito N, Nakamura Y, Mori IC, Murata Y (2009) Nitric oxide functions in both methyl jasmonate signaling and abscisic acid signaling in Arabidopsis guard cells. Plant Signal Behav 4(2):119–120
Sami F, Hayat S (2018) Effect of glucose on the morpho-physiology, photosynthetic efficiency, antioxidant system, and carbohydrate metabolism in *Brassica juncea*. Protoplasma 1–14. https://doi.org/10.1007/s00709-018-1291-4
Saxena I, Srikanth S, Chen Z (2016) Cross talk between H_2O_2 and interacting signal molecules under plant stress response. Front Plant Sci 7:570
Shakirova FM, Sakhabutdinova DR (2003) Changes in the hormonal status of wheat seedlings induced by salicylic acid and salinity. Plant Sci 164:317–322
Shan C, Sun H (2018) Jasmonic acid-induced NO activates MEK1/2 in regulating the metabolism of ascorbate and glutathione in maize leaves. Protoplasma 255:977–983
Shao RX, Xin LF, Guo JM, Zheng HF, Mao J, Han XP, Yang QH (2018) Salicylic acid-induced photosynthetic adaptability of *Zea mays* L. to polyethylene glycol-simulated water deficit is associated with nitric oxide signaling. Photosynthetica 56(4):1370–1377
Si T, Wang X, Zhao C, Huang M, Cai J, Zhou Q, Dai T, Jiang D (2018) The role of hydrogen peroxide in mediating the mechanical wounding-induced freezing tolerance in wheat. Front Plant Sci 9:327
Siboza XI, Bertling I, Odindo AO (2014) Salicylic acid and methyl jasmonate improve chilling tolerance in cold-stored lemon fruit (*Citrus limon*). J Plant Physiol 171:1722–1731
Simontacchi M, García-Mata C, Bartoli CG, Santa-María GE, Lamattina L (2013) Nitric oxide as a key component in hormone-regulated processes. Plant Cell Rep 32:853–866
Singh AP, Dixit G, Mishra S, Dwivedi S, Tiwari M, Mallick S, Pandey V, Trivedi PK, Chakrabarty D, Tripathi RD (2015) Salicylic acid modulates arsenic toxicity by reducing its root to shoot translocation in rice (*Oryza sativa* L.). Front Plant Sci 6:340
Singh AP, Dixit G, Kumar A, Mishra S, Kumar N, Dixit S, Dhankher OP (2017a) A protective role for nitric oxide and salicylic acid for arsenite phytotoxicity in rice (*Oryza sativa* L.). Plant Physiol Biochem 115:163–173
Singh M, Khan MM, Uddin M, Naeem M, Qureshi MI (2017b) Proliferating effect of radiolytically depolymerized carrageenan on physiological attributes, plant water relation parameters, essential oil production and active constituents of *Cymbopogon flexuosus* Steud. under drought stress. PLoS One 12:e0180129
Šírová J, Sedlářová M, Piterková J, Luhová L, Petřivalský M (2011) The role of nitric oxide in the germination of plant seeds and pollen. Plant Sci 181:560–572

Soliman MH, Alayafi AA, El Kelish AA, Abu-Elsaoud AM (2018) Acetylsalicylic acid enhance tolerance of *Phaseolus vulgaris* L. to chilling stress, improving photosynthesis, antioxidants and expression of cold stress responsive genes. Bot Stud 59:6

Song L, Ding W, Zhao M, Sun B, Zhang L (2006) Nitric oxide protects against oxidative stress under heat stress in the calluses from two ecotypes of reed. Plant Sci 171:449–458

Sreenivasulu N, Radchuk V, Alawady A, Borisjuk L, Weier D, Staroske N, Wobus U (2010) De-regulation of abscisic acid contents causes abnormal endosperm development in the barley mutant seg8. Plant J 64:589–603

Sreenivasulu N, Harshavardhan VT, Govind G, Seiler C, Kohli A (2012) Contrapuntal role of ABA: does it mediate stress tolerance or plant growth retardation under long-term drought stress? Gene 506:265–273

Stohr C, Stremlau S (2006) Formation and possible roles of nitric oxide in plant roots. J Exp Bot 57:463–470

Sun LR, Hao FS, Lu BS, Ma LY (2010) AtNOA1 modulates nitric oxide accumulation and stomatal closure induced by salicylic acid in Arabidopsis. Plant Signal Behav 5:1022–1024

Suzuki N, Mittler R (2006) Reactive oxygen species and temperature stresses: a delicate balance between signaling and destruction. Physiol Plant 126:45–51

Tan J, Zhao H, Hong J, Han Y, Li H, Zhao W (2008) Effects of exogenous nitric oxide on photosynthesis, antioxidant capacity and proline accumulation in wheat seedlings subjected to osmotic stress. World J Agric Sci 4:307–313

Tilman D, Balzer C, Hill J, Befort BL (2011) Global food demand and the sustainable intensification of agriculture. Proc Natl Acad Sci 108:20260–20264

Tominaga A, Nagata M, Futsuki A, Abe H, Uchiumi T, Abe M, Kucho K, Hashiguchi M, Akashi R, Hirsch AM, Arima S, Suzuki A (2010) Effect of abscisic acid on symbiotic nitrogen fixation activity in the root nodules of *Lotus japonicas*. Plant Signal Behav 5:440–443

Tossi V, Lamattina L, Cassia R (2009) An increase in the concentration of abscisic acid is critical for nitric oxide-mediated plant adaptive responses to UV-B irradiation. New Phytol 181:871–879

Ueda J, Saniewski M (2006) Methyl jasmonate-induced stimulation of chlorophyll formation in the basal part of tulip bulbs kept under natural light conditions. J Fruit Ornam Plant Res 14:199–210

Verslues PE, Agarwal M, Katiyar-Agarwal S, Zhu J, Zhu JK (2006) Methods and concepts in quantifying resistance to drought, salt and freezing, abiotic stresses that affect plant water status. Plant J 45:523–539

Verstraeten S, Aimo L, Oteiza P (2008) Aluminium and lead: molecular mechanisms of brain toxicity. Arch Toxicol 82:789–802

Walia H, Wilson C, Condamine P, Liu X, Ismail AM, Close TJ (2007) Large-scale expression profiling and physiological characterization of jasmonic acid-mediated adaptation of barley to salinity stress. Plant Cell Environ 30:410–421

Wang G, Xiao Y, Deng X, Zhang H, Li T, Chen H (2018a) Exogenous hydrogen peroxide contributes to heme oxygenase-1 delaying programmed cell death in isolated aleurone layers of rice subjected to drought stress in a cGMP-dependent manner. Front Plant Sci 9:84

Wang W, Wang X, Huang M, Cai J, Zhou Q, Dai T, Cao W, Jiang D (2018b) Hydrogen peroxide and abscisic acid mediate salicylic acid-induced freezing tolerance in wheat. Front Plant Sci 9:1137

Wani SH, Sah SK (2014) Biotechnology and abiotic stress tolerance in rice. J Rice Res 2:e105

Wani SH, Kumar V, Shriram V, Sah SK (2016) Phytohormones and their metabolic engineering for abiotic stress tolerance in crop plants. Crop J 4:162–176

Wani W, Masoodi KZ, Zaid A, Wani SH, Shah F, Meena VS, Mosa KA (2018) Engineering plants for heavy metal stress tolerance. Rend Lincei Sci Fis Nat 29(3):709–723

Wilkinson S, Davies WJ (2010) Drought, ozone, ABA and ethylene: new insights from cell to plant to community. Plant Cell Environ 33:510–525

Wilkinson S, Kudoyarova GR, Veselov DS, Arkhipova TN, Davies WJ (2012) Plant hormone interactions: innovative targets for crop breeding and management. J Exp Bot 63:3499–3509

Wojtyla L, Lechowska K, Kubala S, Garnczarska M (2016) Different modes of hydrogen peroxide action during seed germination. Front Plant Sci 7:731

Yan Z, Zhang W, Chen J, Li X (2015) Methyl jasmonate alleviates cadmium toxicity in *Solanum nigrum* by regulating metal uptake and antioxidative capacity. Biol Plant 59:373–381

Zaid A, Mohammad F (2018) Methyl jasmonate and nitrogen interact to alleviate cadmium stress in mentha arvensis by regulating physio-biochemical damages and ROS detoxification. J Plant Growth Regul 37(4):1331–1348

Zhang WN, Chen WL (2011) Role of salicylic acid in alleviating photochemical damage and autophagic cell death induction of cadmium stress in *Arabidopsis thaliana*. Photochem Photobiol Sci 10:947–955

Zhang J, Jia W, Yang J, Ismail AM (2006) Role of ABA in integrating plant responses to drought and salt stresses. Field Crops Res 97:111–119

Zhang A, Jiang M, Zhang J, Ding H, Xu S, Hu X, Tan M (2007a) Nitric oxide induced by hydrogen peroxide mediates abscisic acid-induced activation of the mitogen-activated protein kinase cascade involved in antioxidant defense in maize leaves. New Phytol 175:36–50

Zhang S, Hu J, Zhang Y, Xie XJ, Knapp A (2007b) Seed priming with brassinolide improves lucerne (*Medicago sativa* L.) seed germination and seedling growth in relation to physiological changes under salinity stress. Aust J Agric Res 58:811–815

Zhang Y, Tan J, Guo Z, Lu S, He S, Shu Z, Zhou B (2009) Increased abscisic acid levels in transgenic tobacco-overexpressing 9 cis-epoxycarotenoid dioxygenase influence H_2O_2 and NO production and antioxidant defences. Plant Cell Environ 32:509–519

Zhang J, Li D, Wei J, Ma W, Kong X, Rengel Z, Chen Q (2018) Melatonin alleviates aluminum-induced root growth inhibition by interfering with nitric oxide production in *Arabidopsis*. Environ Ex Bot. https://doi.org/10.1016/j.envexpbot.2018.08.014

Zhao H, Lin X, Shi A, Chang S (1995) The regulating effects of phenolic compounds on the physiological characteristics and yield of soybeans. Acta Agron Sin 21:351–355

Zhao Y, Dong W, Zhang N, Ai X, Wang M, Huang Z, Xiao L, Xia G (2014) A wheat allene oxide cyclase gene enhances salinity tolerance via jasmonate signaling. Plant Physiol 164:1068–1076

Zhu CQ, Zhang JH, Sun LM, Zhu LF, Abliz B, Hu WJ, Zhong C, Bai ZG, Sajid H, Cao XC, Jin QY (2018) Hydrogen sulfide alleviates aluminum toxicity via decreasing Apoplast and Symplast Al contents in Rice. Front Plant Sci 9:294

Calcium-Mediated Growth Regulation and Abiotic Stress Tolerance in Plants

Khursheda Parvin, Kamrun Nahar, Mirza Hasanuzzaman, M. H. M. Borhannuddin Bhuyan, and Masayuki Fujita

Introduction

Agriculture production is now overwhelmed with different abiotic stresses, such as high temperature, chilling, drought, flooding, salinity, and metal toxicity, which altogether causes reduction in biomass and yield of major crops up to 70% (Thakur et al. 2010; Hasanuzzaman et al. 2012) threatening world food production. This adverse environmental condition is mostly due to direct or indirect effect of

K. Parvin
Laboratory of Plant Stress Responses, Department of Applied Biological Sciences, Faculty of Agriculture, Kagawa University, Takamatsu, Kagawa, Japan

Department of Horticulture, Faculty of Agriculture, Sher-e-Bangla Agricultural University, Dhaka, Bangladesh

K. Nahar
Department of Agricultural Botany, Faculty of Agriculture, Sher-e-Bangla Agricultural University, Dhaka, Bangladesh

M. Hasanuzzaman (✉)
Department of Agronomy, Faculty of Agriculture, Sher-e-Bangla Agricultural University, Dhaka, Bangladesh
e-mail: mirzahasanuzzaman@sau.edu.bd

M. H. M. Borhannuddin Bhuyan
Laboratory of Plant Stress Responses, Department of Applied Biological Sciences, Faculty of Agriculture, Kagawa University, Takamatsu, Kagawa, Japan

Bangladesh Agricultural Research Institute, Joydebpur, Gazipur, Bangladesh

Citrus Research Station, Bangladesh Agricultural Research Institute (BARI), Jaintiapur, Sylhet, Bangladesh

M. Fujita
Laboratory of Plant Stress Responses, Department of Applied Biological Sciences, Faculty of Agriculture, Kagawa University, Takamatsu, Kagawa, Japan

M. Hasanuzzaman et al. (eds.), *Plant Abiotic Stress Tolerance*,
https://doi.org/10.1007/978-3-030-06118-0_13

different anthropogenic activities. Plant faces dehydration stress due to drought, salinity, and high temperature, which limits the plant growth and production (Vorasoot et al. 2003; Thakur et al. 2010). As a negative consequence of abiotic stress, plants suffer from nutritional and hormonal imbalance (Yildiztugay et al. 2014; Ashraf 2009). In addition, plant produces excess amount of reactive oxygen species (ROS), such as singlet oxygen ($^1O_2$), superoxide ($O_2^{\bullet -}$), hydrogen peroxide (H_2O_2), and hydroxyl radicals ($OH^{\bullet}$) at cellular level under stresses resulting in oxidative stress, which potentially damage the intracellular machinery and thus interrupt the antioxidant defense mechanism (Hasanuzzaman et al. 2018; Nahar et al. 2016). Overproduction of ROS induces lipid peroxidation, protein oxidation, disturbance of enzymatic activities, electrolyte leakage (EL), nucleic acid damage as well as cell death (Gill and Tuteja 2010; Hasanuzzaman et al. 2013). Naturally, plants can defend this adverse environmental condition up to certain level for their survival and sustaining next generation. Overcoming environmental stress is not only essential for sustainable increase of food production, but also essential for keeping environmental balance. As a defense mechanism, plants possess different enzymatic and non-enzymatic compounds, which play a significant role in preventing ROS production, check oxidative injury, and other damages. Plant tolerance to abiotic stress is a heterogeneous attribute, which depends on various factors. Thus, a complex stimulus is produced under abiotic stresses causing alteration in genomic expression leading to metabolic changes for making stress tolerance. Genomic expression is occurred due to changes in some physiological and biochemical features in plants under stresses, including changes in morphology, anatomy, water relations, photosynthesis, hormonal balance, ion distribution pattern, and biochemical adaptation (Parida and Das 2005; Hernández et al. 2001; Ashraf and Harris 2013; Acosta-Motos et al. 2015a, b).

Calcium (Ca) is an essential macronutrient and also known as the second messenger, as it acts as signaling molecule in different physiological and biochemical processes in plant to improve stress resistance (White and Broadley 2003; Rahman et al. 2015a, b; Ahmad et al. 2015). Calcium significantly plays its role in cell membrane stabilization, nutrient uptake as well as enzymatic and hormonal regulations to mitigate abiotic stress (Ahmad et al. 2015; Rahman et al. 2015a, b) in plants. Besides oxidative damage, ROS also acts as signal molecules for mediating stress tolerance responses in plant by activating stress-responsive genes governed by Ca^{2+} signaling (Mittler et al. 2004). The relationship between Ca^{2+} signaling and ROS signaling are strongly related, which involve Ca^{2+} and Ca^{2+}-binding proteins, such as calmodulin (Mittler et al. 2004). Thus, under stress condition, cytosolic Ca^{2+} rapidly increases depending on Ca^{2+}-binding protein (Snedden and Fromm 2001; Luan et al. 2002; Sanders et al. 2002) and increased cytosolic Ca^{2+} signal via Ca^{2+}-binding protein to downstream responses involving protection and adjustment of plant with adverse condition. The Ca-dependent protein kinases (CDPKs) activate expression of stress-responsive genes and play an important role in antioxidative stress response (Liu et al. 2006) by regulating physiological responses to abiotic stresses, such as K^+ uptake, gene expression, and stomatal movement (Li et al. 1998; Choi et al.

2005; Yu et al. 2007). Moreover, exogenous Ca^{2+} application increases photosynthesis, enhances stomatal conductance, decreases malondialdehyde (MDA) and proline (Pro) content, and improves catalase (CAT) activity to bring more adaptivity to stress condition (Liang et al. 2009; Tan et al. 2011).

Therefore, in this chapter, we focused the effects of different abiotic stresses on plant growth, development, physiological attributes, and yield. Plants' mechanisms to sustain and tolerate abiotic stress will also be focused. Furthermore, calcium induced abiotic stress tolerance in aspect of growth and physiology and yield improvement will also be reviewed.

Plant Response to Abiotic Stress

Salinity

Salt stress is associated with increasing level of Na^+ and Cl^- ions, which are toxic to plant and causes both ionic and osmotic stresses, which may hamper the membrane integrity, nutrient balance, functions and levels of growth regulators, enzymatic activity, metabolic function, photosynthesis, and other physiological and biochemical activities that ultimately lead to plant death (Mahajan and Tuteja 2005; Hasanuzzaman et al. 2012). Reduced seed germination, retarded vigorous seedling production, obstruction of plant growth and development, restriction in photosynthesis, disruption of dry matter accumulation and partitioning are occurred under salt stress that contributes to lower yield.

Under salinity, as excess salt concentration increases the amount of Na^+ in growth medium, plant uptakes more Na^+ than K^+, increases K^+ efflux that triggers K leakage from cell, and creates an increase of Na/K ratio (Wu and Wang 2012; Parvin et al. 2016; Rahman et al. 2016). Thus, excess Na^+ influx causes membrane depolarization, disturbance of membrane ion channel and nutrient replacement, and further leads to imbalance nutrient uptake and assimilation under salt stress (Nahar et al. 2016; Shabala et al. 2007; Zhao et al. 2007). Nahar et al. (2016) observed reduced amount of Ca, Mg, and Zn in mung bean seedlings leaf due to salt stress.

Besides, salinity is responsible for delayed germination and maturity (Yadav et al. 2011) due to osmotic stress and mineral toxicity, the seed germination process hampered, including restriction of water uptake, solute metabolism and emergence of embryonic organs (Wahid et al. 2014). Salt toxicity also disturbs the seedling emergence, growth and development, and establishment in soil causing lower seedling emergence percentage and increase the mortality (Al-Mutawa 2003; Turhan and Ayaz 2004). Similar result was recorded in *Oryza sativa* (Xu et al. 2011), *Zea mays* (Khodarahmpour et al. 2012) and *Solanum lycopersicum* (Kaveh et al. 2011).

Salt hampers the normal plant growth and development procedure by directly affecting cell expansion, division and differentiation, which results in stunted plant growth (Hasegawa et al. 2000). This phenomenon is due to reduction of soil water potential as well as water uptake ability reduced resulting in slower leaf area expansion. Leaf is the main photosynthetic organ, and thus due to saline-induced leaf area reduction, photosynthesis rate reduced, hence, photosynthate assimilation and translocation also lowered, which causes reduction of plant growth (Mazher et al. 2007).

Plants suffer from low osmotic pressure due to salinity that creates low leaf water potential (Munns 2005; Álvarez et al. 2012). In addition, salinity decreases stomatal conductance and transpiration (Vysotskaya et al. 2010), causing reduced CO_2 assimilation, decrease photosynthesis, as well as lower protein biosynthesis (Parida et al. 2004; Zobel et al. 2007; Mugnai et al. 2009). Reduction of photosynthesis under saline condition depends upon plant species, age of the plant, level and duration of salinity. Salt stress also decreases photosynthetic pigments content, such as, Chl *a* and Chl *b*, and hampered photosynthesis, which was well documented in *O. sativa* (Amirjani 2011), *Brassica juncea* (Ahmad et al. 2012), *Vicia faba* (Azooz et al. 2011) and *Morus alba* (Ahmad et al. 2010).

Under salinity excess amount of different ROS produced in chloroplasts and mitochondria. These ROS induce oxidative damage, like, lipid peroxidation, protein oxidation, denaturation of nucleic acids, imbalance hormones and vital nutrients etc., hence plants faces stress. Biosynthesis of different antioxidant compounds both enzymatic and non-enzymatic are hampered causing reduced scavenging of these ROS, hence plant faces oxidative damage (Nahar et al. 2016; Gill and Tuteja 2010). Many previous reports illustrated the oxidative damage induced by salinity in different crops. Nahar et al. (2016) found higher concentrations of ROS including H_2O_2 and $O_2^{\bullet-}$ with increased lipoxygenase (LOX) activity in *Vigna radiata*, which led to cellular oxidative damage with increased lipid peroxidation. They also found reduced ascorbate (AsA) and dehydroascorbate (DHA) ratio, and altered superoxide dismutase (SOD) and CAT activity, moreover higher methylglyoxal (MG) content was also recorded, which also responsible for oxidative damage. Due to salt stress, higher level of H_2O_2 with higher MDA and overproduced MG were also observed in rice seedling, where LOX and SOD activity increased, while CAT activity decreased (Rahman et al. 2016).

The reduction of yield is the eventual result of changes water relation, stomatal conductance, transpiration, nutritional balance with oxidative damage due to salt stress. Salinity hampers the yield and productivity of crops by changing morpho-physiological and biochemical processes (Kaveh et al. 2011). Yield per plant may reduce up to 62% (Parvin et al. 2017) at 8 dSm^{-1} in *Brassica oleracea* var. *capitata*. Salinity significantly affects reproductive response of crops with yield contributing attributes causing yield loss (Parvin et al. 2015a). Reduction of yield component and yield under salinity was observed in various crop species, such as *V. radiata* (Nahar and Hasanuzzaman 2009), *Solanum lycopersicum* L. (Parvin et al. 2015a), *B. oleracea* var. *capitata* (Parvin et al. 2017), *B. oleracea* var. *italica* (Parvin and Haque 2017).

Drought

As a reaction of global climate changes the frequency and severity of drought has been increasing in an alarming rate (Walter et al. 2011). Due to lack of water or moisture, plants often suffer from drought stress in any growth and developmental stages of life cycle. Shortage of precipitation, lack of groundwater and higher evapotranspiration are primarily responsible for creation of drought (Mishra and Cherkauer 2010). Drought affects plant growth and development by hampering germination, growth and dry matter accumulation. This water shortage causes loss of turgor, reduces energy balance, enzymatic activity, which impairs cell division, elongation and differentiation, and ultimately reduces the productivity of plant (Kiani et al. 2007; Farooq et al. 2009a; Taiz and Zeiger 2010). Adverse effect of drought varies with different plant species, growth stages of plant, and severity and duration of drought. Effect of drought on vegetative stage of plant demonstrated lower economic benefit, while drought stress at reproductive stage is most destructive as it severely affects the yield (Reddy et al. 2003; Yadav et al. 2004). Reduced transpiration by drought stress decreases nutrient absorption and thus plants suffer from nutrient deficiency. Drought-induced disruption of membrane integrity is another reason for reduced water/nutrient uptake. Alteration of enzyme activities and lack of energy dissipation also disturb nutrient accumulation under drought condition (Baligar et al. 2001; Ashraf and Iram 2005). Under drought stress reduced N and P contents were noticed in *L. esculentum* (Subramanian et al. 2006), while decreased N and K were recorded in *Gossypium hirsutum* (McWilliams 2003), yet devastating loss of P was found in *Tagetes erecta* (Asrar and Elhindi 2011).

In addition, water deficit causes alteration in different physiological attributes such as water potential, leaf relative water content, pressure and osmotic potential and transpiration rate (Kirkham 2005). For example, *S. lycopersicum* (Subramanian et al. 2006), *Capparis spinosa* (Ozkur et al. 2009), *Glycine max* (Liu et al. 2004), *Helianthus annuus* (Tezara et al. 2002) showed decreased water and osmotic potential, resulting in low leaf relative water content under drought stress. Subsequently, lower water potential in plant due to drought stress reduces transpiration rate and disrupts nutrients uptake and CO_2 influx. Reduction in transpiration was also reported in *T. aestivum* and *O. sativa* (Siddique et al. 2001), citrumelo (*Poncirus trifoliata* × *Citrus paradisi*) (de Campos et al. 2011).

Upon drought stress, lack of turgidity and subsequent stomatal closure causes reduction in transpiration and CO_2 assimilation, eventually reduce photosynthesis, which was reported in *Phaseolus vulgaris* (Miyashita et al. 2005), *Cocos nucifera* (Gomes et al. 2010), and *H. annuus* (Correia et al. 2006). Moreover, impaired photosynthetic enzyme activities namely ribulose-1,5-bisphosphate carboxylase/oxygenase (Rubisco), pyruvate phosphate dikinase, phosphoenolpyruvate carboxylase, and NADP-malate dehydrogenase also responsible for declining photosynthesis under drought condition (Reddy et al. 2004; Farooq et al. 2009a, b). Drought stress reduces the regeneration of Rubisco causing quick decrease of photosynthesis (Flexas and Medrano 2002; Reddy et al. 2004) and this was observed in different grassland species (Signarbieux and Feller 2011). Drought also decreases the photosynthetic pigment synthesis such as Chl *a* and Chl *b* contents was reduced in *T. erecta* (Asrar and Elhindi 2011).

Water deficit increases photorespiration and declines carbon fixation stimulates oxidative stress by producing H_2O_2 (Noctor et al. 2002; Ghannoum 2009). Drought produces excess amount of ROS, which damage lipid, protein, photosynthesis pigments, nucleic acids and other cellular organelles (Pinheiro and Chaves 2011; Rout and Shaw 2001). Thus, overgenerated ROS reduced stomatal conductance and photosynthesis in *Pinus canariensis* (Tausz 2001), carry out lipid peroxidation in *C. spinosa* (Ozkur et al. 2009) and *Triticum aestivum* (Qiu et al. 2008). In addition, higher level of H_2O_2, $O_2^{\bullet-}$, MDA and LOX activity were observed in *Z. mays*, where activities of SOD, peroxidase (POD) increased with reduction in CAT activity (Anjum et al. 2017). In mung bean, H_2O_2, $O_2^{\bullet-}$, MDA, and MG content were increased under drought, whereas, decreased AsA and increased DHA content resulted in reduction of AsA/DHA ratio (Nahar et al. 2017).

Drought hampers yield contributing characters as well as yield by changing phenological, physiological, and biochemical response of plant. When drought occurs at flowering stage, it increases the pollen sterility, which reduces yield. For instance, declined yield was observed in *H. annuus* (Hussain et al. 2008), *Pennisetum glaucum* L. (Yadav et al. 2004), *Z. mays* (Schussler and Westgate 1995), and *G. max* (Liu et al. 2003) due to drought at flowering period. Nouri-Ganbalani et al. (2009) showed about 50% reduction of average grain yield under water deficit condition in *T. aestivum*. Producing unfilled grain in *O. sativa* was also recorded due to spikelet sterility under drought stress in any growth stage, which results in yield reduction (Shahryari et al. 2008). Similar yield loss was also recorded in *Cicer arietinum* (Mafakheri et al. 2010), *Hordeum vulgare* (Babaeian et al. 2011), *V. radiata* (Ranawake et al. 2011), *L. esculentum* (Pervez et al. 2009), *Abelmoschus esculentus* (Hussein et al. 2011), *Lens culinaris* (Panahyan-e-Kivi et al. 2009), *V. unguiculata* (Zadehbagheri et al. 2012), and *G. max* (Mirakhori et al. 2009).

High Temperature

High temperature (HT) stress is caused by relatively high temperature beyond optimum level for a certain period of time. High temperature stress is one of the most important problems and burning issue for agriculture especially for plant growth and development. Because of rising levels of greenhouse gases, the global temperature is increasing, and it has already been risen by 0.89 ± 0.2 °C, and estimated to be increased by 2.6 ± 4.8 °C between 2081 and 2100 (Stoker et al. 2013). Each degree Celsius of increased temperature may cause up to 17% of crop yield reduction. High temperature stress reduces the plant growth and development, disturbs physiology and biochemical processes such as seed germination, photosynthesis, respiration, transpiration, nutrient uptake, dry matter accumulation and partitioning and thus suffering from nutritional imbalances, $O_2^{\bullet-}$ causes alteration of crop quality as well as yield reduction. This stress also disturbs processes by denaturing of protein, inactivation of enzymes, regeneration of excess ROS, disorganization of membrane structure, etc. and damage of cellular components (Nahar et al. 2015). Types and extent of deteriorative effect of HT stress depend on degree of temperature, time of exposure, and plant species (Hasanuzzaman et al. 2013). High temperature stress hampers seed germination or loses seed vigor, which leads to reduction of seedling emergences and establishment (Hasanuzzaman et al. 2013). Reduction of radicle and plumule growth at the time of seed emergence causes abnormal seedling growth resulting in poor seed germination percentage under heat stress in various crop species (Kumar et al. 2011; Piramila et al. 2012). Grown up plants also suffer from different abnormalities under HT stress such as scorching of leaves and twigs, partial burning of plant parts, leaf senescence and abscission, imbalance of dry matter accumulation and partitioning, hence relatively slower growth rate (Hasanuzzaman et al. 2013). Photosynthesis is negatively affected by HT stress, where chloroplast may injured and alternation of photochemical reactions occurred (Hasanuzzaman et al. 2013; Szymańska et al. 2017). In chloroplast Rubisco activity, PSII activity and ATP synthesis are mostly affected under HT stress (Asthir 2015). Photochemical pigments decreased after exposed to HT stress (Hasanuzzaman et al. 2013; Szymańska et al. 2017). In addition, decreased total chlorophyll (Chl) content including Chl *a* and Chl *b* and imbalance in their ratio are also reported as a negative consequence of HT (Hasanuzzaman et al. 2013). All of these impairs the total photosynthesis and net assimilation rate. Moreover, HT stress induces stomatal closure, reduces the relative water content, and lowers the carbon metabolism (Hasanuzzaman et al. 2013; Szymańska et al. 2017). Reduction of leaf area, relative leaf water potential, and premature leaf senescence also occurred under HT stress (Greer and Weedon 2012; Hasanuzzaman et al. 2013).

Excess amounts of ROS are attributes to HT stress, which causes oxidative damage. Consequently, plant suffers from oxidative stress that induced various physiological damages, such as lipid peroxidation, damage of nucleic acids, protein denaturation, enzyme deactivation, and disruption of cell membrane stability, eventually causes cell death (Qi et al. 2010; Hasanuzzaman et al. 2013; Nahar et al. 2016). Nahar et al. (2017) reported higher content of H_2O_2, $O_2^{\bullet-}$, LOX, and MDA in mung bean when facing HT stress that reduces the ratio of AsA/DHA with increase of SOD activity along with increase of MG content. Heat stress also increases H_2O_2 with increased SOD and CAT activity (Ferreira-Silva et al. 2011).

Ultimate result of HT stress is yield reduction, which imposes a threat to food safety. Negative effect of heat stress on different phenological developmental processes is responsible for yield reduction (Hasanuzzaman et al. 2013). It is well documented that this reduced yield due to HT stress was found in many crops including cereals, pulse, and oil yielding crops (Hasanuzzaman et al. 2013). Reduced yield up to 90% in *O. sativa* was recorded under heat stress due to decreased grain size and weight, including increased spikelet sterility (Suwa et al. 2010). Similarly, reduced seed size and weight are also responsible for yield loss in *Sorghum bicolor* (L.)

Moench (Mohammed and Tarpley 2010). Thus, high temperature induced various changes in morphological, physiological, and yield contributing characters, subsequently reduce yield of different plant species, which is a great threat to crop production worldwide.

Chilling Stress

As high temperature, temperature below optimum level (low temperature) also causes various detrimental effects on plants and thus termed as chilling or cold stress. Low temperature hampers the plant growth, development, productivity, and thus reduces the quality and quantity of yield by affecting different metabolic and physiological processes (Ding et al. 2017; Sheteiwy et al. 2017). Chilling stress causes damage to the germinating seeds and also established plant (Farooq et al. 2017). Poor seed germination percentage, low seed vigor, and slow seedling growth are noticed in plants under chilling stress (Sheteiwy et al. 2017). Delayed seedling emergence is one of the negative consequences of chilling stress and causing poor seedling establishment. Farooq et al. (2017) also observed this phenomenon in chickpea.

Chilling stress reduces water and mineral uptake, stomatal conductance, and photosynthesis causing reduction of plant growth and development (Yadav 2010; Farooq et al. 2017). Chilling stress directly impairs photosynthetic apparatus by damaging thylakoid membranes and also Chl pigments formation (Yadav 2010; Farooq et al. 2017). Due to the imbalanced stomatal movement, CO_2 fixation faces difficulty under chilling stress causing excess sugar accumulation and lipid peroxidation as well as disturbance in the water potential (Allen and Ort 2001; Farooq et al. 2017). Thus, lower CO_2 assimilation rate is observed in plant with reduction of photosynthesis rate (Ding et al. 2017; Farooq et al. 2017). Root hydraulic conductivity is also reduced under chilling stress, which causes reduced leaf water content (Yadav 2010; Farooq et al. 2017). Reduced root growth occurs under chilling stress, resulting in low water and mineral nutrient uptake (Aroca et al. 2003; Farooq et al. 2017). Hence, chilling stress causes major nutrient deficiency in plants, such as N, K, and Ca (Farooq et al. 2017). In addition, chilling stress makes inability of stomatal closing despite having low relative leaf water potential, which ultimately promotes chilling-induced water stress (Yadav 2010; Farooq et al. 2017).

Low temperature stress induces flower and pod abortion, infertility of pollen and ovule, breakdown of pollination and fertilization process, disruption of seed filling and also produces smaller seed, eventually yield loss (Nayyar et al. 2005a; Farooq et al. 2017). Thus, cold stress during reproductive phase has great economic and social consequence (Thakur et al. 2010). Furthermore, chilling stress-induced imbalanced metabolic processes lead to overproduction of ROS and causes oxidative damage to protein, lipid, and DNA (Gururani et al. 2015; Shi et al. 2015; Ding et al. 2017). Ding et al. (2017) observed the increment in H_2O_2 content in tomato plants when exposed to chilling (4 °C) as a reflection of oxidative stress. Banana plant also showed up to 349% increase in H_2O_2 content under chilling exposure (Kang et al. 2003).

On the other hand, chilling-induced imbalanced stomatal regulation leads to lower CO_2 fixation guide to more ATP and reduced $NADP^+$ escorts to the generation of excess ROS (Allen and Ort 2001; Gururani et al. 2015; Mignolet-Spruyt et al. 2016; Ding et al. 2017). Thus, chilling stress inhibits crop yield by damaging plant growth and development including oxidative stress-induced damaged.

Toxic Metals/Metalloids

Various kinds of solutes are required for plant growth and development, which are uptaken from rhizosphere by roots and then distributed to the whole plant body. Uptake of mineral elements with water is very crucial for successful plant life and plays a key role in physiological activity and developmental plasticity in root system (Ovečka and Takáč 2014). Hence, essential ions maintain the structural and physiological functions in plant cells (Conn and Gilliham 2010).

However, despite being essential, the imbalance in relative abundance of these elements including non-physiological concentration in soil can become limiting factors for plant growth through affecting cellular homeostasis (Ovečka and Takáč 2014). Thus, ion toxicity causes irreversible damage through various physiological and developmental changes (Ovečka and Takáč 2014). Despite the plant essential minerals some other metals/metalloids are contaminating the rhizosphere in recent decades due to anthropogenic activities as well as industrialization. These metals/metalloids can be incorporated into plant tissue with plant nutrients at the time of water and nutrient uptake. Metals/metalloids that can be toxic to plants are copper (Cu), iron (Fe), manganese (Mn), zinc (Zn), nickel (Ni), chromium (Cr), aluminum (Al), cobalt (Co), cadmium (Cd), molybdenum (Mo), arsenic (As), and lead (Pb) (Hossain et al. 2012; Dal Corso et al. 2013; Hossain and Komatsu 2013; Ovečka and Takáč 2014).

Agricultural soils are now being contaminated by these metals/metalloids worldwide due to excessive use of phosphatic fertilizers, dust from smelters, bad sewage system, use of sewage sludge, mismanagement of industrial waste (Passariello et al. 2002: Yadav 2010; Lu et al. 2015a, b). These metal ions affect plant growth and metabolism as well as various physiological and biochemical processes, and the degree of negative impacts varies with the metal, soil characters, plant type, etc. (Singh and Agrawal 2010; Sheetal et al. 2016). Excess metals/metalloids causes leaf chlorosis, necrosis, leaf rolling, stunted shoot growth, reduced root growth, alteration of stomatal movement, reduced water potential, reduced membrane integrity, cation efflux, inhibition of photosynthesis, altered enzymatic activities, reduced biomass production, and eventually cause death (Rahman et al. 2015a; Nahar et al. 2016; Sheetal et al. 2016). Comparatively higher amount of toxic metal/metalloid accumulates in roots than shoots (Rahman et al. 2015a) and strongly affects the water and nutrients uptake mechanism by roots (Ovečka and Takáč 2014). Whole process of transpiration and transport among various organs are also inhibited (Ying et al. 2010). Relative leaf water content is also disturbed and causes perturbation of plant—water relationship (Nahar et al. 2016).

Toxic metal causes negative changes in structural and physiological integrity of leaves and then reduces photosynthesis and respiration rate, which has detrimental effect on energy supply and other metabolic processes (Ovečka and Takáč 2014). Consequently, due to changes in the leaves and roots functions, various developmental processes such as embryogenesis, flowering, and seed formation are hampered (Ovečka and Takáč 2014). Reduction of Chl content increased with higher metal content in soil where Chl *a* is more sensitive to metal stress than Chl *b* by showing more reduction (Rahman et al. 2015a). This reduction of Chl has direct effect on inhibition of enzymatic activities and nutrient availability (Van Assche and Clijsters 1990; Sheetal et al. 2016). This reduction of Chl content is also considered for reduction of photosynthesis rate where stomatal conductance also decreased under metal stress (Sheetal et al. 2016). Yield of crop plants also reduced due to metal stress-induced different physiological, biochemical as well as metabolic changes and this yield reduction was recorded (Sheetal et al. 2016). In addition, metal stress also causes cell toxicity by enhancing overproduction of various ROS, which triggers oxidative damage by restricting the antioxidant defense system in plants (Rui et al. 2016; Bücker-Neto et al. 2017) and also reacts with lipids, proteins, and DNA and damages other biomolecules (Rahman et al. 2015a; Nahar et al. 2016).

Mustard (*B. juncea*) showed the increased level of H_2O_2 with lipid peroxidation, EL, as well as enhancement of antioxidant enzymes activities such as SOD, ascorbate peroxidase (APX), and glutathione (GSH), when treated with Cd (Ahmad et al. 2016). Nickel also enhanced the H_2O_2, $OH^•$, and $O_2^{•-}$ content in *O. sativa* (Rajpoot et al. 2016). It was reported that 1 mM concentration of As also enhanced 89% of H_2O_2 along with 51% and 40% decrease of AsA and DHA, respectively, in rice, thus reducing the AsA/DHA ratio where SOD and CAT activity were also increased with higher MG level (Rahman et al. 2015a). Therefore, as like other abiotic stresses, metal stress disrupts the plant morphology, physiology and biochemistry as well as hampers the developmental processes and inhibits plant growth along with yield reduction.

Flooding

Flooding is another form of water stress, which adversely affect the root environment. Most of the plant species are very vulnerable to prolonged submergence or soil waterlogging condition (Loreti et al. 2016). Complete submergence or flash flood is very common in lowland area and seriously hampers the crop survival, leading to extreme yield losses (Gautam et al. 2016). Flooding leads to low oxygen availability to plants causing shift to anaerobic and hypoxic condition, while complete lacking of oxygen (anoxia) is worst for plant survival (Oh and Komatsu 2015; Loreti et al. 2016). Thus, drastic reduction in oxygen concentration leads to physicochemical alteration (Avivi et al. 2016). Flooding stress causes reduced root permeability with initiation of adventitious roots, hampers the mineral and water uptake, and decreases photosynthesis and alteration of hormonal balance (Oh and Komatsu 2015). Flooding affects seed germination, seedling growth, and even established plants. In the first phase of development, seeds and seedlings are more vulnerable to flooding resulting in decreased seed germination as well as seedling establishment (Ferreira et al. 2007; Lal et al. 2015). Submergence affects the plant morphology by reducing shoot, leaf, and root dry weight due to decay, rotting, and death of living tissue and also causes leaf senescence, chlorosis, reduction of stomatal conductance, and photosynthesis leading to poor survival (Gautam et al. 2016). Plant also suffers from dehydration due to reduced leaf relative water content and water potential by decreasing root hydraulic conductivity (Gimeno et al. 2012). The reduction of photosynthesis may be due to decreased stomatal conductance, reduced Chl content, reduced leaf area, and leaf senescence (Avivi et al. 2016). Then, the net CO_2 assimilation rate is also hampered by flooding resulting in lowered yield. Both soluble sugar and starch are decreased in leaf and root due to reduction of CO_2 assimilation and supply of photo-assimilate that in turn is responsible for the reduction of carbohydrate concentration (Gimeno et al. 2012). In addition, plant also faces the nutrient deficiency due to disturbance in nutrient uptake caused by lack of oxygen (Gimeno et al. 2012). Plants also suffer from oxidative damage induced by the overproduction of ROS, which hampers the membrane integrity and induces damage of PSII, thereby reducing photosynthesis (Gautam et al. 2016). These ROS also causes lipid peroxidation, damage of nucleic acid, and some key enzymatic activities (Nahar et al. 2016). Higher amount of H_2O_2 was found in hypoxia-induced *H. vulgare* and *T. aestivum* (Kalashnikov et al. 1994; Biemelt et al. 2000). Waterlogging-stressed maize, pigeon pea, and mung bean also found to be suffered from oxidative damage indicated by ROS overproduction, increased lipid peroxidation, and membrane injury (Yan et al. 1996; Kumutha et al. 2009; Sairam et al. 2011). Hence, flooding causes morphological, physiological, and metabolic changes in plants, where plant growth become stunted, development processes become hampered, eventually plant produces lower yield with poor quality.

Others

Besides the most common abiotic stresses viz., salinity, drought, high temperature, chilling, flooding, and metal/metalloid stresses, plants face other abiotic stresses, such as ultraviolet (UV) radiation, ozone, high and low light, wind, mechanical injury, and herbicidal toxicity. Effects of some of these stresses on plant have been presented below.

Ultraviolet and Other Radiation

Plants are strongly affected by non-ionizing radiation, which reduces plant biomass production, photosynthetic pigments, and also causes severe chloroplast damage resulting in harmful effect for plant growth and development (Stefi et al. 2017). In addition, UV-B radiation causes stress to plant by its detrimental effect on photosynthetic components with degradation of Chl and carotenoid (car) (León-Chan et al. 2017). This inhibition of photosynthesis may be due to higher ROS generation, which causes lipid peroxidation, protein denaturation, nucleic acid damage, and disruption of photosystems (PSI and PSII) and the light harvesting complexes (LHCs) (Nishiyama et al. 2011; Nahar et al. 2016; León-Chan et al. 2017). It is well known that the level of UV-B radiation is higher in tropical and subtropical zone compared to temperate region due to thin ozone layer (Jaakola and Hohtola 2010) that indicate the vulnerability of plants of tropical and subtropical regions to UV-B stress. Ozone, an active form of oxygen, which is formed in troposphere acts as a protectant for plants from the UV radiation damage although it also has some negative impacts depending on its concentration, exposure, and duration ranging from photosynthetic damage to cell death (Chakraborty and Acharya 2017). As an ionizing radiation, gamma radiation also causes damage to DNA and tissue directly or indirectly where direct effect is damage of macromolecules such as lipid peroxidation, inactivation of enzyme, and DNA lesions with breakdown (Esnault et al. 2010). Again with exposure to gamma radiation plant also suffers from oxidative damage through producing excess ROS that leads to cellular damage (van de Walle et al. 2016).

Ozone

Ozone (O_3)—the reactive form of oxygen produced in the troposphere zone, which acts as a protective layer for the entrance of UV radiation to the earth but exerts damaging effects to the living organisms. Anthropogenic activities release air pollutants, such as volatile hydrocarbons and oxides of nitrogen, which can enhance O_3 breakdown through photochemical reactions. Decreased photosynthesis and growth, stomatal regulation, necrosis, surplus excitation energy, and excess generation of ROS have been reported as a common phenomenon of O_3 stress (Dumont et al. 2013; Gottardini et al. 2014; Feng et al. 2016). Ozone can enter through stomata, and depending upon its severity photosynthetic damage and cell death may occur (Oz et al. 2015). Ahlfors et al. (2009) have demonstrated cell death in *Arabidopsis thaliana* due to O_3 stress. Photosynthetic pigments destruction, reduced gas exchange rates, lower Chl *a* fluorescence, decreased antioxidants contents, and altered antioxidative enzyme activity, together with increased lipid per oxidation and accelerated leaf senescence were reported in *T. aestivum* cultivars under O_3 stress (Feng et al. 2016). Combination of drought and O_3 was imposed to different species of oak, which significantly decreased leaf dry weight of *Quercus pubescens* and *Q. cerris* by 70 and 75%, respectively. A decrease in maximum carboxylation rate and an increase in Pro content and MDA were noticed in three different oak species (Cotrozzi et al. 2016). Under O_3 stress, many plant species produce smaller roots. Among 68 different plant species, 22.5% were unaffected, 33.5% had increased stomatal opening, and 44% stomatal closure due to O_3 stress (Mills et al. 2016). The beneficial effect of N on root development was lost at higher levels of O_3 stress (Mills et al. 2016). High O_3 concentration (average 73 ppb) decreased leaf photosynthesis and grain yield of wheat by 20% and 29%, respectively, as compared to plant without O_3 stress (Feng et al. 2008). Ozone-induced changes in Chl *a* fluorescence parameters including reduction in quenching of photochemical efficiency (qP) and quantum yield of PSII (Feng et al. 2011), as well as altered antioxidant defense system (Inada et al. 2012) were noticed to cause physiological disorders. In field condition, continuous exposure to O_3 can decline photosynthesis, increase leaf senescence, and consequently decrease productivity and yield (Feng et al. 2011; Zhu et al. 2011; Burkart et al. 2013).

Nutrient Deficiency Stresses

Macronutrients (N, P, K, Mg, Ca, S), micronutrients (Cl, Fe, B, Mn, Zn, Cu, Ni, Mo), non-mineral (C, H, O) elements are vital elements because these compose structural skeleton and are involved in metabolic, physiological, and developmental processes of plants. Deficiency of these elements negatively affects the plant growth and yield. Differential root elongation pattern in response to N and P deficiencies were reported in *O. sativa* (Sun et al. 2016). Nitrogen deficiency slowed down growth and increased sucrose and trehalose 6-phosphate accumulation in *Arabidopsis* (Nunes et al. 2013). Under N shortage condition, *Z. mays* plant accumulated a number of proteins, which were involved in photosynthesis and metabolism (Nazir et al. 2016). Deficit of P modulates CO_2 assimilation, causes PSII inhibition, and reduces the metabolism in glycolysis and TCA cycle (Hernández and Munne-Bosch 2015). Deficiency of S significantly hampers N assimilation, purine metabolism, increases lipid breakdown, disrupts S assimilation, mitochondrial respiratory chain activity, and causes high photorespiration rate as well as interrupts energy assimilation mechanisms (Victoria et al. 2005; Ostaszewska et al. 2014). *Arabidopsis* plant showed differential responses to S deficiency. Distorted mitochondrial respiratory chain activity, ATP levels, and energy status of cell were noticed under S deficiency stress (Ostaszewska et al. 2014). Decreased gene expression of Rubisco and increased photorespiration were reported in S deficit *Arabidopsis* (Nikiforova et al. 2005). Phosphate fertilizers and fluorine (F) containing pesticides may release F in soil and environment that is one of the most toxic pollutants distressing the environment harshly and limiting plant growth and development (Wahid et al. 2014; Choubisa and Choubisa 2016). Olive plants were treated with NaF (20, 40, and 80 mM, 5 months). Increased F content in plant tissues with a higher level in roots compared to leaves was noticed. High level of H_2O_2, thiobarbituric acid reactive substances (TBRS), lipid peroxidation and electrolyte leakage, reduction of antioxidant enzyme activities (SOD, CAT, and glutathione peroxidase; GPX), and lack of mineral contents have been demonstrated in both roots and leaves especially at the highest dose of NaF (Zouari et al. 2017). Insufficiency of K improved radio caesium (Cs) translocation from stem to old leaves and then to younger organs of rice plant (Nobori et al. 2016). Potassium deficiency decreased stomatal conductance, mesophyll conductance and caused biochemical limitations by 23.9%, 33.0%, and 43.1%, respectively, in *B. napus* (Lu et al. 2016). In *Arabidopsis*, Fe deficiency not only generated H_2O_2 and oxidative stress responses but also this generation of H_2O_2 acted as signal for Fe deficiency responses and up-regulated iron deficiency-induced transcription factor (Le et al. 2016). Reduced roots and shoots growth, leaf necrosis (Hermans et al. 2010), impaired carbon metabolism, and decline of Chl and carbon fixation have been proved through transcriptomic analysis in *A. thaliana* grown under Mg deficit soil (Hermans et al. 2010). Inhibition of photosynthesis, assimilate translocation events including phloem loading, sucrose export, and transport were adversely affected under Mg deficiency stress (Cakmak and Kirkby 2008).

Low Light

Low light also exerts abiotic stress, which causes different morphological, physiological as well as yield contributing characters, yield, and yield quality (Liu et al. 2014); it also hampers the membrane integrity, ATPase activity, and causes photoinhibition of PSII and PSI including decreased Chl content (Meng et al. 2012). Low light may increase plant height and leaf area, which is a kind of plant adaptation to capture more light (Ren et al. 2002; Liu et al. 2009). Low light before the heading stage hampers the fertility of panicle in *O. sativa*. Low light stress after the heading stage decreases net photosynthetic rate, dry matter accumulation, and sink capacity, which decrease the number of filled grain, 1000-grain weight, and the final yield (Kato 1986; Deng et al. 2009; Liu et al. 2009). Low light stress is liable for insufficient supply of assimilates; decreased activity of a soluble starch branching enzymes involved in starch synthesis in grains, which reduces the number of filled grains and increases the percentage of chaffy/unfilled grains (Ren et al. 2003). Effect of low light on growth, yield, pigment composition, photosynthetic efficiency, and antioxidant defense systems of *T. aestivum* cultivars have been studied at different growth stages. Low light stress at the early stage of seedling didn't affect different studied parameters significantly. But at reproductive stage, low light adversely affected different studied parameters. An increase in POD activity was associated with loss of photosynthetic activity and formation of H_2O_2. Harvest index (HI), yield, and nutrient contents were negatively affected, when low light stress was imposed at grain filling stage. No substantial gain was formed under low light at grain filling stage, which affected the final production (Dong et al. 2014).

Herbicide Toxicity

To increase plant productivity, herbicide and pesticide become indispensible, but these chemicals and their residues have become one of the serious ecological evils across the earth (Lu et al. 2015a, b; Yadav et al. 2015). Wheat plants exposed to 0.8–8.0 mg kg^{-1} simetryne (which is the so-called s-triazine herbicide used in agriculture and aquatic environment) for 7 days exhibited suppressed growth and decreased Chl content, overproduction of ROS, injured the membrane lipids with higher activities of antioxidant enzymes, SOD, CAT, POD, APX, glutathione reductase (GR), and glutathione *S*-transferase (GST) (Jiang et al. 2016). Mixtures of herbicides imazethapyr and imazapic or imazapyr and imazapic caused greater reduction in total Chl and car contents in rice plant (Avila et al. 2015). Topramezone is used for post-emergence control of broadleaf and grass weeds in corn field but it was found as detrimental to the green algae *Chlorella vulgaris* growth during the 24–96 h of exposure. Pigments content and relative transcript of photosynthesis-related genes were adversely affected. Membrane permeability damage rate was up to 40.40% due to ROS triggered by topramezone, and morphology was also affected in those green algae (Zhao et al. 2017).

Interactive effects of 24-epibrassinolide (EBL) and imidacloprid (IMI) pesticides on AsA, tocopherol, GSH, polyphenols, and total phenols, *B. juncea* plants were studied and it was revealed that all these antioxidants were significantly increased due to interaction of 24-EBL and IMI (Sharma et al. 2016). Willow (*Salix miyabeana,* cultivar SX64) was exposed to glyphosate (0, 1.4, 2.1, and 2.8 kg ha^{-1}) and aminomethylphosphonic acid (AMPA, the principal glyphosate by-product; 0, 0.28, 1.4, and 2.8 kg ha^{-1}). Decreased Chl pigment contents (and increased Chl degradation), Chl fluorescence, increased oxidative stress markers (higher H_2O_2 content and inhibition of antioxidant enzyme activities) in leaves were evident after 12 h of exposure (Gomes et al. 2016). Root meristems of *Allium cepa* cells showed loss of membrane integrity, increased chromosome aberrations, micronucleus formation, DNA strand breaks, and cell-cycle arrest due to exposure of ZnO nanoparticles. In *Vicia faba* and *Nicotiana tabacum*, toxicity of ZnO nanoparticles was characterized by ROS production, lipid peroxidation, and modulation in activities of antioxidant enzymes (Ghosh et al. 2016). Another popular herbicide paraquat also causes damage to crops. Paraquat is well known for its $O_2^{\bullet-}$ generating capability that resulted in oxidative stress, which was evidenced with higher lipid peroxidation (MDA), H_2O_2 content and $O_2^{\bullet-}$ generation rate, higher LOX activity and MG level. Paraquat toxicity also decreases plant biomass, leaf Chl, and relative water content; disrupts the antioxidant defense by altering the content and activity of antioxidants. Furthermore, activities of glyoxalase system were also disrupted by paraquat in *B. napus* (Hasanuzzaman et al. 2018).

Wind and Mechanical Injury

Almost all of the plants need mechanical support throughout their lifetime. Wind is one of the major mechanical barriers that has an important role on plant growth, morphology, physiology, and ecology (Gardiner et al. 2016). Wind causes leaf scraping, tearing, folding to plant resulting in damage to cuticle, and then induces quick water loss and desiccation as well as allows pathogen entry. Most of the plants including agricultural crops are susceptible to wind damage by lodging (Gardiner et al. 2016). Yield losses under wind stress in *T. aestivum*, *O. sativa*, and *Z. mays* is reported due to insufficient light penetration owing to canopy compactness (Kashiwagi and Ishimaru 2004; Gardiner et al. 2016).

Calcium-Induced Regulation of Growth and Physiology of Plants Under Abiotic Stress

Calcium (Ca) is one of the essential macronutrients for plant growth and development, where Ca plays multifunctional roles in many processes, such as structural and functional integrity of plant membrane, including cell wall stabilization, regulation of ion transportation and selectivity, regulation of ion-exchange behavior, and also involved in activation of different enzyme activities (Rengel 1992; Marschner 2012). It is also called as an intracellular second messenger to engage in extracellular stimuli for intracellular responses and also coordinate a vast range of endogenous actions (Edel et al. 2017). Hence, Ca is not only a mineral nutrient for plant, but also mediates the cell and plant development processes, as well as plant response to different stress conditions by regulating many physiological and cellular aspects (White and Broadley 2003; Hirschi 2004; Tang et al. 2006; Rahman et al. 2015a). Calcium plays its regulatory role on seed germination, growth and development, water relations, photosynthesis, and many more (Table 1). Some of them have been presented below.

Seed Germination

The fundamental phase in plant life cycle is seed germination; it is also a vital phase of growth and development that affect the plants establishment and yield. Seed germination starts with water imbibition, which reactivates the metabolic activity and ends by giving the seedling radicle and plumule emergence from the dry seed (Bewley and Black 1994; Duval et al. 2002). Naturally, seeds contain essential miner nutrient as per requirement for their germination, but availability of these are hampered by stresses like salinity, drought, cold, etc. (Knight and Knight 2001). In this situation, Ca provides stress protection by regulating many physiological and

Table 1 Calcium-induced growth and yield improvement in different crop species under various abiotic stress conditions

Crop species	Abiotic stresses	Exogenous Ca application	Improvements of growth and yield	References
Lycopersicon esculentum Mill.	75 mM NaCl	2.5 mM and 5 mM of Ca as $CaSO_4$	• Increased plant growth and biomass • Improved fruit number, weight, and yield	Tuna et al. (2007)
Phragmites karka Retz.	100, 200, 300, 400, 500 mM NaCl	10 mM $CaCl_2$	• Increased seed germination (80%)	Zehra et al. (2012)
Nicotiana tabacum L.	Heat stress; 30 °C, 35 °C, 40 °C, 45 °C, 47 °C	20 mM $CaCl_2$	• Improved CO_2 assimilation under both stress and recovery • Increased net photosynthesis rate up to 86%	Tan et al. (2011)
Cucumis sativus	50, 100, 150, 200 mM NaCl	CaC_2 at 5, 10, 20, 30 mg plate^{-1}	• Enhanced seed germination through increase ethylene level	Shakar et al. (2016)
Pisum sativum	75 mM or 150 mM NaCl	5.44 mM $CaCl_2$	• Increased root elongation and plant development • Stimulated the remobilizations of seed nutrients stores	Bonilla et al. (2004)
L. esculentum	50, 100, and 150 mM NaCl	100, 200, and 300 mg L^{-1} $CaCl_2$	• Increased leaf number and leaf fresh weight • Increased number of fruit and fruit weight	Lolaei et al. (2012)
L. esculentum	2, 4, 6, 8 dS m^{-1} of NaCl	5, 10 mM of Ca as $CaSO_4.0.5H_2O$	• Reduced the yield reduction rate • Increased flowers and fruit number • Increased fruit size	Parvin et al. (2015b)
Oryza sativa L.	0.5, 1 mM Na_2HAsO_4	10 mM $CaCl_2$	• Decreased As accumulation and restored plant growth and water loss • Lowered Pro content and increased plant dry weight • Increased photosynthetic pigments (Chl *a* and Chl *b*)	Rahman et al. (2015b)

(continued)

Table 1 (continued)

Crop species	Abiotic stresses	Exogenous Ca application	Improvements of growth and yield	References
Cicer arietinum L.	200 μM Cd as $CdSO_4.8H_2O$	50 mM Ca as $CaCl_2$	• Enhanced 67% shoot and 35% root length • Increased biomass production • Pod number and seed yield increased by 29% and 52%, respectively • Reduced Cd accumulation in root about 82% • Chlorophyll and Pro content increased	Ahmad et al. (2016)
Hordeum vulgare	100 μM Co as $CoCl_2.6H_2O$	2 mM $CaCl_2$	• Stimulated growth by increasing length and dry weight • Chlorophyll content was increased up to 39% • Reduced Co concentration in both root and shoot, while translocation rate was also decreased	Lwalaba et al. (2017)
Brassica juncea; Sesbania sesban	50 μM $CdCl_2$	0.2 and 2 mM Ca where 1 mM as CaSO4 and 1 mM as $CaCl_2$	• Resulted in higher shoot biomass production • Increased root length	Eller and Brix (2016)
L. esculentum	150 and 300 μM $NiSO_4$	400 and 700 μM Ca as $CaCl_2$	• Increased shoot and root length • Enhanced leaf RWC; total Chl and car content	Mozafari et al. (2013)
Zoysia japonica	Drought stress; 7 days	5, 10, and 20 mM $CaCl_2$	• Increased above and below ground biomass • Enhanced the Chl content including Chl *a*, *b* and Chl (*a*+*b*) • Exhibited increased photosynthetic rate (Pn)	Xu et al. (2013)
Camellia sinensis L.	Drought; 20 days	50 and 100 μM $CaCl_2$	• Increased leaf dry mass • Improved RWC up to 93%	Upadhyaya et al. (2011)

(continued)

Table 1 (continued)

Crop species	Abiotic stresses	Exogenous Ca application	Improvements of growth and yield	References
Zea mays	Drought; 30% FC	20, 40, and 60 mg L^{-1} Ca^{2+} as $CaCl_2.2H_2O$	• Enhanced photosynthesis (45%) • Resulted increment of 13% in Chl *a*, 20% Chl *b*, and 32% in car • Improved 20% grain yield	Naeem et al. (2017)

cellular responses (Zehra et al. 2012). Under saline condition, Ca restricts the entry of Na^+ and thus reduces the adverse effect of NaCl on seed germination (Marcar 1986; Nayyar 2003; Bonilla et al. 2004; Zehra et al. 2012). Toxicity of salt on germination of *Pisum sativum*, *T. aestivum*, *H. annuus*, *L. esculentum* (Bonilla et al. 2004; Turkmen et al. 2004; Daowei and Moxin 2010; Li-Yun and Ming-You 2010), *Haloxylon ammodendron*, *Artemisia ordosica*, *Aristida adscensionis*, *Bassia dasyphylla*, *Chenopodium album* (Tobe et al. 2004; Yao et al. 2010), and several other halophytic species of Pakistan coast (Gul and Khan 2006) were minimized by Ca application.

Calcium significantly mitigates the toxicity of Na^+ and Mg^{2+} on the germinating seeds of *Kalidium capsicum* (Tobe et al. 1999), *H. vulgare* (Bliss et al. 1986), and *Urochondra setulosa* (Shaikh et al. 2007). Again, chloride and sulfate salts toxicity on seed germination of *K. capsicum* and *H. ammodendron* were alleviated by low concentration of Ca and also reduced the K^+ efflux (Tobe et al. 2004), and thus enhanced seedling growth. Application of $CaCl_2$ (10 mM) improved seed germination up to 80% of *Phragmites karka* under 500 mM of NaCl (Zehra et al. 2012). Use of 50 mM concentration of Ca^{2+} ($CaCl_2$) as seed priming agent showed most beneficial for seed germination regardless of genotypes in *H. vulgare* under drought stress (Kaczmarek et al. 2017). Use of Ca as seed osmopriming is also beneficial for other field and cereal crops such as *O. sativa* (Basra et al. 2004; Farooq et al. 2009a), *T. aestivum* (Nayyar et al. 1995), and *H. annuus* (Kathiresan et al. 1984) by improving germination rate, seedling emergence as well as seedling growth and establishment. Medium or low concentrated Ca^{2+} improved seed germination characteristics of *Koelreuteria paniculata* (Cai et al. 2013). Under waterlogging condition, Ca increased seed germination percentage of rice by making pelleting of seed with CaO_2 as well as improved seedling growth (Mei et al. 2017) because of increased oxygen availability due to increased CaO_2 (Baker and Hatton 1987). Inhibition of germination of *P. karka* was alleviated by Ca^{2+} addition under dark condition (Zehra et al. 2012), which was also observed in *U. setulosa* (Shaikh et al. 2007). Calcium can also partially mitigate the chilling-induced (10 °C) oxidative stress from germinating of seed, such as wheat grains (Nayyar and Kaushal 2002).

Application of Ca causes perturbation in cytoplasmic Ca^{2+} and triggers Ca sensor proteins, for example, calmodulin (CaM), calcenurin B-like proteins (CBLs), and Ca-dependent protein kinases (CDPKs), which are involved in activation of metabolic processes (Zehra et al. 2012). The CDPKs is involved in embryogenesis, seed development, and germination in *Santalum album* (Anil et al. 2000).

Growth

Calcium enhances the plant tolerance to different abiotic stresses by improving the growth and development of crops. Supplemental Ca can enhance the plant dry matter by increasing root and shoot dry weight of *S. lycopersicum* under saline condition (Tuna et al. 2007). Exogenous application of Ca showed the increased plant height, leaf and branch number, leaf area of *S. lycopersicum* under salt-stressed condition (Parvin et al. 2015a). Plant dry weight was also increased with additional Ca application. Improvement of growth- related parameters thus indicates the salinity tolerance behavior of plants supplemented with Ca. Salt-stressed wheat plant showed better plant height, when treated with Ca (Al-Whaibi et al. 2011). Similar result was also found in date palm tree (Jasim et al. 2016). Calcium also improved leaf area under salt-stressed condition in date palm (El-Khawaga 2013; Jasim et al. 2016). Under drought condition, Ca also improved fresh root and shoot biomass along with increased dry weight of *Zoysia japonica,* whereas 10 mM Ca showed a comparative better result (Xu et al. 2013).

Calcium effectively reduces the Cd toxicity, which was observed in *C. arietinum* by improving 66.8% shoot length, 34.79% root length along with increased shoot and root fresh and dry weight (Ahmad et al. 2016). Moreover, Ca may enhance cell elongation and expansion resulting in improvement of plant growth, which was also observed by Hernandez and Almansa (2002) in *P. sativum* and Abdel Latef (2011) in *B. napus*. Again Ca-induced growth promotion was also found in *P. vulgaris* under Cd toxicity (Ismail 2008). Shoot and root biomass were also increased by Ca to Cd-stressed *B. juncea* plant (Eller and Brix 2016). Plant responded under Ni toxicity by showing increased fresh and dry weight of shoot and root with improved leaf area with supplemental Ca application (Mozafari et al. 2013). Application of Ca gave higher dry weight in As-treated rice plant (Rahman et al. 2015a) and also increased fresh and dry weight of Cd-exposed rice seedling (Rahman et al. 2016). Toxic metal accumulation in roots is greater than shoot, which causes cell damage resulting in growth inhibition (Talukdar 2012). Supplemental Ca application in growth medium enhances the rate of Ca accumulation as well as reduces Cd uptake (Tian et al. 2011; Ahmad et al. 2015) resulting in restoration of plant growth.

Water Relations

Leaf relative water content (RWC) was increased by Ca (Mozafari et al. 2013) in Ni-stressed *L. esculentum* plants. Cadmium-treated rice seedling showed higher leaf relative turgidity with the exogenous application of Ca (Rahman et al. 2016). Proline content is related to water status of plant, and its accumulation is increased under different abiotic stress conditions (Hasanuzzaman et al. 2018; Nahar et al. 2015), and shows reduced RWC (Hasanuzzaman and Fujita 2013; Alam et al. 2013; Nahar et al. 2015). Moreover, exogenous Ca improves plant water status by regulating a balanced Pro content (Rahman et al. 2016). Calcium also reduced Pro accumulation in salt-stressed *V. radiata* (Manivannan et al. 2007). Exogenous Ca application on drought-stressed plant showed the lower Pro content, due to increased Pro degrading enzyme (Jaleel et al. 2007). In addition, Ca prevents salt-induced damage from cellular dehydration through regulating osmotic strength of cytoplasm (Arshi et al. 2006). Similar result was also found in Ni-stressed *S. lycopersicum* plants (Mozafari et al. 2013). Improved RWC in *S. lycopersicum* leaves by supplemental Ca under salt stress was also reported (Tuna et al. 2007). Salt-stressed date palm plants showed increased RWC with the supplemental application of Ca (Jasim et al. 2016).

Photosynthesis

Calcium increased the Chl *a*, Chl *b*, total Chl, and car content in chickpea when grown under Cd toxic condition (Ahmad et al. 2016). Thus, exogenous Ca improved photosynthetic pigments in other crops such as *V. faba* (Siddique et al. 2012) and *B. juncea* (Ahmad et al. 2015) under Cd toxicity. It was told that Ca plays its role as second messenger for activating cytokinin activity and thus can improve Chl biosynthesis (Lechowski and Bialczyk 1993).

Leaf Chl content including both Chl *a* and Chl *b* of *L. esculentum* was increased under both Ni-stressed or unstressed condition, and this increment of Chl showed that leaf growth and expansion depend on Ca concentration (Mozafari et al. 2013). Supplementation of Ca improved photosynthetic pigments such as Chl *a*, Chl *b*, and car in Cd-treated rice seedling, where only car had increased up to 44% (Rahman et al. 2016). Increased Chl content was also found in *B. juncea* L. with Ca supplementation under Cd exposure (Ahmad et al. 2015). Exogenous Ca showed increased Chl *a*, Chl *b*, total Chl, and car content in *O. sativa* (Sivasankaramoorthy 2013) and *Withania somnifera* (Jaleel and Azooz 2009). Total Chl content was increased up to 39% in Co-stressed *H. vulgare* plant with exogenous application of Ca (Lwalaba et al. 2017). When plant suffers from loss of Chl under salt stress, Ca rejuvenated Chl in date palm (Jasim et al. 2016). Although Ca successfully increased the Chl content in drought-stressed plant (Xu et al. 2013), this increment in drought-stressed plant indicates the growth promoting behavior of Ca under drought stress (Xu et al. 2013). Photochemical efficiency (Fv/Fm) is known as an indicator for the measurement of photosynthesis efficiency, and this Fv/FM ratio had increased in Ca-treated plant under drought stress (Xu et al. 2013).

Calcium inhibited heat stress-induced loss of Chl possibly by reducing photooxidation or keeping membrane integrity (Coria et al. 1998). Calcium plays its vital role on PSII acceptor side, and oxygen evolving complex (OEC) in heat-stressed plant especially enhances the thermo tolerance and thus maintains photosynthetic capacity (Tan et al. 2011). Exogenous Ca also improves net photosynthetic rate, stomatal conductance, and also photochemical efficiency in HT-stressed *Nicotiana tabacum* plant. Calcium also improved Rubisco activity that enhances photosynthesis as well as CO_2 assimilation rate. Then, reduction of Chl in leaves indicates the reduction of photosynthesis as well as dry matter production, where Ca plays its alleviating role to improve Chl content, as well as photosynthate production under different abiotic-stressed condition.

Ca-Mediated Yield Improvement of Crop Plants Under Abiotic Stress

Ca can regulate the yield improvement of various crops under abiotic stresses such as salinity, drought, flooding, heat, chilling, and heavy metal stress. Calcium supplementation not only improved morphological and physiological features of plant but also significantly increased yield characteristics as well as yield under saline condition (Parvin et al. 2015b).

Salinity at 75 mM concentration was harmful for yield of *S. lycopersicum* by reducing both fruit number and fruit weight per plant, and in this condition exogenous Ca application (5 mM) improved both of these characters and improved yield under salt stress (Tuna et al. 2007). Exogenous Ca also reduced the flower drop, ultimately increased yield under salt stress (Parvin et al. 2015b). Calcium-treated plants showed higher fruit weight per plant, which resulted in higher yield from NaCl-treated tomato plant (Lolaei et al. 2012). Similar result was also observed by Rubio et al. (2009). Calcium supplementation produced higher number of flowers as well as fruit weight of sweet pepper under drought stress condition, and also in control plants (Azam et al. 2016), where 15 mM of Ca gave the best result. Fruit yield of tomato plants was enhanced by Ca as it increased the rate of nitrogen uptake by the fruits. Higher amount of fruit yield production may also be due to the effect of increase in root primordia and results in the greater absorption of nutrients (Siddiq et al. 2009). Supplemental Ca enhanced the pod number and seed yield about 29.42% and 52.45%, respectively, of *C. arietinum* under Cd-stress (Ahmad et al. 2016). Again, application of Ca with combination of K can enhance both pod number and seed yield up to 68.51% and 92.21%, respectively, under Cd-stress. Exogenous application of Ca and K increased yield, which might be due to higher mineral and water uptake, restoration of photosynthetic pigments and rehabilitation of other metabolic processes that were previously affected by Cd. Similar result was also recorded from *Brassica* plant under Cd toxicity where Ca enhanced the biomass yield (Ahmad et al. 2015). Hence, it is evident that the combination of Ca and K supplementation is more effective than Ca solely, and the findings are in conformity with the finding of Siddique et al. (2012) in Cd-stressed *V. faba* (Table 1).

Calcium-Induced Oxidative Damage Mitigation and Up-Regulation of Antioxidant Defense

Upon various abiotic stresses, plant goes through oxidative damage by facing oxidative stress. Being a macronutrient, Ca plays an important role in signal transduction mechanism of plant cell metabolism (Lwalaba et al. 2017). Directly or indirectly Ca regulates the target proteins through Ca-binding proteins, such as CaM to activate protein kinases and other proteins including enzymatic antioxidants (Lwalaba et al. 2017). Calcium keeps membrane more integrated and prevents EL with reduced lipid peroxidation and thus strengthens the plant's tolerance to abiotic stresses (Antosiewicz and Hennig 2004; Hirschi 2004).

Under different abiotic stress situation, there is a net increase in cytosolic Ca^{2+} content to stimulate various biological upstream and downstream responses causing plant adjustment to harmful environment by regulating different antioxidant defense mechanisms and decrease of membrane lipid peroxidation leads to plants survival (Nayyar and Kaushal 2002; Upadhyaya et al. 2011; Zhu et al. 2013).

In metal-stressed condition, Ca hampers metal accumulation through controlling membrane permeability and movements of divalent cations across cell membrane, also competes for transporter site (Farzadfar et al. 2013). Reduction of ROS generation such as H_2O_2 and $O_2^{\bullet-}$ were found in Co-stressed *H. vulgare* plant with the supplementation of Ca. Again the decreased MDA content was recorded as an indicator of reduction of lipid peroxidation resulting in reduction of oxidative stress by Ca under metal stress (Lwalaba et al. 2017). Cobalt-stressed *H. vulgare* seedling showed higher content of GSH, SOD, POD, CAT, and GR naturally to alleviate the oxidative stress, where exogenous use of Ca enhanced all of these antioxidant components compared to stressful condition. Cadmium-exposed rice seedling also showed lower oxidative damage when treated with supplementel Ca by decreasing H_2O_2 and $O_2^{\bullet-}$ including reduced MDA production, together with up-regulated antioxidant defense and glyoxalase system (Rahman et al. 2015a). Moreover, calcium induced lower LOX activity in Cd-stressed rice plant showed comparative relief from oxidative damage (Rahman et al. 2015a). Several other researchers also stated that Ca can effectively reduce the Cd-induced oxidative damage by checking production of ROS in *O. sativa* (Srivastava et al. 2014), *Lens culinaris* (Talukdar 2012), and *Sedum alfredii* (Tian et al. 2011). Arsenic-stressed plant also showed up-regulated antioxidant defense and glyoxalase activity when treated with exogenous Ca (Rahman et al. 2015b). Application of Ca decreased H_2O_2 and MDA by 20.4% and 23.3%, respectively, in Cd-treated *Sesamum indicum* thereby providing cell protection from oxidative damage and up-regulated activities of antioxidant enzymes (Abd_Allah et al. 2017). Moreover, reduction of H_2O_2 and lipid peroxidation due to Ca supplementation in stressed plants were also found in *A. thaliana* (Suzuki 2005), *V. faba* (Siddique et al. 2012), *B. juncea* (Ahmad et al. 2015), and *H. vulgare* (Lwalaba et al. 2017).

Thus, it is well established that exogenous Ca application in growth medium-stressed plants alleviated oxidative stress by decreasing ROS through antioxidative activities. Heat-shocked tobacco plant showed lower MDA content with Ca treatment and Ca-treated plant significantly decreased electrolyte leakage in both normal and high temperatures environment (Tan et al. 2011). Moreover, Ca also increased some antioxidant enzymatic activities and thus reduced the rate of $O_2^{\bullet -}$ production and H_2O_2 content in tobacco leaves under both optimum and high temperature condition.

Chilling stress-induced oxidative damage as evident by lipid peroxidation (MDA) and H_2O_2 was decreased by Ca application in *C. arietinum* seedlings (Nayyar et al. 2005a). Calcium-treated cold-stressed *T. aestivum* plants showed reduction in lipid peroxidation and membrane damage, which indicate the tolerance of plants against cold-stress induced oxidative damage (Nayyar and Kaushal 2002). Calcium supplementation under salt-affected plants increased Pro production, which reflects more tolerance to osmotic stress caused by ROS through acceleration of better enzymatic (POD, CAT) and non-enzymatic antioxidant defense (Jasim et al. 2016). This result is also supported by Al-Whaibi et al. (2011), who studied the salt toxicity alleviation effect of Ca on *T. aestivum* plant. Foliar application of Ca is also very effective in imposing drought tolerance. In favor of this, Naeem et al. (2017) stated that Ca successfully regulates the activities of antioxidant compounds, such as 37% of SOD, 24% of CAT, and 49% of APX along with reduction of lipid peroxidation expressed by 24% decrease of MDA content in drought-exposed *Z. mays* plant. Yet, supplementation of Ca causes improvement of drought tolerance by modifying the water shortage-induced ROS metabolism (Xu et al. 2013; Zhu et al. 2013). Thus, Ca significantly reduces oxidative damage in plants under different abiotic stress condition, making plant more tolerant (Table 2).

Mineral Nutrient Uptake Regulation by Ca Under Abiotic Stress

Calcium is very essential plant nutrient, which plays a role in balancing membrane integrity and ion transport regulations in plants (Sivasankaramoorthy 2013). Supplemental Ca increased K^+, Ca^{2+}, and N in *L. esculentum* plant against salinity (Tuna et al. 2007). Salt-stressed *L. esculentum* plant showed decrease of Na^+ content in leaves along with increased Ca^{2+} and K^+ when treated with exogenous Ca (Parvin et al. 2016), and in this condition 0.25 mM of Ca showed comparatively better result in ion modulation than 0.50 mM concentration, caused increased K^+ content, K^+/Na^+ and Ca^{2+}/Na^+ ratios. Thus, Ca sustains the transport of K^+ and the selectivity of K^+/Na^+ ions. Calcium-treated date palm plants showed increased K^+ content and K^+/Na^+ ration in saline-stressed condition (Jasim et al. 2016). This increase of K^+ content helps to decrease Na^+ uptake under salt-stress, which is required for sustaining osmotic balance (Tuteja and Mahajan 2007); in addition, Ca is also responsible for restoring N, P, and K content in tissue (Grattan and Grieve 1999). Therefore, ion

Table 2 Calcium-induced mitigation of abiotic stress in different crop species

Crops	Stress levels	Ca doses	Effects	References
Lycopersicon esculentum Mill.	75 mM NaCl	2.5 mM and 5 mM of Ca as $CaSO_4$	• Strengthen membrane by decreasing EL • Reduced Na^+ concentration in leaves and increased Ca^{2+}, N, and K^+	Tuna et al. (2007)
Matricaria chamomilla L.	0, 120, and 180 μM $CdCl_2$	0, 0.1, 1, and 5 mM $CaCl_2$	• Notably reduced H_2O_2 and $O_2^{\cdot-}$ • Lowered MDA content • Reduced the CAT, SOD, and POX activity in both shoot and roots	Farzadfar et al. (2013)
Hordeum vulgare L.	100 μM Co as $CoCl_2$. $6H_2O$	2 mM $CaCl_2$	• Reduced Co concentration in both shoots and roots • Decreased H_2O_2 and $O_2^{\cdot-}$ as well as MDA content • Increased GSH content • Enhanced the activities of CAT, POD, SOD, GR	Lwalaba et al. (2017)
Nicotiana tabacum L.	High temperature; 43 °C for 2 h	20 mM $CaCl_2$	• Decreased the content of H_2O_2 and $O_2^{\cdot-}$ • Increased the activity of SOD, CAT, POD, APX, and GR • Reduced both MDA and EL	Tan et al. (2011)
Cicer arietinum L.	Chilling (4 °C); 10 days	1 μM Ca^{2+} as $CaCl_2$	• Decreased 30% of H_2O_2 • Reduced EL up to 30% • Reduced MDA up to 29%	Nayyar et al. (2005b)
Triticum aestivum L.	Low temperature as at 10 °C for 24 h	1 mM $CaCl_2$	• Lowered H_2O_2 and MDA content in embryos • Increased membrane stability and AsA content • Stimulated the activity of CAT, APX, SOD	Nayyar and Kaushal (2002)
T. aestivum	90 mM NaCl	40 mM $CaCl_2$	• Decreased the MDA content • Increased POD, CAT, SOD, GR, and APX activity • Lowered Pro content	Al-Whaibi et al. (2011)
Zea mays	Drought; 30% FC	20, 40, and 60 mg L^{-1} Ca^{2+} as $CaCl_2.2H_2O$	• Improved water potential and turgor potential • Increased stomatal conductance, transpiration rate, and total soluble sugars • Declined H_2O_2 content	Naeem et al. (2017)
Zoysia japonica	Drought; 3, 6, 9, and 16 days	5, 10, and 20 mM $CaCl_2$	• Lowered MDA and Pro contents • Increased SOD, CAT, and POD activities	Xu et al. (2013)

(continued)

Table 2 (continued)

Crops	Stress levels	Ca doses	Effects	References
N. tabacum	0.05 mM $Cd(NO_3)_2$	0.01, 1, 3, 10 mM $Ca(NO_3)_2$	• Accumulated less Cd in roots and thus enhanced Cd tolerance	Antosiewicz and Hennig (2004)
Camellia sinensis L.	Drought; 20 days	50 and 100 μM $CaCl_2$	• Decreased H_2O_2 and lipid peroxidation • Increased the activities of enzymatic antioxidants (SOD, CAT, POX, and GR) • Increased Pro and phenolic content in leaves	Upadhyaya et al. (2011)
Oryza sativa L.	0.25 and 0.5 mM $CdCl_2$	2.5 mM $CaCl_2$	• Reduced Cd uptake upto 64 and 46% in shoot and root, respectively • Increased leaf relative turgidity in Cd-treated plant • Markedly reduced Pro content • Reduced H_2O_2 and $O_2^{\bullet-}$ generations as well as LOX activity • Maintained higher AsA and lower oxidized GSH content • Increased monodehydroascorbate reductase (MDHAR) and dehydroascorbate reductase (DHAR) activity along with 81% increased of GST activity. • GPX, SOD, and CAT activity were also increased • Enhanced glyoxalase I (Gly I) and glyoxalase II (Gly II) activity and resulting declined MG content	Rahman et al. (2015a)
O. sativa	200 mM NaCl	2 mM $CaCl_2$	• Improved ROS and MG detoxification by antioxidant defense and glyoxalase system, respectively • Improved RWC as well as reduced Pro content • Decreased membrane damage indicated by the reduction of 42% MDA content and lowered LOX activity	Rahman et al. (2016)
Catharanthus roseus	10, 15, and 20 days interval drought stress	5 mM $CaCl_2$	• Diminished Pro content • Slightly reduced the H_2O_2 content • Reduced the TBARS content in both root and shoot	Jaleel et al. (2007)

imbalance caused by salinity through substitution of K^+ by Na^+ and decrease of Ca^{2+} can be controlled by the application of exogenous Ca, throughout and restricting Na^+ accumulation and maintaining higher K^+/Na^+ and Ca^{2+}/Na^+ ratios, resulting in enhancement of salt tolerance (Tuna et al. 2007; Wu et al. 2013). Addition of 5 mM Ca caused higher content of Ca^{2+}, N^+, and K^+ in both shoots and roots of *O. sativa* plants against of 50 mM of NaCl (Sivasankaramoorthy 2013). Adequate amount of Ca in growth medium thus improves the K^+/Na^+ selectively by exchanging the uptake ratio in favor of K^+ at the expense of Na^+.

Calcium is also essential for recovery from drought-induced damage by regulating plasma membrane ATPase, which requires to pump back the nutrients that were lost in case of cell damage (Palta 2000).

Calcium treatment caused increase of mineral nutrient uptake, such as K^+ uptake in plant shoots and roots of *L. esculentum* plants under both Ni-stressed and control condition (Mozafari et al. 2013). Moreover, Ca can improve K uptake in Cd-exposed *Z. mays* plants by reducing Cd concentration in roots (Kurtyka et al. 2008). Calcium (50 mM) treatment in Cd-stressed plants appreciably restored mineral elements (S, Mn, Mg, Ca, and K) that were also suppressed in Cd-stressed *C. arietinum* plants in both shoot and root (Ahmad et al. 2016).

Exogenous application of Ca causes an increase of specific utilization rate (SUR) of Ca^{2+}, Mg^{2+}, and K^+ in both stressed and non-stressed plants (Murillo-Amador et al. 2006). Calcium increased the N uptake and N assimilation in *L. esculentum* shoots and pods and responsible for the higher yield (Siddiq et al. 2009).

Conclusion

Now we are under threat of food security, while world population is increasing with decreasing of cropping area. It is essential to increase world food production by making plant more tolerant to different stressful environment. Therefore, it remind us to think about developing some quick actions including use of chemical approaches, especially those are available for farmers and inexpensive compared to molecular approach. Then the use of plant nutrient to enhance crops' tolerance is very promising. Hence, Ca is a very important nutrient element, which also acts in stimulating intracellular responses under environmental deviations. Many researchers already reported about the stress alleviating role of Ca to different kind of abiotic stresses including salinity, drought, heavy metal, heat, chilling, etc. In this chapter, we tried to highlight the available information briefly regarding the physiological responses of plants to supplementation of Ca. However, the mechanisms of stress alleviation by exogenous Ca by enhancing physiological processes demand further research focusing the interactions of Ca with other signaling molecules and phytohormones.

References

Abd_Allah EF, Hashem A, Alqarawi AA, Wirth S, Egamberdieva D (2017) Calcium application enhances growth and alleviates the damaging effects induced by Cd stress in sesame (*Sesamum indicum* L.). J Plant Interact 12:237–243

Abdel Latef AA (2011) Ameliorative effect of calcium chloride on growth, antioxidant enzymes, protein patterns and some metabolic activities of canola (*Brassica napus* L.) under sea water stress. J Plant Nutr 34:1303–1320

Acosta-Motos JR, Díaz-Vivancos P, Álvarez S, Fernández-García N, Sánchez-Blanco MJ, Hernández JA (2015a) NaCl-induced physiological and biochemical adaptative mechanism in the ornamental *Myrtus cummunis* L. plants. J Plant Physiol 183:41–51

Acosta-Motos JR, Díaz-Vivancos P, Álvarez S, Fernández-García N, Sánchez-Blanco MJ, Hernández JA (2015b) Physiological and biochemical mechanisms of the ornamental *Eugenia myrtifolia* L. plants for coping with NaCl stress and recovery. Planta 242:829–846

Ahlfors R, Brosche M, Kollist H, Kangasjärvi J (2009) Nitric oxide modulates ozone induced cell death, hormone biosynthesis and gene expression in *Arabidopsis thaliana*. Plant J 58:1–12

Ahmad P, Jaleel CA, Sharma S (2010) Antioxidative defence system, lipid peroxidation, proline metabolizing enzymes and biochemical activity in two genotypes of *Morus alba* L. subjected to NaCl stress. Russ J Plant Physiol 57:509–517

Ahmad P, Hakeem KR, Kumar A, Ashraf M, Akram NA (2012) Salt-induced changes in photosynthetic activity and oxidative defense system of three cultivars of mustard (*Brassica juncea* L.). Afr J Biotechnol 11:2694–2703

Ahmad P, Sarwat M, Bhat NA, Wani MR, Kazi AG, Tran LP (2015) Alleviation of cadmium toxicity in *Brassica juncea* L. (Czern. & Coss.) by calcium application involves various physiological and biochemical strategies. PLoS One 10:e0114571. https://doi.org/10.1371/journal.pone.0114571

Ahmad P, Abdel Latef AA, Abd_Allah EF, Hashem A, Sarwat M, Anjum NA, Gucel S (2016) Calcium and potassium supplementation enhanced growth, osmolyte secondary metabolite production, and enzymatic antioxidant machinery in cadmium-exposed chickpea (*Cicer arietinum* L.). Front Plant Sci 7:513. https://doi.org/10.3389/fpls.2016.00513

Alam MM, Hasanuzzaman M, Nahar K, Fujita M (2013) Exogenous salicylic acid ameliorates short-term drought stress in mustard (*Brassica juncea* L.) by upregulating the antioxidant defense and glyoxalase system. Aust J Crop Sci 7:1053–1063

Allen DJ, Ort DR (2001) Impact of chilling temperatures on photosynthesis in warm-climate plants. Trends Plant Sci 6:36–42

Al-Mutawa MM (2003) Effect of nitrogen supply on the response of faba bean to salinity. Int J Agric Biol 5:124–128

Al-Whaibi MH, Manzer HS, Mohammed OB (2011) Salicylic acid calcium-induced protection of wheat againsts salinity. Protoplasma 249:769–778

Amirjani MR (2011) Effect of salinity stress on growth, sugar content, pigments and enzyme activity of rice. Int J Bot 7:73–81

Anil VS, Harmon AC, Rao KS (2000) Spatio-temporal accumulation and activity of calcium-dependent protein kinases during embryogenesis, seed development, and germination in sandalwood. Plant Physiol 122:1035–1044

Anjum SA, Ashraf U, Tanveer M, Khan I, Hussain S, Shahzad B, Zohaib A, Abbas F, Saleem MF, Ali I, Wang LC (2017) Drought induced changes in growth, osmolyte accumulation and antioxidant metabolism of three maize hybrids. Front Plant Sci 8:69. https://doi.org/10.3389/fpls.2017.00069

Antosiewicz DM, Hennig J (2004) Overexpression of LCT1 in tobacco enhances the protective action of calcium against cadmium toxicity. Environ Pollut 129:237–245

Aroca R, Vernieri P, Irigoyen JJ, Sanchez-Díaz M, Tognoni F, Pardossi A (2003) Involvement of abscisic acid in leaf and root of maize (*Zea mays* L.) in avoiding chilling-induced water stress. Plant Sci 165:671–679

Arshi A, Abdin MZ, Iqbal M (2006) Effect of $CaCl_2$ on growth performance, photosynthetic efficiency and nitrogen assimilation of *Cichorium intybus* L. grown under NaCl stress. Acta Physiol Plants 28:137–147

Ashraf M (2009) Biotechnological approach of improving plant salt tolerance using antioxidants as markers. Biotechnol Adv 27:84–93

Ashraf M, Harris PJC (2013) Photosynthesis under stressful environments: an overview. Photosynthetica 51:163–190

Ashraf M, Iram A (2005) Drought stress induced changes in some organic substances in nodules and other plant parts of two potential legumes differing in salt tolerance. Flora 200:535–546

Asrar AWA, Elhindi KM (2011) Alleviation of drought stress of marigold (*Tagetes erecta*) plants by using arbuscular mycorrhizal fungi. Saudi J Biol Sci 18:93–98

Asthir B (2015) Protective mechanisms of heat tolerance in crop plants. J Plant Int 10:202–210

Avila LA, Marchezan MG, Camargo ER, Agostinetto D, Schaedler CE, Moraes BS, Langaro AC, Oliveira C (2015) Biochemical alterations of rice in response to plant submersion and herbicides. Procedia Environ Sci 29:73–74

Avivi S, Soeparjono S, Slameto, Ramadhan RA (2016) Physiological characters of sugarcane after flooding stress. Agric Agric Sci Procedia 9:31–39

Azam M, Noman M, Abbasi NA, Ramzan A, Imran M, Hayat A, Abbas H, Akram A (2016) Effect of foliar application of micro-nutrient and soil condition on growth and yield of sweet pepper (*Capsicum annuum* L.). Sci Technol Dev 35:75–81

Azooz MM, Youssef AM, Ahmad P (2011) Evaluation of salicylic acid (SA) application on growth, osmotic solutes and antioxidant enzyme activities on broad bean seedlings grown under diluted seawater. Int J Plant Physiol Biochem 3:253–264

Álvarez S, Gómez-Bellot MJ, Castillo M, Bañón S, Sánchez-Blanco MJ (2012) Osmotic and saline effect on growth, water relations, and ion uptake and translocation in *Phlomis purpurea* plants. Environ Exp Bot 78:138–145

Babaeian M, Esmaeilian Y, Tavassoli A, Asgharzade A, Sadeghi M (2011) The effects of water stress, manure and chemical fertilizer on grain yield and grain nutrient content in barley. Sci Res Essays 6:3697–3701

Baker AM, Hatton W (1987) Calcium peroxide as a seed coating material for padi rice. Plant and Soil 99:379–386

Baligar VC, Fageria NK, He ZL (2001) Nutrient use efficiency in plants. Commun Soil Sci Plant Anal 32:921–950

Basra SMA, Farooq M, Hafeez K, Ahmad N (2004) Osmohardening: a new technique for rice seed invigoration. Inter Rice Res Notes 29:80–81

Bewley JD, Black M (1994) Seeds: physiology of development and germination. Plenum, London

Biemelt S, Keetman U, Mock HP, Grimm B (2000) Expression and activity of isoenzymes of superoxide dismutase in wheat roots in response to hypoxia and anoxia. Plant Cell Environ 23:135–144

Bliss RD, Platt-Aloia KA, Thomson WW (1986) The inhibitory effect of NaCl on barley germination. Plant Cell Environ 9:727–733

Bonilla I, El-Hamdaoui A, Bolanos L (2004) Boron and calcium increase *Pisum sativum* seed germination and seedling development under salt stress. Plant and Soil 267:97–107

Bücker-Neto L, Paiva ALS, Machado RD, Arenhart RA, Margis-Pinheiro M (2017) Interactions between plant hormones and heavy metals responses. Genet Mol Biol 40:373–386

Burkart S, Bender J, Tarkotta B, Faust S, Castagna A, Ranieri A, Weigel HJ (2013) Effects of ozone on leaf senescence, photochemical efficiency and grain yield in two winter wheat cultivars. J Agron Crop Sci 199:275–285

Cai XY, Chen XD, Li CZ, Liu C (2013) Effects of exogenous Ca^{2+} on the seed germination of *Koelreuteria paniculata* in limestone area of Southwest China under drought stress. Zhongguo ke xue yuan Shenyang ying yong sheng tai yan jiu suo zhu ban 24:1341–1346

Cakmak I, Kirkby EA (2008) Role of magnesium in carbon partitioning and alleviating photooxidative damage. Physiol Plant 133:692–704

Chakraborty N, Acharya K (2017) "NO way"! Says the plant to abiotic stress. Plant Gene 11:99–105

Choi HI, Park HJ, Park JH, Kim S, Im MY, Seo HH, Kim YW, Hwang I, Kim SY (2005) *Arabidopsis* calcium dependent protein kinase AtCPK32 interacts with ABF4, a transcriptional regulator of abscisic acid-responsive gene expression, and modulates its activity. Plant Physiol 139:1750–1761

Choubisa SL, Choubisa D (2016) Status of industrial fluoride pollution and its diverse adverse health effects in man and domestic animals in India. Environ Sci Pollut Res 23:7244–7254

Conn S, Gilliham M (2010) Comparative physiology of elemental distributions in plants. Ann Bot 105:1081–1102

Coria NA, Sarquís JI, Peñalosa I, Urzúa M (1998) Heat-induced damage in potato (Solanum tuberosum) tubers: membrane stability, tissue viability, and accumulation of glycoalkaloids. J Agric Food Chem 46:4524–4528

Correia MJ, Osório ML, Osório J, Barrote I, Martins M, David MM (2006) Influence of transient shade periods on the effects of drought on photosynthesis, carbohydrate accumulation and lipid peroxidation in sunflower leaves. Environ Exp Bot 58:75–84

Cotrozzi L, Remorini D, Pellegrini E, Landi M, Massai R, Nali C, Guidi L, Lorenzini G (2016) Variations in physiological and biochemical traits of oak seedlings grown under drought and ozone stress. Physiol Plant 157:69–84

de Campos MKF, de Carvalho K, de Souza FS, Marur CJ, Pereira LFP, Filho JCB, Vieira LGE (2011) Drought tolerance and antioxidant enzymatic activity in transgenic 'Swingle'citrumelo plants over-accumulating proline. Environ Exp Bot 72:242–250

Dal Corso G, Fasani E, Furini A (2013) Recent advances in the analysis of metal hyper accumulation and hyper tolerance in plants using proteomics. Front Plant Sci 4:280. https://doi.org/10.3389/fpls.2013.00280

Daowei Z, Moxin X (2010) Specific ion effects on the seed germination of sunflower. J Plant Nutr 33:255–266

Deng F, Wang L, Yao X, Wang JJ, Ren WJ, Yang WY (2009) Effects of different-growing-stage shading on rice grain-filling and yield. J Sichuan Agric Univ 27:265–269

Ding F, Wang M, Zhang S (2017) Overexpression of a Calvin cycle enzyme SBPase improves tolerance to chilling-induced oxidative stress in tomato plants. Sci Hortic 214:27–33

Dong C, Fu Y, Liu G, Liu H (2014) Low light intensity effects on the growth, photosynthetic characteristics, antioxidant capacity, yield and quality of wheat (*Triticum aestivum* L.) at different growth stages in BLSS. Adv Space Res 53:1557–1566

Dumont J, Spicher F, Montpied P, Dizengremel P, Jolivet Y, Le Thiec D (2013) Effects of ozone on stomatal responses to environmental parameters (blue light, red light, CO_2 and vapour pressure deficit) in three *Populus deltoides* × *Populus nigra* genotypes. Environ Pollut 173:85–96

Duval FD, Renard M, Jaquinod M, Biou V, Montrichard F, Macherel D (2002) Differential expression and functional analysis of three calmodulin isoforms in germinating pea (*Pisum sativum* L.) seeds. Plant J 32:481–493

Edel KH, Marchadier E, Brownlee C, Kudla J, Hetherington AM (2017) The evolution of calcium-based signalling in plants. Curr Biol 27:667–679

El-Khawaga AS (2013) Effect of anti-salinity agents on growth and fruiting of different date palm cultivars. Asian J Crop Sci 5(1):65–80

Eller F, Brix H (2016) Influence of low calcium availability on cadmium uptake and translocation in a fast-growing shrub and a metal accumulating herb. AoB Plant 8:plv143. https://doi.org/10.1093/aobpla/plv143

Esnault MA, Legue F, Chenal C (2010) Molecular mechanisms of ultraviolet radiation-induced DNA damage and repair. Environ Exp Bot 68:231–232

Farooq M, Aziz T, Wahid A, Lee DJ, Siddique KHM (2009a) Chilling tolerance in maize: agronomic and physiological applications. Crop Pasture Sci 60:501–516

Farooq M, Basra SMA, Wahid A, Ahmad N, Saleem BA (2009b) Improving the drought tolerance in rice (*Oryza sativa* L.) by exogenous application of salicylic acid. J Agron Crop Sci 195:237–246

Farooq M, Hussain M, Nawaz A, Dong-Jin Lee DJ, Alghamdi SS, Siddique KHM (2017) Seed priming improves chilling tolerance in chickpea by modulating germination metabolism, trehalose accumulation and carbon assimilation. Plant Physiol Biochem 111:274–283

Farzadfar S, Zarinkamar F, Modarres-Sanavy SAM, Hojati M (2013) Exogenously applied calcium alleviates cadmium toxicity in *Matricaria chamomilla* L. plants. Environ Sci Pollut Res 20:1413–1422

Feng ZZ, Kobayashi K, Ainsworth EA (2008) Impact of elevated ozone concentration on growth, physiology, and yield of wheat (*Triticum aestivum* L.): a meta-analysis. Glob Chang Biol 14:2696–2708

Feng ZZ, Pang J, Kobayashi K, Zhu J, Ort DR (2011) Differential responses in two varieties of winter wheat to elevated ozone concentration under fully open-air field conditions. Glob Chang Biol 17:580–591

Feng Z, Wang L, Pleijel H, Zhu J, Kobayashi K (2016) Differential effects of ozone on photosynthesis of winter wheat among cultivars depend on antioxidative enzymes rather than stomatal conductance. Sci Total Environ 572:404–411

Ferreira CS, Piedade MTF, Junk WJ, Parolin P (2007) Floodplain and upland populations of Amazonian *Himatanthus sucuuba*: effects of flooding on germination, seedling growth and mortality. Environ Exp Bot 60:477–483

Ferreira-Silva SL, Voigt EL, Silva EN, Maia JM, de Vasconcelos Fontenele A, Silveira JAG (2011) High temperature positively modulates oxidative protection in salt-stressed cashew plants. Environ Exp Bot 74:162–170

Flexas J, Medrano H (2002) Drought-inhibition of photosynthesis in C_3 plants: stomatal and non-stomatal limitations revisited. Ann Bot 89(2):183–189

Gardiner B, Berry P, Moulia B (2016) Wind impacts on plant growth, mechanics and damage. Plant Sci 245:94–118

Gautam P, Lal B, Tripathi R, Shahid M, Baig MJ, Raja R, Maharana S, Nayak AK (2016) Role of silica and nitrogen interaction in submergence tolerance of rice. Environ Exp Bot 125:98–109

Ghannoum O (2009) C_4 photosynthesis and water stress. Ann Bot 103:635–644

Ghosh M, Jana A, Sinha S, Jothiramajayam M, Nag A, Chakraborty A, Mukherjee A (2016) Effects of ZnO nanoparticles in plants: cytotoxicity, genotoxicity, deregulation of antioxidant defenses, and cell-cycle arrest. Mutat Res Genet Toxicol Environ Mutagen 807:25–32

Gill SS, Tuteja N (2010) Polyamines and abiotic stress tolerance in plants. Plant Signal Behav 5:26–33

Gimeno V, Syvertsen JP, Simón I, Nieves M, Díaz-López L, Martínez V, García-Sánchez F (2012) Physiological and morphological responses to flooding with fresh or saline water in *Jatropha curcas*. Environ Exp Bot 78:47–55

Gomes FP, Oliva MA, Mielke MS, Almeida AAF, Aquino LA (2010) Osmotic adjustment, proline accumulation and cell membrane stability in leaves of *Cocos nucifera* submitted to drought stress. Sci Hortic 126:379–384

Gomes MP, Le Manac'h SG, Maccario S, Labrecque M, Lucotte M, Juneau P (2016) Differential effects of glyphosate and aminomethylphosphonic acid (AMPA) on photosynthesis and chlorophyll metabolism in willow plants. Pestic Biochem Physiol 130:65–70

Gottardini E, Cristofori A, Cristofolini F, Nali C, Pellegrini E, Bussotti F, Ferretti M (2014) Chlorophyll-related indicators are linked to visible ozone symptoms: evidence from a field study on native *Viburnum lantana* L. plants in northern Italy. Ecol Indic 39:65–74

Grattan SR, Grieve CM (1999) Salinity mineral nutrient relations in horticultural crops. Sci Hortic 78:127–157

Greer DH, Weedon MM (2012) Modelling photosynthetic responses to temperature of grapevine (*Vitis vinifera* cv. Semillon) leaves on vines grown in a hot climate. Plant Cell Environ 35:1050–1064

Gul B, Khan MA (2006) Role of calcium in alleviating salinity effects in coastal halophytes. In: Khan MA, Weber DJ (eds) Ecophysiology of high salinity tolerant plants. Springer, Dordrecht

Gururani MA, Venkatesh J, Tran LP (2015) Regulation of photosynthesis during abiotic stress-induced photoinhibition. Mol Plant 8:1304–1320

Hasanuzzaman M, Hossain MA, Teixeirada Silva JA, Fujita M (2012) Plant response sand tolerance to abiotic oxidative stress: antioxidant defense is a key factor. In: Bandi V, Shanker AK, Shanker C, Mandapaka M (eds) Crop stress and its management: perspectives and strategies. Springer, Berlin, pp 261–316

Hasanuzzaman M, Fujita M (2013) Exogenous sodium nitroprusside alleviates arsenic-induced oxidative stress in wheat (*Triticum aestivum* L.) seedlings by enhancing antioxidant defense and glyoxalase system. Ecotoxicology 22:584–596

Hasanuzzaman K, Nahar K, Alam MM, Roychowdhury R, Fujita M (2013) Physiological, biochemical, and molecular mechanisms of heat stress tolerance in plants. Int J Mol Sci 14:9643–9684

Hasanuzzaman M, Nahar K, Alam MM, Bhuyan MHMB, Oku H, Fujita M (2018) Exogenous nitric oxide pretreatment protects *Brassica napus* L. seedlings from paraquat toxicity through the modulation of antioxidant defense and glyoxalase systems. Plant Physiol Biochem 126:173–186

Hasegawa P, Bressan RA, Zhu JK, Bohnert HJ (2000) Plant cellular and molecular responses to high salinity. Annu Rev Plant Physiol Plant Mol Biol 51:463–499

Hermans C, Vuylsteke M, Coppens F, Craciun A, Inzé D, Verbruggen N (2010) Early transcriptomic changes induced by magnesium deficiency in *Arabidopsis thaliana* reveal the alteration of circadian clock gene expression in roots and the triggering of abscisic acid-responsive genes. New Phytol 187:119–131

Hernandez JA, Almansa MS (2002) Short term effects of salt stress on antioxidant systems and leaf water relations of pea leaves. Physiol Plant 115:251–257

Hernández I, Munne-Bosch S (2015) Linking phosphorous availability with photo-oxidative stress in plants. J Exp Bot 66:2889–2900

Hernández JA, Ferrer MA, Jiménez A, Ros-Barceló A, Sevilla F (2001) Antioxidant systems and O_2/H_2O_2 production in the apoplast of *Pisum sativum* L. leaves: its relation with NaCl-induced necrotic lesions in minor veins. Plant Physiol 127:817–831

Hirschi KD (2004) The calcium conundrum. Both versatile nutrient and specific signal. Plant Physiol 136:2438–2442

Hossain Z, Komatsu S (2013) Contribution of proteomic studies towards understanding plant heavy metal stress response. Front Plant Sci 3:310. https://doi.org/10.3389/fpls.2012.00310

Hossain MA, Piyatida P, da Silva JAT, Fujita M (2012) Molecular mechanism of heavy metal toxicity and tolerance in plants: central role of glutathione in detoxification of reactive oxygen species and methylglyoxal and in heavy metal chelation. J Bot 2010:872875. https://doi.org/10.1155/2012/87287

Hussain M, Malik MA, Farooq M, Ashraf MY, Cheema MA (2008) Improving drought tolerance by exogenous application of glycinebetaine and salicylic acid in sunflower. J Agron Crop Sci 194:193–199

Hussein HA, Metwally AK, Farghaly KA, Bahawirth MA (2011) Effect of irrigation interval (water stress) on vegetative growth and yield in two genotypes of okra. Aust J Basic Appl Sci 5:3024–3032

Inada H, Kondo T, Akhtar N, Hoshino D, Yamaguchi M, Izuta T (2012) Relationship between cultivar difference in the sensitivity of net photosynthesis to ozone and reactive oxygen species scavenging system in Japanese winter wheat (*Triticum aestivum*). Physiol Plant 146:217–227

Ismail MA (2008) Involvement of Ca^{2+} in alleviation of Cd^{2+} toxicity in common bean (*Phaseolus vulgaris* L.) plants. Res J Agric Biol Sci 4:203–209

Jaakola L, Hohtola A (2010) Effect of latitude on flavonoid biosynthesis in plants. Plant Cell Environ 33:1239–1247

Jaleel A, Azooz MM (2009) Exogenous calcium alters pigment composition, α-glutamyl kinase and proline oxidase activities in salt stressed *Withania somnifera*. Plant Omics 2:85–90

Jaleel CA, Manivannan P, Sankar B, Kishorekumar A, Gopi R, Somasundaram R, Panneerselvam R (2007) Water deficit stress mitigation by calcium chloride in *Catharanthus roseus*: effects on oxidative stress, proline metabolism and indole alkaloid accumulation. Colloids Surf B Biointerfaces 60:110–116

Jasim AM, Abbas MF, Shareef HJ (2016) Calcium application mitigates salt stress in date palm (*Phoenix dactylifera* L.) offshoots cultivars of Berhi and Sayer. Acta Agric Slov 107:103–112

Jiang L, Yang Y, Jia LX, Lin JL, Liu Y, Pan B, Lin Y (2016) Biological responses of wheat (*Triticum aestivum*) plants to the herbicide simetryne in soils. Ecotoxicol Environ Saf 127:87–94

Kaczmarek M, Fedorowicz-Strońska O, Głowacka K, Waśkiewicz A, Sadowski J (2017) $CaCl_2$ treatment improves drought stress tolerance in barley (*Hordeum vulgare* L.). Acta Physiol Plant 39:41. https://doi.org/10.1007/s11738-016-2336-y

Kalashnikov YE, Balakhnina TI, Zakrzhevsky DA (1994) Effect of soil hypoxia on activation of oxygen and the system of protection from oxidative destruction in roots and leaves of *Hordeum vulgare*. Russ J Plant Physiol 41:583–588

Kang G, Wang C, Sun G, Wang Z (2003) Salicylic acid changes activities of H_2O_2-metabolizing enzymes and increases the chilling tolerance of banana seedlings. Environ Exp Bot 50(1):9–15

Kashiwagi T, Ishimaru K (2004) Identification and functional analysis of a locus for improvement of lodging resistance in rice. Plant Physiol 134:676–683

Kathiresan K, Kalyani V, Gnanarethinam JL (1984) Effect of seed treatments on field emergence, early growth and some physiological processes of sunflower (Helianthus annuus L.). Field Crops Res 9:215–217

Kato T (1986) Effects of the shading and rachis-branch clipping on the grain-filling process of rice cultivars differing in the grain size. Jpn J Crop Sci 55:252–260

Kaveh H, Nemati H, Farsi M, Jartoodeh SV (2011) How salinity affect germination and emergence of tomato lines. J Biol Environ Sci 5:159–163

Khodarahmpour Z, Ifar M, Motamedi M (2012) Effects of NaCl salinity on maize (*Zea mays* L.) at germination and early seedling stage. Afr J Biotechnol 11:298–304

Kiani SP, Grieu P, Maury P, Hewezi T, Gentzbittel L, Sarrafi A (2007) Genetic variability for physiological traits under drought conditions and differential expression of water stress-associated genes in sunflower L *Helianthus annuus*. Theor Appl Genet 114:193–207

Kirkham MB (2005) Principles of soil and plant water relations. Elsevier, Amsterdam

Knight H, Knight MR (2001) Abiotic stress signaling pathways: specificity and cross-talk. Trends Plant Sci 6:262–267

Kumar S, Kaur R, Kaur N, Bhandhari K, Kaushal N, Gupta K, Bains TS, Nayyar H (2011) Heat-stress induced inhibition in growth and chlorosis in mungbean (*Phaseolus aureus* Roxb.) is partly mitigated by ascorbic acid application and is related to reduction in oxidative stress. Acta Physiol Plant 33:2091–2101

Kumutha D, Ezhilmathi K, Sairam RK, Srivastava GC, Deshmukh PS, Meena RC (2009) Waterlogging induced oxidative stress and antioxidant activity in pigeonpea genotypes. Biol Plant 53:75–84

Kurtyka R, Małkowski E, Kita A, Karcz W (2008) Effect of calcium and cadmium on growth and accumulation of cadmium, calcium, potassium and sodium in maize seedlings. Polish J Environ Stud 17:51–56

Lal B, Gautam P, Rath L, Haldar D, Panda BB, Raja R, Shahid M, Tripathi R, Bhattacharyya P, Mohanty S, Nayak AK (2015) Effect of nutrient application on growth, metabolic and enzymatic activities of rice seedlings during flooding stress and subsequent re-aeration. J Agron Crop Sci 201:138–151

Le CTT, Brumbarova T, Ivanov R, Stoof C, Weber E, Mohrbacher J, Fink-Straube C, Bauer P (2016) Zinc finger of *Arabidopsis thaliana*12 (Zat12) interacts with fer-like iron deficiency-induced transcription factor (Fit) linking iron deficiency and oxidative stress responses. Plant Physiol 170:540–557

Lechowski Z, Bialczyk J (1993) Calcium mediated cytokinin action on chlorophyll synthesis in isolated embryo of Scotspine. Biol Plant 35:53–62

León-Chan RG, López-Meyer M, Osuna-Enciso T, Sañudo-Barajas JA, Heredia JB, León-Félix J (2017) Low temperature and ultraviolet-B radiation affect chlorophyll content and induce the accumulation of UV-B-absorbing and antioxidant compounds in bell pepper (*Capsicum annuum*) plants. Environ Exp Bot 139:143–151

Li J, Lee YR, Assmann SM (1998) Guard cells possess a calcium-dependent protein kinase that phosphorylates the KAT1 potassium channel. Plant Physiol 116:785–795

Liang W, Wang M, Ai X (2009) The role of calcium in regulating photosynthesis and related physiological indexes of cucumber under low light intensity and suboptimal temperature stress. Sci Hortic 123:34–38

Liu F, Jensen CR, Andersen MN (2003) Hydraulic and chemical signals in the control of leaf expansion and stomatal conductance in soybean exposed to drought stress. Funct Plant Biol 30:65–73

Liu F, Jensen CR, Andersen MN (2004) Drought stress effect on carbohydrate concentration in soybean leaves and pods during early reproductive development: its implication in altering pod set. Field Crop Res 86:1–13

Liu F, Yoo BC, Lee JY, Pan W, Harmon AC (2006) Calcium-regulated phosphorylation of soybean serine acetyl transferase in response to oxidative stress. J Biol Chem 281:27405–27415

Liu QH, Zhou XB, Yang LQ, Li T, Zhang JJ (2009) Effects of early growth stage shading on rice flag leaf physiological characters and grain growth at grain-filling stage. Chin J Appl Ecol 20:2135–2141

Liu B, Liu L, Tian L, Cao W, Zhu Y, Asseng S (2014) Post-heading heat stress and yield impact in winter wheat of China. Glob Chang Biol 20:372–381

Li-Yun L, Ming-You W (2010) Effect of $CaCl_2$ on germination of wheat seeds under salt stress. J Henan Agric Sci 1:5–7

Lolaei A, Rezaei MA, Raad MK, Kaviani B (2012) Effect of calcium chloride on growth and yield of tomato under sodium chloride stress. J Ornamental Hortic Plants 2:155–160

Loreti E, Veen HV, Perata P (2016) Plant responses to flooding stress. Curr Opin Plant Biol 33:64–71

Lu YL, Song S, Wang RS, Liu ZY, Meng J, Sweetman AJ, Jenkins A, Ferrier RC, Li H, Luo W, Wang TY (2015a) Impacts of soil and water pollution on food safety and health risks in China. Environ Int 77:5–15

Lu Z, Lu J, Pan Y, Lu P, Li X, Cong R, Ren T (2016) Anatomical variation of mesophyll conductance under potassium deficiency has a vital role in determining leaf photosynthesis. Plant Cell Environ 39:2428–2439

Lu Y, Yao H, Shan D, Jiang Y, Zhang S, Yang J (2015b) Heavy metal residues in soil and accumulation in maize at long- term wastewater irrigation area in Tongliao. China J Chem 2015:628280. https://doi.org/10.1155/2015/628280

Luan S, Kudla J, Rodríguez-Concepción M, Yalovsky S, Gruissem W (2002) Calmodulins and calcineurin B–like proteins: calcium sensors for specific signal response coupling in plants. Plant Cell 14:389–400

Lwalaba JLW, Zvobgo G, Fu L, Zhang X, Mwamba TM, Muhammad N, Mundende RPM, Zhang G (2017) Alleviating effects of calcium on cobalt toxicity in two barley genotypes differing in cobalt tolerance. Ecotoxicol Environ Saf 139:488–495

Mafakheri A, Siosemardeh A, Bahramnejad B, Struik PC, Sohrabi Y (2010) Effect of drought stress on yield, proline and chlorophyll contents in three chickpea cultivars. Aust J Crop Sci 4:580–585

Mahajan S, Tuteja N (2005) Cold, salinity and drought stresses: an overview. Arch Biochem Biophys 444:139–158. https://doi.org/10.1016/j.abb.2005.10.018

Manivannan P, Jaleel CA, Sankar B, Somasundaram R, Murali PV, Sridharan R, Panneerselvam R (2007) Salt stress mitigation by calcium chloride in *Vigna radiata* (L.) wilczek. Acta Biol Cracov Bot 49:105–109

Marcar NE (1986) Effect of the calcium on the salinity tolerance of Wimmera ryegrass (*Lolium rigidum* Gaud. Ev. Wimmera) during germination. Plant and Soil 93:129–132

Marschner H (2012) Mineral nutrition of higher plants. 3rd edn. Academic Press, London

Mazher AA, El-Quesni EF, Farahat MM (2007) Responses of ornamental and woody trees to salinity. World J Agric Sci 3:386–395

McWilliams D (2003) Drought strategies for cotton. Cooperative Extension Service, Circular 582, College of Agriculture and Home Economics. New Mexico State University, USA

Mei J, Wang W, Peng S, Nie L (2017) Seed pelleting with calcium peroxide improves crop establishment of direct-seeded rice under waterlogging conditions. Sci Rep 7:4878. https://doi.org/10.1038/s41598-017-04966-1
Meng Q, Zhou Q, Zheng S, Gao Y (2012) Responses on photosynthesis and variable chlorophyll fluorescence of *Fragaria ananassa* under sound wave. Energy Procedia 16:346–352
Mignolet-Spruyt L, Xu E, Idänheimo N, Hoeberichts FA, Mühlenbock P, Brosché M, Van Breusegem F, Kangasjärvi J (2016) Spreading the news: subcellular and organellar reactive oxygen species production and signalling. J Exp Bot 67:3831–3844
Mills G, Harmens H, Wagg S, Sharps K, Hayes F, Fowler D, Sutton M, Davies W (2016) Ozone impacts on vegetation in a nitrogen enriched and changing climate. Environ Pollut 208:898–908
Mirakhori M, Paknejad F, Moradi F, Ardakani M, Zahedi H, Nazeri P (2009) Effect of drought stress and methanol on yield and yield components of soybean Max (L 17). Am J Biochem Biotechnol 5:162–169
Mishra V, Cherkauer KA (2010) Retrospective droughts in the crop growing season: implications to corn and soybean yield in the midwestern United States. Agric For Meteorol 150:1030–1045
Mittler R, Vanderauwera S, Gollery M, Van Breusegem F (2004) Reactive oxygen gene network of plants. Trends Plant Sci 9:490–498
Miyashita K, Tanakamaru S, Maitani T, Kimura K (2005) Recovery responses of photosynthesis, transpiration, and stomatal conductance in kidney bean following drought stress. Environ Exp Bot 53:205–214
Mohammed AR, Tarpley L (2010) Effects of high night temperature and spikelet position on yield-related parameters of rice (*Oryza sativa* L.) plants. Eur J Agron 33:117–123
Mozafari H, Asrar Z, Pourseyedi FRS, Yaghoobi MM (2013) Calcium and L-histidine interaction on growth improvement of three tomato cultivars under nickel stress. Acta Biol Szegediensis 57:131–144
Mugnai S, Ferrante A, Petrognani L, Serra G, Vernieri P (2009) Stress-induced variation in leaf gas exchange and chlorophyll a fluorescence in *Callistemon* plants. Res J Biol Sci 4:913–921
Munns R (2005) Genes and salt tolerance: bringing them together. Plant Physiol 167:645–663
Murillo-Amador B, Jones HG, Kaya C, Aguilar RL, García-Hernández JL, Troyo-Diéguez E, Ávila-Serrano NY, Rueda-Puente E (2006) Effects of foliar application of calcium nitrate on growth and physiological attributes of cowpea (*Vigna unguiculata* L. Walp.) grown under salt stress. Environ Exp Bot 58:188–196
Naeem M, Naeem MS, Ahmad R, Ahmad R (2017) Foliar-applied calcium induces drought stress tolerance in maize by manipulating osmolyte accumulation and antioxidative responses. Pak J Bot 49:427–434
Nahar K, Hasanuzzaman M (2009) Germination, growth, nodulation and yield performance of three mungbean varieties under different levels of salinity stress. Green Farming 2:825–829
Nahar K, Hasanuzzaman M, Ahamed KU, Hakeem KR, Ozturk M, Fujita M (2015) Plant responses and tolerance to high temperature stress: role of exogenous phytoprotectants. In: Hakeem K (ed) Crop production and global environmental issues. Springer, Cham, pp 385–435
Nahar K, Hasanuzzaman M, Rahman A, Alam MM, Mahmud J, Suzuki T, Fujita M (2016) Polyamines confer salt tolerance in mung bean (*Vigna radiata* L.) by reducing sodium uptake, improving nutrient homeostasis, antioxidant defense, and methylglyoxal detoxification systems. Front Plant Sci 7:1104. https://doi.org/10.3389/fpls.2016.01104
Nahar K, Hasanuzzaman M, Alam MM, Rahman A, Mahmud JA, Suzuki T, Fujita M (2017) Insights into spermine-induced combined high temperature and drought tolerance in mung bean: osmoregulation and roles of antioxidant and glyoxalase system. Protoplasma 254:445–460
Nayyar H (2003) Calcium as environmental sensor in plants. Curr Sci 84:7–10
Nayyar H, Kaushal SK (2002) Chilling induced oxidative stress in germinating wheat grains as affected by water stress and calcium. Biol Plant 45:601–604
Nayyar H, Walia DP, Kaistha BL (1995) Performance of bread wheat (*Triticum aestivum*) seed primed with growth-regulators and inorganic salts. Indian J Agric Sci 65:112–116
Nayyar H, Bains TS, Kumar S, Kaur G (2005a) Chilling effects during seed filling on accumulation of seed reserves and yield of chickpea. J Sci Food Agric 85:1925–1930

Nayyar H, Bains TS, Kumar S (2005b) Chilling stressed chickpea seedlings: effect of cold acclimation, calcium and abscisic acid on cryoprotective solutes and oxidative damage. Environ Exp Bot 54:275–285

Nazir M, Pandey R, Siddiqi T, Ibrahim MM, Qureshi MI, Abraham G, Vengavasi K, Ahmad A (2016) Nitrogen-deficiency stress induces protein expression differentially in low-N tolerant and low-N sensitive maize genotypes. Front Plant Sci 7:298. https://doi.org/10.3389/fpls.2016.00298

Nikiforova VJ, Kopka J, Tolstikov V, Fiehn O, Hopkins L, Hawkesford MJ, Hesse H, Hoefgen R (2005) Systems rebalancing of metabolism in response to sulfur deprivation, as revealed by metabolome analysis of Arabidopsis plants. Plant Physiol 138:304–318

Nishiyama R, Watanabe Y, Fujita Y, Le DT, Kojima M, Werner T, Vankova R, Yamaguchi-Shinozaki K, Shinozaki K, Kakimoto T, Sakakibara H (2011) Analysis of cytokinin mutants and regulation of cytokinin metabolic genes reveals important regulatory roles of cytokinins in drought, salt and abscisic acid responses, and abscisic acid biosynthesis. Plant Cell 23(6):2169–2183. https://doi.org/10.1105/tpc.111.087395

Nobori T, Kobayashi NI, Tanoi K, Nakanishi TM (2016) Alteration in caesium behavior in rice caused by the potassium, phosphorous, and nitrogen deficiency. J Radioanal Nucl Chem 307:1941–1943

Noctor G, Veljovic-Jovanovic S, Driscoll S, Novitskaya L, Foyer C (2002) Drought and oxidative load in the leaves of C_3 plants: a predominant role for photorespiration? Ann Bot 89:841–850

Nouri-Ganbalani A, Nouri-Ganbalani G, Hassanpanah D (2009) Effects of drought stress condition on the yield and yield components of advanced wheat genotypes in Ardabil. Iran J Food Agric Environ 7:228–234

Nunes CM, O'Hara L, Primavesi LF, Delatte TL, Schluepmann H, Somsen GW, Silva AB, Fevereiro PS, Wingler A, Paul MJ (2013) The trehalose 6-phosphate/SnRK1 signalling pathway primes growth recovery following relief of sink limitation. Plant Physiol 162:1720–1732

Oh MW, Komatsu S (2015) Characterization of proteins in soybean roots under flooding and drought stresses. J Proteomics 114:161–181

Ostaszewska M, Juszczuk IM, Kołodziejek I, Rychter AM (2014) Long term sulphur starvation of *Arabidopsis thaliana* modifies mitochondrial ultrastructure and activity and changes tissue energy and redox status. J Plant Physiol 171:549–558

Ovečka M, Takáč T (2014) Managing heavy metal toxicity stress in plants: biological and biotechnological tools. Biotechnol Adv 32:73–86

Oz MT, Eyidogan F, Yucel M, Öktem HA (2015) Functional role of nitric oxide under abiotic stress conditions. In: Khan MN, Mobin M, Mohammad F, Corpas FJ (eds) Nitric oxide action in abiotic stress responses in plants. Springer, Cham, pp 21–41

Ozkur O, Ozdemir F, Bor M, Turkan I (2009) Physiochemical and antioxidant responses of the perennial xerophyte *Capparis ovata* Desf. to drought. Environ Exp Bot 66(3):487–492

Palta JP (2000) Stress interactions at the cellular and membrane levels. Hort Sci 25(11):1377–1381

Panahyan-e-Kivi M, Ebadi A, Tobeh A, Jamaati-e-Somarin S (2009) Evaluation of yield and yield components of lentil genotypes under drought stress. Res J Environ Sci 3:456–460

Parida AK, Das AB (2005) Salt tolerance and salinity effects on plants: a review. Ecotoxicol Environ Saf 60:324–349

Parida AK, Das AB, Mohanty P (2004) Investigations on the antioxidative defense responses to NaCl stress in a mangrove, *Bruguiera parviflora*: differential regulations of isoforms of some antioxidative enzymes. Plant Growth Regul 42:213–226

Parvin K, Haque MN (2017) Protective role of salicylic acid on salt affected broccoli plant. J Agric Ecol Res 10(2). https://doi.org/10.9734/JAERI/2017/30142

Parvin K, Ahmed KU, Islam MM, Haque MN (2015a) Response of tomato plant under salt stress: role of exogenous calcium. J Plant Sci 10(6):222–233

Parvin K, Ahamed KU, Islam MM, Haque MN, Hore PK, Siddik MA, Roy I (2015b) Reproductive behavior of tomato plant under saline condition with exogenous application of calcium. Middle-East J Sci Res 23(12):2920–2926

Parvin K, Ahmed KU, Islam MM, Haque MN (2016) Modulation of ion uptake in tomato (*Lycopersicon esculentum* L.) plants with exogenous application of calcium under salt stress. Poljoprivreda/Agriculture 22(2):40–49

Parvin K, Hasanuzzaman M, Fujita M (2017) Salicylic acid enhances growth and productivity in cabbage (*Brassica oleracea* var. *Capitata* L.) grown under saline condition. Focus Sci 3(1). https://doi.org/10.21859/focsci-030162

Passariello B, Giuliano V, Quaresima S, Barbaro M, Caroli S, Forte G, Garelli G, Iavicoli I (2002) Evaluation of the environmental contamination at an abandoned mining site. Microchem J 73:245–250

Pervez MA, Ayub CM, Khan HA, Shahid MA, Ashraf I (2009) Effect of drought stress on growth, yield and seed quality of tomato (*Lycopersicon esculentum L.*). Pak J Agric Sci 46:174–178

Pinheiro C, Chaves MM (2011) Photosynthesis and drought: can we make metabolic connections from available data? J Exp Bot 62:869–882

Piramila BHM, Prabha AL, Nandagopalan V, Stanley AL (2012) Effect of heat treatment on germination, seedling growth and some biochemical parameters of dry seeds of black gram. Int J Pharm Phytopharmacol Res 1:194–202

Qi Y, Wang H, Zou Y, Liu C, Liu Y, Wang Y, Zhang W (2010) Over-expression of mitochondrial heat shock protein 70 suppresses programmed cell death in rice. FEBS Lett 585:231–239

Qiu GY, Wang L, He X, Zhang X, Chen S, Chen J, Yang Y (2008) Water use efficiency and evapotranspiration of winter wheat and its response to irrigation regime in the north China plain. Agric For Meteorol 148:1848–1859

Rahman A, Mostofa MG, Nahar K, Hasanuzzaman M, Fujita M (2015a) Exogenous calcium alleviates cadmium-induced oxidative stress in rice (*Oryza sativa* L.) seedlings by regulating the antioxidant defense and glyoxalase systems. Braz J Bot 39:393–407

Rahman A, Mostofa MG, Alam MM, Nahar K, Hasanuzzaman M, Fujita M (2015b) Calcium mitigates arsenic toxicity in rice seedlings by reducing arsenic uptake and modulating the antioxidant defense and glyoxalase systems and stress markers. Biomed Res Int 2015:340812. https://doi.org/10.1155/2015/340812

Rahman A, Nahar K, Hasanuzzaman M, Fujita M (2016) Calcium supplementation improves Na^+/K^+ ratio, antioxidant defense and glyoxalase systems in salt-stressed rice seedlings. Front Plant Sci 7:609. https://doi.org/10.3389/fpls.2016.00609

Rajpoot R, Rani A, Srivastava RK, Pandey P, Dubey RS (2016) *Terminalia arjuna* bark extract alleviates nickel toxicity by suppressing its uptake and modulating antioxidative defence in rice seedlings. Protoplasma 253(6):1449–1462

Ranawake AL, Dahanayaka N, Amarasingha UGS, Rodrigo WDRJ, Rodrigo UTD (2011) Effect of water stress on growth and yield of mung bean (*Vigna radiata* L). Trop Agric Res Ext 14:76–79

Reddy TY, Reddy VR, Anbumozhi V (2003) Physiological responses of groundnut (*Arachis hypogea* L.) to drought stress and its amelioration: a critical review. Plant Growth Regul 41(1):75–88

Reddy AR, Chaitanya KV, Vivekanandan M (2004) Drought-induced responses of photosynthesis and antioxidant metabolism in higher plants. J Plant Physiol 161:1189–1202

Ren WJ, Yang WY, Xu JW, Fan GQ, Wang LY, Guan H (2002) Impact of low-light stress on leaves characteristics of rice after heading. J Sichuan Agric Univ 20(3):205–208

Ren WJ, Yang WY, Xu JW, Fan GQ, Ma ZH (2003) Effect of low light on grains growth and quality in rice. Acta Agron Sin 29(5):785–790

Rengel Z (1992) The role of calcium in salt toxicity. Plant Cell Environ 15:625–632

Rout NP, Shaw BP (2001) Salt tolerance in aquatic macrophytes: possible involvement of the antioxidative enzymes. Plant Sci 160(3):415–423

Rubio JS, Garcia-Sanchez F, Rubio F, Martınez V (2009) Yield, blossom-end rot incidence, and fruit quality in pepper plants under moderate salinity are affected by K^+ and Ca^{2+} fertilization. Sci Hortic 119:79–87

Rui H, Chen C, Zhang X, Shen Z, Zhang F (2016) Cd-induced oxidative stress and lignification in the roots of two *Vicia sativa* L. varieties with different Cd tolerances. J Hazard Mater 301:304–313

Sairam RK, Dharmar K, Chinnusamy V, Lekshmy S, Joshi R, Bhattacharya P (2011) NADPH oxidase as the source of ROS produced under waterlogging in roots of mung bean. Biol Plant 55:741–746
Sanders D, Pelloux J, Brownlee C, Harper JF (2002) Calcium at the crossroads of signaling. Plant Cell 14:401–417
Schussler JR, Westgate ME (1995) Assimilate flux determines kernel set at low water potential in maize. Crop Sci 35:1074–1080
Shabala S, Cuin TA, Pottosin I (2007) Polyamines prevent NaCl-induced K^+ efflux from pea mesophyll by blocking non-selective cation channels. FEBS Lett 581:1993–1999
Shahryari R, Gurbanov E, Gadimov A, Hassanpanah D (2008) Tolerance of 42 bread wheat genotypes to drought stress after anthesis. Pak J Biol Sci 11:1330–1335
Shaikh F, Gul B, Li WQ, Jing LX, Khan MA (2007) Effect of calcium and light on the germination of *Urochondra setulosa* under different salts. J Zhejiang Univ Sci B 8:20–26
Shakar M, Yaseen M, Mahmood R, Ahmad I (2016) Calcium carbide induced ethylene modulate biochemical profile of *Cucumis sativus* at seed germination stage to alleviate salt stress. Sci Hortic 213:179–185
Sharma A, Kumar V, Thukral AK, Bhardwaj R (2016) Epibrassinolide-imidacloprid interaction enhances non-enzymatic antioxidants in *Brassica juncea* L. Indian J Plant Physiol 21:70–75
Sheetal KR, Singh SD, Anand A, Prasad S (2016) Heavy metal accumulation and effects on growth, biomass and physiological processes in mustard. Indian J Plant Physiol 21:219–223
Sheteiwy M, Shen H, Xu J, Guan Y, Song W, Hu J (2017) Seed polyamines metabolism induced by seed priming with spermidine and 5-aminolevulinic acid for chilling tolerance improvement in rice (*Oryza sativa* L.) seedlings. Environ Exp Bot 137:58–72
Shi H, Jiang C, Ye T, Tan D, Reiter RJ, Zhang H, Liu R, Chan Z (2015) Comparative physiological, metabolomic, and transcriptomic analyses revealmechanisms of improved abiotic stress resistance in bermudagrass [*cynodon dactylon* (l). pers.] by exogenous melatonin. J Exp Bot 56:1064–1079
Siddiq S, Yaseen M, Mehdi SAR, Khalid A, Kashif SUR (2009) Growth and yield response of tomato (*Lycopersicon esculentum* Mill.) to soil applied calcium carbide and L-methionine. Pak J Bot 41:2455–2464
Siddique KHM, Regan KL, Tennant D, Thomson BD (2001) Water use and water use efficiency of cool season grain legumes in low rainfall Mediterranean-type environments. Eur J Agron 15:267–280
Siddique MH, Al-Whaibi MH, Sakran AH, Basalah MO, Ali HM (2012) Effect of calcium and potassium on antioxidant system of *Vicia faba* L. under cadmium stress. Int J Mol Sci 13:6604–6619
Signarbieux C, Feller U (2011) Non-stomatal limitations of photosynthesis in grassland species under artificial drought in the field. Environ Exp Bot 71:192–197
Singh S, Agrawal M (2010) Effects of municipal waste water irrigation on availability of heavy metals and morpho-physiological characteristics of *Beta vulgaris* L. J Environ Biol 31:727–736
Sivasankaramoorthy S (2013) Effect of supplementary calcium enhances plant growth, photosynthetic pigments and uptake of nutrient in *Oryza sativa* L. under NaCl stress. Int J Chem Life Sci 2:1189–1192
Snedden WA, Fromm H (2001) Calmodulin as a versatile calcium signal transducer in plants. New Phytol 125:35–66
Srivastava RK, Pandey P, Rajpoot R, Rani A, Gautam A, Dubey RS (2014) Exogenous application of calcium and silica alleviates cadmium toxicity by suppressing oxidative damage in rice. Protoplasma 252:959–975
Stefi AL, Margaritis LH, Christodoulakis NS (2017) The effect of the non ionizing radiation on exposed, laboratory cultivated upland cotton (*Gossypium hirsutum* L.) plants. Flora 226:55–64
Stoker T, Dahe Q, Plattner GK (2013) Climate change 2013: the physical science basis, p 12
Subramanian KS, Santhanakrishnan P, Balasubramanian P (2006) Responses of field grown tomato plants to arbuscular mycorrhizal fungal colonization under varying intensities of drought stress. Sci Hortic 107:245–253

Sun H, Bi Y, Tao J, Huang S, Hou M, Xue R, Liang Z, Gu P, Yoneyama K, Xie X, Shen Q, Xu G, Zhang Y (2016) Strigolactones are required for nitric oxide to induce root elongation in response to nitrogen and phosphate deficiencies in rice. Plant Cell Environ 39:1473–1484

Suwa R, Hakata H, Hara H, El-Shemy HA, Adu-Gyamfi JJ, Nguyen NT, Kanai S, Lightfoot DA, Mohapatra PK, Fujita K (2010) High temperature effects on photosynthetic partitioning and sugar metabolism during ear expansion in maize (*Zea mays* L.) genotypes. Plant Physiol Biochem 48:124–130

Suzuki N (2005) Alleviation by calcium of cadmium-induced root growth inhibition in *Arabidopsis* seedlings. Plant Biotech 22:19–25

Szymańska R, Ślesak I, Orzechowska A, Kruk J (2017) Physiological and biochemical responses to high light and temperature stress in plants. Environ Exp Bot 139:165–177

Taiz L, Zeiger E (2010) Plant physiology, 5th edn. Sinauer Associates, Sunderland, MA

Talukdar D (2012) Exogenous calcium alleviates the impact of cadmium induced oxidative stress in *Lens culinaris* Medic. through modulation of antioxidant enzyme activities. J Crop Sci Biotechnol 15:325–334

Tan W, wei Meng Q, Brestic M, Olsovska K, Yang X (2011) Photosynthesis is improved by exogenous calcium in heat-stressed tobacco plants. J Plant Physiol 168:2063–2071

Tang D, Dean WL, Borchman D, Paterson CA (2006) The influence of membrane lipid structure on plasma membrane Ca^{2+}-ATPase activity. Cell Calcium 39:209–216

Tausz M (2001) The role of glutathione in plant response and adaptation to natural stress. In: Grill D, Tausz M, De Kok LJ (eds) Significance of glutathione to plant adaptation to the environment. Springer, Dordrecht, pp 101–122

Tezara W, Mitchell V, Driscoll SP, Lawlor DW (2002) Effects of water deficit and its interaction with CO_2 supply on the biochemistry and physiology of photosynthesis in sunflower. J Exp Bot 53:1781–1791

Thakur P, Kumar S, Malik JA, Berger JD, Nayyar H (2010) Cold stress effects on reproductive develop-ment in grain crops: an overview. Environ Exp Bot 67:429–443

Tian S, Lu L, Zhang J, Wang K, Brown P, He Z, Liang J, Yang X (2011) Calcium protects roots of *Sedum alfredii* H. against cadmium-induced oxidative stress. Chemosphere 84:63–69

Tobe K, Zhang L, Omasa K (1999) Effects of NaCl on seed germination of five non-halophytes species from a Chinese desert environment. Seed Sci Technol 27:851–863

Tobe K, Li X, Omasa K (2004) Effect of five different salts on seed germination and seedling growth of *Haloxylon ammodendron* (Chenopodiaceae). Seed Sci Res 14:345–353

Tuna AL, Kaya C, Ashraf M, Altunlu H, Yokas I, Yagmur B (2007) The effects of calcium sulphate on growth, membrane stability and nutrient uptake of tomato plants grown under salt stress. Environ Exp Bot 59:173–178

Turhan H, Ayaz C (2004) Effect of salinity on seedling emergence and growth of sunflower (*Helianthus annuus* L.) cultivars. Int J Agric Biol 6:149–152

Turkmen O, Dursun A, Turan M, Erdinc C (2004) Calcium and humic acid affect seed germination, growth and nutrient content of tomato (*Lycopersicon esculentum* L.) seedlings under saline soil conditions. Acta Agric Scand B 54:168–174

Tuteja N, Mahajan S (2007) Calcium signaling network in plants. Plant Signal Behav 2:79–85

Upadhyaya H, Panda SK, Dutta BK (2011) $CaCl_2$ improves post-drought recovery potential in *Camellia sinensis* (L) O. Kuntze. Plant Cell Rep 30:495–503

van de Walle J, Horemans N, Saenen E, Van Hees M, Wannijn J, Nauts R, van Gompel A, Vangronsveld J, Vandenhove H, Cuypers A (2016) *Arabidopsis* plants exposed to gamma radiation in two successive generations show a different oxidative stress response. J Environ Radioact 165:270–279

Van Assche F, Clijsters H (1990) Effects of metals on enzyme activity in plants. Plant Cell Environ 13:195–206

Victoria J, Nikiforova VJ, Kopka J, Tolstikov V, Fiehn O, Hopkins L, Hawkesford MJ, Hesse H, Hoefgen R (2005) Systems rebalancing of metabolism in response to sulfur deprivation, as revealed by metabolome analysis of Arabidopsis plant. Plant Physiol 138:304–318

Vorasoot N, Songsri P, Akkasaeng C, Jogloy S, Patanothai A (2003) Effect of water stress on yield and agronomic characters of peanut. Songklanakarin J Sci Technol 25:283–288
Vysotskaya L, Hedley PE, Sharipova G, Veselov D, Kudoyarova G, Morris J, Jones HG (2010) Effect of salinity on water relations of wild barley plants differing in salt tolerance. AoB Plants 2010:plq006. https://doi.org/10.1093/aobpla/plq006
Wahid A, Ahmad SS, Ahmad MN, Khaliq B, Nawaz M, Shah SQ, Shah RU (2014) Assessing the effects of hydrogen fluoride on mango (*Mangifera indica* L.) in the vicinity of a brick kiln field in southern Pakistan. Fluoride 47:307–314
Walter J, Nagy L, Hein R, Rascher U, Beierkuhnlein C, Willner E, Jentsch A (2011) Do plants remember drought? Hints towards a drought-memory in grasses. Environ Exp Bot 71:34–40
White PJ, Broadley MR (2003) Calcium in plants. Ann Bot 92:487–511
Wu GQ, Wang SM (2012) Calcium regulates K^+/Na^+ homeostasis in rice (*Oryza sativa* L.) under saline conditions. Plant Soil Environ 58:121–127
Wu G, Liang N, Feng R-J, Zhang J-J (2013) Evaluation of salinity tolerance in seedlings of sugar beet (*Beta vulgaris* L.) cultivars using proline, soluble sugars and cation accumulation criteria. Acta Physiol Plant 35:2665–2674
Xu S, Hu B, He Z, Ma F, Feng J, Shen W, Yan J (2011) Enhancement of salinity tolerance during rice seed germination by presoaking with hemoglobin. Int J Mol Sci 12:2488–2501
Xu C, Li X, Zhang L (2013) The effect of calcium chloride on growth, photosynthesis, and antioxidant responses of *Zoysia japonica* under drought conditions. PLoS One 8:e68214. https://doi.org/10.1371/journal.pone.0068214
Yadav SK (2010) Cold stress tolerance mechanisms in plants. A review. Agron Sustain Dev 30:515–527
Yadav RS, Hash CT, Bidinger FR, Devos KM, Howarth CJ (2004) Genomic regions associated with grain yield and aspects of post-flowering drought tolerance in pearl millet across stress environments and tester background. Euphytica 136:265–277
Yadav PV, Kumari M, Ahmed Z (2011) Seed priming mediated germination improvement and tolerance to subsequent exposure to cold and salt stress in *Capsicum*. Res J Seed Sci 4:125–136
Yadav IC, Devi NL, Syed JH, Cheng Z, Li J, Zhang G, Jones KC (2015) Current status of persistent organic pesticides residues in air, water, and soil, and their possible effect on neighboring countries: a comprehensive review of India. Sci Total Environ 511:123–137
Yan B, Da Q, Liu X, Huang S, Wang Z (1996) Flooding induced membrane damage, lipid oxidation and activated oxygen generation in corn leaves. Plant and Soil 179:261–268
Yao S, Chen S, Zhao J, Xu D, Lan H, Zhang F (2010) Effect of three salts on germination and seedling survival of dimorphic seeds of *Chenopodium album*. Botany 88:821–828
Yildiztugay E, Ozfidan-Konakci C, Kucukoduk M (2014) The role of antioxidant responses on the tolerance range of extreme halophyte *Salsola crassa* grown under toxic salt concentrations. Ecotoxicol Environ Saf 110:21–30
Ying RR, Qiu RL, Tang YT, Hu PJ, Qiu H, Chen HR, Shi TH, Morel JL (2010) Cadmium tolerance of carbon assimilation enzymes and chloroplast in Zn/Cd hyperaccumulator *Picris divaricata*. J Plant Physiol 167:81–87
Yu XC, Zhu SY, Gao GF, Wang XJ, Zhao R, Zou KQ, Wang XF, Zhang XY, Wu FQ, Peng CC, Zhang DP (2007) Expression of a grape calcium-dependent protein kinase ACPK1 in *Arabidopsis thaliana* promotes plant growth and confers abscisic acid-hypersensitivity in germination, post-germination growth, and stomatal movement. Plant Mol Biol 64:531–538
Zadehbagheri M, Kamelmanesh MM, Javanmardi S, Sharafzadeh S (2012) Effect of drought stress on yield and yield components, relative leaf water content, proline and potassium ion accumulation in different white bean (*Phaseolus vulgaris* L.) genotype. Afr J Agric Res 7:5661–5670
Zehra A, Gul B, Ansari R, Khan MA (2012) Role of calcium in alleviating effect of salinity on germination of *Phragmites karka* seeds. South Afr J Bot 78:122–128
Zhao F, Song CP, He J, Zhu H (2007) Poly amines improve K^+/Na^+ homeostasis in barley seedlings by regulating root ion channel activities. Plant Physiol 145:1061–1072

Zhao F, Xiang Q, Zhou Y, Xu X, Qiu X, Yu Y, Ahmad F (2017) Evaluation of the toxicity of herbicide topramezone to *Chlorella vulgaris*: oxidative stress, cell morphology and photosynthetic activity. Ecotoxicol Environ Saf 143:129–135

Zhu X, Feng Z, Sun T, Liu X, Tang H, Zhu J, Guo W, Kobayashi K (2011) Effects of elevated ozone concentration on yield of four Chinese cultivars of winter wheat under fully open-air field conditions. Glob Chang Biol 17:2697–2706

Zhu X, Feng Y, Liang G, Liu N, Zhu J-K (2013) Aequorin-based luminescence imaging reveals stimulus-and tissue-specific Ca^{2+} dynamics in *Arabidopsis* plants. Mol Plant 6:444–455

Zobel RW, Kinraide TB, Baligar VC (2007) Fine root diameters can change in response to changes in nutrient concentrations. Plant and Soil 297:243–254

Zouari M, Elloumi N, Bellassoued V, Ahmed CB, Krayem M, Delmail D Elfeki A, Rouina BB, Abdallah FB, Labrousse P (2017) Enzymatic antioxidant responses and mineral status in roots and leaves of olive plants subjected to fluoride stress. South Afr J Bot 111:44–49

Silicon: A Sustainable Tool in Abiotic Stress Tolerance in Plants

Chanchal Malhotra and Riti Thapar Kapoor

Introduction

The world's population is approximately seven billion, and it has been predicted to enhance by ten billion in the next 50 years (Glick 2014). Food security is considered as one of the most serious challenges. Recent trends of crop production indicate that fertility of soil is declining due to intensive use of natural resources, indiscriminate use of pesticides, and lack of proper soil management practices (Cakmak 2001). There is a need to use new technologies which could help us to curb the problem of food insecurity. Abiotic stress is known as one of the most important constraints to agricultural production in the world (Meena et al. 2017). Abiotic stresses have become more common and challenging due to the unpredictable climate conditions. The change in climatic conditions has exacerbated the frequency and severity of many abiotic stresses with significant reduction in crop yield (Carmen and Roberto 2011). More than 70% decrease in crop yield has been reported due to land degradation, undesirable effects on agriculture, loss of biodiversity, and abiotic stresses (Veatch-Blohm 2007). The main constraints to agricultural production are change in environmental conditions causing various abiotic stresses such as drought, flood, high and low temperature, salinity, UV-B radiation, inadequate mineral supply, and heavy metal toxicity (Jewell et al. 2010; Shrivastava and Kumar 2015). Hence, there is a need to find the reliable eco-friendly methods to alleviate the impact of abiotic stresses on agricultural system to sustainably meet growing global food demands. The management of the soil with mineral nutrients offers both environmental and economic benefits (Marschner 1995). The positive effects of silicon on the crop yield and quality have been well documented (Liang et al. 2015). The beneficial effects of silicon in stressed plants are more visible than non-stressed plants (Cooke

C. Malhotra · R. T. Kapoor (✉)
Plant Physiology Laboratory, Amity Institute of Biotechnology, Amity University, Noida, Uttar Pradesh, India
e-mail: rkapoor@amity.edu

M. Hasanuzzaman et al. (eds.), *Plant Abiotic Stress Tolerance*,
https://doi.org/10.1007/978-3-030-06118-0_14

and Leishman 2016). The role of silicon in enhancing the resistance of plants to various abiotic stresses has been widely recognized by researchers (Guo et al. 2005; Gunes et al. 2007; Etesami and Beattie 2017; Etesami and Jeong 2018).

Application of silicon in agriculture is a sustainable strategy for the alleviation of abiotic stresses in plants (Cooke and Leishman 2011; Guntzer et al. 2012; Van Bockhaven et al. 2013; Hernandez-Apaolaza 2014; Adrees et al. 2015; Coskun et al. 2016; Imtiaz et al. 2016; Luyckx et al. 2017; Wang et al. 2017; Etesami and Jeong 2018). Silicon fertilizers are safe, eco-friendly, and cost-effective in comparison to other chemical/synthetic fertilizers even for small and marginal plant growers. Hence, silicon can be used as a growth regulator to improve plant growth and resistance under stress conditions.

Occurrence of Silicon in the Soil

Silicon is the second most abundant element in earth's crust in terms of quantity (Mcginnity 2015), and it comprises up to 70% of soil mass (Ma and Yamaji 2008; Ahmed et al. 2014). Silicon forms a major portion of soil as silicate but its availability to plants is low (Zhu and Gong 2014). Balakhnina and Borkowska (2013) reported that most of the sources of silicon are insoluble and in a plant-unavailable form. Silicon exists in the form of silicic acid (H_4SiO_4), a non-charged molecule in soil solution at the concentration 0.1–0.6 mM and pH less than 9 (Epstein 2009). The concentration of silicon in soil solution is equivalent to some of the macroelements such as potassium, calcium, and phosphorus (Epstein 1994). The soil contains 50–400 g silicon/kg of soil but the silicon content in the soil depends on the type of soil as sandy soil contains more silicon in comparison to clay soil (Matichenkov and Calvert 2002). Lovering and Engel (1959) reported that forest in the 1 hectare land can extract about 5000 tonnes of silicon in 5000 years. According to the reports of Food and Agricultural Organization (FAO 1998), approximately 210–224 million tonnes of silicon is removed from the arable soils all around the world and soils become less fertile due to low availability of silicon content. The incorporation of silicon in soil improves the fertility of soil through physical and chemical properties, improved water absorption capacity, and maintenance of nutrients in a soil in the plant available form.

Sources of Silicon

The surface of earth is covered with 27.7% of silicon next to oxygen but the existence of silicon in its pure form is rare (Mitra 2015). Silicon is deposited in the form of quartz (SiO_2), sand and sandstone in the earth crust (Rédei 2008). Agricultural wastes such as silicate slag, steel slag, electric furnace slag, baggase furnace ash, lignite fly ash, and rice straw are rich in silicon and can be used in crop fields to increase the crop yield (Kalra et al. 2003). Calcium silicate obtained as a by-product of phosphorus production is one of the most widely used silicon fertilizer. Potassium silicate is highly soluble and can be used in hydroponic culture for laboratory experiments but it is expensive. The other sources of silicon have been used commercially are silica gel, calcium silicate hydrate, etc.

Silicon: Essential Element for Plants

Earlier researchers reported that silicon is not an essential element for plants but difficulties in growing plants in silicon-free environment made this assumption incorrect and silicon was designated as an essential element for the growth and development of plants. Justus Von Leibig proposed the use of sodium silicate as a silicon fertilizer as early as in the year 1840. Silicon is a beneficial element to higher plants as its effects are frequently linked to morphological, physiological, and molecular aspects of plants (Ma 2004; Lobato et al. 2009). Silicon acts as a quasi-essential element for plants because its deficiency can cause various abnormalities in growth, reproduction, and overall development of plants (Epstein and Bloom 2005). The sand is applied to rice fields at 2–3 tonnes/acre once in 2 or 3 years in southern part of India is considered as a good source of silicon. The farmers in Texas and Florida use certified green sand containing silicates to get higher crop yield. Silicon has been recognized as an agronomically essential element for rice cultivation in Japan because silicon promotes the growth and yield of rice. Silicon nutrition to the plants improves plant protective mechanism against diseases, insect attack, and unfavorable environmental conditions (Guntzer et al. 2011; Dallagnol et al. 2011; Liu et al. 2014).

Silicon Content in Plants

The content of silicon in plants shows a large variability ranging from 0.1 to 10% dry weight (Hodson et al. 2005). According to Marschner (1995), there are three classes of silicon absorbers: (1) silicon accumulator plants as they accumulate large amount of silicon—wheat, rice, millet, and sugarcane; (2) silicon non-accumulator plants—snapdragon plant; (3) silicon-excluder plants—soybean. Silicon accumulators have silicon concentration more than 1% and silicon/calcium ratio > 1 and excluders have silicon concentration below 0.5% and silicon/calcium ratio < 0.5. The plants which do not meet these criteria are considered as intermediate plants. Monocotyledons such as rice, sugarcane, and maize absorb silicon in large quantities due to the presence of silicon transporters in comparison to dicotyledons (Ma et al. 2016). The aerial plant parts accumulate more silicon in comparison to roots. Silicon content in plants increased with the age of crop plants and due to this older leaves are rich in silicon concentration than younger leaves (Henriet et al. 2006).

Silicon Mitigates Abiotic Stresses in Plants

Silicon is known as the only element that is able to increase plant resistance to environmental stresses. Silicon-mediated alleviation of abiotic stresses such as salinity, drought, UV-B radiation, flooding, freezing, lodging, high temperature, and heavy metal toxicity has been reported (Reynolds et al. 2016; Debona et al. 2017; Kim et al. 2017; Etesami and Jeong 2018). The benefits of silicon accumulation in wetland environment have been studied in terms of increased ability to resist water current and allow roots to better penetrate the mud (Struyf and Conley 2008). The most important function of silicon lies in its potential to confer tolerance in plants to multiple stresses (Fig. 1).

Salinity Stress

Salinity is one of the most significant abiotic stresses for crop plants (Chinnusamy et al. 2005). Approximately 20% of the total arable area has been degraded by salinity (Shrivastava and Kumar 2015). Salt interferes with plant growth by ion cytotoxicity (Greenway and Munns 1980), nutrient imbalance (Khan et al. 2000), and oxidative damage (Hernandez et al. 2000). Silicon has been reported to improve salt tolerance in different crops (Liang et al. 2007; Reezi et al. 2009). Silicon can reduce Na^+ uptake by plants under salt stress by (1) reducing membrane permeability of root cells, (2) improving structure and stability of root cells (Exley 2015; Luyckx et al. 2017), (3) mediating stimulation of the root plasma membrane H^+-ATPase, and (4) facilitating Na^+ export from the cell. Exogenous application of silicon improved gas exchange characteristics in many plant species under salinity

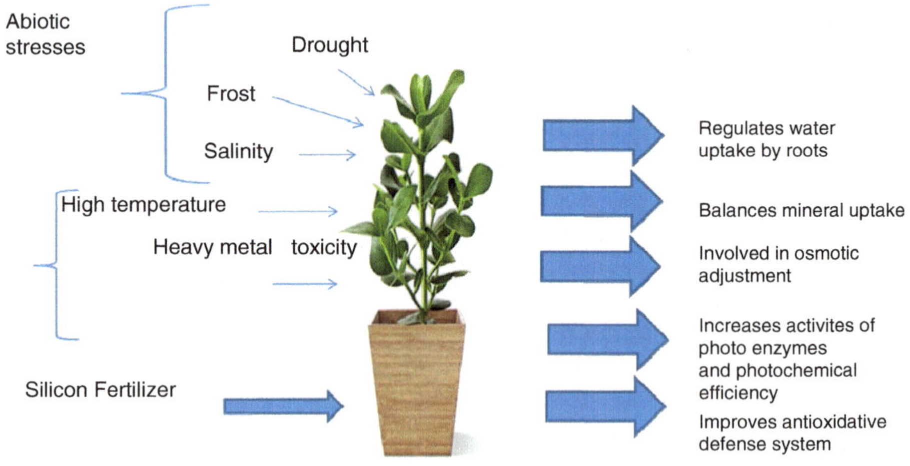

Fig. 1 Effect of silicon fertilizer on abiotic stresses

condition (Etesami and Jeong 2018). The water status of leaf and water use efficiency of crops was increased by silicon application in many salt-stressed plants (Coşkun et al. 2016). Silicon deposited as silica gel in the cell wall of the roots provides binding sites for the salts and reduces their translocation to shoots which consequently improved plant growth and development (Lux et al. 2003).

Silicon deposition and polymerization in leaves limits transpiration and salt accumulation and improved water storage within plant tissues, which allowed higher growth rate, salt dilution within the plant cells, and mitigating salt toxicity effects (Bradbury and Ahmad 1990). Increase in salinity rises Na^+, Cl^-, and SO_4^{2-} contents and decrease in K^+, Ca^{2+}, and Mg^{2+}contents in a number of crop species (John et al. 2003). The addition of silicon to saline environment significantly decreased the Na^+content but increased K^+ concentration (Ashraf et al. 2010). Salinity-induced accumulation of ROS including superoxide radical, hydroxyl radical, and hydrogen peroxide are detrimental to cells at higher concentration because they cause oxidative damage to membrane lipids, proteins, chlorophyll, and nucleic acids (Mittler 2002; Keles et al. 2004; Gunes et al. 2007). The application of silicon under salt stress significantly increased superoxide dismutase, peroxidase, catalase, and glutathione reductase activity in barley roots (Liang et al. 2003) and cucumber (Zhu et al. 2004). The higher efficiency of antioxidant defense enzymes in salt-stressed plants with silicon application coincided with a decrease in their electrolyte leakage and lipid peroxidation level and H_2O_2 content, suggesting that oxidative damage induced by salinity stress can be alleviated by silicon supplementation (Moussa 2006; Wang et al. 2010).

Drought Stress

Drought is one of the major abiotic stresses that adversely affects growth and productivity of agricultural crops (Bodner et al. 2015). Drought or water deficit leads to stomatal closure, reduced transpiration rate, decrease in water potential of plant tissues, decrease in photosynthesis, and ultimately plant growth is inhibited (Yordanov et al. 2003). The ameliorative effect of silicon on drought stress has been related to the reduction in loss of water by transpiration (Romero-Aranda et al. 2006), osmotic adjustment (Trenholm et al. 2004), improved nutrient uptake (Gunes et al. 2008a), and activation of antioxidant defense system (Gunes et al. 2008b). Agarie et al. (1998) reported that deposition of silicon in the cell wall reduced transpiration and increased internal storage of water under drought stress. The mitigating effect of silicon on drought stress could be related to the hydrophilic nature of $SiO_2.nH_2O$ which maintains the water in plant tissues and protects plant tissues from drought (Gong et al. 2005). An important mechanism of drought tolerance is the osmotic adjustment as it helps to retain water in plant tissues despite low water potential (Romero-Aranda et al. 2006). Kaya et al. (2006) found that silicon supply increased Ca concentration which played a vital role in maintaining membrane permeability and stability. Liang (1999) reported that under water stress condition, K^+

concentration increased in the presence of silicon possibly because of the activation of H^+-ATPase in the cell membrane. Potassium is one of the primary osmotic substances which contribute to osmotic adjustment (Ashraf et al. 2001). Silicon-mediated maintenance of higher level of K^+ is essential in achieving better survival with improved growth under water stress (Umar 2002). Drought-induced reduction in chlorophyll content can be alleviated by silicon and hence improved photosynthesis under drought stress (Kaya et al. 2006). The effect of silicon stimulation on root growth may be due to increased root elongation caused by an increase in cell wall extensibility in the growth region as observed in sorghum (Hattori et al. 2003).

Wang et al. (2015) reported significant increase in the root/shoot ratio in silicon-treated plants, and it was due to the silicon-mediated modification in root morphology which increased water uptake ability of silicon-treated plants. The augmented water uptake during the addition of silicon in water deficit condition is the result of improved root hydraulic conductance (Hattori et al. 2008) and root activity (Chen et al. 2011). It has been reported that improved structural stability due to the binding of silicon with cell wall hemicellulose can be beneficial in water shortage condition for the plants (Ma et al. 2015).

Thermal Stress

Temperature is one of the most important environmental factors governing plant growth and development (Monjardino et al. 2005). Thermal stress (cold and heat) may disturb the balance between ROS and antioxidant activity in plants. Application of silicon has been reported to alleviate heat stress in different plant species (Epstein 1999). Heat stress is detrimental to plant growth and dry matter accumulation because it inhibits protein synthesis, enzyme activity, chlorophyll formation, photosynthesis, and increases transpiration (Gibson and Paulsen 1999). Commuri and Jones (2001) reported that extended periods of drought and high temperature lead to poor vigor and shallow rooting. According to Takahashi and Kurata (2007), silicon deposition in the cell wall of xylem vessels prevents the compression of vessels under the conditions of high transpiration caused by heat stress (Hattori et al. 2005; Liang et al. 2007). However, high level of silicon in cell wall reduced transpiration loss caused by higher temperature thus allowing continued metabolic functions under high temperature. According to Epstein (1999), plants wilt less, resist sunburn and are generally more tolerant to heat stress when silicon is applied to plant growth medium. The supply of silicon has been found to produce higher concentrations of RUBP-carboxylase enzyme in leaf tissues which help in CO_2 metabolism in stress environment (Gunes et al. 2008b). Soundararajan et al. (2014) treated *Salvia splendens* with silicon under high temperature and found that the activities of SOD, APX, and GPX were increased.

The chilling and frost stresses can cause irreversible damage to plant cells due to mechanical forces generated by the formation of extracellular ice crystals, cellular dehydration, and increased concentration of intracellular salts (Steponkus 1984). It has been reported that freezing increases the level of ROS in plants and increased lipid peroxidation arising from the accumulation of ROS is the major cause of membrane damage (McKersie et al. 1993). Studies have shown that supply of silicon under low temperature stress regulated the activity of antioxidant defense system and alleviated oxidative damage caused by chilling or frost (Ma et al. 2001). It has been reported that silicon improves leaf and stem strength through deposition in the cuticle and by maintaining cell wall polysaccharide and lignin polymers, thus protected plants from low temperature and frost (Ohyama 1985; Hull 2004). The photosynthetic electron transport, CO_2 fixation, Rubisco activity, and stomatal conductance are the major targets impaired by low temperature stress in plants. Application of silicon under frost conditions resulted in higher concentration of chlorophyll (Schmidt et al. 1999) and RUBP-carboxylase enzyme in leaf tissue. This enzyme regulated the metabolism of CO_2 and enables the plant to make more efficient use of available levels of CO_2 under low temperature stress. The leaves of silicon-treated plants (cucumber, rice, maize, and sunflower) grown hydroponically at low temperature (0–4 °C) were more resistant to cold-induced wilting and root ability to absorb nutrients was higher. Silicon-mediated alleviation of freezing injury may be attributed to enhanced antioxidant defense activities and consequent reduced membrane oxidative damage through better water retention in leaf tissues (Liang et al. 2003; Zhu et al. 2004; Gong et al. 2005).

Radiation

Exogenous application of silicon showed alleviating effects of UV-B stress on soybean, wheat, and maize (Yao et al. 2011; Shen et al. 2014). Many studies have revealed that silicon application can induce resistance to UV-B stress by the modification of physiological and biochemical processes in plants (Schaller et al. 2013; Tripathi et al. 2017). According to Tripathi et al. (2017), UV-B stress was significantly improved with the use of silicon nanoparticles on wheat seedlings.

Wounding Stress

Wounding stress is caused by physical injury in plants, and it may be due to wind or herbivore attack. Such injuries can cause the death of plant tissues and make it vulnerable to pathogenic attack. Wounds initiate oxidative stress thereby damaging the cell membranes. The antioxidant activities of catalase and peroxidase are significantly increased in silicon-treated wounded rice plants compared to control plants and improved mechanical strength to overcome losses from wounding stress (Kim et al. 2014). A similar trend of increased antioxidant enzyme activity was also observed in wheat (Gong et al. 2005), maize, and barley (Liang et al. 2005). The accumulation of antioxidative enzymes in silicon-treated plants functions as a strategy for coping with wounding stress.

Heavy Metal Toxicity

Heavy metal pollution is rapidly increasing, and it originates from natural sources as well as anthropogenic activities including mining, industrial waste, sewage sludge, and excessive use of fertilizer and pesticides (Nagajyoti et al. 2010). The relationship between silicon supplementation and metal tolerance in plants has been widely studied (Shi et al. 2005; da Cunha and do Nascimento 2009; Kaya et al. 2010). Silicon supplementation decreased metal toxicity such as toxicity of aluminum (Wang et al. 2004), boron (Gunes et al. 2007), cadmium (Liang et al. 2005), chromium (Tripathi et al. 2012), copper (Li et al. 2008), and zinc (Neumann and Zur Nieden 2001). The possible mechanism behind the silicon inhibition of metal transport in plants may be due to the thickening of the casparian strips in the endodermis and cell wall of the xylem due to the deposition of silicon in the cell wall (Chen et al. 2000; da Cunha and do Nascimento 2009).

Excessive metal exposure induces leaf chlorosis and necrosis by affecting chlorophyll synthesis and breakdown of the cell membrane. It was reported that silicon reduced the toxic effects of metals in plants by stimulating the production of root exudates (Adrees et al. 2015; Imtiaz et al. 2016). Silicon can reduce metal toxicity in plants by increasing the content and accumulation of macronutrients (Ca, Mg, P, and K) and micronutrients (Zn, Fe, and Mn) (Etesami and Jeong 2018).

The key mechanism of silicon-mediated metal toxicity alleviation in higher plants include: (1) complexation or co-precipitation of toxic metal ions with silicon; (2) immobilization of toxic metal ions in the growth media; (3) stimulation of antioxidant defense system; (4) uptake processes; (5) compartmentation or homogenous distribution of metal ions within plants (Liang et al. 2005); (6) accelerating suberin lamellae, casparian bands, root vascular tissues development, and lignification; (7) enhancing the activity of gas exchange characteristics (net photosynthetic rate, stomatal conductance, transpiration rate, and water use efficiency); (8) change in plant morphological traits (increase in length of leaves, leaf area, number of leaves per plant, root volume, root length, and total root surface area); (9) regulating plant phytohormones; (10) helping the formation of the tertiary endodermal cell wall; (11) localizing in root endodermis, which serve as a barrier to block heavy metal entrance into cells (Adrees et al. 2015; Imtiaz et al. 2016). Silicon has been shown to alleviate the adverse effects of nutritional imbalance stress in plants (Hernandez-Apaolaza 2014; Etesami and Beattie 2017; Etesami and Jeong 2018).

Macronutrients

The excessive use of nitrogen fertilizer causes lodging, mutual shading, and susceptibility of plants to pests and diseases (Ma 2004). The adverse effects due to excessive nitrogen fertilizers can be minimized by the use of silicon as the presence of silicate crystals in plant tissues provide mechanical barrier which hinders feeding of insects and inhibits fungal diseases by inhibiting fungal germ tube penetration of epidermis (Savant et al. 1999). Silicon can alleviate nitrogen deficiency-induced stress in plants by (1) increasing the uptake of nitrogen, (2) improving nodulation and better N_2 fixation in legume plants, (3) increasing N use efficiency, and (4) altering primary metabolism due to stimulating amino acid remobilization.

Phosphorus is essential mineral element but most of phosphorus in soils is in biologically unavailable form. Soundararajan et al. (2016) reported that silicon fertilization increased P availability in various graminaceous species. Silicon can alleviate P shortage-induced stress in plants by (1) improving utilization of P by increasing phosphorylation; (2) decreasing excess Fe, Mn, and Al uptake and (3) increasing water-soluble phosphorus concentration. Owino-Gerroh and Gascho (2005) reported that application of soluble silicon in acid soils can decrease adsorption of phosphorus in soils and increase the amount of bioavailable phosphorus.

Potassium is one of the major macronutrients which has an important role in the growth and development of plants. Silicon can alleviate K-deficiency-induced stress by influencing the availability of K in the soil and plants and modulating antioxidant enzymes to alleviate K-deficiency-induced membrane lipid peroxidation and oxidative stress (Miao et al. 2010; Pei et al. 2010). Silicon can increase calcium and magnesium level in soil and plants. The improved absorption of calcium may be due to an increase in the movement of plasma membrane H^+-ATPase due to the addition of silicon (Etesami and Jeong 2018).

Silicon application influences the nutrient content of sunflower by increasing the accumulation of both macro- and micronutrients (Savić and Marjanović-Jeromela 2013). Silicon can alleviate the effect of micronutrient deficiency stress on the plants grown under nutrient-poor conditions (Pavlovic et al. 2013; Bityutskii et al. 2014; Hernandez-Apaolaza 2014). It has been reported that silicon alleviates chlorosis in plants due to Fe deficiency by (1) forming Fe plaque on root; (2) increasing Fe transport from root to shoot; (3) maintaining the balance of other micronutrients such as Fe/Mn ratio; (4) accumulating Fe-mobilizing compounds such as citrate (in xylem sap and root and shoot tissues) or catechins (in roots); (5) the activation of Fe-deficiency-associated genes and (6) Fe chelate reductase (Etesami and Jeong 2018). Under Fe toxicity conditions, silicon can protect plants from excess of Fe stress by (1) releasing OH^- by roots; (2) increasing the oxidizing capacity of roots, which changes ferrous iron (Fe^{2+}) into ferric iron (Fe^{3+}) and (3) increasing the iron precipitation in the growth media or at root surfaces (iron plaque) as Fe (III)-silicates (You-Qiang et al. 2012).

Silicon is able to mitigate manganese deficiency in plants by enhancing the roots oxidizing capacity, which gives a higher Mn-oxidation rate in the rhizosphere (Li et al. 1999). Silicon can also mitigate the excess stress of Mn by affecting Mn-oxidation state, which promotes its precipitation, regulates the Fenton reaction, and mediates the Mn adsorption on cell walls.

Silicon influences zinc plant nutrition under deficiency stress conditions (Pascual et al. 2016) by (1) remobilizing the Zn pools in the roots, (2) changing Zn distribution in plants, (3) increasing citrate in plants, and (4) activating the Zn-deficiency mechanism. Silicon can also mitigate the excess stress of Zn by avoiding the Zn transport to more sensitive organs of plant by enhancing Zn^{2+} adsorption on the deposits of the silicate.

Si also influences copper plant nutrition in the plants grown under Cu stress conditions by affecting the distribution of Cu within leaves (Frantz et al. 2011). Silicon can mitigate the excess stress of Cu in the plants by the formation of silicon deposits on the cell wall, which increases the Cu-binding sites.

Silicon-Induced Growth Regulation and Abiotic Stress Tolerance in Plants

Plant Growth Improvement

An increase in plant growth and development due to silicon application has been reported in rice (Ma and Yamaji 2008; Kim et al. 2012), cotton (Li et al. 1989), soybean (Hamayun et al. 2010), wheat (Gong et al. 2005), sorghum (Hattori et al. 2005), cucumber (Feng et al. 2009), barley (Savant et al. 1999), bean (Zuccarini 2008), tomato (Al-aghabary et al. 2005; Liang et al. 2007), spinach (Eraslan et al. 2008), and maize (Liang et al. 2005).

Silicon foliar fertilizer is advantageous for sugar beet production and a significant increase has been reported in sugar yield (Artyszak et al. 2014). Silicon has increased yield by 22% in potato (Luz et al. 2008), 30% in rice, and 45% in sugarcane (Kingston 2008). The application of silicon in canola improved yield and oil content and reduced pesticide and fungicide usage (Lynch 2008). Silicon augmentation in soilless cultivation of corn salad improved the yield, quality, and shelf life of corn by the regulation of nutrient acquisition, uptake of nitrate/iron, phenoloxidase gene expression, and protection of chlorophyll pigment from degradation (Gottardi et al. 2012). Silicon inclusion in tissue culture medium resulted in the enhancement of axillary shoot induction, alleviation of hyperhydricity (Soundararajan et al. 2017), callus induction (Islam et al. 2005), and root morphogenesis (Asmar et al. 2013).

Aquaporins are essential transmembrane proteins that maintain the uptake and movement of water molecules across the cell membranes particularly under abiotic stress condition (Boursiac et al. 2005). According to Boursiac et al. (2008), aquaporin activity is susceptible to change in the ROS level as H_2O_2 stimulated by salt stress resulted in the prevention of aquaporin function.

The higher expression of genes related to aquaporin results in the rapid water uptake which also dilutes the excess concentration of Na^+ ions lethal for the plants (Gao et al. 2010). The findings of Sutka et al. (2011) illustrated that the abundance of aquaporin genes in the roots balance the water uptake by the plants even under water deficit conditions. The enhancement of aquaporin-related genes by silicon might improve the water status in plants under salinity and drought stress. The improvement of water status and ion balance aid in the reclamation of plants from abiotic stress. It is known that silicon can increase the uptake of nutrients by increasing root activity (Chen et al. 2011), enhancing water uptake (Sonobe et al. 2010), and improving hydraulic conductance of roots (Hattori et al. 2008).

Increase in Root System of Plants

Abiotic stresses adversely affect the growth and yield of plants by limiting the uptake and translocation of water and essential nutritional ions (Hu and Schmidhalter 2005). Nutrient absorption is linked to root surface region and root length. An increase in root surface area provides more exposed areas for the absorption of dispersed ions (Barber 1995). Silicon can improve root growth and subsequently enhance nutrient uptake, improve nutrient balance, and plant shoot biomass under salinity (Kim et al. 2016; Li et al. 2015) and drought (Chen et al. 2011). It has been reported that silicon facilitates root growth through increasing cell wall extensibility in the growth zone of roots (Vaculík et al. 2009). Silicon has also contributed to stimulation of nutrient uptake by increasing water uptake (Sonobe et al. 2010). The higher root growth increases water acquisition (increased water use efficiency) and nutrient uptake, it subsequently alleviates the adverse stress effects in silicon-treated plants. Water deficit also limits nutrient uptake through roots and subsequent transport to shoots, thereby reducing nutrient availability and metabolism (Farooq et al. 2009). Silicon-mediated selective transport capacity for K^+ over Na^+ and thereby increases in K^+/Na^+, which may be one of the main mechanisms improving plant growth and yield under abiotic stress.

Improvement of Water Relations

Romero-Aranda et al. (2006) stated that silicon helps in water restoration in salt-stressed plants due to its hydrophilic nature. Silicon can decrease salt toxicity for plants and improve plant growth under salinity stress by different mechanism including: (1) increasing water storage in plants which contributes to salt dilution (Tuna et al. 2008); (2) increasing thickness of leaves which results in leaf water content and water potential (Gong et al. 2003) and (3) deposition of silicon in leaves which reduces transpiration from leaf surface and decreases the escape of water molecules from the leaf surface (Keller et al. 2015). It has been recognized that drought stress substantially reduces the leaf water potential and water content in drought-stressed plants. Silicon also influences water relation and improves significantly the photosynthesis and water status in non-irrigated crops (Rizwan et al. 2015). Gao et al. (2005) observed that silicon application enhanced water use efficiency in drought-stressed maize by reducing leaf transpiration rate. According to Rizwan et al. (2015), silicon can influence stomatal movement in plants.

Some of the mechanisms by which silicon can result in higher efficient use of water, enhance root water uptake by roots, and increase drought tolerance in plants under water deficit stress (Zhu and Gong 2014) include: (1) altering transpiration (Farooq et al. 2009); (2) forming a cuticle-silica double layer on leaf epidermal tissue which reduces leaf transpiration and water flow rate in xylem vessels (Zhu and Gong 2014); (3) accumulation of soluble sugars and amino acids (Sonobe et al. 2010); (4) deposition of silicon on root cell wall, which affects the wetting properties of xylem vessels and water/solute transport (Gao et al. 2005); (5) maintaining photosynthetic pigments content (Lobato et al. 2009); (6) improving chloroplast ultrastructure; (7) increasing activities of antioxidant enzymes (Gong et al. 2005); (8) enhancing the activity of phosphoenolpyruvate carboxylase and the concentration of inorganic phosphate in plant leaves (Gong and Chen 2012).

Regulation of Biosynthesis of Compatible Solutes

Under stressful conditions such as salinity and drought, concentration of compatible solutes or osmolytes such as proline (Pei et al. 2010), glycine betaine, and polyols (Parida and Das 2005) increased in plants. Silicon application also increased plant tolerance to salinity and drought stress by modifying the levels of solutes such as proline (Yin et al. 2013), glycine betaine (Torabi et al. 2015), carbohydrates (Ming et al. 2012), polyols, antioxidant compounds (Hashemi et al. 2010), and soluble sugars and free amino acids (Hajiboland et al. 2016). These compatible solutes may increase plant resistance to abiotic stresses by (1) maintaining higher leaf water potential during stress; (2) scavenging free radicals and buffering cellular redox potential, which keep plants protected against oxidative stress and (3) stabilizing subcellular structures (membranes and proteins) (Fahad et al. 2015). It has been found that the compatible solutes may also act as oxygen radical scavengers (An and Liang 2013; Abbas et al. 2015). Silicon alleviated drought stress in the plants by osmotic adjustment inducing the production of soluble sugars and amino acids such as alanine and glutamic acid (Zhu and Gong 2014).

Photosynthesis

It has been reported that silicon delays chlorophyll degradation in plant species under abiotic stresses (Al-aghabary et al. 2004; Feng et al. 2010; Gottardi et al. 2012). Silicon has also been shown to result in higher concentration of chlorophyll per unit area of leaf tissue. There are two possible mechanisms have been reported: The first one was related with the structural protection of the chloroplast membranes by silicon (Al-aghabary et al. 2004; Feng et al. 2010) and second was the effect of silicon on strengthened cell walls which might contribute to a more favorable position of leaves to intercept light and increased the photosynthesis (Ma and Takahashi 2002). Silicon contributes to maintain micronutrients balance such as Fe/Mn ratio (Pich et al. 1994), which is also beneficial in enhancing chlorophyll synthesis and provides a possible explanation for the stimulation in growth of Fe-deficient plants supplied with silicon (Pavlovic et al. 2013; Bityutskii et al. 2014). The increase in citrate concentration in plants has been reported with silicon supply. Citrate is one of the molecules that joined iron on its transport through the xylem (Rellán-Álvarez et al. 2010).

Polyamines Synthesis and Metabolism

Plants with high level of polyamines such as putrescine, spermidine, and spermine reported to possess more resistance against environmental stresses (Quinet et al. 2010; Pottosin and Shabala 2014). The elevated levels of genes responsible for the synthesis of polyamines mitigate the negative effects of oxidative stress (Roy and Wu 2001; Tang et al. 2007). The augmentation of silicon elevated the expression level of S-adenosyl-L-methionine decarboxylase (SAMDC) gene which encodes a vital enzyme involved in the biosynthesis of polyamines. Polyamines are involved in various vital processes such as replication, transcription, translation, stabilization of membranes, and modulation of enzyme activities in addition to stress tolerance. Hence, the regulation of genes involved in polyamine biosynthesis by silicon could not only help in the stress alleviation but also improve the fundamental processes in cells and increased the growth and development of plants.

Antioxidative Defense System

Plants continuously produce several active oxygen species (AOS) or reactive oxygen species (ROS) such as superoxide anion (O_2^-), hydrogen peroxide (H_2O_2), and hydroxyl radical (OH^-) during photosynthesis and respiration processes in different cell organelles such as mitochondria, chloroplast, and peroxisomes. ROS can cause serious oxidative damage to the biomolecules such as protein, membrane lipids, and nucleic acids in the cells (Apel and Hirt 2004; Tripathi et al. 2017). The scavenging of ROS is most important defense mechanism to cope with stress conditions in plants (Baxter et al. 2014). Silicon nutrition in plants enhances the production of antioxidants and enzymes involved in detoxification of the free radicals (Zhu et al. 2004). Plants can maintain homeostasis by two different detoxification mechanisms involving enzymatic and non-enzymatic antioxidants (Mittler 2002; Sytar et al. 2013; Wu et al. 2017). Application of silicon can improve ROS scavenging ability in plants by development of ROS scavenging enzymes such as superoxide dismutase (SOD), peroxidase (POD), catalase (CAT), and ascorbate peroxidase (APX) and non-enzymatic antioxidants such as glutathione, ascorbate, and carotenoids (Crusciol 2009; Shen et al. 2010; Torabi et al. 2015).

Etesami and Jeong (2018) reported that silicon maintained cell membrane permeability and stability of the plants grown under abiotic stresses. Silicon application can also increase the glutathione concentration and suppress the malondialdehyde concentration (Liang et al. 2006). Zhu et al. (2004) reported that silicon is responsible for inhibiting the membrane damage caused by the formation of malondialdehyde, which causes the lipid peroxidation of membranes. By regulating antioxidant defense system, silicon can decrease lipid peroxidation in plants (Kim et al. 2017).

Silicon enhanced abiotic stress tolerance capacity which is linked to accumulation of photorespiratory enzymes (Nwugo and Huerta 2011). Silicon has been reported to prevent the damage of membrane caused by the formation of malondialdehyde (MDA) (Zhu et al. 2004) by regulating antioxidant defense system in plants (Zhu and Gong 2014). Silicon has also been found to decrease the concentration of MDA, the end-product of lipid peroxidation, in salt-stressed barley (Liang et al. 2003), maize (Moussa 2006), and grapevine (Soylemezoglu et al. 2009), thus may help to maintain membrane integrity and decrease membrane permeability.

Ma et al. (2016) found that silicon reduced H_2O_2 accumulation and increased expression of antioxidant enzyme genes in wheat under drought stress. Farooq et al. (2016) observed that silicon treatment increased the antioxidant capacity of rice plants under cadmium stress. Many experiments have assessed the impact of silicon on plants subjected to abiotic stress at the level of gene expression (Liu et al. 2014; Yin et al. 2016), and it is now becoming apparent that silicon may effect primary metabolism in higher plants (Sanglard et al. 2014).

Phytohormone Regulation

Phytohormones are known to play vital role in the ability of plants to acclimatize to different environment by different mechanism (Fahad et al. 2015). It has been reported that silicon application may enhance the plant tolerance to abiotic stresses by adjusting the level of phytohormones (Kim et al. 2014). Gibberellin regulates all aspects of plant life from seed germination to vegetative growth and fruiting (Colebrook et al. 2014). Endogenous bioactive GA1 and GA4 content increased when higher doses of silicon were applied to cucumber plants under salinity and drought stress (Hamayun et al. 2010). This clearly suggests that GAs play significant role in salt and drought stress alleviation. Silicon is known to increase plant growth, which can also be associated with the effect of exogenous gibberellin application to crop plants. It shows that silicon application activates gibberellin biosynthesis in order to maintain growth and impart stress tolerance.

Silicon is essential for life processes such as DNA replication (Okita and Volcani 1978; Martin-Jézéquel et al. 2000). Silicon was found to delay leaf senescence by activation of the cytokinin pathway in both silicon accumulating and non-accumulating plant species (Markovich et al. 2017). Silicon can increase plant tolerance to diseases by increasing the expression of genes involved in plant resistance to pests and diseases, enhancing the production of phenolic compounds, lignin and enzymes of phenylpropanoid pathway and by enhancing the concentration of polyphenol, antimicrobial flavonoids, and anthocyanin (Etesami and Jeong 2018).

Future Perspectives

Based on current knowledge, reviewed here, it can be stated that silicon is noncorrosive, non-pollutive element, and its excessive amount in soil does not pose any harm to plants. Silicon is a modest and major element of soil with enormous benefits to plants especially in the mitigation of abiotic stresses.

We suggest several future avenues of research:

1. Researches are needed to investigate the effect of silicon on plant biochemistry and gene regulation under abiotic stresses.
2. Nanotechnology-related applications can be used to explore possibilities for the formulation of silicon nanoparticles to elevate stress tolerance in plants.
3. The resistance of plants to various environmental stresses can be enhanced by genetic modification of root ability to take up silicon and its accumulation in different plant parts. It will reduce the silicon expenditure in crop fields.
4. Silicon possess multifaceted role in the regulation of genes involved in various physiological mechanism such as photosynthesis, secondary metabolism, polyamine biosynthesis, and transcription. Hence, there is a need of deep investigation related to the molecular level modulations triggered by silicon supplementation for physiological improvement of plant growth under stress conditions.
5. The meta-analysis-related studies on the alleviation of abiotic stresses by silicon highlighted that most studies have focused on single species and single-stress models but there is a lack of studies which looked to compare effects between species or stress types.
6. Farmers and researchers should be aware about the potential application of silicon in agriculture. Farmers should know the correct dose of silicon for particular crop and mode and time of silicon application on the crop plants to increase crop productivity.

Silicon is useful and sustainable fertilizer for crops facing a spectrum of environmental stresses. The application of silicon in the agriculture sector can improve food security by raising crop tolerance to adverse environmental conditions by enhancing adaptability of crops to different stresses. Silicon may be complementary option in silicon-deficient areas as silicon fertilizer can provide economic as well as ecological benefits.

Acknowledgements Authors are thankful to Dr. Chanderdeep Tandon, Director, Amity Institute of Biotechnology, Amity University, Noida, for his persistent encouragement and valuable suggestions.

References

Abbas T, Balal RM, Shahid MA, Pervez MA, Ayyub CM, Aqueel MA, Javaid MM (2015) Silicon-induced alleviation of NaCl toxicity in okra (*Abelmoschus esculentus*) is associated with enhanced photosynthesis, osmoprotectants and antioxidant metabolism. Acta Physiol Plant 37:1–15

Adrees M, Ali S, Rizwan M, Zia-ur-Rehman M, Ibrahim M, Abbas F, Farid M, Qayyum MF, Irshad MK (2015) Mechanisms of silicon-mediated alleviation of heavy metal toxicity in plants: a review. Ecotoxicol Environ Saf 119:186–197

Agarie S, Uchida H, Agata W, Kubota F, Kaufman PB (1998) Effects of silicon on transpiration and leaf conductance in rice plants (*Oryza sativa* L.). Plant Prod Sci 1:89–95

Ahmed M, Asif M, Hassan FU (2014) Augmenting drought tolerance in sorghum by silicon nutrition. Acta Physiol Plant 36:473–483

Al-Aghabary K, Zhu Z, Shi QH (2004) Influence of silicon supply on chloro- phyll content, chlorophyll II fluorescence and antioxidant enzyme activities in tomato plants under salt stress. J Plant Nutr 27:2101–2115

Al-aghabary K, Zhu Z, Shi Q (2005) Influence of silicon supply on chlorophyll content, chlorophyll fluorescence, and antioxidative enzyme activities in tomato plants under salt stress. J Plant Nutr 27:2101–2115

An Y, Liang Z (2013) Drought tolerance of Periploca sepium during seed germination: antioxidant defense and compatible solutes accumulation. Acta Physiol Plant 35:959–967

Apel K, Hirt H (2004) Reactive oxygen species: metabolism, oxidative stress, and signal transduction. Annu Rev Plant Biol 55:373–399

Artyszak A, Gozdowski D, Kucińska K (2014) The effect of foliar fertilization with marine calcite in sugar beet. Plant Soil Environ 60:413–417

Ashraf M, Ahmad A, MeNeilly T (2001) Growth and photosynthetic charac-teristics in the pearl millet under water stress and the different potassium supply. Photosynthetica 39:389–394

Ashraf M, Rahmatullah AM, Ahmad R, Mujeeb F, Sarwar A, Ali L (2010) Alleviation of detrimental effects of NaCl by silicon nutrition in salt-sensitive and salt-tolerant genotypes of sugarcane (*Saccharum officinarum* L). Plant and Soil 326:381–391

Asmar SA, Castro EM, Pasqual M, Pereira FJ, Soares JDR (2013) Changes in leaf anatomy and photosynthesis of micropropagated banana plantlets under different silicon sources. Sci Hortic 161:328–332

Balakhnina T, Borkowska A (2013) Effects of silicon on plant resistanceto environmental stresses: review. Int Agrophys 27:225–232

Barber SA (1995) Soil nutrient bioavailability: a mechanistic approach. Wiley, New York

Baxter A, Mittler R, Suzuki N (2014) ROS as key players in plant stress signalling. J Exp Bot 65:1229–1240

Bityutskii N, Pavlovic J, Yakkonen K, Maksimović V, Nikolic M (2014) Contrasting effect of silicon on iron, zinc and manganese status and accumulation of metal-mobilizing compounds in micronutrient-deficient cucumber. Plant Physiol Biochem 74:205–211

Bodner G, Nakhforoosh A, Kaul HP (2015) Management of crop water under drought: a review. Agron Sustain Dev 35:401–442

Boursiac Y, Chen S, Luu DT, Sorieul M, van den Dries N, Maurel C (2005) Early effects of salinity on water transport in *Arabidopsis* roots. Molecular and cellular features of aquaporin expression. Plant Physiol 139:790–805

Boursiac Y, Prak S, Boudet J, Postaire O, Luu DT, Tournaire-Roux C (2008) The response of *Arabidopsis* root water transport to a challenging environment implicates reactive oxygen species-and phosphorylation-dependent internalization of aquaporins. Plant Signal Behav 3:1096–1098

Bradbury M, Ahmad R (1990) The effect of silicon on the growth of *Prosopis juliflora* growing in saline soil. Plant and Soil 125:71–74

Cakmak I (2001) Plant nutrition research: priorities to meet human needs for food in sustainable ways. Plant and Soil 247:3–24

Carmen B, Roberto D (2011) Soil bacteria support and protect plants against abiotic stresses. In: Shan A (ed) Abiotic stress in plants mechanisms and adaptations. InTech, Croatia, pp 143–170

Chen J, Caldwell RD, Robinson CA, Steinkamp R (2000) Silicon: the estranged medium element. Bulletin 341, Institute of Food and Agricultural Sciences, University of Florida

Chen W, Yao X, Cai K, Chen J (2011) Silicon alleviates drought stress of rice plants by improving plant water status, photosynthesis and mineral nutrient absorption. Biol Trace Elem Res 142:67–76

Chinnusamy V, Jagendorf A, Zhu JK (2005) Understanding and improving salt tolerance in plants. Crop Sci 45:437–448

Colebrook EH, Thomas SG, Phillips AL (2014) The role of gibberellin signalling in plant responses to abiotic stress. J Exp Biol 217:67–75

Commuri PD, Jones RJ (2001) High temperatures during endosperm cell division in maize. A genotypic comparison under in vitro and field conditions. Crop Sci 41:1122–1130

Cooke J, Leishman MR (2011) Silicon concentration and leaf longevity: is silicon a player in the leaf dry mass spectrum? Funct Ecol 25:1181–1188

Cooke J, Leishman MR (2016) Consistent alleviation of abiotic stress with silicon addition: a meta-analysis. Funct Ecol 30:1340–1357

Coskun D, Britto DT, Huynh WQ, Kronzucker HJ (2016) The role of silicon in higher plants under salinity and drought stress. Front Plant Sci 7:1072. https://doi.org/10.3389/fpls.2016.01072

Crusciol CAC (2009) Effects of silicon and drought stress on tuber yield and leaf biochemical characteristics in potato. Crop Sci 49:949–954

da Cunha KPV, do Nascimento CWA (2009). Silicon effects on metal tolerance and structural changes in maize (*Zea mays* L.) grown on a cadmium and zinc enriched soil. Water Air and Soil Pollution. 197: 323–330. https://doi.org/10.1007/s11270-008-9814-9

Dallagnol LJ, Rodrigues FA, DaMatta FM, Mielli MVB, Pereira SC (2011) Deficiency in silicon uptake affects cytological, physiological and biochemical events in the rice–*Bipolaris oryzae* interaction. Phytopathology 101:92–104

Debona D, Rodrigues FA, Datnoff LE (2017) Silicon's role in abiotic and biotic plant stresses. Annu Rev Phytopathol 55:85–107

Epstein E (1994) The anomaly of silicon in plant biology. Proc Natl Acad Sci U S A 91:11–17

Epstein E (1999) Silicon. Annu Rev Plant Physiol Plant Mol Biol 50:641–664

Epstein E (2009) Silicon: its manifold roles in plants. Ann Appl Biol 155:155–160

Epstein E, Bloom AJ (2005) Mineral nutrition of plants: principles and perspectives, 2nd edn. Sinuar Associates, Sunderland, MA, pp 201–240

Eraslan F, Ali I, David J, Gunes A (2008) Interactive effects of salicylic acid and silicon on oxidative damage and antioxidant activity in spinach (*Spinacia oleracea* L cv Matador) grown under boron toxicity and salinity. Plant Growth Regul 55:207–219

Etesami H, Beattie GA (2017) Plant-microbe interactions in adaptation of agricultural crops to abiotic stress conditions. In: Probiotics and plant health. Springer, Berlin, pp 163–200

Etesami H, Jeong BR (2018) Silicon (Si): review and future prospects on the action mechanisms in alleviating biotic and abiotic stresses in plants. Ecotoxicol Environ Saf 147:881–896

Exley C (2015) A possible mechanism of biological silicification in plants. Front Plant Sci 6:853. https://doi.org/10.3389/fpls.2015.00853

Fahad S, Hussain S, Matloob A, Khan FA, Khaliq A, Saud S, Huang J (2015) Phytohormones and plant responses to salinity stress: a review. Plant Growth Regul 75:391–404

FAO (1998) World Agricultural Center, FAOSTAT agricultural statistic data—base gateway
Farooq M, Wahid A, Kobayashi N, Fujita D, Basra SMA (2009) Plant drought stress: effects, mechanisms and management. In: Sustainable agriculture. Springer, Berlin, pp 153–188
Farooq MA, Detterbeck A, Clemens S, Dietz K-J (2016) Silicon-induced reversibility of cadmium toxicity in rice. J Exp Bot 67:3573–3585
Feng JP, Shi QH, Wang XF (2009) Effects of exogenous silicon on photosynthetic capacity and antioxidant enzyme activities in chloroplast of cucumber seedlings under excess manganese. Agric Sci China 8:40–50
Feng J, Shi Q, Wang X, Wei M, Yang F, Xu H (2010) Silicon supplementation ameliorated the inhibition of photosynthesis and nitrate metabolism by cadmium (Cd) toxicity in *Cucumis sativus* L. Sci Hortic 123:521–530
Frantz JM, Khandekar S, Leisner S (2011) Silicon differentially influences copper toxicity response in silicon-accumulator and non-accumulator species. J Am Soc Hort Sci 136:329–338
Gao X, Zou C, Wang L, Zhang F (2005) Silicon improves water use efficiency in maize plants. J Plant Nutr 27:1457–1470
Gao Z, He X, Zhao B, Zhou C, Liang Y, Ge R (2010) Overexpressing a putative aquaporin gene from wheat, TaNIP, enhances salt tolerance in transgenic *Arabidopsis*. Plant Cell Physiol 51:767–775
Gibson LR, Paulsen GM (1999) Yield components of wheat grown under high temperature stress during reproductive growth. Crop Sci 39:1841–1184
Glick BR (2014) Bacteria with ACC deaminase can promote plant growth and help to feed the world. Microbiol Res 169:30–39
Gong H, Chen K (2012) The regulatory role of silicon on water relations, photosynthetic gas exchange and carboxylation activities of wheat leaves in field drought conditions. Acta Physiol Plant 34:1589–1594
Gong HJ, Chen KM, Chen GC, Wang SM, Zhang CI (2003) Effects of silicon on growth of wheat under drought. J Plant Nutr 26:1055–1063
Gong H, Zhu X, Chen K, Wang S, Zhang C (2005) Silicon alleviates oxidative damage of wheat plants in pots under drought. Plant Sci 169:313–321
Gottardi S, Iacuzzo F, Tomasi N, Cortella G, Manzocco L, Pinton R (2012) Beneficial effects of silicon on hydroponically grown corn salad (*Valerianella locusta* L.) plants. Plant Physiol Biochem 56:14–23
Greenway H, Munns R (1980) Mechanism of salt tolerance in nonhalophytes. Annu Rev Plant Physiol 31:149–190
Gunes AA, Bagci IEG, Pilbeam DJ (2007) Silicon-mediated changes of some physiological and enzymatic parameters symptomatic for oxidative stress in spinach and tomato grown in sodic-B toxic soil. Plant and Soil 290:103–114
Gunes A, Kadioglu YK, Pilbeam DJ, Inal A, Coban S, Aksu A (2008a) Influence of silicon on sunflower cultivars under drought stress, II: essential and nonessential element uptake determined by polarized energy dispersive X-ray fluorescence. Commun Soil Sci Plant Anal 39:1904–1927
Gunes A, Pilbeam DJ, Inal A, Coban S (2008b) Influence of silicon on sunflower cultivars under drought stress, I: growth, antioxidant mechanisms, and lipid peroxidation. Commun Soil Sci Plant Anal 39:1885–1903
Guntzer F, Keller C, Meunier JD (2011) Benefits of plant silicon for crops: a review. Agron Sustain Dev 32:201–213
Guntzer F, Keller C, Poulton PR, McGrath SP, Meunier JD (2012) Long-term removal of wheat straw decreases soil amorphous silica at Broadbalk, Rothamsted. Plant and Soil 352:173–184
Guo W, Hou YL, Wang SG, Zhu YG (2005) Effect of silicate on the growth and arsenate uptake by rice (*Oryza sativa* L.) seedlings in solution culture. Plant and Soil 272:173–181
Hajiboland R, Cherghvareh L, Dashtebani F (2016) Effect of silicon supplementation on wheat plants under salt stress. J Plant Proc Funct 5:1–12
Hamayun M, Sohn EY, Khan SA, Shinwari ZK, Khan AL, Lee IJ (2010) Silicon alleviates the adverse effects of salinity and drought stress on growth and endogenous plant growth hormones of soybean (*Glycine max* L.). Pak J Bot 42:1713–1722

Hashemi A, Abdolzadeh A, Sadeghipour HR (2010) Beneficial effects of silicon nutrition in alleviating salinity stress in hydroponically grown canola, *Brassica napus* L. plants. J Soil Sci Plant Nutr 56:244–253

Hattori T, Inanaga S, Tanimoto E, Lux A, Luxová M, Sugimoto Y (2003) Silicon induced changes in viscoelastic properties of sorghum root cell walls. Plant Cell Physiol 44:743–749

Hattori T, Inanaga H, Araki H, An P, Morita S, Luxova M, Lux A (2005) Application of silicon enhanced drought tolerance in Sorghum bicolor. Physiol Plant 123:459–466

Hattori T, Sonobe K, Araki H, Inanaga S, An P, Morita S (2008) Silicon application by sorghum through the alleviation of stress-induced increase in hydraulic resistance. J Plant Nutr 31:1482–1495

Henriet C, Draye X, Oppitz I, Swennen R, Delvaux B (2006) Effects, distribution and uptake of silicon in banana (*Musa* spp.) under controlled conditions. Plant and Soil 287:359–374

Hernandez JA, Jim-enez A, Mullineaux P, Sevilla F (2000) Tolerance of pea (*Pisum sativum* L.) to long-term salt stress is associated with induction of antioxidant defenses. Plant Cell Environ 23:853–862

Hernandez-Apaolaza L (2014) Can silicon partially alleviate micronutrient deficiency in plants? A review. Planta 240:447–458

Hodson MJ, White PJ, Mead A, Broadley MR (2005) Phylogenetic variation in the silicon composition of plants. Ann Bot 96:1027–1046

Hu Y, Schmidhalter U (2005) Drought and salinity: a comparison of their effects on mineral nutrition of plants. J Plant Nutr Soil Sci 168:541–549

Hull RJ (2004) Scientists start to recognize silicon's beneficial effects. Turf Grass Trends 8:69–73

Imtiaz M, Rizwan MS, Mushtaq MA, Ashraf M, Shahzad SM, Yousaf B, Saeed DA, Rizwan M, Nawaz MA, Mehmood S, Tu S (2016) Silicon occurrence, uptake, transport and mechanisms of heavy metals, minerals and salinity enhanced tolerance in plants with future prospects: a review. J Environ Manage 183:521–529

Islam MM, Ahmed M, Mahaldar D (2005) In vitro callus induction andplant regeneration in seed explants of rice (*Oryza sativa* L.). Res J Agric Biol Sci 1:72–75

Jewell MC, Campbell BC, Godwin ID (2010) Transgenic plants for abiotic stress resistance. In: Transgenic crop plants. Springer, Berlin, pp 67–132

John VS, Cabot C, Poschenrieder C, Barcelo J (2003) Efficient leaf ion partitioning, an overriding condition for abscisic acid-controlled stomatal and leaf growth responses to NaCl salinization in two legumes. J Exp Bot 54:2111–2119

Kalra N, Jain MC, Joshi HC, Chaudhary R, Sushilkumar PH, Sharma SK, Kumar V, Kumar R, Harit RC, Khan SA, Hussain MJ (2003) Soil properties and crop productivity as influenced by fly ash incorporation in soil. Environ Monit Assess 87:93–109

Kaya C, Tuna L, Higgs D (2006) Effect of silicon on plant growth and mineral nutrition of maize grown under water-stress conditions. J Plant Nutr 29:1469–1480

Kaya C, Tuna AL, Sonmez O, Ince F, Higgs D (2010) Mitigation effects of silicon on maize plants grown at high zinc. J Plant Nutr 32:1788–1798

Keles Y, Oncel I, Yenice N (2004) Relationship between boron content and antioxidant compounds in Citrus leaves taken from fields with different water source. Plant and Soil 265:343–353

Keller C, Rizwan M, Davidian JC, Pokrovsky OS, Bovet N, Chaurand P, Meunier JD (2015) Effect of silicon on wheat seedlings (*Triticum turgidum* L.) grown in hydroponics and exposed to 0 to 30 μM Cu. Planta 241:847–860

Khan MA, Ungar IA, Showalter AM (2000) Effects of sodium chloride treatments on growth and ion accumulation of the halophyte *Haloxylon recurvum*. Commun Soil Sci Plant Anal 31:2763–2774

Kim YH, Khan AL, Shinwari ZK, Kim DH, Waqas M, Kamran M (2012) Silicon treatment to rice (*Oryza sativa* L. cv. 'Gopumbyeo') plants during different growth periods and its effects on growth and grain yield. Pak J Bot 44:891–897

Kim YH, Khan AL, Waqas M, Shim JK, Kim DH, Lee KY, Lee IJ (2014) Silicon application to rice root zone influenced the phytohormonal and antioxidant responses under salinity stress. J Plant Growth Regul 33:137–149

Kim YH, Khan AL, Waqas M, Shahzad R, Lee IJ (2016). Silicon mediated mitigation of wounding stress acts by up-regulating the rice antioxidant system. Cereal Research Communications. 44: 111–121.

Kim YH, Khan AL, Waqas M, Lee IJ (2017) Silicon regulates antioxidant activities of crop plants under abiotic-induced oxidative stress: a review. Front Plant Sci 8:510. https://doi.org/10.3389/fpls.2017.00510

Kingston G (2008) Silicon fertilisers—requirements and field experiences. In: Silicon in Agriculture 4th International Conference, 31 October, Port Edward, South Africa, p 52

Li YC, Alva AK, Summer ME (1989) Response of cotton cultivars to aluminium in solutions with varying silicon concentrations. J Plant Nutr 12:881–892

Li CH, Chu TD, Liu XB, Yang Q (1999) Silicon nutrition effects and its study and application development in China. In: Proceedings of Symposium of Plant Nutrition, Shaanxi Science and Technology Press, China, pp 329–333

Li J, Leisner SM, Frantz J (2008) Alleviation of copper toxicity in *Arabidopsis thaliana* by silicon addition to hydroponic solutions. J Am Soc Hort Sci 133:670–677

Li H, Zhu Y, Hu Y, Han W, Gong H (2015) Beneficial effects of silicon in alleviating salinity stress of tomato seedlings grown under sand culture. Acta Physiol Plant 37:1–9

Liang YC (1999) Effects of silicon on enzyme activity and sodium, potassium and calcium concentration in barley under salt stress. Plant and Soil 29:217–224

Liang YC, Chen Q, Liu Q, Zhang WH, Ding RX (2003) Exogenous silicon (Si) increases antioxidant enzyme activity and reduces lipid peroxidation in roots of salt-stressed barley (*Hordeum vulgare* L.). J Plant Physiol 160:1157–1164

Liang Y, Si J, Romheld V (2005) Silicon uptake and transport is an active process in *Cucumis sativus*. New Phytol 167:797–804

Liang YC, Hua H, Zhu YG, Zhang J, Cheng C, Romheld V (2006) Importance of plant species and external silicon concentration to active silicon uptake and transport. New Phytol 172:63–67

Liang YC, Sun WC, Zhu YG, Christie P (2007) Mechanisms of silicon-mediated alleviation of abiotic stresses in higher plants: a review. Environ Pollut 147:422–428

Liang Y, Nikolic M, Bélanger R, Gong H, Song A (2015) Effect of silicon on crop growth, yield and quality. In: Liang Y, Nikolic M, Bélanger R, Gong H, Song A (eds) Silicon in agriculture: from theory to practice. Springer, Dordrecht, pp 209–223

Liu P, Yin L, Deng X, Wang S, Tanaka K, Zhang S (2014) Aquaporin mediated increase in root hydraulic conductance is involved in silicon induced improved root water uptake under osmotic stress in *Sorghum bicolor* L. J Exp Bot 65:4747–4756

Lobato AKS, Coimbra GK, Neto MAM, Costa RCL, Santos Filho BG, Oliveira Neto CF, Luz LM, Barreto AGT, Pereira BWF, Alves GAR, Monteiro BS, Marochio CA (2009) Protective action of silicon on relations and photosynthetic pigments in pepper plants induced to water deficit. Res J Agric Biol Sci 4:617–623

Lovering TS, Engel C (1959) Significance of accumulator plants in rock weathering. Bull Geol Soc Am 70:781–800

Lux A, Luxová M, Abe J, Tanimoto E, Hattori T, Inanaga S (2003) The dynamics of silicon deposition in the sorghum root endodermis. New Phytol 158:437–441

Luyckx M, Hausman JF, Lutts S, Guerriero G (2017) Silicon and plants: current knowledge and technological perspectives. Front Plant Sci 8:411. https://doi.org/10.3389/fpls.2017.00411

Luz JMQ, Rodrigues CR, Goncalves MV, Coelho L (2008) The effect of silicate on potatoes in Minas Gerais, Brazil. IV Silicon in Agriculture Conference, 31 October, Universidade Federal de Uberlandia, Amazonas 4C 127 Uberlandia, Brazil, p 67

Lynch M (2008) Silicates in contemporary Australian farming: a 20 year review. IV Silicon in Agriculture Conference, 31 October, South Africa, North Coast Testing Services, Bellingen NSW 2454, Australia, p 49

Ma JF (2004) Role of silicon in enhancing the resistance of plants to biotic and abiotic stresses. Soil Sci Plant Nutr 50:11–18

Ma JF, Takahashi E (2002) Soil, fertilizer and plant silicon research in Japan. Elsevier, Amsterdam

Ma JF, Yamaji N (2008) Functions and transport of silicon in plants. Cell Mol Life Sci 65:3049–3057

Ma JF, Goto S, Tamai K, Ichii M (2001) Role of root hairs and lateral roots in silicon uptake by rice. Plant Physiol 127:1773–1780
Ma J, Cai H, He C, Zhang W, Wang L (2015) Ahemicellulose-bound form of silicon inhibits cadmium ion uptake in rice (*Oryza sativa*) cells. New Phytol 206:1063–1074
Ma D, Sun D, Wang C, Qin H, Ding H, Li Y (2016) Silicon application alleviates drought stress in wheat through transcriptional regulation of multiple antioxidant defense pathways. J Plant Growth Regul 35:1–10
Markovich O, Steiner E, Kouøil S, Tarkowski P, Aharoni A, Elbaum R (2017) Silicon promotes cytokinin biosynthesis and delays senescence in *Arabidopsis* and Sorghum. Plant Cell Environ 40:1189–1196
Marschner H (1995) Mineral nutrition of higher plants. Academic Press, London, p 16
Martin-Jézéquel V, Hildebrand M, Brzezinski MA (2000) Silicon metabolism in diatoms: implications for growth. J Phycol 36:821–840
Matichenkov VV, Calvert DV (2002) Silicon as a beneficial element for sugarcane. J Am Soc Sugar Cane Technol 22:21–30
Mcginnity P (2015) Silicon and its role in crop production. PhD thesis. http://planttuff.com/wpcontentwpcontent/uploads/2015/12/silicon-agricultureiiterature-rvw-1.pdf
McKersie BD, Chen Y, De Beus M, Bowley SR, Bowler C (1993) Superoxide dismutase enhances tolerance of freezing stress in transgenic alfalfa (*Medicago sativa* L). Plant Physiol 103:1155–1163
Meena KK, Sorty AM, Bitla UM, Choudhary K, Gupta P, Pareek A, Singh DP, Prabha R, Sahu PK, Gupta VK, Singh HB, Krishanani KK, Minhas PS (2017) Abiotic stress responses and microbe-mediated mitigation in plants: the omics strategies. Front Plant Sci 8:172. https://doi.org/10.3389/fpls.2017.00172
Miao BH, Han XG, Zhang WH (2010) The ameliorative effect of silicon on soybean seedlings grown in potassium-deficient medium. Ann Bot 105:967–973
Ming DF, Pei ZF, Naeem MS, Gong HJ, Zhou WJ (2012) Silicon alleviates PEG-induced water-deficit stress in upland rice seedlings by enhancing osmotic adjustment. J Agron Crop Sci 198:14–26
Mitra GN (2015) Regulation of nutrient uptake by plants. Springer, New Delhi
Mittler R (2002) Oxidative stress, antioxidants and stress tolerance. Trends Plant Sci 7:405–410
Monjardino P, Smith AG, Jones RJ (2005) Heat stress effects on protein ac-cumulation of maize endosperm. Crop Sci 45:1203–1210
Moussa HR (2006) Influence of exogenous application of silicon on physiological response of salt-stressed maize (*Zea mays* L.). Int J Agric Biol 8:293–297
Nagajyoti PC, Lee KD, Sreekanth TVM (2010) Heavy metals, occurrence and toxicity for plants: a review. Environ Chem Lett 8:199–216
Neumann D, Zur Nieden U (2001) Silicon and heavy metal tolerance of higher plants. Phytochemistry 56:685–692
Nwugo CC, Huerta AJ (2011) The effect of silicon on the leaf proteome of rice (*Oryza sativa* L.) plants under cadmium-stress. J Proteome Res 10:518–528
Ohyama N (1985) Amelioration of cold weather damage of rice by silicate fertilizer application. Agric Hortic J 60:1385–1389
Okita TH, Volcani BE (1978) Role of silicon in diatom metabolism IX. Differential synthesis of DNA polymerases and DNA-binding proteins during silicate starvation and recovery in *Cylindrotheca fusiformis*. Biochim Biophys Acta 519:76–86
Owino-Gerroh C, Gascho G (2005) Effect of silicon on low pH soil phosphorus sorption and on uptake and growth of maize. Commun Soil Sci Plant Anal 35:2369–2378
Parida AK, Das AB (2005) Salt tolerance and salinity effects on plants: a review. Ecotoxicol Environ Saf 60:324–349
Pascual MB, Echevarria V, Gonzalo MJ, Hernández-Apaolaza L (2016) Silicon addition to soybean (*Glycine max* L.) plants alleviates zinc deficiency. Plant Physiol Biochem 108:132–138
Pavlovic J, Samardzic J, Maksimovic V, Timotijevic G, Stevic N, Laursen KH, Hansen TH, Husted S, Schjoerring JK, Liang Y, Nikolic M (2013) Silicon alleviates iron deficiency in cucumber by promoting mobilization of iron in the root apoplast. New Phytol 198:1096–1107

Pei ZF, Ming DF, Liu D, Wan GL, Geng XX, Gong HJ, Zhou WJ (2010) Silicon improves the tolerance of water-deficit stress induced by polyethylene glycol in wheat (*Triticum aestivum* L.) seedlings. J Plant Growth Regul 29:106–115

Pich A, Scholz G, Stephan UW (1994) Iron-dependent changes of heavy metals, nicotianamine, and citrate in different plant organs and in the xylem exudate of two tomato genotypes. Nicotianamine as possible copper translocator. Plant and Soil 165:189–196

Pottosin I, Shabala S (2014) Polyamines control of cation transport across plant membranes: implications for ion homeostasis and abiotic stress signaling. Front Plant Sci 5:154. https://doi.org/10.3389/fpls.2014.00154

Quinet M, Ndayiragije A, Lefèvre I, Lambillotte B, Dupont-Gillain CC, Lutts S (2010) Putrescine differently influences the effect of salt stress on polyamine metabolism and ethylene synthesis in rice cultivars differing in salt resistance. J Exp Bot 61:2719–2733

Rédei GP (2008) Silicon (Si). In: Encyclopedia of genetics, genomics, proteomics, and informatics. Springer, Berlin, p 1817

Reezi S, Babalar M, Kalantari S (2009) Silicon alleviates salt stress, decreases malondialdehyde content and affects petal color of salt stressed cut rose (*Rosa xhybrida* L.) Hot Lady. Afr J Biotechnol 8:1502–1508

Rellán-Álvarez R, Giner-Martínez-Sierra J, Orduna J, Orera I, Rodríguez-Castrillón JÁ, García-Alonso JI, Abadía J, Álvarez-Fernández A (2010) Identification of atri-iron (III), tri-citrate complex in the xylem sap of iron-deficient tomato resupplied with iron: new insights into plant iron long-distance transport. Plant Cell Physiol 51:91–102

Reynolds OL, Padula MP, Zeng R, Gurr GM (2016) Silicon: potential to promote direct and indirect effects on plant defense against arthropod pests in agriculture. Front Plant Sci 7:744. https://doi.org/10.3389/fpls.2016.00744

Rizwan M, Ali S, Ibrahim M, Farid M, Adrees M, Bharwana SA, Zia-ur-Rehman M, Qayyum MF, Abbas F (2015) Mechanisms of silicon-mediated alleviation of drought and salt stress in plants: a review. Environ Sci Pollut Res 22:15416–15431

Romero-Aranda MR, Jurado O, Cuartero J (2006) Silicon alleviates the deleterious salt effect on tomato plant growth by improving plant water status. J Plant Physiol 163:847–855

Roy M, Wu R (2001) Arginine decarboxylase transgene expression and analysis of environmental stress tolerance in transgenic rice. Plant Sci 160:869–875

Sanglard LMVP, Martins SCV, Detmann KC (2014) Silicon nutrition alleviates the negative impacts of arsenic on the photosynthetic apparatus of rice leaves: an analysis of the key limitations of photosynthesis. Physiol Plant 152:355–366

Savant NK, Korndorfer GH, Datnoff LE, Snyder GH (1999) Silicon nutrition and sugarcane production: a review. J Plant Nutr 22:1853–1190

Savić J, Marjanović-Jeromela A (2013) Effect of silicon on sunflower growth and nutrient accumulation under low boron supply. Helia 36:61–68

Schaller J, Brackhage C, B€aucker E, Dudel EG (2013) UV-screening of grasses by plant silica layer? J Biosci 38:413–416

Schmidt RE, Zhang X, Chalmers DR (1999) Response of photosynthesis and superoxide dismutase to silica applied to creeping bent-grass grown under two fertility levels. J Plant Nutr 22:1763–1773

Shen X, ZhouY DL, Li Z, Eneji AE, Li J (2010) Silicon effects on photosynthesis and antioxidant parameters of soybean seedlings under drought and ultraviolet-B radiation. J Plant Physiol 167:1248–1252

Shen X, Li Z, Duan L, Eneji AE, Li J (2014) Silicon effects on the partitioning of mineral elements in soybean seedlings under drought and ultraviolet-B radiation. J Plant Nutr 37:828–836

Shi X, Zhang C, Wang H, Zhang F (2005) Effect of Si on the distribution of cd in rice seedlings. Plant and Soil 272:53–60

Shrivastava P, Kumar R (2015) Soil salinity: a serious environmental issue and plant growth promoting bacteria as one of the tools for its alleviation. Saudi J Biol Sci 22:123–131

Sonobe K, Hattori T, An P, Tsuji W, Eneji AE, Kobayashi S, Kawamura Y, Tanaka K, Inanaga S (2010) Effect of silicon application on sorghum root responses to water stress. J Plant Nutr 34:71–82

Soundararajan P, Sivanesan I, Jana S, Jeong BR (2014) Influence of silicon supplementation on the growth and tolerance to high temperature in *Salvia splendens*. Hortic Environ Biotechnol 55:271–279

Soundararajan P, Manivannan A, Jeong BR (2016) Regulatory mechanisms by silicon to overcome the salinity-induced imbalance of essential nutrient elements, silicon in plants: advances and future prospects. CRC, Boca Raton, FL, pp 47–66

Soundararajan P, Manivannan A, Cho YS, Jeong BR (2017) Exogenous supplementation of silicon improved the recovery of hyperhydric shoots in *Dianthus caryophyllus* L. by stabilizing the physiology and protein expression. Front Plant Sci 8:738. https://doi.org/10.3389/fpls.2017.00738

Soylemezoglu G, Demir K, Inal A, Gunes A (2009) Effect of silicon on antioxidant and stomatal response of two grapevine (*Vitis vinifera* L.) rootstocks grown in boron toxic, saline and boron toxic-saline soil. Sci Hortic 123:240–246

Steponkus PL (1984) Role of the plasma membrane in freezing injury and cold acclimation. Annu Rev Plant Physiol 35:543–584

Struyf E, Conley DJ (2008) Silica: an essential nutrient in wetland biogeochemistry. Front Ecol Environ 7:88–94

Sutka M, Li G, Boudet J, Boursiac Y, Doumas P, Maurel C (2011) Natural variation of root hydraulics in *Arabidopsis* grown in normal and salt-stressed conditions. Plant Physiol 155:1264–1267

Sytar O, Kumar A, Latowski D, Kuczynska P, Strzałka K, Prasad MNV (2013) Heavy metal-induced oxidative damage, defense reactions and detoxification mechanisms in plants. Acta Physiol Plant 35:985–999

Takahashi N, Kurata K (2007) Relationship between transpiration and silica content of the rice panicle under elevated atmospheric carbon dioxide concentration. J Agric Meteorol 63:89–94

Tang W, Newton RJ, Li C, Charles TM (2007) Enhanced stress tolerance in transgenic pine expressing the pepper CaPF1 gene is associated with the polyamine biosynthesis. Plant Cell Rep 26:115–124

Torabi F, Majd A, Enteshari S (2015) The effect of silicon on alleviation of salt stress in borage (*Borago officinalis* L.). Soil Sci Plant Nutr 61:788–798

Trenholm LE, Datnoff LE, Nagara RT (2004) Influence of silicon on drought and shade tolerance of St. Augustine grass. Hort Technol 14:487–490

Tripathi DK, Singh VP, Kumar D, Chauhan DK (2012) Impact of exogenous silicon addition on chromium uptake, growth, mineral elements, oxidative stress, antioxidant capacity, and leaf and root structures in rice seedlings exposed to hexavalent chromium. Acta Physiol Plant 34:279–289

Tripathi DK, Singh S, Singh VP, Prasad SM, Dubey NK, Chauhan DK (2017) Silicon nanoparticles more effectively alleviated UV-B stress than silicon in wheat (*Triticum aestivum*) seedlings. Plant Physiol Biochem 110:70–81

Tuna AL, Kaya C, Higgs D, Murillo-Amador B, Aydemir S, Girgin AR (2008) Silicon improves salinity tolerance in wheat plants. Environ Exp Bot 62:10–16

Umar SM (2002) Genotypic differences in yield and quality of groundnut as affected by potassium nutrition under erratic rainfall conditions. J Plant Nutr 25:1549–1562

Vaculík M, Lux A, Luxová M, Tanimoto E, Lichtscheidl I (2009) Silicon mitigates cadmium inhibitory effects in young maize plants. Environ Exp Bot 67:52–58

Van Bockhaven J, De Vleesschauwer D, Höfte M (2013) Towards establishing broad-spectrum disease resistance in plants: silicon leads the way. J Exp Bot 64:1281–1129

Veatch-Blohm ME (2007) Principles of plant genetics and breeding. Crop Sci 47:1763

Wang YX, Stass A, Horst WJ (2004) Apoplastic binding of aluminum is involved in silicon-induced amelioration of aluminum toxicity in maize. Plant Physiol 136:3762–3770

Wang XD, Ou-yang C, Fan ZR, Gao S, Chen F, Tang L (2010) Effects of exogenous silicon on seed germination and antioxidant enzyme activities of *Momordica charantia* under salt stress. J Anim Plant Sci 6:700–708

Wang S, Liu P, Chen D, Yin L, Li H, Deng X (2015) Silicon enhanced salt tolerance by improving the root water uptake and decreasing the ion toxicity in cucumber. Front Plant Sci 6:759. https://doi.org/10.3389/fpls.2015.00759

Wang M, Gao L, Dong S, Sun Y, Shen Q, Guo S (2017) Role of silicon on plant pathogen interactions. Front Plant Sci 8:701. https://doi.org/10.3389/fpls.2017.00701

Wu X, Yu Y, Baerson SR, Song Y, Liang G, Ding C, Niu J, Pan Z, Zeng R (2017) Interactions between nitrogen and silicon in rice and their effects on resistance toward the brown plant hopper *Nilaparvata lugens*. Front Plant Sci 8:28. https://doi.org/10.3389/fpls.2017.00028

Yao D, Zhang X, Zhao X, Liu C, Wang C, Zhang Z (2011) Transcriptome analysis reveals salt-stressregulated biological processes and key pathways in roots of cotton (*Gossypium hirsutum* L.). Genomics 98:47–55

Yin L, Wang S, Li J, Tanaka K, Oka M (2013) Application of silicon improves salt tolerance through ameliorating osmotic and ionic stresses in the seedling of *Sorghum bicolor*. Acta Physiol Plant 35:3099–3107

Yin L, Wang S, Tanaka K, Fujihara S, Itai A, Den X (2016) Silicon mediated changes in polyamines participate in silicon-induced salt tolerance in *Sorghum bicolor* L. Plant Cell Environ 39:245–258

Yordanov I, Velikova V, Tsonev T (2003) Plant responses to drought and stress tolerance. Bulg J Plant Physiol (Special Issue) 187–206

You-Qiang FU, Hong S, Dao-Ming WU, Kun-Zheng CAI (2012) Silicon-mediated amelioration of Fe^{2+} toxicity in rice (*Oryza sativa* L.) roots. Pedosphere 22:795–802

Zhu Y, Gong H (2014) Beneficial effects of silicon on salt and drought tolerance in plants. Agron Sustain Dev 34:455–472

Zhu ZJ, Wei GQ, Li J, Qian QQ, Yu JQ (2004) Silicon alleviates salt stress and increases antioxidant enzymes activity in leaves of salt-stressed cucumber (*Cucumis sativus* L.). Plant Sci 167:527–533

Zuccarini P (2008) Effects of silicon on photosynthesis, water relations and nutrient uptake of *Phaseolus vulgaris* under NaCl stress. Biol Plant 52:157–160

Responses of *Gerbera jamesonii* Plants to Different Salinity Levels and Leaching Ratios When Grown in Soilless Culture

Özlem Akat Saraçoğlu, M. Kamil Meriç, İ. Hakkı Tüzel, and Yasemin S. Kukul Kurttaş

Introduction

Salinity is one of the important abiotic stress factors limiting plant productivity and development, especially in arid and semiarid parts of the world (Allakhverdiev et al. 2000; Koca et al. 2007). The effects of salinity on plant growth are associated with (1) salt stress (a specific ion effect), (2) water stress (low osmotic potential), (3) nutritional imbalances, and (4) a combination of these factors (Ashraf 1994). The basic elements of crop production can be considered as factors limiting cultivation. In addition, the formulations of basic chemical fertilizers, which are applied together with irrigation water for plant growth and development, also sometimes limit plant growth. All of these factors lead to significant salt accumulation in the plant root zone, which can reduce plant yield and quality, and can cause changes in plant color and shape (Quamme and Stushnoff 1983; Rhoades et al. 1992; Cassaniti et al. 2013). Visible effects of salinity affecting plant physiology and yield emerge with high salinity levels. Sensitivity to salinity varies according to the plant and cultivar (Haman et al. 1997). The ability of plants to withstand root zone salinity (i.e., plant salt tolerance) is affected by the type of salt, the type of plant, the soil, the climate, and their interactions. The relationship between relative yield and salinity (electrical conductivity (EC)) makes it possible to determine this tolerance (Maas and Hoffman 1977; Grattan and Hanson 1993; Grattan 1993; Kotuby-Amacher et al. 1997; Katerji et al. 2003).

Ö. Akat Saraçoğlu (✉)
E.U. Bayindir Vocational School, Ege University, Bayindir-Izmir, Turkey
e-mail: ozlem.akat@ege.edu.tr

M. K. Meriç
E.U. Bergama Vocational School, Ege University, Bergama-Izmir, Turkey

İ. H. Tüzel · Y. S. Kukul Kurttaş
Department of Agricultural Structures and Irrigation, E.U. Faculty of Agriculture, Ege University, Bornova-Izmir, Turkey

M. Hasanuzzaman et al. (eds.), *Plant Abiotic Stress Tolerance*,
https://doi.org/10.1007/978-3-030-06118-0_15

NaCl, which has a harmful effect on plant development, is a salt form commonly found in nature (Quamme and Stushnoff 1983; Munns and Termaat 1986; Rhoades et al. 1992; Grattan 1993; Jacoby 1993). Often, high concentrations of Na and Cl ions are considered synonymous with high salinity. Given that NaCl is the most soluble and widespread salt, it is not surprising that all plants have evolved mechanisms to regulate its accumulation and to preferentially select other nutrients, which are commonly present at low concentrations (Álvarez and Sánchez-Blanco 2013). Cl and Na taken up by plant roots are transported to the shoots and leaves. With regard to comparison of these two ions, it is known that Cl ions are transported to the leaves faster than Na ions and accumulate more in the leaves. Na ions accumulate primarily in the old parts of the root and the plant body. For this reason, the first signs of damage on the leaves of the Cl ion effect, the Na ion, but above a certain limit value of the leaf damage can be observed.

Salinity of irrigation water is more common than salinity of the land (Tyagi 2003). Irrigation water contains a mixture of natural salts such as the cations Na^+, Ca^{+2}, and Mg^{+2} and the anions Cl^-, SO_4^{-2}, HCO_3^-, and CO_3^{-2}. These salts in the irrigation water can accumulate in soil or in substrates; thus, plant growth is adversely affected by excessive salts raising the osmotic pressure of the soil solution and resulting in reduced availability of water for the plants (Ayers and Westcot 1989; Tyagi 2003). In addition to unsuitable irrigation and fertilizer applications in soilless ornamental plant growth, insufficient drainage and limited root volumes result in rapid salt accumulation in the root zone (Quamme and Stushnoff 1983; Sonneveld et al. 2000; Sevgican 2002; Akat 2008). The main purpose of irrigation management in soilless culture is to provide adequate moisture in the root zone, as well as reducing the detrimental effects of salinity, by leaching. In contrast to soil, salt concentrations in substrates can reach excessive levels quickly because of the limited root zone volume and improper management of irrigation, leading to serious problems. As a general approach, to mitigate the harmful effects of salinity, 15–25% leaching (as a ratio of drainage to irrigation) in normal irrigation water conditions and up to 40% in saline irrigation water conditions should be provided. The genotypes of ornamental plants differ greatly in their susceptibility to salt stress (Villarino and Mattson 2011; Cassaniti et al. 2012). With competition for high-quality water, increasing the use of saline water is an alternative for irrigation of salt-tolerant ornamental plants (Cassaniti et al. 2009).

Worldwide, the total land area used for cut flower production is 175,066 ha, and the economic value of this industry was estimated to be €64.7 billion in 2016 (Kazaz 2017). In Turkey, production of cut flowers reached 4858.09 ha with 37 different varieties. Cultivation is spread out among 55 cities located generally in the southern and western parts of the country. More than half of the production is maintained in Izmir (33.4%), Sakarya (22.0%), and Antalya (11.5%). The area used for gerbera production (113.603 ha) is in third place after those used for carnation and cut rose production, and this corresponds to 2.3% of the country's total cut flower production area (TUIK 2017).

The aim of the present study was to determine the evapotranspiration (ET), yield, water use efficiency (WUE), and some quality and growth parameters of gerbera (*Gerbera jamesonii* cv. 'Skylina') grown under different salinity levels and nutrient solution leaching ratios (LRs) in an open soilless culture system. Findings from the first year of this study, including only the ET and the yield, have previously been published by Akat et al. (2009).

Materials and Methods

Test Site, Plant Material, and Growing Conditions

This study was conducted in a tunnel-type heated polyethylene greenhouse located at the Ege University Faculty of Agriculture in Izmir, Turkey (38°27′17.78″N, 27°13′18.82″E), during two consecutive production seasons. The plants were heated by commercial electric heaters only against the risk of frost at night to keep temperature above 0 °C. No heating was needed in the daytime, since the temperature did not drop as low as 0 °C. Insect netting was fastened to the vents of the greenhouse to prevent entry by insects. The shading requirements of the plants were met with a 10% shading net. A misting system was installed over the canopy to regulate indoor humidity until the salinity applications were begun. Brief information concerning the growing seasons is given in Table 1.

Four nutrient solution salinity levels (S_0: control; S_1: S_0 + 1 dS m^{-1}; S_2: S_0 + 2 dS m^{-1}; S_3: S_0 + 3 dS m^{-1}) and two LRs (with drainage volumes of 25% of the applied nutrient solution volume (LR_{25}) and 50% of the applied nutrient solution volume (LR_{50})), in an open soilless culture system, were tested during the study.

The plant material studied was gerbera (*Gerbera jamesonii* cv. 'Skylina'). Some physiological properties of this variety are listed in Table 2 (Schreurs Gerbera Catalogue 1998). Agricultural perlite (with 60% of the particle diameter being 2–5 mm; ETİPER) was chosen as a growing medium because it was available locally and is suitable for soilless culture (Tüzel et al. 1999).

Table 1 Important dates related to growing seasons

Growing season	Length of season (weeks)
First	50
Second	54

Table 2 Physiological properties of *Gerbera jamesonii* cv. 'Skylina'

Property	Value
Average Flower sap length	60 cm
Flower diameter	11–13 cm
Shelf life	12–14 days
Shading requirement	10%

Plastic pots (measuring 75 cm × 23 cm × 16 cm, with a volume of 27.6 L) were filled with 21 L of perlite per pot. Three gerbera seedlings were planted in each pot, so the growing media volume was 7 L per plant and the plant density was eight plants per square meter, with each plant having a growing area of 50 cm × 25 cm. The oldest leaves on the plants were removed regularly during each growing season (Bayçin-Korkut 1998).

Irrigation, Salinity Levels, and Leaching Fractions

Plant water requirements were met by a fogging system; thus, uniform plant growth could be obtained during the first 5 and 10 weeks after planting (WAP) in the first and second growing seasons. Afterward, the water and nutrient requirements of the plants were supplied by Hoagland solution (Table 3). An automated drip irrigation system using pressure-compensated online drippers with a flow rate of 3 L h^{-1} was used to deliver the nutrient solution to the plant root zone. Each plant was irrigated by a dripper. The irrigation was started at 09:00 each day, and the irrigation start times were adjusted according to plant needs by manual observation of root zone moisture and indoor climate conditions. The crops were irrigated a minimum of once a day and a maximum of five times a day during both growing seasons. A timer was used to control irrigation start times and durations, as deemed appropriate, to try and obtain the intended 25% and 50% leaching rates (Smith 1987). The scheduling was reviewed weekly and modified daily, if necessary.

The composition of the nutrient solution was adjusted according to the plant growth period, and the pH was maintained at 5.5–6.5 during the study. A half dose (in which the concentrations of all nutrients were half the full dose) was applied between 5 WAP and 12 WAP in the first growing season, and between 10 WAP and

Table 3 Nutrient solution composition and chemical sources of nutrients

Nutrient	Concentration (ppm)	Chemical source
N	150	NH_4NO_3 (33% N)
P	40	H_3PO_4 (85% P)
K	250	KNO_3 (13%–0%–46%)
Ca	50	$Ca(NO_3)_2$ (15.5%–0%–0%–19%)
Mg	40	$MgSO_4{\cdot}7H_2O$ (9% MgO)
Fe	2	Fe-EDTA (12.5% Fe)
Mn	1.0	$MnSO_4{\cdot}H_2O$ (22–23% Mn)
Zn	0.75	$ZnSO_4{\cdot}7H_2O$ (19–20% Zn)
B	0.2	H_3BO_3
Cu	0.1	$CuSO_4{\cdot}5H_2O$ (20% Cu)
Mo	0.04	$(NH_4)_6Mo_7O_{24}{\cdot}4H_2O$

EDTA ethylenediaminetetraacetic acid

Table 4 Dates of application of nutrient solutions and salinity treatments in the first and second growing seasons

Growing season	Start of half dose[a]	Start of full dose	Start of salinity treatment
First	5th week	12th week	17th week
Second	10th week	16th week	21st week

[a]Freshwater was applied from planting to the half-dose application dates

16 WAP in the second growing season. A full dose was applied in the subsequent weeks. Salinity treatments were started at 17 WAP and 21 WAP in the first and second growing seasons, respectively (Table 4). Salinity levels in the S_1, S_2, and S_3 treatments were adjusted by adding NaCl solution to the same full-dose nutrient solution that was used for the control treatment (S_0), so that 1, 2, and 3 dS m^{-1} increments could be obtained in the S_1, S_2, and S_3 treatments, respectively.

Measurements and Calculations

The indoor air temperature (measured in degrees Celsius) and relative humidity (RH; measured as a percentage) were measured and recorded using a HOBO-Onset data logger (HOBO H08–004-02).

The nutrient solution and drained solution salinity levels were measured regularly using an EC meter (Mettler Toledo model SevenGo Easy).

ET was determined and measured in liters per plant by subtraction of the drained nutrient solution volume from the applied volume in each treatment (Meric et al. 2011).

The water use efficiency of each treatment was calculated according to Eqs. 1 and 2 (Chaves et al. 2004; Gregory 2004; Jones 2004; Howell 2006):

$$\mathrm{WUE_I} = Y/I \tag{1}$$

$$\mathrm{WUE_{ET}} = Y/\mathrm{ET} \tag{2}$$

where $\mathrm{WUE_I}$ and $\mathrm{WUE_{ET}}$ are the water use efficiencies calculated on the basis of the applied nutrient solution and evapotranspiration (measured in kilograms per cubic meter), *Y* is the total yield (i.e., flower weight, measured in kilograms), and *I* and ET are the applied nutrient solution and evapotranspiration (measured in millimeters), respectively.

The plants were harvested when the flowers developed two or three rings of mature stamens. To determine the yield, the cut flowers were counted and weighed (measured in grams) after each harvest. The total fresh weight was separated into the flower and stalk weights. The flower diameter (measured in centimeters), stem length (measured in centimeters), and stem diameter (measured in millimeters) were measured in each harvest to determine the flower quality. The stem diameter measurements were taken from the top (below the receptacle), middle, and lower parts of the stem (above the root collar) (Zeybekoğlu 2000; Yemenici 2000). The relationship between the yield (number of flowers per square meter) and salinity was determined by Eq. 3 (Maas and Hoffman 1977; Hanson 1993; Katerji et al. 2003):

$$Y = 100 - b\,\text{Inner space}\,(\text{0xEF07})\,(EC_e - a) \quad (3)$$

where Y is the yield (measured as a percentage), b is the slope, EC_e is the average root zone salinity (measured in decisiemens (dS) per meter), and a is the salinity threshold value. The EC of the applied nutrient solution was considered instead of the EC_e in this study.

Experimental Design and Statistical Analysis

The adopted experimental design was a randomized split-plot design with three replications. The different salinity levels and LRs were assigned in the main plot and subplots, respectively. Each plot contained nine plants, and a total of 219 gerbera plants was planted in each growing season. The data were evaluated using the TARIST statistical analysis program developed by the Ege University Faculty of Agriculture. The least significant difference (LSD) test was applied to determine significant differences ($p < 0.05$).

Results and Discussion

Greenhouse Climate

The weekly temperature, relative humidity (RH), and vapor pressure deficit (VPD) measured and calculated in the greenhouse are presented in Figs. 1, 2, and 3. While the temperature ranges were 1.8–37.1 °C and 1.4–40.9 °C in the first and second

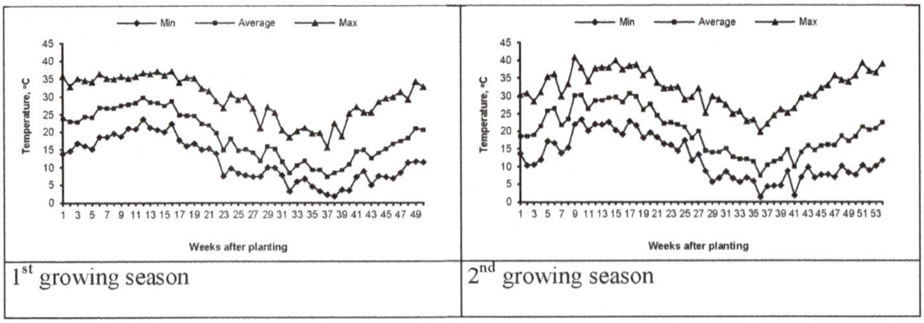

Fig. 1 Weekly indoor temperature in the first and second growing seasons

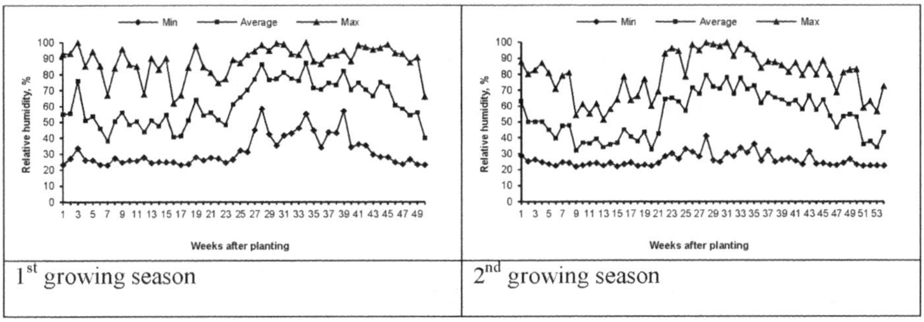

Fig. 2 Weekly indoor relative humidity in the first and second growing seasons

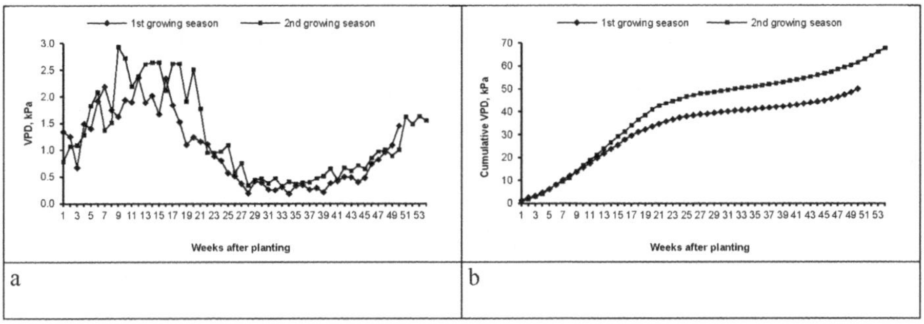

Fig. 3 Weekly indoor vapor pressure deficit (*VPD*) (**a**) and cumulative VPD (**b**) in the first and second growing seasons

growing seasons, respectively, the RH ranges were 23–100% and 21.8–99.8% in the first and second growing seasons, respectively. As a result of the indoor fogging, the RH reached 100% in the weeks during which the system was operated. With regard to the temperature and RH, the VPD ranges in the greenhouse were 0.18–2.36 kPa and 0.32–2.93 kPa in the first and second growing seasons, respectively. The cumulative VPD was higher in the second growing season than in the first growing season because of the relatively higher average temperature and lower average RH.

Applied and Drained Nutrient Solutions, Evapotranspiration, Leaching Ratios, and Electrical Conductivity

The applied and drained nutrient solutions, ET, and LRs obtained from each treatment during the study are presented in Table 5.

The seasonal application of the nutrient solution changed from 41.5–57.7 L per plant in the first growing season to 68.1–100.1 L per plant in the second growing season. More nutrient solution was applied to achieve the intended LRs in the

Table 5 Seasonal applied and drained nutrient solution volumes, evapotranspiration (*ET*), and leaching ratios (*LRs*) of treatments in the first and second growing seasons

Growing season	Treatment		Nutrient solution (liters per plant)	Drained solution (liters per plant)	ET (liters per plant)	LR (%)
First	S_0	LR_{25}	42.6	9.8	32.8	23.0
		LR_{50}	57.7	24.5	33.2	42.5
	S_1	LR_{25}	41.7	11.5	30.2	27.6
		LR_{50}	56.0	22.9	33.1	40.9
	S_2	LR_{25}	41.5	11.9	29.6	28.7
		LR_{50}	57.6	27.7	29.9	48.1
	S_3	LR_{25}	41.8	13.5	28.3	32.3
		LR_{50}	55.0	26.3	28.7	47.8
Second	S_0	LR_{25}	81.6	31.3	50.3	38.4
		LR_{50}	100.1	51.7	48.4	51.6
	S_1	LR_{25}	74.0	27.3	46.7	36.9
		LR_{50}	96.6	51.0	45.6	52.8
	S_2	LR_{25}	72.1	26.3	45.8	36.5
		LR_{50}	93.4	50.3	43.1	53.9
	S_3	LR_{25}	68.1	26.0	42.1	38.2
		LR_{50}	82.2	41.8	40.4	50.9

LR_{25} leaching ratio: drainage volume = 25% of the applied nutrient solution volume, LR_{50} leaching ratio: drainage volume = 50% of the applied nutrient solution volume, S_0 control nutrient solution with no added salt, S_1 control nutrient solution plus salinity increase of 1 dS m^{-1}, S_2 control nutrient solution plus salinity increase of 2 dS m^{-1}, S_3 control nutrient solution plus salinity increase of 3 dS m^{-1}

second growing season than in the first growing season because of the longer growing season and higher evaporative demand (VPD) in the indoor climate (Fig. 3a, b).

The seasonal ET of the plants changed from 33.2–28.3 L per plant in the first growing season to 50.3–40.4 L per plant in the second growing season (Table 5). The total plant water consumption was higher in the second season than in the first season because of the longer growing season and higher VPD in the indoor climate. In both growing seasons, the total ET decreased by 14% and 15%, respectively, with increasing nutrient solution salinity. Although no significant deviations in ET were observed between leaching treatments in both seasons, the effect of leaching on ET was more evident in the S_0 and S_1 treatments in the first growing season, and the ET increased with the increase in LR. In these treatments the LR ranged between 23% and 42.5%. However, in the S_2 and S_3 treatments the increased LR did not decrease the ET, as was observed in S_0 and S_1. In a study on tomatoes in hydroponic conditions, conducted by Pokluda and Kobza (2001), it was shown that climate conditions did not affect the pH of the nutrient solution; the EC levels related to nutrient solution were directly related to the air temperature, RH, and solar radiation; higher temperatures tended to increase the ET of the plants; and increased RH seemed to support the effect of reducing the intake of water and nutrients.

As shown in Fig. 4, the weekly ET in the different treatments showed similar courses during the study. In the first weeks of both seasons, the ET was as low as

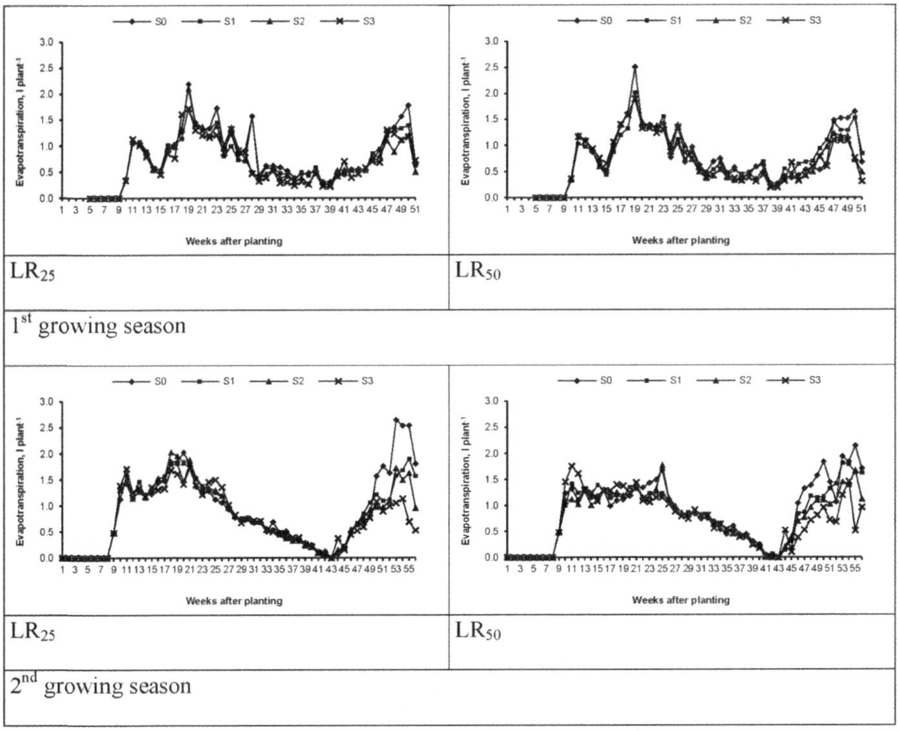

Fig. 4 Weekly evapotranspiration in the first and second growing seasons

0.5 L per plant, and then it increased to 2–2.5 L per plant, related to plant development and an increasing indoor VPD. After that point, ET decreased gradually from 2–2.5 L per plant to 0.5 until the 27th to 43rd weeks, because of the decreasing indoor VPD. The second-highest values were observed after the 47th week, due to the again increasing VPD. In general, ET was highest in the control (S_0) treatment during the peak evaporative demand periods. Since the indoor VPD was higher in the second season than in the first season, the weekly ET was also higher in the second season. In both seasons the highest ET value was achieved in the lowest-salinity (S_0) treatment and the lowest ET value was achieved in the highest-salinity (S_3) treatment. This result was associated with less water intake by the plants in conditions of particularly heavy water stress and salinity levels (Hannah 1998). In a study by Zheng et al. (2005) on gerbera cultivation in a capillary system, the daily plant water consumption was found to be 115 ± 3 mL. In a study using the 'Molina' and 'Lilabella' gerbera cultivars, grown in a hydroponic system in climate-controlled conditions, Syros et al. (2001) found that the daily water consumption varied depending on the season in which the plants were grown, ranging from 800 to 1200 mL per plant. In a study conducted by Savvas and Gizas (2002), using closed-system cultivation, it was reported that 548–563 mL of a nutrient solution was applied to each plant daily, with a drainage rate of 43–45%; accordingly, the daily plant water consumption was 306–315 mL per plant. The daily plant water consumption of the gerbera in that study was lower than the consumption recorded in the present study. These differences were related to the different growing media used, the LRs tested, and the differences in the ecological conditions in which the cultivation was carried out.

The fluctuations in the EC of the applied nutrient solution, starting from the 5th week in the first season and from the 13th week in the second season, are shown in Fig. 5. The EC range was 1–2 dS m^{-1} until the commencement of the salinity treatments in both seasons. From that time onward, the EC of the nutrient solution fluctuated between 1 and 5 dS m^{-1}. Since the EC of freshwater was higher in the second growing season than in the first growing season, the final EC of the full-dose

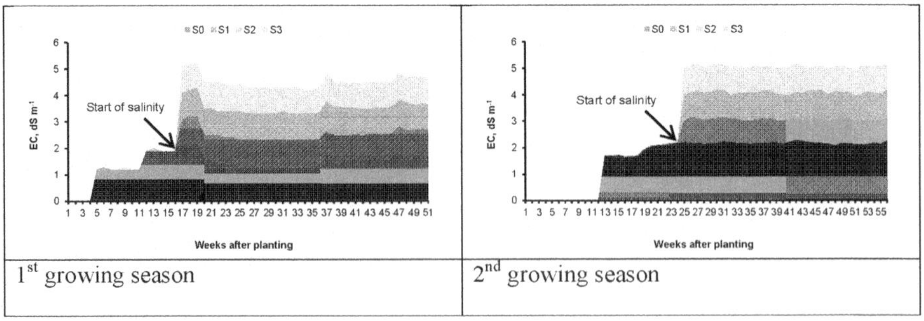

Fig. 5 Electrical conductivity (*EC*) of the applied nutrient solution in the first and second growing seasons

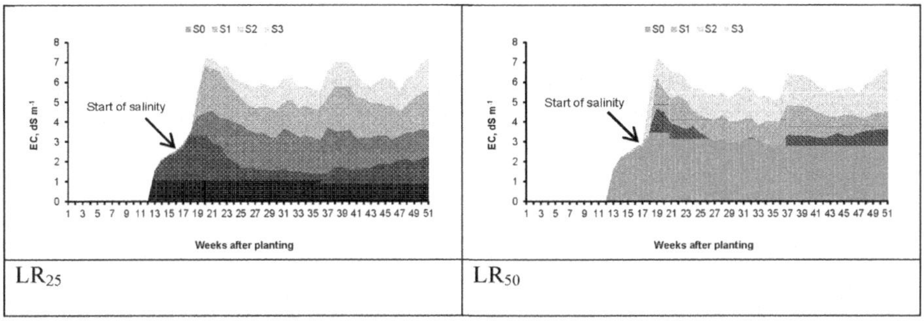

Fig. 6 Electrical conductivity (*EC*) of the drained nutrient solution in the LR_{25} and LR_{50} treatments in the first growing season. LR_{25} leaching ratio: drainage volume = 25% of the applied nutrient solution volume, LR_{50} leaching ratio: drainage volume = 50% of the applied nutrient solution volume

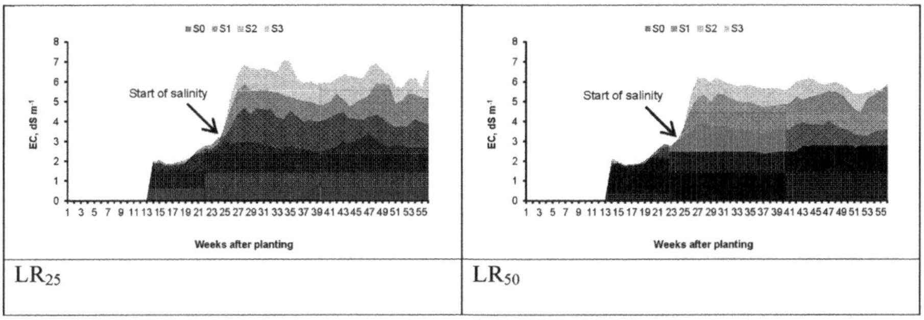

Fig. 7 Electrical conductivity (*EC*) of the drained nutrient solution in the LR_{25} and LR_{50} treatments in the second growing season. LR_{25} leaching ratio: drainage volume = 25% of the applied nutrient solution volume, LR_{50} leaching ratio: drainage volume = 50% of the applied nutrient solution volume

nutrient solution was higher in the second year. In terms of salinity treatments, when the amount of nutrient solution applied to the plants increased, the salinity was decreased.

The EC of the drained nutrient solution ranged between 1 and 7 dS m^{-1} in the first season and between 2 and 7 dS m^{-1} in the second season (Figs. 6 and 7). The average difference in EC between the applied and drained nutrient solutions was greater in the LR_{25} treatment and in the first growing season (Fig. 8). It can be said that more efficient leaching was obtained in the LR_{50} treatment and in the first season, because of the smaller average difference in the EC between the applied and drained nutrient solutions, and the lower EC of the applied nutrient solution due to the lower indoor air temperature and the EC of freshwater. In terms of salinity treatments, leaching was more efficient with the higher-salinity nutrient solution. The results with regard to the LRs applied in both production periods observed reduction in salinity was relative to the leaching ratio applied. As noted by Hanson et al. (1993), the average plant root salinity due to application of high-salinity

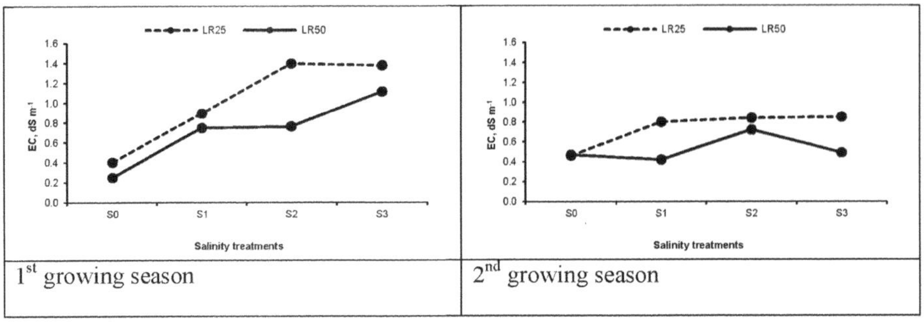

Fig. 8 Efficiency of leaching in terms of the average difference in electrical conductivity (*EC*) between the applied and drained nutrient solutions during salinity treatments in the first and second growing seasons

irrigation water increases under low-LR conditions and decreases under high-LR conditions. The amount of applied nutrient solution was higher in the LR_{50} treatment than in the LR_{25} treatment in both seasons. Since the start times and durations of irrigation were determined by manual observations and the scheduling was not reviewed and modified online, the actual LRs deviated from the predetermined ratios (i.e., 25% and 50%, more or less). Especially in the second season, keeping the LR around the desired ratio was more difficult in the LR_{25} treatment than in the LR_{50} treatment, maybe because of concern about a possible lack of irrigation water. Additionally, the actual LRs were closer to the predefined values in the LR_{25} treatment in the first season and in the LR_{50} treatment in the second season. Giuffrida and Lipari (2003) reported that it was commonly necessary to use a high LR to prevent salt accumulation in the growing media in soilless culture, especially when using nutrient solutions prepared with a low-salt water source. However, the application of a high LR could be associated with problems related to environmental and commercial aspects, necessitating use of low-LR applications. The results showed that there was no significant difference between the LRs with regard to the yield and flower quality parameters in both growing seasons. Thus, reducing the amount of nutrient solution applied in this case showed that water and fertilizer savings, reducing environmental pollution as well as the LRs, could be applied.

Yield and Quality

The plants were harvested 59 and 65 times in the first and second growing seasons, respectively. In general, the yield and quality were negatively affected by increasing salinity. The effects of salinity and leaching treatments on the flower numbers and weight were similar in both seasons (Table 6). Although there were no significant differences between the leaching treatments, the effects of nutrient solution salinity and the LR × nutrient solution salinity interactions on the flower numbers and

Table 6 Effects of treatments on yield and quality parameters of gerbera plants in the first and second growing seasons

Treatment	Number of flowers (per m^2)		Flower weight (g)		Flower diameter (cm)		Stem length (cm)		Stem diameter (cm)					
									First season			Second season		
	First season	Second season	First season	Second season	First season	Second season	First season	Second season	Top	Middle	Bottom	Top	Middle	Bottom
Salinity														
S_0	117.6^{a}	95.8^{a}	24.2^{a}	21.4^{a}	9.4	10.7^{a}	37.9^{a}	38.2^{a}	0.447^{a}	0.591^{a}	0.605^{a}	0.339	0.460	0.473
S_1	100.2^{b}	80.7^{b}	23.9^{a}	20.4^{a}	9.6	10.6^{a}	$37.6^{a,b}$	$36.3^{a,b}$	0.431^{b}	0.572^{b}	0.590^{a}	0.352	0.472	0.485
S_2	89.9^{c}	$69.3^{b,c}$	22.0^{b}	20.0^{a}	9.7	10.6^{a}	$37.3^{a,b}$	$37.3^{a,b}$	$0.424^{b,c}$	0.557^{c}	0.573^{b}	0.342	0.466	0.478
S_3	78.2^{d}	56.8^{c}	20.8^{b}	17.7^{b}	9.6	10.1^{b}	35.5^{b}	35.6^{b}	0.415^{c}	0.542^{d}	0.557^{c}	0.347	0.479	0.491
	*	*	*	*	NS	*	*	*	*	*	*	NS	NS	NS
Leaching														
LR_{25}	94.7	75.5	22.8	19.5	9.7	10.4	37.1	36.5	0.428	0.564	0.580	0.351	0.473	0.486
LR_{50}	98.3	75.8	22.6	20.2	9.4	10.6	37.1	37.3	0.430	0.567	0.583	0.340	0.465	0.478
	NS	NS	NS	NS	NS	NS	NS	NS	NS	NS	NS	NS	NS	NS
LR_{25}														
S_0	118.2^{a}	89.7^{a}	23.9^{a}	21.0^{a}	9.43	10.7^{a}	37.2^{a}	37.3	0.447^{a}	0.589^{a}	0.601^{a}	0.354	0.472	0.486
S_1	96.6^{b}	84.3^{a}	24.2^{a}	$20.7^{a,b}$	9.73	10.6^{a}	38.3^{a}	36.5	0.430^{b}	0.578^{a}	0.595^{a}	0.368	0.490	0.503
S_2	$86.8^{b,c}$	$69.3^{a,b}$	21.5^{b}	$18.8^{b,c}$	9.76	$10.4^{a,b}$	36.8^{a}	36.4	0.419^{b}	0.547^{b}	$0.565^{a,b}$	0.333	0.447	0.461
S_3	77.0^{c}	58.7^{b}	21.7^{b}	$17.8^{b,c}$	9.76	10.1^{b}	35.9^{b}	35.6	0.417^{b}	0.542^{b}	0.558^{b}	0.349	0.483	0.492
	*	*	*	*	NS	*	*	NS	*	*	*	NS	NS	NS
LR_{50}														
S_0	117.0^{a}	102.0^{a}	24.5^{a}	21.8^{a}	9.3	10.8^{a}	38.6^{a}	39.1^{a}	0.447^{a}	0.593^{a}	0.609^{a}	0.324	0.448	0.460
S_1	103.7^{b}	77.0^{b}	23.6^{a}	20.1^{a}	9.5	$10.6^{a,b}$	$36.9^{a,b}$	36.1^{b}	$0.431^{a,b}$	0.567^{b}	0.586^{b}	0.335	0.453	0.466
S_2	93.0^{c}	$69.3^{b,c}$	22.4^{a}	21.2^{a}	9.6	10.7^{a}	$37.7^{a,b}$	$38.2^{a,b}$	$0.429^{b,c}$	0.567^{b}	0.581^{b}	0.354	0.484	0.496
S_3	79.4^{d}	55.0^{c}	19.8^{b}	17.6^{b}	9.3	10.2^{b}	35.0^{b}	35.6^{b}	0.413^{c}	0.541^{c}	0.557^{c}	0.345	0.475	0.490
	*	*	*	*	NS	*	*	*	*	*	*	NS	NS	NS

LR leaching ratio, LR_{25} leaching ratio: drainage volume = 25% of the applied nutrient solution volume, LR_{50} leaching ratio: drainage volume = 50% of the applied nutrient solution volume, *NS* significant, S_0 control nutrient solution with no added salt, S_1 control nutrient solution plus salinity increase of 1 dS m^{-1}, S_2 control nutrient solution plus salinity increase of 2 dS m^{-1}, S_3 control nutrient solution plus salinity increase of 3 dS m^{-1}

*Significant difference ($p < 0.05$) Values followed by different letters (a, b, c and d) are significantly different at $p < 0.05$

weight were found to be significant in both seasons ($p < 0.05$). According to the results, both the number and the weight of the flowers decreased as the salinity of the nutrient solution increased. The flower numbers obtained in the S_3 treatment were 34% and 41% lower than those obtained in the control treatment in the first and second seasons, respectively. Similarly, the flower weights were 14% and 17% lower in the most saline treatment than in the control treatment. In particular, a higher LR in saline conditions provided a greater yield (i.e., greater flower numbers) in both seasons. The plant root zone salinity was kept within certain limits by the addition of leaching water to the irrigation water. As the LR increased, the average root zone salinity decreased, and as a result, the plant yield increased. The effect of the leaching ratio on plant yield increases depends on the tolerance of the plants against soil and irrigation water salinity (Hanson et al. 1993).

In the present study, although it was demonstrated that the effect of the LR applications on the yield was statistically insignificant, this was similar to the finding in the study by Hanson et al. (1993) that plants with high LRs had higher yields. According to the results of the assessment in terms of salinity levels, the different salinity levels we applied were similar to those used in many other studies investigating salt effects (Munns and Termaat 1986; Rhoades et al. 1992; Grattan 1993; Francois and Maas 1994; Hannah 1998; Sivritepe and Eriş 1998; Bass and Van den Berg 2000; Villora et al. 2000; Sonneveld et al. 2000; Picchioni and Graham 2001; Dalla Guda et al. 2001; Shillo et al. 2002). The results of the present study showed that the yield decreased as the EC of the nutrient solution applied in the treatments increased. In addition, the average number of flowers obtained per unit area with the treatment was higher during the first production period than during the second production period. This was related to the effect of the extreme temperatures that were experienced and the recession in plant growth and development in the adaptation cycle after planting. Sonneveld (2001) reported that the maximum yield could be obtained by using nutrient solutions with up to 0.8 dS m^{-1} (1.0 dS m^{-1} in the drained solution) in winter climate conditions and up to 1.5 dS m^{-1} (2.0 dS m^{-1} in the drained solution) in spring climate conditions. The yield results obtained in the present study similarly showed that the highest yield was received from the lowest-EC nutrient and drained solution treatments (S_0). When the results related to the flower quality of gerbera were examined in a similar study on different cut flower species, it was found that the flower shape was not affected by the salinity treatments but the flower size was decreased with increasing salinity, proportionally to the decrease in the flower weight (Sonneveld et al. 2000). The results obtained in the present study were similar in terms of the flower weight reduction with increasing salinity in both production periods, but there was some variation due to the fact that the difference in the flower size with salt application in the first production period did not reach statistical significance. However, the results obtained in the present study with regard to determination of the lowest flower diameter values at the highest salinity level were found to be consistent with the results reported by Sonneveld et al. (2000), showing an association between increasing salinity and decreasing flower quality.

In a similar study conducted by Savvas and Gizas (2002), the effects of recycling the drained solution by adding nutrients to replenish it at three different cation ratios on the yield, flower quality, and nutritional status of soilless-grown gerbera were investigated. The results showed that the three different cation ratios did not affect the flower weight and stem thickness. The total number of flowers per plant ranged between 29.4 and 30.7, the number of marketable flowers per plant ranged between 27.80 and 29.90, the average flower weight ranged between 23.30 and 24.00 g, the flower stem length ranged between 49.10 and 51.30 cm, the flower stem thickness ranged between 0.691 and 0.705 mm, and the flower head diameter ranged between 10.20 and 10.35 cm. In the present study, in the first and second seasons, the flower diameters were 9.4–9.7 cm and 10.1–10.7 cm, respectively; the flower weights were 20.8–24.2 g and 17.7–21.4 g, respectively; the stem lengths were 35.5–37.9 cm and 35.6–38.2 cm, respectively; the flower stem bottom diameters were 0.557–0.605 cm and 0.473–0.491 cm, respectively; the flower stem middle diameters were 0.542–0.591 cm and 0.460–0.479 cm, respectively; and the flower stem top diameters were 0.415–0.447 cm and 0.339–0.352 cm, respectively. These values obtained from the present study are the result of the expected effect of the salt. However, the values of plant quality, which have almost the nearest EC level (1.31–2.29 dS m^{-1}), and even on non-salt-treatment (S_0), are very low compared to the values obtained from cultivation by Savvas and Gizas (2002). This result was related to the physical and chemical differences in the growing media used in cultivation, and to the climate conditions in which the cultivation was carried out—that is, greenhouse conditions without climate control.

Relationships Between Flower Quality and Salinity

The proportional changes in the stem length, flower weight, and number of flowers relative to the average increases in the salinity of the nutrient and drained solutions in the production periods are shown in Figs. 9, 10, and 11, respectively.

When the obtained linear regression graphs are examined, it can be determined that increased salinity reduced the stem length, flower weight, and flower numbers. These reductions in response to an increase of 1 dS m^{-1} in the salinity of the nutrient

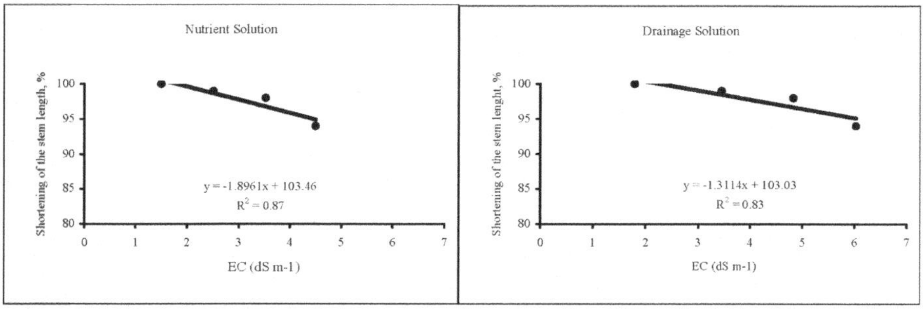

Fig. 9 Proportional changes in the flower stem length relative to the salinity levels of the nutrient and drained solutions during the production periods

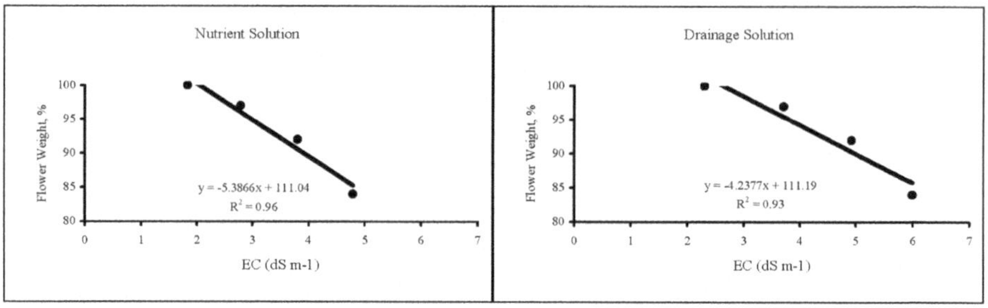

Fig. 10 Proportional changes in the flower weight relative to the salinity levels of the nutrient and drained solutions during the production periods

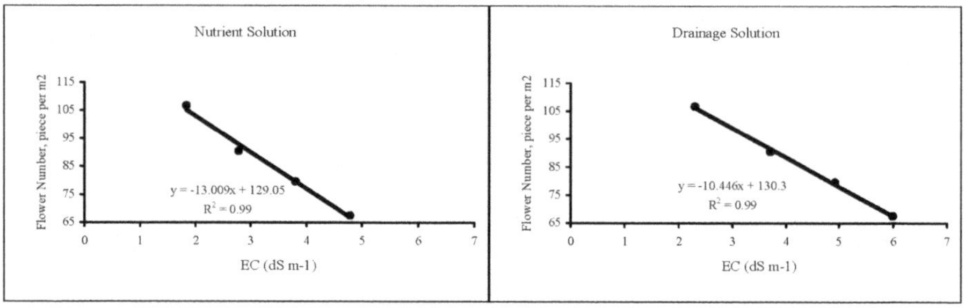

Fig. 11 Proportional changes in the number of flowers relative to the salinity levels of the nutrient and drained solutions during the production periods

Table 7 Decreases in the flower stem length, flower weight, and number of flowers per increase of 1 dS m^{-1} in the salinity of the nutrient and drained solutions (averages per production period)

	Nutrient solution	Drained solution
Decrease in flower stem length (%)	1.9	1.3
Decrease in flower weight (%)	5.4	4.2
Decrease in number of flowers per m^2	13.0	10.5

and drained solutions were 1.9%, 5.4%, and 13.0 flowers per square meter, respectively, for the nutrient solution and 1.3%, 4.2%, and 10.5 flowers per square meter, respectively, for the drained solution (Table 7). However, the reductions in the stem length, flower weight, and flower numbers with increasing salinity of the drained solution were 32%, 22%, and 19% smaller than those seen with increasing salinity of the nutrient solution.

Relationships Between Yield and Salinity

The relative numbers of flowers per square meter (i.e., the yield) corresponding to the EC of the applied and drained nutrient solutions in the different leaching treatments are shown in Figs. 12 and 13, respectively. As shown in Fig. 12 the salinity

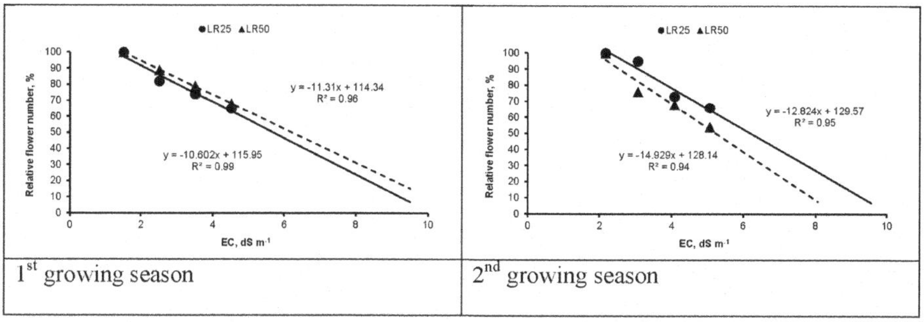

Fig. 12 Linear relationships between the numbers of flowers and the electrical conductivity of the nutrient solution in the different leaching treatments in the first and second growing seasons

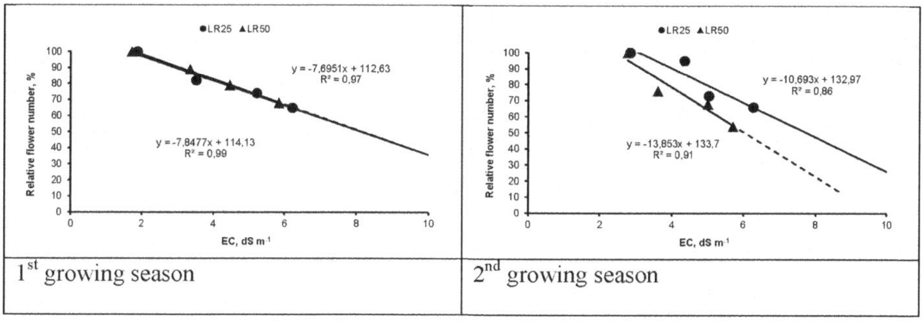

Fig. 13 Linear relationships between the numbers of flowers and the electrical conductivity of the drained solution in the different leaching treatments in the first and second growing seasons

threshold values of the leaching treatments were higher in the second season than in the first season. The relative decrease in flower numbers corresponding to an increase of 1 dS m^{-1} in the nutrient solution salinity was 10.6–14.9% both growing seasons during the study. Because of the higher EC of the freshwater and the higher-than-intended LR in the LR_{25} treatment in the second season, both the salinity threshold value and the EC of the salinity treatments were higher than those in the LR_{50} treatment and in the first season, respectively. Therefore, considering the LR_{25} treatment in the first year and the LR_{50} treatment in the second year, the salinity threshold values were 1.27 and 1.88 dS m^{-1}, respectively, in the LR_{25} and LR_{50} treatments, in terms of the applied nutrient solution salinity. According to Fig. 13, regarding the salinity threshold value determined in the drained solution, the value was higher in the second season than in the first season (as was also the case for the salinity threshold value determined in the applied nutrient solution). The relative flower number decrease corresponding to an increase of 1 dS m^{-1} in the drained solution salinity was 7.7–13.9%. Considering the LR_{25} treatment in the first year and the LR_{50} treatment in the second year, the salinity threshold values were 1.64 and 2.43 dS m^{-1} for the LR_{25} and LR_{50} treatments, respectively, in terms of drained solution salinity. As a result, according to the linear equation giving the salt–yield relationship, in the first season, the average salinity threshold values that started to decrease the yield were determined as 1.4 dS m^{-1} for the nutrient solution and 1.7 dS m^{-1} for the drained solution. During the second season, the average salinity threshold values were found to be 2.1 dS m^{-1} for the nutrient solution and 2.8 dS m^{-1} for the drained solution. In a study by De Kreij et al. (1999), it was expressed that in gerbera cultivation, the calculated salinity threshold values were 1.1 dS m^{-1} for the nutrient solution and 2.1 dS m^{-1} for the drained solution (Adams 2002). In a similar study by Sonneveld et al. (2000), it was found that the threshold value was 2 dS m^{-1}—similar to the values given by De Kreij and Van Os (1989). In another study conducted by De Kreij et al. (1986), two different EC levels were compared for gerbera plant growth in rock wool culture. It was reported that the calculated yield decreases with salinity were 9.7% per increase of 1 dS m^{-1}. The percentage yield decreases with salinity in both seasons in our study were very close to these values. This result showed that the salinity threshold values determined for the two production periods were in agreement with the results of this study.

Water Use Efficiency

The WUE of the treatments in terms of flower numbers and flower weight is shown in Fig. 14.

When the WUE determined on the basis of total flower numbers was evaluated, the WUE values were 0.18–0.35 flowers per liter in the first growing season and ranged from 0.08 to 0.14 flowers per liter in the second season. The highest WUE values were obtained from the lowest salt applications in both seasons. The highest WUE values among the treatments on the basis of the LR were observed with LR_{25} application. According to this, in the first season the WUE values with LR_{25} application in the S_0, S_1, S_2, and S_3 treatments were 0.35, 0.29, 0.26, and 0.23 flowers per liter, respectively; with LR_{50} application, the WUE values were 0.25, 0.23, 0.20, and 0.18 flowers per liter in the S_0, S_1, S_2, and S_3 treatments, respectively. In the second production period, the WUE values with LR_{25} application in the S_0, S_1, S_2, and S_3

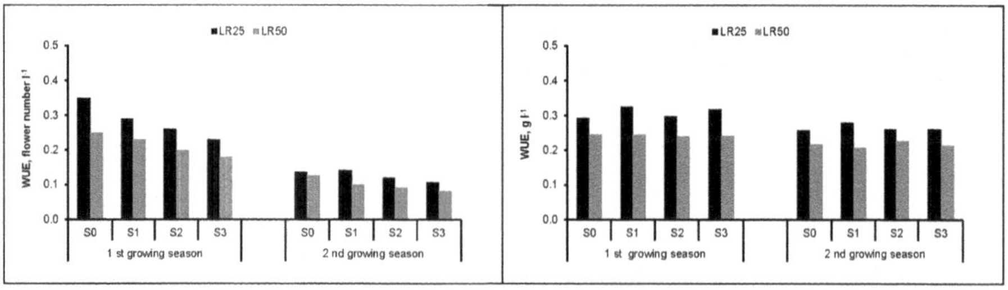

Fig. 14 Water use efficiency of treatments in terms of numbers of flowers and flower weight

treatments were 0.14, 0.14, 0.12, and 0.11 flowers per liter, respectively; with LR_{50} application, the WUE values were 0.13, 0.10, 0.09, and 0.08 flowers per liter in the S_0, S_1, S_2, and S_3 treatments, respectively.

When the WUE determined on the basis of the total flower weight was evaluated, the WUE values were 0.24–0.33 and 0.21–0.28 flower weight L^{-1} in the first and second seasons, respectively. Similarly, the highest WUE values among the leaching treatments were observed in the LR_{25} treatment. No significant numerical difference in WUE between the different salt applications was found. In the first season the WUE values determined in the S_0, S_1, S_2, and S_3 treatments were 0.29, 0.33, 0.30, and 0.32 flower weight L^{-1} in the LR_{25} treatment, and the WUE values for all salinity applications were found to be 0.24 flower weight L^{-1} in the LR_{50} treatment. In the second season the WUE values determined in the S_0, S_1, S_2, and S_3 treatments were 0.26, 0.28, 0.26, and 0.26 flower weight L^{-1}, respectively, in the LR_{25} treatment; and 0.22, 0.21, 0.23, and 0.21 flower weight L^{-1}, respectively, in the LR_{50} treatment. In a study by Martín-Closas and Recasens (2001) to determine WUE in cut rose cultivation, the WUE value was determined as the ratio of the fresh weight of marketable flowers (measured in grams) to the consumed irrigation water (measured in liters). The results showed that the WUE values obtained with perlite were higher than those obtained for tuff (5.6 g L^{-1} for perlite and 3.6 g L^{-1} for tuff). It was reported that the WUE values increased when the air temperature and solar radiation decreased, and the highest values for WUE were determined in the autumn, while the lowest values were determined in summer and spring. The WUE values obtained in the present study were much lower than those observed by Martín-Closas and Recasens (2001). This result was thought to be due to the different plant species (rose versus gerbera) that were investigated. In a study by Tsirogiannis et al. (2010), investigating the effects of high irrigation frequency (HIF) and low irrigation frequency (LIF) on the yield and quality of gerbera plants grown in soilless agricultural conditions, the WUE determined on the basis of flower numbers varied between 0.23 and 0.29 flowers per liter of water. The results showed that the WUE values relative to the salinity levels with both LRs in the first season were similar to those observed in the present study.

Conclusion

When the findings obtained from this study were evaluated together, it was found that when the nutrient and drained solutions were evaluated separately in terms of their salinity levels, on the basis of the percentage decline in yield and the salinity threshold values at which the yield started to decline, the gerbera plant was moderately salt sensitive. In both production periods, the results showed that increasing salinity levels reduced the yield.

The overall results showed that there was no significant difference between the different leaching ratios (LRs) with respect to the yield and flower quality parameters in both growing seasons, whereas the yield and quality did decrease dramatically with increasing salinity. It was found that the highest yield was obtained in the control treatment (S_0) in which no additional salts were applied, while the lowest yield was obtained in the highest-salinity treatment (S_3) in both growing seasons. However, an average increase of 2.1% in the yield (number of flowers) was achieved with application of a high LR (LR_{50}). In comparison with LR_{50} application, LR_{25} application, on average, accounted for 23% of water and equivalent nutrient savings over the growing periods. Since the open system was used for the application of nutrient solution to the plants during the experiment, the drained solution was not re-used in the system after the necessary measurements were made. For this reason, much more solution is taken out than the closed systems where the drained nutrient solution is reused in the system. Based on the results, with LR_{50} application, on average, the amount of removed with the solution was 1.9 times greater than the amount removed with the solution from the closed system. In other words, with LR_{25} application, the water and fertilizer use efficiency was increased significantly, and it was decided that use of a 25% LR is acceptable with use of open-system substrates in soilless culture. These increases were found to be significant in terms of production input. In addition, the absence of a linear increase in the electrical conductivity of the drained solution was another important indicator that this ratio was sufficient to wash salt out of the root zone. Reducing the amount of the nutrient solution applied in this case showed that water and fertilizer savings, as well as a low LR, can be applied to prevent environmental pollution. In addition, when the fertilizers taken from the open system were examined qualitatively and quantitatively, the dimensions of the threats encountered due to environmental pollution were also apparent. In this respect, it is emphasized that some studies carried out in Turkey have increased pollution in groundwater, and some measures need to be taken to address this problem (Eryurt and Sekin 2001; Polat and Yılmaz 2001).

As a result, it is envisaged that cultivation should be managed appropriately, especially to support water and nutrient resources, by using closed systems, which are important in reducing environmental pollution and providing savings. However, appropriate cultivation systems and programs should be developed for other plant species, supported by the results of economic analysis. There is a need for studies in which different irrigation programs, plant nutrition prescriptions, and salinity conditions are jointly assessed in different growing environments. These studies should be carried out not only with the addition of salt solutions, prepared as stock solutions, for addition to the nutrient solution, but also with application of different nutrient solution elements at different rates. In addition, the most important disadvantage of closed systems is the potential for the spread of plant root diseases, for which there is also a need to develop suitable techniques at the manufacturer level.

Acknowledgements This research was supported by the Scientific and Technological Research Council of Turkey (TUBITAK) (Project No TOGTAG-3384).

References

Adams P (2002) Nutritional control in hydroponics. In: Savvas D, Passam H (eds) Hydyroponic production of vegetables and ornamentals. Embryo Publications, Athens, pp 211–261

Akat Ö (2008) Farklı Tuzluluk Düzeyleri ve Yıkama Oranlarının Gerbera Bitkisinde Gelişim, Verim, Kalite ve Su Tüketimi Üzerine Etkileri. EÜ Fen Bil Enst, Doktora Tezi, 231s

Akat O, Tuzel IH, Ozzambak ME (2009) The effects of different salinity levels and leaching fractions on gerbera plants. International Symposium on Strategies Towards Sustainability of Protected Cultivation in Mild Winter Climate. Acta Horticulturae, vol 807, April 8–11, Antalya. pp 233–238

Allakhverdiev SI, Sakamoto A, Nishiyama Y, Inaba M, Murata N (2000) Ionic and osmotic effects of NaCl-induced inactivation of photosystems I and II in *Synechococcus* sp. Plant Physiol 123:1047–1056

Álvarez S, Sánchez-Blanco MJ (2013) Long-term effect of salinity on plant quality, water relations, photosynthetic parameters and ion distribution in *Callistemon citrinus*. Plant Biol 16:757–764

Ashraf M (1994) Organic substances responsible for salt tolerance in *Eruca sativa*. Biol Plant 36:255–259

Ayers RS, Westcot DW (1989) Water quality for agriculture. FAO Irrigation and Drainage Paper 29, Food and Agriculture Organization of the United Nations, Rome

Bass R, Van den Berg D (2000) Sodium accumulation and nutrient discharge in recirculation systems: a case study with roses. Acta Hort 507:157–164

Bayçin-Korkut A (1998) Çiçek Yetiştiriciliği. Hasat Yayıncılık Yayınları, İstabul, 222 s

Cassaniti C, Leonardi C, Flowers TJ (2009) The effect of sodium chloride on ornamental shrubs. Sci Hort 122:586–593

Cassaniti C, Romano D, Flowers TJ (2012) The response of ornamental plants to saline irrigation water, irrigation–water management, pollution and alternative strategies. InTech, Rijeka, pp 131–158

Cassaniti C, Romano D, Hop MECM, Flowers TJ (2013) Growing floricultural crops with brackish water. Environ Exp Bot 92:165–175

Chaves MM, Osorio J, Pereira JS (2004) Water use efficiency and photosynthesis. In: Bacon MA (ed) Water use efficiency in plant biology. Blackwell, Oxford, pp 42–74

Dalla Guda C, Scordo E, Cervelli C, Farina E (2001) Effects of the electrical conductivity of the nutritive solution on *Danae racemosa* growth. In: V International symposium on protected cultivation in mild winter climates: current trends for suistainable technologies, vol 559, pp 627–631

De Kreij C, Van Os PC (1989) Production and quality of gerbera in rockwool as affected by electrical conductivity of the nutrient solution. In: Proceedings of the 7th International Congress on Soilless Culture, pp 255–266

De Kreij C, Van Os PC, Van den Berg TH, Kalkman EC (1986) EC en kalium/calcium-verhouding van voedingsoplossingen bij gerbera. Vakblad voor de Bloemisterij 41:64–65

De Kreij C, Voogt W, Bass R (1999) Nutrient solution and water quality for soilless cultures. Brochure 196. Research Station for Floriculture and Glasshouse Vegetables (PBG), Naaldwijk

Eryurt A, Sekin Y (2001) Manisa Bölgesi Yeraltı Sularında Mevsimsel Değişiklikler Sertlik ve Azotlu Maddeler. Yeraltı Suları ve Çevre Sempozyumu, Çevjeo, 2001, 21–23 Mart, İzmir, s:187–193

Francois LE, Maas EV (1994) Crop response and management on salt-affected soils. In: Mohammad P (ed) Handbook of plant and crop stress. Marcel Dekker, New York, p 696

Giuffrida F, Lipari V (2003) Leaching irrigation to prevent salt accumulation in the substrate. In: Pardossi A, Serra G, Tognoni F (eds) Proceedings of the International Symposium on Managing Greenhouse Crops in Saline Environment. Acta Hort, vol 609, pp 435–440

Grattan SR (1993) How plants responds to salts. In: Hanson B, Grattan SR, Fulton A (eds) Agricultural salinity and drainage. University of California Irrigation Program, University of California, Davis, pp 3–5

Grattan SR, Hanson B (1993) Crop salt tolerance. In: Hanson B, Grattan SR, Fulton A (eds) Agricultural salinity and drainage handbook for water managers. University of California Irrigation Program, University of California, Davis, pp 5–10

Gregory PJ (2004) Agronomic approaches to increasing water use efficiency. In: Bacon MA (ed) Water use efficiency in plant biology. Blackwell, Oxford, pp 42–170

Haman DZ, Capece JC, Smajstrla AG (1997) Irrigating with high salinity water. Electronic Publishing. BUL322/AE091. Gainesville: University of Florida Institute of Food and Agricultural Sciences. http://ufdc.ufl.edu/l/IR00003289/00001. (Last accessed: 14.01.2019) http://edis.ifas.ufl.edu

Hannah M (1998) Cyber Conference Water Quality. http://www.greenbeam.com/cyberconference/woody-plants.1.html

Hanson B (1993) Crop response to leaching fraction and salt distribution. In: Hanson B, Grattan SR, Fulton A (eds) Agricultural salinity and drainage handbook for water managers. University of California Irrigation Program, University of California, Davis, pp 61–65

Hanson B, Grattan SR, Fulton A (1993) Amendments for reclaiming sodic and saline/sodic soils. In: Hanson B, Grattan SR, Fulton A (eds) Agricultural salinity and drainage handbook for water managers. University of California Irrigation Program, University of California, Davis, pp 81–85

Howell TA (2006) Challenges in increasing water use efficiency in irrigated agriculture. In: Yazar A, Gencel B, Tekin S (eds) International symposium on water and land management for sustainable irrigated agriculture. Çukurova University, Adana, pp 53–63

Jacoby B (1993) Mechanisms involved in salt tolerance by plants. In: Pessarakli M (ed) Handbook of plant and crop stress. Marcel Dekker, New York, pp 97–123

Jones H (2004) What's water use efficiency? In: Bacon MA (ed) Water use efficiency in plant biology. Blackwell, Oxford, pp 27–41

Katerji N, Van Hoorn JW, Hamdy A, Mastrorilli M (2003) Salinity effect on crop development and yield analysis of salt tolerance according to several classification methods. Agric Water Manag 62:37–66

Kazaz S (2017) Developments in the crop pattern and in the socio-economic and technological fields in the world ornamental plant sector and the future vision of Turkey. VI. Süs Bitkileri Kongresi. Batı Akdeniz Tarımsal Araştırma Enst. 19–22 Nisan 2016, Antalya (Eds: Özgül Karagüzel, Selma Kösa, M. Uğur Kahraman). Bildiriler kitabı sayfa:3–13

Koca M, Bor M, Ozdemir F, Turkan I (2007) The effect of salt stress on lipid peroxidation, antioxidative enzymes and proline content of sesame cultivars. Environ Exp Bot 60:344–351

Kotuby-Amacher J, Koenig R, Kitchen B (1997) Salinity and plant tolerance. AG-SO-03. Electronic Publishing, Utah State University. http://www.extension.usu.edu/files/agpubs/salini.htm

Maas EV, Hoffman GJ (1977) Crop salt tolerance—current assessment, journal irrigation drain. Div ASCE 103:115–134

Martín-Closas LI, Recasens X (2001) Effect of substrate type (perlite and tuff) in water and nutrient balance of a soilless culture rose production system. International symp on protected cultivation in mild winter climates: current trends for sustainable technologies. Eds: Fernández, Martínez & Castilla. Cartagena, Spain. Acta Hort 559:569–574

Meric MK, Tuzel IH, Tuzel Y, Oztekin GB (2011) Effects of nutrition systems and irrigation programs on tomato in soilless culture. Agric Water Manag 99:19–25

Munns R, Termaat A (1986) Whole plant responses to salinity. Aust J Plant Physiol 13:143–160

Picchioni GA, Graham CJ (2001) Salinity, growth, and ion uptake selectivity of container-grown *Crataegus opaca*. Sci Hort 90:151–166

Pokluda R, Kobza F (2001) Effect of climate conditions on properties of hydroponic nutrient solution. International symposium on protected cultivation in mild winter climates: current trends for sustainable technologies. Eds: Fernández, Martínez & Castilla. Cartagena, Spain. Acta Hort 559: 611–617

Polat M, Yılmaz N (2001) Antalya havzasındaki yeraltı su kaynaklarında kirliliğin belirlenmesi ve koruma stratejileri. Yeraltı Suları ve Çevre Sempozyumu, Çevjeo 2001, 21–23 Mart, İzmir, s.247–253

Quamme HA, Stushnoff C (1983) Resistance to environmental stress. In: Moore JN, Janick J (eds) Methods in fruit breeding. Purdue University Press, West Lafayette, pp 242–266
Rhoades JD, Kandiah A, Mashali AM (1992) Saline waters as resources. The use of saline waters or crop production. FAO Irrigation and Drainage Paper, 48, Food and Agriculture Organization of the United Nations, Rome, 133. pp 5–10
Savvas D, Gizas G (2002) Response of hydroponically grown gerbera to nutrient solution recycling and different nutrient cation rations. Sci Hort 69:267–280
Schreurs Gerbera Catalogue (1998) Schreurs, De Kwakel
Sevgican A (2002) Örtüaltı sebzeciliği (Topraksız Tarım) Cilt – II. E.Ü.Z.F, Yayınları, İzmir
Shannon MC, Noble CI (1990) Genetic approaches for developing economic salt-tolerant crops. In: Tanji K (ed) Agricultural salinity assessment and management. ASCE Manuals and Reports on Engineering Practice. New York, pp 163–185
Shillo R, Ding M, Pasternak D, Zaccai M (2002) Cultivation of cut flower and bulb species with saline water. Sci Hort 92:41–54
Sivritepe N, ve Eriş A (1998) Asmalarda Tuza Dayanım ve Tuza Dayanımda Etkili Bazı Faktörler Üzerinde Araştırmalar. 4. Bağcılık Sempozyumu, 20–24 Ekim 1998, Yalova. Bildiriler Kitabı, s:56–63
Smith DL (1987) Rockwool in horticulture. Grower Books, London, p 153
Sonneveld C (2001) Effects of salinity on substrate grown vegetables and ornamentals in greenhouse horticulture. Thesis, Wageningen University, p 151
Sonneveld C, Baas R, Nijssen HMC, De Hoog J (2000) Effect of salinity substrate grown vegetables and ornamentals in greenhouse horticulture. J Plant Nut 22:1033–1048
Syros T, Economou A, Exarchou E, Schmidt U (2001) Flower and growth evolution of gerbera cultivated oan perlite in an open hydroponic system. In: Maloupa E, Gerasopoulos D (eds) Proceedings of the International Symposium on Growing Media and Hydroponics. Acta Hort 548:625–630
Tsirogiannis I, Katsoulas N, Kittas C (2010) Effect of irrigation scheduling on gerbera flower yield and quality. Hort Sci 45:265–270
Turkish Statistical Institute (TUIK) (2017) Crop production statistics. http://www.turkstat.gov.tr/PreTablo.do?alt_id=1001. Accessed 25 May 2018
Tüzel İH, Tüzel Y, Gül A, Eltez RZ, Altunlu H (1999) Torba Kültürü ile Yapılan Sera Domates Yetiştiriciliğinde Farklı Sulama Programları Ortam ve Ortam Hacimlerinin Verim ve Su Tüketimi Üzerine Etkileri, Türkiye III. Ulusal Bahçe Bitkileri Kongresi, 364–368
Tyagi NK (2003) Managing saline and alkaline water for higher productivity. In: Kijne et al. (eds) Water productivity in agriculture: limits and opportunities for improvement. CABI, Wallingford, pp 60–87
Villarino GH, Mattson NS (2011) Assessing tolerance to sodium chloride salinity in fourteen floriculture species. Hort Technol 21:539–545
Villora G, Moreno A, Pulgar G, Romero L (2000) Yield improvement in zucchini under salt stress: determining micronutrient balance. Sci Hort 86:175–183
Wahome PK (2003) Mechanisms of salt (NaCl) stress tolerance in horticultural crops—a mini review. Acta Hort 609:127–131
Yemenici F (2000) Topraksız gerbera yetiştiriciliğinde bazı atıkların substrat olarak kullanımı üzerine araştırmalar. EÜ Fen Bil Enst, Yüksek Lisans Tezi, 82 s
Zeybekoğlu E (2000) Gerbera Yetiştiriciliğinde Hasat Öncesi ve Sonrası Yapılan Bazı Uygulamaların Verim Kalite ve Vazo Ömrü Üzerine Etkileri. EÜ Fen Bil Enst, Yüksek Lisans Tezi, 71 s
Zheng Y, Graham T, Richard S, Dixon M (2005) Can low nutrient strategies be used for pot gerbera production in closed-loop subirrigation? Acta Hort 691:365–372

Cross Talk of Nitric Oxide and Reactive Oxygen Species in Various Processes of Plant Development: Past and Present

Lekshmy Sathee, Hari Singh Meena, Sandeep B. Adavi, and Shailendra K. Jha

Introduction

Nitric oxide (NO) is a free radical or a reactive oxygen species (ROS) that functions as a diffusible messenger molecule that was first recognized in animals, wherein it plays variable functions ranging from neural transmission and blood circulation to immune system responses and fertilization. NO is a member of the ROS family of molecules, more specifically, falls in the subgroup of molecules that contain nitrogen, also known as reactive nitrogen species (RNS).

It is well established that NO plays an essential role in animal fertilization and embryogenesis (Ignarro 2000). In plants, both NO and other ROS have proven roles in various plant processes ranging from seed germination to reproduction. NO-mediated protein modification plays an important role in seed germination (Zhang et al. 2003). The close interactions of NO, ROS, and free calcium ions control stomatal aperture in plants (Yadav et al. 2013). Another major role of NO is in pathogen defense: Schlicht and Kombrink (2013) reported that NO accumulates at fungal infection sites and proposed that resistance phenotype correlates with NO production, temporally and volumetrically. In addition, Grob et al. (2013) reviewed the cross talk of NO with ROS and antioxidants. The reaction of NO and ROS results in the formation of peroxinitrite ONOO–. The cross talk of ROS and NO has extensive roles in defense, hypersensitive response, programmed cell death (PCD), pollen tube growth and polarity, pollen tube rupture, root growth and self-incompatibility responses (Duan et al. 2007).

L. Sathee (✉) · H. S. Meena · S. B. Adavi
Division of Plant Physiology, ICAR-Indian Agricultural Research Institute, New Delhi, Delhi, India
e-mail: lekshmy@iari.res.in

S. K. Jha
Division of Genetics, ICAR-Indian Agricultural Research Institute, New Delhi, Delhi, India

M. Hasanuzzaman et al. (eds.), *Plant Abiotic Stress Tolerance*,
https://doi.org/10.1007/978-3-030-06118-0_16

NO acts as a messenger in plant defense responses, developmental processes, stress adaptation and symbiosis establishment (Lamattina et al. 2003). Being major cellular antioxidant and precursor for nitrosoglutathione (GSNO), a storage form of NO, glutathione is a connecting node between NO and ROS. NO exerts its function largely by way of S-nitrosylation of target proteins, leading to alteration in enzyme kinetics and transcription of genes (Frungillo et al. 2014). Corpas et al. (2013) suggested that abundance and metabolism of GSNO by the enzymes *S*-nitrosoglutathione reductases (GSNORs) will be an important issue for the future understanding of the role of NO in plant development and stress responses. Other ways of NO signal transduction include protein tyrosine nitration and through extracellular nucleotides (Blume et al. 2013; Salmi et al. 2013). The research facet of NO production and its influences in plants has progressed extensively since the first description in 1998 (Delledonne et al. 1998; Durner et al. 1998). However, the research is far away in deciphering NO biosynthesis by plants cells and how NO signaling mediates umpteen aspects of plant development and responses. This chapter encompasses a brief description on the significance of NO in different plant processes and interaction of ROS and NO in imparting the optimum response. An outline of the biosynthesis, scavenging and detection of NO is also described (Foreman et al. 2003; Monshausen et al. 2007).

Mechanism of NO Production in Plants

NO production in land plants classically involves two main routes. First, a reductive pathway involving both enzymatic and non-enzymatic reduction of nitrite into NO (Gupta et al. 2011). Second, an oxidative pathway requiring a putative nitric oxide synthase (NOS)-like enzyme (Fig. 1). Wildt et al. (1997) demonstrated that in sunflower and sugarcane, increasing the NO_3^- content of the nutrient culture solution enhanced the emission of NO from the plants during nights following the treatment.

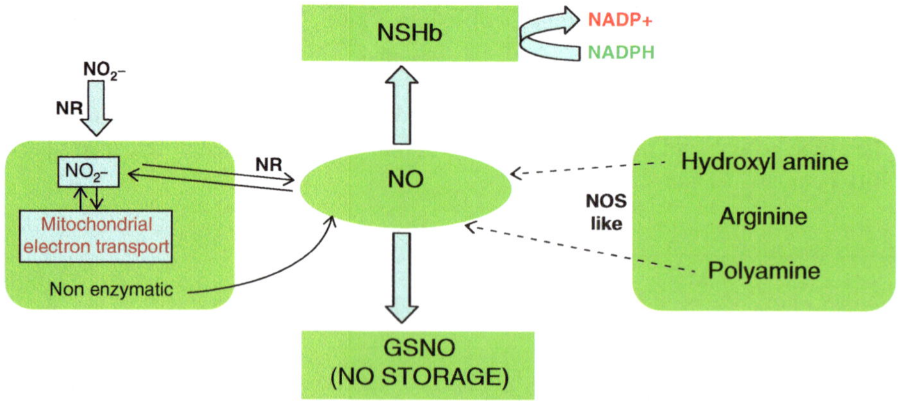

Fig. 1 Mechanism of nitric oxide (NO) homeostasis in plants. *NR* nitrate reductase, *NSHb* non-symbiotic hemoglobin, *NOS* nitric oxide synthase, *GSNO* *S*-nitrosoglutathione

Role of nitrate reductase (NR) in NO production was suspected by low or no NR activity mutants which show no measurable NO. Later *nia1/nia2* double mutants of *Arabidopsis* confirmed the role of NR in the reduction of NO_2^- to NO in Nicotinamide adenine dinucleotide (NADH)-dependent manner under both in vitro and in vivo condition. Under hypoxia/anoxia condition NO_2^- acts as an alternative electron acceptor in mitochondrial electron transport chain allowing mitochondria to oxidize NADH/NADPH and generate a limited amount of ATP together with NO as shown in root mitochondria of several species (Gupta et al. 2011). Both cytochrome c oxidase and AOX (Alternate Oxidase) appear to possess the activity of reducing NO_2^- to NO (Gupta et al. 2011). Finally, several studies also supported the possibility that plant cells synthesize NO through a non-enzymatic reduction of NO_2^-. In particular, a chemical reduction of NO_2^- to NO form was found in the apoplast of barley aleurone layers (Bethke et al. 2004).

The first suggestion that plants do possess mammalian NOS-like enzyme was published in the middle of the 1990s (Yaacov et al. 1998). However, the possibility that such enzyme could catalyze NO synthesis in plants has also been a main controversial issue. First, several studies provided evidence that protein extracts from plant tissues, cultured cells or purified organelles display an NOS-like activity (Durner et al. 1998; Corpas et al. 2009; Du et al. 2015). Many findings suggest effectiveness of animal NOS inhibitors in interfering with NO production in plants. Among the inhibitors, N-Nitro-L-arginine methyl ester (L-NAME), NG-Methyl-L-arginine (L-NMMA), amino guanidine, and S,S0–1,4-phenylene-bis (1,2-ethanediyl)-bis-isothiourea (PBITU) are commonly used (Delledonne et al. 1998; Corpas et al. 2006). In tobacco and *A. thaliana,* enzymes with putative NOS-like activity were recognized: an isoform of glycine decarboxylase complex—P protein (Chandok et al. 2003) and AtNOS/AtNOA1, falling in the guanosine triphosphatase (GTPase) family (Guo et al. 2003).

Experimental evidences further accelerated the notion about the presence of a plant NOS-like enzyme. Initially, Tischner et al. (2007) reported that the L-citrulline-based assay popularly employed to measure an NOS activity is prone to experimental artifacts and is suggestive of cautious analysis. Further, in a proteomic study for identifying the proteins reciprocating to mammalian NOS antibodies did not give any positive results in maize embryonic axes (Butt et al. 2003). Another relevant disagreement regarding plant NOS-like enzyme (Jeandroz et al. 2016) is that despite using transcriptome sequence data of around 1000 species, no mammalian NOS-like sequences were found in plants in species showing NOS-like activities. Further confirming the results shown by Butt et al. (2003) depicting that the maize proteins recognized by mammalian NOS antibodies were unrelated to NOS. Even though in recent past, significant progress has been made in the prehension of the function of NO in plant metabolism, several questions in relation to the biosynthesis remains unanswered. Clearly, land plants do not possess typical mammalian NOS-like proteins (Santolini et al. 2017). Jeandroz et al. (2016) concluded that during evolution plants have developed efficient ways of nitrate assimilation and hence nitrite might serve as a main source for NO.

Physiological Roles of ROS and NO in Plant Development and Environmental Responses

In plants, ROS are mainly produced in chloroplasts, mitochondria, and peroxisomes as a by-product of the metabolism. ROS including hydrogen peroxide (H_2O_2), superoxide anion ($O_2^{•-}$), hydroxyl radicals ($^{•}OH$) and singlet oxygen ($^1O_2$) have all been associated with regulation of different plant biological processes. Mitochondrion wherein the cellular energy metabolism takes place is a main site of ROS production. Mitochondrial alternative oxidase (AOX) also influences ROS and reactive nitrogen species (RNS) generation by the respiratory chain in mitochondria (Vanlerberghe 2013). Peroxisomes are organelles having largely oxidative metabolism and produces superoxide radical ($O_2^{•-}$) as an outcome of their typical metabolism. The chloroplast is also a major location of ROS production in plants (Hideg et al. 2006). The $O_2^{•-}$ and singlet oxygen are produced in chloroplasts by photo-reduction of O_2 and energy transfer from triplet excited chlorophyll to oxygen, respectively. H_2O_2, a ROS of major biological consequences, can be synthesized as a result of the dismutation of superoxide anions by specific enzymes called superoxide dismutase located in various cellular compartments (Noctor et al. 2000; Gechev et al. 2006). An oxidative burst with rapid $O_2^{•-}$ production and its successive dismutation to H_2O_2 in the extracellular space or apoplast is a universal reaction to pathogens, hormonal elicitors, wounding, high temperature, ultra-violet light and ozone (Orozco-Cardenas et al. 2001; Rao and Davis 2001). Besides the oxidative function, H_2O_2 also has a major signaling role in plants (Gechev et al. 2006; Jiang et al. 2011). H_2O_2-mediated signaling induces expression various classes of genes and modulates signaling proteins including protein phosphatases, protein kinases, various transcription factors and plasma membrane calcium channels (Neill et al. 2002; Lin et al. 2012a, b).

Amidst the first roles of NO to be deciphered was the function in defense reaction against bacterial pathogen (Noritake et al. 1996; Delledonne et al. 1998). Accumulating evidences indicated that NO has a role in several physiological and developmental processes: germination, leaf expansion, lateral root development, flowering, stomatal closure, plant hormones signaling, biotic and abiotic stress tolerance (He et al. 2004; Hong et al. 2008; Leitner et al. 2009; Wilkins et al. 2011; Liu et al. 2013; Yadav et al. 2013). Mitochondria and chloroplasts also take part in NO generation in vivo (Galatro et al. 2013; Vanlerberghe 2013). NO also reacts with several targets forming a variety of molecules, such as nitric oxide radicals (NO^-), nitrosonium ion (NO^+), peroxynitrite (ONOO−), *S*-nitrosothiols (SNOs), higher oxides of nitrogen (NOx) and dinitrosyl—iron complexes among others, collectively these NO derivatives are termed as RNS (Di Stasi et al. 2002).

Pollen Tube Growth and Fertilization

NO is recognized as a signaling molecule in plant sexual reproduction also. High amounts of NO was detected in the pollen grain and has been proven to regulate pollen tube growth. Previous studies demonstrated that NO and ROS are produced in a stage and tissue-specific manner in olive (*Olea europaea* L.). Fluorometric assay confirmed that growing pollen tubes produces NO. GSNO a mobile reservoir of NO, produced by the *S*-nitrosylation of reduced glutathione (GSH) was also detected and quantified in pollen tubes. Exogenous NO donors negatively affected both pollen germination and growth; however, these phenomena were partially recovered by NO scavenger c-PTIO. To understand the role of NO on pollen germination, a de novo and a fully annotated reproductive stage transcriptome data from olive was searched for the potential of transcripts amenable for *S*-nitrosylation/Tyr-nitration according to known criteria. Several gene products were found and were classified based on their roles. The significance of *S*-nitrosylation/Tyr-nitration was experimentally validated through Western blotting, immuno-precipitation and proteomic approach (María et al. 2017). There have been attempts to establish the role of NO and ROS during pollen tube growth and signaling in gymnosperms. The pollen tube elongation of Arizona cypress (*Cupressus arizonica* G.) was investigated. NO, ROS and actin were localized by using the particular fluorescent dyes. NO and ROS were found in the generative cell nuclei and also in pollen tube cells. The pollen cell NO content was lowered by NO scavenger and NOS inhibitor. Further, the pollen cell endogenous ROS content was lowered by NADPH oxidase inhibitor. The inhibitor treatments reduced pollen germination and pollen tube growth and induced severe morphological abnormalities. Inhibition of NO and ROS accumulation also severely disrupted the actin cytoskeleton in the pollen tubes.

In the animal system, NO is synthesized by NOS and signaling involves soluble guanylyl cyclases (sGC) that generate cGMP and by phospho diesterases (PDE) that hydrolyze cGMPs (Ignarro 2000). In model plant *Arabidopsis*, homologs of mammalian NOS, sGC and PDEs have not yet been found. However, there are evidences for cGMPs and their action in plants (Penson et al. 1996; Durner et al. 1998). In concurrence with the role of cGMPs, a functional plant guanylyl cyclase was found in *Arabidopsis thaliana* though it displays an unusual protein domain arrangement (Ludidi and Gehring 2003). There are also plenty of reports regarding NO production in plants (Barroso et al. 1999; del Rio et al. 2002). Nitrate reductase (NR) is suggestive of producing NO necessary for stomatal closure (Lamattina et al. 2003). Recent reports describe the presence of new enzymes displaying NOS activity, an inducible NOS showing sequence similarity to glycine decarboxylase (Chandok et al. 2003) and another constitutive NOS form (NOS1) without any homology to mammalian NOS isoforms (Guo et al. 2003). The *Arabidopsis nos1* mutants show impaired flowering and fertility, indicating the role of NO in these developmental events. It can also be speculated that NO synthesized in the peroxisomes (Barroso et al. 1999; Del Rio et al. 2002) could be acting as a negative regulator of pollen tube growth. It can be presumed that the localization of peroxisomes and thereby the site of endogenous NO production ultimately determines the direction into which pollen tube growth resumes. Reports emphasize the role of peroxisomes in the production of NO and ROS that has signaling function. Peroxisomes with their typical and abundant enzymatic machinery participate in an array of developmental processes such as photomorphogenesis in *Arabidopsis* (Hu et al. 2002; Barroso et al. 1999; Del Rio et al. 2002).

The endogenous NO production is correlated with the regulation of pollen tube growth; the in vivo confirmation of this is made difficult by various experimental obstacles. Real-time imaging of pollen tube guidance in vivo would imply the possibility of optically sectioning closed flowers, which implies demanding technical conditions (two photon excitation and water-immersion, long working distance objectives) far from optimized for this specific application. Excitation-derived photo-damage of cells is not an option because it will generate stress-induced bursts of NO production (Lamattina et al. 2003). A possibility for overcoming these obstacles will be either the use of pollen tubes expressing highly fluorescent reporter genes to closely monitor the pollen tube–pistil interaction or otherwise the use of floral mutants with open ovaries and exposed, yet functional ovules. Another problem is related to the high reactivity of NO, with a half-life depending on the redox status of the surrounding environment, namely when ROS are present (Ignarro 2000; Thomas et al. 2001). This makes it difficult to gauge the amount of NO being produced in vivo, so no invasive techniques for NO can easily quantify a putative signal from the female tissue. A self-referencing NO-selective electrode could be used but again tissue accessibility is a limiting factor. Several difficulties arise when interpreting chemical cues identified in different plant species: it can be argued that general mechanisms do not assure species specificity to avoid widespread cross-fertilization (Johnson and Preuss 2002a). One possible explanation could be related to different threshold sensitivities operating for a given molecule from species to species. Otherwise, a different species could use a similar mechanism but with derivative molecules within a single chemical family, which would be transduced into different effects. Given the diversity of molecules shown to have guidance effects on pollen tubes and predicting that more will be uncovered through successive genetic screens it is likely that chemical signaling between the pollen tube and pistil could convey specificity by using universal molecules in various combinations.

The interaction between pollen and stigma is one of the most important stages in the life cycle of a flowering plant because its outcome determines whether fertilization will occur and thus whether seed will be set. This critical cellular communication between the haploid pollen (grain and tube) and the diploid cells of the stigma and style is one of the most precisely adapted activities of the plant—morphologically, physiologically and biochemically (Heslop and Harrison 1978) and has become a paradigm for the study of cell recognition and cell signaling in plants. For fertilization to be achieved pollen must establish molecular congruity/compatibility with the stigma and then following production of a pollen tube with the transmitting tissue of the style and ovary as the pollen tube grows through the pistil to deliver its two sperm cells to an ovule. Thus there must be a continuous exchange of signals both physical and chemical between pollen and pistil from the moment a pollen grain arrives on the stigma to the moment the pollen tube enters the ovule. Identifying these signals and the responses they induce has been the subject of intense research for the past three decades and a picture is emerging of a diverse array of signals that influence pollen germination and pollen tube growth and guidance within the pistil (Franklin and Tong 2002; Johnson and Preuss 2002a, b; Feijo et al. 2004; Dresselhaus 2006). Recently, the animal neurotransmitter, gamma-aminobutyric acid (GABA) was identified as a potential chemo-attractant for pollen tubes in *Arabidopsis* (Palanivelu et al. 2003), while in *Lilium longiflorum*, NO has been implicated in pollen tube guidance as a putative negative regulator of pollen tube growth able to induce tip reorientation (Prado et al. 2004). When a pollen grain lands on a stigma, specific recognition events must take place to establish that: (a) the object that has alighted is a pollen grain and not a fungal spore or bacterium; (b) it is a pollen grain of the correct species, or a closely related species (interspecific hybridization is fairly common in angiosperms); and (c) in most hermaphrodite flowering plants, it is not a pollen grain from the same plant (Heslop and Harrison 1978; Franklin-Tong 2002; Hiscock 2000). While the last of these three recognition events (self-incompatibility) has been studied extensively reviewed in Hiscock and McInnis (2003), relatively little is known about molecular signals and interactions mediating the first two recognition events. The stigma surface is only receptive to pollen for a relatively short period, so the timing of pollination is critical. Pollination either side of this period of optimal female receptivity results in reduced seed set, or no seed set (Herrero 2003). It has long been known that receptive stigmas "ripe" for pollination are characterized by high levels of peroxidase activity (Dupuis and Dumas 1990; McInnis et al. 2006) and tests most widely used to determine pistil receptivity measure stigma peroxidase activity (Dafni and Motte Maues 1998). Nevertheless, the function of these ubiquitous enzymes in stigmas is not known (McInnis et al. 2006). Recently, a stigma-specific peroxidase (SSP) from the ragwort *Senecio squalidus* was identified (McInnis et al. 2005). As part of ongoing work to determine the function of SSP and stigmatic peroxidases, *Senecio* stigmas accumulate high amounts of ROS particularly H_2O_2 in their epidermal cells (papillae) where SSP is localized (McInnis et al. 2006). The presence of such high amounts of ROS/H_2O_2 in the papillae, which receive and discriminate pollen, suggested that ROS/H_2O_2 (and, by potential association, SSP) may be important for stigma function. ROS/H_2O_2 have a variety of roles in cell metabolism but also act as signaling molecules mediating a range of cellular processes from development to defense often in association with NO (Hancock et al. 2006), so it was not unreasonable to speculate that ROS/H_2O_2 might be involved in pollen–stigma interactions.

NO-guidance mechanism would be possible if there were specialized female tissues acting as NO "hot spots," for example at the base of the funiculus where a sharp change in pollen tube growth direction is required or near the embryo sac after fertilization in order to prevent secondary pollen tube from penetrating the micropyle. The fact that *nos1*, the only bona fide NO-producing mutant so far described shows fertility deficiencies is a positive indication that NO may be involved in pollen tube guidance. In past research, pollen tube guidance could not be fully explained by the actions of positive guidance cues. In addition, it remains debatable that tracking down a molecule will overcome questions related to pollen tube path length and thickness (Lush et al. 1998). Yet, a gaseous molecule may overcome these barriers easily. In proposing NO, a diffusible gas as a candidate for pollen tube guidance, it may address a controversial aspect of pollen tube guidance. In *Arabidopsis*, Hulskamp et al. (1995) propose that each ovule guides the pollen tube by chemotatic gradients with ~100 cm range of action at the junction of the ovule with the placenta. However, wild-type *Arabidopsis* pollen tubes make a sharp turn to enter the mycropyle in 10 cm of this area (Shimizu and Okada 2000). The ability of NO to function as a messenger across cell layers and to trigger cellular processes is well established in animals (Ignarro 2000). The negative chemotropism described here for NO is reminiscent of the effects of semaphorins on axon guidance in animals: these proteins function as chemo-repellents, which prompt axons to make right angle turns within an environment that contains both attractants and repellents (Tessier Lavigne and Goodman 1996). Similarly, NO acts as negative effectors on the retinal patterning of the optical lobe in *Drosophila*, where NO prevents further extension of axons beyond their target neurons (Gibbs and Truman 1998). NO function as a guidance cue implies that (1) it is able to form a concentration gradient, (2) it produces a specific response, (3) it remains stable for a given period and (4) it varies in effectiveness with distance to the target (Palanivelu and Preuss 2000). In addition, the response can be prevented if the gradient is perturbed or annihilated by an NO scavenger. The events downstream of NO seem to be at least in part mediated by cGMP. Another tested chemicals sildenafil citrate, a drug that inhibits cGMP-selective PDEs of mammals, facilitated the redirected growth of pollen tubes in response to low doses of NO donors that were themselves ineffective. Previous studies with cyclic nucleotide analogs also suggest that cGMP and cAMP are involved in pollen tube growth control (Moutinho et al. 2001; Elias et al. 2001). A likely target downstream of cGMP is a family of cyclic nucleotide-gated channels (CNGs), also represented in the pollen transcriptome (Becker et al. 2003). Directly or coupled with other transporters, CNGs may regulate the flux of ions such as Ca^{2+}, H^+ and Cl^- that are known to be involved in pollen tube growth control (Feijo et al. 2001; Becker et al. 2003; Feio et al. 1999; Zonia et al. 2002). Cyclic nucleotide balance, modulation of Ca^{2+} channels have recently been linked (Nishiyama et al. 2003).

During the sexual reproduction of flowering plants, pollen grains germinate on receptive stigmas and produce a tip-growing tube with an exceptionally fast growth (Stone et al. 2004), that rapidly penetrates the style to deliver the male gametes to the ovules. To achieve this, pollen tubes have a high energy requirement requiring rapid oxygen uptake (Tadege and Kuhlemeier 1997). Highly polarized pollen tube expansion is dependent on precisely targeted delivery of vesicles containing cell wall material to the tube tip aided by the actin cytoskeleton and a tip-focused intracellular Ca^{2+} ($[Ca^{2+}]$) gradient (Hepler et al. 2001). Pollen tube growth rate and $[Ca^{2+}]$ oscillate with a period of 15–60 s and with a phase shift of 12 s the peak of growth rate precedes and "predicts" the strength of the Ca^{2+} influx maximum (Messerli et al. 2000; Feijó et al. 2001). Oscillations in lily pollen tubes fluorescence at 400 nm (excitation at 360 nm), with the strongest signal at *c*. 30 μm behind the tip, were interpreted as a NAD(P)H reduced coenzyme signal related to mitochondria activity (Cardenas et al. 2006). A complex signaling network is required to regulate this highly localized growth, and pollen tubes provide a model for the analysis of polarized cell development (Hepler et al. 2001). Actin and [Ca^{2+}] are also involved in tip growth of another cell type the root hairs (Hepler et al. 2001). Recently, a requirement for ROS in root-hair growth was demonstrated in *Arabidopsis thaliana* using a loss of function knockout mutant in *AtrbohC/RHD2* (Foreman et al. 2003). *AtrbohC/RHD2* encodes an $O_2^{\bullet-}$ producing NAPDH oxidase NOX, and the mutant had reduced ROS formation at the tip of very short root hairs (Foreman et al. 2003).

Reactive oxygen species are an inevitable consequence of aerobic metabolism but are also generated in a controlled manner and used for a variety of functions including pathogen defense and cell signaling (Halliwell and Gutteridge 1999). The short-lived nature of $O_2^{\bullet-}$ and other ROS makes them ideal signaling molecules or for carrying out other localized biochemical activities. Plasma membrane-localized NOX, which catalyses the extracellular formation of $O_2^{\bullet-}$ from molecular oxygen, using NADPH as an electron donor, is the major source of signaling ROS in mammals (Reeves et al. 2002). Plant NOXs, which are partly homologous to the gp91phox (NOX2) catalytic subunit of the mammalian phagocyte NOX (Keller et al. 1998; Torres et al. 1998), have the same $O_2^{\bullet-}$ generating activity (Sagi and Fluhr 2001). There are 10 NOX (*Atrboh*) genes in *Arabidopsis thaliana* (Foreman et al. 2003). Extracellular $O_2^{\bullet-}$ readily forms other ROS including H_2O_2 and the $OH^{\bullet}$ (Halliwell and Gutteridge 1999). The dismutation of $O_2^{\bullet-}$ to H_2O_2 is spontaneous or catalyzed by cell wall SOD (Halliwell and Gutteridge 1999). H_2O_2 can cross the plant plasma membrane (PM) through aquaporins (Henzler and Steudle 2000). Unlike other ROS, H_2O_2 acts as both an intracellular (Kovtun et al. 2000) and intercellular (Allan and Fluhr 1997) signaling molecule in plants, as it is relatively stable compared with other ROS (Halliwell and Gutteridge 1999). H_2O_2 is also involved in peroxidative cross-linking of cell wall polysaccharides and proteins (Kjellbom et al. 1997; Kerr and Fry 2004). Plant NOXs are implicated in ROS formation during the oxidative burst caused by pathogen infection (Torres et al. 2002) and during abscisic acid-induced stomatal closure (Kwak et al. 2003). NOX activity is also required during normal plant development as transgenic "knockout" plants of *AtrbohF* and the *AtrbohD/F* double mutant are small and have short roots (Torres et al. 2002; Kwak et al. 2003). Furthermore, transgenic tomato plants expressing plant *NOX* RNAi transgenes show a wide range of developmental abnormalities (Sagi et al. 2004). Plant NOXs contain an *N* terminal predicted cytoplasmic domain containing Ca^{2+}-binding EF-hand motifs (Keller et al. 1998; Sagi and Fluhr 2001). In this respect, they are similar to a human gp91phox homolog, NOX5 (Banfi et al. 2001). The activity of both NOX5 and plant NOXs is increased directly by Ca^{2+} in vitro (Banfi et al. 2001; Sagi and Fluhr 2001). These results suggest that [Ca^{2+}] may play a role in the regulation of NOX-mediated $O_2^{\bullet-}$ production in plants.

Programmed Cell Death

Leaf senescence, thought to be another form of plant PCD, is the final stage of leaf development, which is only controlled by organ and also triggered by adverse environmental factors (Pourtau et al. 2004; Munns 2005; Masclaux Daubresse et al. 2007; Jing et al. 2008; Wu et al. 2012). Additionally, phytohormones such as ethylene (ET), Salicylic acid (SA), Jasmonic acid (JA), auxin, ABA and cytokinins all affect leaf senescence (Lim et al. 2007). In *Arabidopsis*, the level of H_2O_2 increases dramatically in leaf tissue during senescence. In addition to its role in oxidizing macromolecules such as proteins and lipids, H_2O_2 has also been proposed to function as a signal to induce the expression of genes involved in the senescence process (Cui et al. 2013). In agreement with its lower antioxidant capacity, senescent leaf tissue was found to contain elevated levels of ROS. In this context, a number of senescence-associated genes (SAGs) characterized from *Arabidopsis* could be induced by ozone (Miller et al. 1999) and the expression of many other SAGs were also induced by ROS (Navabpour et al. 2003), indicating that ROS might function as a signal to promote senescence. Interestingly, senescence-associated NAC genes key regulators of leaf senescence were also found to be rapidly and strongly induced by H_2O_2 treatment in both leaves and roots (Balazadeh et al. 2010). Thus, ROS has a dual role in leaf senescence: to promote the cell death process by directly oxidizing target macromolecules and to drive the expression of senescence-related genes.

Distinct from the positive role of ROS in senescence, NO can both provoke and impede this process, dependent upon its concentration and subcellular location. NO may alleviate the toxicity of ROS and has thus acted as a leaf senescence delaying factor in plants. The NO-deficient mutant *nos1/noa1* showed early leaf senescence (Niu and Guo 2012) and similarly *Arabidopsis* expressing a node grading dioxygenase (NOD) displayed a senescence-like phenotype (Mishina et al. 2007). Furthermore, the level of NO is related with the senescence process and is thought to be an essential component involved in plant senescence signaling cascades. In *Arabidopsis* mutant *dnd1,* which lacks a plasma membrane-localized cation channel (CNGC2), early senescence-associated phenotypes (such as loss of chlorophyll expression level of senescence-associated genes, H_2O_2 generation, lipid peroxidation, tissue necrosis, and SA levels) were all elevated relative to wild type. Basal levels of NO in *dnd1* leaves were lower than wild type, suggesting that the function of CNG C2 may impact downstream "basal" NO production in addition to its role linked to NO signaling. NO generation is therefore thought to act as a negative regulator during plant leaf senescence signaling. The protective effect of NO against ROS-induced cell death can also be linked to the enhanced activity of antioxidant enzymes, as a negative regulator of the chlorophyll catabolic pathway and as drivers for positively maintaining the stability of thylakoid membranes during leaf senescence (Liu et al. 2013). On the other hand, NO can also promote the leaf senescence. *Arabidopsis* AtFer1, one of the best characterized plant ferritin isoforms to date, strongly accumulates on treatment with excess iron, via an NO-mediated pathway. The AtFer1 isoform is functionally involved in events leading to the onset of age-dependent senescence in *Arabidopsis* and its iron detoxification function during senescence is required when ROS accumulates (Murgia et al. 2007). Recently, identification of an NO accrual mutant *nitric oxide excess1* (*noe1*) in rice revealed that *NOE1* encoded a rice catalase (CAT) OsCATC. Interestingly, *noe1* plants exhibited an increase of H_2O_2 in their leaves, which consequently promoted NO production via activation of nitrate reductase. Removal of excess NO reduced cell death in both leaves and suspension cultures derived from *noe1* plants, implicating that NO acts as an important endogenous mediator of H_2O_2-induced leaf cell death. Reduction of intracellular *S*-nitrosylation (SNO) levels, generated by overexpression of OsGSNOR alleviated leaf cell death in *noe1* plants. Thus, *S*-nitrosylation was also involved in light-dependent leaf cell death in *noe1*. Collectively, these data suggest that both NO and SNOs are important mediators in the process of H_2O_2-induced leaf cell death in rice (Lin et al. 2012a, b). OsGSNOR in *noe1* plants reduced SNO levels, consistent with a key role for this enzyme in SNO homeostasis. Moreover, the results show that no change in H_2O_2 content occurred in either *GSNOR*-overexpressing or *GSNOR*-RNAI transgenic lines in the context of *noe1* background, suggesting that NO might function downstream of H_2O_2 in a light-driven leaf cell death in rice. It was found that NO treatment led to rapid cell death and induced H_2O_2 accumulation in maize leaves, and pharmacological studies also suggested that NO-induced cell death is in part mediated via H_2O_2; therefore, H_2O_2 may be involved in NO-induced cell death in maize leaves (Kong et al. 2013). These discrepancies for the role of NO in cell death might be due to the differences in plant species, redox state, and growth conditions. Both NO and H_2O_2 could induce leaf cell senescence.

Abiotic and Biotic Stress Responses

The production of ROS in plants under normal growth conditions is low. However, in response to various environmental stresses, ROS are drastically increased in plants disturbing the normal balance of $O_2^{\bullet -}$, $^{\bullet}OH$ and H_2O_2 in the intracellular environment (Sharma and Dubey 2005). The effects of various environmental stresses such as drought, salinity, chilling, metal toxicity, UV-B radiation and pathogen attack on ROS production are discussed below.

Drought

Under drought stress ROS production is enhanced in several ways. Inhibition of carbon dioxide (CO_2) assimilation coupled with the changes in photosystem activities and photosynthetic transport capacity under drought stress results in the accelerated production of ROS via the chloroplast Mehler reaction (Asada 1999). During drought stress CO_2 fixation is limited due to the stomatal closure which in turn leads to reduced $NADP^+$ regeneration through the Calvin cycle. Due to lack of electron acceptor, over-reduction of the photosynthetic ETC occurs which leads to a higher leakage of electrons to O_2 by the Mehler reaction (Biehler and Fock 1996). Reported 50% more leakage of photosynthetic electrons to the Mehler reaction in drought stressed wheat plants, compared to unstressed plants. Photosynthetic activity is inhibited in plant tissues due to an imbalance between light capture and its utilization under drought stress (Foyer and Noctor 2000). Dissipation of excess light energy in the PSII core and antenna leads to generation of ROS which are potentially dangerous under drought stress conditions (Foyer and Harbinson 1994). Under drought stress, the photorespiratory pathway is also enhanced, especially, when RUBP oxygenation is maximal due to limitation in CO_2 fixation (Noctor et al. 2002). Have estimated that photorespiration is likely to account for over 70% of total H_2O_2 production under drought stress conditions.

Superoxide ($O_2^{\bullet-}$) initiates a chain reaction leading to the production of more toxic radical species, which may cause damage far in excess of the initial reaction products. Under drought stress, one of the real threats towards the chloroplast is the production of the $^{\bullet}OH$ in the thylakoids through "iron-catalyzed reduction of H_2O_2" by both superoxide dismutase (SOD) and AsA. Increased production of ROS leads to oxidative stress in growing plants. Rice seedlings subjected to drought showed an increased concentration of $O_2^{\bullet-}$, increased the level of lipid peroxidation, chlorophyll bleaching, loss of some antioxidants (AsA, GSH, α-tocopherol, and carotenoids), total soluble protein, and thiols (Sharma and Dubey 2005; Boo and Jung 1999). To combat danger posed by ROS, plants possess different scavenging enzymes and metabolites. Enhanced activity of enzymes of antioxidative defense system has been reported under drought stress in several plant species (Sharma and Dubey 2005; Sayfzadeh and Rashidi 2011; Sgherri et al. 2011; Boo and Jung 1999). Comparative study of the antioxidant responses in drought-tolerant and drought-sensitive genotypes revealed higher antioxidant capacity in tolerant genotypes. In contrast to drought-susceptible wheat genotype HD 2329, drought-tolerant wheat genotype C 306 had higher ascorbate peroxidase (APX) and CAT activity, higher AsA content and lower H_2O_2 and MDA content (Sairam et al. 1998). In another study, the drought-tolerant maize genotype Giza 2 was suggested to be comparatively tolerant to water stress compared to drought-sensitive Trihybrid 321 owing to the lower increase in H_2O_2 and MDA content along with higher increase in SOD, CAT, and POX activities (Moussa and Abdel Aziz 2008). Similarly, among two apple rootstocks *M. prunifolia* (drought-tolerant) and *M. hupehensis* (drought-sensitive), *M. hupehensis* was more vulnerable to drought than *M. prunifolia*, resulting in larger increases in the levels of H_2O_2, $O_2^{\bullet-}$, and MDA. The activities of SOD, POD, APX, growth regulator (GR), and DHAR and levels of AsA and glutathione (GSH) increased to a greater extent in *M. prunifolia* than in *M. hupehensis* in response to drought. APX serves as an important component of the antioxidative defense system under drought (Sharma and Dubey 2005). In rice plants, an increase in the capacity of A regeneration system by de novo synthesis of monodehydroascorbate reductase (MDHAR), dehydroascorbate reductase (DHAR), and GR has been shown to be one of the primary responses to water deficit so as to mitigate oxidative stress (Sharma and Dubey 2005; Boo and Jung 1999).

Salinity

Salinity stress results in an excessive generation of ROS (Tanou et al. 2009; Hernandez et al. 2000). High salt concentrations lead to overproduction of the ROS-$O_2^{\bullet-}$, $^{\bullet}OH$, H_2O_2, and $^1O_2$ by impairment of the cellular electron transport within different subcellular compartments such as chloroplasts and mitochondria, as well as from induction of metabolic pathways such as photorespiration. Salt stress can lead to stomatal closure, which reduces CO_2 availability in the leaves and inhibits carbon fixation which, in turn, causes exposure of chloroplasts to excessive excitation energy and over-reduction of photosynthetic electron transport system leading to the enhanced generation of ROS and induced oxidative stress. The low chloroplastic CO_2/O_2 ratio also favors photorespiration leading to increased production of ROS such as H_2O_2. Elevated CO_2 mitigates the oxidative stress caused by salinity, involving lower ROS generation and better maintenance of redox homeostasis as a consequence of higher assimilation rates and lower photorespiration (Perez Lopez et al. 2009). Salinity-induced ROS disrupt normal metabolism through lipid peroxidation, denaturing proteins, and nucleic acids in several plant species (Tanou et al. 2009; Hernandez et al. 2000; KarrayBouraoui et al. 2011). Differential genomic and proteomic screenings carried out in Physcomitrella patens plants showed that they responded to salinity stress by up-regulating a large number of genes involved in antioxidant defense mechanism (Wang et al. 2008). Suggesting that, the antioxidative system may play a crucial role in protecting cells from oxidative damage following exposure to salinity stress in plants. Salinity-induced oxidative stress and the possible relationship between the status of the components of the antioxidative defense system and the salt tolerance in *Indica* rice (*Oryza sativa* L.) genotypes were studied by Mishra et al. (2013). Seedlings of salt-sensitive cultivar showed a substantial increase in the rate of $O_2^{\bullet-}$ production, elevated levels of H_2O_2, malondialdehyde (MDA), declined levels of thiol, AsA, and GSH, and lower activity of antioxidant enzymes compared to salt-tolerant seedlings. It was suggested that a higher status of antioxidants AsA and GSH and a coordinated higher activity of the enzymes SOD, CAT, plant glutathione peroxidase (GPX), APX, and GR can serve as the major determinants in the model for depicting salt tolerance in Indica rice seedlings (Mishra et al. 2013). Similarly, study of immediate responses (enzymatic and non-enzymatic) to salinity-induced oxidative stress in two major rice (*Oryza sativa* L.) cultivars, salt-sensitive Pusa Basmati 1 (PB) and salt-tolerant Pokkali (PK), revealed a lesser extent of membrane damage (lipid peroxidation), lower levels of H_2O_2, higher activity of the ROS scavenging enzyme, CAT, and enhanced levels of antioxidants like ASA and GSH in PK compared to PB (Vaidyanathan et al. 2003). Comparative study using cultivated tomato *Lycopersicon esculentum* L. Mill. cv. M82 (Lem) and its wild salt-tolerant relative L. pennellii (Corr.) D'Arcy accession Atico (Lpa) showed better protection of Lpa roots from salt-induced oxidative damage, at least partially, from the increased activities of the SOD, CAT, APX, MDHAR, and increased contents of AsA and GSH (Shalata et al. 2001). In the salt-stressed root of Lem, a gradual increase in the membrane lipid peroxidation was observed, whereas no change in lipid peroxidation was observed in Lpa.

Salt-tolerant *Plantago maritima* showed a lower level of MDA and a better protection mechanism against oxidative damage caused by salt stress by increasing activities of SOD, CAT, GR, and APX than the salt-sensitive P. media (HediyeSekmen et al. 2007). NADP-dehydrogenases and APX have been suggested as key players in antioxidant defense of olive plants under salt stress conditions (Valderrama et al. 2006; Mittal and Dubey 1991).

Chilling

Chilling stress is a key environmental factor limiting growth and productivity of crop plants. Chilling leads to the overproduction of ROS by exacerbating imbalance between light absorption and light use by inhibiting Calvin-Benson cycle activity (Logan et al. 2006), enhancing photosynthetic electron flux to O_2 and causing over-reduction of respiratory electron transport chain (ETC) (Hu et al. 2008). Chilling stress also causes significant reductions in rbcL and rbcS transcripts, ribulose-1,5-bisphosphate carboxylase (RuBisCO), content and initial RuBisCO activity, leading to higher electron flux to O_2 (Zhou et al. 2006). H_2O_2 accumulation in chloroplast was negatively correlated with the initial RuBisCO activity and photosynthetic rate (Zhou et al. 2006). Chilling-induced oxidative stress evident by increased accumulation of ROS, including H_2O_2 and $O_2^{\bullet-}$, lipid peroxidation, and protein oxidation is a significant factor in relation to chilling injury in plants (Fryer et al. 1998; Prasad 1997; Zhang et al. 2008a, b). Protein carbonyl content, an indication of oxidative damage, was increased two-fold in maize seedlings when exposed to chilling temperatures (Prasad 1997). Lipoxygenase activity, as well as lipid peroxidation, was increased in maize leaves during low temperatures, suggesting that lipoxygenase-mediated peroxidation of membrane lipids contributes to the oxidative damage occurring in chill-stressed maize leaves (Fryer et al. 1998). Responses to chilling-induced oxidative stress include alteration in activities of enzymes of the antioxidant defense system. The activities of antioxidative enzymes APX, MDHAR, DHAR, GR, and SOD increased during chilling periods in maize and strawberry leaves (Fryer et al. 1998; Zhang et al. 2008a, b). However, if the duration of chilling stress is too long, the defense system may not remove overproduced ROS effectively, which may result in severe damage or even cell death (Zhang et al. 2008a, b). Non-enzymic antioxidants (AsA, GSH, carotenoids, and α-tocopherol) also play an important role in cold response. Under cold stress conditions, low-molecular-weight antioxidants, especially, that of reduced AsA, have been suggested to be an important component in plant cell defense (Radyuk et al. 2009). Many comparative studies using chilling-tolerant and sensitive genotypes have shown greater antioxidant capacity in chilling-tolerant species compared to sensitive ones (Jahnke et al. 1991). In rice, higher activities of defense enzymes and higher content of antioxidant under stress were associated with tolerance to chilling (Huang and Guo 2006). The responses of antioxidative system of rice to chilling were investigated in a tolerant cultivar, Xiangnuo-1, and a susceptible cultivar, IR-50. The electrolyte leakage and malondialdehyde content of Xiangnuo-1 were little affected by chilling treatment, but those of IR-50 increased. Activities of SOD, CAT, APX, and GR and AsA content of Xiangnuo-1 remained high, while those of IR-50 decreased under chilling stress. GR activity was also found to increase within 24 h in chilling-tolerant *Zea diploperennis*, but it decreased slightly in chilling-susceptible Z. mays cv. LG11 (Jahnke et al. 1991).

Metal Toxicity

The increasing levels of metals into the environment drastically affect plant growth and metabolism, ultimately, leading to severe losses in crop yields (Salt et al. 1995; Mishra and Dubey 2005). One of the consequences of the presence of the toxic metals within the plant tissues is the formation of ROS, which can be initiated directly or indirectly by the metals and, consequently, leading to oxidative damage to different cell constituents (Shah et al. 2001; Maheshwari and Dubey 2009; Srivastava and Dubey 2011; Sharma and Dubey 2007; Gallego et al. 2002). Under metal stress condition, net photosynthesis (PN) decreases due to damage to photosynthetic metabolism, including photosynthetic electron transport (Phet) (Vinit et al. 2002). For example, copper has been shown to negatively affect components of both the light reactions (e.g., PSII, thylakoid membrane structure, and chlorophyll content) and CO_2 fixation reactions (Vinit et al. 2002; Moustakas et al. 1994). These alterations in photosynthetic metabolism lead to overproduction of ROS such as $O_2^{\bullet-}$, $^{\bullet}OH$, and H_2O_2. The induction of ROS production due to metals (cadmium and zinc) in *Nicotiana tabacum* L. cv. Bright Yellow 2 (TBY-2) cells in suspension cultures showed properties comparable to the elicitor-induced and oxidative burst in other plant cells (Robek Sokolnik et al. 2009). Redox-active metals, such as iron, copper, and chromium, undergo redox cycling producing ROS, whereas redox-inactive metals, such as lead, cadmium, mercury, and others, deplete cells major antioxidants, particularly thiol-containing antioxidants and enzymes (Shah et al. 2001; Maheshwari and Dubey 2009; Srivastava and Dubey 2011; Sharma and Dubey 2007; Gallego et al. 2002; Gallego et al. 1996; Weckx and Clijsters 1996; Yamamoto et al. 1997). If metal-induced production of ROS is not adequately counterbalanced by cellular antioxidants, oxidative damage of lipids, proteins, and nucleic acids ensues (Srivastava and Dubey 2011; Sharma and Dubey 2007; Halliwell and Gutteridge 1989; Dat et al. 2000; Sharma and Dietz 2009; Sandalio et al. 2009). Significant enhancement in lipid peroxidation and a decline in protein thiol contents were observed when rice seedlings were subjected to Al, Ni, and Mn toxicity (Maheshwari and Dubey 2009; Srivastava and Dubey 2011; Sharma and Dubey 2007).

The increased activity of antioxidative enzymes in metal-stressed plants appears to serve as an important component of the antioxidant defense mechanism of plants to combat metal-induced oxidative injury (Shah et al. 2001). Responses of metal exposure to plants vary depending on plant species, tissues, stages of development, type of metal and its concentration. One of the key responses includes triggering of a series of defense mechanisms which involve enzymatic and non-enzymatic components (Shah et al. 2001; Maheshwari and Dubey 2009; Srivastava and Dubey

2011; Sharma and Dubey 2007; Verma and Dubey 2003). Various groups of workers have reported increased activities of antioxidant enzymes like GPX, SOD, APX, MDHAR, DHAR, and GR as well as non-enzymic antioxidants in metal-treated plants and suggested involvement of antioxidant defense system in the adaptive response to metal ions (Shah et al. 2001; Maheshwari and Dubey 2009; Srivastava and Dubey 2011; Sharma and Dubey 2007; Verma and Dubey 2003; Cakmak and Horst 1991). However, results suggest that activation of antioxidant enzymes in response to oxidative stress induced by metals is not enough to confer tolerance to metal accumulation. Comparative study of the antioxidative response of two maize lines differing in Al tolerance suggested that better protection of the Al-tolerant maize roots from Al-induced oxidative damage results, at least partially, from the increased activity of their antioxidative system. After 24 h of Al exposure, a gradual increase in the membrane lipid peroxidation in the Al-stressed root of the susceptible maize line was accompanied by decreased activities of the antioxidant enzymes SOD and peroxidase (POD). In contrast, increased activities of the SOD and POD were found in Al-treated roots of the tolerant maize line, in which the level of membrane lipid peroxidation remained almost unchanged (Giannakoula et al. 2010). Comparative antioxidant profiling of tolerant (TPM-1) and sensitive (TM-4) variety of *Brassica juncea* L. performed after exposure to arsenate [As(V)] and arsenite [As(III)] showed in general, better response of antioxidant enzymes and the level of glutathione in TPM-1 than in TM-4 (Srivastava et al. 2010). These responses presumably allowed TPM-1 to tolerate higher as concentrations as compared with that of TM-4 (Srivastava et al. 2010).

UV-B Radiations

UV-B radiation on plants is now of major concern to plant biologists due to the threat to productivity in global agriculture (Blumthaler and Ambach 1990). Enhanced UV-B significantly inhibits net photosynthetic rate. It has been shown that UV-B treatment results in a decrease in the light-saturated rate of CO_2 assimilation, accompanied by decreases in carboxylation velocity, RuBisCO content and activity (Allen et al. 1997). He et al. (1993) observed marked decrease in the ratios of variable to maximum chlorophyll fluorescence yield and in the quantum yield of photosynthetic O_2 evolution in pea and rice leaves. Limited CO_2 assimilation due to UV-B leads to excessive production of ROS which, in turn, cause oxidative damage in plants (Han et al. 2009; Strid et al. 1994). Rao and coworkers (Strid et al. 1994) suggested that UV-B exposure generates activated O_2 species by increasing NADPH-oxidase activity. Plants must adapt to the deleterious effects of UV-B radiation because they are dependent on sunlight for photosynthesis and, therefore, cannot avoid exposure to UV-B radiation. Plants possess antioxidative enzymatic scavengers SOD, POD, CAT, APX and non-enzymatic antioxidants like GSH, ASA and carotenoids to keep the balance between the production and removal of ROS. In *P. sperata* seedlings although enhanced UV-B (30%) increased the efficiency of antioxidant defense system consisting of UV-B absorbing compounds, carotenoids, and antioxidant enzymes SOD, APX, CAT and GPX (Han et al. 2009), it induced overproduction of ROS and oxidative stress eventually. Peroxidase-related enzymes were found to be preferentially induced by UV-B exposure in *Arabidopsis* (Rao et al. 1996; Gao and Zhang 2008) observed that ASA-deficient mutant vtc1 was more sensitive to supplementary UV-B treatment than wild-type plants and, therefore, suggested that ASA could be considered as an important antioxidant for UV-B radiation.

Biotic Stress

NO and ROS signaling pathways in plant biotic interactions are closely connected. Both can modulate the expression of genes involved in plant stress responses, in primary metabolism or phytohormonal signaling (Grun et al. 2006; Ahlfors et al. 2009; Moreau et al. 2010). Furthermore, NO can react with $O_2{}^{*-}$ to form ONOO– it impacts protein activities through the tyrosine nitration, a post-translational modification (PTM) consisting of the formation of 3-nitrotyrosine residues. NO is also responsible for S-nitrosylation, which consists of the reversible modification of a cysteinyl residue via a nitrosothiol group formation, and metal nitrosylation, which concerns the interaction of an NO moiety with the transition metal of a target-metalloprotein (Besson Bard et al. 2008; Astier et al. 2012; Baudouin 2011; Gaupels et al. 2011). ROS can also promote PTM affecting cysteinyl residues through sulfenylation (Oger et al. 2012).

During plant defense reaction, NO can also react with $O_2{}^{*-}$ to form $ONOO^-$, a common compound of the ROS and NO signaling, which can damage lipids, proteins, DNA and mediate the PTM tyrosine nitration (Astier et al. 2012; Vandelle and Delledonne 2011). The regulation of the $ONOO^-$ concentration in cells is notably achieved through peroxiredoxins (Prx). Interestingly, the peroxidase and $ONOO^-$ reductase activity of the *Arabidopsis thaliana* PrxII E can be inhibited through S-nitrosylation after biotic stress (Romero Puertas et al. 2007) amplifying the deleterious effects of $ONOO^-$.

Another aspect of NO and ROS cross talk during plant defense reaction concerns the NO-dependent modulation of the ROS production. Indeed, the activity of AtRBOHD, a protein responsible for the ROS production observed after pathogen recognition (Suzuki et al. 2011) has been shown to be inhibited by *S*-nitrosylation (Yun et al. 2011). In contrast, NO production has been demonstrated to be required for the full induction of H_2O_2 accumulation following the treatment of *A. thaliana* leaf discs with oligogalacturonides (Rasul et al. 2012). These contradictory results reveal the complex and tight regulation of the NO and ROS signaling during plant defense responses, depending on the model.

Similar to their function in symbiotic interactions and pathogen infection, NO and ROS have also been implicated to play a role in herbivory. To date, their biosynthesis upon herbivore-induced wounding remains contradictory. Increased levels of NO and ROS were reported after herbivore attacks in potato and Lima bean (Bricchi et al. 2010; Lin et al. 2011). However, it was not the case in wounded leaves of tomato where the addition of the NO donor sodium nitroprusside (SNP) blocked H_2O_2 accumulation (Orozco Cardenas and Ryan 2002). Moreover, NO and ROS production differs depending on the wounding method used; in Lima bean herbivore attacks triggered a significantly higher accumulation of NO and H_2O_2 compared to mechanical wounding. This highlights the complexity of plant NO and ROS production during herbivory. The observed variations in NO and ROS production depending on the wounding method are also supported by studies on S-nitrosoglutathione reductase (GSNOR) during plant/herbivore interactions. GSNO the substrate of GSNOR, is one important natural NO reservoir and can be reduced to GSSG and NH_3 (Besson Bard et al. 2008). In *A. thaliana,* GSNOR has emerged as a crucial player in the resistance against pathogens; see Yu et al. (2012) for a detailed description. The important role of GSNOR and GSNO for a successful defense response against herbivores was recently described (Wunsche et al. 2011). Characterization of GSNOR silenced *Nicotiana attenuata* plants suggested that GSNOR activity might be linked to defense phytohormone synthesis, especially JA and ET. Moreover, GSNOR transcript levels are up-regulated after treatment with the oral secretion of *Manduca Sexta,* but interestingly, the corresponding enzymatic activity is suppressed as compared to a mechanical wounded plant. In contrast, mechanically wounded sunflower seedlings displayed a down-regulation of GSNOR transcripts and activity (Chaki et al. 2011). These apparently opposite results likely correlate with the different perception of the plant to an herbivore or a mechanical wounding, resulting in different NO and ROS production patterns. In a further study, GSNO was detected in the local and systemic tissue of mechanically wounded *A. thaliana* indicating that it might be involved in the development of structure–activity relationship (SAR). Furthermore, GSNOR antisense lines showed increased transcript levels of genes involved for JA biosynthesis suggesting that GSNOR functions as a key player during the defense response upon wounding (Espunya et al. 2012). In addition, sweet potato treated with SNP displayed an NO-dependent activation of the antioxidant enzymes Cu/Zn SOD and APX. This activation counteracts the H_2O_2 production triggered by wounding, protecting the cell from a ROS-induced cell death. This example demonstrates once again the interplay of NO and ROS upon wounding. During the last few years, micro-ribonuclease acids (RNAs) were identified as another important feature for post-transcriptional regulation of gene expression (Staiger et al. 2013). These small regulatory RNAs were identified in response to abiotic and biotic stresses (Khraiwesh et al. 2012). In sweet potato, miR828 was up-regulated after wounding and transgenic miR828 overexpressors presented significantly higher H_2O_2 concentrations (Lin et al. 2012a, b). Furthermore, overexpression lines showed increased transcript levels of the defense-related NO inducible phenylalanine ammonia lyase (PAL) gene. In *N. attenuata,* a large number of new herbivory-responsive miRNAs, including miR828, were identified (Bozorov et al. 2012), which could correlate with NO and ROS production. To sum up, it is now widely accepted that NO and ROS are produced during herbivore attacks, and it differs compared to a mechanical wounding event. The regulation of signaling molecules and related proteins like GSNOR and detoxification enzymes like APX and SOD, leading to a successful defense signaling, depends also on the context considered. Further analyses are needed to define the interplay between NO and ROS signaling pathways and their individual roles during plant/herbivore interactions. Some effort could also be put on the determination of the plant NO and ROS signaling control by the herbivore itself.

Nitric Oxide Signaling and Plant Metabolism

Nitric oxide (NO), a redox-active signaling molecule, is an important endogenous signaling molecule involved in many processes like regulating synthesis of the cell wall (Correa-Aragunde et al. 2008; Xiong et al. 2009), ROS metabolism in plants (Delledonne et al. 2001), gene expression and regulation (Bogdan et al. 2001), programmed cell death, maturation and senescence. NO exerts a crucial role in protecting plants against various abiotic stresses. Exogenous NO protects rice leaves from oxidative stress caused by paraquat toxicity by increasing the activities of antioxidant enzymes (Hung et al. 2002). Besides, NO could significantly enhance antioxidative capacity by increasing the activities of CAT, APX and accumulating proline, during wheat seed germination under osmotic stress (Zhang et al. 2003). More recently, NO was found to reduce Al toxicity by preventing oxidative stress in the roots of *Cassia tora* and exogenous NO protects wheat roots from Cd-induced toxicity (Singh et al. 2008).

Nitrate reductase (NR) being the best known pathway for NO production in plants (Rockel et al. 2002) capable of reducing nitrite to NO depending on nitrite accumulation and pH levels; moreover, NO can react reversibly with glutathione (GSH) producing *S*-nitrosoglutathione (GSNO), a reservoir of NO (Sakamoto et al. 2002). GSNO is metabolized by GSNO reductase (GSNOR) which controls NO and nitrosothiol levels, being a key enzyme in most NO-regulated processes, such as pathogen defense, root development, and nitrogen assimilation (Frungillo et al. 2014). It has recently been shown that GSNO inhibits nitrate uptake and its reduction to nitrite which would prevent NR-dependent NO production (Frungillo et al. 2014). Additionally, *S*-nitrosylation of NR under EC with high nitrate supply decreased the NR activity in a same way to that which occur with chilling treatment (Du et al. 2015). Du et al. (2015) showed that regulation of NR under elevated CO_2 depends on nitric oxide synthase-like (NOSl). Moreover, NR activity was shown to be increased by NO under low nitrate condition by interaction with heme and molybdenum centers of NR, which enhances electron transfer during nitrate reduction (Du et al. 2008). It was proposed that NO inhibits the *S*-nitrosoglutathione reductase (GSNOR1) as a result S-nitrosoglutathione content increases, this regulates the nitrate flux and its assimilation (Frungillo et al. 2014).

Conclusions

NO regulates many physiological processes including stress responses, nitrogen metabolism, cell wall biosynthesis, ROS metabolism, gene expression, PCD, etc. It also acts as an elicitor for inducing protection from abiotic and biotic stresses including heavy metal toxicity. The regulatory role of NO and its interaction with ROS signaling and antioxidant defense systems are also well established in model plants. The effective deployment of NO signaling as a stress protectant is explorable for improving crop production in the impending climate change scenario.

References

Ahlfors R, Brosche M, Kollist H, Kangasjarvi J (2009) Nitric oxide modulates ozone induced cell death hormone biosynthesis and gene expression in *Arabidopsis thaliana*. Plant J 58:1–12
Allan AC, Fluhr R (1997) Two distinct sources of elicited reactive oxygen species in tobacco epidermal cells. Plant Cell 9:1559–1572
Allen DJ, Mckee IF, Farage PK, Baker NR (1997) Analysis of limitations to CO_2 assimilation on exposure of leaves of two *Brassica napus* cultivars to UV-B. Plant Cell Environ 20:633–640
Asada K (1999) The water cycle in chloroplasts scavenging of active oxygens and dissipation of excess photons. Annu Rev Plant Biol 50:601–639
Astier J, Kulik A, Koen E, Besson Bard A, Bourque S, Jeandroz S, Lamotte O, Wendehenne D (2012) Protein *S*-nitrosylation what's going on in plants. Free Radic Biol Med 53:1101–1110
Balazadeh S, Siddiqui H, Allu AD, Matallana Ramirez LP, Caldana C, Mehrnia M (2010) A gene regulatory network controlled by the NAC transcription factor ANA C0 92/At NAC 2/ ORE1during salt promoted senescence. Plant J 62:250–264
Banfi B, Molnar G, Maturana A, Steger K, Hegedus B, Demaurex N, Krause KH (2001) A Ca^{2+} activated NADPH oxidase in testis, spleen, and lymph nodes. J Biol Chem 276:37594–37601
Barroso JB, Corpas FJ, Carreras A, Sandalio LM, Valderrama R, Palma JM, Lupianez JA, del Río LA (1999) Localizationof nitric oxide in plant peroxisomes. J Biol Chem 274:36729–36733
Baudouin E (2011) The language of nitric oxide signalling. Plant Biol (Stuttg) 13:233–242
Becker JD, Boavida LC, Carneiro J, Haury M, Feijo JA (2003) Transcriptional profiling of *Arabidopsis* tissues reveals the unique characteristics of the pollen transcriptome. Plant Physiol 133:713–725
Besson Bard A, Pugin A, Wendehenne D (2008) New insights into nitric oxide signaling in plants. Annu Rev Plant Biol 59:21–39
Bethke PC, Gubler F, Jacobsen JV, Jones RL (2004) Dormancy of Arabidopsis seeds and barley grains can be broken by nitric oxide. Planta 219:847–855
Biehler K, Fock H (1996) Evidence for the contribution of the Mehler POD reaction in dissipating excess electrons in drought stressed wheat. Plant Physiol 112:265–272
Blume YB, Krasylenko YA, Demchuk OM, Yemets AI (2013) Tubulin tyrosine nitration regulates microtubule organization in plant cells. Front Plant Sci 4:530–544
Blumthaler M, Ambach W (1990) Indication of increasing solar ultraviolet-B radiation flux in alpine regions. Science 248:206–208
Boo YC, Jung J (1999) Water deficit induced oxidative stress and antioxidative defenses in rice plants. J Plant Physiol 155:255–261
Bogdan C (2001) Nitric oxide and the regulation of gene expression. Trends Cell Biol 11:66–75
Bozorov TA, Baldwin IT, Kim SG (2012) Identification and profiling of miRNAs during herbivory reveals jasmonate dependent and independent patterns of accumulation in Nicotiana attenuata. BMC Plant Biol 12:209
Bricchi I, Leitner M, Foti M, Mithofer A, Boland W, Maffei ME (2010) Robotic mechanical wounding (MecWorm) versus herbivore induced responses early signaling and volatile emission in Lima bean (*Phaseolus lunatus* L.). Planta 232:719–729
Butt Y, Lum J, Lo S (2003) Proteomic identification of plant proteins probed by mammalian nitric oxide synthase antibodies. Planta 216(5):762–771
Cakmak I, Horst WJ (1991) Effect of aluminium on lipid peroxidation, superoxide dismuatse, catalase, and peroxidase activities in root tips of soybean (*Glycine max*). Physiol Plant 83:463–468
Cardenas L, McKenna ST, Kunkel JG, Hepler PK (2006) NAD(P)H oscillates in pollen tubes and is correlated with tip growth. Plant Physiol 142:1460–1468
Chaki M, Valderrama R, Fernandez Ocana AM, Carreras A, Gomez Rodriguez MV, Pedrajas JR, Begara Morales JC, Sanchez Calvo B, Luque F, Leterrier M et al (2011) Mechanical wounding induces a nitrosative stress by down-regulation of GSNO reductase and an increase in S-nitrosothiols in sunflower (*Helianthus annuus*) seedlings. J Exp Bot 62:1803–1813
Chandok MR, Ytterberg AJ, van Wijk KJ, Klessig D (2003) The pathogen inducible nitric oxide synthase (iNOS) in plants is a variant of the p protein of the glycine decarboxylase complex. Cell 113:469–482

Corpas FJ, Alché JDD, Barroso JB (2013) Current overview of S-nitrosoglutathione (GSNO) in higher plants. Front Plant Sci 4:126

Corpas FJ, Barroso JB, Carreras A, Valderrama R, Palma JM, Leon AM et al (2006) Constitutive argininedependent nitric oxide synthase activity in different organs of pea seedlings during plant development. Planta 54:224–246

Corpas FJ, Chaki M, Leterrier M, Barroso JB (2009) Protein tyrosine nitration: a new challenge in plants. Plant Sign Behav 4:920–923

Correa-Aragunde N, Lombardo C, Lamattina L (2008) Nitric oxide: an active nitrogen molecule that modulates cellulose synthesis in tomato roots. New Phytol 179:386–396

Cui MH, Ok SH, Yoo KS, Jung KW, Yoo SD, Shin JS (2013) An *Arabidopsis* cell growth defect factor related protein CRS promotes plants enescence by increasing the production of hydrogen peroxide. Plant Cell Physiol 54:155–167

Dafni A, Motte Maues M (1998) A rapid and simple procedure to determine stigma receptivity. Sexual Plant Rep 11:177–180

Dat J, Vandenabeele S, Vranova E, Van Montagu M, Inze D, Van Breusegem F (2000) Dual action of the active oxygen species during plant stress responses. Cell Mol Life Sci 57:779–795

Del Rio LA, Corpas FJ, Sandalio LM, Palma JM, Gomez M, Barroso JB (2002) Reactive oxygen species antioxidant systems and nitric oxide in peroxisomes. J Exp Bot 53:1255–1272

Delledonne M, Xia Y, Dixon RA, Lamb C (1998) Nitric oxide functions as a signaling plant disease resistance. Nature 394:585–588

Delledonne M, Zeier J, Marocco A, Lamb C (2001) Signal interactions between nitric oxide and reactive oxygen intermediates in the plant hypersensitive disease resistance response. Proc Natl Acad Sci 98:13454–13459

Di Stasi AM, Mallozzi C, Macchia G, Maura G, Petrucci TC, Minetti M (2002) Peroxy nitrite affects exocytosis and SNARE complex formation and induces tyrosine nitration of synaptic proteins. J Neuro Chem 82:420–429

Dresselhaus T (2006) Cell-cell communication during double fertilization. Curr Opin Plant Biol 9:41–47

Du S, Zhang Y, Lin X, Wang Y, Tang C (2008) Regulation of nitrate reductase by nitric oxide in Chinese cabbage pakchoi (*Brassica chinensis* L.). Plant Cell Environ 31(2):195–204

Du S, Zhang R, Zhang P, Liu H, Yan M, Chen N, Ke S (2015) Elevated CO2-induced production of nitric oxide (NO) by NO synthase differentially affects nitrate reductase activity in Arabidopsis plants under different nitrate supplies. J Exp Bot 67(3):893–904

Dupuis I, Dumas C (1990) Biochemical markers of female receptivity in maize (*Zea mays* L.) assessed using *in vitro* fertilization. Plant Sci 70:11–20

Duan YH, Zhang YL, Ye LT, Fan XR, Xu GH, Shen QR (2007) Responses of rice cultivars with different nitrogen use efficiency to partial nitrate nutrition. Ann Bot 99:1153–1160

Durner J, Wendehenne D, Klessig F (1998) Defense gene induction in tobacco by nitric oxide cyclic GMP and cyclic ADP ribose. Proc Natl Acad Sci U S A 95:10328–10333

Elias M, Cvckova F, Obermeyer G, Zarsky V (2001) Microinjection of guanine nucleotide analogue into lily pollen tubes results in isodiametric tip expansion. Plant Biol 3:489–493

Espunya MC, De Michele R, Gomez Cadenas A, Martinez MC (2012) S-Nitrosoglutathione is a component of wound and salicylic acid induced systemic responses in *Arabidopsis thaliana*. J Exp Bot 63:3219–3227

Feijo JA, Sainhas J, Holdaway Clarke T, Cordeiro S, Kunkel JG, Hepler PK (2001) Cellular oscillations and the regulation of growth the pollen tube paradigm. Bioessays 23:86–94

Feijó JA, Sainhas J, Holdaway-Clarke T, Cordeiro MS, Kunkel JG, Hepler PK (2001) Cellular oscillations and the regulation of growth: the pollen tube paradigm. Bioessays 23:86–94

Feijo JA, Costa SS, Prado AM, Becker JD, Certal AC (2004) Signalling by tips. Curr Opin Plant Biol 7:589–598

Feio JA, Sainhas J, Hackett GR, Kunkel JG, Hepler PK (1999) Growing pollen tubes possess a constitutive alkaline band in theclear zone and a growth dependent acidic tip. J Cell Biol 144:483–496

Foreman J, Demidchik V, Bothwell JHF, Mylona P, Miedema H, Torres MA, Linstead P, Costa S, Brownlee C, Jones JDG, Davies JM, Dolan L (2003) Reactive oxygen species produced by NADPH oxidase regulate plant cell growth. Nature 422:442–446

Foyer CH, Harbinson J (1994) Oxygen metabolism and the regulation of photosynthetic electron transport. In: Foyer CH, Mullineaux PM (eds) Causes of photo oxidative stresses and amelioration of defense system in plants. CRC, Boca Raton, FL, pp 1–42

Foyer CH, Noctor G (2000) Oxygen processing in photosynthesis regulation and signaling. New Phytol 146:359–388

Franklin-Tong VE (2002) The difficult question of sex: plants play the mating game. Curr Opin Plant Biol 5:14–18

Frungillo L, Skelly MJ, Loake GJ, Spoel SH, Salgado I (2014) S-nitrosothiols regulate nitric oxide production and storage in plants through the nitrogen assimilation pathway. Nat Commun 5:5401

Fryer MJ, Andrews JR, Oxborough K, Blowers DA, Baker NR (1998) Relationship between CO_2 assimilation, photosynthetic electron transport, and active O^2 metabolism in leaves of maize in the field during periods of low temperature. Plant Physiol 116:571–580

Galatro A, Puntarulo S, Guiamet JJ, Simontacchi M (2013) Chloroplast functionality has a positive effect on nitric oxide level in soybean cotyledons. Plant Physiol Biochem 66:26–33

Gallego SM, Benavides MP, Tomaro ML (1996) Effect of heavy metal ion excess on sunflower leaves evidence for involvement of oxidative stress. Plant Sci 121:151–159

Gallego S, Benavides M, Tomaro M (2002) Involvement of an antioxidant defence system in the adaptive response to heavy metal ions in *Helianthus annuus* L. cells. Plant Growth Regul 36:267–273

Gao Q, Zhang L (2008) Ultraviolet-B-induced oxidative stress and antioxidant defense system responses in ascorbatedeficient*vtc1* mutants of *Arabidopsis thaliana*. J Plant Physiol 165:138–148

Gaupels F, Kuruthukulangarakoola GT, Durner J (2011) Upstream and downstream signals of nitric oxide in pathogen defence. Curr Opin Plant Biol 14:707–714

Gechev TS, Van Breusegem F, Stone JM, Denev I, Laloi C (2006) Reactive oxygen species as signals that modulate plant stress responses and programmed cell death. Bioessays 28:1091–1101

Giannakoula A, Moustakas M, Syros T, Yupsanis T (2010) Aluminum stress induces up regulation of an efficient antioxidant system in the Al tolerant maize line but not in the Al sensitive line. Environ Exp Bot 67:487–494

Gibbs SM, Truman JM (1998) Nitric oxide and cyclic GMP regulate retinal patterning in the optic lobe of *Drosophila*. Neuron 20:83–93

Grob F, Durner J, Gaupels F (2013) Nitric oxide, antioxidants and prooxidants in plant defence responses. Front Plant Sci 4:419

Grun S, Lindermayr C, Sell S, Durner J (2006) Nitric oxide and gene regulation in plants. J Exp Bot 57:507–516

Guo FQ, Okamoto M, Crawford NM (2003) Identification of a plant nitric oxide synthase gene involved in hormonal signalling. Science 302:100–103

Gupta KJ, Fernie AR, Kaiser WM, van Dongen JT (2011) On the origins of nitric oxide. Trends Plant Sci 16:160–168

Halliwell B, Gutteridge JMC (1989) Free radicals in biology and medicine, 2nd edn. Oxford University Press, Oxford

Halliwell B, Gutteridge JMC (1999) Free radicals in biology and medicine. Oxford University Press, Oxford

Han C, Liu Q, Yang Y (2009) Short-term effects of experimental warming and enhanced ultraviolet-B radiation on photosynthesis and antioxidant defense of *Piceaa sperata* seedlings. Plant Growth Regul 58:153–162

Hancock JT, Smirnoff N, Foyer CH (2006) Oxygen metabolism coming up ROSes: a holistic view of the redox metabolism of plant cells. J Exp Bot 57:iv

He J, Huang LK, Chow WS, White cross ML, Anderson JM (1993) Effects of supplementary ultraviolet-B radiation on rice and pea plants. Aust J Plant Physiol 20:129–142

He Y, Tang RH, Hao Y, Stevens RD, Cook CW, Ahn SM et al (2004) Nitric oxide represses the *Arabidopsis* floral transition. Science 305:1968–1971

HediyeSekmen A, Turkan I, Takio S (2007) Differential responses of anti oxidative enzymes and lipid peroxidation to salt stress in salt-tolerant Plantagomaritima and salt sensitive Plantago media. Physiol Plant 131:399–411

Henzler T, Steudle E (2000) Transport and metabolic degradation of hydrogen peroxide in *Chara corallina*: model calculations and measurements with the pressure probe suggest transport of H_2O_2 across water channels. J Exp Bot 51:2053–2066
Hepler PK, Vidali L, Cheung AY (2001) Polarized cell growth in higher plants. Annu Rev Cell Dev Biol 17:159–187
Hern andez JA, Jim enez A, Mullineaux P, Sevilla F (2000) Tolerance of pea (*Pisum sativum* L.) to long-term salt stress is associated with induction of antioxidant defences. Plant Cell Environ 23:853–862
Herrero M (2003) Male and female synchrony and the regulation of mating in flowering plants. Philos Trans R Soc Lond Ser B 358:1019–1024
Heslop-Harrison J (1978) Recognition and response in the pollen–stigma interaction. In: Curtis A (ed) Cell-cell recognition, Society for Experimental Biology Symposium Series, vol 32. Society for Experimental Biology, Cambridge, pp 121–138
Hideg E, Kalai T, Kos PB, Asada K, Hideg K (2006) Singlet oxygen in plants its significance and possible detection with double (fluorescent and spin) indicator reagents. Photo Chem Photo Biol 82:1211–1218
Hiscock SJ (2000) Genetic control of self-incompatibility in *Senecio squalidus* L. (Asteraceae): a successful colonizing species. Heredity 85:10–19
Hiscock SJ, McInnis SM (2003) The diversity of self-incompatibility systems in flowering plants. Plant Biol 5:23–32
Hong JK, Yun BW, Kang JG, Raja MU, Kwon E, Sorhagen K et al (2008) Nitric oxide function and signaling in plant disease resistance. J Exp Bot 59:147–154
Hu J, Aguirre M, Peto C, Alonso J, Ecker J, Chory J (2002) A role for peroxisomes in photo morphogenesis and development of *Arabidopsis*. Science 297:405–410
Hu WH, Song XS, Shi K, Xia XJ, Zhou YH, Yu JQ (2008) Changes in electron transport, superoxide dismutase and ascorbate peroxidase isoenzymes in chloroplasts and mitochondria of cucumber leaves as influenced by chilling. Photosynthetica 46:581–588
Hung KT, Chang CJ, Kao CH (2002) Paraquat toxicity is reduced by nitric oxide in rice leaves. J Plant Physiol 159:159–166
Huang M, Guo Z (2006) Responses of antioxidative system to chilling stress in two rice cultivars differing in sensitivity. Plant Physiol Biochem 44:828–836
Hulskamp M, Schneitz K, Pruitt RE (1995) Genetic evidence for a long range activity that directs pollen tube guidance in *Arabidopsis*. Plant Cell 7:57–64
Ignarro J (2000) In: Ignarro JL (ed) Nitric Oxide Biology and Patho biology. Academic Press, San Diego, CA, pp 3–380
Jahnke LS, Hull MR, Long SP (1991) Chilling stress and oxygen metabolizing enzymes in *Zea mays and Zeadiplo perennis*. Plant Cell Environ 14:97–104
Jeandroz S, Wipf D, Stuehr DJ, Lamattina L, Melkonian M, Tian Z, Zhu Y, Carpenter EJ, Wong GKS, Wendehenne D (2016) Occurrence, structure, and evolution of nitric oxide synthase–like proteins in the plant kingdom. Sci Signal 9(417):re2–re2
Jiang H, Chen Y, Li M, Xu X, Wu G (2011) Overex pression of SGR results in oxidative stress and lesion-mimic cell death in rice seedlings. J Integr Plant Biol 53:375–387
Jing HC, Hebeler R, Oeljeklaus S, Sitek B, Stuhler K, Meyer HE et al (2008) Early leaf senescence is associated with an altered cellular redox balance in *Arabidopsis cpr5/old1* mutants. Plant Biol (Stuttg) 10(Suppl 1):85–98
Johnson MA, Preuss D (2002a) Plotting a course multiple signals guide pollen tubes to their targets. Dev Cell 2:273–281
Johnson MA, Preuss D (2002b) Plotting a course: multiple signals guide pollen tubes to their targets. Dev Cell 2:273–281
KarrayBouraoui N, Harbaoui F, Rabhi M et al (2011) Different antioxidant responses to salt stress in two different provenances of *Carthamus tinctorius* L. Acta Physiol Plant 33:1435–1444
Keller T, Damude HG, Werner D, Doerner P, Dixon RA, Lamb C (1998) A plant homolog of the neutrophil NADPH oxidase gp91phox subunit gene encodes a plasma membrane protein with Ca2+ binding motifs. Plant Cell 10:255–266

Kerr EM, Fry SC (2004) Extracellular cross-linking of xylan and xyloglucan in maize cell-suspension cultures: the role of oxidative phenolic coupling. Planta 219:73–83
Khraiwesh B, Zhu JK, Zhu J (2012) Role of miRNAs and siRNAs in biotic and abiotic stress responses of plants. Biochim Biophys Acta 1819:137–148
Kjellbom P, Snogerup L, Stohr C, Reuzeau C, Mccabe PF, Pennell RI (1997) Oxidative cross-linking of plasma membrane arabinogalactan proteins. Plant J 12:1189–1196
Kong X, Zhang D, Pan J, Zhou Y, Li D (2013) Hydrogen peroxide is involved in nitric oxide induced cell death in maize leaves. Plant Biol (Stuttg) 15:53–59
Kovtun Y, Chiu WL, Tena G, Sheen J (2000) Functional analysis of oxidative stress-activated mitogen-activated protein kinase cascade in plants. Proc Natl Acad Sci U S A 97:2940–2945
Kwak JM, Mori IC, Pei ZM, Leonhardt N, Torres MA, Dangl JL, Bloom RE, Bodde S, Jones JDG, Schroeder JI (2003) NADPH oxidase AtrbohD and AtrbohF genes function in ROS-dependent ABA signaling in *Arabidopsis*. EMBO J 22:2623–2633
Lamattina L, Garcia Mata C, Graziano M, Pagnussat G (2003) Nitric oxide: the versatility of an extensive signal molecule. Annu Rev Plant Biol 54:109–136
Leitner M, Vandelle E, Gaupels F, Bellin D, Delledonne M (2009) NO signal sinthehaze: nitric oxide signaling in plant defence. Curr Opin Plant Biol 12:451–458
Lim PO, Kim HJ, Nam HG (2007) Leaf senescence. Annu Rev Plant Biol 58:115–136
Lin CC, Jih PJ, Lin HH, Lin JS, Chang LL, Shen YH, Jeng ST (2011) Nitric oxide activates superoxide dismutase and ascorbate peroxidase to repress the cell death induced by wounding. Plant Mol Biol 77:235–249
Lin JS, Lin CC, Lin HH, Chen YC, Jeng ST (2012a) MicroR828 regulates lignin and H_2O_2 accumulation in sweet potato on wounding. New Phytol 196:427–440
Lin A, Wang Y, Tang J, Xue P, Li C, Liu L et al (2012b) Nitric oxide and protein *S*-nitrosylation are integral to hydrogen peroxide induced leaf cell death in rice. Plant Physiol 158:451–464
Liu WZ, Kong DD, Gu XX, Gao HB, Wang JZ, Xia M et al (2013) Cytokinin scan act as suppressors of nitricoxide in *Arabidopsis*. Proc Natl Acad Sci U S A 110:1548–1553
Logan BA, Kornyeyev D, Hardison J, Holaday AS (2006) The role of antioxidant enzymes in photo protection. Photosynth Res 88:119–132
Ludidi N, Gehring C (2003) Identification of a novel protein with guanylyl cyclase activity in *Arabidopsis thaliana*. *J Biol Chem* 278:6490–6494
Lush WM, Grieser F, Wolters Arts M (1998) Directional guidance of *Nicotiana alata*pollen tubes *in vitro* and on the stigma. Plant Physiol 118:733–741
Maheshwari R, Dubey RS (2009) Nickel-induced oxidative stress and the role of antioxidant defence in rice seedlings. Plant Growth Regul 59:37–49
María JJQ, Rosario C, Elena LC, Jose AT, Antonio JC, Gonzalo MC, Juan DDA (2017) Generation of nitric oxide by olive (*Olea europaea* L.) pollen during in vitro germination and assessment of the S-nitroso- and nitro-proteomes by computational predictive methods. Nitric Oxide 68:23–37
Masclaux Daubresse C, Purdy S, Lemaitre T, Pourtau N, Taconnat L, Renou JP et al (2007) Genetic variation suggests interaction between cold acclimation and metabolic regulation of leaf senescence. Plant Physiol 143:434–446
McInnis SM, Costa LM, Gutiérrez-Marcos JF, Henderson CA, Hiscock SJ (2005) Isolation and characterization of a polymorphic stigma-specific class III peroxidase gene from *Senecio squalidus* L. (Asteraceae). Plant Mol Biol 57:659–677
McInnis SM, Emery DC, Porter R, Desikan R, Hancock JT, Hiscock SJ (2006) The role of stigma peroxidases in flowering plants: insights from further characterization of a stigma-specific peroxidase (SSP) from *Senecio squalidus* (Asteraceae). J Exp Bot 57:1835–1846
Messerli MA, Creton R, Jaffe LF, Robinson KR (2000) Periodic increases in elongation rate precede increases in cytosolic Ca_{2+} during pollen tube growth. Dev Biol 222:84–98
Miller JD, Arteca RN, Pell EJ (1999) Senescence associated gene expression during ozone induced leaf senescence in *Arabidopsis*. Plant Physiol 120:1015–1024
Mishina TE, Lamb C, Zeier J (2007) Expression of a nitric oxide degrading enzyme induces a senescence programme in *Arabidopsis*. Plant Cell Environ 30:39–52

Mishra S, Dubey RS (2005) Heavy metal toxicity induced alterations in photosynthetic metabolism in plants. In: Pessarakli MED (ed) Handbook of photosynthesis, pp 845–863

Mishra P, Kumari B, Dubey RS (2013) Differential responses of antioxidative defense system to prolonged salinity stress in salt tolerant and salt sensitive Indica rice (*Oryza sativa* L.) seedlings. Protoplasma 250(1):3–19

Mittal R, Dubey RS (1991) Behaviour of peroxidases in rice: changes in enzymatic activity and isoforms in relation to salt tolerance. Plant Physiol Biochem 29:31–40

Monshausen GB, Bibikova TN, Messerli MA, Shi C, Gilroy S (2007) Oscillations in extracellular pH and reactive oxygen species modulate tip growth of Arabidopsis root hairs. Proc Natl Acad Sci USA 104:20996–21001

Moreau M, Lindermayr C, Durner J, Klessig DF (2010) NO synthesis and signaling in plants where do we stand. Physiol Plant 138:372–383

Moussa R, Abdel Aziz SM (2008) Comparative response of drought tolerant and drought sensitive maize genotypes to water stress. Aust J Crop Sci 1:31–36

Moustakas M, Lanaras T, Symeonidis L, Karataglis S (1994) Growth and some photosynthetic characteristics of field grown *Avena sativa* under copper and lead stress. Photosynthetica 30:389–396

Moutinho A, Hussey PJ, Trewavas AT, Malho R (2001) cAMP acts as a second messenger in pollen tube growth and reorientation. Proc Natl Acad Sci U S A 98:10481–10486

Munns R (2005) Genes and salt tolerance: bringing them together. New Phytol 167:645–663

Murgia I, Vazzola V, Tarantino D, Cellier F, Ravet K, Briat JF et al (2007) Knock out off erritin At Fer1 causes earlier onset of age dependent leaf senescence in *Arabidopsis*. Plant Physiol Biochem 45:898–907

Navabpour S, Morris K, Allen R, Harrison EAH, Mackerness S, Buchanan Wollaston V (2003) Expression of senescence enhanced genes in response to oxidative stress. J Exp Bot 54:2285–2292

Neill S, Desikan R, Han- cock J (2002) Hydrogen peroxide signalling. Curr Opin Plant Biol 5:388–395

Nishiyama M, Hoshino A, Tsai L, Henley JR, Goshima Y, Tessier Lavigne M, Poo MM, Hong K (2003) Cyclic AMP/GMP dependent modulation of Ca^{2+} channels sets the polarity of nerve growth cone inducing turning. Nature 423:990–995

Niu YH, Guo FQ (2012) Nitric oxide regulates dark induced leaf senescence through EIN2 in *Arabidopsis*. J Integr Plant Biol 54:516–525

Noctor G, Veljovic-Jovanovic S, Foyer CH (2000) Peroxide process in gin photosynthesis: antioxidant coupling and redox signalling. Philos Trans R Soc Lond B Biol Sci 355:1465–1475

Noctor G, Veljovic Jovanovic S, Driscoll S, Novitskaya L, Foyer CH (2002) Drought and oxidative load in the leaves of C3 plants a predominant role for photorespiration. Ann Bot 89:841–850

Noritake T, Kawakita K, Doke N (1996) Nitric oxide induces phytoalexin accumulation in potato tuber tissues. Plant Cell Physiol 37:113–116

Oger E, Marino D, Guigonis JM, Pauly N, Puppo A (2012) Sulfenylated proteins in the Medicago truncatula Sinorhizobium meliloti symbiosis. J Proteomics 75:4102–4113

Orozco Cardenas ML, Ryan CA (2002) Nitric oxide negatively modulates wound signaling in tomato plants. Plant Physiol 130:487–493

Orozco-Cardenas ML, Narvaez Vasquez J, Ryan CA (2001) Hydrogen peroxide acts as a second messenger for the induction of defense genes in tomato plants in response to wounding, system in, and methyl jasmonate. Plant Cell 13:179–191

Palanivelu R, Preuss D (2000) Pollen tube targeting and axon guidance parallels in tip growth mechanisms. Trends Cell Biol 10:517–524

Palanivelu R, Brass L, Edlund AF, Preuss D (2003) Pollen tube growth and guidance is regulated by POP2, an *Arabidopsis* gene that controls GABA levels. Cell 114:47–59

Penson SP, Schuurink RC, Fath A, Gubler F, Jacobsen JV, Jones RL (1996) cGMP is required for gibberellic acid-induced gene expression in barley aleurone. Plant Cell 8:2325–2333

Perez Lopez U, Robredo A, Lacuesta M et al (2009) The oxidative stress caused by salinity in two barley cultivars is mitigated by elevated CO_2. Physiol Plant 135:29–42

Pourtau N, Mares M, Purdy S, Quentin N, Ruel A, Wingler A (2004) Interactions of abscisic acid and sugar signaling in the regulation of leaf senescence. Planta 219:765–772
Prado AM, Porterfield DM, Feijo JA (2004) Nitric oxide is involved in growth regulation and reorientation of pollen tubes. Development 131:2707–2714
Prasad TK (1997) Role of catalase in inducing chilling tolerance in pre-emergent maize seedlings. Plant Physiol 114:1369–1376
Radyuk MS, Domanskaya IN, Shcherbakov RA, Shalygo NV (2009) Effect of low above zero temperature on the content of low molecular antioxidants and activities of antioxidant enzymes in green barley leaves. Russ J Plant Physiol 56:175–180
Rao MV, Davis KR (2001) The physiology of ozone induced cell death. Planta 213:682–690
Rao MV, Paliyath G, Ormrod DP (1996) Ultraviolet-B and ozone induced biochemical changes in antioxidant enzymes of *Arabidopsis thaliana*. Plant Physiol 110:125–136
Rasul S, DubreuilMaurizi C, Lamotte O, Koen E, Poinssot B, Alcaraz G, Wendehenne D, Jeandroz S (2012) Nitric oxide production mediates oligogalacturonide triggered immunity and resistance to Botrytis cinerea in sexta and is important for jasmonate-elicited responses in Nicotiana attenuata. J Exp Bot 62:4605–4616
Reeves EP, Lu H, Jacobs HL, Messina CGM, Bolsover S, Gabella G, Potma EO, Warley A, Roes J, Segal AW (2002) Killing activity of neutrophils is mediated through activation of proteases by K+ flux. Nature 416:291–297
Robek Sokolnik AZ, Asard H, Gorska Koplinska K, Gorecki RJ (2009) Cadmium and zinc-mediated oxidative burst in tobacco BY-2 cell suspension cultures. Acta Physiol Plant 31:43–49
Rockel P, Strube F, Rockel A, Wildt J, Kaiser WM (2002) Regulation of nitric oxide (NO) production by plant nitrate reductase in vivo and in vitro. J Exp Bot 53:103–110
Romero Puertas MC, Laxa M, Matte A, Zaninotto F, Finkemeier I, Jones AM, Perazzolli M, Vandelle E, Dietz KJ, Delledonne M (2007) S-nitrosylation of peroxiredoxin II E promotes peroxynitrite mediated tyrosine nitration. Plant Cell 19:4120–4130
Sagi M, Fluhr R (2001) Superoxide production by plant homologues of the gp91phox NADPH oxidase. Modulation of activity by calcium and by tobacco mosaic virus infection. Plant Physiol 126:1281–1290
Sagi M, Davydov O, Orazova S, Yesbergenova Z, Ophir R, Stratmann JW, Fluhr R (2004) Plant respiratory burst oxidase homologs impinge on wound responsiveness and development in *Lycopersicon esculentum*. Plant Cell 16:616–628
Sairam RK, Deshmukh PS, Saxena DC (1998) Role of antioxidant systems in wheat genotypes tolerance to water stress. Biol Plant 41:387–394
Sakamoto A, Ueda M, Morikawa H (2002) Arabidopsis glutathione- dependent formaldehyde dehydrogenase is an S-nitrosoglutathione reductase. FEBS Lett 515:20–24
Salt DE, Blaylock M, Kumar NPBA et al (1995) Phytoremediation: a novel strategy for the removal of toxic metals from the environment using plants. Biotechnology 13:468–474
Salmi ML, Clark G, Roux SJ Jr (2013) Current status and proposed roles for nitric oxide as a key mediator of the effects of extracellular nucleotides on plant growth. Front Plant Sci 4:427
Sandalio LM, Rodrıguez Serrano MLA, Rlo d, Romero Puertas MC (2009) Reactive oxygen species and signaling in cadmium toxicity. In: Rio LA, Puppo A (eds) Reactive oxygen species in plant signaling. Springer, Berlin, pp 175–189
Santolini J, Andrè F, Jeandroz S, Wendehenne D (2017) Nitric oxide synthase in plants: where do we stand? Nitric Oxide 63:30–38
Sayfzadeh S, Rashidi M (2011) Response of antioxidant enzymes activities of sugar beet to drought stress. ARPN J Agri Biol Sci 6:27–33
Schlicht M, Kombrink E (2013) The role of nitric oxide in the interaction of Arabidopsis thaliana with the biotrophic fungi, Golovinomyces orontii and Erysiphe pisi. Front Plant Sci 4:351
Sgherri C, Stevanovic B, Navari Izzo F (2011) Role of phenolic acids during dehydration and rehydration of *Ramondaserbica*. Physiol Plant 122:478–485
Shah K, Kumar RG, Verma S, Dubey RS (2001) Effect of cadmium on lipid peroxidation, superoxide anion generation and activities of antioxidant enzymes in growing rice seedlings. Plant Sci 161:1135–1144

Shalata A, Mittova V, Volokita M, Guy M, Tal M (2001) Response of the cultivated tomato and its wild salt tolerant relative *Lycopersiconpennellii*to salt dependent oxidative stress: the root antioxidative system. Physiol Plant 112:487–494

Sharma SS, Dietz KJ (2009) The relationship between metal toxicity and cellular redox imbalance. Trends Plant Sci 14:43–50

Sharma P, Dubey RS (2005) Drought induces oxidative stress and enhances the activities of antioxidant enzymes in growing rice seedlings. Plant Growth Regul 46:209–221

Sharma P, Dubey RS (2007) Involvement of oxidative stress and role of antioxidative defense system in growing rice seedlings exposed to toxic concentrations of aluminum. Plant Cell Rep 26:2027–2038

Shimizu KK, Okada K (2000) Attractive and repulsive interactions between female and male gametophyte in *Arabidopsis* pollen tube guidance. Development 127:4511–4518

Singh HP, Bati DR, Kaur G, Arora K, Kohli RK (2008) Nitric oxide (as sodium nitroprusside) supplementation ameliorates cd toxicity in hydroponically grown wheat roots. Environ Exp Bot 63:158–167

Srivastava S, Dubey RS (2011) Manganese-excess induces oxidative stress, lowers the pool of antioxidants and elevates activities of key antioxidative enzymes in rice seedlings. Plant Growth Regul 64:1–16

Srivastava S, Srivastava AK, Suprasanna P, Dsouza SF (2010) Comparative antioxidant profiling of tolerant and sensitive varieties of *Brassica juncea* L. to arsenate and arsenite exposure. Bull Environ Contam Toxicol 84:342–346

Staiger D, Korneli C, Lummer M, Navarro L (2013) Emerging role for RNA based regulation in plant immunity. New Phytol 197:394–404

Stone LM, Seaton KA, Kuo J, Mccomb JA (2004) Fast pollen tube growth in *Conospermum* species. Ann Bot 93:369–378

Strid A, Chow WS, Anderson JM (1994) UV-B damage and protection at the molecular level in plants. Photosynth Res 39:475–489

Suzuki N, Miller G, Morales J, Shulaev V, Torres MA, Mittler R (2011) Respiratory burst oxidases the engines of ROS signaling. Curr Opin Plant Biol 14:691–699

Tadege M, Kuhlemeier C (1997) Aerobic fermentation during tobacco pollen development. Plant Mol Biol 35:343–354

Tanou G, Molassiotis A, Diamantidis G (2009) Induction of reactive oxygen species and necrotic death-like destruction in strawberry leaves by salinity. Environ Exp Bot 65:270–281

Tessier Lavigne M, Goodman CS (1996) The molecular biology of axon guidance. Science 274:1123–1160

Thomas DD, Liu X, Kantrow SP, Lancaster JR (2001) The biological life time of nitric oxide: implications for the perivascular dynamics of NO and O_2. Proc Natl Acad Sci U S A 98:355–360

Tischner R, Galli M, Heimer YM, Bielefeld S, Okamoto M, Mack A, Crawford NM (2007) Interference with the citrulline-based nitric oxide synthase assay by argininosuccinate lyase activity in Arabidopsis extracts. FEBS J 274(16):4238–4245

Torres MA, Onouchi H, Hamada S, Machida C, Hammond-Kosack KE, Jones JDG (1998) Six *Arabidopsis thaliana* homologues of the human respiratory burst oxidase (gp91phox). Plant J 14:365–370

Torres MA, Dangl JL, Jones JDG (2002) *Arabidopsis* gp91phox homologues AtrbohD and AtrbohF are required for accumulation of reactive oxygen intermediates in the plant defense response. Proc Natl Acad Sci USA 99:517–522

Vaidyanathan H, Sivakumar P, Chakrabarty R, Thomas G (2003) Scavenging of reactive oxygen species in NaCl stressed rice (*Oryza sativa* L.) differential response in salt tolerant and sensitive varieties. Plant Sci 165:1411–1418

Valderrama R, Corpas FJ, Carreras A et al (2006) The dehydrogenase mediated recycling of NADPH is a key antioxidant system against salt induced oxidative stress in olive plants. Plant Cell Environ 29:1449–1459

Vandelle E, Delledonne M (2011) Peroxynitrite formation and function in plants. Plant Sci 181:534–539

Vanlerberghe GC (2013) Alternative oxidase: a mitochondrial respiratory pathway to maintain metabolicand signaling homeostasis during abiotic and biotic stress in plants. Int J Mol Sci 14:6805–6847

Verma S, Dubey RS (2003) Lead toxicity induces lipid peroxidation and alters the activities of antioxidant enzymes in growing rice plants. Plant Sci 164:645–655

Vinit DF, Epron D, Alaoui SB, Badot PM (2002) Effects of copper on growth and on photosynthesis of mature and expanding leaves in cucumber plants. Plant Sci 163:53–58

Wang X, Yang P, Gao Q et al (2008) Proteomic analysis of the response to high-salinity stress in *Physcomitrella patens*. Planta 228:167–177

Wang S, Liang D, Li C, Hao Y, Ma F, Shu H (2012) Influence of drought stress on the cellular ultra structure and antioxidant system in leaves of drought-tolerant and drought-sensitive apple rootstocks. Plant Physiol Biochem 51:81–89

Weckx JEJ, Clijsters HMM (1996) Oxidative damage and defense mechanisms in primary leaves of *Phaseolus vulgaris* as a result of root assimilation of toxic amounts of copper. Physiol Plant 96:506–512

Wildt J, Kley D, Rockel A, Rockel P, Segschneider HJ (1997) Emission of NO from several higher plant species. J Geophys Res Atmos 102(D5):5919–5927

Wilkins KA, Bancroft J, Bosch M, Ings J, Smirnoff N, Franklin-Tong VE (2011) ROS and NO mediate actinre organization and programmed cell death in the self-incompatibility response of *Papaver*. Plant Physiol 156:404–416

Wu XY, Kuai BK, Jia JZ, Jing HC (2012) Regulation of leaf senescence and crop genetic improvement. J Integr Plant Biol 54:936–952

Wunsche H, Baldwin IT, Wu J (2011) S-Nitrosoglutathione reductase (GSNOR) mediates the biosynthesis of jasmonic acid and ethylene induced by feeding of the insect herbivore *Manduca sexta* and is important for jasmonate-elicited responses in *Nicotiana attenuata*. J Exp Bot 62(13):4605–4616

Xiong J, Lu H, Lu K, Duan Y, An L, Zhu C (2009) Cadmium decreases crown root number by decreasing endogenous nitric oxide, which is indispensable for crown root primordia initiation in rice seedlings. Planta 230:599–610

Yaacov YL, Wills RB, Ku VVV (1998) Evidence for the function of the free radical gas nitric oxide (NO•) as an endogenous maturation and senescence regulating factor in higher plants. Plant Physiol Biochem 36:825–833

Yadav S, David A, Baluska F, Bhatla SC (2013) Rapid auxin- induced nitric oxide accumulation and subsequent tyrosine nitration of proteins during adventitious root formation in sunflower hypocotyls. Plant Signal Behav 8(3):e23196

Yamamoto Y, Hachiya A, Matsumoto H (1997) Oxidative damage to membranes by a combination of aluminum and iron in suspension-cultured tobacco cells. Plant Cell Physiol 38:1333–1339

Yu M, Yun BW, Spoel SH, Loake GJ (2012) A sleigh ride through the SNO: regulation of plant immune function by protein Snitrosylation. Curr Opin Plant Biol 15:424–430

Yun BW, Feechan A, Yin M, Saidi NB, Le Bihan T, Yu M, Moore JW, Kang JG, Kwon E, Spoel SH et al (2011) S-nitrosylation of NADPH oxidase regulates cell death in plant immunity. Nature 478:264–268

Zhang H, Shen WB, Xu LL (2003) Effects of nitric oxide on the germination of wheat seeds and its reactive oxygen species metabolisms under osmotic stress. Acta Bot Sin 45:901–905

Zhang Y, Luo Y, Hou YX, Jiang H, Chen Q, Tang RH (2008a) Chilling acclimation induced changes in the distribution of H_2O_2 and antioxidant system of strawberry leaves. Agric J 3:286–291

Zhang Y, Tang HR, Luo Y (2008b) Variation in antioxidant enzyme activities of two strawberry cultivars with shortterm low temperature stress. World J Agric Sci 4:458–462

Zhou YH, Yu JQ, Mao WH, Huang LF, Song XS, Nogues S (2006) Genotypic variation of Rubisco expression, photosynthetic electron flow and antioxidant metabolism in the chloroplasts of chill-exposed cucumber plants. Plant Cell Physiol 47:192–199

Zonia L, Cordeiro S, Tupy J, Feijo JA (2002) Oscillatory chloride efflux at the pollen tube apical tip has a role in growth and osmoregulation and is linked to inositol polyphosphate signaling pathways. Plant Cell 14:2233–2249

Ameliorative Capability of Plant Growth Promoting Rhizobacteria (PGPR) and Arbuscular Mycorrhizal Fungi (AMF) Against Salt Stress in Plant

Tahira Yasmeen, Mohsin Tariq, Shahid Iqbal, Muhammad Saleem Arif, Muhammad Riaz, Sher Muhammad Shahzad, Shafaqat Ali, Muhammad Noman, and Tang Li

Introduction

A saline soil is generally defined as the soil in which the electrical conductivity (EC) of the saturation extract exceeds 4 dS m^{-1} (approximately 40 mM NaCl) in the root zone at 25 °C, and has exchangeable sodium of 15% with pH less than 8.5 (Richards 1954; Munns 2005; Jamil et al. 2011). Soil salinity usually implies the presence of salts including borates (BO_3^{3-}), bicarbonates (HCO_3^-), carbonates (CO_3^{2-}), chlorides

T. Yasmeen (✉)
Department of Environmental Sciences and Engineering, Government College University, Faisalabad, Pakistan

Key Laboratory of Economic Plants and Biotechnology, Kunming Institute of Botany, Chinese Academy of Sciences, Kunming, Yunnan, China
e-mail: tahirayasmeen@gcuf.edu.pk

M. Tariq
Department of Bioinformatics and Biotechnology, Government College University, Faisalabad, Pakistan

Department of Integrative Biology, University of California, Berkeley, CA, USA

S. Iqbal · T. Li
College of Plant Protection, Yunnan Agricultural University, Kunming, China

M. S. Arif · M. Riaz · S. Ali
Department of Environmental Sciences and Engineering, Government College University, Faisalabad, Pakistan

S. M. Shahzad
Department of Soil and Environmental Sciences, University College of Agriculture, University of Sargodha, Sargodha, Pakistan

M. Noman
Department of Bioinformatics and Biotechnology, Government College University, Faisalabad, Pakistan

M. Hasanuzzaman et al. (eds.), *Plant Abiotic Stress Tolerance*,
https://doi.org/10.1007/978-3-030-06118-0_17

(Cl^-), nitrates (NO_3^-) and sulfates (SO_4^{2-}) of sodium (Na), potassium (K), calcium (Ca), magnesium (Mg) and iron (Fe) (Rengasamy 2006). Soil salinization is the process by which a non-saline soil becomes saline due to salt accumulation in terrestrial landscapes. Soil salinization is often a problem in arid regions where ions of soluble salts build up in the soil during the processes of relatively high evaporation (causing seasonal water deficit) and transpiration from leaves of plants. It also implies to the soils where leaching is insufficient to move salts out of the soil profile (Duchaufour 1982; Schofield and Kirkby 2003); ultimately, salts accumulate in the soil to the point that hinders agricultural production (seed germination and plant growth). Consequently, soils become unusable generating environmental health and economic issues. In the beginning, soil organisms are affected due to alteration in metabolic processes resulting in reduced soil fertility which leads to the decline in soil productivity and land transformation into desertified barren land in advanced stages (Tóth et al. 2008; Jones et al. 2012). Globally, about 7% of the total earth's land surface (more than 800 million ha) and 70% of all agricultural soils are categorized as saline soils, which are expected to be responsible of up to 50% arable land loss by the year 2050. Furthermore, annually 10% increase in saline area has also been projected due to various natural and anthropogenic activities (Jamil et al. 2011).

Soil salinization can be categorized into: (1) transient salinization, (2) primary salinization, and (3) secondary salinization. In transient salinization, salt contents change down a soil profile due to different factors including seasonal moisture fluctuations, evapotranspiration and infiltration rate resulting in momentary salinization in the subsoil (Rengasamy 2010). Primary or natural salinization occurs where soil is rich in soluble salts or there is a shallow saline groundwater table and inadequate rain to leach soluble salts from the soil. However, in case of secondary salinization, salts accumulate near the soil surface as a result of rising water tables. This rise in water table may be attributed by implementation of different land management practices such as irrigation or tree clearing (Cisneros et al. 1999; Rengasamy 2006). Inadequate irrigation management leads to secondary salinization that affects 20% of irrigated land worldwide (Glick et al. 2007a, b).

Soil salinization is a subject of increasing concerns in many parts of the world, particularly in arid and semiarid areas (Rengasamy 2006; Parvaiz and Satyawati 2008). Among various environmental stresses affecting agricultural production, soil salinity is one of the most devastating stresses, which causes major reductions in cultivated land area, crop productivity and quality (Yamaguchi and Blumwald 2005; Shahbaz and Ashraf 2013). No accurate recent statistics are available on global extent of salt-affected soils. However, the best available estimates suggest that about 412 million ha area of land is affected by salinity (UNEP 1992). Human-induced salinity, mainly caused by irrigation (without adequate drainage) has affected much smaller area than natural salinity. The extent of human-induced salinity is about 76 million ha of which 52.7 million ha occurs in Asia (FAO 2015).

Salt-affected soils are known to restrain plant growth (Paul 2012) resulting in low agricultural productivity, low economic returns, ecological imbalance, and soil erosions (Hu and Schmidhalter 2002). Plants develop stress-adaptive mechanisms under stress conditions. Halophytes are perceived to be equipped with different

Table 1 Distribution of salt-affected soils

Continent	Saline soils (million ha)
Africa	122.9
South Asia	82.3
North and Central Asia	91.5
Southeast Asia	20.0
South America	69.5
North America	6.2
Mexico/Central America	2.0
Australasia	17.6
World total	412.0

Source: Food and Agriculture Organization of the United Nations and Intergovernmental Technical Panel on Soils (2015)

physiological and biochemical tolerance mechanisms to sustain plant growth in salt-affected soils (Khan et al. 2009). In addition to evolve defense mechanism against salinity stress, plants also adopt symbiotic relationship with plant growth promoting rhizobacteria (PGPR) and arbuscular mycorrhizal fungi to maintain healthy growth process. Plant growth promoting rhizobacteria (PGPR) and arbuscular mycorrhizal fungi (AMF) have been recognized as symbiotic partners of plants to improve plant's tolerance against abiotic stresses such as salinity. Plant growth promoting rhizobacteria (PGPR) not only confer assistance to increased plant growth and reduced susceptibility to diseases but also elicit "induced systemic resistance" (ISR) in plants against different stresses (Kloepper et al. 2004; Van Loon et al. 2004). Mitigation of salt stress by AMF inoculation has also been reported in different plant species (Al-Karaki 2000b; Kohler et al. 2009; Wu et al. 2010).

Considering the severity of the salinization as environmental problem, this book chapter is written to provide a comprehensive review about causes of soil salinization, potential impacts of salinity stress on plants, action mechanisms of plant growth promotion and/or regulation exhibited by PGPR and AMF due to their intrinsic traits that can be helpful to up scale the plant yield and production in stress agriculture (Table 1).

Causes of Soil Salinization

Salinity occurs naturally or induced by human activities (Slinger and Tenison 2007). Natural induced salinity largely occurs in desert and grassland biomes (Wilford et al. 2015). However, human-induced salinity largely occurs in cultivated lands that mainly results by disturbing the soil water balance (Beresford et al. 2001; Rose 2004). The major causes of soil salinization include:

Climate Change

Salinity mainly responds to changes in rainfall and temperature (Schagerl 2016). Therefore, global patterns of salinization changes as climate alter patterns of precipitation and temperature (Schofield and Kirkby 2003). Less rainfall and increasing temperature trend under current scenario in the world could increase salt buildup in soils due to a decreasing ability of precipitation to leach the accumulated salts from soil and an increasing evapotranspiration as a result of increasing temperature. This problem is intensified in arid and semiarid areas due to irregular pattern of rainfall and higher temperature (Parvaiz and Satyawati 2008; Rengasamy 2006). Climate change may increase flood water intrusion and salinity along coastal areas resulting due to sea level rise (Nicholls and Cazenave 2010).

Tree Clearing or Conversion of Woodlands to Agricultural Land

Extensive use of woods for domestic purposes has resulted in the conversion of woodlands into agricultural land and contributed a lot to disturb the hydrological balance (Williamson 1986; George et al. 1999; Runyan and D'Odorico 2010). Uncheck cutting of trees not only increases rate of evaporation from uncovered soils but also causes salt accumulation and water loss (Jamil et al. 2011). Conversion of woodlands to agricultural land has also decreased depth of water table that ultimately causes exclusion of salts from unsaturated soil layer to the root zone (Hatton et al. 2003; Jobbágy and Jackson 2007).

Land Management

Different land management practices like irrigation, fertilization, and summer fallow can help in salinity intrusion.

Irrigation

There is a high competition for fresh water among different sectors (municipal, industrial, and agricultural sectors) due to the current rate of population growth in the world. Consequently, agriculture has limited supply of fresh water (Tilman et al. 2002). Hence pressure has increased to irrigate the land with water of certain salt content (groundwater, drainage water, and treated wastewater). Use of these kinds of water during irrigation aggravates the salinization (Glick et al. 2007a, b). Heavy irrigation under dry climatic conditions particularly in clay texture soils can also cause salt buildup. Moreover, heavy irrigation followed by poor drainage conditions exacerbates the problem (FAO-Unesco 1973).

Fertilization

Commercial fertilizer, compost, and other organic manures contain rich amount of soluble salts (Ca, Mg, Na^+, Cl^-, SO_4^{2-}, K, NH_4^+, NO_3^-, HCO_3^-, and CO_3^{2-}). Over- or imbalanced application of them may cause salinity (Qureshi 2003).

Summer Fallow

Summer fallow management practices may cause increased salinization by increasing the evaporation of soil moisture content (USDA 1998).

Waterlogging

Waterlogging is one of the important causes of salinity. It is the phenomenon of rising of water table. It results from heavy irrigation/rainfall, floods, and seepage from canals or other reservoirs. Rising of water tables remobilizes the salts from the soil layer below the root zone and bring them within <2 m of the soil surface where evaporation leads to salt buildup (Ghassemi et al. 1995; Slinger and Tenison 2007). Floods and seepage from water reservoirs also bring heavy amounts of salts with water and the salinization is aggravated in recharge area (FAO 1997).

Coastal Ecosystems

Coastal ecosystems are highly prone to salinity. Salinization there exacerbated due to rise of sea levels under global warming which causes to increase the sea water intrusion (Mulrennan and Woodroffe 1998). These ecosystems are consisted of tidal marsh zones where salinity is commonly high (Adam 1990). There salinity level is determined by the hydraulic conductivity. The higher hydraulic conductivity lowers the salts aggravation as water rapidly moves out of the soil and washes the soil. Salts aggravation in tidal marshes also depends upon duration of tidal inundation; decreased duration of the tidal inundation allows evapotranspiration to concentrate pore water salinity and salt to accumulate. In addition, wind flow in coastal area transports salts to nearby ecosystems and the salination is aggravated there (Wang et al. 2007).

Weathering of Native Rocks

Physical or chemical weathering of rocks releases salts as a result of minerals (such as carbonate minerals, halite, olivine, and feldspars) breakdown in them, and their subsequent transport can cause salt buildup (Tarbuck and Lutgens 2012).

Impact of Soil Salinity on Plant

Rising water tables in irrigated as well as in non-irrigated areas has significantly affected the agricultural production, water quality, ecological health of in-stream biota, terrestrial biodiversity, and soil fertility of the subjected area. Soil salinity affects plants in two ways: (1) high concentrations of salts in the soil that makes it harder for roots to extract water and (2) high concentrations of salts within the plant which can be toxic for plant growth and development (Munns and Tester 2008). Based on resilience to salinity stress, plants can be categorized into two categories: (1) Halophytes (salt-tolerant plants that can tolerate high internal salt concentration and have capability to absorb salt along with water) and (2) Glycophytes (Salt-resistant plants that cannot tolerate with high internal salt concentration but can survive in certain salt concentration, and have potential to take up water by excluding salts in root zones). Various biochemical pathways facilitate plants to tolerate salt stress including acquisition and retention of water, ion homeostasis, and maintenance of chloroplast functions. Most agricultural plants fall into the salt-resistant category of plants (glycophytes) that can sustain growth in somewhat saline soil by excluding salts at the roots (Greenway and Munns 1980). However, glycophytes are unable to maintain their growth in exceptionally saline soils (DNRQ 1997). Furthermore, the impact of salinity varies with stage of growth, type of plant species, humidity, temperature, light, management practices, and soil fertility level (Munns 2002).

Acclimation of plants to salinized conditions depends upon activation of molecular networks stimulating a chain reaction of stress sensing, signal transduction, and the expression of specific stress-related genes and metabolites. Overwhelming effects of soil salinity can be listed as follows.

Decrease in Plant Growth and Productivity

Under saline soil condition, plants undergo different physiological and morphological changes leading to plant death and/or decreased plant productivity (Allakhverdiev et al. 2000). As plants absorb nutrients through their root systems so roots are very important part of plants. In areas of high salinity, plants absorb large amounts of salt, which not only disrupts their cellular function but also affects branching of root system that causes the release of a stress hormone which ultimately inhibits root growth. Additionally, sodium can cause early onset of leaf death, decreasing the photosynthetic ability of the plant. Sodium and chloride ions become toxic at high levels which ultimately cause burning of leaves, necrotic lesions, and sometimes defoliation (in woody species). High salt concentration also adversely affects seed germination, growth and vigor of seedling, vegetative, flowering and fruiting stages that ultimately causes decreased plant productivity (Sairam and Tyagi 2004).

Physiological Changes

During the commencement and development of salt stress, all major plant process such as protein synthesis, photosynthesis (Iyengar and Reddy 1996), and metabolism (lipid) are affected (Greenway and Munns 1980; Cheeseman 1988; Bohnert et al. 1995). Metabolic imbalances are caused by ion toxicity, osmotic stress, and nutritional deficiency under saline conditions which may lead to oxidative stress (Zhu 2002). Lipids in membranes play imperative role to induce cell resistance under salt stress by altering cell permeability (Bybordi et al. 2010). However, alteration in lipid metabolism and peroxidation occur under high salt stress conditions. Disordering in cohesions of lipids and proteins of membrane has also been observed under salinity stress (Rahdari et al. 2012). Modifications in net photosynthesis and stomatal conductance also occur due to damage in photosynthetic apparatus under saline conditions. This damage may be of varying degree depending upon the exposure time and concentration of salt (Doganlar et al. 2010). Dominance of chlorophyll "a" over chlorophyll "b" also decreases with increasing level of salinity (Mane et al. 2010).

Changes in biochemical pathways also occur, which include compartmentalization of ions at cellular and plant level, selective accumulation and exclusion of ions (Reddy et al. 1992; Iyengar and Reddy 1996; Zhu 2003), alteration in membrane structure and function (Otoch et al. 2001; Wang et al. 2001), modification in the expression of anti-oxidative enzymes, disturbance in plant hormonal balance (Khan and Rizvi 1994), change in the activity of enzymes related to nucleic acid metabolism (Gomes-Filho et al. 2008) and protein metabolism (Dantas et al. 2007). Salinity also induces water stress in plants that influences plant growth by adversely affecting dry matter partitioning, cell division, and cell expansion (Maas and Hoffman 1977). The level of plant hormones such as ABA and cytokinins increases with high salt concentration (Vaidyanathan et al. 1999). The inhibitory effects of salinity on photosynthesis, growth, and translocation of assimilates have been found to be alleviated by ABA (Popova et al. 1995). Dehydration of cell membranes, reduced permeability to carbon dioxide, enhanced senescence, alterations in cytoplasmic structure, and negative feedback are some other considerable phenomenon stimulated under salinity stress in plants (Iyengar and Reddy 1996).

Altered Osmotic Balance and Ion Cytotoxicity

Modification in osmotic balance also happens under saline conditions making extraction of water more difficult from soil. Ability of any plant to adjust osmotic potential in relation to water is an important determinant of growth response (Munns 1993). Reduction in relative leaf water content under salinity stress not only results in loss of turgor but eventually proceeds towards the stomatal closure, limited assimilation of carbon dioxide, and reduced photosynthetic rate (Khan et al. 2000).

Increase in osmotic potential under high salt conditions causes sodium ion leakage into cytosol generating cytotoxicity (Hussain et al. 2008a, b) that inactivate photosynthetic and respiratory electron transport in progressive stages (Allakhverdiev et al. 1999). A considerable decrease in the efficiency of PS II, electron transport chain (ETC), and assimilation rate of CO_2 has been detected under the influence of salinity (Piotr and Grazyna 2005).

The presence of excessive soluble salts in the soil also competes with the mineral nutrient uptake and their metabolism in plants. Increased salt uptake induces specific ion toxicity (like Na^+, Cl^-, and SO_4^{2-}) that decreases the uptake of essential nutrients/ion (e.g., P, K, N, Ca) (Zhu 2001). Most importantly, P concentration in agronomic crops decreases as salinity increases (Qadir and Schubert 2002). Moreover, nutritional disorders in relation to nutrient availability, competitive uptake, transport, and distribution may also result under salinity stress (Rogers et al. 2003; Hu and Schmidhalter 2005). To date, several studies have been conducted to gauge changes occuring in cereals, beans, vegetables, and fruit plants under salinity stress. Some of them are summarized in Table 2.

Ameliorative Strategies to Mitigate Salinity Stress

Strategies to ameliorate the salinity stress can be grouped into management practices, salinity tolerance, and biofertilizers.

Management Practices

Different management practices can be adopted to reclaim the salinization. However, reclaimed soils can revert into saline again unless management practices are continuously followed.

Flooding

Leaching of salts through flooding the soil is an important way to remove excess amounts of the salts. This practice is of great significance where fresh water is abundant (Provin and Pitt 2001). However, it requires time and larger amounts of water to reclaim the salts. There is also the risk of rising water table resulting in waterlogging. Further, the disposal of saline drainage water from salt-affected land is an issue, and it requires recycling before to use it further for irrigation. However, such water upto 4 dS m^{-1} can be used to irrigate the moderately salt-tolerant crops. Water up to 9 dS m^{-1} can also be used for more salt tolerant crops (e.g., sugar beet and cotton) only for limited period of 3 years (Goyal et al. 1999a, b). Consequently, this practice is not always feasible.

Table 2 Response of different plant species to salinity stress

Plant	Impact on plant	Reference
Maize	Decreased plant water potential and water use efficiency; Changed proline contents and photosynthetic ability; Drastically reduced germination rate, radicle length, plumule and seedling length, seed vigor and yield; Minimized toxic damages	Chaum and Kirdmanee (2009), Khodarahmpour et al. (2012), and Carpici et al. (2009)
Rice	Decreased fresh and dry weight of seedlings; Altered osmotic potential, photosynthetic pigments, soluble carbohydrate and proteins; Increased peroxide contents and MDA contents; Decreased leaf area and potassium (K+) content; Increased membrane injury, chlorophyll content and total sugars; Reduced germination; Decreased level of growth and chlorophyll accumulation; Increased proline contents; Increased hydrogen peroxide, peroxidase (POX) activity and anthocyanins; Reduced total dry matter; Reduced chlorophyll *a* and *b* contents of leaves; Increased production of ROS; Elevated ROS-mediated membrane damage, and cellular toxicity; Loss of grain yield	Amirjani (2010), Solangi et al. (2016), Xu et al. (2011), Chunthaburee et al. (2016), Tatar et al. (2010), Senadheera et al. (2012), Hasanuzzaman et al. (2009), Chutipaijit et al. (2011), Dionisio-Sese and Tobita (1998), Linghe and Shannon (2000), and Gain et al. (2004)
Wheat	Significantly reduced leaf growth including leaf area, leaf dry weight, and leaf fresh weight; Induced accumulation of proline and soluble sugars in the leaves; Decreased total carotenoids content; Enhanced accumulation of proline, H_2O_2 and lipid peroxidation; Reduced relative water contents; Reduced germination	Fercha (2011), Hala et al. (2005), Akbarimoghaddam et al. (2011), and Sairam et al. (2002)
Cotton	Altered structural and functional integrity of membranes; Reduced yield	Kurth et al. (1986) and Greenway and Munns (1980)

(continued)

Table 2 (continued)

Plant	Impact on plant	Reference
Barley	Reduced growth due to specific ion toxicity of Na^+ and Cl^-; Reduced growth and photosynthesis; Reduced K^+ and Ca^{2+} uptake; Reduced stomatal conductance; Enhanced chlorophyll degradation; Altered chlorophyll fluorescence (PS II) and function of oxygen evolving complex	Tavakkoli et al. (2011) and Kalaji et al. (2011)
Soybean	Decreased shoot and root weight, total biomass, plant height and leaf number; Reduced grain yield	Dolatabadian et al. (2011) and Greenway and Munns (1980)
Mung bean	Decreased levels of total chlorophyll, chlorophyll *a*, chlorophyll *b*, carotenoids and xanthophylls; Decreased intensity of chlorophyll fluorescence; Reduced yield (numbers of pods per plant, seeds per pod and seed weight)	Saha et al. (2010) and Nahar and Hasanuzzaman (2009)
Cucumber	Decreased total leaf chlorophyll contents	Khan et al. (2013)
Radish	Reduced growth and transpiration; Increased water use efficiency	Marcelis and Van Hooijdonk (1999)
Tomato	Increased production of ROS; Elevated ROS-mediated membrane damage and cellular toxicity	Mittova et al. (2004)
Sugar beet, Cabbage, Amaranth, Pak-choi	Reduced germination rate, root length and fresh weight, shoot length and fresh weight; Reduced yield	Jamil et al. (2006) and Greenway and Munns (1980)
Cabbage, Mustard, Spinach, Canola	Loss of germination percentage, germination rate, seed viability index, seedling and root length, seedling and root fresh weight; Induced changes in photosynthetic (PS II) and electron transport rates, and protein; Increased production of ROS; Elevated ROS-mediated membrane damage and cellular toxicity	Sarker et al. (2014), Ibrar et al. (2003), Ulfat et al. (2007), Bordi (2010), and Mittal et al. (2012)
Strawberry	Reduced the nutrient assimilation, especially of K and Ca; Induced ion imbalances of K, Ca, and Mg	Keutgen and Pawelzik (2009)
Mulberry	Increased production of ROS; Elevated ROS-mediated membrane damage and cellular toxicity	Ahmad et al. (2010)
Citrus	Increased production of ROS; Elevated ROS-mediated membrane damage and cellular toxicity	Gueta-Dahan et al. (1997)

Better Irrigation Practices

Adoption of different irrigation practices (avoiding use of saline water, regulated deficit irrigation/partial root zone drying methodology, and drip/micro jet irrigation) can result in minimizing the salinity effects (Shrivastava and Kumar 2015).

Application of Gypsum

Use of gypsum to replace the sodium with calcium ions followed by leaching through application of excess of water is a quite effective and commonly used method. In saline soils, application of lime (calcium carbonate) is not a good option because these soils sometimes are already rich in carbonate salts and are therefore alkaline (Provin and Pitt 2001; Cucci et al. 2012).

Scraping of Surface Soil

Surface soil with higher salt concentration can be scraped and transported out of the field. This has been practiced around the world (Qureshi et al. 2003).

Pre-Sowing Irrigation with Good Quality Water

To get good seed germination and seedlings establishment, soil is irrigated with good quality water prior to sowing (Goyal et al. 1999a, b).

Ridge/Bed Planting

The salinity stress can be minimized by sowing the seeds on ridges/beds irrigated with furrows (Ahmed et al. 2017). The seeds should be placed on the shoulders of ridges/beds because evaporation from top of the ridge/bed causes salt accumulation there. In the case of alternate furrow irrigation, seeds are placed on the sides of shoulder of ridges to be irrigated. Irrigation through furrows or sprinklers before sowing the seeds is a plus to reduce salt stress (Fischer et al. 2005).

Planting into Standing Water

Field is flooded with good quality water and allowed to percolate which reduce the salts concentrations. When few millimeters of standing water is left, the seeds are dropped over the field, and due to gravity force seeds imbed into the muddy soil surface. This results in good seed germination and seedling establishment. This approach is commonly used in California to grow safflower crop on salt-affected soils (Goyal et al. 1999a, b).

Mulching

Evaporation from surface soil results in salts accumulation and loss of water. To reduce evaporation and conserve water, crop residues and polythene sheets are spread on the soil surface which reduces the upward movement of salts (Al-Dulhli et al. 2010).

Deep Tillage

Surface soil has higher salt concentrations as compared to underlying layers. Deep tillage mix the surface salts in larger volume of soil, thereby reducing its concentration and adverse effects (Araya et al. 2010).

Soil Incorporation of Organic Matter

Green manuring and soil amendments of crop residues and manures are important to improve certain soil properties, i.e., soil tilth, structure, and water infiltration that safeguard against adverse effects of salinity (Diacono and Montemurro 2015).

Selection of Different Vegetative Options

Salt-tolerant crops are better able to grow in saline soils. For example barley, cotton, sugar beet, and canola are highly tolerant, wheat and alfalfa are moderately tolerant while maize and rice are sensitive to salinity (Richards 1969).

Growing of deep rooted perennials instead of shallow rooted annuals or incorporation of deep rooted perennials into current cropping systems results in deep drainage of water that is accumulated during wet season; furthermore, salt accumulation can be avoided through evaporation (Black et al. 1981; Stirzaker et al. 2002). The perfect example is phase farming which involves rotation of herbaceous perennial forage (e.g., alfalfa), i.e., grazed or harvested for hay, with a series of annual crops. The perennial pasture dries the subsoil below the roots of annual crops, thereby creating a buffer zone in which water and nutrients can be held for longer time for the deep rooted perennials in the next phase (Munns 2002).

Salinity Tolerance

Tolerance in plants against the adverse effects of salinity can be increased by different modern approaches.

Breeding for Salt Tolerance

The breeding approach is being utilized to develop tolerance against salinity. This involves screening of collected germplasm for tolerance, crossing of screened germplasms, and then selection of desired plant based upon desired characters (Purty et al. 2008). Based on this technique, some relatively tolerant cultivars have been developed for different crops, i.e., rice, wheat, lucerne, white clover, and citrus (https://www.ars.usda.gov/pacific-west-area/riverside-ca/us-salinity-laboratory/docs/research-databases/). However, there is a great diversity in salinity tolerance within species which is yet needed to explore (Purty et al. 2008).

Through breeding different traits are being developed to control the salinity stress. One of the most important traits related to plant breeding is salt exclusion in which Na^+ or Cl^- accumulates in leaves showing tolerance against adverse conditions of salinity. This trait has a high heritability and has been used to develop tolerance in cultivars of rice, white clover, and lucerne (Munns 2002). Tissue tolerance is another trait which indicates tolerance against high internal Na^+ concentrations. This can be evidenced by an absence of leaf injury under high Na^+ concentrations in leaf. High Na^+ concentrations (>100 mM) inhibit most enzymes. So Na^+ must be compartmentalized in vacuoles. Halophytes have the ability to compartmentalize Na^+ to very high concentrations (about 700 mM) in vacuoles (Flowers et al. 1977). On the other hand, glycophytes can compartmentalize Na^+ in vacuoles up to some extent (200 mM) which is common in photosynthetically active leaves of many species. In wheat leaves, concentrations over 250 mM are potentially toxic because it causes reductions in photosynthesis (James et al. 2002).

Improvement in growth (such as leaf elongation, root elongation, leaf area expansion, and shoot biomass), gas exchange (stomatal conductance, photosynthesis, transpiration efficiency, and chlorophyll fluorescence), turgor and osmotic adjustment, and reduction in leaf injury (chlorophyll content, or electrolyte leakage of cut discs) are other important traits being developed by breeding (James et al. 2002).

Molecular Strategies for Salinity Tolerance

Molecular techniques provide alternative ways to classical plant breeding to achieve salinity tolerance. These techniques benefit the development of salinity-tolerant cultivars based on specific traits. Identification and introduction of genes controlling the desired characters is an important development among molecular strategies. This strategy provided some significant results against salinity tolerance. For example, over-expression of the vacuolar Na^+/H^+ antiporter shows dramatic improvement of vegetative growth and of fruit yield in tomato (Zhang and Blumwald 2001). This antiporter compartmentalize Na^+ in the vacuole, where Na^+ has little chance of toxic effect on metabolism, or to be transported to younger leaves and fruits. However, so far no transgenic has been performed in the field against salinity (Flowers 2004).

This is because salinity tolerance is a multigenic trait, large improvements based on modification of only one gene could only occur if the gene is a transcription factor and regulates a number of genes that control ion transport or some other process involved in salinity tolerance.

Application of Biofertilizers

Application of biofertilizers is one of the important environmental friendly and effective strategies to mitigate the adverse effects of salinity on plant growth. In this regard, the role of arbuscular mycorrhizal fungi (AMF) and plant growth promoting rhizobacteria (PGPR) has enormously been studied (Ahanger et al. 2014a, b; Alqarawi et al. 2014; Hashem et al. 2014). Overall plant growth improves in saline soils when these fertilizers are applied. Both PGPR and AMF help plants to better establish under salinity stress by improving nutrient and water uptake, modifying root structure, improving photosynthetic efficiency, and accumulating osmoregulators (Sheng et al. 2008; Hajiboland et al. 2010; Porcel et al. 2015). Moreover, utilizing AMF and PGPR also improves nutrient cycling in poor soils (Smith and Read 2008). Based on wide scope of AMF and PGPR in saline soils, we entirely focused on them in the next sections of this chapter.

Ameliorative Role of AMF Against Salt Stress

High soil salinity conditions provoke both hyperionic and hyperosmotic stresses which not only lead to plant death but also alter soil texture resulting in reduced soil porosity, aeration, and water conductance (Mahajan and Tuteja 2005). Under natural environmental conditions, plants are colonized with external as well as internal microorganisms. Both plants and microbial commodities of soils are affected by excessive salt concentration which generate toxic and osmotically stressed environment. Consequently, retarded plant and suppressed microbial growth have been observed under salinity stress (Juniper and Abbott 2006). Arbuscular Mycorrhizal Fungi (AMF) being pervasive member of soil microbiota contributes significant role in natural and disturbed ecosystem (Giri et al. 2003; Tang et al. 2009; Navarro et al. 2013). Role of AMF as bio-ameliorators to mitigate plant growth reduction under salinity stress has enormously been reported (Ahanger et al. 2014a; Alqarawi et al. 2014). Studies suggest that the plants colonized with AMF show better growth under salinity stress that may be supported by improved water and nutrients uptake, photosynthetic efficiency, and accumulation of osmoregulators (Sheng et al. 2008; Hajiboland et al. 2010).

AMF colonization brings morphological, nutritional, and physiological changes in colonized plants (Hameed et al. 2014). Modification of root architecture to absorb more water and nutrients (Aroca et al. 2013; Ahanger et al. 2014a; Wu et al. 2014), alteration in physiological status for improved gas exchange capacity, better water use efficiency, and augmented photosynthetic efficiency (Sheng et al. 2008) have been observed in different plants with AMF inoculation. However, progressive development of AMF colonized plant under stress conditions depends upon the type of AMF species involved (Marulanda et al. 2003, 2007; Wu et al. 2007). Ecologically significant key roles of AMF colonization in plants can be abridged as follows: (1) Improvement in soil nutrient cycling especially of low mobile ions (P, N, Cu, Mn, and Fe) particularly in poor soils, (2) Improving water supply to plants under water stress conditions, (3) Improving soil structure and plant establishment, (4) Increasing plant tolerance against biotic and abiotic stresses (Smith and Read 2008).

Some of the apparent influences of AMF colonization on the plants under salinity stress conditions are summarized in Table 3 and described as follows.

Effect on Membrane Stability

Reduced membrane stability index, lipid peroxidation of membrane, and increased production of malondialdehyde (MDA) have been observed in plants under salinity stress in addition to enhanced production of radicals (Alqarawi et al. 2014; Hashem et al. 2016a, b, c). As a result of lipid peroxidation under salinity stress, the ratio of polysaturated fatty acids decreases leading to reduction of membrane integrity and its biological activity that further catalyze leakage of cellular components and disturbance of homeostasis (Alqarawi et al. 2014). However, AMF colonization in the salinity stressed plants improves plants health by maintaining osmotic potential. A reduced production of peroxides and MDA has been observed in AMF-inoculated plants under salinity stress that may be due to upregulated activities of antioxidant enzymes (Hashem et al. 2015a, b).

Effect on Enzymatic and Non-enzymatic Antioxidants Activity

Both enzymatic and non-enzymatic antioxidants are involved in sifting of toxic reactive oxygen species (ROS). Generally, antioxidant enzyme's activities upregulate under stress conditions. ROS produced as a result of stress exposure to the plants are toxic to the plant's metabolic activities posing deleterious impacts on protein, nucleic acid, and lipid molecules. The antioxidant enzymes system comprised of superoxide dismutase (SOD), catalase (CAT), peroxidase (POD), ascorbate peroxidase (APX), and glutathione reductase (GR) are major contributors of defense mechanism that help the stressed plant to counteract the stress-induced oxidative damage (Ahmad et al. 2010; Liu et al. 2014). It has been observed that antioxidant enzymes mediate quick scavenging of ROS to maintain their level below the toxic limits. ROS cause peroxidation of unsaturated lipid component of membranes resulting in the loss of membrane integrity eventually leading to leakage and

Table 3 Ameliorative role of AMF against salinity stress

Plant species	AMF species	Role of AMF under salinity stress	References
Tomato (*Solanum lycopersicum*)	*Glomus fasciculatum*/*Glomus* sp. mixture	Improved growth and mineral (N, P, Mg, Ca, Mn, and Fe) acquisition; Improved dry matter production; Enhanced photosynthesis and productivity of plants; Increased total soluble sugars, polysaccharides, total carbohydrates and total proteins; Increased shoot and root biomass; Increased Chl a & b and PSII activity; Enhanced activity of superoxide dismutase (SOD), catalase (CAT), peroxidase (POD) and ascorbate peroxidase (APX); Inoculation with AMF caused reduction in MDA content	Al-Karaki (2000a, b), Al-Karaki et al. (2001), Hajiboland et al. (2010), Ebrahim and Saleem (2017), Balliu et al. (2015), and Abdel-Latef and Chaoxing (2011)
Lettuce (*Lactuca sativa*)	*G. fasciculatum* *G. mosseae*	Greater root and shoot dry weights; Enhanced transpiration; Enhanced carbon dioxide exchange rate (CER); Enhanced stomatal conductance and water use efficiency (WUE)	Cantrell and Linderman (2001) and Ruiz-Lozano et al. (1996)
Onion *(Allium cepa)*	*Mixture of G. intraradices, G. mosseae, G.aggregatum, Acaulospora trappei, Entrophospora infrequens, Glomus* sp.	Significantly larger onion bulbs at all salt levels than non-AMF-inoculated onions	Cantrell and Linderman (2001)

(continued)

Table 3 (continued)

Plant species	AMF species	Role of AMF under salinity stress	References
Sweet basil (*Ocimum basilicum*)	*G. deserticola*, *G. etunicatum*, *G. mosseae*, *G. intraradices*	Mitigated the reduction of K, P, and Ca uptake due to salinity; Improved balance between K/Na and between Ca/Na in AMF plants; Increased chlorophyll content and water use efficiency; Improved plant growth, photosynthetic efficiency, gas exchange and water use efficiency; Lower extent of lipid peroxidation, higher antioxidant enzyme activities (SOD, APX, and POD); Improved content of lipids, proline, and soluble sugars	Zuccarini and Okurowska (2008), Elhindi et al. (2017) and Hashem et al. (2016c)
Soybean (*Glycine max*)	*G. etunicatum; G. mosseae, G. intraradices*	Improved nutrient uptake including P, K, Zn; Higher fresh and dry weight; Higher root proline; Lower shoot proline and Na concentrations; Improved symbiotic efficiency; Improved nodule formation, leghemoglobin content, nitrogenase activity and auxin synthesis; Enhanced membrane stability; Reduced production of hydrogen peroxide; Reduced lipid peroxidation; Improved expression of auxins; Improved acquisition of nutrient	Sharifi et al. (2007) and Hashem et al. (2016b)
Karna Khatta (*Citrus karna*)	Mixed inoculum of *Glomus* sp. and *Gigaspora* sp.	Improved nutrient uptake (P, N, K); Higher chlorophyll contents; Higher proline contents; Higher sugar contents	Murkute et al. (2006)
Cotton (*Gossypium arboreum*)	*G. mosseae*	Improved nutrient uptake P; Higher root and shoot dry weight	Tian et al. (2004)

(continued)

Table 3 (continued)

Plant species	AMF species	Role of AMF under salinity stress	References
Cotton *(Gossypium hirsutum)*	Indigenous AMF	Promoted leaf proline accumulation in cotton; Higher K+/Na + ratio; Enhanced P uptake; Promoted growth in saline soil	Liu et al. (2016)
Cowpea [*Vigna unguiculata* (L.) Walp.]	*G.mosseae, G.intraradices, G.etunicatum*	Enhanced activity of antioxidant enzymes such as superoxide dismutase (SOD), catalase (CAT), ascorbate peroxidase (APX), peroxidase (POD) and glutathione reductase (GR); Increased proline content; Increased uptake of mineral elements and improved status of osmoregulation	Hashem et al. (2015a, b)
Cleopatra mandarin (*Citrus reshni*); Alemow (*Citrus macrophylla wester*)	*Rhizophagus irregularis, Funneliformis mosseae*	Increased plant growth, Improved P, K, Fe and Cu, Mg uptake	Navarro et al. (2013)
Maize (*Zea mays*)	*G. mosseae*	Increased resistance to soil salinity; Improved nutrient (P) uptake; Higher dry weight of shoot and root; higher relative chlorophyll content; better water status (decreased water saturation deficit, increased water use efficiency, and relative water content); Higher gas exchange capacity (increased photosynthetic rate, stomatal conductance and transpiration rate, and decreased intercellular CO_2 concentration); Higher non-photochemistry and photochemistry efficiency	Feng et al. (2002), Sheng et al. (2008), and Liu et al. (2016)

(continued)

Table 3 (continued)

Plant species	AMF species	Role of AMF under salinity stress	References
Black locust (*R. pseudoacacia*)	*Rhizophagus irregularis*	Improved the net photosynthetic rate, quantum efficiency of photosystem II photochemistry; Improved K^+ content in plants; Reduced Na^+ content; Upregulated the expression of chloroplast genes in leaves; Upregulated gene expression of membrane transport proteins involved in K+/Na + homeostasis in roots	Chen et al. (2017)
Lupine (*Lupinus termis* Forsik)	*G. mosseae, G. intraradices, G. etunicatum*	Increased the activities of antioxidant enzymes (SOD, CAT, APX, and POD); Increased accumulation of osmoregulators (proline, glycine betaine, and sugars); Reduced lipid peroxidation; Increased membrane stability	Hashem et al. (2016a)
Strawberry (*Fragaria ananassa* Duch)	*F. caledonius* *F. mosseae*	Enhanced plant growth, root and shoot mass; Improved salt tolerance and fruit quality	Sinclair et al. (2014)
Wheat	*G. etunicatum, G. mosseae, G. intraradices*	Improved plant dry mass, grain yield	Daeia et al. (2009)
Olive plants (*Olea europaea*)	*G. mosseae, G. intraradices, G. claroideum*	Increased plant growth; Increased plant ability to acquire N, P, and K; Increased shoot and root; Enhanced salt tolerance	Porras-Soriano et al. (2009)
Pearl millet (*Pennisetum glaucum*)	*G. fasciculatum*	Higher total chlorophyll contents; Increased plant growth and nutrient uptake capacity; Improved antioxidant activity; Pronounced accumulation of proline; Higher leaf number, shoot and root length, fresh and dry weight of plant	Borde et al. (2011)

(continued)

Table 3 (continued)

Plant species	AMF species	Role of AMF under salinity stress	References
Cocoa (*Theobroma cacao*)	*Scutellospora, Glomus*	Increased shoot dry weight; Increased P and Ca content in shoot	Chulan and Martin (1992)
Pepper (*Capsicum annuum*)	*G. intraradices*	Maintained greater root and shoot biomass; Higher leaf area, root and shoot dry mass; Lower content of Na and higher content of K and P; Greater cell membrane integrity; Higher proline content in leaves than in roots; Improved plant growth	Beltrano et al. (2013)

desiccation (Ahmad et al. 2010; Rasool et al. 2013). Superoxide radicals are transformed into water and hydrogen peroxide by SOD enzyme. Subsequently, hydrogen peroxide is converted into water and oxygen by CAT or APX enzyme (Mittler 2002). Increased production of H_2O_2 under salinity stress causes disturbances in cellular homeostasis that promotes membrane leakage (Tuna et al. 2008). Increased GR enzyme activity results in enhanced glutathione production that works as electron donor during the conversion of dehydroascorbate into ascorbic acid (Mittler 2002). Upregulation of GR under stress condition reduces formation of superoxide radicals through maintaining the photosynthetic electron transport (Noctor and Foyer 1998; Mittler 2002).

Non-enzymatic defense system includes ascorbic acid, glutathione, phenols, tocopherols, etc. (Mittler 2002; Ahmad et al. 2010). Phenolics (secondary metabolites) are included in non-enzymatic antioxidants, which play important role in defense mechanism by scavenging toxic radicals leading to reduction of oxidative stress and increase of membrane stability (Michalak 2006; Bartwal et al. 2013; Tomar and Agarwal 2013). Ascorbate–glutathione cycle is involved in a series of redox reactions where the net electron flow is from NADPH to H_2O_2 resulting in the conversion of H_2O_2 into water. GR, APX, reduced glutathione, oxidized glutathione and ascorbic acid (ASA) are the important constituents of ascorbate–glutathione pathway which is actively involved in sifting of ROS. ASA acts an electron donor in conversion of H_2O_2 into water and oxygen (Noctor and Foyer 1998; Mittler 2002).

Various studies performed on different plants under salinity stress with or without AMF colonization have proved that the plant inoculated with AMF upregulated antioxidant enzymes activity as compared to un-inoculated plants (Abdel-Latef and Chaoxing 2011; Hashem et al. 2014; Sarwat et al. 2016). Increased activity of enzymatic antioxidants may be attributed towards improved uptake of important nutrients in AMF colonized plants that help to enhance synthesis of biomolecules that serve as cofactor of enzymes. Higher activities of antioxidant enzymes in AMF colonized plant may also be corroborated with improved plant growth under salinity stress (Alguacil et al. 2003; Zhong et al. 2007). Accumulation of phenolic in AMF colonized plants under salinity stress has also been observed; furthermore, the plants maintaining higher contents of phenols are found better adopted to grow under stress conditions as compared to the less accumulating ones (Alqarawi et al. 2014; Hashem et al. 2015a, b).

Effect on Chlorophyll Contents

High salinity impedes different biochemical process such as synthesis of proteins and functioning of pigment protein complex (Levitt 1980; Sultana et al. 1999; El-Tayeb 2005). Inoculation of AMF increases chlorophyll content under normal as well as salt-stressed conditions. However, AMF-inoculated plants maintain higher chlorophyll contents compared to stressed plants (Kaya et al. 2009; Hajiboland et al. 2010; Aroca et al. 2013). The higher chlorophyll contents in AMF-inoculated plants contribute to improve photosynthetic efficiency which maintained plant growth under salinity stress. Enhancement in chlorophyll pigments may be due to colonization of AMF in salt-affected plant roots that supports mineral uptake especially of magnesium (Mg) from soil. Mg is an important component of chlorophyll molecule (Sheng et al. 2008) that helps to build more chlorophyll pigment; resultantly photosynthesis improves to help plant maintenance or establishment under normal and stressed conditions.

Effect on Accumulation of Osmoregulators or Organic Osmolytes

Decrease in the activities of photosynthetic carbon metabolizing enzymes, Rubisco and NADP-dependent malic enzyme (NADP-ME) has also been evident in the salinity stressed plants (Soussi et al. 1998; Fang et al. 2013; Koyro et al. 2013). NADP malic enzymes are supposed to participate in numerous imperative metabolic pathways and plant defense through malate metabolism (Casati et al. 1999). In salt sensitive plants, accumulation of osmoregulators like proline, glycine betaine (GB), sugars, and amino acids decreases when the plants are posed to salinity stress, whereas upregulated expression of osmoregulators for efficient sequestration and compartmentation of toxic ions has been observed in salt tolerance plants (Ahmad and Sharma 2008; Azooz et al. 2011). Proline as well as GB helps to maintain water balance of plants under stress conditions (Ahmad 2010; Ahanger et al. 2014b). In addition, proline regulates stress protective proteins (Thakur and Sharma 2005), whereas GB shields different important components of photosynthetic apparatus, maintains high activity of antioxidant enzymes and lowers the level of H_2O_2 under stress conditions (Hossain and Fujita 2010).

Increased accumulation of GB, proline, and soluble sugars has been recorded in salinity stressed AMF colonized plants (Moghaieb et al. 2004; Khan et al. 2014) that may be ascribed to their role to protect cell membrane from damage and to regulate osmotic pressure as osmoregulators (Hoseini 2010; Masood et al. 2013) that results in buildup of tolerance against salinity stress. However, accumulation of these osmolytes under salt stress varies with host plant as well as AMF species (Rabie and Almadini 2005). The reports inferred on the evidence of enhanced osmolytes concentration in mycorrhizal plants under salt stress (Feng et al. 2002; Al-Garni 2006; Sharifi et al. 2007; Garg and Manchanda 2009) help to assume the role of AMF against ultra-structural damage under salinity stress.

Effect on Nutrient Uptake, Plant Growth and Yield

Reduction in uptake of essential nutrients like potassium (K) and phosphorous (P) also takes place in salinity stressed plant (Al-Karaki 2000b; Kohler et al. 2009). Root colonization with AMF mitigates the stress-induced changes that results in enhanced plant growth especially total biomass as described by Feng et al. (2002) in maize; Abdel-Latef and Chaoxing (2011) in tomato; Giri et al. (2003, 2007) in acacia (*Acacia nilotica*); Sannazzaro et al. (2006) in *Lotus glaber*; and Abd-Allah et al. (2015) in *Sesbania sesban*.

As AMF form mutualistic beneficial relationship with plants, so a bidirectional movement of nutrient is facilitated through this colonization. AMF take up a significant fraction of plant photosynthates (hexose carbon) (Paul and Kucey 1981; Sjöberg 2005; Xie et al. 2014); in return it provides micro- and macronutrient from soil to the plants (Thompson 1990; Hodge et al. 2001; Miransari 2011). The AMF are widely distributed in saline land of terrestrial ecosystem (Yamato et al. 2008) and improve plants performance against salinity stress. AMF induce stress resistance in plants by mediating continuous uptake and assimilation of mineral nutrients that help to maintain plant growth and development under stress conditions (Usha et al. 2005). Recently, many researchers have reported that AMF could improve plants ability to cope with salinity stress (Yano-Melo et al. 2003; Rabie 2005; Jahromi et al. 2008) by improving plant nutrient uptake (Cantrell and Linderman 2001; Asghari et al. 2005) and facilitating water uptake (Berta et al. 1990; Ruiz-Lozano and Azcon 1995).

Effect on Synthesis of Growth Regulators

Endogenous growth hormones including auxins (IAA, IBA), gibberellins (GA3), and dormins (ABA) play very important role as growth regulators. A drastic decline in the synthesis of IAA, IBA, and GA3 under salinity stress occurs, whereas inoculation of plants with AMF improves expression of these hormones especially of IAA and IBA (Waqas et al. 2012; Hashem et al. 2015a, b). AMF also have ability to alter the levels of ABA that leads towards the better adaptability of plants to different environmental stresses including salinity. Biosynthesis of ABA to somewhat lesser extent in AMF colonized plants compared to un-inoculated plants under salinity stress suggests that mycorrhizal plants are relatively less stressed.

Ameliorative Role of PGPR Against Salt Stress

Besides developing protective mechanisms against stressful environment, PGPR can serve as helping hand towards plant tolerance against numerous abiotic stresses including salinity (Dimkpa et al. 2009). Several plant beneficial rhizobacteria belonging to different genera including *Rhizobium, Bradyrhizobium, Bacillus, Pseudomonas, Burkholderia, Pantoea, Achromobacter, Azospirillum, Azobacter,* etc. have been reported to confer resistance to associated plants against abiotic stresses such as salinity, acidity, and alkalinity (Grover et al. 2011; Upadhyay et al. 2009). It has been reported that PGPR inoculated tomato, canola, pepper, lettuce, and bean showed improved growth under saline environment (Barassi et al. 2006; Yildirim and Taylor 2005). The term induced systemic tolerance (IST) is specifically related to PGPR mediated physio-chemical changes resulting in improved resistance against abiotic stress (Yang et al. 2009). PGPR mediated mitigation of salinity may be due to a variety of mechanisms, some of which are described here in the following section and Table 4.

Antioxidant Activity Regulation

The salt stress triggers the increased production of reactive oxygen species (ROS) within plants and causes oxidative damage to plants. Many cellular components especially lipid membranes are vulnerable to ROS. The membrane deterioration caused by ROS leading to cellular toxicity has been described in rice, citrus, and tomato when grown under high salinated conditions (Kim et al. 2005). It has been documented that antioxidant molecules have ROS-scavenging activities and have the potential to counteract the oxidative damage in plants (Spychalla and Desborough 1990). Various antioxidant enzymes such as catalases (EC 1.11.1.6), superoxide dismutases (EC 1.15.1.1), and peroxidases (EC 1.11.1.7) possess ROS degrading activities and thus confer salt tolerance to plants under salt stress (Apel and Hirt 2004; Mittler 2002). The treatment of plants with PGPR showed increased production of numerous oxidative stress resistant enzymes such as superoxide dismutase, peroxidase, polyphenol oxidase, and catalase. It has been investigated that wheat plants inoculated with halotolerant PGPR strain *Dietziana tronolimnaea* STR1 showed elevated gene expression of various antioxidant enzymes including peroxidase, catalase, etc. when cultivated under saline environment than non-inoculated

Table 4 Ameliorative role of PGPR against salinity stress

Plant species	Bacterial species	Effects on plant under salinity stress	References
Solanum lycopersicum (tomato)	*Pseudomonas fluorescens, Pseudomonas aeruginosa, Pseudomonas stutzeri Achromobacter piechaudii* ARV8	Treatment with PGPR strains improved root and shoot growth of tomato by regulating ethylene production through ACC deaminase activity under salinity	Tank and Saraf (2010) and Mayak et al. (2004)
Raphanus sativus L. (radish)	*Staphylococcus kloosii* EY37, *Kocuriaerythromyxa* EY43	Inoculated plants showed better growth, nutrient uptake and reduced Na toxicity under salt stress	Yildirim et al. (2008)
Gossypium hirsutum (cotton)	*Klebsiella oxytoca*	Significant increase in the fresh and dry weight of cotton plant was observed along with enhanced nutrient (Ca, P, N, and K) uptake	Yue et al. (2007)
Triticum aestivum (wheat)	*Paenibacillus* spp., *Bacillus* spp., *Enterobacter* spp., *Pseudomonas putida, Pseudomonas aeruginosa, Dietzia natronolimnaea*	*Bacillus* and *Enterobacter* spp. reduced Na uptake and provided wheat plants with more water and nutrients by producing exopolysaccharides; *Pseudomonas putida* increased the plant height and root length; PGPR treatment also resulted in more biomass production and mitigated salt stress by producing ACC deaminase	Zahir et al. (2009), Nadeem et al. (2010), Upadhyay et al. (2011) and Bharti et al. (2016)
Oryza sativa L. cv. KDML105	*Streptomyces* sp. GMKU336	Inoculation conferred salt tolerance to rice via sequestering ACC by ACC deaminase activity and reduced ethylene production in plants	Jaemsaeng et al. (2018)
Brassica napus L. (canola)	*Pseudomonas species*	Inoculation helped canola plants to mitigate salt stress by ACC deaminase production	Jalili et al. (2009)
Zea mays L. (maize)	*Flavobacterium ferrugineum, Enterobacter aerogenes, Pseudomonas* spp., *Staphylococcus sciuri* SAT-17	ACC deaminase producing rhizobacteria enhanced plant growth by improving water availability and reducing Na uptake	Nadeem et al. (2007) and Akram et al. (2016)
Lactuca sativa (lettuce)	*Bacillus subtilis, Pseudomonas mendocina*	Enhanced shoot biomass and nutrient uptake was observed in plants inoculated with phytohormones and antioxidants producing PGPR in contrast to non-inoculated ones	Arkhipova et al. (2007) and Kohler et al. (2009)

(continued)

Table 4 (continued)

Plant species	Bacterial species	Effects on plant under salinity stress	References
Arachis hypogaea (groundnut)	*Pseudomonas fluorescens TDK1*	Improved plant yield under saline conditions was revealed through ACC deaminase production	Govindasamy et al. (2008)
Spinacia oleracea L. (spinach)	*Bacillus lentimorbus NRRL B-30488*	Helped plants in the amelioration of salinity via antioxidant production	Nautiyal et al. (2008)
Cucumis sativus (cucumber)	*Burkholderia cepacia SE4*, *Promicromonospora sp. SE188*, *Acinetobacter calcoaceticus SE370*	Significant increase in the chlorophyll content and biomass was observed in bacterial strains treated plants as compared to non-treated plants under saline environment. These bacterial strains confer resistance to plants against salinity via phytohormones and antioxidants production	Kang et al. (2014)
Abelmoschus esculentus (Okra)	*Enterobacter* sp. UPMR18	Inoculated plants showed increased germination percentage, chlorophyll content and resistance to oxidative stress when cultivated in salt-affected soil	Habib et al. (2016)

plants (Bharti et al. 2016). Thus these PGPR stimulated ROS-scavenging enzymes can provide plants with resistance against salinity stress by removing hydrogen peroxide (H_2O_2) from salt effected plant rhizosphere (Nautiyal et al. 2008; Tommasi et al. 2001). However, extensive research studies are needed to explore the underlying mechanisms of PGPR-induced oxidative stress tolerance to plant.

Phytohormones Regulation

Phytohormones of rhizobacterial origin induce certain physiological responses in the associated plant. The PGPR improve plant growth by altering root morphology under abiotic stresses by producing various different phytohormones like indole-3-acetic acid (IAA), gibberellic acid (GA), and cytokinins (Kloepper et al. 2007; Paul and Lade 2014). The IAA producing rhizospheric bacteria are thought to be most beneficial in ameliorating salinity stress (Paul and Lade 2014). Sadeghi et al. (2012) reported that wheat plants treated with IAA producing *Streptomyces* strain showed better growth under saline conditions. Similar results were found when potato (Khan and Doty 2009), mung bean, and canola plants (Glick et al. 2007a, b; Kang et al. 2006) were inoculated with IAA synthesizing rhizobacteria. Moreover, the effects of cytokinins and gibberellic acid producing bacterial strains have also been investigated for microbial elicited salinity stress tolerance in plants. It has been observed that auxin, cytokinins, and gibberellic acid producing *Azospirillum* strains enhance plant growth by modifying root architecture (Spaepen et al. 2008). However, the concentrations of phytohormones (specifically auxin) play a key role in plant resistance under different stresses. PGPR producing IAA in less concentration have positive effect on plant growth, whereas bacteria producing IAA in higher concentration negatively affect the root and shoot growth of host plant (Egamberdieva 2009, 2013). Thus, attention is required in screening potential PGPR strains producing phytohormones as bioinoculants to effectively alleviate salt stress.

Exopolysaccharide (EPS) Production

PGPR strains synthesizing exopolysaccharide (EPS) have the potential to improve the soil structure by increasing the macropores volume within soil and aggregation of rhizospheric soil, thus providing plants with more water and nutrients. The EPS producing PGPR strains can chelate cations such as Na^+ (Alami et al. 2000). So it is believed that increasing the number of EPS producing PGPR in the rhizosphere would reduce the Na^+ uptake by plants and thus facilitate plants to grow under saline conditions by mitigating salinity stress (Geddie and Sutherland 1993). Upadhyay et al. (2011) revealed that wheat plants showed improved growth when inoculated with exopolysaccharide producing bacteria belonging to *Bacillus* and *Enterobacter* genera in contrast to non-inoculated plants. However, the mechanism of EPS producing PGPR and their stimulatory effect on plant growth under salinated environment are not well understood yet.

Lowering Electrolytic Leakage

Salt stress fluctuates the ion flux in plants leading to electrolyte imbalance within plants. The higher ionic flux damages the cellular membrane of the plant leading to enhanced permeability causing electrolyte leakage and the leaked electrolyte concentrated around the tissues (Hussain et al. 2008a, b; Kang et al. 2014). Treatment of *Raphanus sativus* L. with PGPR grown under saline conditions showed reduced electrolyte leakage (Yildirim et al. 2008). Similarly, *Pseudomonas* and *Rhizobium* inoculated maize plants depicted low electrolyte leakage when cultivated in salt effected soils (Bano and Fatima 2009; Fazal and Bano 2016). Similar findings were observed by Kang et al. (2014)) in cucumber plants when treated with *Burkholderia cepacia* SE4, *Promicromonospora* sp. SE188, and *Acinetobacter calcoaceticus* SE370. Thus PGPR can maintain the integrity of plant cellular membrane and help to alleviate the harmful effects of salt.

ACC Regulation

The low concentration of ethylene is essential for normal growth and development of plants, but in higher concentration it has detrimental effects on plants resulted in reduced root and shoot proliferation (Bhattacharyya and Jha 2012; Glick et al. 2007a, b). The salinity stress stimulates the high production of 1-aminocyclopropa ne-1-carboxylic acid (ACC) in plant which is directly associated with the synthesis of ethylene resulting in high ethylene level that ultimately harm the plant (Botella et al. 2000). The ACC deaminase producing PGPR is indigenous to various soils and can act as sink for ACC for ensuring the optimum ethylene level required for the normal growth by plants. These bacteria hydrolyze the ACC to ammonia and a-ketobutyrate and re-establish proliferating root system, thus ameliorating the plant salt stress (Belimov et al. 2001; Fahad et al. 2015). The proficient role of ACC deaminase producing bacteria in the growth improvement and salt stress tolerance has been well documented for various crops like groundnut (Govindasamy et al. 2008), tomato (Mayak et al. 2004), canola (Ghosh et al. 2003), pepper (Mayak et al. 2004), etc. Jaemsaeng et al. (2018) described that *Streptomyces* sp. GMKU336 when applied to salt-stressed *Oryza sativa* L. cv. KDML105 promotes plant growth and confers tolerance to salt stress by reducing ethylene level by sequestering ACC via ACC deaminase activity. Similar findings were reported by Sarkar et al. (2018) depicting that halotolerant *Enterobacter* sp. improve the growth of rice plants and help to alleviate salinity stress via ACC deaminase activity. Moreover, ACC deaminase containing *Pseudomonas fluorescens* has a positive impact on the growth of salt-stressed cucumber plant (Nadeem et al. 2016). The rhizospheric bacterial genera containing ACC deaminase activity include *Azospirillum, Bacillus, Pseudomonas, Enterobacter, Alcaligenes, Klebsiella, Variovorax, Acidovorax, Methylobacterium, Achromobacter,* and *Rhizobium* (Duan et al. 2009; Ghosh et al. 2003; Govindasamy et al. 2008; Paul and Lade 2014).

Dual Inoculation of PGPR and AMF

Undoubtedly, PGPR and AMF have potential to ameliorate salinity stress but the capacity can be boosted by dual inoculation of PGPR and AMF to the plants (Gamalero et al. 2010). Inoculation with PGPR and AMF significantly enhances the uptake of essential nutrients particularly P that is mostly limiting factor in alkaline soils (Shirmardi et al. 2010). Positive influence of dual inoculation with PGPR and AMF on water and nutrition absorption, plant growth, and grain yield has also been reported (Najafi et al. 2012). Synergistic relationship of PGPR and AMF not only helps the plants to cope with abiotic stress but also equally beneficial for plants against biotic stresses (plant pathogens). However, the efficacy of co-inoculation is dependent upon both microbial species. In synergistic interactions, combination of certain mechanisms contributed by both bacterial and fungal partner results in positive change on plant morphological, physiological, biochemical, and molecular attributes. However, further investigations are needed to get maximum output from this combination under natural environmental conditions.

Future Prospective

Scientists are engaged to introduce workable strategies against salinity-induced losses of agriculture. Stress-induced damaging impacts can be reduced by exploring hidden potential of naturally occurring microorganisms (bacteria and fungi) and then applying them singly or in different combinations. A number of researchers have already examined positive changes in the plant growth, development, and yield with dual inoculation of bacteria and fungi under stress conditions. Effectiveness of dual inoculation compared to that of individual inoculation has also been reported. However, still there is much more to understand about the underlying mechanisms that are repressed/downregulated under salinity stress or triggered/upregulated by inoculation of bacteria and/or fungi. Understanding about physiological and molecular mechanisms could enable us to use the microbial bioresources as biotechnological tool for amelioration of salinity stress.

References

Abd-Allah EF, Hashem A, Alqarawi AA, Bahkali AH, Alwhibi MS (2015) Enhancing growth performance and systemic acquired resistance of medicinal plant *Sesbania sesban* (L.) Merr using arbuscular mycorrhizal fungi under salt stress. Saudi J Biol Sci 22:274–283

Abdel-Latef AAH, Chaoxing H (2011) Effect of arbuscular mycorrhizal fungi on growth, mineral nutrition, antioxidant enzymes activity and fruit yield of tomato grown under salinity stress. Sci Hortic 127:228–233

Adam P (1990) Saltmarsh ecology. Cambridge University Press, Cambridge

Ahanger MA, Hashem A, Abd Allah EF, Ahmad P (2014a) Arbuscular mycorrhiza in crop improvement under environmental stress. In: Ahmad P, Rasool S (eds) Emerging technologies and management of crop stress tolerance, vol 2, pp 69–95

Ahanger MA, Tyagi SR, Wani MR, Ahmad P (2014b) Drought tolerance: role of organic osmolytes, growth regulators, and mineral nutrients. In: Ahmad P, Wani MR (eds) Physiological mechanisms and adaptation strategies in plants under changing environment. Springer, New York, pp 25–55

Ahmad P (2010) Growth and antioxidant responses in mustard (*Brassica juncea* L.) plants subjected to combined effect of gibberellic acid and salinity. Ach Agron Soil Sci 56:575–588

Ahmad P, Sharma S (2008) Salt stress and phyto-biochemical responses of plants—a review. Plant Soil Environ 54:89–99

Ahmad P, Jaleel CA, Sharma S (2010) Antioxidative defence system, lipid peroxidation, proline metabolizing enzymes and biochemical activity in two genotypes of *Morus alba* L subjected to NaCl stress. Russ J Plant Physiol 57:509–517

Ahmed K, Nawaz MQ, Hussain SS, Rizwan M, Sarfraz M, Wainse GM, Jamil M (2017) Response of onion to different nitrogen levels and method of transplanting in moderately salt affected soil. Acta Agric Slov 109:303–313

Akbarimoghaddam H, Galavi M, Ghanbari A, Panjehkeh N (2011) Salinity effects on seed germination and seedling growth of bread wheat cultivars. Trakia J Sci 9:43–50

Akram M, Shahid M, Tariq M, Azeem M, Javed MT, Saleem S, Riaz S (2016) Deciphering Staphylococcus sciuri SAT-17 mediated anti-oxidative defense mechanisms and growth modulations in salt stressed maize *Zea mays* L. Front Microbiol 7:867

Alami Y, Achouak W, Marol C, Heulin T (2000) Rhizosphere soil aggregation and plant growth promotion of sunflowers by an exopolysaccharide-producing *Rhizobium* sp. strain isolated from sunflower roots. Appl Environ Microbiol 66:3393–3398

Al-Dulhli HS, Al-Rawahy SA, Prathapar S (2010) Effectiveness of mulches to control the soil salinity in sorghum fields irrigated with saline water. A monograph on management of salt-affected soils and water for sustainable agriculture. Sultan Qaboos University, Muscat, pp 41–46

Al-Garni SMS (2006) Increasing NaCl-salt tolerance of a halophytic plant *Phragmites australis* by mycorrhizal symbiosis. Am Eur J Agric Environ Sci 1:119–126

Alguacil MM, Hernández JA, Caravaca F, Portillo B, Roldán A (2003) Antioxidant enzyme activities in shoots from three mycorrhizal shrub species afforested in a degraded semi-arid soil. Plant Physiol 118:562–570

Al-Karaki GN (2000a) Growth, sodium, and potassium uptake and translocation in salt stressed tomato. J Plant Nutr 23:369–379

Al-Karaki GN (2000b) Growth of mycorrhizal tomato and mineral acquisition under salt stress. Mycorrhiza 10:51–54

Al-Karaki GN, Hammad R, Rusan M (2001) Response of two tomato cultivars differing in salt tolerance to inoculation with mycorrhizal fungi under salt stress. Mycorrhiza 11:43–47

Allakhverdiev SI, Nishiyama Y, Suzuki I, Tasaka Y, Sakamoto A, Murata N (1999) Genetic engineering of the unsaturation of fatty acids in membrane lipids alters the tolerance of *Synechocystis* to salt stress. Proc Natl Acad Sci U S A 96:5862–5867

Allakhverdiev SI, Sakamoto A, Nishiyama Y, Inaba M, Murata N (2000) Ionic and osmotic effects of NaCl-induced inactivation of photosystems I and II in *Synechococcus* sp. Plant Physiol 123:1047–1056

Alqarawi AA, Abd Allah EF, Hashem A (2014) Alleviation of salt-induced adverse impact via mycorrhizal fungi in *Ephedra aphylla* Forssk. J Plant Interact 9:802–810

Amirjani MR (2010) Effect of NaCl on some physiological parameters of rice. Environ J Biol Sci 3:6–16

Apel K, Hirt H (2004) Reactive oxygen species: metabolism, oxidative stress, and signal transduction. Annu Rev Plant Biol 55:373–399

Araya GG, Araya K, Zhang H, Ohmiya K, Liu F, Zhang C (2010) Characteristics of salt-affected soils by deep tillage up to 600 mm: disaggregation experiments of soil clods. Eng Agric Environ Food 3:93–99

Arkhipova TN, Prinsen E, Veselov SU, Martinenko EV, Melentiev AI, Kudoyarova GR (2007) Cytokinin producing bacteria enhance plant growth in drying soil. Plant and Soil 292:305–315

Aroca R, Ruiz-Lozano JM, Zamarreno A, Paz JA, Garcia-Mina JM, Pozo MJ, Lopez-Raez JA (2013) Arbuscular mycorrhizal symbiosis influences strigolactone production under salinity and alleviates salt stress in lettuce plants. J Plant Physiol 170:47–55

Asghari HR, Marschner P, Smith SE, Smith FA (2005) Growth response of *Atriplex nummularia* to inoculation with arbuscular mycorrhizal fungi at different salinity levels. Plant and Soil 273:245–256

Azooz MM, Youssef AM, Ahmad P (2011) Evaluation of salicylic acid (SA) application on growth, osmotic solutes and antioxidant enzyme activities on broad bean seedlings grown under diluted seawater. Int J Plant Physiol Biochem 3:253–264

Balliu A, Sallaku G, Rewald B (2015) AMF inoculation enhances growth and improves the nutrient uptake rates of transplanted, salt-stressed tomato seedlings. Sustainability 7:15967–15981

Bano A, Fatima M (2009) Salt tolerance in *Zea mays* (L). following inoculation with *Rhizobium* and *Pseudomonas*. Biol Fertil Soils 45:405–413

Barassi C, Ayrault G, Creus C, Sueldo R, Sobrero M (2006) Seed inoculation with Azospirillum mitigates NaCl effects on lettuce. Sci Hortic 109:8–14

Bartwal A, Mall R, Lohani P, Guru SK, Arora S (2013) Role of secondary metabolites and brassinosteroids in plant defense against environmental stresses. J Plant Growth Regul 32:216–232
Belimov AA, Safronova VI, Sergeyeva TA, Egorova TN, Matveyeva VA, Tsyganov VE, Preisfeld A (2001) Characterization of plant growth promoting rhizobacteria isolated from polluted soils and containing 1-aminocyclopropane-1-carboxylate deaminase. Can J Microbiol 47:642–652
Beltrano J, Ruscitti M, Arango MC, Ronco M (2013) Effects of arbuscular mycorrhiza inoculation on plant growth, biological and physiological parameters and mineral nutrition in pepper grown under different salinity and p levels. J Soil Sci Plant Nutr 13:123–141
Beresford Q, Bekle H, Phillips H, Mulcock J (2001) The salinity crisis: landscapes, communities and politics. University of Western Australia Press, Perth
Berta G, Fusconi A, Trotta A, Scannerini S (1990) Morphogenetic modifications induced by the mycorrhizal fungus *Glomus* strain E3 in the root system of *Allium porrum* L. New Phytol 114:207–215
Bharti N, Pandey SS, Barnawal D, Patel VK, Kalra A (2016) Plant growth promoting rhizobacteria *Dietzia natronolimnaea* modulates the expression of stress responsive genes providing protection of wheat from salinity stress. Sci Rep 6:34768
Bhattacharyya PN, Jha DK (2012) Plant growth-promoting rhizobacteria (PGPR): emergence in agriculture. World J Microbiol Biotechnol 28:1327–1350
Black AL, Brown PL, Halvorson AD, Siddoway FH (1981) Dry land cropping strategies for efficient water-use to control saline seeps in the northern Great Plains, USA. Agric Water Manage 4:295–311
Bohnert HJ, Nelson DE, Jensen RG (1995) Adaptations to environmental stresses. Plant Cell 7:1099–1111
Borde M, Dudhane M, Jite P (2011) Growth photosynthetic activity and antioxidant responses of mycorrhizal and non-mycorrhizal bajra (*Pennisetum glaucum*) crop under salinity stress condition. Crop Prot 30:265–271
Bordi A (2010) The influence of salt stress on seed germination, growth and yield of canola cultivars. Not Bot Horti Agrobo 38:128–133
Botella M, Del Amor F, Amorós A, Serrano M, Martínez V, Cerdá A (2000) Polyamine, ethylene and other physico-chemical parameters in tomato (*Lycopersicon esculentum*) fruits as affected by salinity. Physiol Plant 109:428–434
Bybordi A, Tabatabaei SJ, Ahmedov A (2010) Effects of salinity stress on fatty acids composition of Canola (*Brassica napus* L.). Food Agric J 8:113–115
Cantrell IC, Linderman RG (2001) Preinoculation of lettuce and onion with VA mycorrhizal fungi reduces deleterious effects of soil salinity. Plant and Soil 233:269–281
Carpici EB, Celik N, Bayram G (2009) Effects of salt stress on germination of somemaize (*Zea mays* L.) cultivars. Afr J Biotechnol 8:4918–4922
Casati P, Drincovich MF, Edwards GE, Andreo CS (1999) Malate metabolism by NADP-malic enzyme in plant defense. Photosynth Res 61:99–105
Chaum S, Kirdmanee C (2009) Effect of salt stress on proline accumulation, photosynthetic ability and growth characters in two maize cultivars. Pak J Bot 41:87–98
Cheeseman JM (1988) Mechanism of salinity tolerance in plants. Plant Physiol 87:547–550
Chen J, Zhang H, Zhang X, Tang M (2017) Arbuscular mycorrhizal symbiosis alleviates salt stress in black locust through improved photosynthesis, water status, and K^+/Na^+ homeostasis. Front Plant Sci 8:1739
Chulan HA, Martin K (1992) The vesicular-arbuscular (VA) mycorrhiza and its effect on growth of vegetatively propagated *Theobroma cacao* L. Plant and Soil 144:227–233
Chunthaburee S, Dongsansuk A, Sanitchon J, Pattanagul W, Theerakulpisut P (2016) Physiological and biochemical parameters for evaluation and clustering of rice cultivars differing in salt tolerance at seedling stage. Saudi J Biol Sci 23:467–477
Chutipaijit S, Cha-um S, Sompornpailin K (2011) High contents of proline and anthocyanin increase protective response to salinity in *Oryza sativa* L. spp. indica. Aust J Crop Sci 5:1191–1198

Cisneros JM, Cantero JJ, Cantero A (1999) Vegetation, soil hydrophysical properties, and grazing relationships in saline-sodic soils of Central Argentina. Can J Soil Sci 79:399–409
Cucci G, Lacolla G, Pallara M, Laviano R (2012) Reclamation of saline and saline-sodic soils using gypsum and leaching water. Afr J Agric Res 7:6508–6514
Daeia G, Ardekania MR, Rejalic F, Teimurib S, Miransari M (2009) Alleviation of salinity stress on wheat yield, yield components, and nutrient uptake using arbuscular mycorrhizal fungi under field conditions. J Plant Physiol 166:617–625
Dantas BF, De Sa RL, Aragao CA (2007) Germination, initial growth and cotyledon protein content of bean cultivars under salinity stress. Rev Bras Sementes 29:106–110
Diacono M, Montemurro F (2015) Effectiveness of organic wastes as fertilizers and amendments in salt-affected soils. Agriculture 5:221–230
Dimkpa C, Weinand T, Asch F (2009) Plant–rhizobacteria interactions alleviate abiotic stress conditions. Plant Cell Environ 32:1682–1694
Dionisio-Sese ML, Tobita S (1998) Antioxidant responses of rice seedlings to salinity stress. Plant Sci 135:1–9
DNRQ (1997) Salinity management handbook. Department of Natural Resources, Queensland. ISBN 0 7242 7412 X
Doganlar ZB, Demir K, Basak H, Gul I (2010) Effects of salt stress on pigment and total soluble protein contents of the three different tomato cultivars. Afr J Agric 5:2056–2065
Dolatabadian A, Modarressanavy SAM, Ghanati F (2011) Effect of salinity on growth, xylem structure and anatomical characteristics of soybean. Not Sci Bio 3:41–45
Duan J, Müller KM, Charles TC, Vesely S, Glick BR (2009) 1-aminocyclopropane-1-carboxylate (ACC) deaminase genes in rhizobia from southern Saskatchewan. Microb Ecol 57:423–436
Duchaufour P (1982) Pedology: pedogenesis and classification. Allen and Unwin, London
Ebrahim MKH, Saleem AR (2017) Alleviating salt stress in tomato inoculated with mycorrhizae: photosynthetic performance and enzymatic antioxidants. J Taibah Uni Sci 11:850–860
Egamberdieva D (2009) Alleviation of salt stress by plant growth regulators and IAA producing bacteria in wheat. Acta Physiol Plant 31:861–864
Egamberdieva D (2013) The role of phytohormone producing bacteria in alleviating salt stress in crop plants. Biotechnological techniques of stress tolerance in plants. Studium, Houston, TX, pp 21–39
Elhindi KM, El-Din AS, Elgorban AM (2017) The impact of arbuscular mycorrhizal fungi in mitigating salt-induced adverse effects in sweet basil (*Ocimum basilicum* L). Saudi J Biol Sci 24:170–179
El-Tayeb MA (2005) Response of barley grains to the interactive effect of salinity and salicylic acid. Plant Growth Regul 45:215–224
Fahad S, Hussain S, Bano A, Saud S, Hassan S, Shan D, Khan FA, Khan F, Chen Y, Wu C, Tabassum MA, Chun MX, Afzal M, Jan A, Jan MT, Huang J (2015) Potential role of phytohormones and plant growth-promoting rhizobacteria in abiotic stresses: consequences for changing environment. Environ Sci Pollut Res 22:4907–4921
Fang XJ, Wen ZX, Hao YN, Zhou Z, Li MS, Tao DM, Dong XU, Lai MJ (2013) Activities of principal photosynthetic enzymes in green macroalga *Ulva linza*: functional implication of C_4 pathway in CO_2 assimilation. Sci China Life Sci 56:571–580
FAO (1997) Irrigation potential in Africa: a basin approach. FAO Land and Water Bulletin 4, Rome
FAO (2015) FAO and ITPS: status of the world's soil resources (SWSR)—main report. Food and Agriculture Organization of the United Nations and Intergovernmental Technical Panel on Soils, Rome
FAO-Unesco (1973) Irrigation, drainage and salinity. An International Sourcebook. UNESCO/Hutchinson, Paris/London, p 510
Fazal A, Bano A (2016) Role of plant growth-promoting rhizobacteria (pgpr), biochar, and chemical fertilizer under salinity stress. Comm Soil Sci Plant Anal 47:1985–1993
Feng G, Zhang FS, Li XL, Tian CY, Tang C, Rengel Z (2002) Improved tolerance of maize plants to salt stress by arbuscular mycorrhiza is related to higher accumulation of soluble sugars in roots. Mycorrhiza 12:185–190

Fercha A (2011) Some physiological and biochemical effects of NaCl salinity on durum wheat (*Triticum durum* Desf.). Adv Biol Res 5:315–322
Fischer RA, Sayre K, Monasterio IO (2005) The effect of raised bed planting on irrigated wheat yield as influenced by variety and row spacing. In Proceedings of Workshop, Evaluation and performance of permanent raised bed cropping systems in Asia, Australia and Mexico held in Griffith, NSW, Australia
Flowers TJ (2004) Improving crop salt tolerance. J Exp Bot 55:307–319
Flowers TJ, Troke PF, Yeo AR (1977) The mechanism of salt tolerance in halophytes. Annu Rev Plant Physiol 28:89–121
Gain P, Mannan MA, Pal PS, Hossain MM, Parvin S (2004) Effect of salinity on some yield attributes of rice. Pak J Biol Sci 7:760–762
Gamalero E, Berta G, Massa N, Glick BR, Lingua G (2010) Interactions between *Pseudomonas putida* UW4 and *Gigaspora rosea* BEG9 and their consequences on the growth of cucumber under salt stress conditions. J Appl Microbiol 108:236–245
Garg N, Manchanda G (2009) Role of arbuscular mycorrhizae in the alleviation of ionic, osmotic and oxidative stresses induced by Mycorrhiza salinity in *Cajanus cajan* (L.) Millsp. (Pigeonpea). J Agron Crop Sci 195:110–123
Geddie JL, Sutherland I (1993) Uptake of metals by bacterial polysaccharides. J Appl Microbiol 74:467–472
George RJ, Nulsen RA, Ferdowsian R, Raper GP (1999) Interactions between trees and groundwaters in recharge and discharge areas—a survey of West Australian sites. Agric Water Manag 39:91–113
Ghassemi F, Jakeman AJ, Nix HA (1995) Salinisation of land and water resources: human causes, extent, management and case studies. UNSW Press, Sydney
Ghosh S, Penterman JN, Little RD, Chavez R, Glick BR (2003) Three newly isolated plant growth-promoting bacilli facilitate the seedling growth of canola, Brassica campestris. Plant Physiol Biochem 41:277–281
Giri B, Kapoor R, Mukerji KG (2003) Influence of arbuscular mycorrhizal fungi and salinity on growth, biomass, and mineral nutrition of *Acacia auriculiformis*. Biol Fertil Soils 38:170–175
Giri B, Kapoor R, Mukerji KG (2007) Improved tolerance of *Acacia nilotica* to salt stress by arbuscular mycorrhiza, *Glomus fasciculatum* may be partly related to elevated K/Na ratios in root and shoot tissues. Microb Ecol 54:753–760
Glick BR, Todorovic B, Czarny J, Cheng Z, Duan J, McConkey B (2007a) Promotion of plant growth by bacterial ACC deaminase. Crit Rev Plant Sci 26:227–242
Glick BR, Cheng Z, Czarny J, Duan J (2007b) Promotion of plant growth by ACC deaminase-producing soil bacteria. Eur J Plant Pathol 119:329–339
Gomes-Filho E, Lima CRFM, Costa JH, da Silva AC, daGuia SLM, de Lacerda CF, Prisco JT (2008) Cowpea ribonuclease: properties and effect of NaCl-salinity on its activation during seed germination and seedling establishment. Plant Cell Rep 27:147–157
Govindasamy V, Senthilkumar M, Gaikwad K, Annapurna K (2008) Isolation and characterization of ACC deaminase gene from two plant growth-promoting rhizobacteria. Curr Microbiol 57:312–317
Goyal SS, Sharma SK, Rainsa DW, Lauchli A (1999a) Long term reuse of drainage waters of varying salinities for crop irrigation in a cotton-safflower rotation system in the San Joaquin Valley of California—a nine year study: I. Cotton (*Gossypium hirsutum* L.). J Crop Prod 2:181–213
Goyal SS, Sharma SK, Rains DW, Lauchli A (1999b) Long term reuse of drainage waters of varying salinities for crop irrigation in a cotton-safflower rotation system in the San Joaquin Valley of California—a nine year study: II. Safflower (*Carthamus tinctorius* L.). J Crop Prod 2:215–227
Greenway H, Munns R (1980) Mechanisms of salt tolerance in non-halophytes. Annu Rev Plant Physiol 31:149–190
Grover M, Ali SZ, Sandhya V, Rasul A, Venkateswarlu B (2011) Role of microorganisms in adaptation of agriculture crops to abiotic stresses. World J Microbiol Biotechnol 27:1231–1240

Gueta-Dahan Y, Yaniv Z, Zilinskas BA, Ben-Hayyim G (1997) Salt and oxidative stress: similar and specific responses and their relation to salt tolerance in citrus. Planta 204:460–469

Habib SH, Kausar H, Saud HM (2016) Plant growth-promoting rhizobacteria enhance salinity stress tolerance in okra through ROS-scavenging enzymes. Biomed Res Int 2016:1–10

Hajiboland R, Aliasgharzadeh N, Laiegh SF, Poschenrieder C (2010) Colonization with arbuscular mycorrhizal fungi improves salinity tolerance of tomato (*Solanum lycopersicum* L.) plants. Plant and Soil 331:313–327

Hala M, El-Bassiouny S, Bekheta MA (2005) Effect of salt stress on relative water content, lipid peroxidation, polyamines, amino acids and ethylene of two wheat cultivars. Int J Agric Biol 3:363–368

Hameed A, Egamberdieva D, Abd_Allah EF, Hashem A, Kumar A, Ahmad P (2014) Salinity stress and arbuscular mycorrhizal symbiosis in plants. In: Miransari M (ed) Use of microbes for the alleviation of soil stresses. Springer, New York, pp 139–159

Hasanuzzaman M, Fujita M, Islam MN, Ahamed KU, Nahar K (2009) Performance of four irrigated rice varieties under different levels of salinity stress. Int J Integr Biol 6:85–90

Hashem A, Abd_Allah EF, Alqarawi AA, El-Didamony G, Mona SA, Egamberdieva D (2014) Alleviation of adverse impact of salinity on faba bean (*Vicia faba* L.) by arbuscular mycorrhizal fungi. Pak J Bot 46:2003–2013

Hashem A, Abd_Allah EF, Alqarawi AA, Mona SA, Alenazi MM, Egamberdieva D, Ahmad P (2015a) Arbuscular mycorrhizal fungi mitigates NaCl induced adverse effects on *Solanum lycopersicum* L. Pak J Bot 47:327–340

Hashem A, Abd-Allah EF, Alqarawi AA, Egamberdieva D (2015b) Induction of salt stress tolerance in cowpea [*Vigna unguiculata* (L.) Walp.] by arbuscular mycorrhizal fungi. Legume Res 38:579–588

Hashem A, Abd-Allah EF, Alqarawi AA, Wirth S, Egamberdieva D (2016a) Arbuscular mycorrhizal fungi alleviate salt stress in Lupine (*Lupinus termis* Forsik) through modulation of antioxidant defense systems and physiological traits. Legume Res 39:198–207

Hashem A, Abd-Allah EF, Alqarawi AA, Wirth S, Egamberdieva D (2016b) Comparing symbiotic performance and physiological responses of two soybean cultivars to arbuscular mycorrhizal fungi under salt stress. Saudi J Biol Sci. https://doi.org/10.1016/j.sjbs.2016.11.015

Hashem A, Salwa AA, Alqarawi AA, Abdullah EF, Egamberdieva D (2016c) Arbuscular mycorrhizal fungi enhance basil tolerance to salt stress through improved physiological and nutritional status. Pak J Bot 48:37–45

Hatton TJ, Ruprecht J, George R (2003) Preclearing hydrology of the Western Australia wheatbelt: target for the future? Plant and Soil 257:341–356

Hodge A, Campbell CD, Fitter AH (2001) An arbuscular mycorrhizal fungus accelerates decomposition and acquires nitrogen directly from organic material. Nature 413:297–299

Hoseini SM (2010) Studying effects of salinity stress on germination, proline and carbohydrate content in Thyme (*Thymus vulgaris* L) seedlings. Int J Agric Crop Sci 2:34–38

Hossain MA, Fujita M (2010) Evidence for a role of exogenous glycine betaine and proline in antioxidant defense and methyl glyoxal detoxification systems in mung bean seedlings under salt stress. Physiol Mol Biol Plants 16:19–29

Hu Y, Schmidhalter U (2002) Limitation of salt stress to plant growth. In: Hock B, Elstner CF (eds) Plant toxicology. Marcel Dekker Inc, New York, pp 91–224

Hu Y, Schmidhalter U (2005) Drought and salinity: a comparison of their effects on mineral nutrition of plants. J Plant Nutr Soil Sci 168:541–549

Hussain TM, Hazara M, Sultan Z, Saleh BK, Gopal GR (2008a) Recent advances in salt stress biology a review. Biotechnol Mol Biol Rev 3:8–13

Hussain TM, Chandrasekhar T, Hazara M, Sultan Z, Saleh B, Gopa GR (2008b) Recent advances in salt stress biology. Biotechnol J 3:1008–1013

Ibrar M, Jabeen M, Tabassum J, Hussain F, Ilahi I (2003) Salt tolerance potential of *Brassica juncea* Linn. J Sci Technol Uni Peshawar 27:79–84

Iyengar ERR, Reddy MP (1996) Photosynthesis in highly salt-tolerant plants. In: Pessaraki M (ed) Handbook of hotosynthesis. Marcel Dekker, New York, pp 897–909

Jaemsaeng R, Jantasuriyarat C, Thamchaipenet A (2018) Molecular interaction of 1-aminocyclopropane-1-carboxylate deaminase (ACCD)-producing endophytic Streptomyces sp. GMKU 336 towards salt-stress resistance of Oryza sativa L. cv. KDML105. Sci Rep 8:1950

Jahromi F, Aroca R, Porcel R, Ruiz-Lozano JM (2008) Influence of salinity on the in vitro development of *Glomus intraradices* and on the in vivo physiological and molecular responses of mycorrhizal lettuce plants. Microb Ecol 55:45–53

Jalili F, Khavazi K, Pazira E, Nejati A, Rahmani HA, Sadaghiani HR, Miransari M (2009) Isolation and characterization of ACC deaminase-producing fluorescent pseudomonads, to alleviate salinity stress on canola (*Brassica napus* L.) growth. J Plant Physiol 166:667–674

James RA, Rivelli AR, Munns R, von Caemmerer S (2002) Factors affecting CO_2 assimilation, leaf injury and growth in salt-stressed durum wheat. Funct Plant Biol 29:1393–1403

Jamil M, Lee DB, Jung KY, Ashraf M, Lee SC, Rh ES (2006) Effect of salt (NaCl) stress on germination and early seedling growth of four vegetables species. J Central Eur Agric 7:273–281

Jamil A, Riaz S, Ashraf M, Foolad MR (2011) Gene expression profiling of plants under salt stress. Crit Rev Plant Sci 30:435–458

Jobbágy EG, Jackson RB (2007) Groundwater and soil chemical changes under phreatophytic tree plantations. J Geophys Res 112:G02013. https://doi.org/10.1029/2006JG000246

Jones A, Panagos P, Barcelo S, Bouraoui F, Bosco C, Dewitte O, Gardi C, Hervás J, Hiederer R, Jeffery S (2012) The state of soil in Europe—a contribution of the JRC to the European Environment Agency's Environment State and Outlook R-SOER 2010

Juniper S, Abbott LK (2006) Soil salinity delays germination and limits growth of hyphae from propagules of arbuscular mycorrhizal fungi. Mycorrhiza 16:371–379

Kalaji HM, Govindjee BK, Koscielniakd J, Zük-Gołaszewska K (2011) Effects of salt stress on photosystem II efficiency and CO_2 assimilation of two Syrian barley landraces. Environ Exp Bot 73:64–72

Kang BR, Yang KY, Cho BH, Han TH, Kim IS, Lee MC, Anderson AJ, Kim YC (2006) Production of indole-3-acetic acid in the plant-beneficial strain Pseudomonas chlororaphis O6 is negatively regulated by the global sensor kinase GacS. Curr Microbiol 52:473–476

Kang SM, Khan AL, Waqas M, You YH, Kim JH, Kim JG, Hamayun M, Lee IJ (2014) Plant growth-promoting rhizobacteria reduce adverse effects of salinity and osmotic stress by regulating phytohormones and antioxidants in Cucumis sativus. J Plant Interact 9:673–682

Kaya C, Ashraf M, Sonmez O, Aydemir S, Tuna AL, Cullu MA (2009) The influence of arbuscular mycorrhizal colonization on key growth parameters and fruit yield of pepper plants grown at high salinity. Sci Hortic 121:1–6

Keutgen AJ, Pawelzik E (2009) Impacts of NaCl stress on plant growth and mineral nutrient assimilation in two cultivars of strawberry. Environ Exp Bot 65:170–176

Khan Z, Doty SL (2009) Characterization of bacterial endophytes of sweet potato plants. Plant and Soil 322:197–207

Khan MA, Rizvi Y (1994) Effect of salinity, temperature and growth regulators on the germination and early seedling growth of *Atriplex griffithii* var. Stocksii. Can J Bot 72:475–479

Khan MA, Irwin A, Allan MS (2000) The effect of salinity on the growth, water status and ion content of a leaf succulent perennial halophyte, *Suaeda fruticosa* L. Arid Environ J 45:73–84

Khan MA, Ansari R, Ali H, Gul B, Nielsen BL (2009) *Panicum turgidum*, a potentially sustainable cattle feed alternative tomaize for saline areas. Agric Ecosyst Environ 129:542–546

Khan MM, Al-Mas'oudi RSM, Al-Said F, Khan I (2013) Salinity effects on growth, electrolyte leakage, chlorophyll content and lipid peroxidationin cucumber (*Cucumis sativus* L.). In: Int Conf Food Agric Sci IPCBEE, vol 55. IACSIT Press, Singapore. https://doi.org/10.7763/IPCBEE.2013.V55.6

Khan MIR, Asgher M, Khan NA (2014) Alleviation of salt-induced photosynthesis and growth inhibition by salicylic acid involves glycine betaine and ethylene in mungbean (*Vigna radiata* L.). Plant Physiol Biochem 80:67–74

Khodarahmpour Z, Ifara M, Motamedi M (2012) Effects of NaCl salinity on maize (*Zea mays* L.) at germination and early seedling stage. Afr J Biotechnol 11:298–304

Kim SY, Lim JH, Park MR, Kim YJ, Park TI, Seo YW, Choi KG, Yun SJ (2005) Enhanced antioxidant enzymes are associated with reduced hydrogen peroxide in barley roots under saline stress. BMB Rep 38:218–224

Kloepper JW, Ryu CM, Zhang S (2004) Induced systemic resistance and promotion of plant growth by *Bacillus* species. Phytopathology 94:1259–1266

Kloepper J, Gutierrez-Estrada A, McInroy J (2007) Photoperiod regulates elicitation of growth promotion but not induced resistance by plant growth-promoting rhizobacteria. Can J Microbiol 53:159–167

Kohler J, Hernandez JA, Caravaca F, Roldan A (2009) Induction of antioxidant enzymes is involved in the greater effectiveness of a PGPR versus AM fungi with respect to increasing the tolerance of lettuce to severe salt stress. Environ Exp Bot 65:245–252

Koyro HW, Hussain T, Huchzermeyer B, Khan MA (2013) Photosynthetic and growth responses of a perennial halophytic grass *Panicum turgidum* to increasing NaCl concentrations. Environ Exp Bot 91:22–29

Kurth E, Cramer GR, Lauchli A, Epstein E (1986) Effects of NaCl and $CaCl_2$ on cell enlargement and cell production in cotton roots. Plant Physiol 82:1102–1106

Levitt J (1980) Responses of plants to environmental stresses, vol 2. Academic, New York

Linghe Z, Shannon MC (2000) Salinity effects on seedling growth and yield components of rice. Crop Sci 40:996–1003

Liu L, Sun H, Chen J, Zhang Y, Li D, Li C (2014) Effects of cadmium (Cd) on seedling growth traits and photosynthesis parameters in cotton. Plant Omics J 7:284–290

Liu S, Guo X, Feng G, Maimaitiaili B, Fan J, He X (2016) Indigenous arbuscular mycorrhizal fungi can alleviate salt stress and promote growth of cotton and maize in saline fields. Plant and Soil 398:195–206

Maas EV, Hoffman GJ (1977) Crop salt tolerance—current assessment. Proc Am Soc Civil Eng 103:115–134

Mahajan S, Tuteja N (2005) Cold, salinity and drought stresses: an overview. Arch Biochem Biophys 444:139–158

Mane AV, Karadge BA, Samant JS (2010) Salinity induced changes in photosynthetic pigments and polyphenols of *Cymbopogon nardus* (L.) Rendle. J Chem Pharm Res 2:338–347

Marcelis LFM, Van Hooijdonk J (1999) Effect of salinity on growth, water use and nutrient use in radish (*Raphanus sativus* L.). Plant and Soil 215:57–64

Marulanda A, Azcón R, Ruiz-Lozano JM (2003) Contribution of six arbuscular mycorrhizal fungal isolates to water uptake by *Lactuca sativa* plants under drought stress. Physiol Plant 119:526–533

Marulanda A, Porcel R, Barea JM, Azcón R (2007) Drought tolerance and antioxidant activities in lavender plants colonized by native drought-tolerant or drought-sensitive *Glomus* species. Microb Ecol 54:543–552

Masood A, Iqbal N, Asgher M, Khan MIR, Fatma M, Khan NA (2013) Variation in carbohydrate accumulation in two cultivars of mustard and its association with salt tolerance. J Funct Environ Bot 3:94–102

Mayak S, Tirosh T, Glick BR (2004) Plant growth-promoting bacteria confer resistance in tomato plants to salt stress. Plant Physiol Biochem 2:565–572

Michalak A (2006) Phenolic compounds and their antioxidant activity in plants growing under heavy metal stress. Polish J Environ Stud 15:523–530

Miransari M (2011) Arbuscular mycorrhizal fungi and nitrogen uptake. Review article. Arch Microbiol 193:77–81

Mittler R (2002) Oxidative stress, antioxidants and stress tolerance. Trends Plant Sci 7:405–410

Mittal S, Kumari N, Sharma V (2012) Differential response of salt stress on *Brassica juncea*: photosynthetic performance, pigment, proline, D1 and antioxidant enzymes. Plant Physiol Biochem 54:17–26

Mittova V, Guy M, Tal M, Volokita M (2004) Salinity upregulates the antioxidative system in root mitochondria and peroxisomes of the wild salt-tolerant tomato species *Lycopersicon pennellii*. J Exp Bot 55:1105–1113

Moghaieb REA, Saneoka H, Fujita K (2004) Effect of salinity on osmotic adjustment, glycine betaine accumulation and the betaine aldehyde dehydrogenase gene expression in two halophytic plants, *Salicornia europaea* and *Suaeda maritime*. Plant Sci 166:1345–1349
Mulrennan ME, Woodroffe CD (1998) Saltwater intrusion into the coastal plains of the Lower Mary River, Northern Territory, Australia. J Environ Manage 54:169–188
Munns R (1993) Physiological processes limiting plant growth in saline soil: some dogmas and hypotheses. Plant Cell Environ 16:15–24
Munns R (2002) Comparative physiology of salt and water stress. Plant Cell Environ 25:239–250
Munns R (2005) Genes and salt tolerance: bringing them together. New Phytol 167:645–663
Munns R, Tester M (2008) Mechanisms of salinity tolerance. Plant Biol J 59:651–681
Murkute AA, Sharma S, Singh SK (2006) Studies on salt stress tolerance of citrus rootstock genotypes with arbuscular mycorrhizal fungi. Hort Sci 33:70–76
Nadeem SM, Zahir ZA, Naveed M, Arshad M (2007) Preliminary investigations on inducing salt tolerance in maize through inoculation with rhizobacteria containing ACC deaminase activity. Can J Microbiol 53:1141–1149
Nadeem SM, Zahir ZA, Naveed M, Asghar HN, Arshad M (2010) Rhizobacteria capable of producing ACC-deaminase may mitigate salt stress in wheat. Soil Sci Soc Am J 74:533–542
Nadeem SM, Ahmad M, Naveed M, Imran M, Zahir ZA, Crowley DE (2016) Relationship between in vitro characterization and comparative efficacy of plant growth-promoting rhizobacteria for improving cucumber salt tolerance. Arch Microbiol 198:379–387
Nahar K, Hasanuzzaman M (2009) Germination, growth, nodulation and yield performance of three mung bean varieties under different levels of salinity stress. Green Farming 2:825–829
Najafi A, Ardakani MR, Rejali F, Sajedi N (2012) Response of winter barley to co-inoculation with *Azotobacter* and Mycorrhiza fungi influenced by plant growth promoting rhizobacteria. Ann Biol Res 3:4002–4006
Nautiyal CS, Govindarajan R, Lavania M, Pushpangadan P (2008) Novel mechanism of modulating natural antioxidants in functional foods: involvement of plant growth promoting rhizobacteria NRRL B-30488. J Agric Food Chem 56:4474–4481
Navarro JM, Perez-Tornero O, Morte A (2013) Alleviation of salt stress in citrus seedlings inoculated with arbuscular mycorrhizal fungi depends on the rootstock salt tolerance. J Plant Physiol 171:76–85
Nicholls RJ, Cazenave A (2010) Sea-level rise and its impact on coastal zones. Science 328:1517–1520
Noctor G, Foyer CH (1998) Simultaneous measurement of foliar glutathione, γ-glutamyl cysteine and amino acids by high-performance liquid chromatography: comparison with two other assay methods for glutathione. Anal Biochem 264:98–110
Otoch MDL, Sobreira ACM, deAragao MEF, Orellano EG, Lima MDS, deMelo DF (2001) Salt modulation of vacuolar H^+-ATPase and H^+-pyrophosphatase activities in *Vigna unguiculata*. J Plant Physiol 158:545–551
Parvaiz A, Satyawati S (2008) Salt stress and phyto biochemical responses of plants. Plant Soil J 54:89–99
Paul D (2012) Osmotic stress adaptations in rhizobacteria. J Basic Microbiol 52:1–10
Paul EA, Kucey RMN (1981) Carbon flow in plant microbial associations photosynthesis. Science 213:473–474
Paul D, Lade H (2014) Plant-growth-promoting rhizobacteria to improve crop growth in saline soils: a review. Agron Sustain Dev 34:737–752
Piotr S, Grazyna K (2005) Antioxidant defense in the leaves of C_3 and C_4 plants under salinity stress. Physiol Plant 125:31–40
Popova LP, Stoinova ZG, Maslenkova LT (1995) Involvement of abscisic acid in photosynthetic process in *Hordeum vulgare* L. during salinity stress. J Plant Growth Regul 14:211–218
Porcel R, Redondo-Gómez S, Mateos-Naranjo E, Aroca R, Garcia R, Ruiz-Lozano JM (2015) Arbuscular mycorrhizal symbiosis ameliorates the optimum quantum yield of photosystem II and reduces non-photochemical quenching in rice plants subjected to salt stress. J Plant Physiol 185:75–83.

Porras-Soriano A, Soriano-Martín ML, Porras-Piedra A, Azcón R (2009) Arbuscular mycorrhizal-fungi increased growth, nutrient uptake and tolerance to salinity in olive trees under nursery conditions. J Plant Physiol 166:1350–1359

Provin TL, Pitt JL (2001) Managing soil salinity. Texas AgriLife Extension Service publication E-60. Texas A&M University Publication, College Station, TX

Purty RS, Kumar G, Singla-Pareek SL, Pareek A (2008) Towards salinity tolerance in *Brassica*: an overview. Physiol Mol Biol Plants 14:39–49

Qadir M, Schubert S (2002) Degradation processes and nutrient constraints in sodic soils. Land Degrad Dev 13:275–294

Qureshi RH (2003) Impact of agriculture practices on environmental sustainability in South Asia. In: Proceedings of better agricultural practices for environmental sustainability. Asian Productivity Organization (APO), Tokyo

Qureshi RH, Aslam M, Akhtar J (2003) Productivity enhancement in the salt-affected lands of Joint Satiana Pilot Project area of Pakistan. In: Goyal SS, Sharma SK, Rains DW (eds) Crop production in saline environments: Global and integrative perspectives. The Food Products Press, Binghamton, New York, pp 277–297

Rabie GH (2005) Influence of arbuscular mycorrhizal fungi and kinetin on the response of mungbean plants to irrigation with seawater. Mycorrhiza 15:225–230

Rabie GH, Almadini AM (2005) Role of bioinoculants in development of salt-tolerance of *Vicia faba* plants under salinity stress. Afr J Biotechnol 4:210–220

Rahdari P, Tavakoli S, Hosseini SM (2012) Studying of salinity stress effect on germination, proline, sugar, protein, lipid and chlorophyll content in Purslane (*Portulaca oleraceae* L.) leaves. Stress Physiol Biol J 8:182–193

Rasool S, Ahmad A, Siddiqi TO, Ahmad P (2013) Changes in growth, lipid peroxidation and some key antioxidant enzymes in chickpea genotypes under salt stress. Acta Physiol Planta 35:1039–1050

Reddy MP, Sanish S, Iyengar ERR (1992) Photosynthetic studies and compartmentation of ions in different tissues of *Salicornia brachiata* Roxb. under saline conditions. Photosynthetica 26:173–179

Rengasamy P (2006) World salinization with emphasis on Australia. J Exp Bot 57:1017–1023

Rengasamy P (2010) Soil processes affecting crop production in salt-affected soils. Funct Plant Biol 37:613–620

Richards LA (1954) Diagnosis and improvement of saline and alkali soils. USDA, Washington, DC

Richards LA (1969) Diagnosis and improvement of saline and alkali soils. Agriculture Handbook No. 60. United States Department of Agriculture, Washington DC

Rogers ME, Grieve CM, Shannon MC (2003) Plant growth and ion relations in lucerne (*Medicago sativa* L.) in response to the combined effects of NaCl and P. Plant and Soil 253:187–194

Rose C (2004) An introduction to the environmental physics of soil, water and watersheds. Cambridge University Press, Cambridge

Ruiz-Lozano JM, Azcon R (1995) Hyphal contribution to water uptake in mycorrhizal plants as affected by the fungal species and water status. Physiol Plant 95:472–478

Ruiz-Lozano JM, Azcón R, Gómez M (1996) Alleviation of salt stress by arbuscular-mycorrhizal *Glomus* species in *Lactuca sativa* plants. Physiol Plant 98:767–772

Runyan CW, D'Odorico P (2010) Ecohydrological feedbacks between salt accumulation and vegetation dynamics: role of vegetation-groundwater interactions. Water Resour Res 46:W11561

Sadeghi A, Karimi E, Dahaji PA, Javid MG, Dalvand Y, Askari H (2012) Plant growth promoting activity of an auxin and siderophore producing isolate of Streptomyces under saline soil conditions. World J Microbiol Biotechnol 28:1503–1509

Saha P, Chatterjee P, Biswas AK (2010) NaCl pretreatment alleviates salt stress by enhancement of antioxidant defense system and osmolyte accumulation in mungbean (*Vigna radiata* L. Wilczek). Indian J Exp Biol 48:593–600

Sairam RK, Tyagi A (2004) Physiology and molecular biology of salinity stress tolerance in plants. Curr Sci J 86:407–421

Sairam RK, Roa KV, Srivastava GC (2002) Differential response of wheat genotypes to long term salinity stress in relation to oxidative stress, antioxidant activity and osmolyte concentration. Plant Sci 163:1037–1046

Sannazzaro AI, Ruíz OA, Alberto EO, Menendez AB (2006) Alleviation of salt stress in *Lotusglaber* by *Glomus intraradices*. Plant and Soil 285:279–287

Sarkar A, Ghosh PK, Pramanik K, Mitra S, Soren T, Pandey S, Mondale MH, Maitia TK (2018) A halotolerant Enterobacter sp. displaying ACC deaminase activity promotes rice seedling growth under salt stress. Res Microbiol 169:20–32

Sarker A, Hossain I, Kashem A (2014) Salinity (NaCl) tolerance of four vegetable crops during germination and early seedling growth. Int J Latest Res Sci Technol 3:91–95

Sarwat M, Hashem A, Ahanger MA, Abd-Allah EF, Alqarawi AA, Alyemeni MN, Ahmad P, Gucel S (2016) Mitigation of NaCl Stress by arbuscular mycorrhizal fungi through the modulation of osmolytes, antioxidants and secondary metabolites in mustard (*Brassica juncea* L.) plants. Front Plant Sci 7:869

Schagerl M (2016) Soda lakes of East Africa. Springer, Cham

Schofield RV, Kirkby MJ (2003) Application of salinization indicators and initial development of potential global soil salinization scenario under climatic change. Global Biogeochem Cycles 17:1078

Senadheera P, Tirimanne S, Maathuis FJM (2012) Long term salinity stress reveals variety specific differences in root oxidative stress response. Ric Sci 19:36–43

Shahbaz M, Ashraf M (2013) Improving salanity tolerance in cereals. Cri Rev Plant Sci 32:237–249

Sharifi M, Ghorbanli M, Ebrahimzadeh H (2007) Improved growth of salinity-stressed soybean after inoculation with salt pre-treated mycorrhizal fungi. J Plant Physiol 164:1144–1151

Sheng M, Tang M, Chen H, Yang B, Zhang F, Huang Y (2008) Influence of arbuscular mycorrhizae on photosynthesis and water status of maize plants under salt stress. Mycorrhiza 18:287–296

Shirmardi M, Savaghebi GR, Khavazi K, Akbarzadeh A, Farahbakhsh M, Rejali F, Sadat A (2010) Effect of microbial inoculants on uptake of nutrient elements in two cultivars of sunflower (*Helianthus annuus* L.) in saline soils. Not Sci Bio 2:57–66

Shrivastava P, Kumar R (2015) Soil salinity: a serious environmental issue and plant growth promoting bacteria as one of the tools for its alleviation. Saudi J Biol Sci 22:123–131

Sinclair G, Charest C, Dalpe Y, Khanizadeh S (2014) Influence of colonization by arbuscular mycorrhizal fungi on three strawberry cultivars under salty conditions. Agric Food Sci 23:146–158

Sjöberg J (2005) Arbuscular mycorrhizal fungi–occurrence in Sweden and interaction with a plant pathogenic fungus in barley. PhD Thesis. Swedish University of Agricultural Sciences, Uppsala, pp 1–55

Slinger D, Tenison K (2007) Salinity Glove Box Guide: NSW Murray and Murrumbidgee Catchments, NSW Department of Primary Industries

Smith SE, Read DJ (2008) Mycorrhizal symbiosis. Academic, San Diego, CA

Solangi SB, Chachar QI, Chachar SD, Solangi AB, Solangi JA (2016) Effect of salinity (NaCl) stress on physiological characteristics of rice (*Oryza sativa* L.) at early seedling stage. J Agric Technol 12:263–279

Soussi M, Ocana A, Lluch C (1998) Effects of salt stress on growth, photosynthesis and nitrogen fixation in chick-pea (*Cicer arietinum* L.). J Exp Bot 49:1329–1337

Spaepen S, Dobbelaere S, Croonenborghs A, Vanderleyden J (2008) Effects of *Azospirillum brasilense* indole-3-acetic acid production on inoculated wheat plants. Plant and Soil 312:15–23

Spychalla JP, Desborough SL (1990) Superoxide dismutase, catalase, and α-tocopherol content of stored potato tubers. Plant Physiol 94:1214–1218

Stirzaker RJ, Vertessy RA, Sarre A (2002) Trees, water and salt: an Australian guide to using trees for healthy catchments and productive farms. Rural Industries Research and Development Corporation, Canberra

Sultana N, Keda T, Itoh R (1999) Effect of NaCl salinity on photosynthesis and dry matter accumulation in developing rice grains. Environ Exp Bot 42:211–220

Tang M, Chen H, Huang JC, Tian ZQ (2009) AM fungi effects on the growth and physiology of *Zea mays* seedlings under diesel stress. Soil Biol Biochem 41:936–940
Tank N, Saraf M (2010) Salinity-resistant plant growth promoting rhizobacteria ameliorates sodium chloride stress on tomato plants. J Plant Interact 5:51–58
Tarbuck EJ, Lutgens FK (2012) Earth science. Prentice Hall/Pearson, Upper Saddle River
Tatar O, Brueck H, Gevreka MN, Asch F (2010) Physiological responses of two Turkish rice (*Oryza sativa* L.) varieties to salinity. Turk J Agric For 34:451–459
Tavakkoli E, Fatehi F, Coventry S, Rengasamy P, McDonald GK (2011) Additive effects of Na^+ and Cl^- ions on barley growth under salinity stress. J Exp Bot 62:2189–2203
Thakur M, Sharma AD (2005) Salt stress induced proline accumulation in germinating embryos: evidence suggesting a role of proline in seed germination. J Arid Environ 62:517–523
Thompson JP (1990) Soil sterilization methods to show VA-mycorrhizae aid P and Zn nutrition of wheat in vertisols. Soil Biol Biochem 22:229–240
Tian CY, Feng G, Li XL, Zhang FS (2004) Different effects of arbuscular mycorrhizal fungal isolates from saline or non-saline soil on salinity tolerance of plants. Appl Soil Ecol 26:143–148
Tilman D, Cassman KG, Matson PA, Naylor R, Polasky S (2002) Agricultural sustainability and intensive production practices. Nature 418:671–677
Tomar NS, Agarwal RM (2013) Influence of treatment of *Jatropha curcas* L. leachates and potassium on growth and phytochemical constituents of wheat (*Triticum aestivum* L.). Am J Plant Sci 4:1134–1150
Tommasi F, Paciolla C, de Pinto MC, Gara LD (2001) A comparative study of glutathione and ascorbate metabolism during germination of *Pinus pinea* L. seeds. J Exp Bot 52:1647–1654
Tuna AL, Kaya C, Dikilitas M, Higgs D (2008) The combined effects of gibberellic acid and salinity on some antioxidant enzyme activities, plant growth parameters and nutritional status in maize plants. Environ Exp Bot 62:1–9
Tóth G, Montanarella L, Rusco E (2008) Threats to soil quality in Europe. Institute for Environment and Sustainability, Land Management and Natural Hazards Unit, Office for the Official Publications of the European Communities, Luxembourg, EUR 23438 EN, pp 162
Ulfat M, Athar H, Ashraf M, Akram NA, Jamil A (2007) Appraisal of physiological and biochemical selection criteria for evaluation of salt tolerance in canola (*Brassica napus* L.). Pak J Bot 39:1593–1608
UNEP (1992) Proceedings of the Ad-hoc Expert Group Meeting to Discuss Global Soil Databases and Appraisal of GLASOD/SOTER, February 24–28. Nairobi, UNEP
Upadhyay SK, Singh DP, Saikia R (2009) Genetic diversity of plant growth promoting rhizobacteria isolated from rhizospheric soil of wheat under saline condition. Curr Microbiol 59:489–496
Upadhyay SK, Singhd S, Singh P (2011) Exopolysaccharide-producing plant growth-promoting rhizobacteria under salinity condition. Pedosphere 21:214–222
USDA (1998) Soil Quality Resource Concerns: Salinization. Natural Resources Conservation Service, Washington, DC
Usha K, Mathew R, Singh B (2005) Effect of three species of arbuscular mycorrhiza on bud sprout and ripening in grapevine (*Vitis vinifera* L.) cv. Perlette. Biol Agric Hort 23:73–83
Vaidyanathan R, Kuruvilla S, Thomas G (1999) Characterization and expression pattern of an abscisic acid and osmotic stress responsive gene from rice. Plant Sci 140:21–30
Van Loon LC, Bakker PA, Pieterse CM (2004) Systemic resistance induced by rhizosphere bacteria. Annu Rev Phytopathol 36:453–483
Wang BS, Luttge U, Ratajczak R (2001) Effects of salt treatment and osmotic stress on V-ATPase and V-PPase in leaves of the halophyte *Suaeda salsa*. J Exp Bot 52:2355–2365
Wang H, Hsieh YP, Harwell MA, Huang W (2007) Modeling soil salinity distribution along topographic gradients in tidal salt marshes in Atlantic and Gulf coastal regions. Ecol Model 201:429–439
Waqas M, Khan AL, Kamran M, Hamayun M, Kang SM, Kim YH, Lee IJ (2012) Endophytic fungi produce gibberellins and indole acetic acid and promotes host-plant growth during stress. Molecules 17:10754–10,773

Wilford J, de Caritat P, Bui E (2015) Modelling the abundance of soil calcium carbonate across Australia using geochemical survey data and environmental predictors. Geoderma 259:81–92
Williamson DR (1986) The hydrology of salt affected soils in Australia. Reclam Reveg Res 5:181–196
Wu QS, Zou YN, Xia RX, Wang MY (2007) Five *Glomus* species affect water relations of *Citrus tangerine* during drought stress. Bot Stud 48:147–154
Wu QS, Zou YN, He XH (2010) Contributions of arbuscular mycorrhizal fungi to growth, photosynthesis, root morphology and ionic balance of citrus seedlings under salt stress. Acta Physiol Plant 32:297–304
Wu QS, Zou YN, Abd_Allah EF (2014) Mycorrhizal association and ROS in plants. In: Ahmad P (ed) Oxidative damage to plants. Elsevier, Amsterdam, pp 453–475
Xie X, Weng B, Cai B, Dong Y, Yan C (2014) Effects of arbuscular mycorrhizal inoculation and phosphorus supply on the growth and nutrient uptake of *Kandelia obovata* (Sheue, Liu & Yong) seedlings in autoclaved soil. Appl Soil Ecol 75:162–171
Xu S, Hu B, He Z, Ma F, Feng J, Shen W, Yan J (2011) Enhancement of salinity tolerance during rice seed germination by presoaking with hemoglobin. Int J Mol Sci 12:2488–2501
Yamaguchi T, Blumwald E (2005) Developing salt-tolerant crop plants: challanges and opportunities. Trends Plant Sci 10(12):615–620
Yamato M, Ikeda S, Iwase K (2008) Community of arbuscular mycorrhizal fungi in coastal vegetation on Okinawa Island and effect of the isolated fungi on growth of sorghum under salt-treated conditions. Mycorrhiza 18:241–249
Yang J, Kloepper JW, Ryu CM (2009) Rhizosphere bacteria help plants tolerate abiotic stress. Trends Plant Sci 14:1–4
Yano-Melo AM, Saggin OJ, Maia LC (2003) Tolerance of mycorrhized banana (*Musa* sp. cv. Pacovan) plantlets to saline stress. Agric Ecosyst Environ 95:343–348
Yildirim E, Taylor A (2005) Effect of biological treatments on growth of bean plants under salt stress. Science 123:1
Yildirim E, Turan M, Donmez MF (2008) Mitigation of salt stress in radish (*Raphanus sativus* L.) by plant growth promoting rhizobacteria. Roumanian Biotechnol Lett 13:3933–3943
Yue H, Mo W, Li C, Zheng Y, Li H (2007) The salt stress relief and growth promotion effect of Rs-5 on cotton. Plant and Soil 297:139–145
Zahir ZA, Ghani U, Naveed M, Nadeem SM, Asghar HN (2009) Comparative effectiveness of Pseudomonas and Serratia sp. containing ACC-deaminase for improving growth and yield of wheat (*Triticum aestivum* L.) under salt-stressed conditions. Arch Microbiol 191:415–424
Zhang HX, Blumwald E (2001) Transgenic salt-tolerant tomato plants accumulate salt in foliage but not in fruit. Nat Biotechnol 19:765–768
Zhong QH, Chao XH, Zhibin Z, Zhirong Z, Huai SW (2007) Changes in antioxidative enzymes and cell membrane osmosis in tomato colonized by arbuscular mycorrhizae under NaCl stress. Colloids Surf B Biointerfaces 59:128–133
Zhu JK (2001) Plant salt tolerance. Trends Plant Sci 6:66–71
Zhu JK (2002) Salt and drought stress signal transduction in plants. Plant Biol J 53:247–273
Zhu JK (2003) Regulation of ion homeostasis under salt stress. Curr Opin Plant Biol 6:441–445
Zuccarini P, Okurowska P (2008) Effects of mycorrhizal colonization and fertilization on growth and photosynthesis of sweet basil under salt stress. J Plant Nutr 31:497–513

Plant MIRnome: miRNA Biogenesis and Abiotic Stress Response

Deepu Pandita

Introduction

Abiotic stress is defined as the negative impact of non-living ecological/environmental factors on the living organisms in a specific environment. Abiotic stress is the most detrimental factor regarding the growth and productivity of crops globally. To prevaricate environmental/abiotic stresses, plant life exhibits tolerance or avoidance by acclimation and adaptation evolved through natural selection (Yamaguchi-Shinozaki and Shinozaki 2006). Tolerance to abiotic stresses is a very intricate phenomenon involving interactions between stress-responsive elements and molecular and biochemical factors influencing plant growth and development. Today, the microRNAs (miRNAs/miRs)-mediated stress-responsive molecular mechanism has been identified. miRNAs are 21 to 25-nucleotide double-stranded, endogenous, non-coding RNAs, which regulate the gene expression at transcriptional and post-transcriptional levels through translational repression of target mRNA in plants, mRNA cleavage, chromatin remodeling, and/or DNA methylation. The stress-induced miRNAs up-regulation brings about down-regulation of their target mRNAs, while their down-regulation leads to up-regulation, buildup, and function of positive regulators (Chinnusamy et al. 2007). The investigations were formerly aimed to identify abiotic stress-responsive plant miRNAs, their expression profiling and functions tolerance and stress responses such as drought, salinity, extreme temperatures, nutrient deprivation, and heavy metals. Several plant species subjected to abiotic stress conditions reported changes in miRNAs expressions connected with their growth and development. Several protein-coding genes and miRNAs have been identified for regulating abiotic stresses responses to plant, but still the regulatory mechanisms at molecular levels are unreported. The need of the future is the transformative tools to adapt crops to inconsiderate surroundings (Zhang and Wang

D. Pandita (✉)
Government Department of School Education, Jammu, Jammu and Kashmir, India

M. Hasanuzzaman et al. (eds.), *Plant Abiotic Stress Tolerance*,
https://doi.org/10.1007/978-3-030-06118-0_18

2015), for which miRNAs act as prospective targets for genetic engineering and manipulations for producing abiotic stress-tolerant crop plants. Diverse abiotic stresses-responsive microRNAs have been found in crop plants like rice, barley, wheat, sugarcane, legumes, tomato, potato, and in many other species (Barciszewska-Pacak et al. 2015). miRNAs play a fundamental role as negative regulators of gene expression in the cellular, biological, and developmental processes and abiotic stress responses, such as drought (Zhou et al. 2010), salinity (Li et al. 2013a, b, c), cold (Zhang et al. 2009a, b), heat (Zhao et al. 2016), nutritional deficiency (Kulcheski et al. 2015; Kumar et al. 2017), oxidative stress (Zhang 2015), and mechanical stress (Kateryna et al. 2016; Zhang 2015). Abiotic response elements were found in the promoter region of many microRNA genes and the use of bioinformatics tools suggests their transcriptional regulation (Zhao et al. 2013). In view of this, this chapter summarizes the herein miRNAs, their biogenesis, and recent updates on regulatory role of plant miRNAs in abiotic stress responses.

Yesteryears of miRNA

The first miRNA identified was lin-4 in the nematode *Caenorhabditis elegans* (Lee et al. 1993), considered as small temporal RNAs (stRNAs). Gene *lin-4* was not undergoing translation but only transcription into a pair of petite RNAs-a shorter RNA 22 nucleotides long and a longer one roughly 61 nucleotides in length (stem-loop precursor). The stem-loop structured longer RNA was envisaged as the precursor of the 21-mer RNA. This shorter *lin-4* RNA was actually the founding member of the abundantly available class of small regulatory RNAs known today as miRNA. The *lin-4* RNAs had partial antisense complementarity with manifold sites in 3′-untranslated region (3′ UTR) of *lin-14* gene transcript, in that way inhibiting the translation of *lin-14* without reducing *lin-14* transcript levels (Lee et al. 1993; Wightman et al. 1993). It was only after year 2000 that this small regulatory RNA became complex when *let-7* (22-nt regulatory RNA) was also discovered in *C. elegans*. Gene *let-7* promotes the transition from late-larval to fully developed cell in *C. elegans* (Reinhart et al. 2000; Vella et al. 2004). The *let-7* miRNA is ubiquitous in animal phyla with its homologs in the genomes of human, Drosophila, and bilateral animals, thus ascertaining biological significance of miRNAs (Pasquinelli et al. 2000). >100 additional genes were identified in various organisms, for instance, flies, worms, and human cells in <1 year (Lagos-Quintana et al. 2001; Lau et al. 2001; Lee and Ambros 2001). As a consequence of its developmental stage transition-specific regulation, *lin-4* and *let-7* (originator members) were reported as small temporal RNAs (stRNAs). Afterwards the name "microRNA" came into reality, including all other small RNAs with analogous characters and regulatory functions (Vella et al. 2004). From various biological organisms including plants, miRNAs have been cloned later and exposed their ubiquitous nature in plants. The first plant miRNA (miR156, miR159, miR164, miR171, etc.) was discovered in *Arabidopsis thaliana* in 2000, 13 years after their discovery in animals (Reinhart et al. 2002). Llave et al. (2002b) cloned a huge number of miRNAs of predominantly 21-nt to 24-nt length from *Arabidopsis thaliana*. Amid 125 miRNAs from the *Arabidopsis* genome reported, about 90% were intergenic in nature, few from intragenic regions of the protein-coding genes and transposable element derived (Mette et al. 2002). Differential expression profiling of 16 *Arabidopsis thaliana* miRNAs was performed, out of which half were conserved in *Oryza sativa*. *Arabidopsis thaliana* Dicer (RNase III family enzyme) homolog CARPEL FACTORY, when mutated, stopped the miRNA accumulation concluding that an analogous mechanism also exists in plant kingdom which direct miRNA processing as in animal world (Park et al. 2002; Reinhart et al. 2002). Regardless of some similarity between the plant and animal miRNAs, the variations between them were more conspicuous. It includes the size range of the mature miRNAs, miRNA gene structure, 5′ nucleotide predilections, and most notably absence of any genetic conservation between the two groups and the differences in the modes of miRNA biogenesis and functions. These facts suggest that plant and animal kingdoms had twofold origins in their evolutionary lineages. To date, a total of 48,885 mature miRNAs sequences from 271 different species are registered in the miRBase database (Release 22, March 2018) and 8433 miRNAs from 121 plant species are listed in PMRD-plant miRNA database (Pandita 2018).

Biogenesis of miRNAs and Mode of Action

Transcription, miRNA precursor processing, maturation/modification, and finally execution to RISC (a multiprotein complex) are the steps in miRNA biogenesis.

MIR Transcription

The miRNAs are transcribed from non-coding nuclear miRNA genes (MIR), a majority of which are located mainly in intergenic and intragenic (intron) regions, very few of them are located in intragenic (exon) regions and in 5′-UTR or 3′-UTR regions. MIR gene has a transcription unit comprising of its own promoter, transcribing region and terminator (Lee et al. 2004). Plant miRNAs are usually transcribed by RNA polymerase II. A transcriptional coactivator (mediator) helps recruit RNA pol II to MIR loci (Kim et al. 2011). NEGATIVE ON TATA LESS2 (NOT2), CELL DIVISION CYCLE 5 (CDC5), and the Elongator complex are other factors of MIR transcription which interact with RNA pol II and the dicing complex entailing their functions in bridging RNA pol II transcription and pri-miRNA processing (Zhang et al. 2013; Fang et al. 2015). POWERDRESS factor supports the transcription of MIR172a, MIR172b, and MIR172c by enhancing RNA pol II occupancy at their promoter region but has no effect on MIR172d or MIR172e (Baek et al. 2013). DNA-dependent RNA polymerase II transcribes a miRNA gene (MIR) into nascent stem-loop double-stranded primary miRNA transcripts (pri-miRNA) with a 5′ capping and 3′ polyadenylation.

miRNA Precursor Processing

pri-miRNAs are processed by the dicing complex with three core components DICER-LIKE1 (DCL1), HYPONASTIC LEAVES1 (HYL1), and SERRATE (SE) to mature miRNA/miRNA* duplexes (Fukudome and Fukuhara 2017). Primary miRNA transcripts (pri-miRNAs) are then processed into the hairpin structured precursor miRNA (pre-miRNA) of roughly 70 nucleotides by a predominant DCL RNAse III endonuclease, DICER-LIKE 1 (DCL 1) in plants except miR822 and miR839 which are generated by DCL4 (Rajagopalan et al. 2006) and Drosha enzyme in animals (Kurihara et al. 2006; Reinhart et al. 2002) in the Dicing bodies (D bodies). A part of the pri-miRNA folds to compose a stem-loop structure which is sliced by the DCL1 enzyme. DCL1-mediated processing occurs as a rule at the lower stem or base, while in minority cases the terminal loop controls the processing. Amid long-stem or terminal loop-branched pri-miRNAs, the miRNA processing might be bidirectional, i.e., from base to loop and loop to base, resulting in productive and abortive processing of miRNAs, respectively. DCL1 complexes generally cut pri-miRNAs at a distance of 16–17 bp from the single-strand–double-strand junction (Zhu et al. 2013). Double-stranded pre-miRNA, miRNA-miRNA* duplex is formed by the endonucleolytic activity of DCL1. DCL requires the proteins HYPONASTIC LEAVES1 (HYL1-DOUBLE-STRANDED RNA-BINDING PROTEINS (DRBs)) to facilitate accurate miRNA precursor processing and guide strand selection during AGO loading (Manavella et al. 2012) and SERRATE (SE-Zn finger protein) for processing (Han et al. 2004; Vazquez et al. 2004; Lobbes et al. 2006; Yang et al. 2006a). Together with battery of other proteins, CAP-BINDING protein COMPLEX (CBC) (CBP80) (CBP20) (Gonatopoulos-Pournatzis and Cowling 2014), STABILIZED1 (STA1) (Ben Chaabane et al. 2013), THO1/HPR1/EMU and THO2 (Furumizu et al. 2010; Francisco-Mangilet et al. 2015), SICKLE (Zhan et al. 2012), TOUGH (TGH) (Ren et al. 2012b), PSR1-INTERACTING PROTEIN 1 (PINP1) (Qiao et al. 2015), and MODIFIER OF SNC1, 2 (MOS2) (Wu et al. 2013) CDC5, NOT2, Elongator, PRL1 (PROTEIN PLEIOTROPIC REGULATORY LOCUS 1), and DDL (DAWDLE) (Zhang et al. 2013, 2014; Fang et al. 2015) proteins promote miRNA processing and biogenesis.

miRNA Stabilization and RISC Formation

Pre-miRNA is unstable in nucleus and the stability of the miRNA–miRNA* duplex is brought about by methylation at 2′ OH of 3′ terminal nucleotides by an *S*-adenosyl methionine-dependent methyltransferase protein HUA ENHANCER 1 (HEN1) (Yu et al. 2005). Methyl groups positioned on the 3′ terminal nucleotides of each strand avert their uridylation and succeeding degradation by the SMALL RNA DEGRADING NUCLEASE (SDN) class of exonucleases (Li et al. 2005; Yang et al. 2006b). Then export factor HASTY5 (HST5) which is an ortholog of exportin-5 protein exports the tailored miRNA miRNA* duplex from nucleus to cytoplasm (Park et al. 2005) wherein it is further processed into mature double-stranded miRNA.

Arabidopsis thaliana RISC loading at specific subcellular locations takes place in four steps:

1. Formation of complex of AGO1 and HEAT SHOCK PROTEIN 90 (HSP90) dimer.
2. Binding of ATP to HSP90 which causes conformational change of AGO1 so that small RNA duplex is incorporated into the AGO1–HSP90 protein complex.
3. ATP hydrolysis for dissociation of AGO1 from HSP90.
4. AGO1 conformational change caused by HSP90 dissociation causes passenger strand removal and formation of mature RISC (Iki et al. 2010).

The miRNA-miRNA* duplex is unwound in cytoplasm and one strand of duplex is loaded into AGO1 protein to form miRISC (Baumberger and Baulcombe 2005; Qi et al. 2005). The miRNA-programmed silencing complex is known as the RNA-induced silencing complex (RISC) (Hammond et al. 2000; Llave et al. 2002a; Tang et al. 2003). How the guide strand is selected is affected by miRNA precursor processing factors, structure of the small RNA duplex, and the nature of 5′ end nucleotide. HYL1 and HYL1 phosphatase CPL1 factors assist in guide strand selection (Manavella et al. 2012). The nature of the 5′ nucleotides and the bulges in miRNA/miRNA* duplex directs AGO loading. The majority of miRNA guide strands initiates with a 5′-terminal uridine and are incorporated into AGO1. AGO1 also prefers miRNA/miRNA* duplexes with central mismatches whereas AGO2 prefers miRNA duplexes without central mismatches (Ren et al. 2014). Cloning and expression data signify that the miRNA strand (guide strand) of miRNA miRNA* duplex accumulates to a great deal of higher levels in vivo in comparison to miRNA* or the passenger strand (Reinhart et al. 2002). This asymmetric accretion of guide and passenger strand is accomplished by preferential loading of the guide (miRNA) strand into the RNA-induced silencing complex (RISC), where it is prevented from degradation, whereas the passenger (miRNA*) strand is preferentially kept out of the silencing complex and is thus degraded. Most miRNA–miRNA* duplexes also have asymmetry in terms of energy; the 5′ ends of most miRNAs are not that stably paired than those of the corresponding miRNA*s (Khvorova et al. 2003). Consequently, at the end of the miRNA biogenesis pathway, a small single-stranded RNA is integrated into the silencing complex.

A central component of the RNA-induced silencing complex (RISC) is a member of the Argonaute protein family. Argonaute proteins have two conserved domains, the PAZ and PIWI (Carmell et al. 2002). PAZ domain is an RNA-binding domain (Lingel et al. 2003) while PIWI domain is structurally and functionally analogous to RNase H enzymes (Song et al. 2004). Many biological organisms contain different members of Argonaute family so is an evidence for their functional diversification. *Arabidopsis thaliana* AGO1 binds miRNAs and causes target cleavage in vitro (Baumberger and Baulcombe 2005; Qi et al. 2005) and *ago1* mutants have enhanced levels of miRNA targets in vivo (Vaucheret et al. 2004). The target mRNA cleavage takes place at the tenth or 11th nucleotide of the 5′ end of the miRNA, despite of the miRNA length and the base pairing of the 5′ end of the miRNA with the target mRNA is required (Floyd and Bowman 2004). The consequential cleavage products have 3′ hydroxyl and 5′ phosphate groups, analogous to other enzyme products with "slicer" activity, for instance, RNase H (Song et al. 2004). Plant miRNAs are typically loaded with U at the 5′ end and are accordingly sorted in the AGO1-mediated RISC complexes (Voinnet 2009). A few *Arabidopsis* miRNAs have U to A mutation at the 5′ end which is sorted in AGO2 complexes in vivo leading to loss of their silencing potential (Mi et al. 2008). Bioinformatics analysis of the next-generation sequencing (NGS) data showed that there is a wide spectrum of variations in the incidence of U at the 5′ position in monocots (45–85%) and dicots (45–70%). In potato, U and A rule are present at equal instances at the 5′ position (Takeda et al. 2008). Few plant miRNAs show undergo modifications at the 5′ and 3′ ends whereas the 3′ end modifications are more frequent (Ameres and Zamore 2013). The modifications are additions or deletions and occur either due to faulty processing of DCL1 or could be the effect of post-maturation processes of miRNAs (Voinnet 2009). Over 80% of the modified *Arabidopsis thaliana* and *Oryza sativa* miRNAs are found in the AGO1 and AGO4 complexes signifying that the modified miRNAs have a function.

Mode of Action

Mature miRNA target genes at the post-transcriptional level either by Transcript Cleavage or Translation Inhibition/Repression (Rogers and Chen 2013). Plant miRNAs and their target mRNAs have nearly perfect complementarity, so transcript cleavage was considered to be the principal mode of action (Chen 2009; Voinnet 2009). But a high degree of sequence complementarity is not refractory to translational repression (Yang et al. 2012; Li et al. 2013a, b, c). So sequence complementarity does not decide the plant miRNAs mode of action. Plants miRNAs mainly function at the post-transcriptional gene silencing (PTGS) level and guide the PIWI domain of AGO proteins with endonuclease activity to cleave the target mRNA between positions 10 and 11 (Zhu et al. 2011). The cleavage fragments are subsequently degraded by exonucleases. In plants, translation repression is less often observed than transcript cleavage and occurs on membrane-bound polysomes (MBPs) and requires KATANIN 1 (KTN1) (Brodersen et al. 2008), VARICOSE (VCS) (Brodersen et al. 2008), SUO (Yang et al. 2012), and endoplasmic reticulum (ER)-localized ALTERED MERISTEM PROGRAM1 (AMP1) (Li et al. 2013a, b, c). The exact way of translational inhibition is still ambiguous, but during translation, miRNAs do not target mRNA by cleavage but stop translation by blocking read-through of the ribosome (Wang et al. 2008) or hinder the movement of ribosomes and/or inhibit translation initiation (Iwakawa and Tomari 2013).

miRNAs in Abiotic Stress Response

Abiotic stresses negatively impinge on growth, productivity, morphology, physiology, biochemistry, and molecular machinery in plants. So appreciating the molecular basis of genetic interactions during abiotic stresses such as drought, salinity, heat, and cold is obligatory to develop transgenics or modified plants adaptable to upcoming climate changes. Along with alterations in the expression of TFs, signaling components and genes, miRNAs profile has also been reported to vary during various abiotic stress conditions (Zhang 2015). The miRNAs are fundamental players in plant abiotic stress response and intercede the abiotic responses by modulating the transcript level of themselves, their mRNA targets, or the activity/mode of action of miRNA–protein complexes. Thus, results in manipulation of the timing of protein expression, protein location, and protein quantity expressed from other genes when plant is subjected to harsh environment of stress. Thus, a big role for small miRNAs has been ascertained in modification of the gene expression in reaction to diverse environmental stresses and diseases (Sunkar and Zhu 2004; Ruiz-Ferrer and Voinnet 2009; Zhou et al. 2010; Meng et al. 2011; Yu et al. 2012; Mittal et al. 2013). Some 1511 miRs are identified in various abiotic stresses of plant species (Zhang et al. 2013b). The gene regulation by miRNAs in response to abiotic stresses in plants cops up mainly at the post-transcriptional level (Ding et al. 2013; Feng et al. 2013; Ozhuner et al. 2013; Wang et al. 2014a, b). The miRNAs regulate the expression of specific target genes (Sunkar et al. 2012) and because of being conserved across plant kingdom; it is probable of their regulating similar targets of the entire plants. The targets for the conserved miRNAs comprise several TFs, for instance, MYB, NAC1, and homeodomain-leucine zipper protein (HD-ZIP) (Jones-Rhoades and Bartel 2004), which are mainly stress-responsive factors in plants (Fang et al. 2006; Xu et al. 2008).

Drought Stress

Drought is a customary recurring climate characteristic worldwide which declines crop productivity to a great extent. However, nature has endowed plants with a defense system to survive under harsh climatic conditions. One of the defense mechanisms at molecular level is the reprogramming of gene expression by microRNAs (miRNAs). Expression profile of microRNAs (miRNAs) and many genes and metabolites like vacuolar acid invertase, glutathione S-transferase (GST), abscisic acid (ABA)-inducible genes (LEA, RAB, COR, Rubisco), helicase, proline, dehydrins, and carbohydrates (Nezhadahmadi et al. 2013; Sunkar and Zhu 2004; Kulcheski et al. 2011; Sunkar et al. 2012) changes during drought stress. Plant miRNAs as chief gene regulators are anticipated to regulate the drought-responsive genes and play an imperative position in water stress regulatory networks (Khraiwesh et al. 2012). The stress-responsive miRNAs identified in diverse plant species are too numerous. Drought-responsive miRNAs have been documented in several plant species such as *Arabidopsis thaliana* (Sunkar and Zhu 2004; Liu et al. 2008), *Oryza sativa* (Zhou et al. 2010), cowpea (Barrera-Figueroa et al. 2011), tobacco (Frazier et al. 2011), soybean (Kulcheski et al. 2011), *Phaseolus vulgaris* (Arenas-Huertero et al. 2009), sugarcane (Gentile et al. 2015) and list is escalating and have been summarized in Table 1. During drought stress in *Arabidopsis thaliana* (Liu et al. 2008), miR156, miR159, miR167, miR168, miR171, miR172, miR319, miR393, miR394a, miR395c, miR395e, miR396, and miR397 are up-regulated, while miR161, miR168a, miR168b, miR169, miR171a, and miR319c are down-regulated. The miR169 because of being down-regulated leads to profusion of its target nuclear factor YA5, which in switches on various drought-responsive genes. Zhou et al. (2010) reported differential expressions of 30 drought-regulated miRNAs during drought stress, out of which 19 were novel to *Arabidopsis thaliana*. Genome-wide expression analysis in rice revealed that a set of 16 miRNAs were down-regulated and 14 miRNAs were significantly up-regulated during drought stress (Zhou et al. 2010). The *miR156, miR159, miR168, miR170, miR171, miR172, miR319, miR396, miR397, miR408, miR529, miR896, miR1030, miR1035, miR1050, miR1088*, and *miR1126* were down-regulated by drought stress while 14 miRNAs, *miR159, miR169, miR171, miR319, miR395, miR474, miR845, miR851, miR854, miR896, miR901, miR903, miR1026*, and *miR1125* were unregulated. Notably, *miR171, miR319*, and *miR896* were both up-regulated and down-regulated (Zhou et al. 2010). The targets of these miRNAs were mainly TFs (Zhou et al. 2010). Trindade et al. (2010) reported strong inverse correlation for miR398a/b and miR408 and their targets, copper superoxide dismutase (CSD1/2), mitochondrial cytochrome c oxidase, and plastocyanin, confirming a vital connection in adaptation to drought and copper homoeostasis. The knock-down of microRNA166 bestows *Oryza sativa* with drought resistance (Zhang et al. 2018). The comparative expression profiles of drought-tolerant rice variety N22 and drought-sensitive rice variety Pusa Basmati 1, showed miRNAs with variety-specific expression patterns during drought stress (Kansal et al. 2015). The leaf and root tissues of *Triticum turgidum* showed drought-induced expression in 438 miRNAs in comparison to 205 miRNAs under controlled conditions while 13 miRNAs of drought-tolerant wild emmer wheat showed differentially expressed (Kantar et al. 2010). The sensitive and tolerant soybean cultivars showed differential expression in miR166-5p, miR169f-3p, miR1513c, miR397ab, miR-Seq13, and miR166f which can facilitate crop improvement research studies (Kulcheski et al. 2011). Recently in sugarcane, the expression pattern of miRNA was observed to be dependent on the species, type of stress, tissue type, and growth condition. *MiR396* and *miR171* were differentially expressed in the majority of cases (Gentile et al. 2015).

Table 1 Drought-responsive miRNAs in different plant species

miRNA	Plant species	Target gene	Reference
miR156	*Triticum dicoccoides* *Arabidopsis thaliana* *Hordeum vulgare* *Oryza sativa*	SBP family of transcription factors	Eldem et al. (2012) Kantar et al. (2011) Liu et al. (2008) Ren et al. (2012)
miR157	*Prunus persica*	SBP family of transcription factors	Eldem et al. (2012)
miR159	*Prunus persica* *Oryza sativa* *Arabidopsis thaliana*	MYB and TCP transcription factors	Arenas-Huertero et al. (2009) Eldem et al. (2012) Jones-Rhoades and Bartel (2004)
miR160	*Populus trichocarpa* *Prunus persica* *Populus tomentosa*	Auxin response factors ARF 10, ARF 16, and ARF 17	Eldem et al. (2012) Liu et al. (2007) Jones-Rhoades and Bartel (2004)
miR162	*Populus tomentosa*	Dicer Like1	Ren et al. (2012)
miR164	*Brachypodium distachyon* *Medicago truncatula* *Populus trichocarpa*	NAC domain TF	Shuai et al. (2013) Wang et al. (2011)
miR165	*Prunus persica*	HD-ZIPIII transcription factor	Eldem et al. (2012)
miR166	*Glycine max* *Triticum dicoccoides*	HD-ZIPIII transcription factor	Kantar et al. (2011) Li et al. (2011a, b)
miR167	*Populus tomentosa* *Arabidopsis thaliana* *Prunus persica*	Auxin response factors ARF6 and ARF8	Eldem et al. (2012) Liu et al. (2008) Ren et al. (2012)
miR168	*Arabidopsis thaliana* *Oryza sativa* *Zea mays*	ARGONAUTE1, MAPK	Liu et al. (2008) Wei et al. (2009) Zhou et al. (2010)
miR169	*Arabidopsis thaliana* *Oryza sativa* *Medicago truncatula* *Prunus persica* *Glycine max* *Populus tomentosa* *Lycopersicon esculentum*	NF-YA transcription factor subunit A-3, NF-YA transcription factor subunit A-10, SlMRP1	Eldem et al. (2012) Li et al. (2008) Li et al. (2011a, b) Qin et al. (2011) Ren et al. (2012) Trindade et al. (2010) Wang et al. (2011) Zhang et al. (2011) Zhao et al. (2007) Zhou et al. (2010)

(continued)

Table 1 (continued)

miRNA	Plant species	Target gene	Reference
miR170	*Arabidopsis thaliana* *Oryza sativa*	SCL transcription factor	Sun (2012) Zhou et al. (2010)
miR171	*Arabidopsis thaliana* *Oryza sativa* *Medicago truncatula* *Prunus persica* *Populus tomentosa* *Triticum dicoccoides*	GRAS transcription factors	Eldem et al. (2012) Kantar et al. (2011) Llave et al. (2002a, b) Liu et al. (2008) Ren et al. (2012) Wang et al. (2011)
miR172	*Arabidopsis thaliana* *Oryza sativa* *Populus tomentosa*	Floral homeotic protein APETALA2 bZIP transcription factor family protein	Jones-Rhoades and Bartel (2004) Ren et al. (2012) Zhou et al. (2010)
miR393	*Prunus persica* *Arabidopsis thaliana*	TIR1 and AFB2 and AFB3	Liu et al. (2008) Navarro et al. (2006) Eldem et al. (2012)
miR394	*Populus tomentosa* *Populus trichocarpa* *Glycine max*	Dehydration-responsive protein and F-box proteins	Li et al. (2011a, b) Ren et al. (2012) Shuai et al. (2013)
miR395	*Oryza sativa* *Populus tomentosa* *Prunus persica*	Sulfate transporter	Eldem et al. (2012) Liang et al. (2010) Ren et al. (2012) Zhou et al. (2010)
miR396	*Arabidopsis thaliana* *Oryza sativa* *Prunus persica* *Medicago truncatula*	GRL transcription factors; ceramidase genes	Eldem et al. (2012) Kantar et al. (2011) Liu et al. (2008) Liu and Yu (2009) Zhou et al. (2010)
miR397	*Arabidopsis thaliana* *Oryza sativa* *Prunus persica* *Populus tomentosa*	Laccases	Abdel-Ghany and Pilon (2008) Ding and Zhu (2009) Eldem et al. (2012) Ren et al. (2012) Sunkar and Zhu (2004) Zhou et al. (2010)
miR399	*Medicago truncatula* *Populus tomentosa*	Phosphate transporter	Bari et al. (2006) Ren et al. (2012) Wang et al. (2011) Jones-Rhoades and Bartel (2004)

(continued)

Table 1 (continued)

miRNA	Plant species	Target gene	Reference
miR474	*Oryza sativa* *Triticum dicoccoides* *Zea mays*	Proline dehydrogenase Kinesin, a pentatricopeptide repeat (PPR) family protein	Kantar et al. (2011) Lu et al. (2005) Zhou et al. (2010)
miR528	*Zea mays*	POD, peroxidase	Wei et al. (2009)
miR1432	*Triticum dicoccoides*	Poly (ADP-ribose) polymerase; calcium binding, EF hand domains	Kantar et al. (2011) Zhang et al. (2009a, b)
miR2118	*Medicago truncatula*	TIR-NBS-LRR domain-protein	Jagadeeswaran et al. (2009)

Salinity Stress

Salt stress is caused by an amalgamation of the level, duration, and timing of salinity exposure and it negatively impinges on the quantity and quality of crop production (Blumwald and Grover 2006; Gepstein et al. 2006). Some 217 miRNAs are reported till date involved in salinity stress in various plant species like *Arabidopsis*, *Oryza sativa*, *Zea mays*, *Triticum aestivum*, *Glycine max*, *Glycine soja*, *Medicago truncatula*, *Nicotiana tabacum*, *Gossypium hirsutum*, *Panicum virgatum*, *Populus euphratica*, *Saccharum officinarum,* and *Phaseolus vulgaris*. In rice, expression patterns of 41 miRNAs was identified in response to drought, salt, cold, or ABA treatments (Shen et al. 2010) and 23 novel miRNAs were cloned from salt-stressed basmati rice (Sanan-Mishra et al. 2009). Drought and salinity stress-induced miR393, which targets an auxin transporter gene (OsAUX1) and a rice tiller inhibitor gene (OsTIR1) when introduced to rice plants increases their tiller numbers, causes early flowering and reduced tolerance to salt and hypersensitiveness to auxin (Xia et al. 2012). The salt stress decreases Osa-miR396c expression in an ABA-dependent manner and its overexpression causes decreased salt stress tolerance (Gao et al. 2011). miR169g, miR169n, and miR169o are three salt-inducible members of miR-169 family members (17) (Zhao et al. 2009), which target NF-YA gene transcripts. But Osa-miR169g only gets up-regulated in drought stress (Zhao et al. 2007). In *Raphanus sativus*, Sun et al. (2015) reported 49 already known and 22 novel salt-responsive miRNAs which targeted the genes of signaling, ion- homeostasis, and plant growth. The salt-tolerant (NC286) and salt-sensitive (Huangzao4) maize lines divulged the differential expression of various miRNAs throughout salinity stress (Ding et al. 2009). In *Arabidopsis thaliana*, miR156, miR158, miR159, miR165, miR167, miR168, miR169, miR171, miR319, miR393, miR394, miR396, and miR397 were up-regulated during salinity stress while miR398 was down-regulated (Liu et al. 2008). The salt-induced up-regulation of miR393 is known in both *Oryza sativa* and *Arabidopsis* under while miR167 is under salt stress regulation *Arabidopsis* but not in *Oryza sativa* (Sunkar and Zhu 2004; Lv et al. 2010). Overexpression of *osa-miR393* lead to enhanced salt tolerance in *Arabidopsis* (Gao et al. 2011). The plants showing salt-responsive miRNAs are summarized in Table 2.

Table 2 Salt-responsive miRNAs in different plant species

miRNA	Plant species	Target gene	Reference
miR16	*Oryza sativa*	Germin-like protein; ethylene-insensitive3 (EIN3)-like 1 protein	Sanan-Mishra et al. (2009)
miR29	*Oryza sativa*	Strictosidine synthase precursor	Barrera-Figueroa et al. (2012)
miR2001	*Oryza sativa*	Protein GPR107 precursor	Jian et al. (2010)
miR2003	*Oryza sativa*	HEAT repeat family protein; ribosomal protein S11; NAC domain-protein 90	Jian et al. (2010)
miR2005	*Oryza sativa*	Nitrate and chloride transporter; phosphate carrier protein	Jian et al. (2010)
miR156	*Panicum virgatum* *Populus trichocarpa* *Vigna unguiculata* *Arabidopsis thaliana* *Populus euphratica* *Zea mays*	Squamosa promoter-binding like-like TF POPTR_0007s01030 SPL-binding protein Squamosa promoter-binding like-binding protein Cationic amino acid transporter SPL-like transcription factor	Sun et al. (2012) Li et al. (2013a, b, c) Paul et al. (2011) Liu et al. (2008) Qin et al. (2011) Ding et al. (2009)
miR159	*Phaseolus vulgaris* *Panicum virgatum* *Arabidopsis thaliana* *Nicotiana tabacum*	MYB TF MYB TF MYB TF MYB TF	Jones-Rhoades and Bartel (2004) Sun et al. (2012) Chen et al. (2012) Frazier et al. (2011)
miR160	*Populus trichocarpa* *Setaria italica* *Vigna unguiculata* *Triticum aestivum* *Gossypium raimondii*	POPTR_0002s09050 Auxin response factor Auxin response factor Auxin response factor Auxin response factor	Li et al. (2013a, b, c) Khan et al. (2014) Paul et al. (2011) Lu et al. (2011) Xie et al. (2014)
miR162	*Setaria italica* *Zea mays* *Panicum virgatum*	DICER-LIKE 1 (DCL1) DICER-LIKE 1 (DCL1) DICER-LIKE 1 (DCL1)	Khan et al. (2014) Ding et al. (2009) Sun et al. (2012)
miR164	*Populus trichocarpa* *Zea mays* *Arabidopsis thaliana* *Oryza sativa*	POPTR_0007s08420 NAC family gene NAC family gene CUC2 no apical meristem (NAM) protein; NAC domain-containing protein; helicase	Li et al. (2013a, b, c) Ding et al. (2009) Amor et al. (2009) Macovei and Tuteja (2012)
miR165	*Arabidopsis thaliana*	Class III HD-ZIP TFs	Liu et al. (2008)
miR166	*Glycine max* *Zea mays* *Arabidopsis thaliana*	SBP-like TFs SBP-like TFs SBP-like TFs	Li et al. (2011b) Kong et al. (2010) Amor et al. (2009)
miR167	*Triticum aestivum* *Arabidopsis thaliana* *Nicotiana tabacum*	Auxin response factor Auxin response factor 8 Auxin response factor	Lu et al. (2011) Kinoshita et al. (2012) Frazier et al. (2011)
miR168	*Zea mays* *Arabidopsis thaliana* *Populus euphratica*	ARGONAUTE1 ARGONAUTE1 MYB TF	Ding et al. (2009) Liu et al. (2008) Qin et al. (2011)

(continued)

Table 2 (continued)

miRNA	Plant species	Target gene	Reference
miR169	*Populus euphratica* *Glycine max* *Vigna unguiculata* *Arabidopsis thaliana* *Nicotiana tabacum* *Zea mays* *Oryza sativa*	CCAAT-binding TF CBF HAP2-like factor CCAAT-binding TF CCAAT-binding TFs CBF HAP2-like factor NFY-A T CBF HAP2-like factor	Qin et al. (2011) Li et al. (2011b) Paul et al. (2011) Liu et al. (2008) Frazier et al. (2011) Luan et al. (2015) Zhao et al. (2009)
miR171	*Setaria italica* *Arabidopsis thaliana* *Populus trichocarpa*	Scarecrow-like TFs Scarecrow-like TF POPTR_0001s00480	Khan et al. (2014) Liu et al. (2008) Li et al. (2013a, b, c)
miR172	*Gossypium raimondii* *Nicotiana tabacum*	APETALA2-like factor APETALA2-like factor	Xie et al. (2014) Frazier et al. (2011)
miR319	*Arabidopsis thaliana*	TCP TFs	Liu et al. (2008)
miR393	*Arabidopsis thaliana* *Oryza sativa*	F-box protein Phytosulfokine receptor precursor; GRF-interacting factor (GIF)	Sunkar and Zhu (2004) Gao et al. (2011)
miR394	*Arabidopsis thaliana* *Glycine max*	F-box protein F-box protein	Liu et al. (2008) Li et al. (2011b)
miR395	*Zea mays* *Nicotiana tabacum* *Panicum virgatum* *Arabidopsis thaliana*	ATP sulfurylase ATP sulfurylase ATP sulfurylase ATP sulfurylase	Ding et al. (2009) Frazier et al. (2011) Sun et al. (2012) Kim et al. (2010b)
miR396	*Populus trichocarpa* *Nicotiana tabacum* *Zea mays* *Arabidopsis thaliana* *Oryza sativa*	GRL TFs GRL TFs GRL TFs GRL TFs GRL transcription factors; Rhodanese-like protein; kinesin-like protein B	Frazier et al. (2011) Ding et al. (2009) Zhou et al. (2012) Liu et al. (2008) Gao et al. (2010)
miR397	*Panicum virgatum* *Arabidopsis thaliana*	cDNA L-ascorbate oxidase precursor Laccases	Sun et al. (2012) Sunkar and Zhu (2004)
miR398	*Arabidopsis thaliana*	Cu/Zn superoxide dismutase	Jagadeeswaran et al. (2009)
miR399	*Medicago truncatula*	ATP-dependent RNA helicase	Lelandais-Briere et al. (2009)
miR402	*Arabidopsis thaliana*	DEMETER-LIKE protein 3	Kim et al. (2010a)
miR417	*Arabidopsis thaliana*	C2-domain containing and SNF7 family protein	Jung and Kang (2007)
miR474c	*Populus trichocarpa*	Protein kinase; kinesin	Zhou et al. (2012)
miR482	*Populus trichocarpa*	TIR-NBS-LRR resistance protein	Lu et al. (2008)
miR530a	*Populus trichocarpa*	F-box domain-protein	Lu et al. (2008)
miR1446	*Populus euphratica*	Gibberellin response modulator-like protein	Lu et al. (2008)
miR1447	*Populus euphratica*	ATP-binding transport protein	Lu et al. (2008)

Heat Stress

Temperature fluctuations during day and different seasons in the surrounding environment of a plant require a reprogramming of gene expression pattern to adjust to such notable shifts in temperature. The climatic change and global warming may adversely impact plant's physiology including seed maturation and grain filling. Several heat-responsive miRNAs have been identified in plant species (Cao et al. 2014). Nine heat stress-responsive miRNAs were identified in wheat, eight being conserved across plants. miR156, miR159, miR160, miR166, miR168, miR169, miR393, and miR827 were up-regulated and only miR172 showed down-regulation after heat stress (Xin et al. 2010). Heat stress-tolerant and susceptible cultivars of wheat exhibit differential miRNAs gene expression pattern at 40 °C (Xin et al. 2011). In *Populus trichocarpa* and *Arabidopsis thaliana* miR168 shows up-regulated expression (Lu et al. 2008; Liu et al. 2008) and down-regulation in rice (Lv et al. 2010) and expression of miR171 was vice versa in *Arabidopsis thaliana* and *Oryza sativa* (Liu et al. 2008; Lv et al. 2010) thus confirming species-specific role of miRNAs. Heat stress up-regulates miR398 and down-regulates target gene mRNA such as CSD1, CSD2 (copper/zinc superoxide dismutase), and CCS (a gene encoding a copper chaperone for both CSD1 and CSD2). Transgenic plants of miR398-resistant forms of CSD1, CSD2, and CCS being sensitive resulted into increased damage to the plant and flowers under heat stress as compared to transgenics with normal genes of CSD1, CSD2 or CCS, while *csd1, csd2,* and *ccs* mutant plants were extra heat tolerant than the wild type. In *Arabidopsis thaliana*, HSF genes encoding heat stress transcription factors and HSP genes encoding heat shock proteins showed decreased expression in heat-sensitive transgenic plants expressing miR398-resistant forms of CSD1, CSD2, or CCS and increased expression in the heat-tolerant *csd1, csd2*, and *ccs* plants, thus suggesting role of miR398 in activation of a regulatory loop which is decisive for thermo tolerance in *Arabidopsis* (Guan et al. 2013). Stief et al. (2014) reported that miR156 regulates the persistent heat stress memory in *Arabidopsis* and the plants were found to be extra tolerant to heat stress. Rice plants overexpressing miR159 were extra sensitive to heat stress (Wang et al. 2012).

Cold Stress

Cold stress is a major abiotic stress factor that restricts the agricultural productivity in hilly areas. Tolerance to chilling (0–15 °C) and freezing (< 0 °C) temperatures varies in plants. Temperate vegetation is tolerant to chilling while tropical and subtropical plants and crops, for instance, rice, maize, and tomato are sensitive to chilling and deficient in the competence for cold acclimation. Cold-responsive gene regulatory networks are controlled by various miRNAs. Comparative profiling of miRNA expression in *Arabidopsis*, *Brachypodium*, and *Populus trichocarpa* during cold stress revealed similarity plus disparity in miRNA regulation. *Arabidopsis* miR393 is the first plant defense miRNA discovered which down-regulated its target gene, ubiquitin E3 ligase, and thus enhances the quantity of cold-induced proteins (Navarro et al. 2006; Sunkar et al. 2004; Samuel et al. 2008; Liu et al. 2008). The miR397 and miR169 were up-regulated in *Arabidopsis*, *Brachypodium*, and *Populus trichocarpa*, while miR172 was up-regulated in *Arabidopsis* and *Brachypodium* but not in *Populus trichocarpa* (Zhang et al. 2009a, b). Cold stress induces up-regulation of miR168 levels in *Populus trichocarpa* and *Arabidopsis* (Lu et al. 2008; Liu et al. 2008) and down-regulation in *Oryza sativa* (Lv et al. 2010). In *Oryza sativa* during cold stress, miR171 was down-regulated (Lv et al. 2010) and up-regulated in *Arabidopsis* (Liu et al. 2008). According to Sunkar and Zhu (2004), during cold stress miR393, miR397b, miR402, and miR319c is up-regulated in *Arabidopsis*, while wide transcriptome analyses revealed that miR165/166, miR169, miR172, miR393, miR396, miR397, and miR408 are notably up-regulated while miR156/157, miR159/319, miR164, miR394, and miR398 are only transitorily or slightly regulated during cold stress (Zhou et al. 2008). *Arabidopsis* miR165/166, miR393, miR396, and miR408 play main part in the regulatory network of cold-responsive genes (Sunkar and Zhu 2004; Liu et al. 2008). In *Oryza sativa*, 18 cold-responsive down-regulated miRNAs, miR156k, miR166k, miR166m, miR167a/b/c, miR168b, miR169e, miR169f, miR169 h, miR171a, miR535, miR319a/b, miR1884b, miR444a.1, miR1850, miR1868, miR1320, miR1435, and miR1876 have been identified which target genes of the SPL, SCL, MADS box, or TCP families, which show increased expression in response to cold stress (Lu and Huang 2008; Lv et al. 2010). Overexpression of *Osa*-miR319 in *Oryza sativa* caused enhanced cold tolerance (4 °C) after chilling acclimation (12 °C) in transgenic rice seedlings. Notably, under 4 and 12 °C, *Osa*-miR319a and *Osa*-miR319b were down-regulated and the expression of miR319-targeted genes, *OsPCF5* and *OsPCF8,* was induced. Moreover, down-regulating *OsPCF5* and *OsPCF8* expression in RNA interference (RNAi) plants cause enhanced cold tolerance suggesting role of miR319 in cold tolerance in *Oryza sativa* (Yang et al. 2013). According to Wang et al. (2014a, b), *Osa-miR319b* up-regulation and targeting of *OsPCF6* and *OsTCP21* also leads to cold stress tolerance. In two *B. rapa* L. varieties, "Longyou 7" (cold-tolerant) and "Tianyou 4" (cold-sensitive), 353 cold-responsive miRNAs (84 putative novel and 269 conserved miRNAs), were identified from the leaves and roots of under cold stress of −4 °C for 8 h, out of which miR166h-3p-1, miR398b-3p, miR398b-3p-1, miR408d, miR156a-5p, miR396h, miR845a-1, and miR166u were conserved and Bra-novel-miR3153-5p and Bra-novel-miR3172-5p were novel miRNAs indicative of miRNA-mediated cold tolerance in this plant (Zeng et al. 2018).

Nutrient Homeostasis

The macronutrient sulfur is an important component of amino acid residue cysteine, which is further changed into sulfur-containing defense compounds, i.e., glutathione, phytoalexins, and glucosinolates. The first ever report signifying the correlation between plant miRNAs and stress response regulation was given by Jones-Rhoades and Bartel (2004) in *Arabidopsis*, when they observed several novel miRNAs, including miR395 to be up-regulated during sulfate starvation. Under sulfate deprivation, miR395 activates gene families, low-affinity sulfate transporter (SULTR2; 1/AST68), and 3 ATP sulfurylase family members (APS1, APS3, and APS4) (Jones-Rhoades and Bartel 2004; Allen et al. 2005) and also remobilizes the sulfate between leaves in *Arabidopsis* (Liang et al. 2010). The miR395 supports the plant in sulfate paucity to acclimatize growth and development. Phosphate is an indispensable plant nutrient with role in DNA replication, phospholipid bilayer formation, and ATP synthesis. When Pi is in less quantity, expression of miR156, miR399, miR778, miR827, and miR2111 is up-regulated while miR169, miR395, and miR398 get down-regulated (Hsieh et al. 2009). Pi deficiency leads to induction of a transcription factor phosphate starvation response 1 (PHR1) which is the master regulator of the PHR regulation pathway in *Arabidopsis thaliana*; and up-regulates miR399 which consecutively down-regulates PHO2 (E2 ubiquitin-conjugating enzyme) predominantly in roots (Aung et al. 2006; Bari et al. 2006). The phosphate deficiency-induced expression of miR399 with a corresponding decrease in its target ubiquitin-conjugating enzyme (UBC) in *Arabidopsis* is the second discovery of the interlink of miRNAs with abiotic stress (Fujii et al. 2005; Aung et al. 2006; Bari et al. 2006). Moreover, miR827 and miR2111 also get up-regulated and target E3 ligases (At1g02860 and At1g63010, respectively), indicating phosphate homeostasis regulation by a miRNA-controlled ubiquitination-mediated pathway (Fujii et al. 2005; Chiou et al. 2006). The transgenic rapeseed plants overexpressing miR395 show noteworthy augmentation of tolerance to cadmium stress (Zhang et al. 2013a).

Oxidative Stress and Hypoxia

Oxidative damage of the cell is caused by the toxic compounds called Reactive Oxygen Species (ROS) produced as by-products of photosynthesis and respiration (Mittler et al. 2004). ROS is a secondary messenger molecule in stress-associated signal transduction pathways (Vandenabeele et al. 2003; Mittler et al. 2004) and also plays a function in the biological process regulation. In *Arabidopsis*, oxidative stress down-regulates miR398 which otherwise suppresses expression of two Cu/Zn superoxide dismutases (CSD1 and CSD2/SODs) (Bonnet et al. 2004; Jones-Rhoades and Bartel 2004) authenticating a correlation between miRNAs and reactive oxygen species (ROS) pathway. Transgenic lines of miR398 resistant (with high quantity of CSD2) are resistant to high light intensity, heavy metals, and other oxidative stressors (Sunkar et al. 2006). Hypoxia, a low oxygen condition, causes switching from aerobic to anaerobic respiration (Bailey-Serres and Voesenek 2008). With the high-throughput microarray technology in submerged maize roots, Zm-miR166, Zm-miR167, Zm-miR171, Os-miR396, Zm-miR399, Zm-miR159, At-miR395, Pt-miR474, and Os-miR528 were identified (Zhang et al. 2008) and by using next-generation sequencing in *Arabidopsis* roots, 19 hypoxia-responsive miRNA families were identified, which target other transcripts (Moldovan et al. 2009).

Mechanical Stress

Mechanical stress to plants means bending of branches and stems by gravity, wind, or other peripheral abiotic agents. Transcript levels of miRNAs in tension-stressed, compression-stressed xylem and unstressed xylem show reduced expression of some miR156, miR162, miR164, miR475, miR480, and miR481 and up-regulates miR408 under stress. In compression stress, miR160 and miR172 were down-regulated and miR168 was up-regulated in the tension-created tissue (Lu et al. 2005). Wound stress up-regulates miR828 in miR828 overexpressing lines in sweet potato causing higher lignin biosynthesis and H_2O_2 production which further plays a function in defense mechanisms (Lin et al. 2012).

Conclusions and Future Perspectives

Many miRNA families play important functions in abiotic stress conditions, for instance, cold, heat, drought, salt, etc. (Sunkar et al. 2012). MiRNA-mediated post-transcriptional gene regulation is the most vital mechanism of abiotic stress tolerance in plants, causing cleavage or translational repression of the target genes. High-throughput technologies of sequencing and miRNA microarray have confirmed many stress-responsive miRNAs. Up-regulation of miRNAs during abiotic stress can down-regulate their target gene transcripts, thus act as a negative regulator. Alternatively, down-regulated miRNAs bring about amassing of their target gene transcripts so positively regulate stress conditions. Regulatory networks of miRNAs have yet to be understood completely. Unraveling the regulatory roles of miRNAs may unlock new possibilities of crop improvement for enhanced food production to quench the hunger of the rising populace. MicroRNA interference (miRNAi) technology endows with an efficient podium for functional genomics investigations to better understand the basic underlying plant developmental mechanisms, plant abiotic stress responses, and agricultural applications. The miRNAs are reported as the crucial regulators of plant abiotic stress responses. How the plants work under severe ecological conditions can be understood from the way miRNAs regulate stress-responsive genes and further, manipulation and modification of the gene expression can help in producing stress-tolerant plants and crop improvement on one side and secondly by breeding improved crop cultivars with enhanced agronomic traits on other side.

References

Abdel-Ghany SE, Pilon M (2008) MicroRNA-mediated systemic down-regulation of copper protein expression in response to low copper availability in *Arabidopsis*. J Biol Chem 283:15932–15945

Allen E, Xie Z, Gustafson AM, Carrington JC (2005) MicroRNA-directed phasing during trans-acting siRNA biogenesis in plants. Cell 121:207–221

Ameres SL, Zamore PD (2013) Diversifying microRNA sequence and function. Nat Rev Mol Cell Biol 14:475–488

Amor BB, Wirth S, Merchan F, Laporte P, d'Aubenton- Carafa Y, Hirsch J, Maizel A, Mallory A, Lucas A, Deragon JM, Vaucheret H, Thermes C, Crespi M (2009) Novel long non-protein coding RNAs involved in *Arabidopsis* differentiation and stress responses. Genome Res 19:57–69

Arenas-Huertero C, Perez B, Rabanal F, Blanco-Melo D, De la Rosa C, Estrada-Navarrete G, Sanchez F, Alicia Covarrubias A, Luis Reyes J (2009) Conserved and novel miRNAs in the legume *Phaseolus vulgaris* in response to stress. Plant Mol Biol 70:385–401

Aung K, Lin SI, Wu CC, Huang YT, Su CL, Chiou TJ (2006) pho2, a phosphate over accumulator, is caused by a nonsense mutation in a microRNA399 target gene. Plant Physiol 141:1000–1011

Baek D, Kim MC, Chun HJ, Kang S, Park HC, Shin G, Park J, Shen M, Hong H, Kim WY et al (2013) Regulation of miR399f transcription by AtMYB2 affects phosphate starvation responses in Arabidopsis. Plant Physiol 161:362–373

Bailey-Serres J, Voesenek LA (2008) Flooding stress: acclimations and genetic diversity. Annu Rev Plant Biol 59:313–339

Barciszewska-Pacak M, Milanowska K, Knop K, Bielewicz D, Nuc P, Plewka P, Pacak AM, Vazquez F, Karlowski W, Jarmolowski A, Szweykowska-Kulinska Z (2015) Arabidopsis microRNA expression regulation in a wide range of abiotic stress responses. Front Plant Sci 6:410

Bari R, Datt Pant B, Stitt M, Scheible WR (2006) PHO2 MicroRNA399 and PHR1 define a phosphate-signaling pathway in plants. Plant Physiol 141:988–999

Barrera-Figueroa BE, Gao L, Diop NN, Wu Z, Ehlers JD, Roberts PA, Close TJ, Zhu JK, Liu R (2011) Identification and comparative analysis of drought-associated microRNAs in two cowpea genotypes. BMC Plant Biol 11:127

Barrera-Figueroa BE, Gao L, Wu Z, Zhou X, Zhu J, Jin H, Liu R, Zhu JK (2012) High throughput sequencing reveals novel and abiotic stress-regulated microRNAs in the inflorescences of rice. BMC Plant Biol 12:132

Baumberger N, Baulcombe DC (2005) Arabidopsis ARGONAUTE1 is an RNA slicer that selectively recruits microRNAs and short interfering RNAs. Proc Natl Acad Sci U S A 102:11928–11933

Ben Chaabane S, Liu R, Chinnusamy V, Kwon Y, Park JH, Kim SY, Zhu JK, Yang SW, Lee BH (2013) STA1, an *Arabidopsis* pre-mRNA processing factor 6 homolog, is a new player involved in miRNA biogenesis. Nucleic Acids Res 41:1984–1997

Blumwald E, Grover A (2006) Salt tolerance. In: Halford NG (ed) Plant biotechnology: current and future uses of genetically modified crops. Wiley, London, pp 206–224

Bonnet E, Wuyts J, Rouze P, Van de PY (2004) Detection of 91 potential conserved plant microRNAs in *Arabidopsis thaliana* and *Oryza sativa* identifies important target genes. Proc Nat Acad Sci U S A 101:11511–11516

Brodersen P, Sakvarelidze-Achard L, Bruun-Rasmussen M, Dunoyer P, Yamamoto YY, Sieburth L, Voinnet O (2008) Widespread translational inhibition by plant miRNAs and siRNAs. Science 320:1185–1190

Cao X, Wu Z, Jiang F, Zhou R, Yang Z (2014) Identification of chilling stress-responsive tomato microRNAs and their target genes by high- throughput sequencing and degradome analysis. BMC Genomics 15:11–30

Carmell MA, Xuan Z, Zhang MQ, Hannon GJ (2002) The Argonaute family: tentacles that reach into RNAi, developmental control, stem cell maintenance, and tumorigenesis. Genes Dev 16:2733–2742

Chinnusamy V, Zhu J, Zhu JK (2007) Cold stress regulation of gene expression in plants. Trends Plant Sci 12:444–451
Chiou T, Aung K, Lin S, Wu C, Chiang S, Su C (2006) Regulation of phosphate homeostasis by microRNA in *Arabidopsis*. Plant Cell 18:412–421
Chen X (2009) Small RNAs and their roles in plant development. Annual Review of Cell and Developmental Biology 25:21–44
Chen L, Wang T, Zhao M, Tian Q, Zhang WH (2012) Identification of aluminum-responsive microRNAs in Medicago truncatula by genome-wide highthroughput sequencing. Planta 235:375–386
Ding YF, Zhu C (2009) The role of microRNAs in copper and cadmium homeostasis. Biochem Biophys Res Commun 386:6–10
Ding D, Zhang L, Wang H, Liu Z, Zhang Z, Zheng Y (2009) Differential expression of miRNAs in response to salt stress in maize roots. Ann Bot 103:29–38
Ding Y, Tao Y, Zhu C (2013) Emerging roles of microRNAs in the mediation of drought stress response in plants. J Exp Bot 64:3077–3086
Eldem V, Akcay UC, Ozhuner E, Bakir Y, Uranbey S, Unver T (2012) Genome-wide identification of miRNAs responsive to drought in peach (*Prunus persica*) by high-throughput deep sequencing. PLoS One 7:e50298
Fang Q, Xu Z, Song R (2006) Cloning, characterization and genetic engineering of FLC homolog in *Thellungiella halophila*. Biochem Biophys Res Commun 347:707–714
Fang X, Cui Y, Li Y, Qi Y (2015) Transcription and processing of primary microRNAs are coupled by Elongator complex in Arabidopsis. Nat Plants 1:15075
Feng H, Zhang Q, Wang Q, Wang X, Liu J, Li M, Huang L, Kang Z (2013) Target of tae-miR408, a chemocyanin-like protein gene (TaCLP1), plays positive roles in wheat response to high-salinity, heavy cupric stress and stripe rust. Plant Mol Biol 83:433–443
Floyd SK, Bowman JL (2004) Gene regulation: ancient microRNA target sequences in plants. Nature 428:485–486
Francisco-Mangilet AG, Karlsson P, Kim MH, Eo HJ, Oh SA, Kim JH, Kulcheski FR, Park SK, Manavella PA (2015) THO2, a core member of the THO/TREX complex, is required for microRNA production in *Arabidopsis*. Plant J 82:1018–1029
Frazier TP, Sun G, Burklew CE, Zhang B (2011) Salt and drought stresses induce the aberrant expression of microRNA genes in tobacco. Mol Biotechnol 49:159–165
Fujii H, Chiou TJ, Lin SI, Aung K, Zhu JK (2005) A miRNA involved in phosphate-starvation response in *Arabidopsis*. Curr Biol 15:2038–2043
Fukudome A, Fukuhara T (2017) Plant dicer-like proteins: double-stranded RNA cleaving enzymes for small RNA biogenesis. J Plant Res 130:33–44
Furumizu C, Tsukaya H, Komeda Y (2010) Characterization of *EMU*, the *Arabidopsis* homolog of the yeast THO complex member *HPR*1. RNA 16:1809–1817
Gao P, Bai X, Yang L, Lv D, Li Y, Cai H, JiW GD, Zhu Y (2010) Over-expression of osa-MIR396c decreases salt and alkali stress tolerance. Planta 231:991–1001
Gao P, Bai X, Yang L, Lv D, Pan X, Li Y, Cai H, Ji W, Chen Q, Zhu Y (2011) Osa-MIR393: a salinity and alkaline stress-related microRNA gene. Mol Biol Rep 38:237–242
Gentile A, Dias LI, Mattos RS, Ferreira TH, Menossi M (2015) MicroRNAs and drought responses in sugarcane. Front Plant Sci 6:58
Gepstein S, Grover A, Blumwald E (2006) Producing biopharmaceuticals in the desert: building an abiotic stress tolerance in plants for salt, heat, and drought. In: Knablein J, Muller RHW (eds) Modern biopharmaceuticals: design, development and optimization. Wiley-VCH Verlag GmbH, Weinhaum, pp 967–994
Gonatopoulos-Pournatzis T, Cowling VH (2014) Cap-binding complex (CBC). Biochem J 457:231–242
Guan Q, Lu X, Zeng H, Zhang Y, Zhu J (2013) Heat stress induction of *miR398* triggers a regulatory loop that is critical for thermo tolerance in *Arabidopsis*. Plant J 74:840–851
Hammond SC, Bernstein E, Beach D, Hannon GJ (2000) An RNA-directed nuclease mediates posttranscriptional gene silencing in Drosophila cells. Nature 404:293–296

Han MH, Goud S, Song L, Fedoroff N (2004) The Arabidopsis double-stranded RNA-binding protein HYL1plays a role in microRNA-mediated gene regulation. Proc Natl Acad Sci U S A 101:1093–1098

Hsieh LC, Lin SI, Shih AC, Chen JW, Lin WY, Tseng CY, Li WH, Chiou TJ (2009) Uncovering small RNA-mediated responses to phosphate deficiency in *Arabidopsis* by deep sequencing. Plant Physiol 151:2120–2132

Iki T, Yoshikawa M, Nishikiori M, Jaudal MC, Matsumoto-Yokoyama E, Mitsuhara I, Meshi T, Ishikawa M (2010) In vitro assembly of plant RNA induced silencing complexes facilitated by molecular chaperone HSP90. Mol Cell 39:282–291

Iwakawa HO, Tomari Y (2013) Molecular insights into microRNA-mediated translational repression in plants. Mol Cell 52:591–601

Jagadeeswaran G, Zheng Y, Li YF, Shukla LI, Matts J, Hoyt P, Macmil SL, Wiley GB, Roe BA, Zhang W, Sunkar R (2009) Cloning and characterization of small RNAs from *Medicago truncatula* reveals four novel legume-specific microRNA families. New Phytol 184:85–98

Jian X, Zhang L, Li G, Zhang L, Wang X, Cao X, Fang X, Zha FC (2010) Identification of novel stress-regulated microRNAs from *Oryza sativa* L. Genomics 95:47–50

Jones-Rhoades MW, Bartel DP (2004) Computational identification of plant microRNAs and their targets, including a stress-induced miRNA. Mol Cell 14:787–799

Jung HJ, Kang H (2007) Expression and functional analyses of microRNA417 in *Arabidopsis thaliana* under stress conditions. Plant Physiol Biochem 45:805–811

Kansal S, Devi RM, Balyan SC, Arora MK, Singh AK, Mathur S, Raghuvanshi S (2015) Unique miRNome during anthesis in drought-tolerant indica rice var. Nagina 22. Planta 241:1543–1559

Kantar M, Unver T, Budak H (2010) Regulation of barley miRNAs upon dehydration stress correlated with target gene expression. Funct Integr Genomics 10:493–507

Kantar M, Lucas S, Budak H (2011) miRNA expression patterns of *Triticum dicoccoides* in response to shock drought stress. Planta 233:471–484

Kateryna F, Benoit L, Olivier H (2016) Interplay between miRNA regulation and mechanical stress for *CUC* gene expression at the shoot apical meristem. Plant Signal Behav 11:3

Khan Y, Yadav A, Bonthala VS, Muthamilarasan M, Yadav CB, Prasad M (2014) Comprehensive genome-wide identification and expression profiling of foxtail millet [*Setaria italica* (L.)] miRNAs in response to abiotic stress and development of miRNA database. Plant Cell Tiss Org Cult 118:279–292

Khraiwesh B, Zhu JK, Zhu J (2012) Role of miRNAs and siRNAs in biotic and abiotic stress responses of plants. Biochim Biophys Acta 1819:137–148

Khvorova A, Reynolds A, Jayasena SD (2003) Functional siRNAs and miRNAs exhibit strand bias. Cell 115:209–216

Kim JL, Kwak KJ, Jung HJ, Lee HJ, Kang H (2010a) MicroRNA402 affects seed germination of *Arabidopsis thaliana* under stress conditions via targeting DEMETER-LIKE Protein3 mRNA. Plant Cell Physiol 51:1079–1083

Kim JY, Lee HJ, Jung HJ, Maruyama K, Suzuki N, Kang H (2010b) Overexpression of microRNA395c or 395e affects differently the seed germination of *Arabidopsis thaliana* under stress conditions. Planta 232:1447–1454

Kim YJ, Zheng B, Yu Y, Won SY, Mo B, Chen X (2011) The role of mediator in small and long noncoding RNA production in *Arabidopsis thaliana*. EMBO J 30:814–822

Kulcheski FR, de Oliveira LF, Molina LG, Almerao MP, Rodrigues FA, Marcolino J, Barbosa JF, Stolf-Moreira R, Nepomuceno AL, Marcelino Guimaraes FC, Abdelnoor RV, Nascimento LC, Carazzolle MF, Pereira GA, Margis R (2011) Identification of novel soybean microRNAs involved in abiotic and biotic stresses. BMC Genomics 12:307

Kinoshita N, Wang H, Kasahara H, Liu J, Macpherson C, Machida Y, Kamiya Y, Hannah MA, Chua NH (2012) IAA-Ala resistant3, an evolutionarily conserved target of miR167, mediates Arabidopsis root architecture changes during high osmotic stress. The Plant Cell 24:3590–3602

Kulcheski FR, Côrrea R, Gomes IA, de Lima JC and Margis R (2015) NPK macronutrients and microRNA homeostasis. Front. Plant Sci. 6:451

Kong Y, Elling A, Chen B, Deng X (2010) Differential expression of microRNAs in maize inbred and hybrid lines during salt and drought stress. American Journal of Plant Sciences 1:69–76

Kumar S, Verma S, Trivedi PK (2017) Involvement of small RNAs in phosphorus and Sulfur sensing, Signaling and stress: current update. Front Plant Sci 8:285

Kurihara Y, Takashi Y, Watanabe Y (2006) The interaction between DCL1 and HYL1 is important for efficient and precise processing of pri-miRNA in plant microRNA biogenesis. RNA 12:206–212

Lagos-Quintana M, Rauhut R, Lendeckel W, Tuschl T (2001) Identification of novel genes coding for small expressed RNAs. Science 294:853–858

Lau NC, Lim LP, Weinstein EG, Bartel DP (2001) An abundant class of tiny RNAs with probable regulatory roles in *Caenorhabditis elegans*. Science 294:858–862

Lee RC, Ambros V (2001) An extensive class of small RNAs in *Caenorhabditis elegans*. Science 294:862–864

Lee RC, Feinbaum RL, Ambros V (1993) The *C. elegans* heterochronic gene *lin-4* encodes small RNAs with antisense complementarity to *lin-14*. Cell 75:843–854

Lee Y, Kim M, Han J, Yeom K, Lee S, Baek SH et al (2004) MicroRNA genes are transcribed by RNA polymerase II. EMBO J 23:4051–4060

Lelandais-Briere C, Naya L, Sallet E, Calenge F, Frugier F, Hartmann C, Gouzy J, Crespi M (2009) Genome-wide *Medicago truncatula* small RNA analysis revealed novel microRNAs and isoforms differentially regulated in roots and nodules. Plant Cell 21:2780–2796

Li J, Yang Z, Yu B, Liu J, Chen X (2005) Methylation protects miRNAs and siRNAs from a 3′-end uridylation activity in *Arabidopsis*. Curr Biol 15:1501–1507

Li WX, Oono Y, Zhu J, He XJ, Wu JM, Iida K, Lu XY, Cui X, Jin H, Zhu JK (2008) The *Arabidopsis* NFYA5 transcription factor is regulated transcriptionally and posttranscriptionally to promote drought resistance. Plant Cell 20:2238–2251

Li H, Dong Y, Yin H, Wang N, Yang J, Liu X, Wang Y, Wu J, Li X (2011a) Characterization of the stress associated microRNAs in *Glycine max* by deep sequencing. BMC Plant Biol 11:170

Li T, Li H, Zhang YX, Liu JY (2011b) Identification and analysis of seven H2O2-responsive miRNAs and 32 new miRNAs in the seedlings of rice (*Oryza sativa* L. ssp. indica). Nucleic Acids Res 39:2821–2833

Li B, Duan H, Li J, Deng XW, Yin W, Xia X (2013a) Global identification of miRNAs and targets in *Populus euphratica* under salt stress. Plant Mol Biol 81:525–539

Li S, Liu L, Zhuang X, Yu Y, Liu X, Cui X, Ji L, Pan Z, Cao X, Mo B et al (2013b) MicroRNAs inhibit the translation of target mRNAs on the endoplasmic reticulum in *Arabidopsis*. Cell 153:562–574

Li Y, Zhang Y, Shi D, Liu X, Qin J, Ge Q et al (2013c) Spatial-temporal analysis of zinc homeostasis reveals the response mechanisms to acute zinc deficiency in *Sorghum bicolor*. New Phytol 200:1102–1115

Liang G, Yang F, Yu D (2010) MicroRNA395 mediates regulation of sulfate accumulation and allocation in *Arabidopsis thaliana*. Plant J 62:1046–1057

Lin JS, Lin CC, Lin HH, Chen YC, Jeng ST (2012) MicroR828 regulates lignin and H2O2 accumulation in sweet potato on wounding. New Phytol 196:427–440

Lingel A, Simon B, Izaurralde E, Sattler M (2003) Structure and nucleic-acid binding of the *Drosophila* Argonaute 2 PAZ domain. Nature 426:465–469

Liu PP, Montgomery TA, Fahlgren N, Kasschau KD, Nonogaki H, Carrington JC (2007) Repression of AUXIN RESPONSE FACTOR10 by microRNA160 is critical for seed germination and post-germination stages. The Plant Journal 52:133–146

Liu D, Yu D (2009) MicroRNA (miR396) negatively regulates expression of ceramidase-like genes in *Arabidopsis*. Prog Nat Sci 19:781–785

Liu HH, Tian X, Li YJ, Wu CA, Zheng CC (2008) Microarray-based analysis of stress-regulated microRNAs in *Arabidopsis thaliana*. RNA 14:836–843

Llave C, Kasschau KD, Rector MA, Carrington JC (2002a) Endogenous and silencing-associated small RNAs in plants. Plant Cell 14:1605–1619

Llave C, Xie Z, Kasschau KD, Carrington JC (2002b) Cleavage of *Scarecrow-like* mRNA targets directed by a class of *Arabidopsis* miRNA. Science 297:2053–2056

Lobbes D, Rallapalli G, Schmidt DD, Martin C, Clarke J (2006) SERRATE: a new player on the plant microRNA scene. EMBO Rep 7:1052–1058

Lu XY, Huang XL (2008) Plant miRNAs and abiotic stress responses. Biochem Biophys Res Commun 368:458–462

Lu SF, Sun YH, Shi R, Clark C, Li LG, Chiang VL (2005) Novel and mechanical stress-responsive micro-RNAs in *Populus trichocarpa* that are absent from *Arabidopsis*. Plant Cell 17:2186–2203

Lu S, Sun YH, Chiang VL (2008) Stress-responsive microRNAs in *Populus*. Plant J 55:131–151

Lu W, Li J, Liu F, Gu J, Guo C, Xu L, Zhang H, Xiao K (2011) Expression pattern of wheat miRNAs under salinity stress and prediction of salt-inducible miRNAs targets. Front Agric China 5:413–422

Luan M, Xu M, Lu Y, Zhang L, Fan Y, Wang L (2015) Expression of zma-miR169 miRNAs and their target ZmNF-YA genes in response to abiotic stress in maize leaves. Gene 555:178–185

Lv DK, Bai X, Li Y, Ding XD, Ge Y, Cai H, Ji W, Wu N, Zhu YM (2010) Profiling of cold-stress-responsive miRNAs in rice by microarrays. Gene 459:39–47

Macovei A, Tuteja N (2012) microRNAs targeting DEAD-box helicases are involved in salinity stress response in rice (*Oryza sativa* L.). BMC Plant Biol 12:183

Manavella PA, Hagmann J, Ott F, Laubinger S, Franz M, Macek B, Weigel D (2012) Fast-forward genetics identifies plant CPL phosphatases as regulators of miRNA processing factor HYL1. Cell 151:859–870

Meng Y, Shao C, Gou L, Jin Y, Chen M (2011) Construction of microRNA and microRNA*-mediated regulatory networks in plants. RNA Biol 8:1124–1148

Mette MF, van der Winden J, Matzke M, Matzke AJ (2002) Short RNAs can identify new candidate transposable element families in Arabidopsis. Plant Physiol 130:6–9

Mi S, Cai T, Hu Y, Chen Y, Hodges E, Ni F, Wu L, Li S, Zhou H, Long C, Chen S, Hannon GJ, Qi Y (2008) Sorting of small RNAs into *Arabidopsis* argonaute complexes is directed by the 5′ terminal nucleotide. Cell 133:116–127

Mittal D, Mukherjee SK, Vasudevan M, Sanan-Mishra N (2013) Identification of tissue-preferential expression patterns of rice miRNAs. J Cell Biochem 114:2071–2081

Mittler R, Vanderauwera S, Gollery M, Van Breusegem F (2004) Reactive oxygen gene network of plants. Trends Plant Sci 9:490–498

Moldovan D, Spriggs A, Yang J, Pogson BJ, Dennis ES, Wilson IW (2009) Hypoxia-responsive microRNAs and trans-acting small interfering RNAs in *Arabidopsis*. J Exp Bot 61:165–177

Navarro L, Dunoyer P, Jay F, Arnold B, Dharmasiri N, Estelle M, Voinnet O, Jones JD (2006) A plant miRNA contributes to antibacterial resistance by repressing auxin signaling. Science 312:436–443

Nezhadahmadi A, Prodhan ZH, Faruq G (2013) Drought tolerance in wheat. Sci World J:1–12

Ozhuner E, Eldem V, Ipek A, Okay S, Sakcali S, Zhang B, Boke H, Unver T (2013) Boron stress responsive microRNAs and their targets in barley. PLoS One 8:e59543

Pandita D (2018) Plant miRNAs: micro structure and macro character. Res Rev J Agric Allied Sci 7:83–84

Park W, Li J, Song R, Messing J, Chen X (2002) CARPEL FACTORY, a dicer homolog, and HEN1, a novel protein, act in microRNA metabolism in *Arabidopsis thaliana*. Curr Biol 12:1484–1495

Park MY, Wu G, Gonzalez-Sulser A, Vaucheret H, Poethig RS (2005) Nuclear processing and export of microRNAs in *Arabidopsis*. Proc Natl Acad Sci U S A 102:3691–3696

Pasquinelli A, Reinhart B, Slack F, Maller B, Ruvkun G (2000) Conservation across animal phylogeny of the sequence and temporal regulation of the 21 nucleotide *C. elegans let-7* heterochronic regulatory RNA. Nature 408:86–89

Paul S, Kundu A, Pal A (2011) Identification and validation of conserved microRNAs along with their differential expression in roots of *Vigna unguiculata* grown under salt stress. Plant Cell Tiss Org Cult 105:233–242

Qi Y, Denli AM, Hannon GJ (2005) Biochemical specialization within *Arabidopsis* RNA silencing pathways. Mol Cell 19:421–428

Qiao Y, Shi J, Zhai Y, Hou Y, Ma W (2015) *Phytophthora* effector targets a novel component of small RNA pathway in plants to promote infection. Proc Natl Acad Sci U S A 112:5850–5855

Qin Y, Duan Z, Xia X, Yin W (2011) Expression profiles of precursor and mature microRNAs under dehydration and high salinity shock in *Populus euphratica*. Plant Cell Rep 30:1893–1907

Rajagopalan R, Vaucheret H, Trejo J, Bartel DP (2006) A diverse and evolutionarily fluid set of microRNAs in *Arabidopsis thaliana*. Genes Dev 20:3407–3425

Reinhart B, Slack F, Basson M, Pasquinelli A, Bettinger J, Rougvie A, Horvitz R, Ruvkun G (2000) The 21 nucleotide *let-7* RNA regulates *C. elegans* developmental timing. Nature 403:901–906

Reinhart BJ, Weinstein EG, Rhoades MW, Bartel B, Bartel DP (2002) MicroRNAs in plants. Genes Dev 16:1616–1626

Ren Y, Chen L, Zhang Y, Kang X, Zhang Z, Wang Y (2012) Identification of novel and conserved *Populus tomentosa* microRNA as components of a response to water stress. Funct Integr Genomics 12:327–339

Ren G, Xie M, Dou Y, Zhang S, Zhang C, Yu B (2012b) Regulation of miRNA abundance by RNA binding protein TOUGH in *Arabidopsis*. Proc Natl Acad Sci U S A 109:12817–12821

Ren G, Xie M, Zhang S, Vinovskis C, Chen X, Yu B (2014) Methylation protects microRNAs from an AGO1-associated activity that uridylates 50 RNA fragments generated by AGO1cleavage. Proc Natl Acad Sci U S A 111:6365–6370

Rogers K, Chen X (2013) Biogenesis, turnover, mode of action of plant microRNAs. Plant Cell 25:2383–2399

Ruiz-Ferrer V, Voinnet O (2009) Roles of plant small RNAs in biotic stress responses. Annu Rev Plant Biol 60:485–510

Samuel MA, Mudgil Y, Salt JN, Delmas F, Ramachandran S, Chilelli A, Goring DR (2008) Interactions between the S-domain receptor kinases and at PUB-ARM E3 ubiquitin ligases suggest a conserved signaling pathway in *Arabidopsis*. Plant Physiol 147:2084–2095

Sanan-Mishra N, Kumar V, Sopory SK, Mukherjee SK (2009) Cloning and validation of novel miRNA from basmati rice indicates cross talk between abiotic and biotic stresses. Mol Genet Genom 282:463–474

Shen J, Xie K, Xiong L (2010) Global expression profiling of rice microRNAs by one-tube stem-loop reverse transcription quantitative PCR revealed important roles of microRNAs in abiotic stress responses. Mol Genet Genom 284:477–488

Shuai P, Liang D, Zhang Z, Yin W, Xia X (2013) Identification of drought-responsive and novel *Populus trichocarpa* microRNAs by high throughput sequencing and their targets using degradome analysis. BMC Genomics 14:233

Song JJ, Smith SK, Hannon GJ, Joshua T (2004) Crystal structure of Argonaute and its implications for RISC slicer activity. Science 305:1434–1437

Stief A, Altmann S, Hoffmann K, Pant BD, Scheible WR, Baurle I (2014) Arabidopsis miR156 regulates tolerance to recurring environmental stress through SPL transcription factors. Plant Cell 26:1792–1807

Sun G (2012) MicroRNAs and their diverse functions in plants. Plant Mol Biol 80:17–36

Sun G, Stewart CN, Xiao P, Zhang B (2012) MicroRNA expression analysis in the cellulosic biofuel crop switchgrass (*Panicum virgatum*) under abiotic stress. PLoS One 7:e32017

Sun X, Xu L, Wang Y, Yu R, Zhu X, Luo X et al (2015) Identification of novel and salt-responsive miRNAs to explore miRNA-mediated regulatory network of salt stress response in radish (*Raphanus sativus* L.). BMC Genomics 16:197

Sunkar R, Zhu JK (2004) Novel and stress-regulated microRNAs and other small RNAs from *Arabidopsis*. Plant Cell 16:2001–2019

Sunkar R, Kapoor A, Zhu J (2006) Posttranscriptional induction of two cu/Zn superoxide dismutase genes in *Arabidopsis* is mediated by down-regulation of miR398 and important for oxidative stress tolerance. Plant Cell 18:2051–2065

Sunkar R, Li YF, Jagadeeswaran G (2012) Functions of microRNAs in plant stress responses. Trends Plant Sci 17:196–203

Takeda A, Iwasaki S, Watanabe T, Utsumi M, Watanabe Y (2008) The mechanism selecting the guide strand from small RNA duplexes is different among argonaute proteins. Plant Cell Physiol 49:493–500

Tang G, Reinhart BJ, Bartel DP, Zamore PD (2003) A biochemical framework for RNA silencing in plants. Genes Dev 17:49–63

Trindade I, Capitao C, Dalmay T, Fevereiro MP, dos Santos DM (2010) miR398 and miR408 are up-regulated in response to water deficit in *Medicago truncatula*. Planta 231:705–716

Vandenabeele S, Van-Der-Kelen K, Dat J, Gadjev I, Boonefaes T, Morsa S, Rottiers P, Slooten L, Van-Montagu M, Zabeau M, Inze D, Van-Breusegem F (2003) A comprehensive analysis of hydrogen peroxide-induced gene expression in tobacco. Proc Natl Acad Sci U S A 100:16113–16118

Vaucheret H, Vazquez F, Crete P, Bartel DP (2004) The action of ARGONAUTE1 in the miRNA pathway and its regulation by the miRNA pathway are crucial for plant development. Genes Dev 18:1187–1197

Vazquez F, Gasciolli V, Crete P, Vaucheret H (2004) The nuclear dsRNA binding protein HYL1 is required for microRNA accumulation and plant development, but not posttranscriptional transgene silencing. Curr Biol 14:346–351

Vella MC, Choi EY, Lin SY, Reinert K, Slack FJ (2004) The *C. elegans* microRNA *let-7* binds to imperfect *let-7* complementary sites from the *lin-41* 3′UTR. Genes Dev 18:132–137

Voinnet O (2009) Origin, biogenesis, and activity of plant microRNAs. Cell 136:669–687

Wang B, Malik R, Nigg EA, Korner R (2008) Evaluation of the low-specificity protease elastase for large-scale phosphoproteome analysis. Anal Chem 80:9526–9533

Wang T, Chen L, Zhao M, Tian Q, Zhang WH (2011) Identification of drought-responsive microRNAs in *Medicago truncatula* by genome-wide high-throughput sequencing. BMC Genomics 12:1–11

Wang Y, Sun F, Cao H, Peng H, Ni Z, Sun Q, Yao Y (2012) TamiR159 directed wheat TaGAMYB cleavage and its involvement in anther development and heat response. PLoS One 7:e48445

Wang S, Sun X, Hoshino Y, Yu Y, Jia B, Sun Z, Zhu Y (2014a) *MicroRNA319* positively regulates cold tolerance by targeting *OsPCF6* and *OsTCP21* in rice (*Oryza sativa* L.). PLoS One 9:e91357

Wang B, Sun YF, Song N, Wei JP, Wang XJ, Feng H, Yin ZY, Kang ZS (2014b) MicroRNAS involving in cold, wounding and salt stresses in *Triticum aestivum* L. Plant Physiol Biochem 80:90–96

Wei L, Zhang D, Xiang F, Zhang Z (2009) Differentially expressed miRNAs potentially involved in the regulation of defense mechanism to drought stress in maize seedlings. Int J Plant Sci 170:979–989

Wightman B, Ha I, Ruvkun G (1993) Posttranscriptional regulation of the heterochronic gene *lin-14* by *lin-4*mediates temporal pattern formation in *C. elegans*. Cell 75:855–862

Wu X, Shi Y, Li J, Xu L, Fang Y, Li X, Qi Y (2013) A role for the RNA-binding protein MOS2 in microRNA maturation in *Arabidopsis*. Cell Res 23:645–657

Xia K, Wang R, Ou X, Fang Z, Tian C, Duan J (2012) OsTIR1 and OsAFB2 down regulation via OsmiR393 over expression leads to more tillers, early flowering and less tolerance to salt and drought in rice. PLoS One 7:e30039

Xie F, Wang Q, Sun R, Zhang B (2014) Deep sequencing reveals important roles of microRNAs in response to drought and salinity stress in cotton. J Exp Bot 66:789–804

Xin M, Wang Y, Yao Y, Xie C, Peng H, Ni Z et al (2010) Diverse set of microRNAs are responsive to powdery mildew infection and heat stress in wheat (*Triticum aestivum* L.). BMC Plant Biol 10:123

Xin M, Wang Y, Yao Y, Song N, Hu Z, Qin D et al (2011) Identification and characterization of wheat long non-protein coding RNAs responsive to powdery mildew infection and heat stress by using microarray analysis and SBS sequencing. BMC Plant Biol 11:61

Xu DQ, Huang J, Guo SQ, Yang X, Bao YM, Tang HJ, Zhang HS (2008) Over expression of a TFIIIA-type zinc finger protein gene ZFP252 enhances drought and salt tolerance in rice (*Oryza sativa* L.). FEBS Lett 582:1037–1043

Yamaguchi-Shinozaki K, Shinozaki K (2006) Transcriptional regulatory networks in cellular responses and tolerance to dehydration and cold stresses. Annu Rev Plant Biol 57:781–803

Yang L, Liu Z, Lu F, Dong A, Huang H (2006a) SERRATE is a novel nuclear regulator in primary microRNA processing in *Arabidopsis*. Plant J 47:841–850

Yang W, Chendrimada TP, Wang Q et al (2006b) Modulation of microRNA processing and expression through RNA editing by ADAR deaminases. Nat Struct Mol Biol 13:13–21

Yang L, Wu G, Poethig RS (2012) Mutations in the GW-repeat protein SUO reveal a developmental function for microRNA-mediated translational repression in *Arabidopsis*. Proc Natl Acad Sci U S A 109:315–320

Yang CH, Li DY, Mao DH, Liu X, Ji CJ et al (2013) Over expression of microRNA319 impacts leaf morphogenesis and leads to enhanced cold tolerance in rice (*Oryza sativa* L.). Plant Cell Environ 36:2207–2218

Yu B, Yang Z, Li J, Minakhina S, Yang M, Padgett RW, Steward R, Chen X (2005) Methylation as a crucial step in plant microRNA biogenesis. Science 307:932–935

Yu X, Wang H, Lu YZ, Ruiter M, Cariaso M, Prins M, Van TA, He YK (2012) Identification of conserved and novel microRNAs that are responsive to heat stress in *Brassica rapa*. J Exp Bot 63:1025–1038

Zeng X, Xu Y, Jiang J, Zhang F, Ma L, Wu D, Wang Y, Sun W (2018) Identification of cold stress responsive microRNAs in two winter turnip rape (Brassica rapa L.) by high throughput sequencing. BMC Plant Biol. 18:52

Zhan X, Wang B, Li H, Liu R, Kalia RK, Zhu JK, Chinnusamy V (2012) *Arabidopsis* proline-rich protein important for development and abiotic stress tolerance is involved in microRNAs biogenesis. Proc Natl Acad Sci U S A 109:18198–18203

Zhang B (2015) MicroRNA: a new target for improving plant tolerance to abiotic stress. J Exp Bot 66:1749–1761

Zhang B, Wang Q (2015) MicroRNA-based biotechnology for plant improvement. J Cell Physiol 230:1–15

Zhang JF, Yuan LJ, Shao Y, Du W, Yan DW, Lu YT (2008) The disturbance of small RNA pathways enhanced abscisic acid response and multiple stress responses in *Arabidopsis*. Plant Cell Environ 31:562–574

Zhang L, Chia JM, Kumari S, Stein JC, Liu Z, Narechania A, Maher CA, Guill K, McMullen MD, Ware D (2009a) A genome-wide characterization of microRNA genes in maize. PLoS Genet 5:e1000716

Zhang J, Xu Y, Huan Q, Chong K (2009b) Deep sequencing of *Brachypodium* small RNAs at the global genome level identifies microRNAs involved in cold stress response. BMC Genomics 10:449

Zhang X, Zou Z, Gong P, Zhang J, Ziaf K, Li H, Xiao F, Ye Z (2011) Over-expression of microRNA169 confers enhanced drought tolerance to tomato. Biotechnol Lett 33:403–409

Zhang S, Xie M, Ren G, Yu B (2013) CDC5, a DNA binding protein, positively regulates posttranscriptional processing and/or transcription of primary microRNA transcripts. Proc Natl Acad Sci U S A 110:17588–17593

Zhang LW, Song JB, Shu XX, Zhang Y, Yang ZM (2013a) miR395 is involved in detoxification of cadmium in *Brassica napus*. J Hazard Mater 15:204–211

Zhang S, Yue Y, Sheng L, Wu Y, Fan G, Li A, Hu X, Shang Guan M, Wei C (2013b) PASmiR: a literature curated database for miRNA molecular regulation in plant response to abiotic stress. BMC Plant Biol 13:33

Zhang S, Liu Y, Yu B (2014) PRL1, an RNA-binding protein, positively regulates the accumulation of miRNAs and siRNAs in *Arabidopsis*. PLoS Genet 10:e1004841

Zhang J, Zhang H , Srivastava AK, Pan Y, Bai J, Fang J, Shi H, Zhu JK (2018) Knockdown of Rice MicroRNA166 Confers Drought Resistance by Causing Leaf Rolling and Altering Stem Xylem Development. Plant Physiology 176(3):2082–2094

Zhao B, Liang R, Ge L, Li W, Xiao H, Lin H, Ruan K, Jin Y (2007) Identification of drought-induced microRNAs in rice. Biochem Biophys Res Commun 354:585–590

Zhao B, Ge L, Liang R, Li W, Ruan K, Lin H, Jin Y (2009) Members of miR-169 family are induced by high salinity and transiently inhibit the NF-YA transcription factor. BMC Mol Biol 10:29

Zhao X, Liu X, Guo C, Gu J, Xiao K (2013) Identification and characterization of microRNAs from wheat (*Triticum aestivum* L.) under phosphorus deprivation. J Plant Biochem Biotechnol 22:113–123

Zhao J, He Q, Chen G, Wang L, Jin B (2016) Regulation of non-coding RNAs in heat stress responses of plants. Front Plant Sci 7:1213

Zhou ZS, Wang SJ, Yang ZM (2008) Biological detection and analysis of mercury toxicity to alfalfa (*Medicago sativa*) plants. Chemosphere 70:1500–1509

Zhou L, Liu Y, Liu Z, Kong D, Duan M, Luo L (2010) Genome-wide identification and analysis of drought-responsive microRNAs in *Oryza sativa*. J Exp Bot 61:4157–4168

Zhou J, Liu M, Jiang J, Qiao G, Lin S, Li H, Xie L, Zhuo R (2012) Expression profile of miRNAs in *Populus cathayana* L. and *Salix matsudana* Koidz under salt stress. Mol Biol Rep 39:8645–8654

Zhu H, Hu F, Wang R, Zhou X, Sze SH, Liou LW, Barefoot A, Dickman M, Zhang X (2011) *Arabidopsis* Argonaute10 specifically sequesters miR166/165 to regulate shoot apical meristem development. Cell 145:242–256

Index

M. Hasanuzzaman et al. (eds.), *Plant Abiotic Stress Tolerance*,
https://doi.org/10.1007/978-3-030-06118-0

N

Printed by Printforce, the Netherlands